S0-CDN-315

MOUNTAIN VIEW PUBLIC LIBRARY
1000184299

MOUNTAIN VIEW
PUBLIC LIBRARY
Mountain View, Calif.

JANE'S

SPACEFLIGHT DIRECTORY

REGINALD TURNILL

FOREWORD BY DEKE SLAYTON

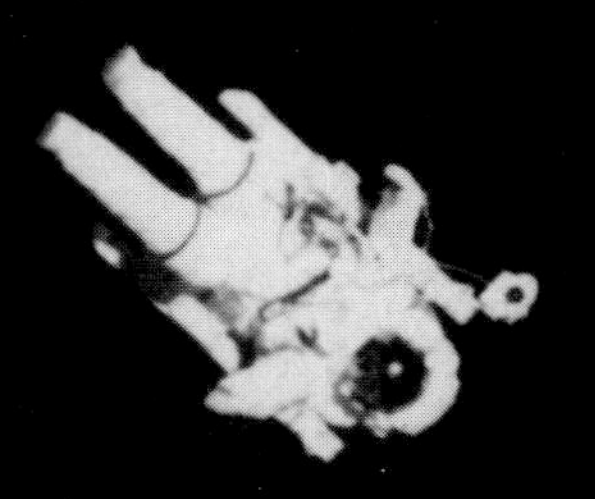

JANE'S

FOREWORD

First published in the United Kingdom in 1984 by
Jane's Publishing Company Limited,
238 City Road, London EC1V 2PU

ISBN 0 7106 0208 1

Distributed in the Philippines and the USA and its dependencies by
Jane's Publishing Inc,
135 West 50th Street,
New York, NY 10020

Typeset by Method Limited,
Woodford Green, Essex

Printed by Biddles Ltd,
Guildford, Surrey

Jane's Spaceflight Directory is the most comprehensive compilation of data on space operations I have seen anywhere. Reg Turnill has been a true scholar of the subject for 25 years and we have had the pleasure of his company at manned launches since their inception. His unmatched coverage is marked by an authority based on a high degree of personal experience.

The first quarter-century of the Earth's space age has seen an exponential expansion in technology and in our ability to explore the solar system. We must all be proud and amazed that mankind made the jump from the first unmanned satellite to a first step on the Moon in a mere dozen years. That feat will for a long time to come undoubtedly be seen in the history books as the most significant space event. However, the second dozen years have been almost equally spectacular. They have been marked by the large-scale exploitation of Earth orbits for commercial and scientific applications. The development and very successful early flights of the Space Shuttle will generate a boom in the scientific and commercial utilisation of space. And who in the world could not be impressed by the spectacular Jupiter and Saturn flypasts by the Voyager series? Or by the fantastic photography and data returned from Mars by the Vikings and from Venus by the Veneras?

We have been blessed with the chance to participate in and observe the most technologically productive 25 years in recorded history. Let us hope we are wise enough to apply this technology for the betterment of mankind for many centuries to come. Reg Turnill has accurately and thoroughly documented space history to date, and without a doubt future historians will find this reference work extremely valuable.

Deke Slayton
(Director, NASA Flight Crew Operations, 1963–72)

Below: The author introducing Dr Wernher von Braun, creator of the Saturn 5 Moon rocket, during a BBC TV programme shortly before von Braun's death in 1977. ***(Margaret Turnill)***

Right: Deke Slayton floats from Apollo into Soyuz during the 1975 Soviet-American mission. Last of the original Mercury Seven to fly, he said it was worth the 16yr wait. As Flight Crew Director during those years, he was largely responsible for deciding which astronauts went to the Moon. ***(NASA)***

CONTENTS

INTRODUCTION

SATELLITE LAUNCHES BY COUNTRY

	1980	1957- 1981	1982	1983	Total 1957- 1983
USSR	1,312	94	101	98	1,605
USA	682	15	13	17	727
Japan	17	3	1	3	24
China	8	1	1	1	11
France	9	0	0	0	9
Europe	1	1	0	1	3
India	1	1	0	1	3
UK	1	0	0	0	1
USA/Intelsat	25	2	2	1	30
USSR/ Intercosmos	202	0	0	0	22
USA/Europe	13	0	0	1	14
USA/UK	11	1	0	0	12
USA/Canada	9	0	2	0	11
USSR/France	5	1	0	0	6
USA/Italy	5	0	0	0	5
USA/W Germany	5	0	0	0	5
USA/NATO	5	0	0	0	5
USA/France	4	0	0	0	4
USSR/India	2	1	0	0	3
USA/Japan	3	0	0	0	3
USA/Australia	2	0	0	0	2
USA/Indonesia	2	0	0	0	2
France/W Germany	1	0	0	0	1
USA/Netherlands	1	0	0	0	1
USA/Spain	1	0	0	0	1
Europe/India	0	1	0	0	1
USA/India	0	0	1	1	2
Europe/ Intelsat	0	0	0	1	1
USA/Canada/ Indonesia	0	0	0	1	1
USA/Netherlands/ UK	0	0	0	1	1
Total launches	2,145	123	121	127	2,516

The *Table of Earth Satellites*, continuously maintained by Britain's Royal Aircraft Establishment since Sputnik 1 in 1957 and reproduced here with RAE permission, provides a unique record of what is happening in space. Listed first are the countries which have launched their own satellites with their own rockets; then come the launching countries with the number of launches they have performed for other countries. Thus it shows that France, in addition to launching nine satellites with her own rockets, has had six others launched by the USSR and four by the US, and has also launched a W German satellite with a French rocket.

The satellite table above summarises what this directory is about. By the end of 1983 the number of launches into Earth orbit or deep space had risen to 2,516. Because many of the rockets carried multiple payloads, the total of satellites was 3,446. In addition, there were over 300 unsuccessful launch attempts. *Jane's Spaceflight Directory* seeks to cover and explain all of those launches.

This is how the late President Brezhnev of the Soviet Union summed up the first 25 years of the Space Age in a message to the UN conference on the study and use of space for peaceful purposes, held in Vienna in August 1982: "Cosmonautics has become a reliable assistant of the geologist and the sailor, the agronomist and the meteorologist, the communications specialist and the doctor, the cartographer and the forestry worker." Russia's steady progress with long-duration manned spaceflight, and America's successes with the Space Shuttle, following the brilliant Viking and Voyager visits to Mars, Jupiter and Saturn, are among the many positive achievements recorded in this directory. But both at Vienna and at the 33rd International Astronautical Federation Congress at Paris the following month increasing anxiety was expressed about the way man is using what the late President Kennedy so aptly described as "this new ocean". The geosynchronous orbit is already becoming overcrowded as the launcher nations fill the maximum of 180 available slots with a mixture of communications, weather, scientific, broadcast and military satellites for themselves and other nations rich enough to pay for them. (Unique among the poorer nations is India, with the foresight to leap into space technology despite the cost in order to ensure that she shares in its benefits.) The collision hazard is already so great that in the West warnings of imminent collisions are issued so that active satellites can be commanded to take evasive action.

The alarming increase in space debris – consisting not only of dead satellites, rocket upper stages and discarded protective panels, but also of thousands of fragments resulting from over 70 explosions (some accidental, others deliberate) – greatly occupied the delegates of the more than 100 countries represented at the two conferences. There is evidence that collisions have already destroyed some unmanned satellites, and the threat to men in space and their hopes of establishing permanent orbiting platforms steadily increases. Suggestions for ways of meeting this threat include an international clearing house for orbital information, a ban on space explosions, and a specially allotted orbit for discarded space material; this might be at 800-1,000km altitude, where most of the debris already circulates. Special international satellites entirely devoted to debris tracking may soon become necessary. Fuller details of this problem are given under "Monitoring" in the Military section of this book.

Space war inevitable?

The length of the Military section reflects the growing concern about the increase in military space activities, which are creating much of the debris. The belief among US defence chiefs that in the next 25 years contests in space are not only possible but almost inevitable have led to the setting up of the new USAF Space Command. For many years past Russia's space launches, though not fully understood, have been more than 70% military in purpose, with another 15% being partly military. And as this directory went to press NASA's tradition of an "open" space programme was being steadily eroded by the addition of secret control rooms at the Kennedy and Johnson space centres, and by pressure for a separate military fleet of Space Shuttles. The USAF would like a more advanced version, launchable from mobile platforms; Boeing proposes a hydrogen-powered 747 for the latter task.

A new breed of military astronauts was being quietly selected too, recalling the brief period from 1967 when the USAF had 16 astronauts of its own until the Manned Orbiting Laboratory (MOL) was cancelled and seven of them transferred to NASA's present 77-strong team. All have taken an active part in the early Shuttle missions, with Richard Truly becoming head of the US Navy's new Space Command. At first the new military astronauts will work as mission specialists alongside NASA's Shuttle pilots; the first, still unnamed, was due to have flown on the cancelled STS-10. But the long-term aim is undoubtedly to form them into a separate team who will nurse their secrets as they ferry to and from the inevitable manned military reconnaissance platforms.

Theoretically the battle for space-based laser weapons is settling down to be grimmer and much more expensive than the race to the Moon 20 years ago. The arguments

Artist's concept of reusable Air Launched Sortie Vehicle, designed by Boeing for the USAF. It must use all its 9 engines to achieve low Earth orbit. The 747 launcher needs an additional hydrogen-powered engine in the tail. ***(Boeing)***

ebb and flow, and current thinking is that the technology is unlikely to be available until the turn of the century. US defence philosophy used to be that the man on the high ground - the Moon - could dominate the world. That proved to be nonsense, and now the argument is that whoever can produce an operational laser satellite can control access to space. Some US intelligence estimates still insist that Russia is likely to be able to place high-energy lasers in orbit in the 1984-86 period. At present America expects to have operational space lasers by 1989, though their development could be speeded up to match the Soviet deployment if more money were provided. Cost estimates are vague, ranging from $12 billion - less than the Apollo Moon-landing programme 20 years ago - to around $200 billion. That's one reason why both politicians and defence experts in the Pentagon are suggesting that technically advanced "Western" allies like West Germany and Japan should be asked to co-operate. And Russia, it is pointed out, is spending three to five times as much as America on space laser development. Politically, some argue, there is no reason why defensive "laser battle stations" should not be developed since the Salt and anti-ballistic missile defence treaties permit their use.

The first laser satellites would be used to defend America against bomber attacks and to destroy tankers, shutting off the enemy's oil supplies. But that would mean directing their beams through the earth's atmosphere, which quickly reduces their range and effectiveness. Unimpeded in space, the beams could be deadly over ranges of thousands of kilometres.

But the true importance of space-based laser battle stations lies in the fact that in the long run they would transform defence strategy as dramatically as did the deployment of operational nuclear weapons in the 1950s. They would be able to detect and destroy nuclear-tipped ballistic missiles a few minutes after they had left their launchpads, thus providing an effective defence not only for the United States but for her allies as well. And with countries like Libya, Iraq, India and Pakistan now moving towards development of their own nuclear weapons, such a defence will become increasingly urgent. Even more important, according to Senator Malcolm Wallop (Republican, Wyoming), is the prospect that American space-based lasers would make Russia's huge investment in offensive missiles obsolete overnight. It would probably take the Soviet Union another decade to design laser-resistant nuclear missiles, retool its factories, and produce a sixth generation of such weapons. That would mean another decade during which the West could feel safe against the possibility of surprise nuclear attack, another ten years of negotiating time - and another ten years during which the Soviet leaders might move towards less imperialistic and more democratic policies.

Space station studies

The Soviet Union is currently estimated to be spending an annual $18 billion on space - $3-4 billion more than the US. And with the US military space effort getting $8·5 billion in FY 1983 compared with NASA's $6·4 billion to cover all aspects of manned and unmanned spaceflight, the civilian agency's real hope of funding a permanently manned platform for such activities as materials processing would probably be to share one with the military - if the latter would permit it! An indication of the financial pressures on NASA in recent years was the agency's decision to send engineers to Washington's National Air and Space Museum to recover components from the unused Skylab on display there, and to use them on Shuttle missions.

Encouraged by NASA, the European Space Agency is doing its own space station studies, as are individual countries like Britain, Canada, France, Germany, Italy and Japan, awake at last to the need to join in if they are to obtain a share of the technology and commercial benefits available within the next decade. But even while Britain and Germany are competing with alternative designs for free-flying platforms based on Britain's Shuttle pallets and Germany's SPAS, some Americans are becoming increasingly concerned at the way other countries are seeking to move in on US-originated space technology. As a result, NASA has given Boeing a study contract for yet another pallet system.

The Reagan Administration's long-awaited space policy document, issued in July 1982, supported the Space Shuttle in general terms but made little reference to future projects. It was left to James Beggs, the current NASA Administrator, to deal with this in separate speeches and statements. He pointed out that NASA had the ability to develop, construct and place in orbit a permanent manned space station by 1990 at a cost of about $3·5 billion, and that this would "dramatically improve" US abilities to operate in space for both civil and military purposes.

The emphasis laid on increasing the joint civil/defence management of space programmes in the President's space policy document represented a significant change. There would be launch priority for national security missions, and separate but strongly interacting defence and civil programmes.

The Government aims to create a climate which would make possible and encourage private-sector investment and involvement in space activities. (Disputes over the application of this policy to Landsat are currently threatening its future operation.) In an apparent reference to some attempts by Third World and other nations to claim sovereignty over geosynchronous orbit slots above their territory, the document stated: "The US rejects any claims to sovereignty by any nation over space or celestial bodies. . . . Purposeful interference with space systems shall be viewed as an infringement upon sovereign rights."

NASA's uncertainties were at last dispelled with President Reagan's direction in Jan 1984 to begin development over 8yr and at a cost of $8 billion of a

One of many current Space Station designs. Vertical control module with solar arrays would be launched first. 4 main modules would follow, with manufacturing, satellite replacement and repair facilities and a launch platform for higher-altitude missions being added as required. ***(Lockheed)***

permanently manned space station capable of housing 6-8 astronauts. He called for a bold and imaginative programme to maintain US leadership in space well into the 21st century. The fight for funding lay ahead, but after vigorously opposing the Space Station DoD has apparently agreed to remain neutral so long as Shuttle development is not affected by the programme. As we go to press, NASA is seeking foreign participation worth up to $3 billion. Anything less than $1 billion-worth would not be considered worthwhile, and ESA and other interested parties are being told that they must decide by the end of 1984 if they are to join in the 2yr design phase.

Aware of European dissatisfaction that the agreement under which ESA contributed Spacelab to the Shuttle programme provides no continuing access to this facility, NASA is ready to agree formally that those contributing to the station will have long-term user rights. This is however sure to run into the recurring US industry claims for priority in the exploitation of American-developed space technology.

The Soviet lead

While the US is undoubtedly well ahead in space technology, the Russians have been far more energetic and successful in applying the technology that they possess. As a result, the US cannot hope to catch up with the Soviet Union in space station operations, from crew rotation to materials processing, until well into the 1990s. A study of the Soyuz flights and of the manning of Salyut 6 for nearly two of its five years in orbit reveals just how much practical work has been done. It is now being continued aboard Salyut 7 even though, as the Soviets themselves accept, this station incorporates only minor advances on its predecessor. More forthcoming nowadays about their future plans, Soviet space leaders have indicated that assembly of a large, modular space station is not likely to begin before the mid-1980s, though a great deal of preparatory work is being done. During their 185 days in Salyut 6 cosmonauts Popov and Ryumin devoted a lot of time to developing assembly and repair techniques. They found that it was just about possible to melt gold, silver, copper and other materials and apply them as protective coatings to metal plates, as if repairing micrometeorite damage. Since then there have been hundreds of experiments during the 375 days of occupation of Salyut 7. In addition, the Soyuz T-5 crew tried out new tools for assembly work in space, while the T-9 crew used them effectively to fix additional solar panels to the station.

Like America, the Soviet Union talks of sending raw materials into orbit for fabrication there. For some years the US manufacturers have been offering a variety of systems for extruding miles of lightweight beams from Shuttle loads of material. Now Russia talks of delivering metals and plastics which could be converted in orbit into thin-walled elements to serve as building blocks for space stations.

Increased Soviet confidence in their space competence has been expressed in several ways over the last few years. There has been a welcome increase in attendance and participation at international conferences, and early in 1982 Russia was reported to have delivered to the UN Economic Commission the first comprehensive geological map series of Africa, covering 35 million square miles and providing information on 79 different minerals.

And at last the cosmonauts have been allowed to become human. We learned that Valeri Ryumin, who was later to spend a whole year in orbit, went though the most miserable period of his life, worrying what he might have done wrong after failing to dock on Soyuz 25, his first trip into space. During his flight with Leonid Popov they spilled 2½lit while attempting to water their plants. The resulting gigantic weightless globule threatened to contaminate equipment and cause short-circuits. Placing themselves on each side of it, with a great effort they drank it all between them. More recently we had the visit of Svetlana Savitskaya to Salyut 7, during which she firmly rejected light-hearted suggestions by her four male companions that she should do the cooking.

Techniques for assembling a space platform 3 times the size of a football field have already been developed. After automatic deployment of the platform from the Shuttle cargo bay an astronaut would check the latching mechanisms. *(Lockheed)*

The importance of space prestige

Savitskaya's mission was the latest example of the importance attached by Russia to space prestige. While America did not take the first foreigner into space until Spacelab 1 at the end of 1983, Russia has in the last four years improved her relations with the scientific communities in ten countries (nine Soviet-bloc plus France) by taking their representatives on seven-day missions. An Indian will be the next to go.

But women have been a different matter. While America has for some time had eight women poised to take their place alongside the men in orbit, veteran Soviet cosmonauts Generals Shatalov and Nikolayev (the latter was married to Valentina Tereshkova, the world's first spacewoman) have been positively patronising when explaining why they had no female cosmonauts under training. "Nowadays we keep our women here on Earth," said Nikolayev. "We love them very much; we spare them as much as possible. However, in future they will surely work on board space stations as doctors, geologists, astronomers - and of course as stewardesses." But within weeks of America's announcement that Dr Sally Ride would fly on STS-7 in April 1983, the Soviets declared that they had two women, a pilot-engineer and a flight engineer, under training. Svetlana, instead of Ride, thus became the first woman to make a working trip into orbit, as compared with Tereshkova's three-day test flight nearly 20 years ago.

As we went to press, Dr Kathryn Sullivan was due to become the first woman spacewalker, on the 14th Shuttle mission (41-G in the new designation system) in Aug 1984 - though there was still time for a Soviet woman to beat her to it. All 6 women selected in 1978 were included in the 53 flight assignments for 1984, with 2 flying together on 41-G, since Sally Ride gets her second flight on that mission. But although America's astronauts now include 3 married couples (Anna and William Fisher, Rhea Seddon and Robert Gibson, and Sally Ride and Stephen Hawley) they have so far only been assigned to fly with other people's spouses. However, we should see the first married couple in space together in the next year or two.

1982 was a bad year for aerospace insurers. India's failed Insat was insured for $65 million, while Ford Aerospace had also insured itself against the possible loss of performance bonuses amounting to another $5·3 million.

The loss of Ariane L-5 cost the insurers much less, for ESA had covered Marecs B for only $20 million, mainly because they could not afford more but also because they felt confident of success. The true cost of this failure can be judged from the fact that Marecs A had been covered for a total of $96 million against the possibility of having to buy another satellite, another launcher and more insurance. As a result Third World countries keen to order satellites are increasingly finding that they cannot get international funding unless they are willing to insure their launches.

1983, characterised by an unbroken run of success, was followed by the double loss on the 10th Shuttle flight in Feb 1984 of Westar 6 and Palapa B2 when their PAM-D upper stages failed. Claims are being made for $105 million by Western Union and $75 million by the Indonesian Government, and the 5·5% premiums they paid are likely to rise to 10-12% for future insurers.

The steady accumulation of space debris is leading to the need for "third party" insurance, and Lloyds is currently offering for $100,000 coverage of the launch and early years of operation. The major risk covered is the possibility of a launch failure leading to debris falling on land. But the policy also covers any damage done by the satellite colliding - perhaps long after its useful life is over - with an active spacecraft.

Halley's Comet missions

Following the outstanding success of the US Pioneer, Viking and Voyager programmes, which have filled in so many gaps in man's knowledge of the solar system, NASA is finding it difficult to justify the cost of continuing these activities. Consequently, it will be left to ESA, the USSR and Japan to intercept and examine Halley's Comet when it reappears in 1986. For several years NASA will be preoccupied with building up successful Space Shuttle test flights into a commercially viable programme. "Reimbursable" (paying) launches will be interspersed with the Defence Department flights described in the Military section. The occasional, exciting scientific payloads started with Spacelab 1 in 1983 will be highlighted by the Space Telescope in 1986. It is hoped that this at least will provide US scientists with some useful observations of Halley's Comet. And in a brilliant afterthought, ISEE-3 has been diverted by lunar gravity into an orbit which should take it through the tail of Comet Giacobini-Zinner on 11 Sep 1985. This is a cheap way, it is hoped, of snatching for America the prestige of man's first comet interception and inspection.

ESA's Giotto spacecraft, to be launched from Kourou in July 1985, should intercept Halley's Comet in March 1986. 4 more interceptors will be sent by the Soviet Union and Japan. ***(British Aerospace)***

The Soviet Union's space shuttle still seems to be some years off, and the long-awaited Saturn V-class launcher is still said to be under development for launch about 1986. Academician Vladimir Kotelnikov, chairman of Intercosmos, told the author at the end of 1982 that Russia's priorities in solar system exploration were still Venus and Mars, in that order. Venus is seen as important because of the need to ensure that the Earth's atmosphere does not go the same way as that of this hothouse planet. Mars would be visited by robot soil-samplers before long. Kotelnikov would make no prophecy about the prospects of a manned flight to Mars, though there is some evidence that Soviet scientists have not abandoned their Martian ambitions. A 400-day window for this mission will occur in 1986–7, allowing 140 days to reach Mars, 30 days orbiting the Red Planet (with or without a landing) and a 230-day return flight. Round trips to the Moon and Venus would seem likely as rehearsals beforehand.

Talk of solar power stations has inevitably receded during this era of temporarily plentiful fuel supplies, though design work is quietly continuing in both the East and West. Some Soviet scientists talk of establishing a solar power or "helio station" not in near-Earth space but near Mercury, where solar activity is hundreds of times greater. They claim to know already how to use the "super high-frequency range" to transmit the energy to Earth, though they admit that a system of relaying devices would be needed.

On the American side perhaps the most significant proposal is Boeing's Air Launched Sortie Vehicle, a mini-Shuttle and External Tank launched from the top of a Boeing 747 powered by hydrogen-fuelled engines. The concept, which envisages a first mission in 1988, would give Boeing the valuable bonus of being the first to develop a hydrogen-powered aircraft.

Jane's Spaceflight Directory is international in outlook, but since it is published in Britain perhaps one can end with a personal note of regret that Britain is increasingly being left behind in the space race. Among European nations, West Germany has achieved notable successes in the past year not only as the major contributor to ESA's Spacelab but also with her independently designed and funded SPAS-01 satellite. Italy is rapidly moving into

Lunar mining facility, designed for JSC by Eagle Engineering of Houston, could produce water, hydrogen and oxygen from ilmenite-rich soil. Workers live in 2 half-buried External Tanks at top centre. Top left, lunar transporter takes liquid oxygen to orbiting way station. ***(NASA)***

space technology, and France of course continues to pursue a growing national space programme alongside her ESA contributions. With British industry and politicians still sceptical and unbelievably ignorant about the commercial potential of microgravity production, the chances are that West Germany will emerge as NASA's major partner in this area. Meanwhile, although the British production line of communications satellites is worth a healthy £300 million, it is facing increasingly effective competition from the French.

STS-9/Spacelab 1: first lift-off for the European Space Agency's Spacelab, on 28 November 1983. Crew for the flight was a record 6 men. ***(NASA)***

ACKNOWLEDGEMENTS

Jane's Spaceflight Directory is a completely rewritten and updated version of *The Observer's Spaceflight Directory*, first published by Frederick Warne in 1978. It is the result of 26 years of first-hand space coverage by the author. The thanks of the author and publisher must go first to NASA (the US National Aeronautics and Space Administration) for the unfailing help of its officers throughout this period. The NASA news departments at Washington, Cape Canaveral, Houston, Huntsville and Pasadena are always a pleasure to visit, and their press kits and news reference books remain unequalled - except possibly by those of the US space contractors, beginning with Rockwell International, prime contractor for both Apollo and the Space Shuttle. Equally important among the major contractors are Boeing, Chrysler, Grumman, Hughes, IBM, Lockheed, Martin Marietta, McDonnell Douglas, RCA and TRW. Their information, photographs and other illustrations form the backbone of the book.

Other sources of information and photographs gratefully acknowledged include the European Space Agency, Japan's National Space Development Agency, British Aerospace, France's CNES and Aérospatiale, and other European companies named in this book.

The Royal Aircraft Establishment's *Table of Earth Satellites* is an essential and unique source of information, and thanks are due to Dr D.G. King-Hele for permission to continue reproducing its latest census of space vehicles. Equally, thanks are due to the late Dr Charles Sheldon and his successor, Ms Marcia Smith, for permission to use information and tables from their annual surveys of international space activities for the US Library of Congress. Magazines from which additional information is gleaned must as always be headed by *Aviation Week and Space Technology*; others providing a steady flow of information are *Flight International*, *Air et Cosmos*, (US) *Air Force Magazine*, *Comsat* and the British Interplanetary Society's *Spaceflight*.

In addition to contributing to the Launchers and Military sections, Robert D. Christy provided the satellite launch tables for 1982 and 1983, identifying the purpose of the satellites as far as possible. Jonathan McDowell of Cambridge University contributed to the Solar System section.

Novosti, Tass and Graham Turnill provided some Soviet pictures, and Associated Press and Popperfoto a number of other pictures. Chartwell Illustrators, C.P. Vick and D.R. Woods are responsible for the line drawings. Grateful thanks too to Margaret Turnill, whose research, typing and checking have matched the author's output for many years, and to Brendan Gallagher of Jane's for the all-important final editing.

METRIC SYSTEM

Spacemen have a disconcerting way of expressing weights and measurements indiscriminately in a mixture of metric and Imperial figures. In this book the metric system is used throughout, with only an occasional exception; one such is atmospheric pressure, which is still more familiar to most people when expressed in pounds per square inch. Thrust, nowadays expressed in newtons, is also more meaningful to most people when given alternatively in pounds rather than kilogrammes. For convenience, NASA's conversion table follows:

	Multiply	*By*	*To obtain*
Distance	feet	0·3048	metres
	metres	3·281	feet
	kilometres	3281	feet
	kilometres	0·6214	statute miles
	statute miles	1·609	kilometres
	nautical miles	1·852	kilometres
	nautical miles	1·1508	statute miles
	statute miles	0·8689	nautical miles
	statute miles	1760	yards
Velocity	feet/sec	0·3048	metres/sec
	metres/sec	3·281	feet/sec
	metres/sec	2·237	statute mph
	feet/sec	0·6818	statute miles/hr
	feet/sec	0·5925	nautical miles/hr
	statute miles/hr	1·609	km/hr
	nautical miles/hr (knots)	1·852	km/hr
	km/hr	0·6214	statute miles/hr
Liquid measure, weight	gallons	3·785	litres
	litres	0·2642	gallons
	pounds	0·4536	kilogrammes
	kilogrammes	2·205	pounds
	metric tonnes	1000	kilogrammes
	short tons	907·2	kilogrammes
Volume	cubic feet	0·02832	cubic metres
Pressure	pounds/sq in	70·31	grammes/sq cm
Thrust	pounds	4·448	newtons
	newtons	0·225	pounds
Temperature	Centigrade	1·8, add 32	Fahrenheit

A newton is the force which will accelerate 1kg of mass 1m per second per second. 1kg of force = 9·807 newtons.

SPACE LOGS

TABLE OF SATELLITE LAUNCHES 1982

Designation	Name	Date	Site	Vehicle	Perigree (km)	Apogee (km)	Period (min)	Inclination (deg)	Notes
1A	Cosmos 1331	7 Jan	PL	C-1	773	810	100·7	74·0	Military tactical communications using store/dump technique
2A	Cosmos 1332	12 Jan	PL	A-2	210	248	89·1	82·3	Photo-reconnaissance, recovered after 13 days
3A	Cosmos 1333	14 Jan	PL	C-1	970	1,016	105·0	82·9	Navigation satellite
4A	RCA Satcom 4	16 Jan	ER	Delta 3910	geosynchronous above 83°W				Commercial comsat serving US cable TV networks
5A	Cosmos 1334	23 Feb	PL	A-2	226	288	89·7	72·9	Photo-reconnaissance, recovered after 14 days
6A	OPS-2849	21 Jan	WR	Titan 3B	560	644	96·8	97·3	Military satellite, possibly reconnaissance
7A	Cosmos 1335	29 Jan	PL	C-1	479	514	94·6	74·1	Possibly electronic intelligence-gathering
8A	Cosmos 1336	30 Jan	TT	A-2	169	356	89·8	70·3	Photo-reconnaissance, re-entered after 27 days
9A	Ekran 8	5 Feb	TT	D-1-e	geosynchronous above 99°E				Television comsat serving remote communities
10A	Cosmos 1337	11 Feb	TT	F-1	428	446	93·3	65·1	Electronic intelligence-gathering over ocean areas
11A	Cosmos 1338	16 Feb	PL	A-2	357	414	92·3	72·9	Photo-reconnaissance, recovered after 14 days
12A	Cosmos 1339	17 Feb	PL	C-1	954	1,016	104·9	82·9	Navigation satellite
13A	Cosmos 1340	19 Feb	PL	A-1	628	650	97·6	81·2	Electronic intelligence-gathering
14A	Westar 4	26 Feb	ER	Delta 3910	geosynchronous above 99°W				Private comsat
15A	Molniya 1 (53)	26 Feb	PL	A-2-e	474	39,892	718·0	62·9	Comsat serving the Orbita system
16A	Cosmos 1341	3 Mar	PL	A-2-e	630	39,684	717·0	62·9	Missile early-warning satellite
17A	Intelsat 5 (F-4)	5 Mar	ER	Atlas-Centaur	geosynchronous above 63°E				Commercial comsat
18A	Cosmos 1342	5 Mar	PL	A-2	227	288	89·7	72·9	Photo-reconnaissance, recovered after 14 days
19A	OPS-8701	6 Mar	ER	Titan 3C	geosynchronous above 68°W				Missile early-warning satellite
20A	Gorizont 5	15 Mar	TT	D-1-e	geosynchronous above 54°E				Comsat serving the Moskva international system
21A	Cosmos 1343	17 Mar	PL	A-2	225	282	89·6	72·8	Photo-reconnaissance, recovered after 14 days
22A	STS-3	22 Mar	ER	*Columbia*	244	255	89·4	38·0	Third test of *Columbia*, crew Lousma and Fullerton. Landed White Sands, New Mexico, after 8 days
23A	Molniya 3 (18)	24 Mar	PL	A-2-e	627	39,768	718·6	62·8	Comsat serving the Orbita system within the USSR and abroad
24A	Cosmos 1344	24 Mar	PL	C-1	967	1,011	105·0	82·9	Navigation satellite
25A	Meteor 2 (8)	25 Mar	PL	F-1	941	961	104·1	82·5	Meteorological satellite
26A	Cosmos 1345	31 Mar	PL	C-1	504	545	95·2	74·0	Possibly electronic intelligence-gathering
27A	Cosmos 1346	31 Mar	PL	A-1	621	660	97·6	81·2	Electronic intelligence-gathering
28A	Cosmos 1347	2 Apr	TT	A-2	172	340	89·7	70·4	Photo-reconnaissance, re-entered after 50 days

1 Launch sites: **TT** Tyuratam, USSR; **PL** Plesetsk, USSR; **KY** Kapustin Yar, USSR; **ER** Kennedy Space Centre and Cape Canaveral AFB, USA; **WR** Vandenberg AFB, USA; **KO** Kourou, French Guiana; **SH** Shuang Cheng Tse, Peoples' Republic of China; **SR** Sriharikota Island, India; **TA** Tanegashima Space Centre, Japan; **KA** Kagoshima Space Centre, Japan.
2 Soviet launchers are designated under the system devised by the late Dr Charles S. Sheldon.

Designation	Name	Date	Site	Vehicle	Perigree (km)	Apogee (km)	Period (min)	Inclination (deg)	Notes
29A	Cosmos 1348	7 Apr	PL	A-2-e	596	39,751	717·6	62·9	Missile early-warning satellite
30A	Cosmos 1349	8 Apr	PL	C-1	968	1,012	105·0	82·9	Navigation satellite
31A	Insat 1A	10 Apr	ER	Delta 3910	geosynchronous above 74°E				Indian combined weather-reporting and communications satellite, failed in orbit
32A	Cosmos 1350	15 Apr	PL	A-2	171	357	89·8	67·2	Photo-reconnaissance, de-orbited after 31 days
33A	Salyut 7	19 Apr	TT	D-1	343	356	91·4	51·6	Manned space laboratory
34A	Cosmos 1351	21 Apr	KY	C-1	348	547	93·5	50·7	Possibly electronic intelligence-gathering
35A	Cosmos 1352	21 Apr	TT	A-2	349	415	92·2	70·4	Photo-reconnaissance, recovered after 14 days
36A	Cosmos 1353	23 Apr	PL	A-2	211	241	89·1	82·3	Earth-resources photography, recovered after 13 days
37A	Cosmos 1354	28 Apr	PL	C-1	794	807	100·9	74·0	Military tactical communications using store/dump technique
38A	Cosmos 1355	29 Apr	TT	F-1	428	446	93·3	65·1	Electronic intelligence-gathering over ocean areas
39A	Cosmos 1356	5 May	PL	A-1	631	669	97·8	81·2	Electronic intelligence-gathering
40A–H	Cosmos 1357–64	6 May	PL	C-1	1,402 1,473	1,479 to 1,526	114·7 116·0	74·0 74·0	Single launch of 8 satellites for short-range tactical communications
41A	OPS-5642	11 May	WR	Titan 3D	174	258	88·9	96·4	Big Bird reconnaissance satellite, de-orbited after 208 days
41C					699	705	98·9	96·0	Electronic intelligence-gathering, launched piggyback with Big Bird
42A	Soyuz T-5	13 May	TT	A-2	192 (later) 343	231 356	88·7 91·4	51·6 51·6	Carried crew of Berezovoi and Lebedev to Salyut 7. Landed after 107 days with the Soyuz T-7 crew. Berezovoi and Lebedev flew for 211 days
43A	Cosmos 1365	14 May	TT	F-1	248	265	89·7	65·0	Ocean surveillance using radar. Nuclear reactor boosted to 900km orbit after 136 days
33C	Iskra 2	17 May	—	—	336	342	91·3	51·6	Amateur radio relay satellite released from Salyut 7
44A	Cosmos 1366	17 May	TT	D-1-e	geosynchronous above 90°E				Experimental comsat
45A	Cosmos 1367	20 May	PL	A-2-e	585	39,390	717·8	62·8	Missile early-warning satellite
46A	Cosmos 1368	21 May	TT	A-2	233	260	89·5	70·4	Photo-reconnaissance, recovered after 13 days
47A	Progress 13	23 May	TT	A-2	191 (later) 338	278 343	88·9 91·3	51·6 51·6	Unmanned supply vessel serving Salyut 7, de-orbited after 14 days
48A	Cosmos 1369	25 May	PL	A-2	261	283	90·0	82·3	Earth-resources photography, recovered after 14 days
49A	Cosmos 1370	28 May	TT	A-2	197	275	89·2	64·9	Photo-reconnaissance, re-entered after 44 days
50A	Molniya 1 (54)	28 May	PL	A-2-e	631	39,732	717·9	62·8	Comsat serving the Orbita system
51A	Cosmos 1371	1 Jun	PL	C-1	791	811	100·9	74·0	Military tactical comsat using store/dump technique
52A	Cosmos 1372	1 Jun	TT	F-1	250	264	89·7	65·0	Ocean surveillance using radar. Nuclear reactor boosted to 900km orbit after 71 days

Designation	Name	Date	Site	Vehicle	Perigree (km)	Apogee (km)	Period (min)	Inclination (deg)	Notes
53A	Cosmos 1373	2 Jun	TT	A-2	365	448	92·7	70·4	Photo-reconnaissance, recovered after 14 days
54A	Cosmos 1374	5 Jun	KY	C-1	225	225	89·0	50·7	Single-orbit recovery test of unmanned scale model of Shuttle-type vehicle
55A	Cosmos 1375	6 Jun	PL	C-1	980	1,012	105·0	65·8	Target for satellite interception test
56A	Cosmos 1376	8 Jun	PL	A-2	253	278	89·9	82·4	Earth-resources photography, recovered after 14 days
57A	Cosmos 1377	8 Jun	TT	A-2	172	343	89·7	64·9	Photo-reconnaissance, re-entered after 44 days
58A	Westar 5	9 Jun	ER	Delta 3910	geosynchronous above 140°W				Private comsat
59A	Cosmos 1378	10 Jun	PL	F-1	634	663	97·8	82·5	Possibly electronic intelligence-gathering
60A	Cosmos 1379	18 Jun	TT	F-1	507	1,020	100·1	65·9	Satellite interceptor using Cosmos 1375 as a target
61A	Cosmos 1380	18 Jun	PL	C-1	140	725	93·3	82·9	Intended navigation satellite, launcher shut down early
62A	Cosmos 1381	18 Jun	TT	A-2	376	436	92·7	70·4	Photo-reconnaissance, recovered after 13 days
63A	Soyuz T-6	24 Jun	TT	A-2	189	233	88·7	51·7	Carried Dzhanibekov, Ivanchenkov and Chrétien to Salyut 7, landed after 8 days
					(later) 343	356	91·4	51·6	
64A	Cosmos 1382	25 Jun	PL	A-2-e	597	39,737	717·4	62·8	Missile early-warning satellite
65A	STS-4	27 Jun	ER	*Columbia*	295	302	90·3	28·5	Fourth flight of *Columbia*, crew Mattingly and Hartsfield, landed Edwards AFB after 7 days
66A	Cosmos 1383	29 Jun	PL	C-1	989	1,028	105·4	82·9	Navigation satellite, also carrying Cospas search and rescue package to monitor emergency radio frequencies
67A	Cosmos 1384	30 Jun	PL	A-2	170	354	89·8	67·2	Photo-reconnaissance, re-entered after 30 days
68A	Cosmos 1385	6 Jul	PL	A-2	186	236	88·8	82·3	Earth-resources photography, recovered after 14 days
69A	Cosmos 1386	7 Jul	PL	C-1	954	1,009	104·8	83·0	Navigation satellite
70A	Progress 14	10 Jul	TT	A-2	186	242	88·7	51·6	Unmanned supply vessel serving Salyut 7, de-orbited after 34 days
					(later) 301	325	90·7	51·6	
71A	Cosmos 1387	13 Jul	PL	A-2	212	234	89·1	82·3	Earth-resources photography, recovered after 13 days
72A	Landsat 4	16 Jul	WR	Delta 3920	680	700	98·6	98·3	Long-term Earth-resources satellite
73A–H	Cosmos 1388–95	21 Jul	PL	C-1	1,395	1,476	114·6	74·0	Single launch of 8 satellites for short-range tactical military communications
						to			
					1,476	1,517	115·9	74·0	
74A	Molniya 1 (55)	21 Jul	PL	A-2-e	603	39,750	717·7	63·1	Comsat serving the Orbita system
75A	Cosmos 1396	27 Jul	PL	A-2	229	292	89·8	72·9	Photo-reconnaissance, recovered after 14 days
76A	Cosmos 1397	29 Jul	KY	C-1	343	541	93·4	50·7	Possibly electronic intelligence-gathering
77A	Cosmos 1398	3 Aug	PL	A-2	216	234	89·0	82·4	Earth-resources photography, recovered after 10 days
78A	Cosmos 1399	4 Aug	TT	A-2	170	345	89·7	64·9	Photo-reconnaissance, re-entered after 43 days
79A	Cosmos 1400	5 Aug	PL	A-1	630	654	97·6	81·2	Electronic intelligence-gathering

Designation	Name	Date	Site	Vehicle	Perigree (km)	Apogee (km)	Period (min)	Inclination (deg)	Notes
80A	Soyuz T-7	19 Aug	TT	A-2	228 (later) 292	280 301	89·5 90·2	51·6 51·6	Carried Serebrov, Popov and Savitskaya to Salyut 7. Crew landed in Soyuz T-5 after 8 days. Soyuz T-7 landed after 113 days with the Soyuz T-5 crew
81A	Cosmos 1401	20 Aug	PL	A-2	261	274	89·9	82·3	Earth-resources photography, recovered after 14 days
82A	Telesat 5	26 Aug	ER	Delta 3920	geosynchronous above 104°W				Canadian comsat
83A	Molniya 3 (19)	27 Aug	PL	A-2-e	450	39,910	717·8	62·9	Comsat serving the Orbita system within the USSR and abroad
84A	Cosmos 1402	30 Aug	TT	F-1	251	264	89·6	65·0	Ocean surveillance using radar. Nuclear power pack fell back into the atmosphere in Jan 1983
85A	Cosmos 1403	1 Sep	TT	A-2	354	416	92·3	70·4	Photo-reconnaissance, recovered after 14 days
86A	Cosmos 1404	1 Sep	PL	A-2	358	414	92·3	72·9	Photo-reconnaissance, recovered after 14 days
87A	Kiku 4	3 Sep	TA	N-1	965	1,228	107·1	44·6	Japanese engineering test satellite
88A	Cosmos 1405	4 Sep	TT	F-1	430	444	93·3	65·0	Electronic intelligence-gathering over ocean areas
89A	Cosmos 1406	8 Sep	PL	A-2	212	220	88·8	82·3	Earth-resources photography, recovered after 12 days
90A	China 12	9 Sep	SH	Long March 2	175	384	90·1	63·0	Reconnaissance satellite test, recovered after 5 days
91A	Cosmos 1407	15 Sep	PL	A-2	174	340	89·6	67·2	Photo-reconnaissance, re-entered after 31 days
92A	Cosmos 1408	16 Sep	PL	F-1	635	669	97·8	82·6	Possibly electronic intelligence-gathering
93A	Ekran 9	16 Sep	TT	D-1-e	geosynchronous above 99°E				Television comsat serving remote communities
94A	Progress 15	18 Sep	TT	A-2	190 (later) 313	230 333	89·0 91·0	51·6 51·6	Unmanned supply vessel serving Salyut 7, de-orbited after 29 days
95A	Cosmos 1409	22 Sep	PL	A-2-e	612	39,688	716·7	63·1	Missile early-warning satellite
96A	Cosmos 1410	24 Sep	PL	F-1	1,495	1,502	116·0	82·6	Possibly a geodetic satellite
97A	Intelsat 5 (F-5)	28 Sep	ER	Atlas-Centaur	geosynchronous above 63°E				Commercial comsat
98A	Cosmos 1411	30 Sep	PL	A-2	357	414	92·3	72·9	Photo-reconnaissance, recovered after 14 days
99A	Cosmos 1412	2 Oct	TT	F-1	251	266	89·7	65·0	Ocean surveillance using radar. Nuclear reactor separated and boosted to 900km orbit after 39 days
100A,D,E	Cosmos 1413–15	12 Oct	TT	D-1-e	19,065	19,080	673·4	64·8	Navigation satellites. 1st Glonass launch
101A	Cosmos 1416	14 Oct	TT	A-2	231	278	89·6	70·4	Photo-reconnaissance, recovered after 14 days
102A	Cosmos 1417	19 Oct	PL	C-1	963	1,012	104·9	83·0	Navigation satellite
103A	Gorizont 6	20 Oct	TT	D-1-e	geosynchronous above 85°E				Comsat serving the Moskva international system
104A	Cosmos 1418	21 Oct	KY	C-1	370	414	92·3	50·7	Possibly electronic intelligence-gathering
105A	RCA Satcom 5	28 Oct	ER	Delta 3924	geosynchronous above 140°W				Commercial communications satellite
106A, B	DSCS	30 Oct	ER	Titan 34D	geosynchronous above 14°W and 135°W				Single launch of two military comsats

Designation	Name	Date	Site	Vehicle	Perigree (km)	Apogee (km)	Period (min)	Inclination (deg)	Notes
107A	Progress 16	31 Oct	TT	A-2	186 (later) 355	245 364	88·8 91·7	51·6 51·6	Unmanned supply vessel serving Salyut 7, de-orbited after 44 days
108A	Cosmos 1419	2 Nov	TT	A-2	228	285	89·6	70·3	Photo-reconnaissance, recovered after 14 days
109A	Cosmos 1420	11 Nov	PL	C-1	780	810	100·8	74·0	Military tactical comsat using store/dump technique
110A	STS-5	11 Nov	ER	*Columbia*	294	317	90·5	28·5	Fifth flight of *Columbia*. Carried crew of Brand, Overmyer, Allen and Lenoir, landed Edwards AFB after 5 days
110B	SBS-3	11 Nov	—	PAM	geosynchronous above 94°W				Business comsat launched from Shuttle
110C	Telesat 6	12 Nov	—	PAM	geosynchronous above 113°W				Canadian comsat launched from Shuttle
111A	OPS-9627	17 Nov	WR	Titan 3D	281	518	92·6	97·0	KH11-type reconnaissance satellite
112A	Cosmos 1421	18 Nov	TT	A-2	230	280	89·6	70·3	Photo-reconnaissance, recovered after 14 days
33D	Iskra 3	18 Nov	—	—	350	365	91·5	51·6	Amateur radio relay satellite released from Salyut 7
113A	Raduga 11	26 Nov	TT	D-1-e	geosynchronous above 35°E				Comsat serving the Orbita system
114A	Cosmos 1422	3 Dec	PL	A-2	228	287	89·7	72·9	Photo-reconnaissance, recovered after 14 days
115A	Cosmos 1423	8 Dec	TT	A-2-e	405	515	93·8	62·8	Molniya 1 comsat, last rocket stage exploded on ignition
116A	Meteor 2 (9)	14 Dec	PL	A-1	810	895	102·0	81·3	Meteorological satellite
117A	Cosmos 1424	16 Dec	TT	A-2	170	350	89·7	64·9	Photo-reconnaissance, re-entered after 43 days
118A	OPS-9845	21 Dec	WR	Thor-Burner 2	816	826	101·4	98·7	USAF meteorological satellite
119A	Cosmos 1425	23 Dec	TT	A-2	348	416	92·2	70·0	Photo-reconnaissance, recovered after 14 days
120A	Cosmos 1426	28 Dec	TT	A-2	205	340	89·9	50·5	Photo-reconnaissance, possibly carrying out parallel studies with Salyut 7. Recovered after 67 days
121A	Cosmos 1427	29 Dec	PL	C-1	443	500	94·0	65·8	Possibly carrying out electronic intelligence-gathering

TABLE OF SATELLITE LAUNCHES 1983

Designation	Name	Date	Site	Vehicle	Perigree (km)	Apogee (km)	Period (min)	Inclination (deg)	Notes
1A	Cosmos 1428	12 Jan	PL	C-1	995	1,007	104·7	82·9	Navigation satellite
2A–H	Cosmos 1429–36	19 Jan	PL	C-1	1,402 1,468	1,468 to 1,521	114·6 115·9	74·0 74·0	Single launch of 8 satellites for short-range tactical communications
3A	Cosmos 1437	20 Jan	PL	A-1	630	657	97·6	81·2	Electronic intelligence-gathering
4A	IRAS	26 Jan	WR	Delta 3910	896	913	103·1	99·1	Infra-Red Astronomy Satellite, joint venture by the US, UK and Netherlands
5A	Cosmos 1438	27 Jan	TT	A-2	175	293	89·2	70·4	Photo-reconnaissance, recovered after 11 days
6A	Sakura 2A	4 Feb	TA	N-2	geosynchronous above 130°E				Comsat, Japanese-built and launched
7A	Cosmos 1439	6 Feb	TT	A-2	160	295	89·1	70·4	Photo-reconnaissance, recovered after 16 days
8A, B, E, F and H	OPS-0252 (5 payloads)	9 Feb	WR	Atlas F	1,050	1,170	107·5	63·4	Military ocean surveillance using radar

1 Launch sites: **TT** Tyuratam, USSR; **PL** Plesetsk, USSR; **KY** Kapustin Yar, USSR; **ER** Kennedy Space Centre and Cape Canaveral AFB, USA; **WR** Vandenberg AFB, USA; **KO** Kourou, French Guiana; **SH** Shuang Cheng Tse, Peoples' Republic of China; **SR** Sriharikota Island, India; **TA** Tanegashima Space Centre, Japan; **KA** Kagoshima Space Centre, Japan.
2 Soviet launchers are designated under the system devised by the late Dr Charles S. Sheldon.

Designation	Name	Date	Site	Vehicle	Perigree (km)	Apogee (km)	Period (min)	Inclination (deg)	Notes
9A	Cosmos 1440	10 Feb	PL	A-2	260	275	89·9	82·3	Earth-resources photography, recovered after 14 days
10A	Cosmos 1441	16 Feb	PL	A-1	630	643	97·4	81·2	Electronic intelligence-gathering
11A	Tenma	20 Feb	KA	Mu 3S	488	503	94·4	31·5	Scientific satellite, carrying out X-ray observations of the sky
12A	Cosmos 1442	25 Feb	PL	A-2	169	360	89·8	67·2	Photo-reconnaissance, re-entered after 45 days
13A	Cosmos 1443	2 Mar	TT	D-1	195	235	88·7	51·6	Large module delivering supplies to and providing extra space aboard Salyut 7. Cargo-return module landed 23 Aug. Main craft de-orbited 19 Sep
14A	Cosmos 1444	2 Mar	PL	A-2	358	416	92·3	72·8	Photo-reconnaissance, recovered after 14 days
15A	Molniya 3 (20)	11 Mar	PL	A-2-e	439	39,917	717·8	62·8	Comsat serving the Orbita system, used for both domestic USSR and international communications
16A	Ekran 10	12 Mar	TT	D-1-e	geosynchronous above 99°E				Television comsat serving remote communities
17A	Cosmos 1445	15 Mar	KY	C-1	200	225	88·0	50·7	Single-orbit recovery test of unmanned scale model of a Shuttle-type vehicle
18A	Cosmos 1446	16 Mar	TT	A-2	222	241	89·1	69·9	Photo-reconnaissance, recovered after 14 days
19A	Molniya 1 (56)	16 Mar	PL	A-2-e	448	39,903	717·7	62·8	Comsat serving the Orbita system
20A	Astron	23 Mar	TT	D-1-e	2,000	200,000	5,880	51·5	Large astronomical satellite carrying ultra-violet telescope and sensors
21A	Cosmos 1447	24 Mar	PL	C-1	960	1,015	104·9	83·0	Navigation satellite, also carrying Cospas search and rescue package to monitor emergency radio frequencies
22A	NOAA-8	28 Mar	WR	Atlas F	808	830	101·3	98·8	Meteorological satellite, also carrying Sarsat package (similar to Cospas on Cosmos 1447)
23A	Cosmos 1448	30 Mar	PL	C-1	965	1,005	104·8	83·0	Navigation satellite
24A	Cosmos 1449	31 Mar	PL	A-2	356	417	92·3	72·9	Photo-reconnaissance, recovered after 15 days
25A	Molniya 1 (57)	2 Apr	TT	A-2-e	507	39,843	717·1	63·0	Comsat serving the Orbita system
26A	STS-6	4 Apr	ER	*Challenger*	280	290	90·0	28·5	First flight of *Challenger*, with crew of Weitz, Bobko, Peterson and Musgrave. Landed at Edwards AFB after 5 days
26B	TDRS-1	5 Apr	—	IUS	geosynchronous above 41°W				Launched from *Challenger*'s cargo bay. Satellite to improve spacecraft-to-ground communications
27A	Cosmos 1450	6 Apr	PL	C-1	470	515	94·4	65·9	Possibly electronic intelligence-gathering
28A	Raduga 12	8 Apr	TT	D-1-e	geosynchronous above 85°E				Comsat using Orbita system
29A	Cosmos 1451	8 Apr	PL	A-2	227	323	90·0	82·4	Photo-reconnaissance, recovered after 14 days
30A	RCA Satcom 6	11 Apr	ER	Delta 3924	geosynchronous above 128°W				Commercial comsat
31A	Cosmos 1452	12 Apr	PL	C-1	758	810	100·8	74·0	Tactical military communications using store/dump technique
32A	OPS-2925	15 Apr	WR	Titan 3B/ Agena D	135	298	88·9	96·5	Photo-reconnaissance, re-entered after 4 months

Designation	Name	Date	Site	Vehicle	Perigree (km)	Apogee (km)	Period (min)	Inclination (deg)	Notes
33A	Rohini 3	17 Apr	SR	SLV-3	388	852	97·0	46·6	Indian scientific satellite
34A	Cosmos 1453	19 Apr	PL	C-1	470	517	94·5	74·0	Possibly electronic intelligence-gathering
35A	Soyuz T-8	20 Apr	TT	A-2	180	228	88·5	51·6	Intended to carry Titov, Strekalov and Serebrov to Salyut 7. A problem with the rendezvous radar forced cancellation of the docking and a return to Earth after 2 days
36A	Cosmos 1454	22 Apr	PL	A-2	170	343	89·6	67·1	Photo-reconnaissance, re-entered after 30 days
37A	Cosmos 1455	23 Apr	PL	F-1	635	665	97·8	82·5	Possibly electronic reconnaissance
38A	Cosmos 1456	25 Apr	PL	A-2-e	620	39,170	717·3	63·0	Missile early-warning satellite
39A	Cosmos 1457	26 Apr	TT	A-2	171	350	89·7	70·4	Photo-reconnaissance, re-entered after 43 days
40A	Cosmos 1458	28 Apr	PL	A-2	212	245	89·1	82·3	Earth-resources photography, recovered after 13 days
41A	GOES-6	28 Apr	ER	Delta 3914	geosynchronous above 135°W				Meteorological satellite
42A	Cosmos 1459	6 May	PL	C-1	946	1,020	104·8	83·0	Navigation satellite
43A	Cosmos 1460	6 May	TT	A-2	350	417	92·2	70·3	Photo-reconnaissance, recovered after 14 days
44A	Cosmos 1461	7 May	TT	F-1	429	444	93·3	65·0	Electronic reconnaissance over ocean areas
45A	Cosmos 1462	17 May	PL	A-2	259	277	89·9	82·3	Earth-resources photography, recovered after 14 days
46A	Cosmos 1463	19 May	PL	C-1	300	1,550	103·6	82·9	Possibly ionospheric research
47A	Intelsat 5 (F-6)	19 May	ER	Atlas-Centaur	geosynchronous above the Atlantic Ocean				Commercial comsat, including a maritime communications package
48A	Cosmos 1464	24 May	PL	C-1	968	1,011	104·9	82·9	Navigation satellite
49A	Cosmos 1465	26 May	KY	C-1	349	543	93·4	50·7	Possibly electronic intelligence-gathering
50A	Cosmos 1466	26 May	TT	A-2	174	345	89·7	64·9	Photo-reconnaissance, re-entered after 14 days
51A	Exosat	26 May	ER	Delta 3914	356	191,581	5,431	72·5	European-built X-ray observatory
52A	Cosmos 1467	31 May	PL	A-2	357	417	92·3	72·9	Photo-reconnaissance, recovered after 12 days
53A	Venus 15	2 Jun	TT	D-1-e	—	—	—	—	Radar-carrying Venus orbiter, entered planetary orbit 10 Oct
54A	Venus 16	7 Jun	TT	D-1-e	—	—	—	—	Radar-carrying Venus orbiter, entered planetary orbit 14 Oct
55A	Cosmos 1468	7 Jun	PL	A-2	252	277	89·9	82·3	Earth-resources photography, recovered after 14 days
56A, C and D	OPS-6432 (3 payloads)	10 Jun	WR	Atlas F	1,045	1,165	107·4	63·3	Military ocean surveillance using radar
57A	Cosmos 1469	14 Jun	PL	A-2	231	342	90·29	72·8	Photo-reconnaissance, recovered after 10 days
58A	ECS-1	16 Jun	KO	Ariane (L-6)	geosynchronous above 10°E				Communications satellite, European-built and launched
58B	Oscar 10				3,919	35,531	699·3	26·4	Amateur radio relay
59A	STS-7	18 Jun	ER	*Challenger*	295	302	90·3	28·5	Second flight of *Challenger*, with crew of Crippen, Hauck, Fabian and Ride. Tested RMS, landed at Edwards AFB after 6 days
59B	Telesat 7	18 Jun	—	PAM	geosynchronous above 113°W				Canadian comsat launched from Shuttle
59C	Palapa 3	19 Jun	—	PAM	geosynchronous above 105°E				Indonesian comsat launched from Shuttle

Designation	Name	Date	Site	Vehicle	Perigree (km)	Apogee (km)	Period (min)	Inclination (deg)	Notes
59F	SPAS-01	22 Jun	—	—	295	302	90·3	28·5	Test of West German-designed satellite pallet, released from and recovered by Shuttle
60A	OPS-0721	20 Jun	WR	Titan 3D	159	259	88·8	96·5	Big Bird-type photo-reconnaissance satellite
61A	Cosmos 1470	22 Jun	PL	F-1	633	667	97·8	82·5	Possibly carrying out electronic reconnaissance
62A	Soyuz T-9	27 Jun	TT	A-2	197 (later) 324	227 337	88·8 91·2	51·6 51·6	Carried crew of Lyakhov and Alexandrov to the Salyut 7/Cosmos 1443 complex. Spacecraft and crew recovered after 150 days
63A	Hilat	27 Jun	WR	Scout	771	836	101·0	82·0	Modified Transit-type satellite researching the ionosphere at high latitudes
64A	Cosmos 1471	28 Jun	PL	A-2	122	345	89·7	67·1	Photo-reconnaissance, re-entered after 30 days
65A	Galaxy 1	28 Jun	ER	Delta 3920	geosynchronous above 134°W				Commercial comsat offering channels for lease or purchase
66A	Gorizont 7	30 Jun	TT	D-1-e	geosynchronous above 13°W				Comsat serving the Moskva international system
67A	Prognoz 9	1 Jul	TT	A-2-e	380	720,000	26·7 days	65·5	Scientific satellite exploring the magnetosphere and deep space
68A	Cosmos 1472	5 Jul	PL	A-2	336	360	91·6	82·4	Earth-resources photography, recovered after 14 days
69A–H	Cosmos 1473–1480	6 Jul	PL	C-1	1,394 1,463	1,463 to 1,520	114·5 115·8	74·0 74·0	Single launch of 8 satellites for short-range tactical communications
70A	Cosmos 1481	8 Jul	PL	A-2-e	642	39,199	707·4	62·9	Missile early-warning satellite. Desired 718min orbit not achieved
71A	Cosmos 1482	13 Jul	TT	A-2	351	411	92·2	70·0	Photo-reconnaissance, recovered after 14 days
72A	Navstar 8	14 Jul	WR	Atlas F	19,921	20,435	717·8	62·8	Navigation satellite
73A	Molniya 1 (58)	19 Jul	TT	A-2-e	457	39,882	717·5	62·9	Communications satellite serving the Orbita system
74A	Cosmos 1483	20 Jul	PL	A-2	258	273	89·9	82·3	Earth-resources photography, recovered after 14 days
75A	Cosmos 1484	24 Jul	TT	A-1	592	661	97·3	98·0	Earth-resources satellite
76A	Cosmos 1485	26 Jul	PL	A-2	356	414	92·3	72·9	Photo-reconnaissance, recovered after 14 days
77A	Telstar 3A	28 Jul	ER	Delta 3920	geosynchronous above 66°W				Communications satellite
78A	SDS-8	31 Jul	WR	Titan 3B/ Agena D	500	39,000	718	63	Military communications in north polar regions
79A	Cosmos 1486	3 Aug	PL	C-1	783	804	100·8	74·1	Military tactical communications using store/dump technique
80A	Cosmos 1487	5 Aug	PL	A-2	258	273	89·9	82·3	Earth-resources photography, recovered after 14 days
81A	Sakura 2B	5 Aug	TA	N-2	geosynchronous above 135°E				Japanese communications satellite
82A	Cosmos 1488	9 Aug	PL	A-2	355	415	92·3	72·9	Photo-reconnaissance, recovered after 14 days
83A	Cosmos 1489	10 Aug	TT	A-2	175	301	89·3	64·7	Photo-reconnaissance, recovered after 14 days
84A–C	Cosmos 1490–1492	10 Aug	TT	D-1-e	19,103	19,156	675·7	64·9	Navigation satellites. 2nd Glonass launch

Designation	*Name*	*Date*	*Site*	*Vehicle*	*Perigree (km)*	*Apogee (km)*	*Period (min)*	*Inclination (deg)*	*Notes*
85A	Progress 17	17 Aug	TT	A-2	189 (later) 334	241 355	88·8 91·4	51·6 51·6	Unmanned supply vessel serving Salyut 7, de-orbited after 32 days
86A	China 13	19 Aug	SH	Long March 2	171	388	90·1	63·3	Possibly a test of satellite recovery technique, though no recovery made
87A	Cosmos 1493	23 Aug	PL	A-2	358	412	92·3	72·8	Photo-reconnaissance, recovered after 14 days
88A	Raduga 13	25 Aug	TT	D-1-e	geosynchronous above 46°E				Comsat serving the Orbita system
89A	STS-8	30 Aug	ER	*Columbia*	295	302	90·3	28·5	Sixth flight of *Columbia*. Carried crew of Truly, Brandenstein, Thornton, Bluford and Gardner, landed at Edwards AFB after 6 days
89B	Insat 1B	31 Aug		PAM	geosynchronous above 46°E				Indian combined weather-reporting and communications satellite
90A	Molniya 3 (21)	30 Aug	PL	A-2-e	426	39,953	718·3	62·9	Comsat serving the Orbita system within the USSR and abroad
91A	Cosmos 1494	31 Aug	KY	C-1	345	550	93·5	50·7	Possibly electronic intelligence-gathering
92A	Cosmos 1495	3 Sep	PL	A-2	213	235	89·0	82·3	Earth-resources photography, recovered after 13 days
93A	Cosmos 1496	7 Sep	TT	A-2	168	340	89·6	67·2	Photo-reconnaissance, re-entered after 42 days
94A	RCA Satcom 7	8 Sep	ER	Delta 3924	geosynchronous above 72°W				Commercial communications satellite
95A	Cosmos 1497	9 Sep	PL	A-2	356	413	92·3	72·9	Photo-reconnaissance, recovered after 14 days
96A	Cosmos 1498	14 Sep	PL	A-2	259	272	89·9	82·3	Earth-resources photography, recovered after 14 days
97A	Cosmos 1499	17 Sep	PL	A-2	356	414	92·3	72·9	Photo-reconnaissance, recovered after 14 days
98A	Galaxy 2	22 Sep	ER	Delta 3920	geosynchronous above 74°W				Commercial comsat offering channels for lease or purchase
99A	Cosmos 1500	28 Sep	PL	F-1	633	665	97·8	82·5	Specialised ocean-survey satellite
100A	Ekran 11	29 Sep	TT	D-1-e	geosynchronous above 99°E				Television comsat serving remote communities
101A	Cosmos 1501	30 Sep	PL	C-1	466	512	94·5	82·9	Possibly electronic intelligence-gathering
102A	Cosmos 1502	5 Oct	PL	C-1	368	411	92·4	65·9	Possibly electronic intelligence-gathering
103A	Cosmos 1503	12 Oct	PL	C-1	788	808	100·9	74·1	Military tactical comsat using store/dump technique
104A	Cosmos 1504	14 Oct	TT	A-2	171	306	89·3	64·9	Photo-reconnaissance, re-entered after 53 days
105A	Intelsat 5 (F-7)	19 Oct	KO	Ariane (L-7)	geosynchronous above 16°E				Commercial comsat, including maritime communications package
106A	Progress 18	20 Oct	TT	A-2	193 (later) 328	242 345	88·8 91·3	51·6 51·6	Unmanned supply vessel serving Salyut 7, de-orbited after 27 days
107A	Cosmos 1505	21 Oct	PL	A-2	356	414	92·3	72·9	Photo-reconnaissance, recovered after 14 days
108A	Cosmos 1506	26 Oct	PL	C-1	950	1,014	104·8	83·0	Navigation satellite
109A	Meteor 2 (10)	28 Oct	PL	A-1	752	888	101·4	81·2	Meteorological satellite
110A	Cosmos 1507	29 Oct	TT	F-1	433	442	93·4	65·1	Electronic intelligence-gathering over ocean areas
111A	Cosmos 1508	11 Nov	PL	C-1	397	1,964	109·1	82·9	Possibly ionospheric research
112A	Cosmos 1509	17 Nov	PL	A-2	225	290	89·7	72·9	Photo-reconnaissance, recovered after 14 days
113A	OPS-1294	18 Nov	WR	Atlas F	814	831	101·4	98·7	USAF meteorological satellite

Designation	Name	Date	Site	Vehicle	Perigree (km)	Apogee (km)	Period (min)	Inclination (deg)	Notes
114A	Molniya 1 (59)	18 Nov	PL	A-2-e	464	39,894	717·9	62·8	Comsat serving the Orbita system
115A	Cosmos 1510	24 Nov	PL	C-1	1,479	1,524	116·0	73·6	Possibly a geodetic satellite
116A	STS-9	28 Nov	ER	*Columbia*	239	251	89·4	57·0	First Spacelab mission. Carried crew of Young, Shaw, Garriott, Parker, Lichtenberg and Merbold, landed Edwards AFB after 10 days
117A	Cosmos 1511	30 Nov	PL	A-2	167	338	89·6	67·1	Photo-reconnaissance, recovered after 44 days
118A	Gorizont 8	30 Nov	TT	D-1-e	geosynchronous above 85°E				Comsat serving the Moskva international system
119A	Cosmos 1512	7 Dec	PL	A-2	355	416	92·3	72·9	Photo-reconnaissance, recovered after 14 days
120A	Cosmos 1513	8 Dec	PL	C-1	961	1,016	104·9	82·9	Navigation satellite
121A	Cosmos 1514	14 Dec	PL	A-2	214	259	89·3	82·3	Biological research satellite under the Intercosmos programme and including US experiments, recovered after 5 days
122A	Cosmos 1515	15 Dec	PL	F-1	636	663	97·7	82·5	Possibly electronic intelligence-gathering
123A	Molniya 3 (22)	21 Dec	PL	A-2-e	616	39,748	718·0	62·8	Comsat serving the Orbita system within the USSR and abroad
124A	Cosmos 1516	27 Dec	TT	A-2	205	270	89·3	64·9	Photo-reconaissance, re-entered after 44 days
125A	Cosmos 1517	27 Dec	KY	C-1	181	221	88·5	50·7	Single-orbit recovery test of unmanned scale model of shuttle-type vehicle
126A	Cosmos 1518	28 Dec	PL	A-2-e	586	39,695	716·3	63·0	Missile early-warning satellite
127A-C	Cosmos 1519-21	29 Dec	TT	D-1-e	19,015	19,139	673·6	64·8	Navigation satellites for ships and aircraft; 3rd Glonass launch

US MANNED FLIGHTS

Spacecraft	Launch date	Crew	Revolutions	Duration (days, hr, min)	Details
Mercury 3	05.05.61	Alan Shepard	Sub-orbital	00.00.15	1st American in space
Mercury 4	21.07.61	Virgil Grissom	Sub-orbital	00.00.16	Capsule sank
Mercury 6	20.02.62	John Glenn	3	00.04.55	1st American in orbit
Mercury 7	24.05.62	M. Scott Carpenter	3	00.04.56	Landed 402km from target
Mercury 8	03.10.62	Walter Schirra	6	00.09.13	Landed 8km from target
Mercury 9	15.05.63	L. Gordon Cooper	22	01.10.20	1st long flight by an American
Gemini 3	23.03.65	Virgil Grissom John Young	3	00.04.53	1st manned orbital manoeuvres
Gemini 4	03.06.65	James McDivitt Edward White	62	04.01.56	21min spacewalk (White)
Gemini 5	21.08.65	L. Gordon Cooper Charles Conrad	120	07.22.56	1st extended manned flight
Gemini 7	04.12.65	Frank Borman James Lovell	206	13.18.35	Longest US flight for 8 years
Gemini 6	15.12.65	Walter Schirra Thomas Stafford	16	01.01.51	Rendezvous to within 1·8m of Gemini 7
Gemini 8	16.03.66	Neil Armstrong David Scott	6	00.10.41	1st docking; emergency splashdown
Gemini 9	03.06.66	Thomas Stafford Eugene Cernan	45	03.00.21	2hr spacewalk (Cernan)

Spacecraft	*Launch date*	*Crew*	*Revolutions*	*Duration (days, hr, min)*	*Details*
Gemini 10	18.07.66	John Young Michael Collins	43	02.22.47	Rendezvous with 2 targets; Agena package retrieved
Gemini 11	12.09.66	Charles Conrad Richard Gordon	44	02.23.17	Rendezvous and docking
Gemini 12	11.11.66	James Lovell Edwin Aldrin	59	03.22.34	Dockings; 3 spacewalks
Apollo 7	11.10.68	Walter Schirra Donn Eisele Walter Cunningham	163	10.20.09	1st manned Apollo flight
Apollo 8	21.12.68	Frank Borman James Lovell William Anders	10 lunar revs	06.03.00	1st manned flight around Moon
Apollo 9	03.03.69	James McDivitt David Scott Russell Schweickart	151	10.01.01	Docking with Lunar Module
Apollo 10	18.05.69	Thomas Stafford Eugene Cernan John Young	31 lunar revs	08.00.03	Descent to within 14km of Moon
Apollo 11	16.07.69	Neil Armstrong Edward Aldrin Michael Collins	31 CML revs	08.03.18	Armstrong and Aldrin land on Moon; 20kg samples
Apollo 12	14.11.69	Charles Conrad Richard Gordon Alan Bean	49 CML revs	10.04.36	2 EVAs, total 7hr 46min; 34kg samples
Apollo 13	11.04.70	James Lovell John Swigert Fred Haise	—	05.22.55	Mission aborted following oxygen tank explosion
Apollo 14	31.01.71	Alan Shepard Stuart Roosa Edgar Mitchell	34 CML revs	09.00.42	2 EVAs, total 9hr 25min; 44kg samples
Apollo 15	26.07.71	David Scott James Irwin Alfred Worden	74 CML revs	12.07.12	3 EVAs, total 19hr 08min; 78kg samples
Apollo 16	16.04.72	John Young Thomas Mattingly Charles Duke	64 CML revs	11.01.51	3 EVAs, total 20hr 14min; 97·5kg samples
Apollo 17	06.12.72	Eugene Cernan Ronald Evans Harrison Schmitt	75 CML revs	12.13.51	3 EVAs, total 22hr 04min; 113kg samples
Skylab 1	14.05.73	Unmanned	34,981	—	Vibration damage during lift-off. Re-entered 11.07.79
Skylab 2	25.05.73	Charles Conrad Joseph Kerwin Paul Weitz	404	28.00.50	Exceeded Soviet duration record. EVAs repaired damage
Skylab 3	28.07.73	Alan Bean Owen Garriott Jack Lousma	858	59.11.09	Rescue mission prepared but not needed
Skylab 4	16.11.73	Gerald Carr Edward Gibson William Pogue	1,214	84.01.15	Total Skylab EVAs 82hr 47min
ASTP	15.07.73	Thomas Stafford Vance Brand Donald Slayton	138	09.01.28	1st US–Soviet joint flight; 2 days' docked activities

CML revs = Command Module lunar revolutions. **Note** The above manned flights totalled 189,985,890km (118,077,569 miles) of space travel.

US MANNED FLIGHTS: SPACE SHUTTLE (as at Jun 1984)

Flight	*Launch date Orbiter*	*Crew*	*Duration*	*Details*
STS-1	12.04.81 102	Young, Crippen	02.06.21	Near-perfect flight
STS-2	12.11.81 102	Engle, Truly	02.06.13	5-day mission halved by fuel cell fault
STS-3	22.03.82 102	Lousma, Fullerton	08.00.05	Extra day because of storm at Northrup landing strip
STS-4	27.07.82 102	Mattingly, Hartsfield	07.01.10	1st concrete landing. SRBs lost, DoD package failed
STS-5	11.11.82 102	Brand, Overmyer, Allen, Lenoir	05.02.14	1st operational flight; SBS-3, Anik C3 deployed; EVA failed
STS-6	05.04.83 099	Weitz, Bobko, Peterson, Musgrave	05.00.23	Delayed 3 months by engine leaks, etc. TDRS-1, 1st EVA
STS-7	18.06.83 099	Crippen, Hauck, Fabian, Ride	06.02.24	1st US woman; 1st satellite retrieval; Anik C2, Palapa B
STS-8	30.08.83 099	Truly, Brandenstein, Bluford, Gardner, Thornton	06.00.07	1st night launch/landing; Insat B

Note Cancellations of STS-10 and 12, and other delays, led to new flight designation system: 1st figure = FY; 2nd fig = KSC (1) or VAFB (2); final letter = sequence.

Flight	*Launch date Orbiter*	*Crew*	*Duration*	*Details*
STS-9 41-A	28.11.83 102	Young, Shaw, Garriott, Parker, Lichtenberg (PS), Merbold (PS)	10.07.47	Spacelab 1; 3yr late. Science 90% successful

Flight No Designation	*Launch date Orbiter*	*Crew*	*Duration*	*Details*
10 41-B	03.02.84 099	Brand, Gibson, McCandless, Stewart, McNair	07.23.17	1st MMU flights. Westar 6 and Palapa B2 lost. SMM rehearsal
11 41-C	04.04.84 099	Crippen, Scobee, Nelson, van Hoften, Hart	06.23.40	LDEF. SMM repair. 1st orbital retrieval and repair
12 41-D	Jul 84 103	Hartsfield, Coats, Resnik, Hawley, Mullane, C. Walker	07	OAST-1, LFC-1, Syncom 4-1. 1st commercial PS 2 launch failures
41-E	Cancelled 099	Mattingly, Shriver, Onizuka, Buchli, USAF PS	—	2nd cancellation due to IUS failure
13 41-F	29.08.84 103	Bobko, Williams, Seddon, Hoffman, Griggs	07	Spartan 1, SBS-4, Telstar 3C, Syncom 4-2
14 41-G	01.10.84 099	Crippen, McBride, Sullivan, Ride, Leestma, Canadian PS	08	OSTA-3, ERBS
41-H	Nov 84 103	Hauck, D. Walker, A. Fisher, Gardner, Allen, C. Walker	?	Possible Palapa B2 rescue by Allen and Gardner. Would take over 51-A payload
15 51-A	02.11.84 103	Brandenstein, Creighton, Lucid, Fabian, Nagel	06	MSL-1, Telesat 8. Palapa rescue alternative
16 51-C	09.12.84 099	?	?	DoD
17 51-B	17.01.85 103	Overmyer, Gregory, Lind, Thagard, Thornton, Berg, Wang	07	Spacelab 3
18 51-E	12.02.85 099	Engle, Covey, Buchli, Lounge, W. Fisher, Baudry (France)	05	Telesat 9, TDRS-2, Echograph (France)
19 51-D	18.03.85 103	Shaw, O'Connor, Cleave, Spring, Ross, PS (Hughes)	06	Syncom 4-3. LDEF retrieval
21 51-F	17.04.85 099	Fullerton, Griggs, Musgrave, England, Henize, Johnston, Trinh	07	Spacelab 2. 1st woman PS
22 51-G	30.05.85 102	5	07	EASE/Access, Telstar 3D, Morelos A, Arabsat A

Flight No *Designation*	*Launch date* *Orbiter*	*Crew*	*Duration*	*Details*
23 51-L	02.07.85 102	6	07	EOS-1, TDRS-3
24 51-I	08.08.85 099	5	07	MSL-2, Syncom 4-4, Aussat 1, ASC-1
25 51-J	09.09.85	?	?	DoD
26 61-A	14.10.85 102	7	07	Spacelab D1 (Germany)
27 62-A	15.10.85 103	?	?	DoD 86-1V. 1st Vandenberg launch
28 61-B	06.11.85 099	5	07	Morelos B, Satcom Ku-1, Aussat 2, MSI-3
29 51-H	27.11.85 104	Brand, Smith, Springer, Garriott, Lampton, Lichtenberg, Nicollier	07	EOM-1, Spacelab reflight. 1st *Atlantis* flight
30 61-C	23.12.85 102	5	07	Spartan 2, Westar 7, Satcom Ku-2, GStar 3
31 61-D	22.01.86 099	Cdr, Fabian, Bagian, Seddon, PSs	07	Spacelab 4
32 61-E	06.03.86 102	Cdr, Plt, Parker, Leestma, Hoffman	07	Intelsat 6-1, Astro 1
33 61-F	15.05.86 099	4	02	ISPM
34 61-G	21.05.86 104	4	02	Galileo
35 61-H	27.06.86 102	5	07	EOS-2, Insat 1C, Skynet 4B, STC DBS-1
36 61-I	16.07.86 099	5	07	Intelsat 6-2, MSL-4, Sunlab 1
37 61-J	13.08.86 104		03	Space Telescope (could be earlier)
38 61-K	21.08.86 102	?	?	DoD
39 61-L	12.09.86 099	?	?	Payload opportunity
40 62-B	29.09.86 104	?	?	DoD
41 71-A	01.10.86 102	4	07	MSL-5, ASC-2, STC DBS-2, DoD PAM-1
42 71-B	23.10.86 102	4	07	Satcom Ku-3, CRRES, Payload opportunity
43 71-C	06.11.86 099	?	?	DoD
44 71-D	26.11.86 102	6	07	Astro 2, DoD PAM-2, Usat 1
45 71-E	17.12.86 102	4	07	EOM-2, DoD PAM-3, Galaxy Ku-2
46 72-A	02.01.87 103	5	07	Landsat 4 repair
47 71-F	14.01.87 099	4	07	OAST-2, MSL-6, Intelsat 6-3
48 71-G	28.01.87 104	4	07	Payload opportunity
49 71-H	Feb 87 102	4	05	STC DBS-3 Payload opportunity
50 71-I	Mar 87 099	4	07	MSL-7, Usat 2, DoD PAM-4, Orion 1
51 71-J	Mar 87 104	4	07	C2 Spacelines
52 71-K	Apr 87 102	4	07	DoD PAM-5, Satcom Ku-4, Unisat 1
53 72-B	Apr 87 103	4	07	OSTA-5, Payload opportunity
54 71-L	May 87 099	4	07	Spartan 3, Satcom 1, STC DBS-4, Payload opportunity

Flight No Designation	*Launch date Orbiter*	*Crew*	*Duration*	*Details*
55 71-M	May 87 104	6	07	IML-1 (Spacelab 8)
56 71-N	Jun 87 102	4	07	MSL-8, CFMF-1, Intelsat 6-4
57 71-O	Jul 87 099	4	07	Unisat 2, SBS-6, DoD PAM-6, Spacenet 3
58 71-P	Jul 87 104	?	?	DoD
59 71-Q	Jul 87 102	6	07	Astro 3, Orion 2, DoD PAM-7
60 71-R	Aug 87 099	4	07	OAST-3, LDEF-2, DoD PAM-7
61 71-S	Aug 87 104	4	07	Sunlab 2, MSL-9, Rosat
62 71-T	Sep 87 102	?	?	DoD
63 82-A	Oct 87 103	?	?	DoD
64 81-A	Oct 87 099	4	07	Eureca, STS DBS-5, DoD PAM-8, RCA DBS-4
65 81-B	Oct 87 104	4	07	MSL-10, EUVE, Intelsat 6-5
66 81-C	Nov 87 102	4	07	EOM-3, ASC-3, DoD PAM-9, Spacenet 4
67 81-D	Nov 87 099	?	?	DoD
68 82-B	Dec 87 103	4	07	OSTA-7, COBE
69 81-E	Dec 87 104	4	07	TSS-1, Lageos 2, DoD PAM-10, Unisat 3
70 81-F	Jan 88 102	4	07	CFMF-2, DBS Lux 1, DoD PAM-11, Fordsat 1
71 81-G	Jan 88 099	6	07	Spacelab J
72 81-H	Feb 88 104	4	07	MSL-11, Westar 8, DoD PAM-12, Fordsat 2
73 81-I	Mar 88 102	5	07	SHEAL-1, Orion 3, Usat 4, Westar A
74 81-J	Mar 88 099	?	?	DoD
75 81-K	Apr 88 104	4	03	VRM
76 81-L	May 88 102	4	07	GRO, Eureca retrieval, RCA DBS-5
77 81-M	May 88 099	4	07	CFMF-3, Orion 4, Telesat 11, DoD PAM-13
78 81-N	Jun 88 104	?	?	DoD
79 82-C	Jun 88 103	4	07	SP Plasma 1, Payload opportunity
80 81-O	Jun 88 102	6	07	LS Lab 3
81 81-P	Jul 88 099	4	07	Telstar 3B, Aussat 3, DoD PAM-14, MSL-12
82 81-Q	Jul 88 104	4	07	DoD PAM-15, Fordsat 3, USSB-1
83 81-R	Aug 88 102	4	07	Leasecraft 101, Italsat 1
84 81-S	Aug 88 099	4	07	MSL-13, DoD PAM-16, Westar B, RCA DBS-2
85 82-D	Sep 88 103	?	?	DoD

Shuttle Orbiters: 099 *Challenger*, 102 *Columbia*, 103 *Discovery*, 104 *Atlantis*. **Crew composition:** 1st name is cdr; 2nd is plt; 3rd, 4th and 5th MS; subsequent PS. **Payloads:** Flights 41-D to 71-G are carrying payloads for commercial customers who have made progress payments, budgeted NASA programmes and signed DoD missions. Flights 71-H to 91-T list payloads for commercial customers who have made "earnest" but no progress payments yet, budgeted NASA programmes and signed DoD missions.

SOVIET MANNED FLIGHTS

Spacecraft	Launch date	Crew	Revolutions	Duration (days, hr, min)	Details
Vostok 1	12.04.61	Yuri Gagarin	1	00.01.48	1st man in space
Vostok 2	06.08.61	Gherman Titov	17	01.01.18	1st day in space
Vostok 3	11.08.62	Andrian Nikolayev	64	03.22.27	1st double flight
Vostok 4	12.08.62	Pavel Popovich	48	02.22.29	
Vostok 5	14.06.63	Valeri Bykovsky	81	04.23.06	2nd double flight
Vostok 6	16.06.63	Valentina Tereshkova	48	02.22.50	1st woman in space
Voskhod 1	12.10.64	Vladimir Komarov Konstantin Feoktistov Boris Yegorov	16	01.00.17	1st 3-man craft
Voskhod 2	18.03.65	Pavel Belyayev Alexei Leonov	17	01.02.02	1st spacewalk (10min), by Leonov
Soyuz 1	23.04.67	Vladimir Komarov	18	01.02.45	Re-entry parachute snarled, Komarov killed
Soyuz 3	26.10.68	Georgi Beregovoi	64	03.22.51	Rendezvous practice with unmanned Soyuz 2
Soyuz 4	14.01.69	Vladimir Shatalov	48	02.23.14	1st docking of 2 manned craft; Khrunov and Yeliseyev returned in Soyuz 4
Soyuz 5	15.01.69	Boris Volynov Yevgeni Khrunov Alexei Yeliseyev	50	03.00.46	
Soyuz 6	11.10.69	Georgi Shonin Valeri Kubasov	80	04.22.42	
Soyuz 7	12.10.69	Anatoli Filipchenko Vladislav Volkov Viktor Gorbatko	80	04.22.41	30 manual control manoeuvres, 1st space welding
Soyuz 8	13.10.69	Vladimir Shatalov Alexei Yeliseyev	80	04.22.41	
Soyuz 9	01.06.70	Andrian Nikolayev Vitali Sevastyanov	285	17.16.59	Endurance record
Salyut 1	19.04.71	Orbital science station	2,800		Destroyed 11 Oct 1971
Souyz 10	23.04.71	Vladimir Shatalov Alexei Yeliseyev Nikolai Rukavishnikov	30	02.00.45	Docked with Salyut but no entry made
Soyuz 11	06.06.71	Georgi Dobrovolsky Vladislav Volkov Viktor Patsayev	380	23.17.40	23 days spent in Salyut; crew killed on re-entry
Soyuz 12	27.09.73	Vasili Lazarev Oleg Makarov	32	02.47.16	Tested post-disaster modifications
Soyuz 13	18.12.73	Pyotr Klimuk Valentin Lebedev	128	07.20.55	Soviet and US spacemen in orbit together for 1st time
Salyut 3	25.06.74	Unmanned	3,430		Destroyed 24 Jan 1975
Soyuz 14	03.07.74	Pavel Popovich Yuri Artyukhin	252	15.17.30	Docked with Salyut 3
Soyuz 15	26.08.74	Gennadi Sarafanov Lev Demin	32	02.00.12	Failed to dock with Salyut 3
Soyuz 16	2.12.74	Anatoli Filipchenko Nikolai Rukavishnikov	96	05.22.24	ASTP rehearsal
Salyut 4	26.12.74	Unmanned	12,000		Occupied by 2 crews for 93 days; destroyed 3 Feb 1977
Soyuz 17	11.01.75	Alexei Gubarev Georgi Grechko	467	29.14.40	New Soviet duration record in Salyut 4
Soyuz 18A	05.04.75	Vasili Lazarev Oleg Makarov	—	—	1st abort between launch and insertion
Soyuz 18B	24.05.75	Pyotr Klimuk Vitali Sevastyanov	993	62.23.20	Soviet duration record increased
Soyuz 19 (ASTP)	15.07.75	Alexei Leonov Valeri Kubasov	96	05.22.31	1st US/USSR joint flight
Soyuz 20	17.11.75	Unmanned			Resupply rehearsal to Salyut 4
Salyut 5	22.06.76	Unmanned	6,630		Occupied by 2 crews for 65 days. Destroyed 8 Aug 1977
Soyuz 21	06.07.76	Boris Volynov Vitali Zholobov	789	49.05.24	48 days in Salyut 5
Soyuz 22	15.09.76	Valeri Bykovsky Vladimir Aksyonov	127	07.21.54	Probably 1st manned spy satellite
Soyuz 23	14.10.76	Vyacheslav Zudov Valeri Rozhdestvensky	32	02.00.06	1st Soviet splashdown after emergency return

Spacecraft	Launch date	Crew	Revolutions	Duration (days, hr, min)	Details
Soyuz 24	07.02.77	Viktor Gorbatko Yuri Glazkov	286	17.16.08	Docked with Salyut 5
Salyut 6	29.09.77	Unmanned			
Soyuz 25	09.10.77	Vladimir Kovalyonok Valeri Ryumin	32	02.00.46	Failed to dock with Salyut 6
Soyuz 26/27	10.12.77	Yuri Romanenko Georgi Grechko	1,520	96.10.00	New world duration record
Soyuz 27/26	10.01.78	Vladimir Dzhanibekov Oleg Makarov	96	06.00.04	1st double docking
Soyuz 28	02.03.78	Alexei Gubarev Vladimir Remek (Czech)	124	07.22.17	1st international flight
Soyuz 29/31	15.06.78	Vladimir Kovalyonok Alexander Ivanchenkov	2,233	139.14.18	1st 100+ days
Soyuz 30	27.06.78	Pyotr Klimuk Miroslaw Hermaszewski	126	07.22.04	1st Pole
Soyuz 31/29	26.08.78	Valeri Bykovsky Sigmund Jahn	125	07.20.49	1st East German
Soyuz 32/34	25.02.79	Vladimir Lyakhov Valeri Ryumin	2,800	175.00.36	Duration record
Soyuz 33	10.04.79	Nikolai Rukavishnikov Georgi Ivanov	31	01.23.01	Failed to dock with Salyut 6; 1st Bulgarian
Soyuz 34	06.06.79	Unmanned	1,180	73.18.24	Used for S32 crew return
Soyuz T-1	16.12.79	Unmanned	1,606	100.09.20	Docked with Salyut 6
Soyuz 35/37	09.04.80	Leonid Popov Valeri Ryumin	2,957	184.19.02	Duration record
Soyuz 36/35	25.05.80	Valeri Kubasov Bertalan Farkas	126	07.20.46	1st Hungarian
Soyuz T-2	05.06.80	Yuri Malyshev Vladimir Aksyonov	62	03.22.41	1st manned T flight
Soyuz 37/36	23.07.80	Viktor Gorbatko Pham Tuan	126	07.21.42	1st Vietnamese
Soyuz 38	18.09.80	Yuri Romanenko Arnaldo Tamayo Mendez	125	07.20.34	1st Cuban/black
Soyuz T-3	27.11.80	Leonid Kizim Oleg Makarov Gennadi Strekalov	205	12.19.08	Resumption of 3-crew flights
Soyuz T-4	13.03.81	Vladimir Kovalyonok Viktor Savinykh	1,196	74.18.43	50th Soviet/ 100th spaceman
Soyuz 39	22.03.81	Vladimir Dzhanibekov Jugderdemidiyn Gurragcha	126	07.20.43	1st Mongolian
Soyuz 40	14.05.81	Leonid Popov Dumitru Prunariu	125	07.20.38	1st Romanian
Salyut 7	19.04.82	Unmanned			
Soyuz T-5/T-7	13.05.82	Anatoli Berezovoi Valentin Lebedev	3,381	211.08.05	1st mission to Salyut 7
Soyuz T-6	24.06.82	Vladimir Dzhanibekov Alexander Ivanchenkov Jean-Loup Chrétien	126	07.21.51	1st Frenchman
Soyuz T-7	19.08.82	Leonid Popov Alexander Serebrov Svetlana Savitskaya	126	07.21.52	2nd spacewoman
Soyuz T-8	20.04.83	Vladimir Titov Gennadi Strekalov Alexander Serebrov	32	02.00.18	Failed to dock with Salyut 7
Soyuz T-9	27.06.83	Vladimir Lyakhov Alexander Alexandrov	2,390	149.09.46	2 EVAs added 2 solar panels
Soyuz T-10A	27.09.83	Vladimir Titov Gennadi Strekalov	—	—	Launchpad fire, 1st use of escape tower
Soyuz T-10B	08.02.84	Leonid Kizim Vladimir Solovyev Oleg Atkov		150+	1st long-stay triple crew. 5 EVAs: 17hr 50min record total
Soyuz T-11	03.04.84	Yuri Malyshev Gennadi Strekalov Rakesh Sharma	126	07.21.51	1st Indian. 6 crew aboard Salyut 7

Note Earth revolution figures are approximate. A revolution is 6min longer than an orbit, and there is frequent confusion between the two. USSR totals are based on official figures when available, and estimates when they are not. NASA's own official figures tend to vary; the US has been credited with two Earth revolutions for each Apollo Moon flight.

MAJOR UNMANNED FLIGHTS

Date	Spacecraft	Achievement
04.10.57	Sputnik 1	1st artificial satellite
31.01.58	Explorer 1	Discovered Van Allen radiation belts
17.03.58	Vanguard 1	Plotted Earth's shape
02.01.59	Luna 1	Earth-escape spacecraft
07.08.59	Explorer 6	TV pictures from space
12.09.59	Luna 2	Lunar impact
04.10.59	Luna 3	Lunar far-side picture
01.04.60	Tiros 1	Weather satellite
24.05.60	Midas 2	Missile detection satellite
22.06.60	Transit/Solrad	1st multiple payload
10.08.60	Discoverer 13	Payload recovered
12.02.61	Sputnik 8	Orbital platform launch
12.02.61	Venus 1	Venus fly-by launched by Sputnik 8
27.08.62	Mariner 2	Data from Venus
01.11.62	Mars 1	Mars fly-by
26.07.63	Syncom 2	Synchronous satellite
17.10.63	Vela 1	Satellite to detect nuclear explosions
28.07.64	Ranger 7	1st close-up Moon pictures
28.11.64	Mariner 4	Mars pictures
03.04.65	Snapshot 1	Nuclear reactor in orbit
06.04.65	Early Bird (Intelsat 1)	Commercial TV communications
16.07.65	Proton 1	Cosmic ray measurements
16.11.65	Venus 3	Venus impact
31.01.66	Luna 9	Soft-landed on Moon; pictures from surface
31.03.66	Luna 10	Lunar orbiter
10.10.66	Lunar Orbiter 1	Pictures from lunar orbit
21.12.66	Luna 13	Tested density of lunar surface
25.01.67	Cosmos 139	Fractional orbit bombardment satellite (FOBS)
17.04.67	Surveyor 3	Dug lunar trench
12.06.67	Venus 4	Investigated Venusian atmosphere
01.07.67	Dodge 1	1st full-face colour picture of Earth
08.09.67	Surveyor 5	Chemical analysis of lunar soil
30.10.67	Cosmos 186/188	Automatic docking
21.09.68	Zond 5	Circumlunar flight and recovery
20.10.68	Cosmos 249	Inspection/destruction satellite
10.11.68	Zond 6	Skip re-entry after lunar flight
17.08.70	Venus 7	Soft-landed on Venus
12.09.70	Luna 16	Automatic return of lunar soil
10.11.70	Luna 17	Lunar robot
19.05.71	Mars 2	Impacted on Mars
28.05.71	Mars 3	Soft-landed on Mars
30.05.71	Mariner 9	Orbited Mars
03.03.72	Pioneer 10	Launched to Jupiter; 300 pictures of Jupiter, Ganymede, Callisto and Europa
27.03.72	Venus 8	Analysed Venusian soil
23.07.72	Landsat 1	1st Earth-resources satellite
08.06.75	Venus 9	Venus surface pictures
20.08.75	Viking 1	Mars surface pictures
05.09.77	Voyager 1	Jupiter/Saturn moon pictures
20.01.78	Progress 1	Robot fuel transfer
27.06.78	Seasat 1	Sea resources satellite
12.08.78	ISEE-3	Libration halo orbit
02.06.83	Venera 15	1st high-resolution Venusian surface pictures

Bruce McCandless, his feet anchored in the foot restraint, is moved around on the end of the RMS, being operated by Terry Hart from the flight deck. The empty cradle for the lost Westar 6 satellite can be seen in the payload bay. *(NASA)*

NATIONAL SPACE PROGRAMMES

ARGENTINA

Argentina was due to decide in 1984 on a proposal to launch two Argensat satellites into stationary orbits to supply 300 Earth stations with telephone, TV, educational and medical services. Cost is estimated at about $240 million, with launch by Shuttle or Ariane in 1987/88.

AUSTRALIA

History Australia's direct involvement in space activities inevitably diminished with the run-down of the Woomera rocket range north of Adelaide, which was set up jointly with Britain at the end of the war to test military missiles and the Blue Streak, Black Knight and Black Arrow launchers. The provision of the Woomera facilities enabled Australia to become a member of ELDO, the now defunct European space organisation. But when that was replaced by ESA launch activities were transferred to the more suitable equatorial site at Kourou, French Guiana. Australia has however continued to provide vital facilities for NASA's Deep Space Network; the 64m and 34m antennas already in use at Tidbinbilla are being joined by a second 34m antenna (replacing the former 26m unit at Honeysuckle), with a fourth to be installed later. The Orroral and Yarragadee stations in Western and Eastern Australia provided critical support for the OMS burns in the early stages of the Shuttle programme. But the long-term future of such stations was threatened by the development of the TDRS system, which enables the US to operate with far fewer overseas stations.

A government-owned company called Aussat has ordered three Hughes Type 376 satellites costing a total of about $180 million to improve domestic communications and supply radio and TV to sparsely populated areas. There will be two in orbit and one ground spare, and they will work in the 12-14GHz band. Having orginally reserved launch positions on both Ariane and Shuttle, Australia announced in Oct 1982 that the spacecraft would be launched by Shuttle in Jul and Oct 1985. It was subsequently revealed that an Australian payload specialist would fly on one of the missions, making Australia the first country to take up this NASA offer to major payload sponsors.

The space logs credit Australia with the two small test satellites detailed below:

Wresat L 29 Nov 1967 by Sparta from Woomera. Wt 45kg. Orbit 106 × 777km. Incl 83°. Australia's first satellite was launched by a modified American Redstone rocket with additional 2nd and 3rd-stage solid-propellant motors. Wresat (from Weapons Research Establishment, Australia, which developed it) transmitted data for 5 days and re-entered after 42 days.

Oscar 5 L 23 Jan 1970 by Thor-Delta from Vandenberg. Wt 18kg. Launched piggyback on ITOS-1 into a 1,435 × 1,481km orbit with 102° incl, this was the 5th amateur radio satellite and transmitted for 46 days. Orbital life 10,000yr (see US - Oscar).

BRAZIL

History Brazil took the lead among developing nations in making use of satellites for both economic and educational purposes. Pending completion of a long-term programme to develop her own launch system, she is planning a pair of domestic comsats, SBTS-1 and 2, based on a $130 million proposal from Spar/Hughes, for launch by Ariane in Feb and Aug 1985. In the 1960s two separate institutes were created: INPE, the Institute for Space Research, to handle civilian space research; and IAE, the Institute for Space Activities, responsible for rocketry development. Their headquarters are at Sao Jose dos Campos in central Brazil. Since Apr 1973 that has been the site of a Landsat Earth receiving station, and INPE's director has said that for internal development "Landsat is a better form of aid than dollars." Its data are being used to prepare maps of Brazil showing forest areas, geological structures, inland lakes and waterways. Settlers can also be watched via Landsat pictures to ensure that they clear only the agreed one third of their areas for agricultural use, preserving the remainder as forest. Landsat data have been used to identify potential breeding areas of a snail known to be the carrier of an intestinal disease which affects 10% of the Brazilian population.

Following an experiment with NASA's ATS-6 involving educational TV relays to 500 schools and 15,000 students in north-eastern Brazil in 1975, INPE has an ambitious plan to use geostationary satellites to provide a standard curriculum for the 24 million pupils in Brazil's primary schools.

IAE has produced three sounding rockets. Sonda 1, a two-stage solid-propellant rocket, has been used more than 200 times and carries 4·5kg payloads to altitudes of 100km. The single-stage Sonda 2, used more than 30 times, can carry 62kg to 180km. Sonda 3, first launched in 1976, uses Sonda 2 as its second stage and can carry 50kg to 500km. Further developments are intended to lead to a three-stage rocket capable of use as a satellite launcher. Launch centre is near Natal (see Space Centres).

BRITAIN

History With orders worth over £300 million at the end of 1983, British Aerospace Dynamics could claim to be the largest manufacturer of communications satellites outside the US and USSR. BAe is currently responsible for five ECS, two Marecs, and Olympus (L-Sat) for ESA, as well as building the UK's £70 million Skynet 4. The Government's financial contributions to Intelsat, Inmarsat and ESA had thus secured a reasonable outlet for the scientific talent which in the early space years had tended to drain away to Canada and the US, where it played a major role in both manned and unmanned projects.

But traditional political anxieties as to whether costs outweigh the possible merits of a space programme persist. A proposal to create a national space agency to match that of France was shelved yet again in 1981, and by the end of 1983 BAe's hopes that its Spacelab pallet would be used as the basis of Eureca, ESA's projected free-flying scientific platform, had been demolished by Germany, whose 60% investment in Spacelab (compared with Britain's 6%) inevitably won the day. Worse still,

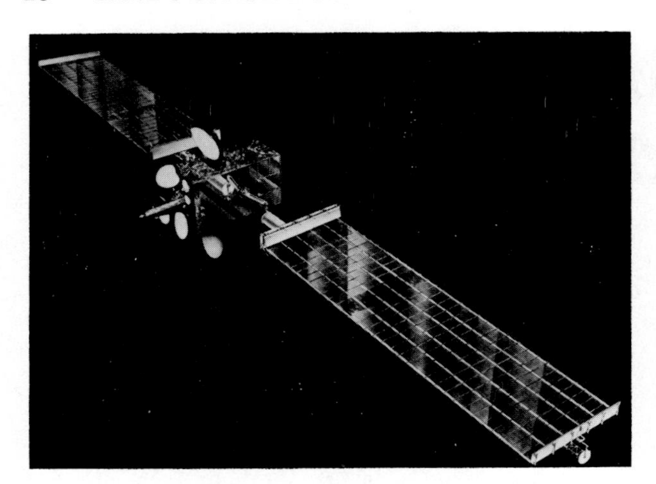

Olympus, a large multi-purpose telecommunications platform, for which ESA has appointed British Aerospace and Matra as prime contractors. A market for 150 Olympus is forecast. *(ESA)*

Britain's Black Arrow launcher lifts off from Woomera with the Prospero test satellite in October 1971 on its one and only flight into orbit. *(RAE)*

British indifference to space activities is not confined to the highest levels: a national schools competition seeking ideas for sponsored scientific experiments to be flown as Getaway Specials in the Shuttle had to be abandoned for lack of interest.

Apart from communications satellites, Britain's biggest recent space production programme covered the Spacelab pallet systems. All 20 of the pallets ordered by ESA in return for Britain's 6% contribution to Spacelab have now been delivered; they brought BAe a £420,000 contract in Nov 1982 for the cradle system for Hughes' Leasat, five of which will be launched by Shuttle for US Navy use. This partnership provides the main prospect for the immediate future, with BAe and Hughes bidding for a £500 million Inmarsat contract.

BAe is prime contractor for the £34 million Giotto mission to Halley's Comet, as well as contributing to ISPM and ERS-1. In response to a forecast that a world remote-sensing market worth £250 million a year would develop by 1990 (see US - Landsat), Britain is investing £45 million during 1983-6 in a joint science-industry programme covering data-acquisition, processing, dissemination, interpretation and forecasting. The RAE National Remote Sensing Centre at Farnborough, established in 1980, already has much experience in handling Landsat data via Kiruna, Sweden. This continued research is intended to maintain Britain's competitive position in the face of French plans with Spot. At the same time, however, Britain gave the company Spot Imaging its first distribution rights within the UK in 1983.

Following up her success with the Ariel series, Britain provided the ground control for the three-nation Infra-red Astronomical Telescope, L 25 Jan 1983, which detected another solar system and five comets during its 10-month life (see International - IRAS).

Britain has launched and orbited one completely national satellite and paid the US to launch 12 more, nine of which were British-built. But Britain's first nationally launched satellite was also her last. Like the orginal US and Soviet programmes, the British space effort depended initially on the conversion of a ballistic missile into a launcher. This was the Hawker Siddeley Blue Streak, finally cancelled as a missile in 1960 though continued until 1973 as a launcher first stage for the European ELDO programme. A more modest attempt at a national launch vehicle was announced in Sep 1964. It was decided to develop a small three-stage vehicle from the successful Westland Black Knight research rocket. Construction and supply of three launchers, to be called Black Arrow, with the first two stages directly evolved from Black Knight, was ordered in Mar 1967. The purpose was to develop and test in space new components for communications satellites, and to develop a tool for space research. The successful accomplishment of those objectives in Oct 1971 was soured by the fact that in Jul the Government had announced that the programme would be cancelled after the final launch. But with this one satellite Britain became the sixth nation to achieve orbital capability - after Russia, America, France, Japan and China.

By early 1984 NASA, anxious to stimulate British participation in the Space Station, had agreed to carry a UK payload specialist on each of the Skynet 4 Shuttle launches due in Dec 1985 and Jun 1986. There was also some possibility that a British scientist-astronaut could be fitted in even earlier. The 2 selected to fly will be chosen from 4 candidates (see Spacemen section) who are being given 6 months' training at an "expenses only" cost to Britain of about $150,000 each.

National launches

Black Arrow/Prospero Test flights and the final launch result were as follows:

X-1 L 27 Jun 1969 from Woomera, Aus. Veered off course, and had to be destroyed 50sec after lift-off. Intended to provide data on 1st and 2nd-stage performance, with separation of dummy 3rd stage.

R-1 L 4 Mar 1970. Completely successful test of 1st and 2nd-stage engines, and of 3rd-stage systems (propulsion test not included), and spin-up of dummy payload. 2nd stage and payload impacted as planned, 15min after launch, in Indian Ocean, 3,050km NW of Woomera.

R-2 L Sep 1970. Intended to prove 3rd-stage propulsion system and inject into orbit X-2 development payload. 1st objective achieved but orbit not reached due to failure to maintain 2nd-stage (HTP) tank pressurisation.

R-3 Prospero L 28 Oct 1971 by Black Arrow from Woomera. Wt 66kg. Orbit 547 × 1,582km. Incl 82°. Design aim of 1yr life more than achieved; 812 passes were monitored from total of 4,960 orbits in first year, and on-board tape recorder replayed 697 times. It failed on 24 May 1973 after 730 replays, but after 10yr in orbit the Lasham, England, telemetry station was still able to reactivate it periodically to acquire real-time data. Orbital life 150yr.

US launches

Ariel The first international satellite programme evolving from a 1959 NASA offer to launch foreign payloads when such experiments were of mutual scientific interest.

Ariel 1 L 26 Apr 1962 by Delta from C Canaveral (orbit 389 × 1,214km, wt 60kg, incl 54°), transmitted ionospheric and X-ray data until Nov 1964 and had a 15yr life.

Ariel 2 L 27 Mar 1964 by Scout from Wallops I (orbit 285 × 1,362km, wt 68kg, incl 52°), also returned data until Nov 1964 and re-entered Dec 18 1967.

Ariel 3 L 5 May 1967 by Scout from Vandenberg (orbit 498 × 606km, wt 90kg, incl 80°), was first all-British satellite and measured radio noise from lightning and from galactic sources at frequencies too low to penetrate the atmosphere. Spin-stabilised, it was very successful, operating for over 2yr; re-entered in Dec 1970.

Ariel 4 L 11 Dec 1971 by Scout from Vandenberg (orbit 485 × 599km, wt 100kg, incl 83°), continued electron density, radio noise and other experiments until Apr 1973, when it was decided enough data had been obtained. It was then switched on occasionally to obtain data in support of sounding rocket experiments. Re-entered 12 Dec 1978.

Ariel 5 L 15 Oct 1974 by Scout from San Marco (orbit 504 × 549km, wt 129kg, incl 3°), carried 6 experiments to study X-ray sources, and at the end of its 2nd year British experimenters announced that a new sort of star, called a "rapid burster," had been identified. Working with the Dutch ANS satellite and America's SAS-1, they had identified about 12 rapid bursters on the opposite side of the galaxy. These sources are a type of black hole which emits bursts of X-rays lasting around 10sec at intervals of several hours. At that point Ariel 5 had discovered 60 new sources of X-ray signals.

After 5yr of successful operation Ariel 5 re-entered on 14 Mar 1980.

Ariel 6 L 2 Jun 1979 by Scout from Wallops I. Wt 154kg. Orbit 600 × 654km. Incl 55°. Built by Marconi. By the time it was shut down in Feb 1982, having run out of attitude-control gas, Ariel 6 had studied 30 X-ray objects in detail, with excellent data from several black hole candidates. Its low-energy X-ray telescopes showed that Cygnus X2 contained a white dwarf. Observations of ultra-heavy cosmic ray particles showed a striking overabundance of elements with charges between 58 and 72. Orbital life 20yr.

Skynet Military satellites designed to provide exclusive and highly secure voice, telegraph and facsimile links for both strategic and tactical communications between UK military headquarters and ships and bases in Middle and Far East. Drum-shaped cylinders 137cm dia × 81cm high, they are spin-stabilised but with a despun antenna. They had an unsatisfactory early history.

Skynet 1 L 21 Nov 1969 from C Kennedy by Delta (wt 422kg), was successfully placed in geosynchronous orbit over the Indian Ocean but probably operated for less than a year.

Skynet 1B L 19 Aug 1970 by Thor-Delta from C Kennedy (wt 285kg), had to be abandoned in its transfer orbit of 270 × 36,058km due to the failure of the apogee motor. Orbital life unknown.

Skynet 2A L 19 Jan 1974 by Thor-Delta from C Canaveral (wt 422kg), was built by Marconi in England and was the first communications satellite built outside the US or USSR. 3 times as powerful as Skynet 1s. A launch failure placed it in a tumbling 120 × 1,857km orbit and it re-entered after 6 days.

Skynet 2B L 23 Nov 1974 by Delta from C Canaveral, was successfully placed in synchronous orbit over the Indian Ocean. With a life of at least 3yr, it provided communications from the UK to W Australia and was probably of limited use during the 1982 Falklands War. The original Skynet system included 17 ground stations, with 12·8m master terminal at Oakhanger, air-transportable terminals for use on land, and 1m shipborne terminals. By the time the system was operational successive defence cuts and British withdrawals from the Far and Near East had made it largely unnecessary. Its cost was never disclosed.

Skynet 4 The need to provide the Royal Navy and other UK forces with secure and exclusive communications was recognised in 1981 with the award to BAe of a £70 million contract for 2 1,250kg spacecraft capable of relaying voice, data and telex to and from naval vessels and fixed and mobile land terminals. With communications equipment provided by Marconi, they brought to 10 the production line based on OTS. Canadian Astronautics of Ottawa is designing and building the UHF antenna. Skynet 4A will be launched in late 1985 and Skynet 4B in 1986, both by Shuttle despite strong French pressure to use Ariane. Determining factors in the choice of launcher were British hopes that Skynet 4 would be adopted as the standard NATO system, and a NASA offer to carry a British payload specialist when the launches are carried out.

Skynet 2B, Britain's 2nd-generation military communications satellite, placed over the Indian Ocean in 1974, is seen here on top of its Delta launcher. *(NASA)*

X series The only satellite in this projected series, X-4 (re-named Miranda in orbit, in the Shakespearian tradition of the UK series), was launched 9 Mar 1974 by Scout from Vandenberg. Wt 93kg. Orbit 703 × 918km. Incl 97·81°, which made it near Sun-synchronous. British-built by Hawker Siddeley for the Dept of Trade and Industry, its main purpose was to test a new type of 3-axis attitude control as an alternative to the spin-stabilised systems used on previous UK spacecraft. Despite minor control problems, good data were said to have been obtained. Orbital life 150yr. Total cost was over £5 million, plus about £1·5 million launch costs.

Uosat (Oscar 9) L 6 Oct 1981 by Delta 2310 from Vandenberg. Wt 60kg. Orbit 554km polar, Sun-synchronous. Launched with NASA's SME, Uosat was designed and constructed by the University of Surrey in conjunction with AMSAT (the international Amateur Radio Satellite Corporation). Intended to encourage radio amateurs and educational establishments to become involved in space alongside professional scientists, it carries instruments intended to increase understanding of radio wave propagation from the HF to microwave regions. Its HF, phase-related beacons transmit on 7, 14, 21 and 29MHz. Other instruments include an Earth-imaging experiment and a speech synthesiser for schools experiments. Tumbling and other technical problems took over a year to sort out, but by Feb 1983 it was possible to pick up the voice synthesiser and it was hoped that the TV capability would also become operational. Orbital life 5yr.

Uosat 2 Due for L Feb 1984 with Landsat D-Prime, this 2nd University of Surrey amateur satellite was to include a small TV camera to relay Earth and auroral display pictures to home and school receivers from a 560km polar orbit at the upper edge of the ionosphere. The project was subject to NASA approval.

Skylark Originally developed by the Royal Aircraft Establishment and British Aircraft Corporation for the 1957/8 International Geophysical Year, Skylark sounding rockets continue to provide an invaluable space research tool with their 6min of microgravity experiment time. They cost about £100,000 each, with an average length of 9m and 1,850kg launch wt. The original altitude of 160km with 45kg payload had been increased by the 3-stage Skylark 12 series to 135kg to 800km and 200kg to lower altitudes. The 250th launch, carried out in 1976 from Woomera, was an unsuccessful joint effort with NASA to produce an energy-level map of X-radiation from the exploded Puppis-A supernova. The total of 386 launches by the end of 1982 included 81 for ESRO/ESA, and many successful Earth-resources investigations for Germany, Argentina and other countries. Since 1981 a continuing series has been launched from Kiruna, Sweden, with 350kg payloads under Germany's Texus programme in preparation for future Spacelab materials-processing activities. The Swedish Space Corporation also participates in these experiments, which concern metallurgy, optical glasses, ceramics, fluid dynamics, and physics and chemistry generally.

UKS Britain is providing a 74kg, 1m-dia sub-satellite as a contribution to AMPTE (see US - Explorers). A joint effort with the US and Germany to study how energy carried by the solar wind is intercepted and stored in the magnetic fields that form the Earth's radiation belts and other parts of the comet-shaped magnetosphere out to 100,000km, AMPTE is a 3-satellite mission due for Thor-Delta launch in Aug 1984. The Rutherford-Appleton Laboratory will collect data for 4hr on each 44hr orbit for about 1yr.

Unisat BAe, Marconi and British Telecom have formed Satcom International to provide a national broadcasting and telecommunications system with 2 enhanced ECS satellites called Unisat. Each has a lift-off wt of 1,380kg and offers 3,000 equivalent voice circuits or 2 TV channels, and 4 repeaters for mixed business services in TDMA. Cost was estimated at £150 million with 1 ground spare. The British Government has approved 2 BBC satellite TV channels, for which the BBC would pay Satcom about £12 million per year. Britain's full 5-channel DBS entitlement would require a larger craft such as Olympus. Proposals include a transatlantic service, which would have to be negotiated with Intelsat. By Dec 1983 the BBC was having second thoughts about the viability of the project, on which some £50 million had already been spent.

CANADA

History Canada's total of 12 satellites by 1983, all launched by NASA, had played a major role in pioneering domestic satellite communications. Telesat Canada, jointly owned by the government and telephone and telegraph companies, became the world's first operator of domestic satellites with the launch of Anik A1 in 1972. With Anik C2, launched by STS-7 in 1983, Telesat Canada pioneered five-channel TV direct to US subscribers' homes in Nov 1983. Some 20 years of close collaboration with NASA, and the development at a cost of $C100 million of the Remote Manipulator System for the Space Shuttle, have made Canada the largest national contributor to the Space Transportation System outside the US. In return, in Apr 1980 NASA ordered three more RMS sets for the next three Orbiters at a cost of $63 million. Canada has been an observer member of ESA since 1975, and in 1978 signed a co-operation agreement under which Canada will pay 1% of the net fixed common costs of ESA's general budget. During 1981-85 Canada is spending a planned total of $C476 million on space activities, including six Anik launches: five by NASA (one by Delta in 1982 at $25 million; three by Shuttle in 1983–4 at $10 million each; one by Shuttle in 1985 at $19 million), and one by Ariane 3 in 1985. Aniks C1-3 were bought from Hughes under a 1978 contract worth $53·6 million plus $13·7 million incentive awards.

NASA has recognised Canada's contributions to the Shuttle by providing 3 flights for payload specialists. The 1st will fly on Mission 51-A, due in Nov 1984, to supervise the launch of Anik C1 and a Getaway Special experiment designed by 2 Canadian students. The 2nd will fly in Nov 1985 to test a space vision system aimed at increasing the usefulness of the RMS. The 3rd will take part in spacesickness research in 1966, covering the disorientation experienced when waking from sleep in zero g. The National Research Council of Canada has reduced 4,300 astronaut applicants to a team of 6.

Canada's immensely successful robot arm is seen here at work during STS-7. Note the TV camera on the elbow. Picture was taken from Germany's SPAS-01, which the arm had placed in free flight above *Challenger*. (NASA)

Alouette Canada's first satellite, Alouette 1, L 29 Sep 1962 by Thor-Agena B from Vandenberg into a 997 × 1,026km orbit with 80° incl, was the first satellite to return useful data - on the ionosphere - for a period exceeding 6yr. It has a 2,000yr life. Alouette 2, L 29 Nov 1965 (orbit 505 × 2,987km; incl 80°; orbital life 500yr), was the first in a series of three craft in the joint Canadian-NASA International Satellites for Ionospheric Studies (ISIS) programme. This also transmitted useful data for over 6yr.

ISIS A continuation of Alouette studies, ISIS-1, L 30 Jan 1969 by TAID (Thor-Delta) from Vandenberg into a 578 × 3,526km orbit inclined at 88°, weighed 241kg. It carried eight Canadian and four US experiments to measure the daily and seasonal fluctuations in the electron density of the upper atmosphere, study radio and cosmic noise emissions, and measure energetic particles interacting with the ionosphere. Orbital life 250yr. The studies were continued by ISIS-2, L 1 Apr 1971 by Thor-Delta from Vandenberg into a 1,358 × 1,429km orbit inclined at 88°. Wt 264kg. Orbital life 8,000yr.

Anik Named "brother" in the Eskimo language, and also known as Telesat since it is operated by Telesat Canada from its satellite control centre in Ottawa, this series established the world's first domestic commercial communications system.

Anik A With launch wt of 562kg and in-orbit wt of 295kg, these spacecraft were spin-stabilised with despun antenna and feeds, and had a 7yr station-keeping capability. Each could handle 1 colour TV channel or 960 1-way phone links. They started commercial communications over Canada's sprawling 10 million sq km, with some circuits being leased to US carriers for private voice, data and video transmissions to Alaska. Anik A1, L 9 Nov 1972 by Thor-Delta from C Canaveral and placed at 114°W, retired Jul 1982; Anik A2, L 20 Apr 1973 and placed at 109°W, retired Oct 1982; Anik A3, L 7 May 1975 and placed at 119°W, upgraded to handle 10 colour TV channels or 9,600 phone circuits.

Anik B With 7yr design life, this followed up CTS in the 12-14GHz band with experiments in education, telemedicine, conferencing and Eskimo broadcasting. B1, L 16 Dec 1978 by Delta from C Canaveral (474kg in-orbit wt), replaced A2.

Anik C and D These spacecraft are being used to establish a 5-satellite system providing all Canada with telephone links, live coverage of Parliament, and cable and educational TV. Cylindrical and spin-stabilised, the Cs are operating at 14 and 12GHz. They offer 16 channels each, equivalent to 32 TV programmes or 21,504 voice links. In-orbit wt 567kg. Life 8-10yr. Spot-beam antenna systems focus radio energy into 4 regional patterns covering most of Canada from east to west. C3 and C2 L (in reverse order) by STS-5 and STS-7 on 11 Nov 1982 and 18 Jun 1983. Placed at 117·5°W and 112·5°W. Leased facilities from C2 were expected to enable United Satellite Communications Inc of New York to introduce 5-channel TV direct to domestic dish antennas in Nov 1983. C1 to follow in 1984. The 2 Ds are based on the Hughes 376 with 24 channels and 620kg in-orbit wt. Anik D1 (also known as Telesat 5), L 26 Aug 1982 by Delta/PAM from C Canaveral (660kg in-orbit wt), placed at 104°W.

CTS The Communications Technology Satellite (L 17 Jan 1976 by Delta from C Canaveral; launch wt 675kg; synchronous orbit at 116°W) was a joint US/Canadian project designed to prove that powerful satellites could bring low-cost TV to remote areas anywhere on the globe. An advance on ATS-6, CTS was Canadian-designed and built, and equipped with a TRW transmitter operating for the first time in the 12-14GHz band and at 10 times the power of previous comsats. ESA contributed its 16·5m solar sails. In 3½yr of operations 160 US experiments ranged from business teleconferences and meetings between Congressional members and their constituents several states away to emergency use in the Johnstown flood disaster of 1977, when all commercial lines were out of service.

In 1978 pictures and sound from a UN conference in Buenos Aires were relayed to UN HQ at New York, where simultaneous translation into 5 languages was carried out and fed back into the conferees' earphones. CTS proved that new frequency bands could be successfully exploited.

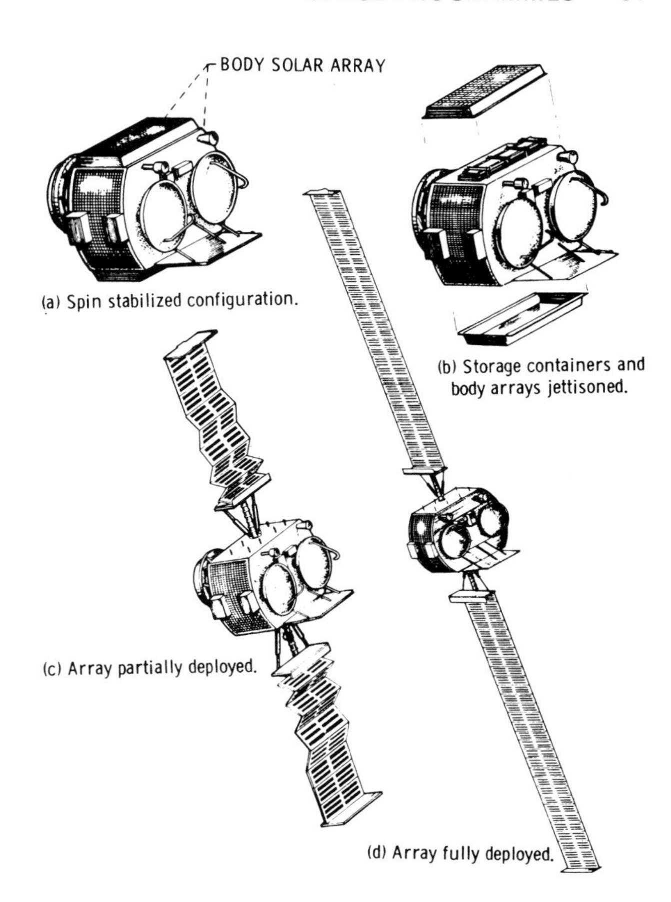

Actuation sequence of the CTS extendible solar array.

RMS The Remote Manipulator System, developed by Spar Aerospace Ltd of Toronto, is Canada's $100 million contribution to the US Space Shuttle. It consists of a 15m-long mechanical arm with joints similar to the human shoulder, elbow and wrist. Operated by an astronaut from the portside crew station of the Orbiter flight deck, it will be used to deploy from the cargo bay satellites of up to 29,480kg wt, 18·28m length and 4·57m dia, and to retrieve already orbiting satellites for onboard servicing or return to Earth. With the largest spacecraft there will be a clearance of only 7·6cm on each side. First RMS was delivered in time for STS-2 in Oct 1981; a further 3 systems were to be delivered for installation on the rest of the Orbiter fleet. A second arm may be installed on the opposite side of the cargo bay for more complicated operations such as space station assembly. Canada's National Research Council is funding the system partly because the technology developed for it is applicable to Arctic and ocean mineral and petroleum exploration, and to operations with the heavy-water type of atomic reactors used in Canada. At the time of STS-2 Spar management was concerned that the 1st RMS could be used for only 2-10 flights without danger of failure, as against the specified 100-mission life. They regarded this figure as unrealistic, and considered that NASA had set hardware and software specifications more stringent than for most other Shuttle systems because the US agency was not itself paying for RMS. Development started in Aug 1974.

CHINA

History China's space scientists admit to slow progress since launching their first satellite in Apr 1970. By the end of 1983 only 13 had been placed in orbit, many delays having been caused by political and economic changes. But a quantum leap forward came in early 1984 with the successful use of a Chinese launcher to place a Chinese-built TV satellite (China 15) in synchronous orbit. Preparations for this had included a 1983 agreement with Italy to use that country's Sirio 1 to help China to develop geostationary communications technology, and a Mar

1984 $20 million contract to Canada to supply and equip 26 Earth stations. Despite this national success, however, in May 1984 the Chinese Broadcasting Satellite Corporation invited bids from US and European companies for 3 direct-broadcast satellites, 2 for orbit and 1 as a ground spare. Each is intended to cover the whole of China with 1 high-power channel. Reservations, each covered by a non-refundable $100,000 deposit, have been made with Arianespace for 1987 and 1988 launches, and back-up reservations on the US Shuttle were also expected. But plans to put men in space have been postponed indefinitely. In Jun 1981 Prof Shen Yuan, president of the Beijing Institute of Aeronautics and Astronautics, told the author that "no serious training" was being given to any Chinese astronauts. While a comprehensive space industry was taking shape, "China's aerospace technology is still rather backward compared with the advanced countries of the world." During scientific exchanges with the US, Japan and ESA countries in recent years it has become clear that China's space scientists think their most urgent need is for defence, Earth-resources, communications and broadcast satellites.

Development of China's space boosters and long-range military rockets is largely credited to Dr Tsien Wei-Ch'ang, who was expelled from the US in 1949 as a result of the late Senator Joseph McCarthy's anti-communist campaigns. Tsien had taken a PhD at Toronto University and then worked as a research engineer at the Jet Propulsion Laboratory, Pasadena, until ordered to return to mainland China. In Nov 1972 he revisited the US as leader of a team of Chinese scientists.

China 1 L 24 Apr 1970. Wt 173kg. Orbit 441 × 2,386km. Incl 68·4°. China's first satellite made her the 5th space country. It carried a transmitter which broadcast "The East is Red," a Chinese song paying tribute to Chairman Mao, and announced the times it passed over various parts of the world. Probably spheroid, with 1m dia, it transmitted until Jun 1970; orbital life 100yr.

China 2 L 3 Mar 1971. Wt 221kg. Orbit 265 × 1,825km. Incl 69·9°. Powered by solar cells, transmitted data for 12 days before the batteries apparently failed. Orbital life 8yr.

China 3 L 26 Jul 1975. Wt 3,500kg. Orbit 184 × 461km. Incl 69°. During its 50-day life this craft passed regularly over Russia, the US and Eastern Europe. No details of weight, capability or duration of transmissions were given. Decayed 14 Sep 1975.

China 4 L 26 Nov 1975. Wt 3,500kg. Orbit 173 × 483km. Incl 63°. Part of this satellite was apparently successfully recovered on 2 Dec after 7 days in orbit. This was a major step forward for China's space programme, indicating the ability of her rockets to place craft in precise orbits from which recovery within Chinese territory is possible. The satellite's recovery was front-paged on Chinese newspapers alongside reports of the first meeting in Peking between President Ford and Chairman Mao. It was not however revealed whether recovery was made by aircraft snatch during descent, on land or at sea, nor whether the contents were undamaged. The remainder of the satellite continued to transmit from a 175 × 399km orbit until re-entry on 29 Dec 1975.

China 5 L 16 Dec 1975. Wt 3,500kg. Orbit 186 × 387km. Incl 69°. With a similar orbit to China 3, this test vehicle decayed after 42 days.

China 6 L 30 Aug 1976. Wt ? 270kg. Orbit 195 × 2,145km. Incl 69°. Similar to earlier small satellites, with an orbital life of 6yr. Decayed after 817 days.

China 7 L 7 Dec 1976. Wt ? 3,600kg. Orbit 172 × 479km. Incl 59°. Recovered about 10 Dec, leaving a residual object, possibly the booster, weighing 1,200kg in orbit for a further 23 days. A fragment which decayed after only 2 days might have been a spacecraft.

China 8 L 26 Jan 1978. Wt 3,600kg. Orbit 169 × 479kg. Incl 57°. Recovered, possibly with dog inside, after 12 days.

China 9-11 L 19 Sep 1981 by CSL-2. 1st multiple launch, placing triple payload in 235 × 1,600km orbits with 59·5° incl. Described as space physics satellites. The 1st was a balloon (? 1·2m dia) + sphere with 7 days' life. 2nd and 3rd were a truncated cone and octagonal prism + 4 vanes; 382 and 332 days' life respectively.

China 12 L 9 Sep 1982 by Long March 2 from Shuang Cheng Tse. Wt ? 3,600kg. Orbit 172 × 393km. Incl 63°. Remote-sensing satellite. Re-entry capsule recovered 5 days later, suggesting reconnaissance satellite test. Main body re-entered after 12 days.

China 13 L 19 Aug 1983. Orbit 170 × 400km. Incl 63°. Re-entry vehicle recovered 24 Aug; possible operational reconnaissance flight.

China 14 L 29 Jan 1984 by CZ-3 from new launch site in Szechuan Province initially into 309 × 451km orbit, then after 12 revs boosted to 359 × 6,479km at 36° incl. First thought to be unsuccessful attempt at achieving synchronous orbit; later believed to be successful 1st test of Long March 3 launcher.

Future projects The Chinese Academy of Sciences undertook to pay the US $200,000 a year from 1982 to receive Landsat images at a ground station being built near Beijing. By mid-1981 aerial photography had covered only 5% of China's 9·6 million sq km, and the Chinese Communications Satellite Corp had proposed to buy a US communications satellite for service in 1981/2; a similar date was originally given for their own experimental comsat. The Central Meteorological Bureau, responsible for both civil and military systems, had also hoped for a polar-orbiting, Sun-synchronous, 3-axis-stabilised metsat at 900km altitude in 1982, followed by a geostationary metsat in 1985. China's economic problems led to the postponement of the US buy and postponement of her own launches, but there is no reason to doubt that the projects are being continued. In 1982 China again gave economic problems as the reason for suspending plans for Germany's MBB to develop a Chinese TV system by the late 1980s.

CZECHOSLOVAKIA

History Czechoslovakia has contributed instruments and experiments to every one of the Soviet Union's Intercosmos satellites, and to many others as well. Under the agreement providing for Intercosmos member countries to participate in manned spaceflights, Vladimir Remek, a Czech military pilot, became the first man outside America or Russia to go into space. As cosmonaut-researcher on the Soyuz 28 mission in 1978 he spent a week in Salyut 6.

Magion, the first satellite designed and built in Czechoslovakia (by the Geophysical Institute, Czechoslovak Academy of Sciences), was launched with Intercosmos 18 on 24 Oct 1978. Weighing 15kg, it was placed in a near-polar 404 × 772km orbit with 83° inclination. It was planned to study low-frequency electromagnetic fields by obtaining data from the pair of slowly diverging satellites. But a power malfunction meant that except for a brief period data-acquisition was sub-standard. Magion decayed on 11 Sep 1981.

Czech instruments for Intercosmos have included a laser reflector to measure the distance between the satellite and the Earth, an instrument to measure the mass and speed of meteorites, and a silicon detector to measure cosmic radiation. Prognoz has been provided with instruments for the study of solar flares and rays. Czechoslovakia is also contributing to the Vega satellite due to intercept Halley's Comet in 1986.

FRANCE

History France is the world's third space nation. She consistently spends more on both satellites and launchers than any other country except Russia and America. And her unique ability to co-operate with both the major space powers simultaneously has probably enabled her to become more knowledgeable than any other nation about the potential uses of space for commercial and possibly military purposes.

In a striking example of successful French collaboration with the Soviets, two fighter pilots, Maj Patrick Baudry, 35, and Lt-Col Jean-Loup Chrétien, 42, began cosmonaut training at Star City in Sep 1980. This culminated in Chrétien's 7-day mission in Salyut 7 in Jun 1982.

By the end of 1982 France had been responsible for the successful launch of 20 satellites: nine were entirely national, while four had been launched by NASA and six by Russia. France herself had launched one for West Germany. Her ninth successful national satellite was also her last, however, since the national Diamant launcher was abandoned in favour of the European (though largely French) Ariane. Examples of the way French scientists regularly contribute to both Soviet and US major projects include the laser mirror on Russia's Lunokhod vehicles, experiments on the 1973 Mars probes, and the ultra-violet telescopes on the 1973 US Skylab and the 1983 Soviet Astron. Contributions to the Salyut 7 mission included an optical helmet for use in balance experiments. Paying tribute to 15 years of Franco-Soviet collaboration on over 30 experiments, Academician Sagdeyev, director of the USSR Space Research Institute, spoke of the resulting mutual benefits and said that their joint study of cosmic radiation had gained wide international recognition.

France became the third nation to achieve orbital capability when the 18-tonne, three-stage Diamant placed the A-1 satellite in orbit on 26 Nov 1965, eight years after Sputnik 1. There was no loss of resolve when 1973's final collapse of ELDO, in which France had been the driving force, coincided with a run of misfortunes in her national programme. The future of the ambitious space centre at Kourou in French Guiana was also threatened, only to be secured by the re-emergence of the European Space Agency, with its headquarters in Paris and France once again the leading member. France's space programme has been firmly built around the Centre National d'Etudes Spatiales (CNES), founded in 1962.

The CNES budget - Fr 4,762 million in 1984, up 34% on 1983 - is a good indication of future trends. Spending on ESA came down from 60·9% in 1979 to 37% in 1983, while investment in bilateral programmes - mainly with Russia, America and Germany - rose from 4·4% to 12·58%. These figures reflect a change of policy in 1980, when following the success of Symphonie it was decided to concentrate much more upon applications projects such as direct-broadcast satellites (see International - Eutelsat), Telecom 1 (a two-satellite system for national communications due to be launched in 1984) and Spot (a Landsat-type Earth-resources satellite aimed at European and African markets and due for launch in 1985). With a potential market of more than 100,000 Spot images a year, CNES is considering charging $500–1,000 per scene, presented in digital form on computer-compatible tapes. Although this is 3-5 times more than the US Landsat rate, CNES argues that it still compares very favourably with high-altitude photography. This would yield an income of $50-100 million a year, enough to cover the entire cost of the system. It seems however that this approach may have to be revised in the light of the May 1983 report to the US Government to the effect that Landsat operations cannot be successfully commercialised because the predicted "huge potential market" does not exist. Spot technology will no doubt also be used for SAMRO, a military observation and reconnaissance satellite which is also to be launched around 1984-5.

Jean-Loup Chrétien, 2nd from left, aboard Russia's Salyut 7 in 1982. His back-up then, Patrick Baudry, is now due to fly on the Shuttle in 1986. *(CNES)*

CNES is co-operating in a number of future Soviet scientific missions, including Vega, which involves balloon-carrying Venus landers. A follow-on mission releasing Montgolfier-type balloons in the planet's atmosphere is planned for the 1990s. Intercosmos has invited France to propose experiments for a Soviet 100km lunar polar orbiter in 1987-1990. The two countries are considering a submillimetric orbital telescope to study the galaxies, and a dual-spacecraft mission called Interball to study Earth's magnetosphere. A space dust collector and a French detector to measure oscillations of the Sun are other projects. But French hopes of sending spationaute Patrick Baudry on a long-duration Salyut mission have been dashed by official political coolness over Afghanistan; this may have stimulated the recent eagerness to collaborate with NASA, which has agreed to fly the Echograph system on a 1984 Shuttle flight. It will not be accompanied by a French astronaut, however, though it is hoped that an improved version, accompanied by Baudry, will be carried on a follow-on mission in 1985-6. Joint missions to the asteroids and outer planets have also been discussed with NASA.

The first Ariane launch in December 1979. This vehicle is largely French-funded and the result of French insistence that Europe must have an independent alternative to US and Soviet launchers. *(ESA)*

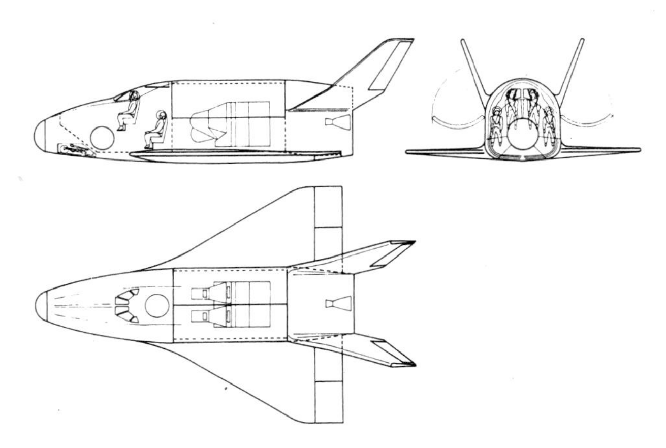

Hermes, a French proposal for a 4-crew mini-shuttle for launch by Ariane 5 in the 1990s, has recently found favour with both ESA and NASA. NASA's Administrator thinks it could be used for changing crews on the US Space Station. *(CNES)*

Diamant launcher Diamant A achieved the first 3 orbital launchings outside America and Russia. Diamant B, 23·54m long and weighing 24,600kg at launch, developed 35,000kg (77,000lb) thrust for 112sec at lift-off. First launching was at Kourou on 10 Mar 1970, when France's first foreign payload was placed in orbit. Of 12 launches, 10 were successful. Much useful technology was transferred to ESA's Ariane.

National launches

A-1 (Asterix) L 26 Nov 1965 by Diamant from Hammaguir. Wt 41·7kg. Orbit 528 × 1,768km. Incl 34°. First French satellite. Transmitted for 2 days. Orbital life 200yr.

D-1A (Diapason) L 17 Feb 1966 by Diamant from Hammaguir. Wt 20kg. Orbit 504 × 2,753km. Incl 34°. 3rd French satellite and 2nd national launch. Carried out geodetic research. Orbital life 200yr.

D-1C (Diademe 1) L 8 Feb 1967 by Diamant from Hammaguir. Wt 22·6kg. Orbit 580 × 1,340km. Incl 40°. Geodetic satellite, operational in spite of low apogee. Orbital life 100yr.

D-1D (Diademe 2) L 15 Feb 1967 by Diamant from Hammaguir. Wt 22·6kg. Orbit 592 × 1,886km. Incl 39°. Provided laser and Doppler data for 3 months; orbital life 200yr.

Wika/Mika L 10 Mar 1970 by Diamant B from Kourou. Wt Wika 50kg, Mika 64·8kg. Orbits 307 × 1,700 and 1,746km. Incl 5°. The W German Wika was the first foreign payload to be orbited by France; it carried out 30 days of geocorona and upper-atmosphere investigation, and had a 5yr orbital life. Mika's task was to monitor the performance of the Diamant B launcher, which had 15,420kg thrust, liquid 1st stage, solid-propellant 2nd stage and 3rd-stage apogee motor to spin payload into orbit.

Peole L 12 Dec 1970 by Diamant B from Kourou. Wt 69·7kg. Orbit 580 × 747km. Incl 15°. Test flight to qualify the Eole meteorology satellite, successfully launched by Scout B from Wallops I on 16 Aug 1971. This project, conducted in collaboration with NASA, involved use of a satellite to track hundreds of 4m pressurised meteorological balloons carrying capsules on cables and measuring ambient temperatures as they drifted at a constant height.

D-2A (Tournesol 1) L 15 Apr 1971 by Diamant B from Kourou. Wt 90kg. Orbit 456 × 703km. Incl 46°. Carried 5 experiments to study solar radiation and ultra-violet range. Orbital life 6yr.

D-2A L 5 Dec 1971 by Diamant B from Kourou. Wt 97kg. Intended for a higher orbit than the previous satellite to continue solar radiation studies, but 2nd stage of launcher failed.

D-2A L 21 May 1973 by Diamant B from Kourou. Twin satellites, Castor and Pollux, fell into the sea; the launcher apparently failed to produce sufficent thrust.

Starlette L 6 Feb 1975 by Diamant B from Kourou. Wt 68kg. Orbit 804 × 1,138km. Incl 50°. A tiny, 10in-dia, spheroid, this was an early laser reflector for geodetic studies. Orbital life 2,000yr.

D-5A (Pollux) and D-5B (Castor) L 17 May 1975 by Diamant BP-4 from Kourou. Twin satellites, wts 35kg and 77kg, placed in 270 × 1,270km orbits with 30° incl. Pollux carried microrocket experiments and Castor an accelerometer. Orbital lives 80 days and 3yr.

D-2B (Aura) L 27 Sep 1975 by Diamant BP-4 from Kourou. Wt 110kg. Orbit 499 × 723km. Incl 97°. France's last national launch carried experiments to study ultra-violet radiation from the Sun and hot new stars. Orbital life 15yr.

Telecom Two satellites, with a 3rd ground spare, were approved in 1979 to provide France and her overseas territories with TV, phone, digital data and military communications. Prime contractor is Matra. Stationed at 10°W and 7°W, Telecoms will have 650kg in-orbit wt and 7yr life. Communications for France will be supplied by 6 active and 3 standby repeaters operating at 12-14GHz; data links with overseas territories will be handled by 4 active and 2 standby 4-6GHz repeaters with 1,000 phone or 1 TV channel capacity each. Military data will be handled by 2 active and 1 standby 7-8GHz repeaters serving the Syracuse network, which links military installations, ships and mobile ground stations. Launch of TDF-1 due Nov 1985; TDF-2 undecided.

France's Spot Earth-resources satellite is intended to provide Europe and Africa with images more detailed than Landsat's. *(CNES)*

Spot This Earth-resources system will be used for oil and mineral exploration, crop and environment monitoring, coastal zone studies and land-use mapping. The company Spot Image – with $4·1 million capital allocated by public institutions and private companies in France, Belgium and Sweden, headquarters at Toulouse and a chairman who is also Director of Applications Programmes at CNES – will attempt to market multispectral and panchromatic (black-and-white) images worldwide in competition with Landsat. Spot 1, due for launch by Ariane L-15 in Jan 1985, will be based on a multimission platform design and will be capable of 20m resolution in the multispectral mode and 10m with black-and-white imaging. Stereo images will be made possible by ground processing of 2 images taken on successive passes over the same area. By the end of 1983 marketing contracts had been signed with Britain, Norway, Finland and Nepal.

Athos This experimental telecommunications satellite will be used to study the 30/20GHz frequencies, at which the successors to the current ECS and Telecom satellites could operate. With transfer-orbit wt of 1,450kg and 877kg in-orbit wt, Athos will be launched on Ariane 4's test flight in 1985. Matra is prime contractor for the French National Telecommunications Centre.

US launches

FR-1 L 6 Dec 1965 by Scout from Vandenberg into circular 780km orbit to study ionosphere.

Eole L 16 Aug 1971 by Scout from Wallops I. Orbit 678 × 903km. Meteorological satellite interrogating 500 balloons drifting in atmosphere at 12,000m.

Symphonie A Franco-German project under which Symphonie 1 was launched 19 Dec 1974 and Symphonie 2 27 Aug 1975 by Thor-Deltas from Cape Canaveral into synchronous orbits at 11·5°W. With 237kg orbital wt, they each provided 2 TV channels or 300 two-way phone circuits covering Europe, the Middle East, Africa,

France's TDF-1 and Germany's TV-Sat are being jointly developed around this design to provide both countries with direct-to-home TV. *(Aérospatiale)*

S America and part of N America. The experiment was a merging of the French SAROS and German Olympia projects, with ground stations in each country using a small 16m antenna to experiment with multiple access. The Symphonies were successfully used for experiments, demonstrations and operational services, including educational broadcasts to Africa, commercial TV, as a "hot line" (from 1 Feb 1978) linking the heads of government of France, Germany and Iran, and for UN and UNESCO communications. A direct result was the Franco-German agreement to develop the TDF-1 and TV-Sat direct-broadcast satellites for launch in 1984/85.

USSR launches

Aureole L 21 Dec 1971 by C1 from Plesetsk. Orbit 410 × 2,500km. To study upper atmosphere and polar lights.

SRET-1 L 4 Apr 1972, probably by C1 from Plesetsk. Orbit 460 × 39,248km. To study radiation effects on solar cells.

SRET-2 L 5 Jun 1975 piggyback with Molniya from Plesetsk to test Meteosat cooling system.

Signe 3 L 17 Jun 1977 from Kapustin Yar. Wt 102kg. Orbit 459 × 519km. Incl 50·6°. Studied gamma radiation in the galaxies and carried out star cataloguing for 733 days.

GERMAN FEDERAL REPUBLIC

History It was Germany's pre-war rocket team, led by Dr Wernher von Braun and Dr Kurt Debus after they had transferred their allegiance to the US at the end of the Second World War, which ensured the building of the great Saturn V rocket and the Apollo Moon landings. Inevitably, German interest in space continues at a high level, encouraged by a series of successes in 1983. This included MBB's SPAS-01, which became the world's first satellite to be deployed and retrieved (see US, Space Shuttle - STS-7), and the long-awaited STS-9/Spacelab 1 mission. Until then second in Europe to that of France, Germany's space programme may be about to take the lead with the ambitious German/Italian space station proposal known as Columbus and described under Italy.

Six national satellites, five of them launched by NASA and one by France, culminated in the pair of Helios solar explorers, which are yielding much new knowledge of the Sun. Germany also joined France in the Symphonie project. In Feb 1976, however, the German Government decided that there would be no more national or even bilateral satellite projects such as Symphonie; two-thirds of the 1976–79 space budget of DM2,200 million (£425 million) would be allocated to the European Space Agency, and only one-third to national space efforts.

While France decided to put her main effort into the Ariane launcher, Germany became the largest contributor (53·3%) to Spacelab, developed by prime contractor ERNO and companies from ten European countries for the US Space Shuttle. In addition, DM40 million was being spent over an eight-year period to 1984 on contributions by Dornier and MBB to Ariane. The 1983 budget totalled DM770 million.

Germany's SPAS-01 satellite, photographed from *Challenger's* flight deck as it was deployed and retrieved by the robot arm on STS-7. *(NASA)*

The increasingly ambitious nature of Germany's future projects, with a return to national satellites and the selection in Dec 1982 of two national astronauts, reflects some disillusionment with ESA's co-operative efforts. Germany has already paid deposits to NASA for two national Spacelab missions, designated Spacelabs D1 and D4 and provisionally assigned to 61-A and 91-C, due in 1985 and 1988. The first of these, costing $65 million, will be devoted to materials processing and life sciences, the second to astrophysics. German research into the use of micro-gravity conditions for industrial production began in 1978 with Texus, a programme using six high-altitude sounding rockets giving 6min of micro-gravity. It continued with Maus, automatically controlled materials experiments carried first in the Shuttle cargo bay and then on the SPAS reusable satellite. The programme should be completed with the D1 mission, though there may already be operational units by then. In another significant development, the government-run aerospace agency, DFLVR, has renewed its links with the once unpopular Otrag launch company, which is still working towards a privately funded launcher.

There are plans for a German X-ray satellite, Rosat (see International), to be launched by Shuttle in 1987, and private and scientific organisations have booked 25 small self-contained Shuttle payloads on a space-available basis. Two other major collaborations with NASA involve participation in the Galileo mission to Jupiter (MBB is developing the propulsion module), and the Active Magnetospheric Particle Tracer Explorers (AMPTE), a two-spacecraft project due to fly in 1984. The Max Planck Institute will provide the ion release module, and the US Charge Composition Explorer will detect seven ion releases by the German craft. MBB is also engaged in a Franco-German programme leading to the complementary TV-Sat and TDF-1 satellites for launch in 1985. Each will transmit three TV programmes to every home receiver in France and Germany from geostationary orbit; they will have a seven-year life and will be followed by later national satellites, each with a capacity of five channels.

Germany's Amsat was being given government help to plan a small ion-drive satellite for the international Oscar series. It would be used to explore the asteroids and possibly orbit three of them.

Azur L 8 Nov 1969 by Scout from Vandenberg. Wt 71kg. Orbit 387 × 3,150km. Incl 103°. The first in a projected series of German-designed and built research satellites. The solar-synchronous orbit enabled the 7 experiments to observe Earth's radiation belts, solar particle flows, etc, in a co-operative programme with the US to augment data from US Explorer and OGO satellites. Life 100yr.

Dial L 10 Mar 1970 by Diamant B from Kourou, the name combining the words Diamant launcher and *Allemand* (German). Wt 63kg. Orbit 301 × 1,631km. France's first foreign launch, it was intended to study hydrogen geocorona and the ionosphere but excessive vibration during launch made data evaluation difficult. Re-entered 5 Oct 1978.

Aeros 1 L 16 Dec 1972 by Scout from Vandenberg. Wt 127kg. Orbit 223 × 867km. Incl 97°. With 5 experiments (4 German, 1 US), Aeros 1 dipped into the Earth's upper atmosphere 3,844 times, measuring short-wave ultraviolet radiation from the Sun, which is dangerous to man, before re-entering on 22 Aug 1973.

Aeros 2 L 16 Jul 1974 by Scout from Vandenberg. Wt 127kg. Orbit 224 × 869km. Incl 97°. Intended to continue upper-atmosphere research. Data were said to be "acceptable," although the intended orbit to continue the work of Aeros 1 had been 230 × 900km. Re-entry 25 Sep 1975 after 436 days.

Helios 1 L 10 Dec 1974 by Titan-Centaur from C Canaveral. Wt 370kg. Heliocentric orbit. First of 2 spool-shaped spacecraft placed in solar orbits taking them nearer the Sun than any previous craft. On 15 Mar 1975 Helios 1 passed within 48 million km at 238,000km/hr, survived temperatures high enough to melt lead, and made the first close-in observations of the Sun's surface and solar wind. The craft rotated once every second to distribute evenly the heat coming from the Sun, 90% of which was deflected by optical surface mirrors. Heat from the central compartment was radiated to space via louvre systems, keeping the internal temperature below 30°C compared with an outside temperature of 370°C. Helios 1 detected 15 times more micrometeorites close to the Sun than there are near Earth. It is not known whether they are transported by the Sun's corona after being pulled in by the Sun's gravity from elsewhere in space.

Helios 2 (L 15 Jan 1976 by Titan-Centaur) was placed in a similar orbit taking it 3 million km closer to the Sun and subjecting it to 10% more heat; it was hoped it would reveal more about the unexpected discoveries of Helios 1. Named after the Sun god of ancient Greece, Helios was the largest bilateral project in which NASA had participated up to that time, with Germany paying $180 million of the total $260 million costs. The spacecraft were controlled from the German space centre near Munich, and data were correlated with results from IMP Explorers 47 and 50 in Earth orbit, the Pioneer solar orbiters, and Pioneers 10 and 11 in the outer solar system.

SPAS-01 Developed by MBB (wt 1,800kg, 4·3m long, 1·4m wide), Space Pallet Satellite became the 1st spacecraft to be deployed and retrieved by the Shuttle RMS, on STS-7. Made of 0·7m-long, 6cm-dia carbon-fibre tubes forming a truss bridge, it has 12 compressed-nitrogen attitude-control thrusters, a data-processing system, radio, radar and a 28V battery. Total cost is about $13 million, including 1st launch, plus $8-10 million for 8 experiment packages. Follow-on versions will be able to fly free for months before retrieval; first of them will be ESA's Eureca. MBB says that SPAS platforms could be flown on single-discipline missions lasting 6 months at a cost of about $25 million. With an added rocket motor they could be placed in geostationary orbits.

Postsats Also known as DFS, these craft will be derived from ECS to provide Germany's 1st national communications satellite services. Weighing about 1,400kg each, 2 will be orbited by Ariane 3 or 4 starting in late 1986, with a 3rd as ground spare. Prime contractor is Siemens.

GIRL German Infra-Red Laboratory is a 40cm-aperture telescope cooled by super-liquid helium at 1·6°K and designed to study origins of stars and active galaxies. MBB/ERNO is prime contractor. 1st flight of this reusable craft is likely to be on a 14-day Shuttle/Spacelab mission in 1987.

Two Helios spacecraft, launched in 1974 and 1976, survived temperatures high enough to melt lead when they made the first "close" observations of the Sun's surface. ***(NASA)***

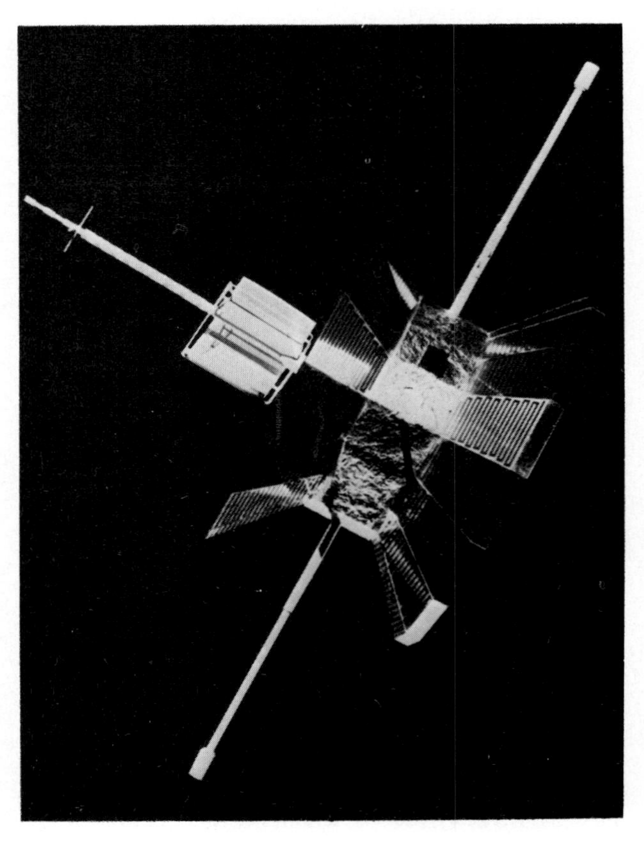

INDIA

History Despite the inevitable setbacks, India's vigorous space programme enabled her to become the 7th nation to achieve orbital launch capability. After six years of planning and production the Indian Space Research Organisation's SLV-3 launcher orbited a 45·3kg scientific satellite in Jul 1980. Prime Minister Mrs Gandhi said that the satellite and launcher had cost R200 million (about $12 million). Further improvements are planned, culminating in the Polar SLV able to launch 1,000kg into polar orbit by 1988. Like France, India has benefited from simultaneous co-operation with Russia, America and ESA. After 2 years' training Maj Rakesh Sharma of the Indian Air Force took part in the very successful 7-day Soyuz T-11/Salyut 7 mission in Apr 1984. Of her seven satellites up to end-1983, three had been launched by Russia, three nationally and one by the US. The latter was the first of two multi-purpose Insats from Ford Aerospace, following up the successful educational broadcasts to Indian villages by ATS-6 in 1975 with permanent domestic communications, direct TV and meteorological services. Plans include a fourth launch by Russia, of an Indian Remote Sensing (IRS) craft in 1986. India will have to pay for this launch, having been given the first three.

By 1983 ISRO employed 10,000 people compared with ESA's 1,400, and had a 10-year programme embracing launchers and sounding rockets for use from a national launch site, and communications and remote-sensing satellites.

India's Insat 1B being launched by *Challenger* on STS-8 to replace the failed Insat 1A. It transmits TV to rural communities via 100,000 simple Earth terminals. *(NASA)*

National launches

Rohini 1B L 18 Jul 1980 by SLV-3 from Sriharikota. Wt 35kg. Orbit 305 × 919km. Incl 44·7°. India's 1st successful launch, this experimental satellite followed the failure of Rohini 1A on 10 Aug 1979. Function was to monitor the launcher and itself; 1·2yr life.

Rohini 2 L 31 May 1981 by SLV-3 from Sriharikota. Wt ? 32kg. Orbit 186 × 418km instead of planned 296 × 834km. Incl 46°. Re-entered in 8 days.

Rohini 3 L 17 Apr 1983 by SLV-3 from Sriharikota. Wt 41·5kg. Orbit 371 × 861km. Incl 46·6°. Carries 2 cameras and L-band beacon, and was soon sending back pictures of Indian landmarks. Orbital life 5yr.

Other launches

Aryabhata L 19 Apr 1975 by Skean + Restart stage from Kapustin Yar. Wt ? 300kg. Orbit 563 × 619km. Named after the 5th-century astronomer and mathematician and intended for X-ray astronomy. Indian scientists prepared it "under the scientific supervision of Soviet engineers". It was designed for 6 months' operations and an orbital life of 10yr, and control was to have been handed over to Indian scientists on the 3rd day. But the Indian-designed transformer failed and only a few days' useful life was obtained.

Bhaskara 1 L 7 Jun 1979 from ? Kapustin Yar. Wt 442kg. Orbit 519 × 541kg. Incl 50·6°. Named after 7th-century Indian astronomer. With 2 TV cameras and microwave radiometers, it cost R65 million and was to spend 1yr studying India's earth and water resources. Useful ocean and land surface data were received but the TV cameras malfunctioned.

Bhaskara 2 L 20 Nov 1981 from Kapustin Yar. Wt 444kg. Orbit 541 × 557km. Incl 50·7°. Declared operational by ISRO following receipt of more than 300 TV images of Indian sub-continent. 9yr orbital life.

Apple (Ariane Passenger Payload Experiment) L 19 Jun 1981 by Ariane from Kourou. Wt 670kg. Geostationary orbit. India's first indigenous test communications satellite, Apple was launched with ESA's Meteosat 2. Indian TV programmes and educational teleconferences were relayed despite a jammed solar panel.

Insat 1A L 10 Apr 1982 by Delta 3910/PAM from C Canaveral. Wt 1,152kg. Synchronous direct TV/meteorological satellite to be placed at 74°E. Jammed C-band antenna released on 22 Apr. In addition to providing nationwide direct TV to rural areas, it was intended to keep a half-hourly weather watch for cyclones, etc. After emerging from the solar eclipse affecting all geosynchronous satellites, however, it was found to have run out of attitude-control fuel for some reason and had to be abandoned on 4 Sep 1982 instead of operating for the expected 7yr.

Insat 1B L 30 Aug 1983 by STS-8 (wt 1,152kg), Insat 1B almost suffered the same fate as its predecessor. Video recordings suggested that it might have been struck by Orbiter debris just after launch, though this could not be confirmed. It was not until mid-Sep that Ford and Indian controllers at India's Hassan Satellite Control Centre succeeded in deploying its solar array. By then the satellite had been stationed at 74°E in place of Insat 1A. Full operation was achieved in Oct. Described as the most complex civil satellite so far, Insat 1B will use its 12 transponders for the next 7yr to provide India with communications and direct nationwide TV services to thousands of remote villages, plus a comprehensive weather and disaster-warning service. A total 100,000 Indian-built, receive-only, 3-3·6m-dia Earth terminals are already in place to supply rural communities with social and education programmes. Total cost of Insat 1B and the back-up Insat 1C, including launch with 2 PAM-Ds, was approximately $140 million.

INDONESIA

History Indonesia's Palapa ("fruit of effort") satellite system is enabling the country's 6,064 inhabited islands, with their 147 million population and 5,000km span, to develop a national communications system. This would have been impossible to achieve by terrestrial means.

Palapa A1 and A2 L 8 Jul 1976 and 10 Mar 1977 by Delta from C Canaveral. Wt 575kg. Ht 3·7m, dia 1·9m. Geostationary at 83°E and 77°E. Still operating 7yr after launch. Identical to Canada's Anik and Western Union's Westars except for a modified 1·5m-dia parabolic dish antenna giving maximum illumination of the Indonesian landmass, these craft have 12 transponders providing 4,000 voice circuits or 12 simultaneous TV channels. They transmit to 125 Earth stations.

Palapa B First in series, B1, L 18 Jun 1983 by STS-6 from C Canaveral and placed at 108°E. In-orbit wt 628kg, ht 6·83m, dia 2·16m. Twice as big and four times as powerful as the As, these replacements have an 8yr life. Built by Hughes, they have 24 transponders with 10W output, and they will also serve the Philippines, Thailand, Malaysia and Singapore. Cost of the Palapa B contract in 1979 was $74·5 million, plus $11 million B1 launch costs paid to NASA. Palapa B2 was L 6 Feb 1984, 3 days late, on Shuttle's 10th mission. Following loss of Westar 6 on 1st day, Palapa B2 was also lost. Similar malfunctions of their PAM-D upper stages placed both in useless 321 × 1,100km orbits instead of transfer orbits. The Indonesian Government decided to claim the $75 million insurance and order a replacement.

Landsat In 1981 Indonesia became the 11th country to contract with NASA to build a ground station to receive, process, archive and disseminate Landsat data for uses such as crop, resources and pollution monitoring. NASA makes the information available in return for an annual access fee of $200,000.

TERS The Tropical Earth Resources Satellite, to be built jointly with the Netherlands, is planned to enter service in the early 1990s.

Indonesia's Palapa B1 in the Shuttle payload bay alongside the similar Westar before launch on STS-6. Unlike their successors on Mission 10, these were both successfully orbited. *(NASA)*

ITALY

History In addition to her increasingly effective contributions to ESA, totalling 9% of that budget, Italy has embarked on a major space programme involving the launch of six satellites during 1983-89. One of these, the ambitious Shuttle Tethered Satellite in 1987, will probably call for a flight by an Italian astronaut. Italy's increased interest in space activities was rewarded in Oct 1983 with a NASA contract worth $1·3 million for the design and construction of the transporter needed to carry Orbiters horizontally between launchpad and runway at Vandenberg.

Early in 1984 Italy joined W Germany in proposing that ESA should authorise development of the Columbus space station facility to operate either independently for Europe or in conjunction with the US Space Station. Using Spacelab and German Eureca/SPAS technology, it would include pressurised manned modules, unmanned free-flying platforms, support vehicles and service modules. A 1st mission in 1993 would dock a pressurised module to the US Space Station; launch would be by Shuttle or Ariane, depending upon ESA and/or French participation. It would lead to a manned, free-flying Columbus space laboratory by 1999. NASA has welcomed the initiative.

A small but effective programme built around Italy's international platform in the Indian Ocean off San Marco, Kenya, was carried out during 1964-74. A total of four San Marco satellites were launched to study atmosphere density; they were due to be followed up with another San Marco launch in co-operation with NASA. Italy's fifth national satellite, Sirio 1, was being repositioned in 1983 under a little noticed co-operative agreement with China which enables the latter to start tests related to her planned geostationary communications satellites. Iris, Italy's upper stage now under development, is described in the Launcher section.

San Marco 1 L 15 Dec 1964 by Scout from Wallops I. Wt 24kg. Orbit 194 × 697km. Incl 38°. Measured atmospheric density until its decay on 13 Sep 1965.

San Marco 2 L 26 Apr 1967 by Scout from San Marco. Wt 129kg. Orbit 185×211km. Incl 3°. The first launch from the San Marco platform, it was the first of 3 spacecraft ingeniously designed to measure atmospheric density; they consisted of 2 concentric spheres of different mass joined by 3 flexible arms, so that the drag on the outer sphere could be measured as it dipped into the atmosphere. Re-entered 14 Oct 1967.

San Marco 3 L 24 Apr 1971 by Scout from San Marco. Wt 164kg. Orbit 222 × 718km. Incl 3°. Dec Nov 29 1971.

San Marco 4 L 18 Feb 1974 by Scout from San Marco. Wt 164kg. Orbit 231 × 910km. Incl 3°. With Explorer 51 taking measurements in the auroral zone and San Marco 4 on the equator, it was possible to compare the different responses to energy coming from the Sun, and to study the theory that most solar energy enters the atmosphere at the poles. San Marco 4 transmitted for 806 days until decay on 4 May 1976.

San Marco D/L Due for launch by Scout from San Marco into a 290 × 800km orbit. A 1m sphere weighing 240kg, San Marco D/L will continue the programme's atmospheric studies for about 1½yr. NASA and Britain are co-operating.

Sirio 1 L 25 Aug 1977 by Delta from C Canaveral. Wt 398kg; 220kg in orbit. Geosynchronous orbit at 15°W. Italy's first experimental comsat, Sirio was developed by the company now called Compagnia Nazionale Aerospaziale to study super-high-frequency (SHF) propagation in rain, snow and fog. Controlled from Fucino, Italy, it was to be followed by ESA's Sirio 2, but that craft was lost in the Ariane L-5 failure in Sep 1982. Following the 1983 space co-operation agreement with China, Sirio 1 was being moved to 65°E to enable China to develop geostationary communications in the 18GHz range.

Future projects The Shuttle Tethered Satellite, designed to continue Italy's atmospheric studies, is a joint project with NASA. This small craft will be trolled out from the Shuttle and down into the atmosphere at the end of a super-strong cord measuring as much as 97km long. Planned to fly in Feb 1987, this system would overcome the disadvantages of both sounding rockets, which last for only minutes, and low-altitude satellites, which decay within a few hours, in gathering atmospheric, magnetospheric and gravity data in the Earth's upper atmosphere. SAX, an X-ray astronomy satellite, is planned to make the 2nd use of Italy's Iris upper stage in a Shuttle launch in late 1987. Italsat, for domestic communications, is scheduled for Shuttle launch in Nov 1987. Sarit, for domestic TV, and Sicral, a comsat for Italy's military and civil services, are both planned to enter service in about 1989.

JAPAN

Development Japan has by far the most ambitious long-term national programme after those of the US and the Soviet Union, and it probably surpasses even that of the 11-nation ESA. 1978 saw the announcement of a 15-year plan costing $14 billion and involving the launch of 80 spacecraft on 76 boosters by 1992. By 1983 this had been only slightly modified to a total of 78 satellites during 1984-2000. They will include survey missions to the Moon, Mars and Venus, a Halley's Comet interceptor (abandoned as too costly by the US), and regular flights by Japanese science-astronauts in the US Shuttle to operate materials-processing and life sciences experiments. The first Japanese science-astronauts are now due to fly aboard the Shuttle in 1987. Following that, the plan is for between one and three Japanese to fly every year. Selection of the first three or four payload specialists was due to begin in Apr 1983.

With her usual commercial prescience Japan had by Mar 1980 selected 45 priority items for materials processing, and another 17 in the life sciences field. The choice had been simplified by twice-yearly launches of the Nissan Motors TT-500A two-stage sounding rocket; there were occasional setbacks on the programme, however, with two rockets being lost in Jan and Aug 1981.

While the project is officially described as "just a dream," the National Space Development Agency of Japan (NASDA) continues with design work on a four-man mini-shuttle. 1987 is the target date for a first flight, government funding and production of the H-2 rocket permitting (see Launcher section). Progress is being made on developing a thermal protection system, and it is hoped that three re-entry capsules will have been recovered by 1986. It has already been decided that Japan's shuttle, unlike America's, would have two jet engines to permit it to make a powered landing.

History Japan was the fourth country after the Soviet Union, US and France to achieve national satellite capability. After many setbacks and disappointments her first test satellite, Osumi, was finally launched on 11 Feb 1970. By Sep 1983 26 Japanese satellites had been launched, all nationally achieved apart from three placed in geostationary orbits by NASA. After two failures, in Feb 1977 Kiku 2 established Japan as the third country to achieve geostationary capability.

Despite the fact that Osumi was 2½ years later than originally planned, Japan's space budget over the 16-year period that led to final success totalled only $189 million. The first operational satellite, Shinsei, came in Sep 1971. The ambitions of Japanese scientists to become the world's third space nation, almost achieved, had been foiled partly because there were two rival space programmes, with Tokyo University working on solid-fuel rockets and the government's Science and Technology Agency on liquid-fuel rockets. This was resolved by the setting up on 1 Oct 1969 of NASDA, which is responsible for the development and launch of applications satellites from the Tanegashima Space Centre (with launchpads at Osaki and Takesaki on the south-eastern tip of the island). The Institute of Space and Astronautical Science (ISAS), University of Tokyo, remains responsible for both the scientific satellites and development of their launch vehicles at the original Kagoshima Space Centre, construction of which began in 1962. The 1983 budget totalled Y1,143 billion, of which Y906·5 billion was allotted to NASDA and Y156·7 billion to ISAS.

Each NASDA satellite is named after the flower which happens to be blooming at the time of its launch. But there is one complication: if the launch is delayed there has to be a hurried change of name!

National launches

Osumi L 11 Feb 1970 by Lambda 4S-5 from Kagoshima. Wt 24kg. Orbit 340 × 5,140km. Incl 31°. Test satellite was 4th stage. Transmitted 17hr. Orbital life 80yr.

Tansei Test vehicles, named Tansei ("Light Blue") after Tokyo University colours.

Tansei 1 L 16 Feb 1971 by M-4S-2 from Kagoshima. Wt 63kg. Orbit 990 × 1,110km. Incl 30°. Demonstrated launcher's capability. Operated for full lifetime of 1wk. Orbital life 1,000yr.

Tansei 2 L 16 Feb 1974 by M-3C-1 from Kagoshima. Wt 56kg. Orbit 290 × 3,240km. Incl 31°. Confirmed launcher performance and satellite attitude controls. Orbital life 9yr.

Tansei 3 L 19 Feb 1977 by M-3H-1 from Kagoshima. Wt 129kg. Orbit 790 × 3,810km. Incl 66°. Tested satellite attitude stabilisation. Orbital life 2,000yr.

Tansei 4 L 17 Feb 1980 by M-3C-1 from Kagoshima. Tested future scientific satellite sub-systems.

Japan's Marine Observation Satellite (MOS-1), for launch in late 1986, will use 3 types of sensor to monitor sea states. ***(NASDA)***

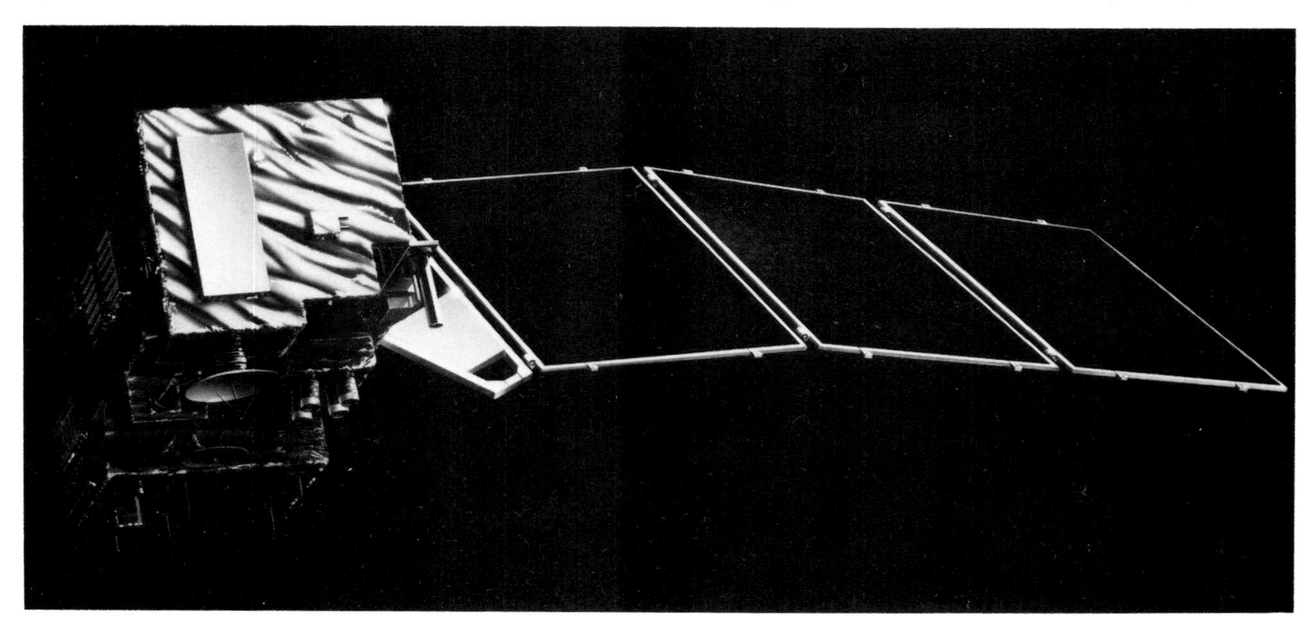

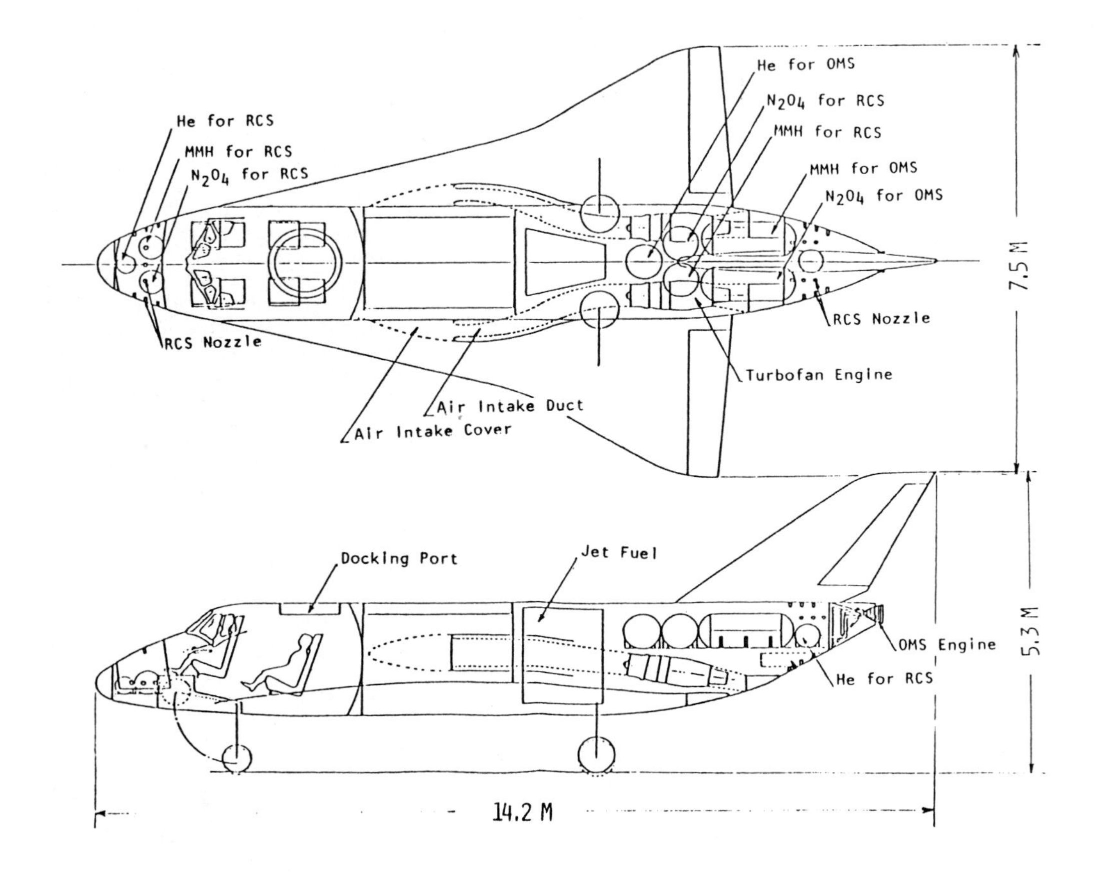

Japan's proposed mini-shuttle would have a 4-man crew and jet engines for powered landings. *(NASDA)*

ETS-3 pioneered for Japan the technologies needed for large satellites with high power requirements. *(NASDA)*

Scientific satellites

SS-1 (Shinsei, "New Star") L 28 Sep 1971 by M-4S-3 from Kagoshima. Wt 65kg. Orbit 870 × 1,870km. Incl 32°. Observed solar HF radio emissions, plasma and cosmic rays. Data recorder failed after 4 months but real-time data continued. Orbital life 5,000yr.

SS-2 (Denpa) L 19 Aug 1972 by M-4S-4 from Kagoshima. Wt 75kg. Orbit 250 × 6,750km. Incl 31°. Observed plasma waves and density, electron flux and geomagnetism. Dec 19 May 1980.

SS-3 (Taiyo, "The Sun") L 24 Feb 1975 by M-3C-2 from Kagoshima. Wt 86kg. Orbit 255 × 3,135km. Incl 31°. Intended for 250 × 2,000km orbit to observe solar radiation; in spite of incorrect orbit succeeded in monitoring solar X-rays, ultra-violet radiation, etc. Dec 29 Jun 1980.

SS-5 (Kyokko, Exos A) L 4 Feb 1978 by M-3H from Kagoshima. Wt 126kg. Orbit 640 × 3,980km. Incl 65°. Took 1sec "snapshots" showing auroral activity, etc, and transmitted them by slow-scan TV. Orbital life 300yr.

SS-6 (Jikiken, Exos B) L 16 Sep 1978 by M-3H. Wt 91kg. Orbit 230 × 30,050km. Incl 31°. Studied magnetosphere and plasma resonance and echo phenomena for over 3yr. Orbital life 20yr.

SS-4 (Hakucho, Corsa B) L 21 Feb 1979 by M-3C. Wt 96kg. Orbit 550 × 580km. Incl 30°. Followed Corsa A, which failed 4 Feb 1976 to conduct panoramic survey of X-ray bursts. Orbital life 10yr.

SS-7 (Hinotori, Astro A) L 21 Feb 1981 by M-3S. Wt 190kg. Orbit 580 × 640km. Incl 31°. Observing spectrum of solar hard X-ray flares. Orbital life 30yr.

SS-8 (Tenma, Astro B) L 20 Feb 1983 by M-3S from Kagoshima. Wt 180kg. Orbit 350 × 600km. Incl 31°. Observing X-ray sources in the core of active galaxies. SS-9, due in 1984, will be Exos C. SS-10 will be Planet A, due for L 1985 for ultra-violet imaging of Halley's Comet. SS-11 Astro C, at 400kg Japan's largest scientific satellite, will observe galaxies and X-ray sources following L in 1987.

Engineering test satellites

ETS-1 (Kiku, "Chrysanthemum") L 9 Sep 1975 by N-1 from Tanegashima. Wt 83kg. Orbit 980 × 1,098km. Incl 47°. First launch for both NASDA and Tanegashima. Test of launch, satellite tracking and antenna extension technology. Orbital life 800yr.

ETS-2 (Kiku 2) L 23 Feb 1977 by N-1 from Tanegashima. Wt 130kg. 1st geostationary satellite, manoeuvred to 130°E.

ETS-4 (Kiku 3) L 11 Feb 1981 by N-2 from Tanegashima. Wt 640kg. Orbit 220 × 35,082km. Incl 29°.

ETS-3 (Kiku 4) L 3 Sep 1982 by the last N-1 from Tanegashima into a 1,000km circular orbit. Wt 375kg. Incl 45°. Designed for 1yr of testing 3-axis control, deployable solar paddles, and ion engines for future space propulsion. Future ETS include 1 each for communications experiments with moving vehicles in 1987, work on multiple beams and digital 3-axis controls in 1992, and attitude-control and lightweight structure research.

Ume ("Japanese Apricot") Japan produced the world's first ionospheric critical-frequency distribution chart with the second of these 2 satellites, which was also used for co-operative observations with Canada.

Ume 1 L 29 Feb 1976 by N-1 from Tanegashima. Wt 135kg. Orbit 994 × 1,013km. Incl 70°. Power failure after one month.

Ume 2 Back-up L 16 Feb 1978 by N-1. Wt 141kg. Orbit 980 × 1,220km. Incl 69°. Still operating in 1981, studying sources of interference with short-wave radio communications.

GMS-1 (Himawari 1) L 14 Jul 1977 by Delta from C Canaveral. Wt 315kg. Orbit geostationary at 140°E. Japan's contribution to the Global Atmospheric Research Programme. From Apr 1978 provided cloud pictures 14 times per day for TV weather forecasts in Japan and some other SE Asian countries.

GMS-2 (Himawari 2) L 10 Aug 1981 by N-2 from Tanegashima. Wt 670kg full; 281kg empty. Replaced GMS-1, but by mid-1984 both were beginning to fail.

GMS-3A and 3B have been ordered from Hughes; their launches may be advanced.

Communications

CS-1 (Sakura, "Cherry") L 15 Dec 1977 by Delta from C Canaveral. Wt 340kg. Orbit geostationary at 135°E. Carried out the world's first experiments on domestic communications in the quasi-millimetric band and 4yr later was still being used for tests with small mobile Earth stations.

BS-1 (Yuri) Experimental broadcasting satellite L 8 Apr 1978 by Delta from C Canaveral. Wt 355kg. Orbit geostationary at 110°E. Used for 2yr on investigations of direct home TV with 1m antennas.

Ayame 1 and 2 L 6 Feb 1979 and 22 Feb 1980 by N-1 from Tanegashima. Wt 130kg. Experimental communications satellites for millimetric-wave space communications. Both failed to enter geostationary orbit.

CS-2a (Sakura 2a) L 4 Feb 1983 by N-2 from Tanegashima and placed at 132°E. Japan's 1st commercial communications satellite, providing routine and emergency links for remote islands. It used a submillimetre bandwidth (20-30GHz), for the first time in the world, to counter the effects of microwave disturbances on Earth.

CS-2b L 6 Aug 1983 by N-2 from Tanegashima to 135°E. Designed to provide phone and telex services to both mainland and islands. Users include police and firemen. Built by Ford and Mitsubishi, it is spin-stabilised and has a 3yr life.

BS-2a L 23 Jan 1984 by N-2 to station at 110°E to improve TV in outlying islands, failed by mid-1984. BS-4 will operate in the 22-27GHz band from 1996.

Future projects In addition to the scientific, engineering and communications projects already mentioned, Japan is working on the 750kg Marine Observations Satellite (MOS-1), for launch in 1986 by N-2 into Sun-synchronous orbit, and the Geodetic Satellite (GS-1), balloon-shaped so that it will reflect both solar light and laser beams. GS-1 will be used to determine, among other things, the exact location of remote islands around Japan. Experiments with particle accelerators were contributed to Spacelab 1 in 1983.

Japan started commercial communications with the CS-2 series in 1983. They have a 5yr mission life. *(NASDA)*

BS-2 broadcasting satellites, intended to provide the Japanese archipelago with 2-channel TV, have run into trouble because of malfunctioning US components. *(NASDA)*

NETHERLANDS

History Up to 1970 the Netherlands' space activities were carried out mainly within the European organisations. In 1971, however, the government decided to fund two Astronomical Netherlands Satellites (ANS) at a cost of G76 million with NASA help. Dutch laboratories have also carried out scientific experiments on NASA and Japanese sounding rockets, and contributed X-ray and ultra-violet experiments to four European satellites and to America's OGO-5.

As a 2·7% participant in the European Space Agency the Netherlands is currently developing a recovery system for the Ariane rocket first stage. Fokker is prime contractor, and the parachute system was due to have its first operational test on the 7th Ariane mission at the end of 1983. The Netherlands also has an 11·8% share in ESA's L-Sat, and is building the satellite's service, propulsion and communications modules.

ANS-1 L 30 Aug 1974 by Scout from Vandenberg. Wt 129kg. Orbit 258 × 1,173km. Incl 98°. Designed and built by a Dutch industrial consortium, the Netherlands' first satellite was intended for a near-polar, Sun-synchronous circular orbit of 500km. Its studies of the ultra-violet spectrum of young, hot stars and both soft and hard X-rays from cosmic sources were controlled from the European Space Agency's ESOC centre at Darmstadt, Germany. ANS-1 re-entered 14 Jun 1977.

TERS In 1983 the Netherlands and Indonesia began studies aimed at launching a Tropical Earth Resources Satellite into equatorial orbit at 1,680km altitude and about 45° incl. To be launched in the 1990s, the 750kg TERS would be 3-axis-stabilised and equipped with sensors which would reject images of cloudy areas before they were sent. It would be able to observe tropical regions from 10°N to 10°S, with 8-20m spatial resolution.

SPAIN

History Spain contributed 2·1% of the European Space Agency's budget in 1982 and is becoming increasingly active in space research. Her only national satellite so far was the 20kg Intasat 1, launched 15 Nov 1974 (with NOAA-4 and Oscar 7) by Delta from Vandenberg into a 1,442 × 1,462km orbit inclined at 115°. It was an ionospheric beacon, with two-year operational and 10,000-year orbital life. Spain provides a control and monitoring station at Villafranca for ESA's Marecs satellites.

SWEDEN

History Sweden has been a member of the European Space Agency since 1973, and in 1982 contributed 1·6% of the budget. She has also made substantial contributions to Soviet space projects, including Prognoz and Intercosmos. In Apr 1979 the government decided to embark upon a national space programme to enable its industries to be more competitive in the international space communications market. Viking, to become Sweden's first satellite, and Tele-X, the Scandinavian direct-broadcast satellite, are the first results.

Viking Due for tandem launch with France's Spot in 1985, it is primarily intended to gain experience in system development and management. Launch wt is 550kg, and a positioning motor will place it in an 822 × 15,000km polar orbit, where it will provide data to follow up Sweden's sounding-rocket studies of the interaction of the solar wind and the Earth's magnetosphere, and the behaviour of the resulting Aurora Borealis. The Swedish Space Corporation awarded a $25·2 million contract to Saab-Scania in Sep 1980. The spacecraft platform, to be supplied by Boeing, is based on the USAF Small Scientific Satellites and NASA Atmospheric Explorers. Boeing is offering the platform to other countries seeking to launch small satellites which will fit neatly beneath the main payload on many Ariane launches; it can also be used for military reconnaissance. Viking lifetime will be 8 months.

Tele-X The Swedish Space Corporation and Eurosatellite signed an agreement in Sep 1983 covering development and manufacture of Tele-X as a Scandinavian DBS. Based on the Franco-German TV-Sat/TDF-1, it will cost over £100 million and is due for launch by Ariane (costing £33·5 million) in 1986. 3 TV channels will operate in the 12GHz band, with 2 more for data and video transmissions. Norway has taken a 15% share in the project, and Finland will probably take 4-5%. Other possible participants are Iceland and Denmark. Aérospatiale is project manager, assisted by Saab-Scania.

SWITZERLAND

History Switzerland is a founder member of the European Space Agency, contributing 1·4% of the 1982 budget, and also of Star, a European consortium of companies specialising in communications and applications satellites. So far her space projects have all been on a collaborative basis with Europe and NASA, though she has built the Zenit sounding rocket, which has been launched at least a dozen times since 1970. Her most famous contributions were the instrument packages provided by Berne Physics Institute for the Apollo and Skylab programmes. These included the simple but effective sheet of aluminium set up by the Apollo astronauts to register the composition of the solar wind as it reaches the surface of the Moon.

Sweden's Viking spacecraft, due for Ariane launch in 1985, will study high-latitude communications interference caused by the solar wind. ***(Boeing)***

UNITED STATES

ABC

Advanced Business Communications is planning 3 ABC satellites, including 1 ground spare, for launch in Dec 1986 and Feb 1987 to 83°W and 130°W.

ACTS

The Advanced Communications Technology Satellite (ACTS) was being developed by RCA Astro Electronics Division under contract from NASA's Lewis Research Centre for Shuttle launch in 1988, to operate in the Ka band. The 1,040kg experimental satellite would have cost about $260 million of public funding, and a total programme starting with a 2,780kg commercial satellite in 1990 about $500 million. Early in 1984 Hughes Communications Galaxy Inc, planning a competing project, successfully challenged the use of public funds for ACTS and NASA dropped the project.

Aeros

Space Services Inc of Houston, American Science & Technology Corp of Bethesda, Md, and Aeros Data Corp of Bethesda announced the formation of Space America in Sep 1983. Using the Conestoga-2 launcher, for which they are also responsible, they are planning a 3-satellite commercial remote-sensing system called Aeros. Launches, probably from Cat Island off the Mississippi coast, are due in 1986, 1988 and 1990. Ball Aerospace will build the 1st satellite and Honeywell the sensors. Operating at 917km altitude, it will have 6 visible bands capable of 80m resolution and 2 rated at 43m, plus, possibly, 2 short-wave infra-red bands rated at 80m. The 2nd satellite will incorporate 20m and 40m sensors providing 6 and 12-day repeat coverage. The system will focus on the 3 key resources - food, energy and water - aiming to provide much cheaper services than either France's Spot or the US Landsat. Shuttle launches would be considered if they were cheaper than Conestoga.

Amersat

American Satellite Co gave RCA a $100 million contract in early 1983 for 3 communications satellites based on Spacenet. 1 will be a ground spare. To be Shuttle-launched to 81°W and 128°W in Sep 1985 and Mar 1986, they will have 6 Ku-band and 12 C-band transponders, and a 10yr life.

Apollo

History The 3-man Apollo spacecraft was first proposed by NASA in July 1960 for Earth orbital and circumlunar flights, to be launched by a Saturn 1-type rocket with 6,672kN (680,400kg) thrust. On 25 May 1961 President John F. Kennedy proposed that the US should establish as a national goal a manned landing on the Moon by the end of the decade, and this in turn led to the development of Saturn 5, with 33,360kN (3·40 million kg) of thrust. North American Rockwell was selected as prime contractor for the Apollo Command and Service Modules, and Grumman Aircraft for the Lunar Module. On 28 May 1964 a Saturn I placed the first Apollo Command Module in orbit from Cape Kennedy; parallel development of the spacecraft and rockets was making remarkable progress until, on 27 Jan 1967, an electrical arc from wiring in a spacecraft being ground-tested at Cape Kennedy started a fire which became catastrophic in the 100% oxygen atmosphere. Astronauts Grissom, White and Chaffee were burned to death, and the first manned flight was delayed for 18 months. On 9 Nov 1967 the first unmanned test of the combined Apollo/Saturn 5 vehicles - designated Apollo 4 - was successfully accomplished. Apollo 5, in Jan 1968, successfully tested the Lunar Module systems, including firings in Earth orbit of both the ascent and descent propulsion systems; in Apr 1968 Apollo 6, the 2nd unmanned test of the combined Apollo/Saturn 5, was only partially successful, with "pogo" vertical oscillations affecting the first stage. Nevertheless, the first manned flight, Apollo 7, went ahead in Oct 1968, and the first lunar landing, by Apollo 11, was brilliantly achieved only 9 months later.

APOLLO MISSION PROFILE

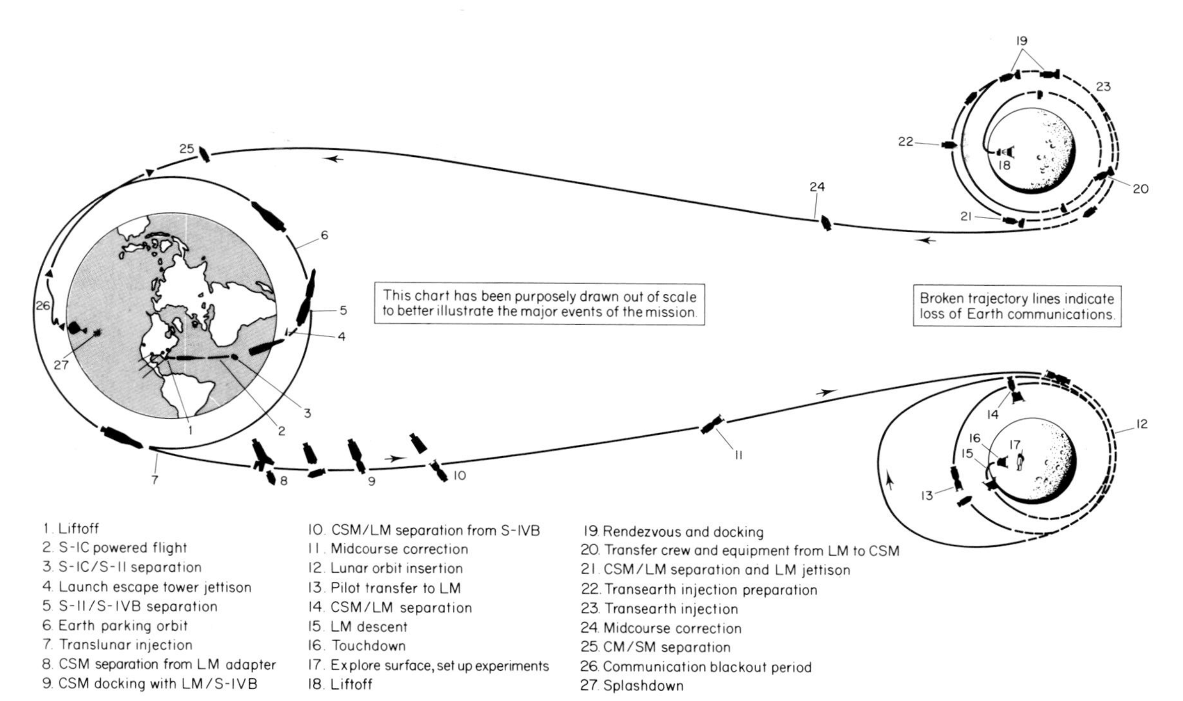

Spacecraft description The Apollo spacecraft included the Command Module (CM), providing control centre and living quarters for the 3-man crew, with total lift-off weight of about 5,800kg and splashdown weight of about 5,300kg; the Service Module (SM), containing the 91·18kN (9,300kg) thrust Service Propulsion System (SPS) engine, its propellant and most of the mission's consumables (with a total weight of up to 25,000kg, it could not be entered by the astronauts and was jettisoned before re-entry); and the Lunar Module (LM), a 2-stage vehicle with descent and ascent stages, used to convey 2 astronauts between lunar orbit and the Moon's surface, and which provided living quarters on the Moon. Total weight rose to 16,600kg on the last flights. The mechanism mounted in the CM docking tunnel enabled the CM and LM to be connected twice during a normal lunar mission: at the beginning of translunar flight, when the CM had to turn around, dock with and withdraw the LM from its launch position inside the 3rd-stage Saturn 4B rocket; and when the LM ascent stage returned from lunar orbit. The Lunar Rover Vehicle (LRV) was used with great success on Apollos 15-17. A manned spacecraft on wheels, with Earth weight of 209kg, it gave the astronauts with their equipment a 10km radius of exploration from their touchdown point on the lunar surface. Its remotely controlled TV unit, with umbrella-like antenna, enabled Mission Control at Houston to watch what the astronauts were doing and to select their own lunar views for scientific study without troubling the astronauts.

The first Moon landing. Neil Armstrong took this picture of Ed Aldrin soon after Apollo 11's 1969 touchdown. ***(NASA)***

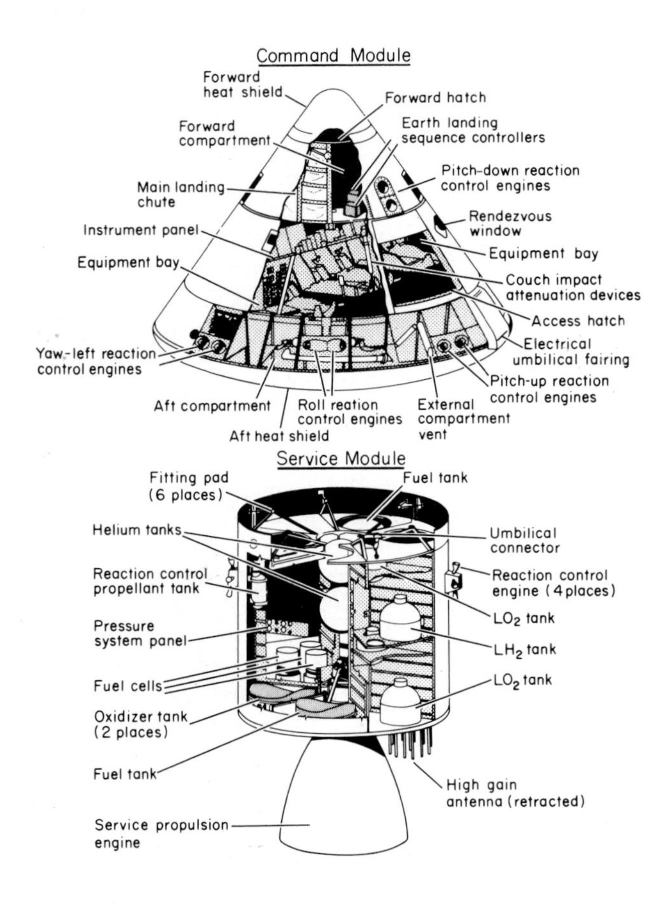

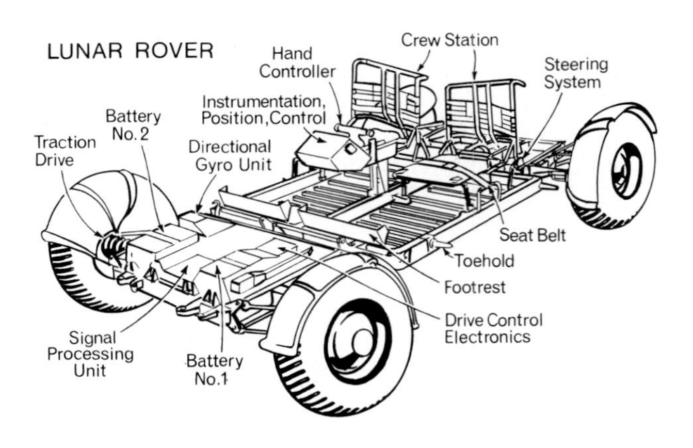

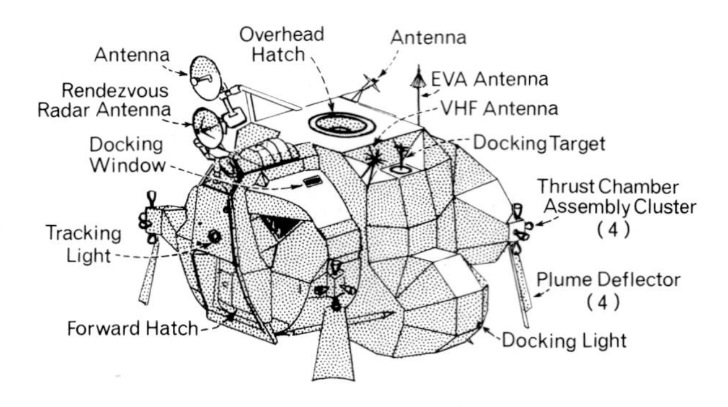

LUNAR MODULE - ASCENT STAGE

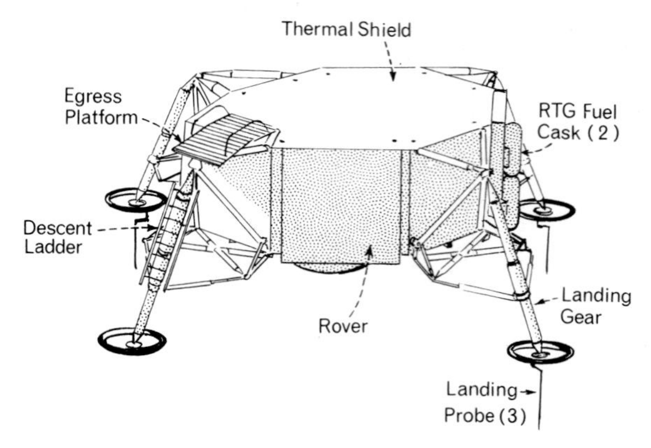

LUNAR MODULE - DESCENT STAGE

Apollo 1 Scheduled for launch in Mar 1967, this mission was tragically delayed when on 27 Jan 1967 a short-circuit in the electrical systems set fire to the all-oxygen atmosphere while the crew, Virgil Grissom, Edward White and Roger Chaffee, were carrying out a full launch rehearsal on Pad 34. White, reaching back over his head, was unable to open the hatch in the few seconds before the crew was overwhelmed. At that time it was known that there were many things wrong with the Apollo hardware, and in retrospect the 18-month delay caused by the disaster and the resulting complete redesign are seen as contributing significantly to the smooth progress of the programme when flights did begin. The death of the three astronauts also changed the whole crew sequence, directly affecting the choice of the men who would become the first to land on the Moon. Grissom was on record as saying: "If we die, we want people to accept it. We hope that if anything happens to us it will not delay the programme. The conquest of space is worth the risk of life." So, when his widow brought a lawsuit against North American Rockwell and NASA two years after the fire, some astronauts thought it was a breach of faith with Grissom. In 1971 North American Rockwell quietly settled the lawsuit out of court with a payment of $350,000, thereby establishing a cash value for an astronaut's life. On the company's initiative similar settlements were bestowed on the widows of Chaffee and White.

Apollo 7 Oct 1968. The first manned flight of the Command and Service Modules conducted in Earth orbit, lasting nearly 11 days (260hr and 163 orbits). Walter Schirra, Donn Eisele and Walter Cunningham successfully tested the operational qualities of the space vehicle; measured the performance of the Service Module's main propulsion engine by firing it both automatically and manually on eight occasions; simulated the all-important manoeuvre of extracting the Lunar Module (which was not in fact carried) from the S4B 3rd-stage rocket; and checked out the performance of the heatshield during re-entry. It set the pattern of success which carried the Apollo project right through to its successful Moon landings.

Apollo 8 Dec 1968. Man's first flight around the Moon. Frank Borman, James Lovell and William Anders were the first men to be launched by the 3,000-tonne Saturn 5. The critical lunar orbit insertion (LOI) manoeuvre was achieved by firing the SPS engine behind the Moon on 24 Dec. After that the astronauts spent 20hr circling the Moon, filming and photographing the far side, never before seen by man, and obtaining paictures of craters, rilles and potential landing sites on the near side. It was at this time that Frank Borman took the famous "Earthrise" picture, showing the Earth rising above the Moon's horizon, and on Christmas morning stirred the whole world with his reading of the "In the beginning" passage from Genesis. Splashdown in the Pacific at the end of the 147hr mission was just 11sec earlier than the time, computed months before, given in the flightplan.

Apollo 9 Mar 1969. The first manned flight with the Lunar Module. James McDivitt, David Scott and Russell Schweickart were launched after a 2-day postponement because of sore throats and nasal congestion. Although the 24hr flight was confined to Earth orbit, separation of the Lunar Module from the Command Module, followed by rendezvous and docking, were carried out in conditions simulating those which would be encountered on the Moon-landing missions. On the 4th day, with McDivitt and Schweickart in the LM and Scott in the CM, the 2 craft were separated. When 182km apart the LM's descent stage was jettisoned and the ascent stage, simulating a lift-off from the Moon, successfully fired its 15·5kN (1,590kg) thrust rocket engine to rendezvous and dock with the CM. The ascent stage, jettisoned before re-entry, did not decay until 26 Oct 1981.

Apollo 10 May 1969. This 8-day (192hr) flight by Thomas Stafford, Eugene Cernan and John Young was the successful dress rehearsal for the actual Moon landing 2 months later. The first time the complete Apollo spacecraft had orbited the Moon, it took man to within 14·5km of the lunar surface. The mission closely followed the Apollo 11 flightplan. The CM remained in a 111km lunar orbit for nearly 32 revolutions, with Young in control, while Stafford and Cernan undocked and made a simulated landing in the LM by twice descending to within 14·5km of the surface. The spacecraft successfully docked after 8hr of separation.

Apollo 8 lifting off for man's first flight around the Moon. *(NASA)*

The Apollo 1 crew patch, prepared in December 1966, was withheld following the launchpad fire in January 1967 and has rarely been published. *(NASA)*

Man's first view of Earthrise from the Moon. Picture signed for the author by Apollo 8 commander Frank Borman. *(NASA)*

David Scott, in the open Command Module hatch, inspects the Lunar Module during Apollo 9, the lunar lander's first flight test. *(NASA)*

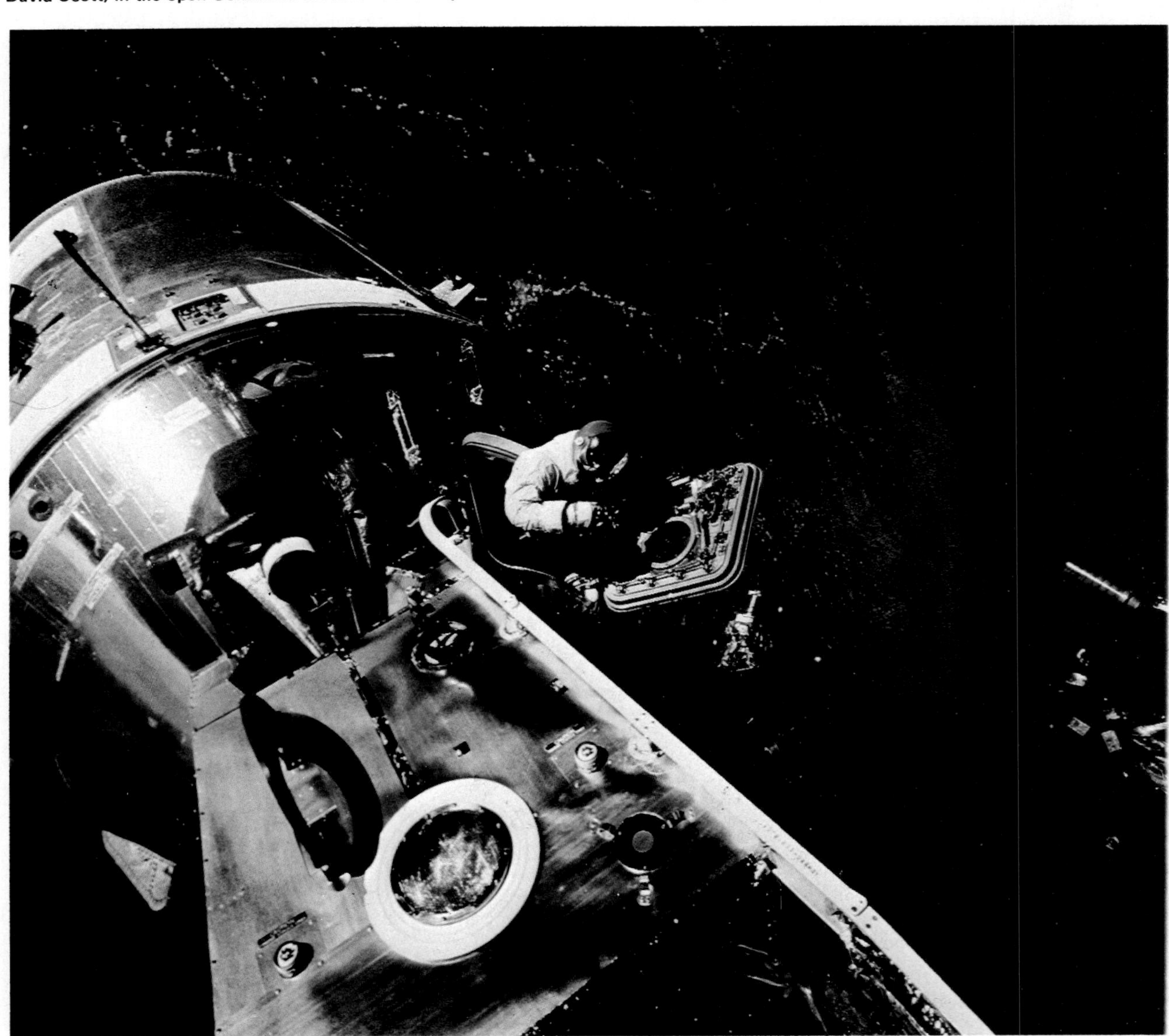

Top: Apollo 10 view of Triesnecker crater, which has a diameter of 31km. *(NASA)*

Above: Stafford and Young in the Lunar Module took this picture of the Command Module as they re-docked after the final Moon landing rehearsal. *(NASA)*

Apollo 11 July 1969. The historic flight by Neil Armstrong, Edwin Aldrin and Michael Collins which culminated in Armstrong and Aldrin becoming the first men to step on the Moon. Launch from Cape Kennedy was achieved without any technical problems on Wed 16 Jul. Lunar orbit insertion was successfully achieved on Sat 19 Jul, and the spacecraft was placed in a 100 × 121km orbit. On Sun 20 Jul, while Collins remained in control of the CM (code-named *Columbia*), Armstrong and Aldrin entered the LM (code-named *Eagle*). The 2 spacecraft separated on the 13th lunar orbit, and *Eagle*'s descent engine was fired behind the Moon. As he neared the surface, Armstrong decided to take over manual control because the spacecraft was approaching an area in the Sea of Tranquillity strewn with boulders.

At that stage the mission was almost aborted. Responsible for saving it was Steve Bales, a young engineer in Mission Control charged with monitoring the performance of *Eagle*'s computers. Persistent malfunction alarms were called by Aldrin and confirmed by Capcom, indictating that one of the computers was overloading. The unit in question was a vital part of the radar system which calculated the LM's altitude and rate of descent, and on which the life of the crew depended. Steve Bales had to decide whether the computer would clear itself or to recommend an abort. His quiet "Go, Flight" in a hushed Mission Control cleared *Eagle* to continue the landing from 2,000m. (Later he was to stand beside the Apollo 11 crew at the White House to receive the Medal of Freedom in recognition of his cool judgement.)

At 21·17 BST (16·17 EDT) on Sun 20 Jul, with less than 2% of descent propellant remaining, Armstrong told Mission Control: "Contact light, OK, engine stop ... Houston, Tranquillity Base here. The Eagle has landed." Mission Control replied: "Roger, Tranquillity. We copy you on the ground. You've got a bunch of guys about to turn blue. We're breathing again. Thanks a lot." After checking the LM's systems in case an emergency lift-off was required, Armstrong and Aldrin elected not to take the 4hr rest period scheduled in the flightplan, but to go straight ahead with preparations for their 2¼hr of activity on the Moon's surface. The arduous process of donning spacesuits and backpacks, and depressurising *Eagle*, took a good deal longer than expected. But finally Armstrong, on hands and knees, backed carefully out of the spacecraft on to the small platform or porch outside and descended the 3·05m ladder attached to the landing leg.

At 03·56 BST on Mon 21 Jul (22·56 EDT on Sun 20 Jul) he became the first man to step on the Moon. As he placed his left foot on the dusty surface, a breathless world could just distinguish his words: "That's one small step for a man; one giant leap for mankind." Aldrin joined him on the surface 18min later, and the 2 astronauts practised the best way of moving about in one-sixth gravity by walking, running and trying out a "kangaroo hop". They erected a 1·52m US flag extended on a wire frame since there is no wind on the Moon, saluted it, and received a congratulatory telephone call from President Nixon. A plaque fixed to the spacecraft's ladder was unveiled. It read: "Here men from the planet Earth first set foot upon the Moon, July 1969 A.D. We came in peace for all mankind" and bore the signatures of the 3 Apollo 11 astronauts and of President Nixon. Armstrong and Aldrin collected 21·75kg of rock and soil samples, and placed on the surface of the Moon a laser reflector to enable scientists to measure Earth-Moon distances by reflecting light beams from it, and a seismometer to measure meteorite impacts and moonquakes.

Apollo 11: instruments placed on the Moon included a seismometer and laser reflector. ***(NASA)***

Below: This stereo view of a 3in-square patch of the lunar surface was taken by the Apollo 11 crew with a 35mm close-up camera. Shiny, spherical particles enabled scientists to deduce the processes that shaped and modified the surface. ***(NASA)***

Bottom: Pre-flight picture of Apollo 11 crew Neil Armstrong, Michael Collins and Edwin Aldrin. Collins stayed in lunar orbit while the others explored the Sea of Tranquillity. ***(NASA)***

The launching of the ascent stage went as planned, and docking was achieved 3½hr later. An uneventful return journey ended in splashdown at 17·50 BST on Thur 24 Jul (12·50 EDT). After donning biological isolation garments the astronauts were taken by helicopter to the recovery ship, where they immediately entered the Mobile Quarantine Facility (a modified trailer), in which they travelled to the Lunar Receiving Laboratory at Houston, Texas. Thus they were kept in isolation – a precaution which proved to be unnecessary – for 21 days after lift-off from the lunar surface to ensure that no lunar contamination was brought back to Earth. During their 8-day mission the Apollo 11 astronauts had travelled 1,533,225km.

Apollo 12 Nov 1969. This 2nd lunar landing mission, by Charles Conrad, Richard Gordon and Alan Bean, started sensationally, for as the launch was being made through a rain squall the Saturn 5 rocket was struck by lightning. The spacecraft's electrical system was briefly put out of action and for the first time during a manned launch they were very close to an abort. When the all-Navy crew separated in lunar orbit the Command and Lunar Modules were code-named *Yankee Clipper* and *Intrepid*, with Conrad and Bean making the landing. Because Apollo 11 had touched down 6·5km beyond the target point extra course corrections were made, and a soft undocking carried out.

At the end of the 579,365km journey *Intrepid* touched down as planned, only 183m from the Surveyor 3 spacecraft which had been soft-landed in the Ocean of Storms 31 months earlier. Conrad and Bean were on the Moon for 31½hr, during which they made 2 excursions totalling 7½hr on the surface. In addition to collecting over 34kg of rocks and soil samples, they deployed the first Apollo Lunar Surface Experiment Package (ALSEP), 6 scientific experiments with a nuclear-powered battery giving the equipment an operational life of at least a year, and brought back the TV camera and other parts of the Surveyor spacecraft so that scientists could discover the effects on them of long-term exposure to the solar wind, and of the extreme variations of temperature in vacuum conditions. Although the astronauts were again kept in quarantine for 3 weeks after leaving the lunar surface as a precaution against bringing back any possible contamination to Earth, it was again found to be unnecessary.

Apollo 13 Apr 1970. This 3rd Moon landing mission was intended to be the first of 3 landings devoted to geological research on the lunar surface. Instead, an explosion on board when the spacecraft was 329,915km from Earth all but cost the lives of the crew and turned the mission into a 3½-day rescue drama surpassing any space fiction story as tens of thousands of technicians in the US aerospace industry worked to bring the crippled spacecraft safely home.

The commander, Jim Lovell, about to make his 4th spaceflight, command pilot Thomas Mattingly and lunar pilot Fred Haise, neither of whom had flown before, had been in contact with German measles. Finally it was decided that Mattingly had no immunity to the disease

Apollo 12, top, touched down only 200m from Surveyor 3, which had landed on the Ocean of Storms 2½yr earlier. Astronaut Pete Conrad walked across and brought its TV camera back to Earth for examination. *(NASA)*

and he was replaced with the back-up command pilot, Jack Swigert. Progress towards the Moon became so uneventful and routine that public interest in the expedition rapidly waned. But on 13 Apr, at 55hr 54min into the flight, came the explosion. It was 9pm at Mission Control at Houston, 4am in Britain; the astronauts had just been congratulated on a routine but successful TV transmission and were being advised on how to locate Comet Bennett, when Swigert interrupted sharply: "Hey, we've got a problem here." There had been "a pretty large bang", followed by a "Main B Bus undervolt". (The spacecraft's major electrical harnesses are known as A and B, and called "buses" by engineers.) Only later was it established that the module had been ripped open by an explosion amid its oxygen tanks; but 10min after, amid mounting tension as oxygen readings in the spacecraft fell to zero, the crew reported they could see "a gas of some sort" venting into space.

Only 80min after the crisis began, journalists were told that the Moon landing had been abandoned and that the aim must be to swing the spacecraft round the Moon and use the Lunar Module's power systems and supplies to bring it back to Earth. 10min after that Mission Control told the crew: "We're starting to think about the LM lifeboat." Swigert replied: "That's something we're thinking about too" - the only spoken acknowledgement at any time of the acute danger they faced. With only 15min of power left in the CM, Mission Control instructed the crew to make their way into the LM. At first it was thought that, even by using the LM as a lifeboat to tow the crippled spacecraft home, only 38hr of power, water and oxygen were available - about half as much as would be needed to get the crew home. But as technicians brought their computers and simulators into use, techniques were devised for powering down the systems so that, although it meant increasingly cold and uncomfortable conditions, there would be an ample margin for the return. 4 firings of the LM's descent engine were carried out in an elaborate series of manoeuvres - accompanied by heart-stopping moments on 2 occasions when the astronauts, tired and chilled, made mistakes.

The CM, code-named *Odyssey*, jettisoned the SM 3½hr before splashdown, and as it moved away there were exclamations from the astronauts as they saw and photographed for the first time the extent of the damage. Finally the LM, code-named *Aquarius*, was jettisoned 1hr before re-entry began. At that time no one knew for certain whether the explosion had damaged the heatshield. Evidence that it had not was dramatically provided when the spacecraft, descending on its 3 parachutes, suddenly filled the TV screens of the world's hundreds of millions of viewers. Apollo 13 ended as a triumphant failure.

It was 5 years later, in NASA's official Apollo history, before the full details of what went wrong were revealed by Jim Lovell himself. The CSM's No 2 oxygen tank

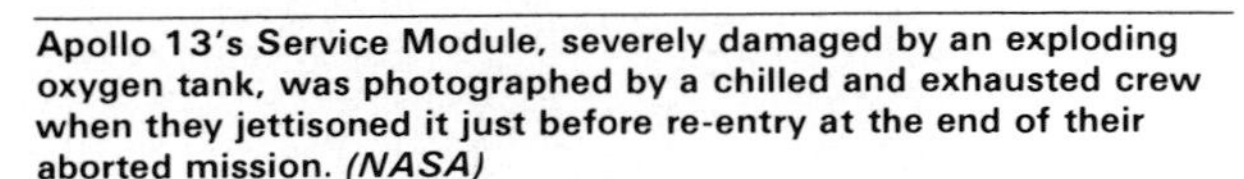

Apollo 13's Service Module, severely damaged by an exploding oxygen tank, was photographed by a chilled and exhausted crew when they jettisoned it just before re-entry at the end of their aborted mission. ***(NASA)***

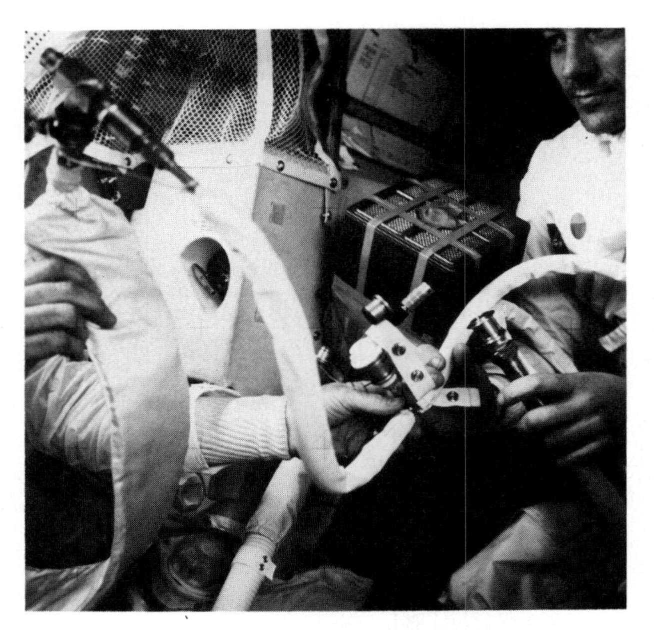

The Apollo 13 crew survived an uncomfortable journey home in their Lunar Module "lifeboat," using the jury-rigged arrangement shown here to purge carbon dioxide from their atmosphere. Swigert is seen on right. ***(NASA)***

overheated and blew up (putting No 1 out of action as well) because its heater switches had welded shut when subjected to excessive pre-launch electric currents. Originally Tank No 2 had been installed in Apollo 10; there was trouble with it then, and it had been removed for modification and damaged during the removal. This is how Lovell concluded his description of this famous drama: "Furthermore other warning signs during testing went unheeded, and the tank, damaged from 8hr overheating, was a potential bomb the next time it was filled with oxygen. That bomb exploded on 13 Apr 1970 - 200,000 miles from Earth".

Apollo 14 Feb 1971. The 3rd successful Moon landing expedition, commanded by Alan Shepard, America's first man in space and the only one of the original 7 Mercury astonauts finally to set foot on the Moon. Shepard's leadership, together with the immense physical endurance he displayed, proved that criticisms of his selection as commander at the age of 47 and after recovery from serious ear trouble were completely unjustified. He had flown only one 15min space "lob" and his companions, CM pilot Stuart Roosa and LM pilot Edgar Mitchell, had had no previous spaceflight experience at all, so this was easily the least experienced of the Apollo crews.

For the first time in the programme the CSM was placed in a very low lunar orbit, with an apocynthion of 112km and a low point of only 15,250m. This meant that the LM would have more fuel available for hovering and selecting a safe landing point. After separation on the 12th lunar orbit Roosa took *Kitty Hawk*, the CSM, up to a 112km circular orbit. Shepard brought the spacecraft to a safe touchdown only 26·5m from the target at 10·18 GMT on 5 Feb. Although it was 75sec late, there were 60sec of fuel left, compared with Armstrong's 20sec on Apollo 11. When Shepard stepped on to the Moon he commented: "It's been a long way but we're here". The first EVA, from depress to repress, lasted 4hr 44min; Mitchell's "thumper" device, which should have fired small explosive charges on the surface of the Moon to generate readings on the seismometer, was not completely successful, with only 14 of the 21 charges going off. Next day Shepard and Mitchell loaded up their 2-wheel trolley and towing it set off for the rim of Cone Crater, nearly 1·6km away and 122m above the touchdown point. After 2hr 10min they were 50min behind schedule and tiring. Shepard's heart rate reached 150, Mitchell's 128; they finally turned back short of the rim and one of the highlights of the mission, rolling stones down the inside of 38m-deep Cone Crater, had to be abandoned.

Before lift-off from the lunar surface Shepard produced 2 golfballs and used the handle of the contingency sample equipment to tee off. TV viewers were amused by the first lunar golfer, but for years after the incident was exploited by critics of the space programme, who bemoaned the cost of "sending a man to play golf on the Moon". The second EVA lasted 4hr 41min. A record 44kg of lunar rocks and soil was brought back to Earth. The crew were the last lunar landers to have to endure the 3-week quarantine procedure after their return.

Apollo 15 July 1971. The first of 3 "J Series" missions intended to exploit the scientific potential of the Apollo hardware. Lasting 12 days 7hr, it was 2 days longer than any previous Apollo flight, and such an unqualified success that one estimate was that David Scott, the commander, Alfred Worden, the CMP, and James Irwin, LMP, had brought back as much scientific informtion as Charles Darwin acquired during his 5-year voyage in *Beagle* in 1831-6. The all-US Air Force crew had in fact named the LM *Falcon* after the Air Force mascot, and the CSM *Endeavour* after the ship that carried James Cook on his 18th-centry scientific voyages.

The crew donned their spacesuits when, a few hours before entering lunar orbit, they blew a 2·90m × 1·52m panel off the SM to expose 8 scientific experiments in the bay containing the Scientific Instrument Module (SIMBAY). The experiments included mapping and panoramic cameras, and spectrometers, 2 of them on extendable booms more than 6·1m long, for assessing the composition of the lunar surface, solar X-ray interaction, and particle emissions while the CSM spent 6 days orbiting the Moon. The SIMBAY also carried a 35·4kg subsatellite which was successfully ejected following the conclusion of the lunar landing to spend a year studying the Moon's mascons and other phenomena.

Falcon had about 3,050m clearance over the top of the 3,960m Apennine Mountains as it dropped down to land in a basin near Hadley Rille, a 366m chasm named after the 18th-century English mathematician. *Falcon* came to rest at 22·16 GMT on 30 Jul within a few score metres of the target, but with a 10° tilt because one footpad was in a 1·52m crater. This caused Scott and Irwin much inconvenience with the swinging hatch door when they were crawling in and out in their cumbersome space suits for the 3 EVAs which gave them a total of 19hr 08min on the surface.

They had some difficulty deploying the first Lunar Rover. Though they found that when driving in one-sixth g the vehicle was a "real bucking bronco," making seat belts essential, Scott and Irwin were delighted with it. When they reached the rim of Hadley Rille, 4km from the LM, the world shared their view when they turned on the TV camera mounted on the Rover and deployed the umbrella-like antenna to transmit excellent pictures direct to Earth. On this first EVA, however, energy expended and thus oxygen consumption were 17% greater than expected, and Mission Control warned that activities would have to be curtailed. Before it ended, however, the 3rd ALSEP was deployed. It was an exciting moment for geologists, who were confident that by using the Apollo 12, 14 and 15 ALSEPs they could finally pinpoint and even predict the monthly quakes occurring in 11 areas (mostly in the Copernicus region) each time the Moon passes at its closest point to Earth.

The 2nd EVA proved to be the most rewarding period so far on the Moon. Scott excitedly pronounced that he thought they had discovered what they went to find: a 10cm piece of anorthosite rock believed to be part of the pristine Moon and quickly dubbed the "Genesis rock". (Later, geologists decided it was 4,150 million years old, 150 million years older than any previous sample recovered, but it was *not* as old as the Moon itself.) The astronauts finally returned to *Falcon* after driving a total of 28km and collecting 78kg of rock samples.

The TV camera on the Rover was carefully set by Scott so that lift-off from the Moon could be televised for the first time. The result was an astonishing 2sec of TV when the screen burst into red and green as the ascent stage rushed upwards to rejoin Worden, who had had a busy 73hr alone in *Endeavour* conducting the orbital science experiments. Satellite ejection was followed by trans-Earth injection (TEI) on the 74th lunar orbit. Then came Worden's planned 19min spacewalk 321,870km from Earth to retrieve 2 film cassettes from the SIMBAY. It was the first spacewalk ever made for a practical, working purpose. After re-entry one of the three 25·3m parachutes failed to open. The recovery carrier warned the crew to "stand by for a hard landing" and they emerged none the worse for their dramatic splashdown.

Right: A panel was blown off Apollo 15's Command Module to expose cameras and instruments designed to study the lunar surface while the CM waited for the Lunar Module's return. The CM is seen here over the Sea of Fertility. *(NASA)*

Apollo 15: commander David Scott drove the Lunar Rover to the very edge of Hadley Rille. Photo taken by astronaut Jim Irwin. *(NASA)*

The reason for the parachute failure was never conclusively established; most likely causes were either faulty metal links connecting the suspension lines, or residual RCS fuel burning through the shroud lines while being dumped during the final descent. On the subsequent missions the links were strengthened and the fuel retained on board.

Details of some of the crew's activities during their 3rd day on the Moon were only revealed some years later. In Irwin's words: "We played post office and cancelled the first stamps of a new issue commemorating US achievements in space. With our own cancellation device – which worked in a vacuum – we imprinted August 2, 1971, which was the first day of issue." The crew had taken 400 unauthorised postal covers or envelopes, in addition to about 250 official ones, as part of their "personal preference kits". Some were to be sold in a complicated deal involving a German businessman, and the 3 crew members expected to use the $8,000 they would each get to set up a trust fund for the education of their children. The story leaked out when the sale of the envelopes at $1,500 each began earlier than the astronauts expected. After NASA and Congressional inquiries they were reprimanded and removed from their assignment as back-up crew for Apollo 17. It effectively ended their astronaut careers. But in 1983, with NASA short of a payload on STS-8 and planning to send up 2 tons of stamped envelopes in a deal with the US Post Office, Worden sued the agency for the return of the envelopes they had confiscated. NASA was apparently advised that it would lose the action, and in August 1983 returned the envelopes. Some estimates put their current value at around $500,000.

Apollo 16 Apr 1972. The 5th Moon landing was a triumphant success, achieved despite the fact that at one point the mission appeared to be a repeat of Apollo 13, with the possibility of an emergency return to Earth. It lasted 11 days 1hr 51min, 24hr 45min less than the original flightplan. Total lift-off weight was 2,914,000kg; lift-off thrust was 34,355kN (3,503,405kg). Spacecraft weight was 48,602kg.

A dozen technical problems kept John Young, commander, Thomas (Ken) Mattingly, CMP, and Charles Duke, LMP, troubleshooting for much of the outward journey. The major crisis came after the undocking on the 12th lunar orbit. Mattingly, having delivered LM *Orion* into a 107 × 19km orbit, had to fire CSM *Caspar*'s main SPS engine to circularise his orbit and be available for a possible rescue manoeuvre before *Orion* attempted its lunar landing. He did not do so because of indications of yaw oscillations in the back-up system for firing the engine. With only minutes to go before starting their final descent Young and Duke were given the first "space wave-off". Both craft were ordered to continue orbiting and to reduce the gap between them, ready to re-dock, while Mission Control studied the problem and even talked of once more using the Lunar Module as a liferaft to tow home a disabled Command Module. But on the 15th orbit Christopher Kraft, Houston's director, gave a "go" for landing. Mattingly, using the primary system, successfully fired the SPS engine to circularise his orbit. Young and Duke finally settled 6hr late on the Cayley Plains of the Descartes landing site at 02·23 GMT on 21 Apr, 200m N and 150m W of their Double Spot landmark. They were only about 3m from a crater some 7·6m deep.

Young had become the first man to orbit the Moon twice; he and Duke, the 9th and 10th men on the Moon, became the most enthusiastic, vocal and efficient of them all. But after their exhausting experience during the approach they elected to sleep before making the first EVA. Duke, who had complained earlier about his spacesuit being tight, had no difficulty at the start of the first, 7hr 11min, EVA. The one major disappointment on the Moon came when Young backed into and broke the cable connecting the top-priority heat-flow experiment to the ALSEP powerplant, thus destroying the scientists' hopes of checking the discovery by the Apollo 15 crew that heat was flowing from the lunar interior at a far higher rate than expected. Although at a later stage a rear mudguard fell off the Lunar Rover, causing the astronauts to be sprayed with soil, it performed well. At the end of EVA 1 Young, filmed by Duke, carried out the planned "Grand Prix," driving the Rover flat-out in circles and skidding it to test wheel grip.

Apollo 16: John Young, already on his 4th flight, tries out the Rover on the Descartes landing site. His flimsy vehicle reached a top speed of 17km downhill. ***(NASA)***

EVA 2 dramatically demonstrated the improved quality of the TV pictures from the Moon. When the astronauts had driven 4·1km to Stone Mountain and 232m up it they could be seen working on ridges and crater rims. Geologists on the ground used the TV camera completely independently to look around and zoom in for close-ups of boulders and other features of interest. The geologists were embarrassed by the lack of evidence of volcanic activity.

EVA 3 was shortened to 5hr 40min to meet the revised lift-off time. It was devoted to exploring North Ray crater, a large depression 4·9km from *Orion*. Before re-entering for the last time Young and Duke competed to see how high they could jump. Duke, who had tumbled many times, fell heavily backwards, causing much concern at Mission Control, but escaped with no personal or spacesuit damage. Ascent and docking went perfectly, but a wrongly positioned switch caused *Orion* to start tumbling immediately after undocking. *Caspar* had to make a hurried evasive manoeuvre, and *Orion* had to be left in lunar orbit instead of being crashed on the lunar surface. In view of the slightly suspect back-up system for firing the SPS, the journey home started as soon as the 40kg subsatellite had been ejected. (The planned plane change to place the subsatellite at a more suitable inclination was cancelled to avoid another SPS burn. As a result, the subsatellite lasted only a few weeks; after 425 revolutions it crashed into mountains on the far side on 29 May 1972.) The highlight of a relatively quiet return flight was Mattingly's EVA 274,000km from Earth to recover film cassettes from the SIMBAY. All 3 parachutes deployed perfectly after re-entry and were recovered for examination because of the Apollo 15 failure. The CM, however, remained upside-down in the water until the buoyancy balloons were released. Residual propellants were retained on board for the first time because of fears that they might have damaged Apollo 15's parachute lines. But the wisdom of venting them on previous missions was demonstrated when they exploded after the spacecraft had been taken back to North American at Downey, California, causing damage and injuring a number of technicians.

A special diet of food and orange-juice laced with potassium made this crew the first to survive a mission without any irregular heartbeats, removing doubts about the practicability of the 56-day flights planned for Skylab. Young and Duke clocked up record totals of 20hr 40min on the lunar surface, drove 27km and brought back an estimated 97·5kg of lunar samples.

Apollo 17 Dec 1972. The final Moon landing proved to be the most perfect of the Apollo expeditions. The crew, Eugene Cernan, commander, making his 3rd flight, Ronald Evans, CMP, and Dr Harrison (Jack) Schmitt, LMP, the first trained geologist to visit the Moon, stayed longer, travelled further and brought back more lunar samples, 113kg, than any previous crew. America's first night launch saw night turned into day by pure-gold flame as Saturn 5 extended its unbroken success record by placing the spacecraft accurately on course. Total lift-off weight was 2,923,383kg; lift-off thrust was 34,099kN (3,476,794kg). Spacecraft weight totalled 52,744kg: CM *America* 30,320kg, LM *Challenger* 16,448kg.

On their 12th revolution in lunar orbit Cernan and Schmitt separated *Challenger* from *America*. When Evans had safely established the CM for a possible rescue manoeuvre in a 100 × 125km orbit Cernan swept the LM down between 2 2,130m-high mountain ranges to touch down between the Taurus Mountains and Littrow Crater on the edge of the Sea of Serenity. The area had been chosen by scientists because Alfred Worden, orbiting in the Apollo 15 CM, had seen conical mounds similar to those formed on Earth by volcanic debris accumulating around a vent. The surrounding area ranged from the most ancient highlands to young mantling material, or dust, overlaying the lowest areas. Co-ordinates of the 6th and final Apollo landing, at 19·55 GMT on 11 Dec, were 20°12′16″N, 30° 45′ 0″E. During the ALSEP work the Earth audience enjoyed both panoramic and close-up views showing with perfect clarity the smooth-topped mountains which surrounded them and, at one place, an avalanche of light material.

Cernan, after some initial difficulty which caused him to use more of his oxygen than expected, succeeded in drilling both of the 1m holes required for the heat-flow experiment; after the Apollo 16 failure this had been given top priority. Because Cernan had inadvertently knocked off the Lunar Rover's right rear mudguard the astronauts and their equipment were showered with Moondust during their first drive. They drove south to Sterno Crater and gathered 21·7kg of samples. Back inside *Challenger*, they were advised from Houston by John Young on how to improvise a makeshift mudguard by taping together 4 of their lunar maps and using emergency lighting clips to fasten them to the Rover at the start of the 2nd EVA. Then came a full hour's drive taking them over 6·4km W from the LM to the foot of the 2,130m-high mountain range. On the way back Cernan kicked up orange soil and for 20min, with Schmitt scrambling excitedly around the rim of Shorty Crater, scientists were riveted to the TV screens. The remotely operated zoom lens clearly showed the rust-coloured soil. There was general agreement that the astronauts had found a volcanic vent, and that there had been, and perhaps still was, water on the Moon. But 3 months later, following analysis of the orange soil, the US Geological Survey announced that it consisted of tiny, orange-coloured glass beads formed by the heat generated by an ancient meteor impact. By 1975, however, it was decided that the beads, unusually rich in lead, zinc and sulphur, were of volcanic origin and had been thrown out from perhaps 300km inside the lunar interior during a period when the lunar mantle had partially melted. With this still in the future, Cernan and Schmitt re-entered the LM after travelling a record 19km and setting an EVA

Apollo 17: during the final EVA at Taurus-Littrow, Schmitt studied a bus-sized boulder and brought back football-sized samples. ***(NASA)***

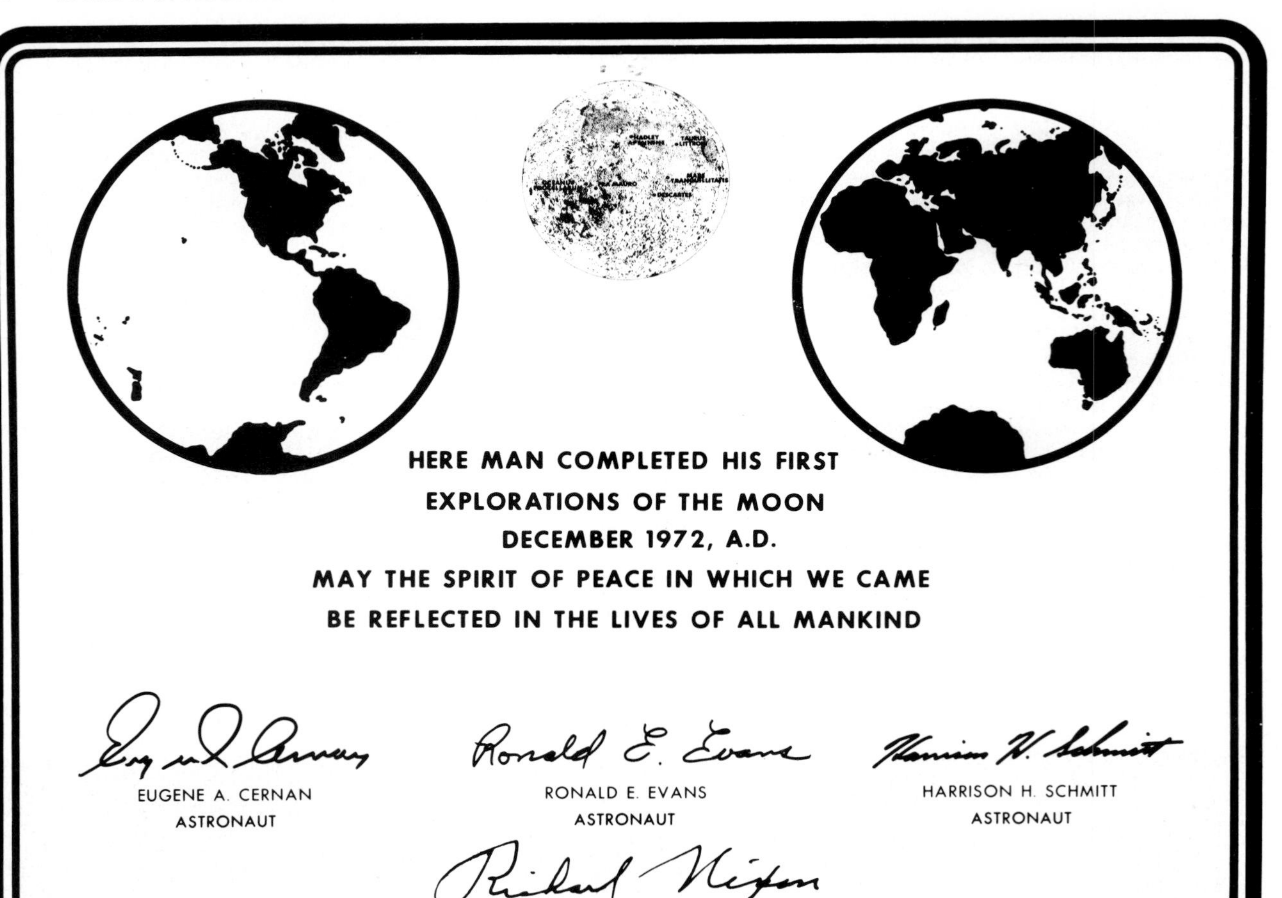

The Apollo 17 crew, the 6th and last to visit the Moon, left this plaque on their Lunar Rover to show the position of all the landing sites. *(NASA)*

LM *Challenger* pictured by Ron Evans just before the final docking in lunar orbit. *(NASA)*

record of 7hr 37min 21sec. On the final EVA there was dramatic TV of Schmitt scaling boulders the size of London buses to collect samples. The Lunar Rover was taken up mountain sides so steep that its wire-mesh wheels were dented. During their 3 EVAs Cernan and Schmitt had travelled a total of 35km in the Rover and spent 22hr 4min outside *Challenger*. They had been on the Moon's surface for 74hr 59min.

Its record load of 113kg of Moon rocks meant that the LM was 18kg overweight. Nevertheless, lift-off, vividly seen via the TV camera carefully pre-positioned on the abandoned Rover, went smoothly. The extra weight was offset by reorganising the engine firing to burn more fuel more quickly. During Evans' 60min cislunar spacewalk 290,000km from Earth he retrieved film and data canisters from the SIMBAY without difficulty. Splashdown in the Pacific, with the CM settling right way up for once, was 6·4km from the recovery ship. During the 12-day 13hr 51min flight the CM (and Evans) had travelled 2,390,984km.

Summary and conclusions The total costs of Project Apollo, including the development and production of all the hardware, the facilities, tracking networks and launch costs, were $24 billion (£10,000 million). The main disappointment was that, after Apollo 17, man seemed as far as ever from understanding the origin of the Moon. The first Moon landings, the technical mastery of space travel and the final scientific expeditions to the lunar surface must nevertheless rank as the most significant step so far in man's development.

NASA itself admits that, "since the Apollo landing sites were highly restricted by operational and safety limitations, most of the Moon's surface, containing many intensely interesting and mysterious features, remains untouched, awaiting mankind's next series of contacts". The total of almost 385kg of lunar materials amassed by the conclusion of Apollo 17 should enable techniques to be developed for making lunar bases self-supporting. By the end of Apollo 14 it had already been established that both water and oxygen could be produced from the lunar soil, since it contains a high percentage of iron oxide.

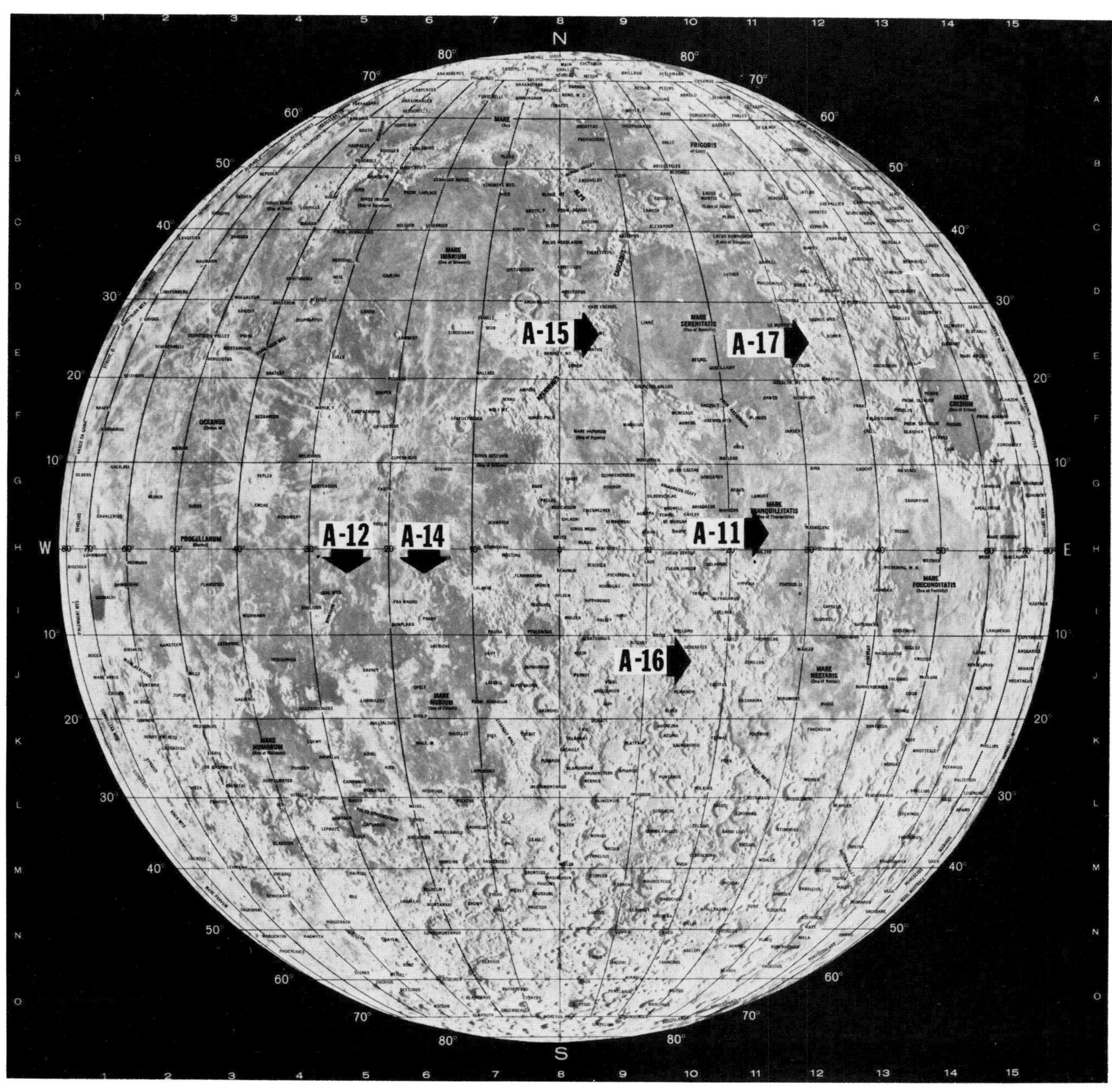

This map shows how widely spread were the six Moon landing sites. But the astronauts' hopes of exploring the more rugged terrain towards the South Pole were frustrated by cancellation of Apollos 18-20. *(NASA)*

By using a solar furnace and introducing hydrogen, 6·35kg of water could be produced from 45·4kg of iron oxide, and the water then separated into oxygen and hydrogen. Thus life could be supported and fuel provided for space vehicles.

There was much scientific criticism when, on 1 Oct 1977, NASA finally shut down the Houston Control Centre, which had been receiving the stream of information from the 5 ALSEPs (Apollo Lunar Scientific Experiment Packages) left on the Moon's surface by the crews of Apollos 12, 14, 15, 16 and 17. It was a purely economic measure, designed to save the $2 million per year cost of the Centre. From that day, although the nuclear-powered ALSEPs were all in perfect working order (apart from some occasional mysterious lapses by Apollo 14), their continuing signals about moonquakes, temperatures, etc, went unrecorded. But Apollo 12's ALSEP was in its 8th year, and Apollo 17's in its 5th, and NASA had accumulated 29 years of operating time compared with the cumulative total of 6 years for which the equipment was designed. Even so, there were still insufficient data for the geologists to be certain whether the Moon has a molten core. To determine that, they needed at least one large "event" (greater than 10^{19} ergs) on the far side. It never came before closedown, although the 10,000 moonquakes and 2,000 meteorite impacts that had been recorded included major quakes on the near side on 13 May 1972, and on the far side (10^{18} ergs) on 19 Sep 1973.

Astro

This project calls for 3 7-day flights of a 3-telescope ultraviolet astronomy instrument package mounted on 2 Spacelab pallets with the Instrument Pointing System being developed for Spacelab 2. Scheduled for Shuttle missions 32, 44 and 59 in 1986-7, the flights will enable scientists to sample objects around the celestial sphere, providing new opportunities for the study of hot stars, galaxies, quasars, supernova remnants and the interstellar medium. Halley's Comet will also be studied during the 1st flight. In addition to 2 mission specialists, 2 career astronomers will fly as payload specialists to operate the instruments.

Applications Technology Satellites

History Arthur C. Clarke, who first conceived the idea of synchronous satellites, described ATS-6 as "the most ambitious and important educational experiment in history". It was launched in 1974 to beam direct TV programmes into village homes in underdeveloped countries; its achievements since are described below. Originally ATS was conceived as a 5-satellite series, starting in Dec 1966, to test components and techniques for communications, meteorological and navigation satellites. ATS-1 and 3 proved that predictions that such vehicles would be able to provide fixes to within 10m of any point on the surface of the Earth were correct. Before the ATS programme NASA tried out communications satellites at low, intermediate and synchronous orbits; they included passive reflectors such as the 33m-dia Echo balloons and active repeaters such as Relay 1 (L 13 Dec 1962; wt 78kg; orbit 1,322 × 7,439km; lifetime 100,000yr) and Syncom (see International - Intelsat). The ATS programme, based on the results of these experiments, included Syncom techniques such as spin stabilisation and complicated synchronous orbit manoeuvres. From their geosynchronous orbits 35,680km above the equator ATS satellites match Earth's rotational period and thus remain stationary in relation to a selected point on the Earth's surface. They view about 45% of the globe, compared with the 3% of a spacecraft in a 320km orbit.

ATS-1 L 7 Dec 1966 by Atlas-Agena D from C Canaveral. Wt 352kg. Soon after lift-off this satellite, 1·42m dia and 1·34m high, with 23,870 solar cells, was successfully manoeuvred into its permanent position over the Pacific; in 1982 it was still operational - making it the oldest working geostationary satellite - though transmitting only on command. In addition to many scientific and technological experiments it provided a wide variety of services. The capabilities of such satellites were vividly demonstrated when ATS-1 provided Australia with 10hr of TV from Canada's Expo '67. On 12 Oct 1967 the Japanese were able to see their Prime Minister's visit to Australia, and later that year the world saw President Johnson and other heads of state attending the funeral of Australian Prime Minister Holt. In the same year many lives and millions of dollars were saved in Alaska when ATS-1 provided emergency communications during the great flood there. In 1969, when the Apollo 11 astronauts returned from the first Moon landing, ATS-1 provided the main communications link between President Nixon and USS *Hornet*, the recovery ship.

ATS-2 L 6 Apr 1967 by Atlas-Agena D from C Canaveral. Wt 370kg. Intended to be gravity-gradient-stabilised and placed in a medium, 11,100km, equatorial orbit, but a failure in the Agena rocket prevented the booster from re-igniting at a crucial point in the trajectory. As a result, the satellite remained in a 185 × 11,177km orbit at 28° incl. Severe decelerations in this irregular orbit overcame the stabilisation system, and only limited data were obtained from the systems and experiments as ATS-2 slowly tumbled and rotated. It re-entered on 2 Sep 1969.

ATS-3 L 6 Nov 1967 by Atlas-Agena D from C Canaveral. Wt (after apogee motor burnout) 365kg. A perfect launch placed this satellite in a synchronous equatorial orbit of 35,765 × 35,807km over the Atlantic Ocean, where it was still operational and transmitting on command 15yr later. The most notable of many achievements was the sending back of the first colour photograph of Earth from space; it also obtained many cloud pictures and was used to monitor severe storms. On 21 Nov 1967 it provided the first ground-to-spacecraft-to-aircraft communications link over the Atlantic during a Pan American flight. Transmissions to and from the aircraft via the satellite were monitored not only in America but in London, Hamburg, Frankfurt and Buenos Aires. Later ATS-3 provided support for the Apollo Moon landings and TV coverage of Pope Paul's visit to Bogota, Colombia, in 1968. In a series of maritime experiments covering ship location and ship-to-shore communications it demonstrated that such satellites could bring major improvements to the management of shipping fleets. At the end of 1974 it was being used by the US National Oceanic and Atmospheric Administration (NOAA) to predict and measure rainfall in remote areas by the comparison of satellite pictures of the brightness and size of clouds with radar scans of their density. This system will provide flood warnings to farm populations in mountain areas. During the 1980 Mount St Helens eruption ATS-3 was used to relay communications from a jeep at the disaster site. In 1983, stationed at 105°W, it was being used for remote site communications by users of a 2-suitcase terminal taking 2min to set up.

ATS-4 L 10 Aug 1968 by Atlas-Centaur from C Canaveral. Wt 392kg. Intended to be a gravity-gradient spacecraft, but a Centaur failure left it in a 217 × 772km orbit. Although its systems were turned on, few data were obtained. Dec 17 Oct 1968.

ATS-5 L 12 Aug 1969 by Atlas-Centaur from C Canaveral. Wt (after apogee motor jettison) 433kg. Another gravity-gradient spacecraft, ATS-5 was still providing useful information on command at the end of 1976. From the start, however, only 9 of its 14 experiments were operational because the spacecraft began to wobble on its spin axis shortly after separation. The hydrazine attitude-control jets had to fire 15 times faster than normal to reduce the wobble to required limits. The satellite was allowed to drift to its planned geostationary location at the equator, 1,770km W of Quito, Ecuador, taking 20 days instead of 11hr if the apogee motor had been fired a 2nd time. ATS-5 concluded its flight upright but wrongly spinning counter-clockwise; this meant that its gravity-gradient experiment booms could not be deployed. However, its auroral particle experiments sent back a large quantity of useful data until the end of 1974. The US tanker *Manhattan* was used to demonstrate the ability of ATS-5 to locate ships precisely by L-band ranging signals from the Mojave (Calif) ground station, and to exchange teletype messages with unattended equipment on board ship. Other tests included transmissions for the Federal Aviation Administration to a Boeing aircraft, and investigations of the effects of rain on signal fading.

ATS-6 L 30 May 1974 by Titan 3C from C Canaveral. Wt 1,402kg. Orbit 35,781 × 35,791km. Incl 1·6°. The world's first educational satellite, intended to bring the benefits of space technology to backward areas by lessening ignorance and poverty. 14hr after launch it was perfectly stationed over the equator above the Galapagos Islands, giving 40% global coverage and capable of being relocated E or W. ATS-6 is 8·51m high, topped by a 9m-dia reflector antenna, with 2 combined VHF antennas and solar panels above that on the ends of a 16m boom. It radiated 200,000W of effective radio energy, compared with the 6,400W produced by Intelsat 4. Combined with the ability to point the antenna within one-tenth of a degree of arc, this meant that it could beam colour TV direct to thousands of small ground receivers. Essentially a full-size ground transmitter placed in orbit, it began its 20 experiments by providing the first satellite educational course on 2 Jul 1974; more than 600 elementary school teachers in 8 Appalachian states took a graduate-level course by means of ATS-6 transmissions. After a year of providing remote schools in the Rocky Mountains with special courses, and 2-way communications for medical diagnoses in Alaska, the ATS ground station at Rosman, N Carolina, began the complicated operation of moving ATS-6 a distance of 12,800km so that it could be used for the promised year-long Indian educational experiment. The 2-storey spacecraft was set in motion by 2 burns of its onboard rocket motor; the 2nd, lasting 5hr 37min 17sec, was the longest to date by a chemical rocket in space. 3 separate braking manoeuvres spread over a week brought ATS-6 to a halt at its new station above Lake Victoria in E Africa 6 weeks later. There it played a major part in the Apollo-Soyuz flight of Jul 1975: for 50min of each ASTP orbit Apollo communications were passed to Houston via ATS-6. Its presence resolved a disagreement between America and Russia as to where the docking should take place. Both countries wanted it to be within range of their own tracking stations; ATS-6 permitted it to take place over Europe, within range of Soviet stations, while at the same time transmissions to and from Houston were relayed by the satellite. On 2 Aug, only 9 days after the end of Apollo-Soyuz, educational transmissions to 2,400 villages in 7 states in India began. Half the villages were equipped with 3m-dia antennas made of chicken wire, plus a TV set, costing a total of $1,000. India's government-operated TV network (normally able to transmit to only 4 cities) had devised a series of simple programmes demonstrating, for instance, how to plant rice correctly by keeping the shoots a fist's width apart, and how to cook using only the primitive utensils available in an Indian village. Another lesson was aimed at persuading Indian mothers to feed commonly available solid food to their babies instead of exclusively breast-feeding them until the age of 3. From a ground station at Ahmedabad, N of Bombay, the Indian Space Research Organisation then transmitted the programmes via ATS-6, which NASA made available for 4hr per day for 1yr. In some villages there was anti-climax when their screens failed to work for the opening speech by the Prime Minister, Mrs Gandhi. But in many others hundreds of people, including naked children, crowded round single sets to watch the programmes. Other achievements by ATS-6 included the first aircraft-to-ship communications relay via satellite; the first direct flight control of an aircraft by an ocean air traffic controller using a satellite; and the first direction of a search-and-rescue operation (simulated) by satellite. By the end of 1974, however, one of its experiments had failed. This was the Very High Resolution Radiometer, a camera-like instrument designed to test new ways of gathering meteorological information from synchronous orbit. It did however send back over 1,000 images before the failure.

By May 1979 3 of ATS-6's 4 stationkeeping thrusters had failed and the reliability of the 4th was questionable. So on 30 Jun 1979 this thruster was used to boost ATS-6 several hundred km above stationary orbit, out of the way of future occupants of its slot. It had done 5yr work instead of the planned 2; during that time proposals to launch its back-up as ATS-7 had been dropped as an economy measure. The series was built by Hughes Aircraft. Cost of the ATS-6 mission was $180 million, plus $25 million for the launch.

ATS 6: Indian technicians near Hyderabad prepare to receive their first transmissions. ***(NASA)***

ATS-6, which pioneered direct-broadcast TV, had 16m solar paddles and a 9m-diameter reflector. The Earth-viewing module is seen below.

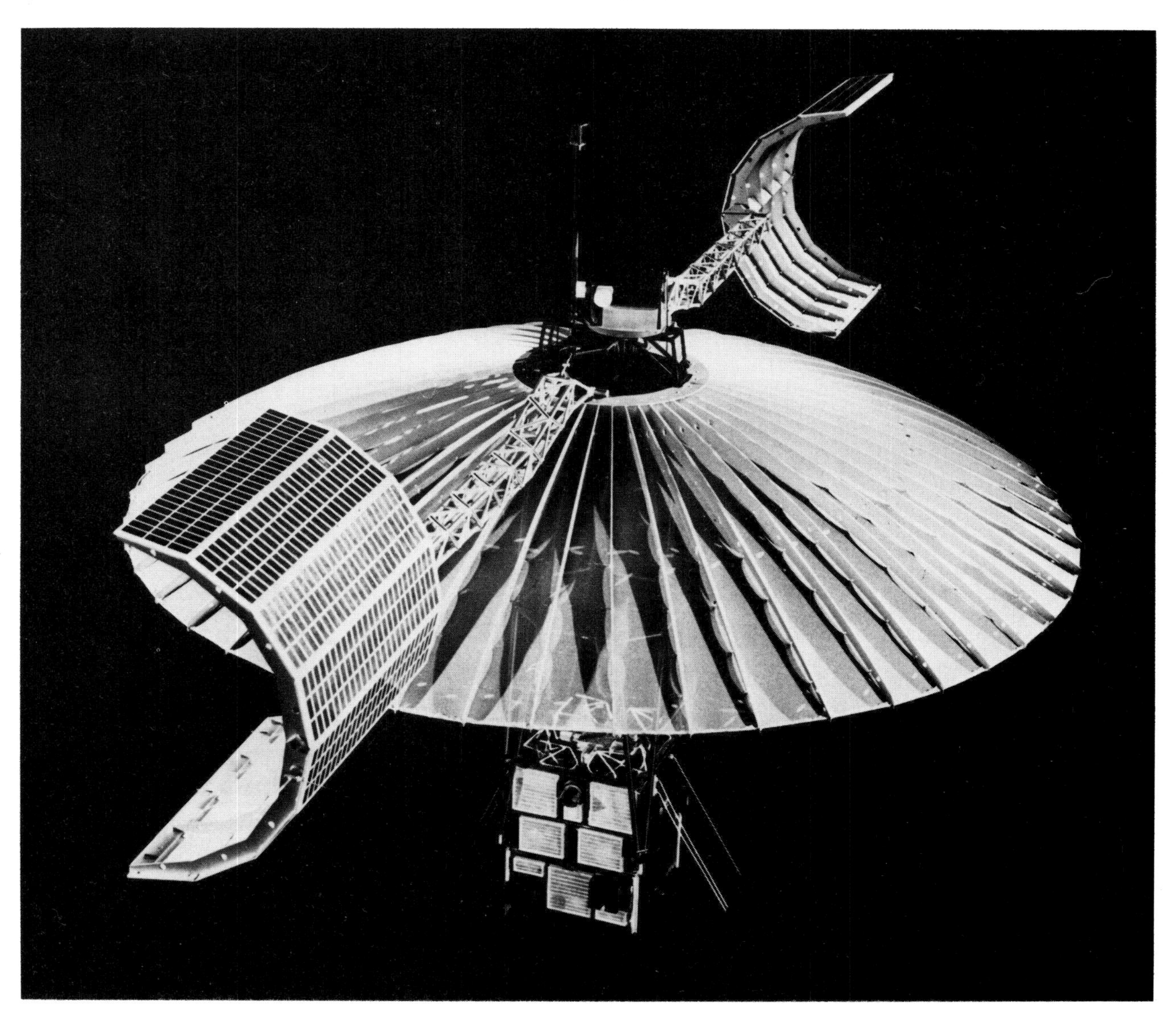

AXAF

The Advanced X-Ray Astrophysics Facility, due for launch by Shuttle into a 500km orbit in the early 1990s, will resume the work started by Uhuru in 1970 and continued by HEAO-2 until its re-entry in 1981. International scientists were asked in Oct 1983 for proposals for the use of this 10-tonne, 4·2 × 13m facility, which is designed to study X-ray emissions from all known astronomical objects. It will be maintained in orbit for 15yr.

Biosatellite

History America's least successful space programme, it was originally intended to investigate the prolonged effects of weightlessness and radioactivity on living organisms. There were to be 6 flights, the first 2 with plants and organisms, the next 2 with pig-tailed monkeys, and the last 2 with rodents. Bonny, the first of the pig-tailed monkeys, was sent up while the world's space correspondents were gathered at the Cape for the first Moon landing. The principle of using animals in this way was heavily criticised before the flight. Astronauts themselves took the view that it was pointless to use animals, which were unable to cope with technical troubles , and that men were willing and able to take part in such experiments. Then, when Bonny was launched and had a bad time, there was so much criticism from the assembled correspondents that the last 3 flights were cancelled on the grounds of "economy". Apart from OFO-1 (Orbiting Frog Otolith) in 1970, in which 2 male bullfrogs were sent up to test effects on ears and balance, it was not until 6yr later that NASA was able to resume such biological experiments - and then only by joining in on a Soviet biological satellite, Cosmos 782. The first unmanned Soviet-American satellite, this was a follow-up to the successful Apollo-Soyuz flight.

Bios 1 L 14 Dec 1966 by TAD from C Canaveral. Wt 426kg. Orbit 307 × 316km. Incl 33·5°. Retro-rocket failure led to the recovery capsule, a covered aluminium bowl of 79cm dia with 0·1699 cu m of payload space, being stranded in orbit. Spacecraft re-entered 2 months later, on 15 Feb 1967.

Bios 2 L 7 Sep 1967 by TAD from C Canaveral. Wt 508kg. Orbit 302 × 326km. Incl 33·5°. Similar to Bios 1, this carried 13 experiments, with a Strontium 85 radiation source to subject the weightless plant and animal specimens to regular doses of radiation. Onboard watering systems, with lights and cameras, enabled the specimens to be photographed every 10min and a total of 268 times during the flight. The 3-day flight was cut short at 45hr - 17 orbits earlier than planned - because of communications problems and a tropical storm in the Pacific recovery area. Subsequent pictures of an originally healthy pepper plant showed it looking very dejected indeed after 17hr 40min of weightlessness. Recovery was however successfully achieved, and showed that animal cells were least affected by weightlessness.

Bios 3 L 28 Jun 1969 by LTTA-Delta N from C Canaveral. Wt 697kg. Orbit 355 × 386km. Incl 33·5°. The mission cost $96 million. To monitor his progress Bonny the pig-tailed monkey had 24 sensors planted in his body, including 6 in the brain and one in each eye. It was these pre-flight preparations which led to criticism. He had to earn his food twice a day by pushing various buttons for 15min at a time. But though he had enjoyed doing the exercises on the ground Bonny proved to be lonely and unhappy in space, and refused to work or drink. After 9 days of what was intended to be a 30-day flight he was brought back, only to die shortly after being recovered. The official report recorded: "Inflight experiment data satisfactory for brief mission; post-flight experiments minimal due to premature post-recovery death of primate" (see also International - Biosatellites).

Comsat

History Comsat (Communications Satellite Corporation) is privately owned and was set up in Feb 1963 to carry out a US Government mandate to establish a global commercial communications satellite system in co-operation with other countries. It was designated as the US participant in Intelsat, which it operates on behalf of the 105 national members. By 1 Mar 1980 Comsat had a 22·6% investment in Intelsat; national contributions are adjusted annually to match each country's total use of the system in the previous 6 months. Comsat also participates in Comstar, Marisat, Inmarsat and SBS.

Comsat General

Marisat After several years' international controversy as to who should provide satellite communications for ships and aircraft, Comsat, which operates the Intelsat organisation, set up a subsidiary called Comsat General. This made a start in Feb 1976 by providing the $100 million Marisat 2-satellite experimental system for commercial shipping and offshore installations. The US Navy assisted by contracting to use much of its capacity for up to 3yr. Norwegian cruise ships and Esso tankers were among the 14 vessels initially fitted with receivers; their owners had agreed to pay $9,600 per ship for a year's service (see International - Marisat, Inmarsat).

Comstar

Comstars 1, 2 and 3 (L 13 May 1976, 22 Jul 1976 and 29 Jun 1978 by Atlas-Centaur) were placed in stationary orbit by Comsat General. Each weighs 1,516kg, has about 18,000 circuits and is leased for its 7yr lifetime to AT & T at $1·3 million per month. Comstar 4, intended as a back-up to conserve the power of the spacecraft already in service, was added on 21 Feb 1981. Built by Hughes, they are supplying communications services to major US metropolitan areas and Hawaii. Comsat General expects a return of $340 million on the $197 million investment in this programme.

Discoverer

History An early programme of major importance. Its tests of orbital manoeuvring and re-entry techniques not only played a large part in enabling the first US manned flights to be carried out in Project Mercury, but also led to the spy satellite systems now in regular use. Between 1959 and 1962, after which this type of work was classified, there were 38 launches, all by the US Air Force from the Western Test Range at Vandenberg. Project objectives were "military space research; development of capsule recovery techniques; and biological research". The series began 13 months after the successful launch of Explorer 1, America's first satellite. But for 18 months after that, as the list of the early flights illustrates, it was dogged by failures. The plan was for a modified Thor IRBM to boost a 2nd-stage Agena A rocket to near-orbit altitude; after separation the Agena's 6,800kg-thrust engine was fired to place the whole stage in orbit. There its gas-jet orientation system turned the vehicle through 180° so that the ballistic nosecone, containing a 136kg ejection capsule, was pointing rearwards and tilted down. It thus met the requirement that for a satellite to drop a bomb or return an object to Earth, the object must separate and use retro-rockets fired in the opposite direction to the satellite's path and at a downward angle to ensure that it re-enters the

Launch of a Discoverer spy satellite by Thor-Agena from Vandenberg. *(NASA)*

atmosphere. At first ejections were often at the wrong angle, and sometimes upwards. This resulted in a false alarm in Feb 1960, when the US Navy's early detection system found an unknown satellite, believed to be Russian, in a near-polar orbit. It was ultimately established that it was an early Discoverer capsule which had been ejected upwards instead of down. Ejection was achieved by explosive bolts on either orbit 17 or 33. At 15,240m a parachute pulled the recovery package clear of the heatshield, which fell into the sea, usually the Pacific near Hawaii. Radio and light beacons, and even radar chaff, were used to assist location as the equipment package descended into the target area. At first C-119 transport aircraft, towing trapeze-like frameworks at a height of 2,400m, patrolled the area in some numbers in vain efforts to develop the technique of "snatching" the parachute harness with the package attached before it reached the sea. A sea recovery was finally achieved on Discoverer 13, and the first mid-air capture on the following mission.

It was Discoverer 14 which almost certainly provided America with her first space photographs of the Soviet Union. Beginning with Discoverer 16 - which failed to achieve orbit - these satellites were launched with the more powerful Agena B. 8m longer and carrying more propellants for a new throttlable engine, it could orbit 950kg. Beginning with Discoverer 30 in Sep 1961, C-130s were used for mid-air recovery.

Amid much public concern that America's military capacity in space was probably far behind Russia's, the programme was pursued with almost frantic haste. There were 12 launches in 1961, with the last 2 officially designated Discoverers being launched in 1962. Looking at the large percentage of failures, it is easy to forget that space computer techniques, as well as hardware such as re-entry nosecones, were still being developed. But in the 21 launches between the first mid-air success and Discoverer 38 there were only 7 more successful mid-air recoveries; 4 sea recoveries could be counted partial successes; 11 launches were admitted failures. The high failure rate attracted much public comment; Russia meanwhile was able to pursue her parallel programme in complete secrecy, merely announcing the Cosmos numbers of her launches. Inevitably, from 22 Nov 1961 the US Department of Defence decided to classify all military spaceflights. Since then it has been possible to follow such activities only indirectly by studying the *Tables of Earth Satellites* compiled by Britain's Royal Aircraft Establishment. These show 27 capsule ejections in orbit during 1963-71; undesignated launchings are listed under the name of the rocket, Atlas-Agenas B and D until 1967, when the more powerful Thor-Agena and Titan-Agena began to come into use. These launches have long since passed from the experimental to the operational stage. The Military section contains further details of the regular passes now being made over Russia, China, the Middle East and other countries, whether or not they happen to be US allies, by satellites able to return data either directly or by means of ejected film packs. Below is a list of the early Discoverers and, for comparison, details of the last of them:

Discoverer 1 L 28 Feb 1959 by Thor-Agena A from Vandenberg. Wt 590kg (including 111kg of instruments). Orbit 159 × 974km. Incl 90°. First satellite in polar orbit. Objective was to test propulsion, guidance, staging and communications, but tumbling made accurate tracking impossible. Re-entered after 5 days.

Discoverer 2 L 13 Apr 1959 by Thor-Agena A from Vandenberg. Wt 730kg. Orbit 229 × 354km. Incl 90°. Objective was to eject and recover a 88kg hemispherical capsule containing a working life-support system. Capsule ejected on orbit 17 but was lost in Arctic. Satellite re-entered after 13 days.

Discoverer 3 L 3 Jun 1959 by Thor-Agena A from Vandenberg. Similar to Discoverer 2 but carrying 4 black mice in recoverable capsule. Failed to orbit after 2nd-stage failure.

Discoverer 4 L 25 Jun 1959. Similar to Discoverer 2. Failed to orbit, again because of 2nd-stage failure.

Discoverer 5 L 13 Aug 1959. Similar to Discoverer 2. Orbit 219 × 724km. Incl 80°. Capsule ejected but not recovered. Satellite dec 11 Feb 1961.

Discoverer 6 L 19 Aug 1959. Orbit 210 × 850km. Incl 84°. Capsule again ejected on orbit 17 but recovery failed.

Discoverer 7 L 7 Nov 1959. Orbit 159 × 835km. Incl 81°. Poor stabilisation, and capsule was not ejected. Dec after 19 days.

Discoverer 8 L 20 Nov 1959. Orbit 193 × 1,660km. Incl 80°. Capsule ejected on orbit 15 but overshot recovery area. Dec 8 Mar 1960.

Discoverer 9 L 4 Feb 1960. Failed to orbit; premature 1st-stage cut-off.

Discoverer 10 L 19 Feb 1960. Failed to orbit; destroyed by range safety officer when it went off course.

Discoverer 11 L 15 Apr 1960. Orbit 165 × 603km. Incl 80°. Capsule was ejected on orbit 17 but recovery failed. Dec after 11 days.

Discoverer 12 L 29 Jun 1960. Failed to orbit following 2nd-stage attitude instability.

Discoverer 13 L 10 Aug 1960. Orbit 253 × 694km. Incl 82°. Success at last; capsule ejected on orbit 17 and recovered from sea. Satellite dec 14 Dec 1960.

Discoverer 14 L 18 Aug 1960. Orbit 182 × 808km. Incl 79°. After ejection on orbit 17 capsule was captured for the first time in mid-air by patrolling aircraft. Dec after 28 days.

Discoverer 38 L 27 Feb 1962 by Thor-Agena B from Vandenberg. Wt 952kg. Orbit 334 × 495km. Incl 82°. Last in the series to be officially announced; capsule was successfully ejected and recovered in mid-air after 65 orbits. Satellite dec 21 Mar 1962.

ERBS

Earth Radiation Budget Satellite (ERBS), due for launch by Shuttle in Aug 1984, will work with NOAA-6 and 7, using scanning and non-scanning radiometers to measure solar radiation received and emitted by different Earth regions. More solar energy is absorbed in some regions and more thermal energy emitted in others; the resulting temperature differentials set wind and ocean currents in motion as heat is transferred to cooled areas. Thus Earth's "radiation budget" is perhaps the weather's main driving force. ERBS will be placed in a 600km circular orbit at 46° incl, and will communicate via TDRSS. In Nov 1980 NASA gave Ball Aerospace Systems of Boulder, Colorado, a $21 million contract to integrate the government-supplied instruments with the satellite bus and provide pre and post-launch support.

Explorer

History Explorer 1 was the United States' first satellite, and the world's third. Now numbering over 60, the Explorer series is comparable to the non-military part of Russia's Cosmos series. It is NASA's principal means of conducting long-term automated investigations of the Earth, the interplanetary medium and Sun-Earth relationships, and astronomical studies of the Sun, stars and galaxies which do not require more complex observatories like OSO, OAO and HEAO.

The series began before NASA's creation, and in a flurry of haste following the successful Soviet launch of Sputnik 1 on 4 Oct 1957. While the Russians had fulfilled their undertaking, made 2yr earlier, to launch a satellite for meteorological purposes as part of the International Geophysical Year of 1957-58, America's own efforts, concentrated on the Vanguard project, had been a dismal failure. It was at this point that Washington turned at last to Dr Wernher von Braun's group at the Army Ballistic Missile Agency at Huntsville, whose satellite proposals had been repeatedly turned down. They had produced a rocket called Juno 1, a 4-stage development of Jupiter C, which itself had been developed from Redstone, the 21·3m rocket which von Braun had evolved for America from his wartime German V2. In a plan submitted in April 1957 ABMA had recommended a programme to launch six 7·7kg satellites, the first of which would orbit in Sep 1957. This, as it turned out, would have given America a one-month lead over Russia in what was to become the major event of the 20th century, the Space Race. 3 weeks after Sputnik 1, ABMA was given authority to go ahead with plans to launch 2 satellites, with a target date of 30 Jan 1958 for the first. Explorer 1 was successfully launched 1 day behind this schedule. To this technical success can be added its major contribution to the International Geophysical Year – confirmation of the existence of radiation belts around the Earth, forecast by and named after Dr James Van Allen. While the early Explorers were tiny compared with Russia's Sputniks, their miniaturised instruments gathered data of extreme scientific value. They began a trend which ultimately put America far ahead in space techniques, and ensured, in spite of Russia's apparent lead in the early years, that the first men on the Moon were Americans.

After the US Army had developed and launched 5 Explorers, 3 successfully, the series was taken over by NASA when that agency was formed in Oct 1958. The major missions are listed below, grouped where possible:

Explorer 1 L 31 Jun 1958 by Jupiter C from C Canaveral. Wt (including integral last-stage motor) 14kg. Orbit 360 × 2,532km. Incl 65°. Total length, with rocket case, 2m. Explorer 1 carried 8kg of instruments designed to gather and transmit data on cosmic rays, meteorites and orbital temperatures. Confirmed the existence of belt of intense radiation beginning 965km above the Earth. Continued transmissions until 23 May 1958. Remained in orbit over 12yr, re-entering over S Pacific after 58,000 revolutions on 31 Mar 1970.

Explorer 2 L 5 Mar 1958 by Jupiter C from C Canaveral. Wt 14kg. Failed to orbit due to unsuccessful 4th-stage ignition.

Explorer 3 L 26 Mar 1958 by Jupiter C from C Canaveral. Wt 14kg. Orbit 195 × 2,810km. Incl 31°. Similar to Explorer 1 but with addition of small magnetic tape recorder able to release 2hr of stored data on cosmic-ray bombardment in 5sec as satellite passed over ground station. Transmitted until 16 Jun 1958; dec 28 Jun 1958.

Explorer 4 L 26 Jul 1958 by Jupiter C from C Canaveral. Wt 17kg. Orbit 262 × 2,210km. Incl 50°. Mapped Project Argus radiation until 6 Oct 1958; dec 23 Oct 1959.

Explorer 5 L 24 Aug 1958 by Jupiter C from C Canaveral. Wt 17kg. Failed to orbit; upper stage fired in wrong direction, leading to collision of rocket and instrument section.

Explorer S1 L 16 Jul 1959 by Juno 2 from C Canaveral. Wt 42kg. Failed to orbit; destroyed by range safety officer. The last of 5 successive failures (others were 2 Vanguards and 2 Discoverers).

Explorer 6 L 7 Aug 1959 by Thor-Able from C Canaveral. Wt 64·4kg. Incl 47°. "Paddlewheel" satellite 0·66m in dia, with 4 solar panels 0·5m sq for recharging the batteries. Instruments observed behaviour of radio waves in ionosphere, mapped Earth's magnetic field and cloud cover, etc. It also sent back first photograph of Earth. Transmitted until 6 Oct 1959; dec Jul 1961.

Explorer 7 L 13 Oct 1959 by Juno 2 from C Canaveral. Wt 41kg. Orbit 557 × 1,088km. Incl 50°. Returned data on Earth's magnetic field and solar flares until 24 Jul 1961. Orbital life 70yr.

Explorer S46 L 23 Mar 1960 by Juno 2 from C Canaveral. Wt 16kg. Failed to orbit; upper stage apparently failed to ignite.

Explorers 8, 20, 22 and 27 Ionosphere Explorers (L 3 Nov 1960, 25 Aug 1964, 10 Oct 1964, 29 Apr 1965) designed to measure electron distribution in space and time between satellite ht (approx 1,000km) and ht of max electron density (approx 350km). Also involved in this programme were 9 international launches: Ariel 1, Alouette 1, Isis X, FR-1, Esro 1, Isis A, Esro 1B, Isis B and Ariel 4.

Explorers 9, 19, 24/25 and 39/40 Air Density Explorers (L 16 Feb 1961, 19 Dec 1963, 21 Nov 1964, 8 Aug 1968) designed to determine the effect of thin air on satellite motion. 3·7m-dia inflatable spheres placed in approx 300 × 2,500km orbits, they observed the upper thermosphere and lower exosphere over the entire globe as a function of latitude, season and local solar time. Explorers 25 and 40 were dual-launched magnetosphere satellites which attempted to correlate atmospheric density readings with corpuscular energy measurements.

Explorers 18, 21, 28, 33, 34, 35, 41, 43, 47 and 50 These 10 Interplanetary Monitoring Platforms (IMPs) investigated Earth's radiation environment during a complete 11yr cycle of solar activity. They provided a vast range of new information about what goes on in space between the Earth and Moon and beyond the Moon. They discovered that the solar wind includes a double ion stream; the 2 streams penetrate one another and have a speed difference of over 300,000km/hr. While the first IMP (Exp 18, L 26 Nov 1963; orbit 125,000 × 202,000km) weighed only 62kg with a 35W power supply, Exp 50 (L 26 Oct 1973; orbit 94,697 × 238,989km) weighed 81kg with double power supply for spacecraft and experiments of 120W and 42W. Exp 35 (L 19 Jul 1967) was placed in lunar orbit and studied the Moon's magnetic field and radiation belt, as well as Earth's magnetospheric "tail". IMP spacecraft provided continuous warning of solar flare radiation during the Apollo and Skylab missions, and were used as survey baselines for the Pioneer 10 and 11 missions to Jupiter and the Mariner missions to Venus and Mercury in 1973-74. Their achievements include defining the nature and extent of the magnetosphere, the enormous, teardrop-shaped envelope surrounding Earth

Explorer 35, launched in 1967, was one of 10 Interplanetary Monitoring Platforms which made notable discoveries about the solar wind and the Earth's magnetospheric "tail". *(NASA)*

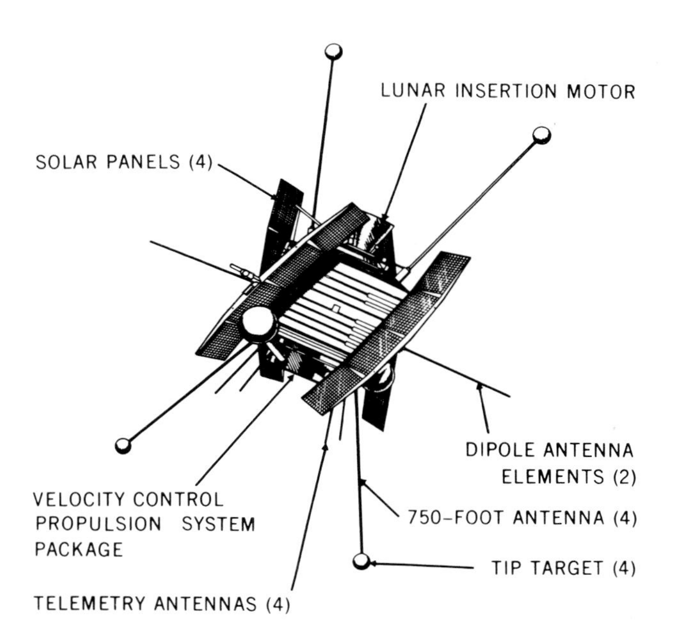

Above right: Explorer 49 was the 2nd of 2 Radio Astronomy Explorers which spent 4yr measuring galactic radio sources. *(NASA)*

which is formed by the particles of the solar wind as they enter the Earth's magnetic field. Launches were by Delta; orbital life is mostly over 1 million yr.

ISEE-1-3 was a follow-on project in 1977-8 (see below).

Explorers 29, 30 and 36 Geodetic Explorers (L 6 Nov 1965, 18 Nov 1965, 11 Jan 1968) launched by the US Navy to monitor solar ultra-violet rays and carry out geometric geodesy (the study of the size and shape of the Earth).

Explorers 38 and 49 Radio Astronomy Explorers. Exp 38 (L 4 Jul 1968; orbit 5,851 × 5,862km) spent 4yr listening to galactic radio sources, though for about one-third of its observing time its receivers were swamped with noise (radio broadcasting, storms, etc) coming from Earth. For that reason Exp 49 (L 10 Jun 1973) was placed in a 1,109 × 1,120km lunar orbit so that when it was on the far side the Moon would shield its radio experiments from Earthly interference. Once in lunar orbit, its 4 radio antennas, 458m between tips, formed a huge "X".

Explorers 42, 48 and 53 These Small Astronomy Satellites built up a sky map of gamma rays, the electromagnetic radiations which cannot be detected on Earth because they are absorbed in its atmosphere. Exp 42 (SAS-1), L 12 Dec 1970 by Scout from San Marco (orbit 521 × 563km; wt 306kg; incl 3°), was named Uhuru in honour of Kenya's independence day. It catalogued over 200 celestial X-ray sources and found the first evidence for the existence of a possible black hole in Cygnus. Exp 48 (SAS-2), L 15 Nov 1972 from San Marco, and Exp 53 (SAS-3), L 7 May 1975, entered similar orbits and continued to explore the dynamics of the Milky Way. Exp 53, with greater pointing accuracy, followed up the UK Ariel 5's discovery of an X-ray source 5 times more intense than any previous source, and identified a neutron star (Vela X-1) with a mass 1·7 times that of the Sun. This series was also supported by ANS, L 30 Aug 1974.

Explorers 51, 54 and 55 Atmosphere Explorers. This series, following up the aeronomy satellites Exp 17 and 32 of 1963 and 1966, continued to study Earth's atmosphere, particularly pollution of the upper layers (22-25km altitude) by, for example, freon gas emitted by aerosols. Exp 51 (L 16 Dec 1973, orbit 158 × 4,303km) had its orbit lowered to 120km every few weeks to take low-altitude samples, before being raised again after a few days to prevent decay as a result of atmospheric drag. Exp 54 (L 6 Oct 1975; orbit 155 × 3,816km) sampled nitric oxide, one of the controlling agents of ozone production and depletion. Exp 55, later known as AE-5 (L 20 Nov 1975; orbit 156 × 2,983km), dipped into the upper layers of the atmosphere to study changes in the total ozone field, the Earth's heat balance and energy conservation mechanisms for 5½yr and 31,268 orbits. Onboard propulsion prevented early decay until it finally re-entered on 10 Jun 1981.

ISEE-1, 2 and 3 International Sun-Earth Explorers, part of the International Magnetosphere Study, followed up the IMP series. ISEE-1 and 3 were provided by NASA, and ISEE-2 by ESA; for details of ISEE-1 and 2 (L jointly 22 Oct 1977) see ESA - ISEE. ISEE-3, L 12 Aug 1978 by Delta from C Canaveral (wt 469kg), had to be launched during a window lasting only 5min in order to achieve, 3 months later, the first "halo orbit" around the Sun-Earth/Moon libration point (L1). Located 1·6 million km from Earth, this is where the gravitational forces of the Sun and Earth-Moon system balance each other out. Boosted by 8 hydrazine thruster firings per year, the satellite rises and falls every 6 months 150,000km above and below the ecliptic plane so that solar radio emissions do not interfere with tracking and data acquisition. In this position it provides approx 1hr warning of solar activity before it reaches ISEE-1 and 2. In this way scientists tell

In 1978 ISEE-3 became the first satellite to be placed in "halo orbit" around Libration Point One, where the gravitational forces of Sun, Earth and Moon balance out. *(NASA)*

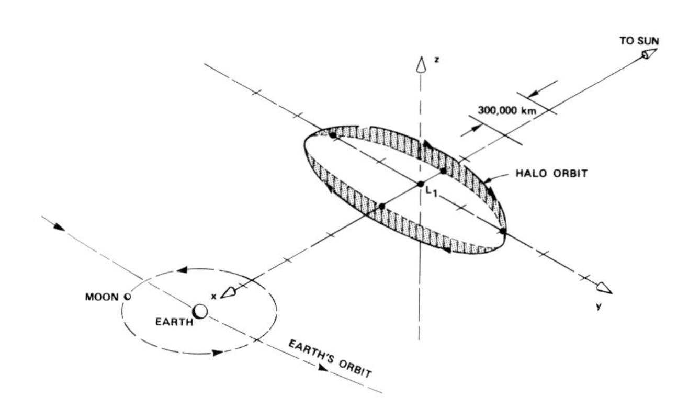

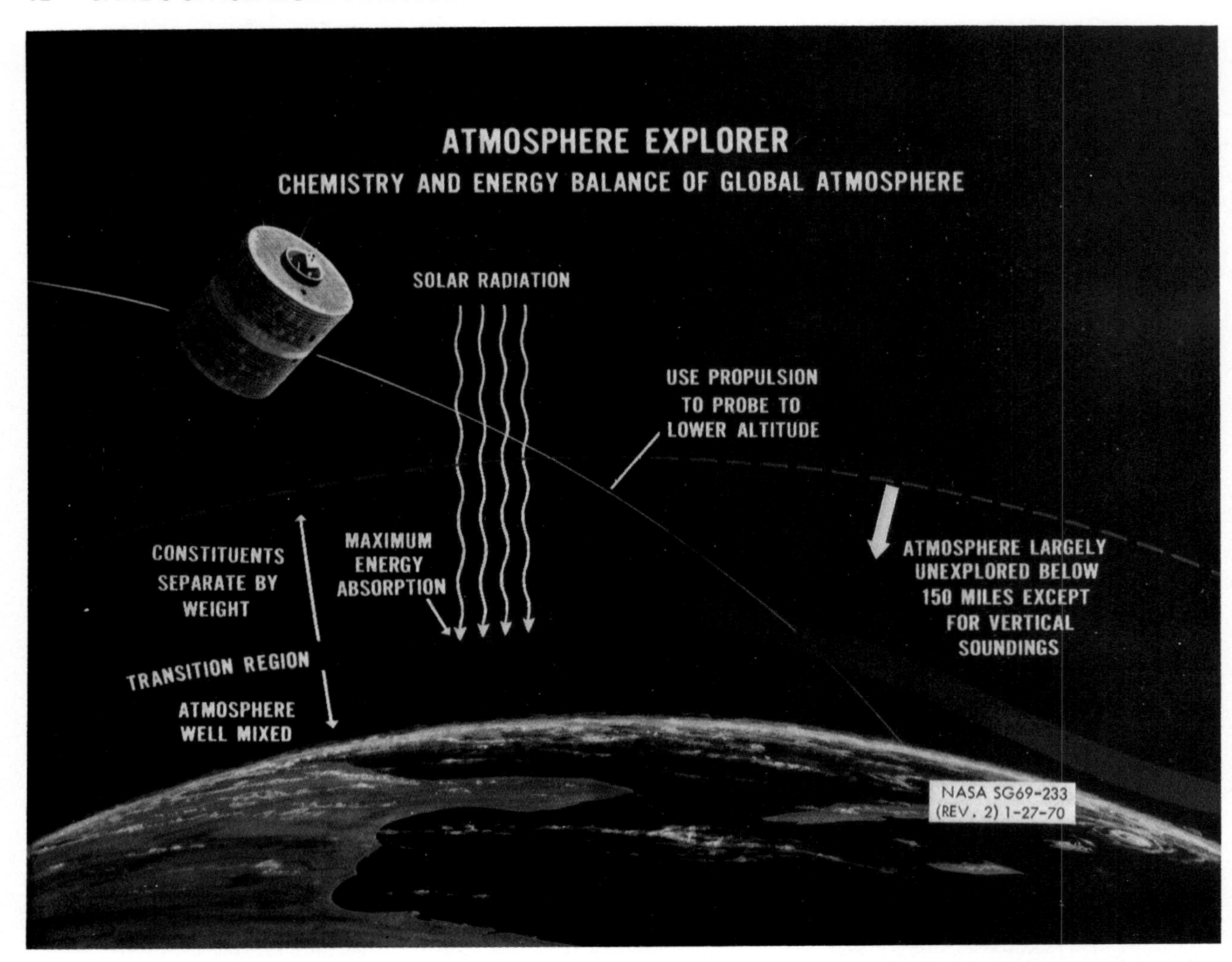

whether phenomena are caused by the Sun or by activities within the magnetosphere. With a 3yr design life, ISEE-1 and 2 may be functional for 6yr. ISEE-3, originally due to remain in this role for 10yr, was diverted back towards Earth in late 1982 so that it could be directed into a series of lunar swingbys – the last in Dec 1983, as close as 100km – to throw it towards man's first encounter with a comet. The aim is that it should pass through the tail, about 3,000km from the head, of Giacobini-Zinner on 11 Sep 1985. This will happen 6 months before Soviet, European and Japanese spacecraft intercept Halley's Comet. Through ISEE-3 can take no pictures it should return data on plasma densities, flow speeds, temperatures and heavy ions in the tail.

Having successfully increased its speed from 4,679 to 8,278km/hr by sweeping behind the Moon on 22 Dec, ISEE-3 was renamed the International Cometary Explorer by NASA. It had survived 28min without solar power (its battery having failed earlier in 1983) and had also been turned off for that period for the 1st time in its 5yr 4 months in orbit. After intercepting the comet 70·7 million km from earth, it may now be swung back towards Earth in the hope that a space tug may be available to pick it up, with its coating of cometary dust, between the years 2012 and 2015.

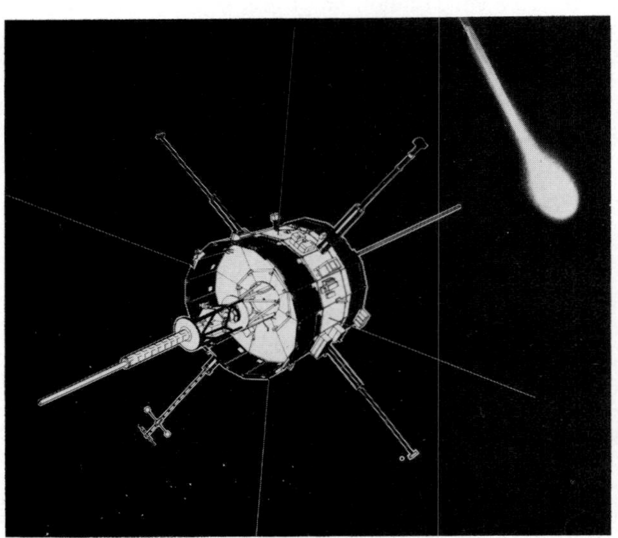

Artist's concept of ISEE-3 leaving Comet Giacobini-Zinner after flight through its tail in September 1985. It should be man's first comet interception. ***(NASA)***

IUE International Ultra-violet Explorer, L 26 Jan 1978 by Delta from C Canaveral. Wt 1,479kg (669kg in orbit). Modified synchronous orbit, 45,588 × 26,005km at 28·3°. The first astronomical satellite in synchronous orbit, IUE was a joint venture by NASA, ESA and the UK. Within 1yr it had provided 200 scientists from 17 countries with over 9,000 images of astronomical objects unobtainable from ground-based instruments to help them research basic questions about the composition of stars, nebulae and galaxies and how they develop. IUE is operated 16hr a day by the US from Goddard Space Centre, and by ESA and the UK for 8hr from Villafranca, Spain, where over 20,000 images have now been acquired and archived. In Oct 1978 IUE studied Comet Seargent and identified several molecular species thrown off as it receded from the Sun. Other results include the first ultra-violet observation of a supernova, and the first high-resolution ultra-violet spectrum of a star in another galaxy. While studying globular clusters in our galaxy IUE carried out observations of 10-20 bright stars which "may well be orbiting a massive black hole the size or mass of 1,000 solar systems". In 1979 it also discovered that the Milky Way galaxy is surrounded by a gaseous galactic corona, a hot gas envelope with a temperature of around 100,000°C and extending out as far as 25,000 light-years. Highlight of 1981 activities was the observation of "a target of opportunity," a supernova in the galaxy NGC 4536.

In late 1983 the international research team announced that the quasar-like heat of the galaxy known as NGC 4151 was powered by a black hole 100 million times heavier than the Sun. It was the 1st time the centre of a quasar had been weighed; to do so they had investigated gas clouds found to be moving at 50 million km/hr very close to the galaxy's core. NGC 4151, 50 million light-years away, is a spiral galaxy similar to our own Milky Way.

Applications Explorers AEM-1 (also known as HCMM, for Heat Capacity Mapping Mission) L 26 Apr 1978 by Scout from Wallops. Wt 134kg. Orbit 560 × 641km. Incl 97·6°. Orbital life 60yr. Thermal infra-red data were combined with Landsat data by German scientists working with the Goddard Centre to produce vividly enhanced images of such areas as Germany, Egypt and the Sinai Desert which were of value in oil and mineral exploration.

AEM-2 (SAGE, for Stratospheric Aerosol and Gas Experiment) L 18 Feb 1979 by Scout from Wallops. Wt 147kg. Orbit 549 × 661km. Incl 55°. Orbital life 40yr. Onboard photometer provides data on ozone and aerosol content by looking through the atmosphere during 15 sunrises and sunsets each day. The results are then compared with ground measurements made in US, Japan and Europe and with Nimbus 7 data. By 1982 AEM-2 observations of the 1980 Mount St Helen eruptions had helped to establish that such events did not seem to be related to the prolonged drought and heatwave in the US SW, nor to the cold, wet summer of N Europe in 1980.

Dynamics Explorers Complementing the 3 Atmosphere Explorers and the 3 ISEEs, DE-1 and DE-2 were launched together on 3 Aug 1981 by Delta from Vandenberg to study the flow of solar energy and matter from space through the Earth's magnetic field and into the upper atmosphere. Because the 2nd stage of the Delta was 113kg short of propellant due to a ground error, the high points of the elliptical 570 × 22,500km and 309 × 1,012km orbits were 2,340km and 286km lower than intended, reducing the 2-1/2yr operational life of DE-2 by about 6 months. DE-1 is providing images showing auroral storms, and among other things is conducting the first search from space for marine bioluminescence. RCA built the spacecraft, and their cost and the launch totalled $77 million. The 3 programmes are intended to lead on to the 4-satellite OPEN (Origin of Plasmas in the Earth's Neighbourhood) mission in the mid-1980s.

Dual Dynamic Explorers followed up the ISEE trio, and made the first search for marine bioluminescence. *(NASA)*

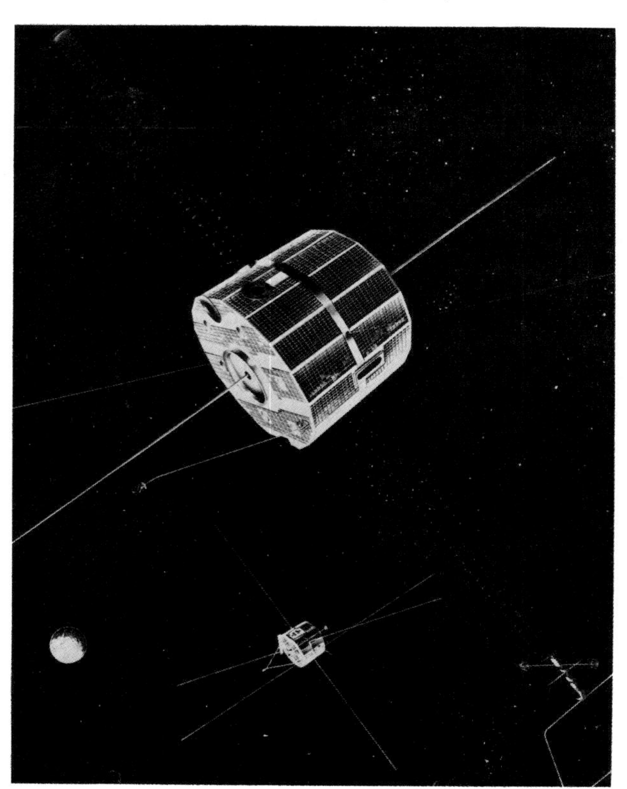

SME ORBIT GEOMETRY

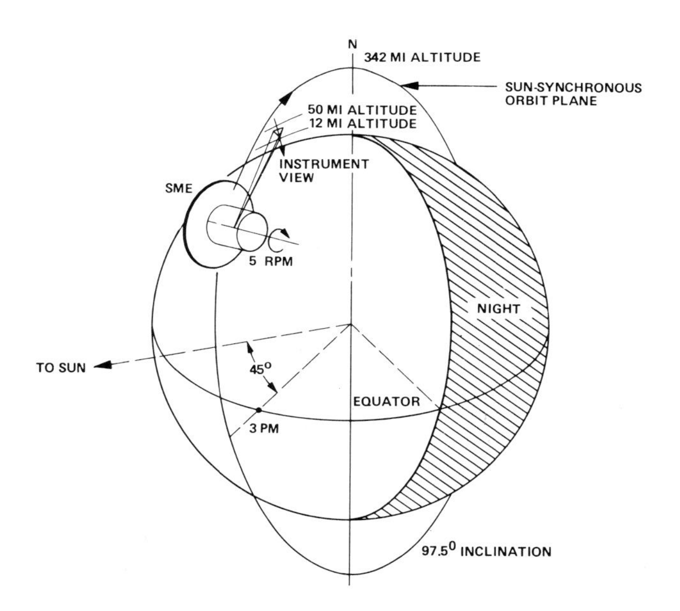

SME Solar Mesosphere Explorer L 6 Oct 1981 by Delta from Vandenberg. Wt 437kg. Orbit Sun-synchronous, 538 × 542km. Incl 97°. SME was designed to spend at least 1yr studying atmospheric ozone and the processes that form and destroy it. The mesosphere is the region lying above the stratosphere and below the ionosphere at an altitude of 30-50km. 5 SME instruments simultaneously monitor ozone and minor atmospheric-constituent quantities, water-vapour abundance and temperature, and the amount of incoming solar radiation to determine the role each plays in ozone production and distribution throughout the mesosphere. Ball Aerospace provided the spacecraft for JPL.

AMPTE A joint NASA/German (Max Planck Institute) project, the Active Magnetospheric Particle Tracer Explorers, planned for launch in 1984, are a follow-on to Firewheel, lost in the 1980 Ariane L-02 launch failure. A triple-payload Thor/Delta will place a 700kg German spacecraft, a 70kg UK satellite developed by the Rutherford-Appleton Laboratories, and a US satellite developed by Johns Hopkins University in an initial 50 × 50,000km orbit. The German and British craft will then be boosted into 500 × 130,000km orbits. Over an 8-month period the German craft will make 7 releases of barium and lithium ions into the solar wind and into the Earth's magnetic tail; the first release, in late 1984, is expected to result in the formation of a comet-type phenomenon visible from Earth. The US craft will remain in the magnetosphere at a significant distance to make tracing observations of the ions; Earth-based stations and aircraft will also study the solar wind reaction.

Fordsat

Based on Ford's Intelsat 5 series, 4 Fordsats have been ordered by Ford Aerospace and Communications to provide a domestic network from 1988, when Fordsats 1-3 are to be Shuttle-launched. With 1,540kg launch weight and 10yr life, they are claimed to be the 1st craft combining C and Ku-band transponders. They will have integrated propulsion, eliminating the need for an upper stage to take them from the Shuttle to geosynchronous altitude. These features, it is claimed, will result in a 40% saving compared with competing systems. Ford Aerospace has also offered to develop Fordsat into a "Supersat" family of communications craft capable of competing with current Hughes systems.

Galaxy

Galaxy 1, L 28 Jun 1983 by Delta/PAM-D from C Canaveral (in-orbit wt 520kg) and placed at 134°W, was the 1st cable TV statellite. Owned and operated by Hughes Communications Inc, a subsidiary of Hughes Aircraft, it was the 1st of 3. Galaxy has 18 primary transponders guaranteed for 9yr, plus 6 secondaries, all operating at 4/6GHz. Capacity is being sold to cable TV programmers such as Home Box Office Inc, which supplies 7 million subscribers with uninterrupted films, sport, etc; other users supply news programmes. Future applications are expected to include shopping by TV, burglar alarms operated by sensors, and voting facilities. Galaxy 2, L 22 Sep 1983 by Delta from C Canaveral, is stationed at 74°W. With Galaxy 3 due to enter service at 93·5°W in Jul 1984, 50-state coverage is expected.

In May 1983 Hughes filed for authorisation for a follow-on system of 3 more Galaxies operating in the Ku band and able to supply smaller, low-cost Earth terminals with voice, data and teleconferencing services. Launch would be by either Shuttle or Ariane, and the system would become operational in 1987 with a 10yr life.

Galileo

Plans to follow up the Voyager missions with a Jupiter orbiter and probe mission started in 1978, and by 1981 $300 million had been invested in the project. In Nov 1980 Hughes Aircraft was awarded a $40 million contract to develop the probe-carrier spacecraft, to be launched by Shuttle in Mar 1984 and to reach the planet 1,200 days (3·2yr) later in Jul 1987. The probe would separate 100 days before planet arrival and enter Jupiter's atmosphere on the sunlit side at 160,000km/hr, slowing in a few seconds to a few hundred km/hr, when a parachute would open. A vertical profile of the atmosphere, believed to consist of the original material from which the stars were formed, would explain Jovian weather and provide insights into Earth weather mechanisms. The carrier, which would relay the probe's data back to Earth, would go into an orbit similar to that of Ganymede, 1,002,400km from Jupiter, and use the Jovian satellite's gravity to skew its orbit at each near approach so that in 11 passes in 20 months it would cover all regions of Jupiter. Powered by 2 more advanced versions of the ERDA-developed nuclear power sources used on previous outer-planet missions, Galileo would have a 5m-dia furlable antenna. It would send back pictures of the larger moons with a resolution of 30-50m, and, among other studies, would even be able to identify materials on the surface of the moons. Over the years budget cuts have led to many delays and changes in the original plan, which aimed at a Dec 1981 launch. The abandonment of a 3-stage IUS in favour of a Centaur upper stage modifed for the Shuttle was another major cause of delay, followed in turn by cancellation of the "wide-body" Centaur in favour of a 2-stage IUS. By 1982, however, launch had been fixed for STS 61-G on 21 May 1986. A propulsion stage was to be added to the Galileo spacecraft, with an Earth swingby to pick up enough energy for the trajectory to Jupiter. This meant that flight time would be over 4 years. Before 1982 was out, however, wide-body Centaur had been reinstated once more; the latest plan is that Galileo will be launched in May 1986 on a 2yr journey, with greater fuel margins making possible additional encounters within the Jovian system. By 1983 total cost was estimated at $630 million, and it was being proposed that for another $100 million spare parts already acquired could be used for an additional Galileo/Saturn orbiter. This would be launched in 1987-8, using Earth gravity and having an 8yr journey time.

How Galileo should release its atmospheric probe (right) before going into orbit for a 20-month survey of the giant planet and its satellites. *(NASA)*

Gamma Ray Observatory

Designed to explore space sources of gamma rays - the most energetic form of radiation known - the Gamma Ray Observatory is intended to follow up the work of HEAO-3, SAS-2 and ESA's COS-B. Conceived in 1978 for launch in 1984, GRO is now due to be launched by Shuttle into a 400km circular orbit with 28° incl in 1988, and to operate for 2yr. At 10,432kg wt, 7·6m length and 3·8m dia, it will be one of the largest observatories ever launched. Its instruments and telescopes are expected to provide a deeper understanding of supernovas, pulsars, quasars and radio galaxies; illuminate the character of the universe shortly after its creation; and indicate the possible existence of anti-matter in the universe.

Gemini

History Successor to Mercury, the Gemini spacecraft was twice the weight and drew heavily on proven Mercury technology. But it was also far more advanced, complex and versatile, and achieved far more than its original aim, which was to bridge the gap in US manned spaceflight between the end of Mercury and the start of Apollo. As it was, 2 years elapsed between the last Mercury mission and the first Gemini flight. But once the programme did begin, 10 missions, Gemini 3-12, were flown at a breathtaking rate of 5 a year in 1965-6. Invaluable experience was obtained in long-duration flight, rendezvous, docking, use of target-vehicle propulsion for orbital manoeuvres, extravehicular activity (EVA, or space-walking), and guided re-entry. In those 2 years, the US put 20 men into space, while the Soviet Union achieved only the 2-man Voskhod 2 flight. Gemini 8 achieved the first space docking and established the technical lead that finally enabled the US to land the first men on the Moon. The project achieved 9 dockings with a target, and 7 different ways of docking were worked out.

Gemini was the first spacecraft to use fuel cells (a British development), which generate electrical power through the chemical reaction of oxygen and hydrogen; each cell produced 0·57lit of drinking water per hour as a by-product. The first 2 flights were unmanned tests; the first 7 in the series, launched by Titan 2 rockets, were code-named GT; the last 5, which involved an Agena rocket being placed in orbit to act as a docking target, were designated GTA. Total cost was $1,283·4 million.

Gemini 3 First manned flight in the series, a 3-orbit mission in March 1965. The first computer was carried into space. Weighing 22·7kg and capable of making 7,000 calculations a sec, it enabled mission commander Grisson to compute the thrust needed to change his orbit. From that time, instead of being carried helplessly round and round the world on a fixed orbital path, man had a genuine ability to "fly" in space.

Gemini 4 On this flight, 3 months later, the computer failed and Jim McDivitt had to make a manual re-entry. He was 1sec late punching the button to fire the retro-rockets, and the spacecraft came down 64·4km off course. All the same, it was a triumphant flight, lasting 4 days to become the first US long-duration mission. 3 months after Leonov had first tried it, Ed White ventured outside for a 21min spacewalk. Experimentally, and somewhat uncertainly, he manoeuvred himself with a hand-held oxygen thruster.

Gemini 5 An 8-day flight by Cooper and Conrad in Aug 1965 which proved that men could withstand weightlessness for as long as it would take to fly to the Moon, visit it briefly and return.

Gemini 7 A 14-day flight in Dec 1965 by Borman and Lovell. It held the long-duration record for 5 years until finally overtaken by Russia's 18-day Soyuz 9 in 1970.

Gemini 6 This mission, postponed in Oct 1965 at T - 42min when the Agena rocket intended as a rendezvous target was lost, used Gemini 7 as the target instead. On Gemini 7's 11th day in orbit Schirra and Stafford went up to join them for 1 day. Schirra brought his craft within 2m of Gemini 7 to achieve the first real rendezvous of men in space. The manoeuvre required more than 35,000 individual thruster firings.

The 2-man Gemini spacecraft. *(NASA)*

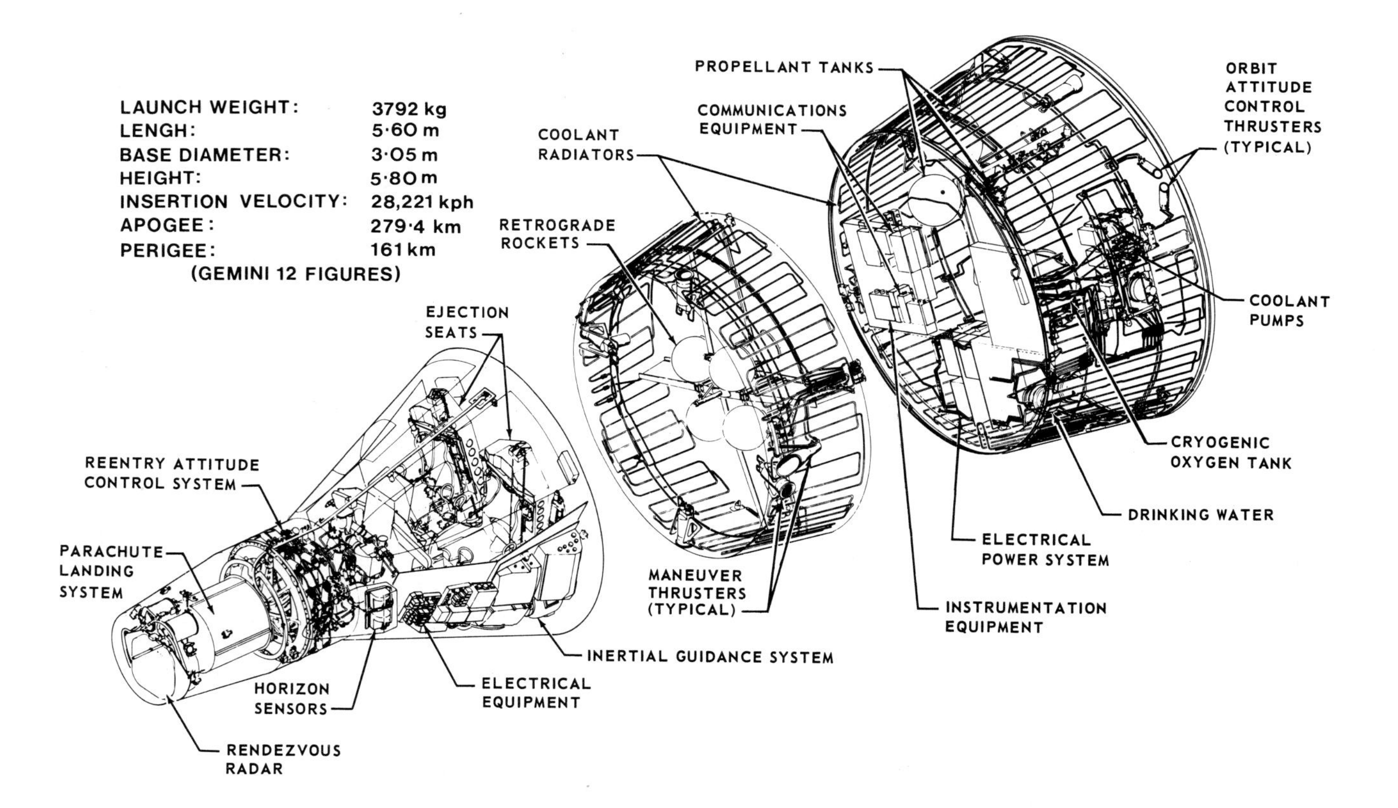

Gemini 8 On 16 Mar 1966 Armstrong and Scott achieved the first docking, linking their spacecraft with an Agena target rocket placed in orbit 101min earlier. Shortly afterwards the linked vehicles began tumbling and spinning out of control as a result of a jammed thruster. The astronauts escaped only by firing their retro-rockets, and had to return to Earth 2 days early. They survived man's first space emergency, and the drama completely obscured the fact that it was on this mission that the US finally overtook the Soviet Union in space technology.

Gemini 9 The unlucky flight. Its original crew, See and Bassett, were killed when trying to land a jet fighter in bad weather at the McDonnell works at St Louis, crashing through the factory roof only yards from the spacecraft. The mission was finally flown by the back-up crew, Stafford and Cernan, in June 1966 after 2 delays: first, the Agena target was lost; then, when a reserve target was launched, Gemini 9 missed its window. But when it did take place the flight included a 2hr spacewalk by Cernan, and splashdown was less than 0·80km from the recovery ship.

Gemini 10 A 3-day mission in July 1966 during which Young and Collins first docked with an Agena target rocket, then fired its engine to boost themselves into a 761km orbit. There they separated and rendezvoused with Gemini 8's Agena target, which had been left in a parking orbit for this purpose. Collins spacewalked across to it and retrieved a dust-collecting device.

Gemini 11 In Sept 1966 Conrad and Gordon made a direct-ascent rendezvous and docked with an Agena target while still on their first revolution. During a 44min spacewalk Gordon connected the 2 craft with a 30m tether. When undocked the Gemini's thrusters were used to put the craft into a cartwheel motion, producing 0·00015g in the world's first attempt to create artificial gravity. Automatic, computer-steered re-entry was achieved for the first time.

Gemini 8: the Agena target rocket with which Armstrong and Scott achieved the first docking — only to lose control and be forced into an emergency re-entry. *(NASA)*

Gemini 4: Ed White making the first US spacewalk. He died in the Apollo 1 fire in 1967. *(NASA)*

The world's first space "rendezvous": Gemini 7 as seen from only 6m by the Gemini 6 crew. *(NASA)*

Gemini 12 The last mission, flown by Lovell and Aldrin in Nov 1966, included 3 spacewalks totalling 5½ hr by Aldrin. These were the first fully successful attempts at EVA. Once again the spacecraft docked with an Agena target, artificial-gravity experiments were carried out, and a fully automatic re-entry performed.

GStar

3 GStar satellites have been ordered from RCA at a total cost of $100 million by GTE Satellite Corp to provide digital data, voice and video services to all 50 US states starting in 1984. Intended for both private and government use, the spacecraft will be stationed at 106° W and 103° W, with the 3rd in reserve. They will operate in the 12/14GHz band and be able to handle 30,000 simultaneous phone calls or 300 2-way video conferences. GStars 1A and 1B were due for Ariane launch in Sep 1984 and Jan 1985 at a total cost of $50 million.

HEAO

History Conceived in the 1960s and several times postponed, 3 High Energy Astronomical Observatories were finally launched in 1977, 1978 and 1979, and dramatically changed scientific views of the universe. Their instruments were developed in a series of balloon flights to altitudes of over 36km in 1974-5. According to Dr Riccardo Giacconi of the Harvard-Smithsonian Centre for Astrophysics, HEAO-2 (named *Einstein* to mark the 100th anniversary of the great theorist's birth) replaced the concept of majestically rotating galaxies, evolving over billions of years, with one of a dynamic, explosive universe in constant turmoil. At the time of writing, the evidence was thought to favour a universe that would keep expanding for ever, though the mass of stored data which have yet to be analysed may contain more surprises about what another of the principal scientists called the "black beasts" lying in outer space. HEAO is to be followed by an 11-tonne Gamma Ray Observatory, due for launch in 1988 and capable with Space Shuttle support of carrying out a 15yr study from a 480km orbit. Control of HEAO spacecraft was directed by the Marshall Space Flight Centre, with TRW as prime contractor. Total cost was about $246 million.

HEAO-1 L 12 Aug 1977 by Atlas-Centaur from C Canaveral. Wt 2,720kg. Orbit 428 × 447km. Incl 22·7°. In its first 6 months HEAO-1 mapped the X-ray sky, increasing the total of known X-ray sources from 350 to nearly 1,500. It then studied more than 300 stellar sources of particular interest, located a new black hole candidate (bringing the total to 4), and detected hot thermal plasma distributed throughout space. Re-entered 15 Mar 1979.

HEAO-2 L 13 Nov 1978 by Atlas-Centaur from C Canveral. Wt 2,720kg. Orbit 520 × 541km. Incl 23·5°. Named *Einstein*, operated for 2yr 5 months apart from a 4-month gap in 1980 due to a faulty gyro which inexplicably later returned to full life. It carried the largest X-ray telescope built and the first capable of producing images of X-ray objects other than the Sun, enabling it to study the most powerful emitting objects yet observed. These are quasars estimated to lie more than 10 billion light-years from Earth. A new class of very hot and massive young stars – "O" stars – not previously known as X-ray sources were identified, some of them in the Eta Carinae nebula. The first images of an X-ray "burster," an object which suddenly gives off intense bursts of energy, were obtained. In Jun 1980 HEAO-2 found that Jupiter and Earth emit X-rays, making them the only planets in our solar system known to do so. HEAO-2 re-entered 25 Mar 1982; it is to be succeeded in the 1990s by AXAF.

The 3 HEAOs, seen here against a background of some of the galaxies they examined, changed our concept of the universe. *(TRW)*

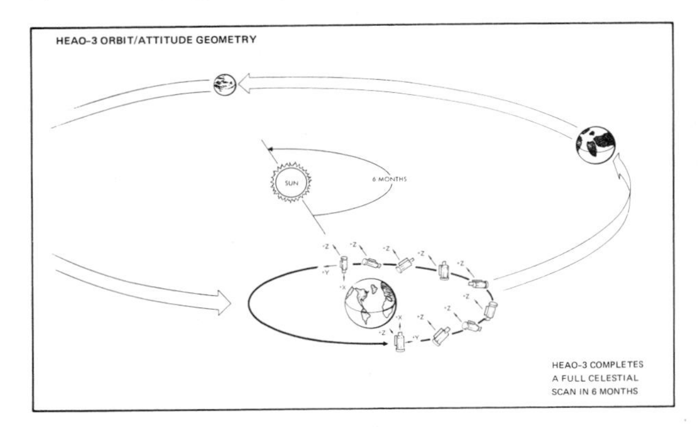

HEAO-3 L 20 Sep 1979 by Atlas-Centaur from C Canaveral. Wt 2,720kg. Orbit 485 × 501km. Incl 43·61°. After operating for 20 months compared with its 1yr design life HEAO-3 ran out of attitude-control propellant on 20 May 1981 and was powered down. Its survey of gamma and cosmic rays, supplementing those of its predecessors, brought more surprises - notably by casting doubt on the widely held belief that cosmic rays came from supernovae explosions. HEAO-3's observations suggested instead that cosmic rays derive their enormous energy from heating and cooling processes in the interstellar medium. Orbital life 4yr. Its work is to be followed up in 1988 by the Gamma Ray Observatory.

Lageos

History A tool to help predict earthquakes, the Laser Geodynamics Satellite, L 4 May 1976 by Delta 2913 from Vandenberg into a near-circular 5,900km orbit with 110° incl, was NASA's first to be devoted wholly to laser ranging. With a practical life of 50yr and orbital life of about 8 million yr, it was designed to act as a permanent reference point so that Earth's progress could be tracked relative to the satellite, in contrast to the traditional system of tracking satellites relative to the Earth. It should ultimately make it possible to track movements as small as 2cm on Earth's surface. The US Geological Survey, responsible for earthquake research and prediction, will use Lageos to measure continental drift. By paying particular attention to earthquake areas such as the San Andreas Fault in California, it should be possible to provide early warning of impending earthquakes.

The first 4yr until 1980 were devoted to determining Lageos' precise orbit and to building up a global network of 14 Earth stations supported by a mobile, truck-mounted laser-ranging system. By accurately measuring the time for a laser pulse to travel to the satellite and return, the position of the laser system could be determined to about 10cm (4in). Under NASA's Crustal Dynamics Project, started in 1979, 56 scientific investigators from 12 countries (France, West Germany, Netherlands, Switzerland, Spain, Sweden, Australia, New Zealand, Venezuela, Canada, Britain and US) are until 1986 making repeated measurements between their locations and Lageos; to lunar reflectors planted on the Moon by Apollos 14, 15 and 16 and 2 unmanned Lunas; and by very long-baseline microwave interferometry using extragalactic radio sources. By these methods crustal movements as small as 1cm per year can be determined. It is currently estimated that such movements average 1-20cm per year. By the end of 1981 Lageos results had showed that the San Andreas Fault was moving 50% faster than expected, and that earthquake activity there could be regarded as "overdue". Lageos 2 was due to be launched by Shuttle in Nov 1986.

Spacecraft description With 60cm dia and 411kg wt, Lageos is an aluminium sphere with brass core. Its 426 prisms, called cube-corner reflectors, give it the dimpled appearance of a golf ball. The 3-dimensional prisms reflect laser beams back to their source, regardless of the angle from which they come. The satellite had to be big enough to accommodate a large number of retroflectors, but small enough not to be affected by solar radiation drag. Aluminium construction would have been too light, and brass too heavy; finally 2 aluminium hemispheres were bolted together around a brass core to provide a large mass/surface ratio. Materials were also selected to reduce interaction between the satellite and Earth's magnetic field. Each end of the bolt connecting the hemispheres carries a copy of a message prepared by Dr Carl Sagan. It includes 3 maps of Earth's surface showing the period 225 million yr ago when, it is believed, the land masses were one super-continent (sometimes called "Pangaea"); the position of the continents as they are now; and their estimated position 8·4 million yr from now when the satellite, so solid it will survive re-entry, falls back to Earth.

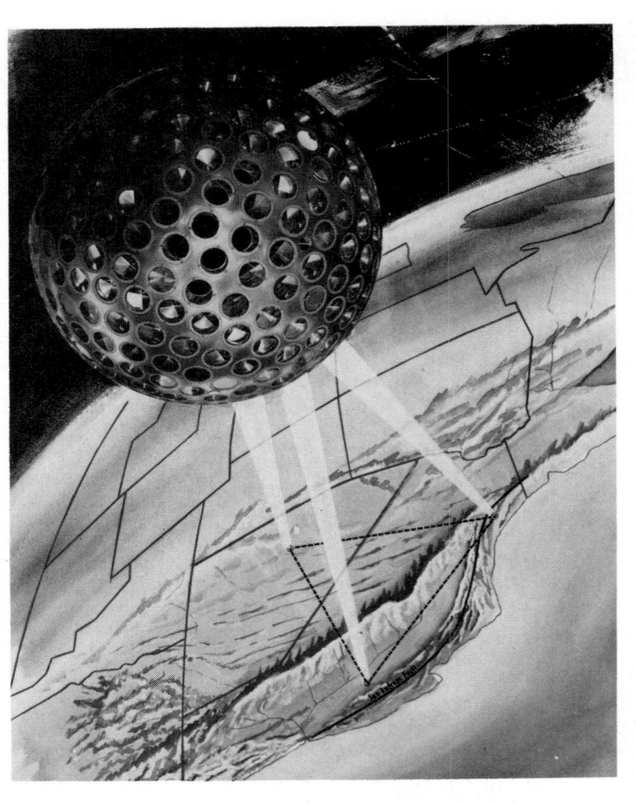

Lageos is being used to measure continental drift in the hope that it will be possible to predict earthquakes, with particular attention being paid to California's San Andreas Fault. ***(NASA)***

Landsat

History Landsat data, which include about 1·5 million stored views of Earth, are now used by scientists and local officials in more than 100 countries for making crop inventories and flood assessments, and identifying likely areas for mineral exploration. 11 nations receive and process their own data. Originally called Earth Resources Technology Satellites (ERTS), Landsats now routinely provide accurate assessments of likely crop yields around the world. This information has great political and economic significance. For instance, the US Agriculture Department buys images of US, Soviet and Chinese grain crops, with the result that no large country, like the Soviet Union in the past, can now quietly buy up the produce of other countries when faced with a poor yield of its own.

Landsats 1-3, made for NASA by General Electric, were improved and enlarged versions of the Nimbus weather satellites, which they resembled in appearance. The original plan was to launch 2 in successive years, 1972 and 1973, as an experiment in systematically surveying the Earth's surface to study the health of its crops and the potential use and development of its land and oceans. The dramatic flow of vivid, revealing photographs (actually "false-colour images") sent back from the moment Landsat 1 became operational were regarded as sensational by the 300 investigators in the 50 participating countries. Their quality led to the information being put to immediate practical use. It became clear that they were directly relevant to the management of the world's 3 major resources, food, energy and the environment, and when Landsat 2 was launched the then NASA Administrator, Dr James Fletcher, said he if had to pick one space-age development to save the world it would be Landsat and its successor spacecraft. Before launch NASA decreed that all ERTS data, including 9,000 pictures per week, should be unclassified and made available to the public. The pictures are sent by the NASA ground stations to the US Deptartment of the Interior's Earth Resources Observations System Data Centre in Sioux Falls, SD, where they could be bought for prices starting at $1·25. An unexpected result was that for the first time the public were able to obtain and study detailed pictures of Soviet and Chinese launch centres and missile sites; such photographs continue to be highly classified when obtained by spy satellites. The immediate success of the project caused the processing and distribution systems to be swamped with requests. A number of countries have since installed or are installing their own ground receiving stations; they include South Africa, Canada, Brazil, Italy, Sweden, Japan, Thailand, India, Australia, Argentina, Chile, Zaire, Iran and China. They pay construction costs of $4-7 million, $1·2 million per yr operational costs, and an annual service fee of $200,000.

But despite all this success, with Landsats 2 and 3 still partly operational despite being long time-expired, by 1982 the long-term Landsat programme had run into both technical and financial troubles. Launch of Landsat 4 had slipped to Jul 1982. General Electric, whose prime contract for Landsat D and D-Prime (as Landsat 4 and its back-up were known before launch) was worth $505 million, had run into cost and technical problems with both the flight and ground systems. Hughes had hit similar trouble with the Thematic Mapper, a highly advanced instrument designed to provide data for forecasting world food production, to give early warning of adverse agricultural conditions, and to help in locating mineral and petroleum deposits. However, when it finally got under way the much delayed mission looked like being immensely successful, and likely to fulfil plans to produce 10-12 scenes per day by Oct 1983 and 50 per day by Jan 1985. But by Aug 1983 multiple failures due to what NASA described as "a dumb design" had resulted in an NOAA request to launch the back-up Landsat D-Prime at an extra cost of $25 million, and as soon as possible if Landsat's 11yr imaging capability was to be maintained. The programme's technical and political troubles were at the same time compounded by the IUS failure which damaged and placed TDRS-1 in the wrong orbit following the STS-6 launch.

NASA costs for the first 3 Landsats totalled $251 million, including $149 million for the spacecraft and $14 million for the launchers. By Jul 1981 it had become clear that the series would represent a total US investment of $1 billion. The Reagan Administration decided at that stage that the Earth remote-sensing programme must either die or be taken over by the private sector. Comsat, Hughes and several other companies expressed interest subject to Government guarantees of a minimum level of data purchase. But in May 1983 a report prepared for the US Commerce Department confirmed earlier suggestions that there was no evidence to support the predicted "huge potential market" for Landsat data, which could not therefore be successfully commercialised.

Landsat 1 L 23 Jul 1972 by Thor-Delta from Point Arguello. Wt 891kg. Orbit 901 × 920km. Incl 99°. The near-polar orbit enabled the satellite to circle Earth 14 times a day, and to view any point on Earth except for small areas around the poles. Because the orbit is also Sun-synchronous, every 18 days the Landsat cameras could view the same spot at the same time of day; each pass enabled the cameras to view a swathe 185km wide with some overlap on each pass; it was back in its original position after 252 passes. By 30 Mar 1973, when faults occurred in the videotape recorder, the N American continent had been photographed 10 times; the total of over 34,000 images taken since launch included all the world's major land masses at least once. The tape-recorder fault meant that images could no longer be stored for transmission, though when the satellite was over one of NASA's 3 ground receiving stations in California, Alaska and Maryland it was still possible to transmit live pictures of the entire N American continent. An example of the ability of such satellites to make available to the public pictures of launch sites was one shot of southern Russia showing 7 major ABM complexes near the Chinese border (Landsat No E-1049-05260-7).

By the time Landsat 1 had to be retired on 16 Jan 1978, due to general wear and tear after 5½yr of operation, it had returned 300,000 pictures and proved the potential of remote sensing in the fields of geology, oceanography, agriculture, forestry, hydrology, urban planning, crop prediction and many other disciplines. Orbital life of the Landsats is 70-100yr.

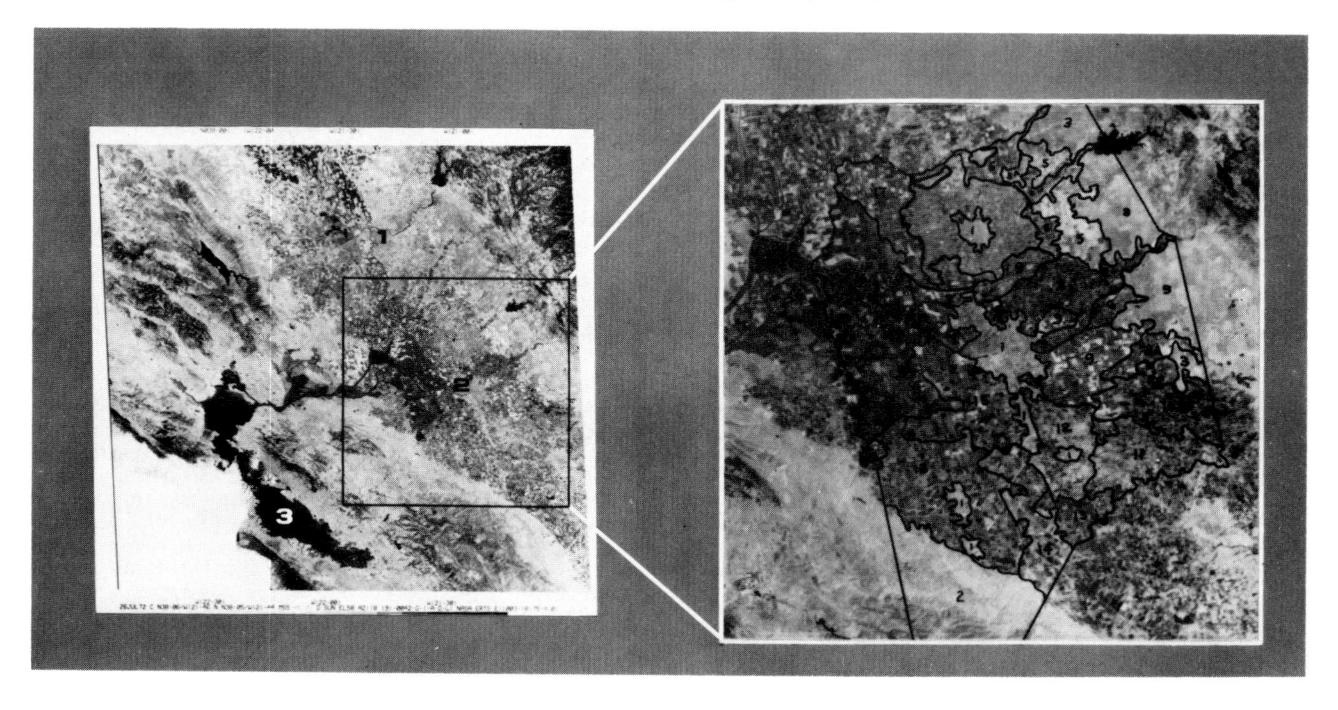

ERTS-1 image produced on 27 July 1973. (1) Sacramento; (2) Stockton; (3) San Francisco Bay. Smaller square encloses San Joaquin County, detailed right.

Land use map of San Joaquin County. 17 distinct land uses were identified from ERTS image. Examples: (1) Urban and commercially developed areas; (2) Rangeland; (5) Irrigated pasture, dry crops; (11) Orchards and vineyards; (16) Corn, alfalfa, sugar beet. *(NASA)*

Mosaic of Italy compiled from 46 Landsat 1 infra-red pictures. To obtain a cloud-free result the pictures had to be taken during all 4 seasons in 1972-3. *(NASA)*

Landsat 2 L 22 Jan 1975 by Thor-Delta from Vandenberg. Wt 816kg. Orbit 907 × 918km. Incl 99°. This Landsat incorporated a 14kg digital computer with 4,096-word memory which was able to handle 55 separate commands from ground stations and carry out routine operations for up to 24hr – a vital function since the satellite was out of tracking station range for 80% of the time. Landsat 2 was intended to supplement and later replace Landsat 1, and the 2 were able to work together until 1978, giving repetitive Earth coverage every 9 instead of 18 days. Their data were used to measure crop yields so that shortages and gluts could be forecast and avoided; to map mountain snows so that the amount of run-off for irrigation and generating electric power would be known in advance; to help cities, regional authorities and developing nations to plan use of their land resources; to monitor potential earthquake zones and monitor offshore sewage and industrial wastes; and to assess disasters like the major oil well blowout in the Gulf of Mexico and volcanic eruptions. Retired on its 5th birthday in 1980 due to wear of the attitude-control system, Landsat 2 was revived 6 months later by engineers of the Goddard Space Flight Centre, who developed a new magnetic compensation system to keep the spacecraft stabilised. Though its tape recorders had ceased working it was still operating in the direct read-out mode in 1981, providing Earth images to supplement the limited activities of Landsat 3. Landsat 2 also provided a picture (E-2101-07085-5) giving the public full details for the first time of Russia's Kapustin Yar launch centre.

Landsat 3 L 5 Mar 1978 by Delta from Pt Arguello. Wt 960kg. Orbit 900 × 918km. Incl 99°. Further sensor refinements included the Multispectral Scanner Subsystem, able to detect temperatures differences in vegetation, bodies of water and urban areas by day and night; and an improved Return Beam Vidicon offering a 50% resolution increase and enabling areas as small as half an acre to be identified. One of its most interesting pictures, taken soon after launch, showed C Canaveral; every launchpad, both NASA and Defence Department, was visible in detail. The photograph gave a vivid insight into the amount of information being obtained by more sophisticated, but classified, spy satellites about Soviet launch and missile sites, ports and airfields. The 3yr Large Area Crop Inventory Experiment (LACIE), concluded in Nov 1978, showed that Landsat's multispectral scanners revealed the Soviet Union's wheat crop to be 91·4 million tonnes, less than 1% below the official Soviet figure of 92 million tonnes. There was less success with predictions of wheat yields in Canada and the US, where the long, narrow fields can easily be confused with adjoining spring-planted crops. But improvements in Landsat 4's instruments were intended to remedy this, making it impossible for countries like Russia to repeat past successes in buying up world wheat surpluses cheaply when anticipating a poor crop of her own. Since wheat is the world's most important grain crop, Australia, China, Brazil and Argentina were also included in LACIE. As a follow-on it was decided to monitor wheat, corn, rice, barley, sorghum, soya beans and sunflower crops in Russia, China, Brazil, Mexico, Australia, Canada and the US. In Jan 1981, after 2yr of operations, Landsat 3's multispectral scanner malfunctioned. But the Return Beam Vidicon continued to complement Landsat 2's still operational Multispectral Scanner.

Landsat 4 L 16 Jul 1982 by Delta 3920 from Vandenberg. Wt 1,941kg. Orbit 689 × 696km. Incl 98·3°. The Sun-synchronous orbit enables it to image the same 185km swath of Earth's surface every 16 days; any part of the globe, except for a small area around the poles, can be covered. First pictures sent back by the new Hughes Thematic Mapper, designed to provide data in 7 spectral bands, showed the Detroit area to 30m resolution compared with Landsat 3's 80m. Project officials described data quality as "beyond expections"; ships could be seen in the Detroit River, airport taxiway intersections could be identified, and houses and factories were shown in great detail. Landsat 4 was designed to obtain orbital position information from GPS, and to relay its information via TDRSS. NOAA was due to take over spacecraft control from NASA on 31 Jan 1983, with responsibility for distributing data through the Dept of Interior's Earth Resources Observation System (EROS) in Sioux Falls, SD, while NASA undertook responsiblity for data-processing until Jan 1985.

By Aug 1983, however, Landsat 4 was a dying spacecraft, suffering from multiple malfunctions. Wires carrying power from the solar panels were breaking as a result of day/night temperature variations; 50% power had been lost, with further loss likely. The Thematic Mapper ground link had failed, and hopes that TDRS-1 would replace this were dashed by initial problems with that satellite. Finally, the prime command and data-handling computer had failed. NOAA, faced with a revenue loss of about $600,000 a month, requested launch of Landsat D-Prime and the Senate agreed to $20·4 million being spent on a Feb 1984 launch. Meanwhile, NASA began studying a Landsat 4 repair mission. It was due to be retrieved in any case by STS-2V in Jun 1986, when its altitude would be down to 315km, but could not be reached by the Shuttle any earlier than Oct 1985, when the Vandenberg launch site becomes available. A decision had to be made on whether Landsat 4's fuel should be used to bring it down to about 160km before control was lost; in such a case Shuttle astronauts would have to refuel as well as repair it so that it could be manoeuvred back to its correct height. Meanwhile, a Delta launch from Vandenberg to replace it with Landsat D-Prime was due in Mar 1984.

Landsat 5 L 1 Mar 1984 by Delta 3910 from Vandenberg. Wt 1,941kg. Orbit 705km circular, Sun-synchronous. Incl 98·2°. Britain's Uosat 2 (Oscar 11) was secondary payload. Landsat 5, modified to prevent a repetition of Landsat 4's solar array and X-band transponder failures, again carries the Multispectral Scanner and the advanced, high-resolution Thematic Mapper, covering 7 spectral bands with greatly improved spatial and radiometric resolution. NOAA has operating responsibility, but NASA retains Thematic Mapper control until 1985. NOAA plans a 2-satellite operation for as long as Landsat 4's remote-sensing instruments continue to function.

Spacecraft description Butterfly-shaped. Ht 3m. Width 3·4m with solar paddles deployed. Sensors and electronics are housed inside a 1·5m-dia sensory ring below the paddles. Sensors as follows. MSS (Multispectral Scanner Subsystem): this collects data by continually scanning the ground directly below in 4 spectral bands, 2 in the visible spectrum and 2 in the infra-red. RBV (Return Beam Vidicon Subsystem): this views the same 185km swathe as the MSS. 3 4,125-line cameras are reshuttered simultaneously every 25sec to produce 185 × 185km overlapping images of the ground along the direction of satellite motion. WBVTR (Wide Band Video Tape Re-

Landsat 2: composite multispectral scanner picture of the Dead Sea region. *(ESA)*

corder): because Landsat is not within range of a ground station for much of its time, 2 tape recorder systems can store images, and later transmit them, simultaneously. Data Collection System: this collects measurements as it passes over remote platforms installed on icebergs, etc, to send up readings of soil, water and other conditions for later retransmission and comparison with data collected from the other sensors.

Remote sensing The basis of Landsat is the fact that all objects, living or inanimate, transmit or reflect visible and invisible light, and thus have a distinctive "signature". All energy coming to Earth from the Sun is reflected, transmitted or absorbed in objects on the ground, each in its own way. Remote sensing - assessing the characteristics of objects from a distance - goes back to aerial photography in the 1930s. It began to develop more rapidly with military reconnaissance in the Second World War, followed by the use of sounding rockets and satellites after that. A camera is a remote sensor in that it records the shape and colour of an object by its reflected light without touching the object. The human eye is the simplest example of remote sensing, though most of the information reflected or radiated by the Earth cannot be detected by the human eye. Near-infra-red light lies just beyond our vision. Healthy green vegetation is even brighter in the infra-red than in the visible, and this information is particularly valuable to farmers because it can provide early warning if their crops are sick. Just as the eye can cover only a minute portion of the total electromagnetic spectrum, so no single instrument is capable of sensing and measuring the radiations from objects with different physical and chemical properties. One of the objects of Landsat is to determine what sort of sensors, and what combination of them, will yield the most useful information.

Picture reading and results The energy reflections described above are converted by the Landsat scanners into electrical signals in 4 selected bands. (In this case, Bands 1 and 2 are in the visible wavelengths of 0·5 - 0·6 and 0·6 - 0·7 micrometres; Bands 3 and 4, which are not visible to the human eye, are in the near-infra-red portion of the spectrum, with wavelengths of 0·7 - 0·8 and 0·8 - 1·1 micrometres.) These reflections are processed into digital bits, transmitted to a receiving station, and then reprocessed into either black-and-white reproductions of what was seen on Earth, or as false-colour images by projecting the data of 3 of the 4 bands through blue, red and green filters. The colours assigned are in the same order as the primary colours of the visible spectrum, but result in an "exchanged" colour: what we seen as green in Band 1 is shown as blue; what we see as red in Band 2 appears as green; and Band 3, which normally we cannot see, appears as red. Band 4 may be used instead of Band 3, and also appears red.

The result of this is that clear water appears black in Bands 3 and 4, because water almost totally absorbs radiant energy, sending back hardly any reflections; water carrying silt, or otherwise polluted, appears blue. Trees and plants appear bright red because of the very high reflectivity of chlorophyll-bearing leaves in the near-infra-red. Vegetation brightness depends on such things as the size of leaves, big leaves showing up as brighter than small leaves, with the effect that hardwood trees register as brighter than pine trees. The big leaves of tobacco plants are brighter than wheat. Crop brightness depends on plant health, so that healthy crops are much brighter than diseased vegetation. But crop disease is difficult or impossible to detect on a single photograph; abnormal changes, suggesting disease, only show up when successive pictures are compared.

The amount of knowledge which can be extracted from the data and pictures increases as the scientists learn the technique of reading them. But at an early stage observers were able to detect geologic faults and water-bearing rock areas in Nebraska, Illinois and New York State which had been unknown before; areas of clear and polluted waters in Chesapeake Bay were readily discernible. From Brazil came reports that Landsat pictures had revealed that villages and towns were sometimes wrongly located on maps by tens of kilometres, and that lagoons shown as 20km long were in reality over 100km long. Ghana reported that locust control was being attempted because pictures had shown vegetation at the edge of deserts which attracted breeding locusts. Iran reported that it had become apparent that lowering of the Caspian Sea by evaporation had changed the shape of the Bandar Shah peninsula. Pictures of Britain disclose, among other things, that a fault starts near Harwich on the E coast and runs right through London to Land's End.

LDEF

The Long Duration Exposure Facility (LDEF) is a NASA-developed free-flying, 12-sided, 9·14m-long, 4·27m-dia open-grid structure weighing 3,636kg. Designed to be placed in orbit by the Shuttle every 18 months, it will be retrieved after exposure periods of 10-12 months. 1st deployment, carrying 57 scientific, applications and technology experiments, was due on Mission 11 in Apr 1984, with retrieval in Feb 1985. Deployment and recovery are carried out with the RMS. Having been deployed on its 1st mission into a gravity-gradient-stabilised attitude in a 463km circular orbit with 28·5° incl, LDEF will be left passive, without manoeuvring or communications facilities. It offers vibration-free, low-acceleration exposure to the space environment for simple, long-term experiments. It is regarded as a prototype for free-flyers likely to accompany the Space Station, and would seem to perform many of the roles ESA had foreseen for Spacelab hardware.

Leasecraft

Based on the Fairchild Multi-Mission Spacecraft (MMS) used for Landsat 4 and the Solar Maximum Mission, this joint Fairchild/NASA 5,400kg spacecraft bus will be placed in a parking orbit to await payloads from government or commercial customers wishing to fly materials-processing or scientific packages for periods longer than normal Shuttle missions. Fairchild will organise payload installation, maintenance and recovery, and expects to have flight-qualified Shuttle payload specialists on the payroll to carry out these operations. Control and data-recovery transmissions will be handled by TDRSS. Leasecraft will contain a centrally mounted propulsion module for transfer from and to the Shuttle orbit height, up to 1,000km. McDonnell Douglas Astronautics is considering using the facility for its electrophoresis pharmaceutical processing system, weighing 4,500kg at launch. NASA's proposed 1986 Advanced X-ray Astronomy Facility is another potential user. Fairchild hopes to be operating 12 Leasecraft by 1990, though progress will be governed by the ability of Shuttle astronauts to work with orbiting satellites, as demonstrated by the planned SMM repair mission. In Sep 1983 Fairchild and NASA signed a agreement under which Fairchild will spend $200 million building Leasecraft 1 for launch by Shuttle in 1987, although a commercial customer had still to be found. Hire charges of $2-4 million per month for primary payloads and $½-1 million for secondaries are proposed.

Lunar Orbiter

History The second of 3 unmanned exploration projects, carried out in parallel with the 3 manned programmes aimed at getting men on the Moon before 1970. Following the successful Ranger flights, which yielded the world's first TV pictures of the Moon's surface, 5 Lunar Orbiters were launched within a year, starting on 10 Aug 1966, to help select Apollo landing sites in equatorial regions from 43°E to 56°W. Other objectives were to study variations in lunar gravity, radiation and micrometeoroids. Orbiter 1 was placed in a 191 × 1,867km lunar orbit with 12° inclination. Its pictures, covering 5·18 million sq km of the Moon, included 41,440 sq km of potential Apollo landing areas. Perturbations in its orbit provided the first knowledge of what became known as "mascons". At least a dozen in number, these mass concentrations, usually associated with the lunar "seas," have a powerful gravitational effect on spacecraft remaining for lengthy periods in lunar orbit. All 5 Orbiters were immensely successful, and it proved possible to manoeuvre them by Earth commands into orbits descending as low as 40km. Objects as small as 1m across were photographed, and their pictures permitted the creation of the first complete lunar atlas. The first 4 Orbiters provided between them experience of several thousand lunar orbits before each was deliberately crashed on to the surface with the last of its attitude-control gas to ensure that there was no radio-frequency interference with later missions. Orbiter 4 provided the first pictures of the lunar south pole. Orbiter 5, launched on 1 Aug 1967, was retained after it had completed its photography as a target for NASA's Manned Spaceflight Network until its final controlled impact on 31 Jan 1968. By then the 3rd unmanned lunar exploration project, Surveyor, was also nearing completion.

Spacecraft description With a truncated cone structure, Lunar Orbiters were folded for launch. When deployed, the 4 windmill-like solar panels and antennas provided a maximum span of 5·6m, and 1·6m ht. Total wt of 390kg included a photographic laboratory weighing only 65·8kg but carrying 2 cameras for wide-angle and telephoto coverage, and film processing and photo readout (scanning) systems. These viewed the Moon through a quartz window protected by a mechanical flap. Launcher: Atlas-Agena D.

Mariner

History One of NASA's 4 major planetary exploration programmes. By the end of the remarkable Mariner 10 mission the secrets of the smallest planet, Mercury, had been revealed and over half of its surface photographed. Much had been learned about Mars and Venus, with the promise of more revelations ahead. The Mariner programme had complemented and leapfrogged Pioneer, and built up the technology for the Viking landings in 1976. The first 9 Mariner launches, spread over 10yr, included 3 intended for Venus and 6 for Mars. Their success enabled the 10th to be aimed at Mercury, passing Venus on the way, with the following pair intended to fly past Jupiter and Saturn. The latter replaced the more expensive "Grand Tour" missions planned for 1977-79, when all 5 outer planets were lined up in such a way that their gravity could be used to swing spacecraft past each in turn; this occurs only once in 180yr. The revised plan cost only $320 million compared with the Grand Tour's $900 million. The 2-yearly launch windows (when Earth and Mars come within about 56 million km of each other) were used by Mariner 4 in 1967 to obtain man's first close look at another planet; and by Mariners 6 and 7 in 1969 to follow up with much better pictures. In 1971 Mariner 9 became man's first planetary orbiter, providing a complete map of the Red Planet. The secrets of Venus, still largely concealed beneath her dense cloud cover despite Russia's landings, were finally exposed by the Pioneer Venus orbiter and probes launched in 1978. What were to have been Mariners 11 and 12, renamed Voyagers 1 and 2, went on to explore Jupiter, Saturn and the outermost planets. The later Mariner missions are described in some detail to illustrate the developing technology which transformed man's knowledge and understanding of the solar system.

Spacecraft description The early Mariners, 3·04m long and 1·52m across the base, were very similar to the Ranger spacecraft used for lunar impact flights: a tubular centre was attached to a hexagonal base from which a dish antenna and solar cell panels were extended. The weight of Mariners 3 and 4 was increased to 260kg from the 202kg of the first two craft. An octagonal magnesium centrebody had 4 rectangular solar cell panels to power its computer and sequencer, TV camera, cosmic-ray telescope, etc, and a hydrazine-fuelled main engine. Mariners 6 and 7, twice as large again, had an 8-sided magnesium framework with 8 compartments containing electronics, TV assembly, etc, and 4 2·13m-long rectangular solar panels with attitude-control jets at their tips. Each had 2 TV cameras (wide and narrow-angle) mounted on a rotating platform and able to resolve objects down to 275m. Mariners 8 and 9 retained the same basic design, but the need for a 136kg-thrust retro-engine to inject them into Mars orbit again increased total wt, to 1,031kg; this included 454kg of fuel. The narrow-angle TV camera could resolve features down to 100m. Other instruments for investigating the atmosphere and surface included an infra-red radiometer, ultra-violet spectrometer, and an infra-red interferometer spectrometer.

With Mariner 10 proving for the first time the practicability of using the gravitational pull of one planet to reach another, it was possible to halve the weight of the succeeding Voyager spacecraft.

Mariner 1 L 22 Jul 1962 by Atlas-Agena B from C Canaveral. Wt 202kg. This first attempt at a Venus flyby failed because of an error in the flight guidance equation; the rocket went off course immediately after launch and had to be blown up. The object had been to obtain details of the Venusian atmosphere, cloud cover, magnetic field, etc.

Mariner 2 L 27 Aug 1962 by Atlas-Agena B from C Canaveral. Wt 202kg. The first successful planetary flyby, it was fired into a Venusian trajectory from Earth parking orbit. After a 109-day journey it flew past the planet at a distance of 34,830km, providing 35min of instrument scanning. Surface temperatures registered as 428°C, above the melting point of lead and far higher than expected. The atomosphere appeared to contain no water vapour. The cloud layer was unbroken, with one spot near the southern end of the terminator 11°C cooler than the rest, possibly due to a mountain range. It was also established that, unlike Earth, Venus did not have a strong magnetic field and radiation belt.

Mariner 3 L 5 Nov 1964 by Atlas-Agena D from C Canaveral. Wt 261kg. Intended to take 21 TV pictures as it passed Mars at a distance of 13,840km, but failed to achieve the necessary speed of 41,228km/hr when fired from Earth parking orbit. Although Mariner 3 went into solar orbit, Mars was missed by a wide margin.

Mariner 4 L 28 Nov 1964 by Atlas-Agena D from C Canaveral. Wt 261kg. Following Mariner 3's failure, launch of the 2nd of the pair of vehicles was delayed till the end of the Martian window. Injection was successful and problems resulting from the instruments locking on to the wrong stars were overcome. Canopus was acquired 2 days after launch, and after 228 days and a flight of 523 million km Mars was passed at a distance of 9,844km on 14 Jul 1965. During the next 10 days 21 TV pictures and 22 lines of a 22nd photograph were received at NASA's Jet Propulsion Laboratory at Pasadena, California. Man's first close-range pictures of another planet showed that Mars was heavily cratered, more Moon-like than Earth-like, very dry, with no trace of surface water, and certainly not possessing any of the canals imagined by astronomers. Although the possibility of some form of life could not be ruled out, the very thin atmosphere, coupled with evidence that there might never have been enough water for oceans or streams, made any advanced life forms seem most unlikely. Long after passing Mars, Mariner 4, by then in solar orbit, provided convincing evidence that such vehicles could be operated for many years. 2½yr later, with Mariner 4 90 million km from Earth, a JPL command turned on the TV equipment and fired the spacecraft engine for 70sec.

Mariner 5 L 14 Jun 1967 by Atlas-Agena D from C Canaveral. Wt 425kg. Originally the back-up vehicle for Mariner 4, this was modified for flight towards the Sun and Venus, instead of away from the Sun to Mars. Solar panels were reversed and reduced in size, and a thermal shield added. A flight of 349 million km resulted in Mariner 5 passing only 3,990km ahead of Venus in its orbit around the Sun on 19 Oct 1967. Using more advanced instruments, surface temperatures of about 267°C were recorded. Measurements of the magnetic field ranged between zero and 1/300th of Earth's. An electrified ionosphere was identified at the top of the atmosphere.

Mariner 6 and 7 L 24 Feb and 27 Mar 1969 by Atlas-Centaur from C Canaveral. Wt 413kg. Intended to study the atmosphere and surface of Mars as part of the search for extraterrestrial life, and to develop technology for later Mars missions, these flights were immensely successful. The author, watching the 201 TV pictures flowing back to JPL at Pasadena, found it even more exciting and dramatic than the first Apollo Moon landing a few days earlier. The spacecraft passed Mars at distances of 3,412 and 3,524km. Mariner 6 had flown 387·8 million km in 156 days to arrive on 31 Jul for encounter at 95·7 million km, about 5½ light-min, from Earth. Mariner 7 flew 316·9 million km in 130 days for encounter on 5 Aug at 99·4 million km. Mariner 7 was probably struck by a small meteoroid a few days before arriving: after loss of signal, commands to switch antennas were successful both in restoring communications and establishing that it had been damaged, losing some of its telemetry channels. A slight velocity change caused it to arrive 10sec late. The spacecraft began sending back far-encounter pictures from distances of up to 1,126,540km, but more interesting still were the 24 near-encounter pictures sent back by Mariner 6 during its 68min of closest approach, and the 33 near-encounter pictures provided by Mariner 7's 74min. Mariner 6, concentrating on the equatorial region, dramatically established that Nix Olympica, at first thought to be a gigantic crater, was a 24km-high volcano with a 64km-wide crater at the top. Mariner 7, concentrating on the southern hemisphere and part of the south polar ice cap, confirmed that this was largely solid carbon dioxide (dry ice), with perhaps a little water. Mars proved to be heavily cratered, with a thin atmosphere consisting of at least 98% carbon dioxide. Its craters differed from those on the Moon as a result of being worn down by winds and dust. It seemed that any advanced form of life could be ruled out, but one scientist speculated on the possibility of some form of life which had no need for liquid water. One of the Mariner 7 pictures showed a minute, potato-shaped speck which proved to be one of the 2 Martian moons, Phobos. Full details of the moons and of the Red Planet itself were to be finally obtained by Mariner 9 during the next launch window. Cost of this twin mission was $148 million.

This wide-angle photograph taken from 3,600km by Mariner 6 showed that some Martian regions are very similar to the Moon. Covering an area about 1,000km across, it revealed craters ranging from 128km to 5km in dia, some fresh, some barely discernible. ***(NASA)***

Mariners 6 and 7, which began the search for life on Mars, arrived just after the first Apollo Moon landing in 1969. Attitude-control gas jets were mounted on the tips of the 4 solar panels.

Mariner 8 L 8 May 1971 by Atlas-Centaur from C Canaveral. Wt 1,031kg. Intended to be the first of a pair of Martian orbiters, but an autopilot fault sent the 2nd stage off course and it fell into the Atlantic 1,450km SE of Cape Kennedy.

Mariner 9 L 30 May 1971 by Atlas-Centaur from C Canaveral. Wt 1,031kg. Intended to map 70% of Mars during 90 days in orbit around it, Mariner remained operational 349 days before it was shut down following exhaustion of its attitude-control nitrogen gas. By then it had circled the Red Planet 698 times, mapped the whole of it, and transmitted 7,329 TV pictures, including detailed photographs of both Phobos and Deimos. Following the loss of Mariner 8, plans were revised so that Mariner 9 could cover both missions. The spacecraft arrived at Mars on 13 Nov 1971 at the end of a 167-day flight covering 397 million km. A 15min firing of its 136kg-thrust liquid-propellant engine reduced the approach speed relative to Mars from 18,000km/hr to 12,500km/hr and placed it (after a later trim manoeuvre) in a 12hr, 1,387 × 17,140km Martian orbit. It thus became the first man-made object to orbit another planet. (Russia's Mars 2 and 3 followed later in 1971.) The braking burn reduced the spacecraft's wt to 590kg. As the spacecraft was approaching Mars in mid-Nov it took 3 series of pictures of a violent dust-storm which astronomers had been watching envelop the entire Martian globe during a 2-month period. While this delayed Mariner 9's mapping work for 6 weeks, it provided a unique opportunity for its instruments to peer down into the most extensive dust storm to occur on Mars since 1924. It reached an altitude of 50-60km. Only the bright, waning ice cap at the S pole and 4 dark mountain peaks (one of them Nix Olympica) were visible through the haze, which had the effect of cooling the surface and warming the atmosphere. When the dust storm subsided Mariner 9 was able to observe the changing seasons below for more than half a Martian year. By the end of its mission Mars was known to be a geologically active planet different from both Earth and Moon, with volcanic mountains and calderas (craters) larger than any on Earth. There is a vast equatorial crevasse which would dwarf America's Grand Canyon, 4,000km long and plunging to a depth of 6,096m. The Martian canals were an illusion, yet this gigantic rift was never suspected, showing up on Earth-based telescopes only as dark markings. One theory is that the dark trough is warmed by the Sun at one end while it is still dark at the other, resulting in violent winds each day. Contrary to earlier conclusions, it is now thought that free-flowing water may have existed on Mars at one time, and that dust storms and cloudiness account for much of the variability of appearance that has puzzled astronomers for centuries. Because of a previously unknown gravity-field variation in Mars' equatorial plane, the Mariner's orbital period was found to be too short to permit satisfactory tracking from Earth. So after the dust storm cleared, the spacecraft's engine was fired to raise the periapsis, or low point, to 1,625km, so

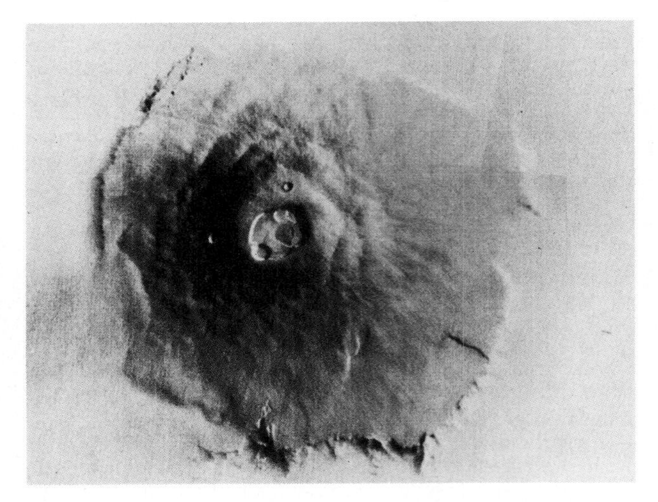

Mars' Nix Olympica, a gigantic volcanic mountain 500km across at the base (twice as broad as any similar Earth feature), was photographed by Mariner 9 in 1972. *(NASA)*

keeping Mariner 9 visible to the Goldstone, California, station. With the Martian surface clear at last, the mapping cameras looked down on a shrinking S polar cap; sinuous channels which appeared to be dried-up river beds cut by water; chaotic, bouldery terrain first glimpsed by Mariners 6 and 7; and huge impact craters, their floors covered with wind-blown dunes. Surface temperatures ranged from 81°F on the equator to - 189°F at the poles; the N pole was much colder than Earth's coldest spot, a point in Antarctica at which - 125°F has been recorded. Several localised dust storms were seen after the main storm cleared. Variable cloud patterns were observed, mainly in the north but also over large volcanoes. They were believed to contain water ice, though if large quantities of water exist they seem certain to be locked in the permanent polar ice caps. Atmospheric winds were measured at up to 185km/hr. Nix Olympica, 500km across at the base, is the Red Planet's highest spot. Reaching at least 17km above the surrounding plain, it is far higher than Everest on a planet half the size of Earth. The tiny Martian moons were both studied: Deimos, orbiting at 20,070km, has a 16km equator and is 9·6km from N to S. Phobos, orbiting at only 5,986km, has a 27·3km equator and is 19·3km from N to S. Both moons are heavily cratered, apparently from meteorite impacts. Gravity is so low on Phobos that a man could throw a cricket ball into orbit around it. From 2 Apr to 4 Jun 1972 the Mariner's instruments were turned off while its orbit took it into Mars' shadow during each twice-a-day revolution. After they had been successfully turned on again it became possible to study the N polar region and to look for potential landing sites for the Viking 1975 project. The last of 45,960 commands to Mariner 9 turned off its radio transmitter; the craft is expected to remain in Martian orbit for at least 50yr. The costs of Mariners 8 and 9 totalled $136·4 million.

Mariner 10 L 3 Nov 1973 by Atlas-Centaur from C Canaveral. Wt 503kg. This was the first dual-planet mission, and the first designed to use the gravitational attraction of one planet to reach another. Despite a series of technical failures soon after launch - such as TV heaters failing to turn on - its remarkable success provided man with his first detailed knowledge of Mercury. Course corrections were successfully performed on 13 Nov and 21 Jan 1974 (the latter a 3·8sec burn) to ensure passing Venus at a distance of 5,760km, thus obtaining the gravity-assisted bending of the trajectory needed to enable the onboard rocket engine to achieve a Mercury encounter. Closest approach to Venus was on 5 Feb, and during the flyby 3,500 pictures were sent back by its twin TV cameras. Although both US and Soviet spacecraft had visited Venus before, this was the first mission to carry cameras equipped with ultra-violet filters able to send back pictures of the global circulation of the Venusian atmosphere. These pictures confirmed radar measurements made by NASA's Pasadena laboratory in 1962 and suggesting that the upper atmosphere hurtles round in 4 days compared with the 243-day rotation of Venus itself. The pictures showed striking details of circulation patterns swirling from the equator; those of the edge of the planet's disc showed dense, lower layers below the fast-moving upper cloud deck. Because Venus has no detectable magnetic field the solar wind acts directly upon the very dense atmosphere, forming an ionosphere which in turn generates a bow shock as the planet orbits the Sun (see Venus entry in Soviet section for more information).

On 16 Mar 1974 a 3rd trajectory correction placed Mariner 10 on a course which took it past Mercury at a distance of only 271km on 29 Mar. In an 11-day period - 6 days before closest approach and 5 days after - a total of 2,300 TV pictures of truly remarkable clarity came back to Earth. They revealed a highly cratered, lunar-like surface. Impact craters, valleys, and features resembling the mare regions of the Moon were recorded down to resolutions as low as 100m. Numerous impact craters had central peaks like those on the Moon; overlapping craters suggested a fall of meteorites striking the surface in quick succession. Lava-filled craters indicated surface activity after the main impact patterns had been formed. The first Mercury feature to be given a name - a large, multiple-rayed crater in the centre of a half-disc picture built up from 18 photographs - was named after the late Gerard Kuiper, principal investigator on Ranger 7, which sent back the first lunar photos. Kuiper was a member of the Mariner 10 team until his death in 1973. The spacecraft's infra-red radiometer recorded temperatures ranging from 370°F on the planet's day side to minus 280°F on the night side at local midnight. It was estimated that surface temperatures on the day side could reach 560 - 800°F, depending on the planet's distance from the Sun. This total temperature range, as great as 1,000°F, suggests that the planet is covered by a thin blanket of porous material with a very low thermal conductivity. But while Mercury's crust is Moonlike, the interior is much more like Earth's. A minor disappointment came when what was thought to be a small moon perhaps 26km in dia was later found to be a star in the background. On 9 and 10 May Mariner 10 was placed on course for its 2nd Mercury flyby with a 2-part burn, performed that way to avoid overheating the rocket engine. A coincidence of celestial mechanics - the fact that Mariner's 176-day orbit round the Sun was in phase with Mercury's 88-day orbit - made a 3rd encounter possible. But correction of overheating problems, and getting the spacecraft out of "search-roll mode" after it had drifted off Canopus, caused unexpectedly high consumption of the nitrogen gas used by the reaction-control system.

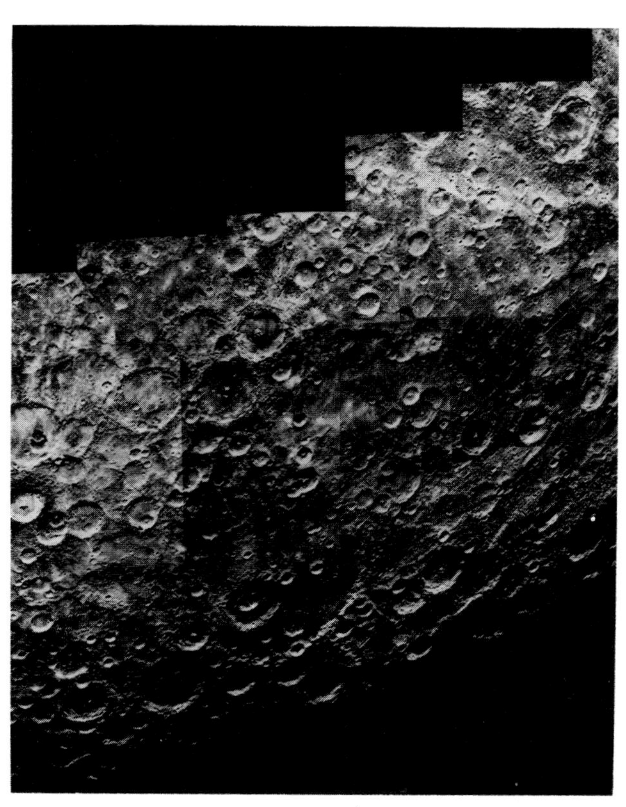

Man's first look at Mercury. Mariner 10 showed the south polar region to be thickly covered with both old and new craters. Note the fresh crater and ray system at right. *(NASA)*

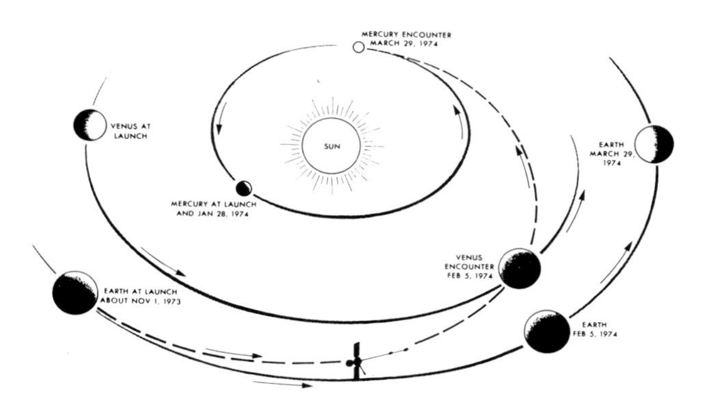

Mariner 10's Mercury-encounter trajectory.

Mariner 10 completed its 2nd sweep past Mercury on 21 Sep 1974, with closest approach at 48,000km. With the spacecraft 170 million km from Earth (19 million km further than on the first pass), Mariner 10 sent back about 1,000 TV pictures. The closest, taken over the S pole with resolutions down to 1km, showed that Mercury was once struck with such force in that area that a crater 1,290km wide was left, and the opposite side of the planet was scarred by the vibrations. The colliding object could have been as big as 96km in dia. The crater was named the Caloris Basin since it becomes extremely hot because of the planet's proximity to the Sun. Scientists studying the pictures decided that extensive thrust faults and scarp formations were due to buckling effects caused by 2 slabs of the surface moving towards each other. These contrasted with the tension features on both Mars and the Moon, formed when sections of the surface crust pulled apart. It became clear, too, that Mercury's interior differs from that of the Moon, with an iron, or high-iron-content, core accounting for about half the planet's volume.

Mariner 10's 3rd and final flight past Mercury, with nearest approach of 319km on 16 Mar 1975 and another 1,000 pictures of the surface, completed a triumphant mission for the Jet Propulsion Laboratory's guidance and control team. With its orginal 3·6kg of attitude-control nitrogen down to only 0·13kg, the controllers developed a new "solar sailing" technique to conserve fuel during the 6-month flight around the Sun following the 2nd encounter. The pressure of solar radiation on the movable high-gain antenna and the solar panels was used to stabilise the spacecraft and to keep it on course between trajectory-correction manoeuvres. The last of 8 course corrections – the most ever made on a single mission – was achieved with a 3sec burst of the orbital propulsion system. This directed the spacecraft towards Mercury from the sunlit side so that it then looped behind in relation to the Sun. TV pictures were taken during the inbound and outbound periods. Despite a technical fault at Canberra, Australia, which put the big receiving dish out of action, pictures showing surface details as small as 450-900m were obtained. The final batch of pictures meant that 57% of Mercury's visible surface had been photographed. The pictures suffered from some blurring, however, because of the spacecraft's speed of 40,000km/hr as it swept past. Altogether Mariner 10 sent back more than 10,000 pictures of Mercury. The primary aim of the 3rd flyby, however, was not pictures but to probe the planet's magnetosphere, streaming out into space on the anti-Sun side. The 3rd pass confirmed the finding of the spacecraft's magnetometers a year earlier: that Mercury is one of the few magnetised planets in the solar system. Its magnetic field, scientists decided, was not just the effect of the solar wind on Mercury's surface. It was caused by the planet's interior, though they could not decide immediately whether this was due to permanently magnetised rocks or to an "active dynamo mechanism" in a fluid core. On 24 Mar 1975, following a signal indicating that the control gas was exhausted, the transmitter was switched off before this historic spacecraft drifted out of communication with Earth. Having penetrated the mysteries of Mercury in just one flight, Mariner 10 is now orbiting the Sun and will eventually fall into it. Excluding the booster, launch costs and NASA tracking charges, the first flyby cost $96·87 million. The subsequent 2 flybys added $2·98 million, making a total of $99·85 million, cheap indeed for such a brilliantly productive mission.

Future Mariners Four Mariner Mark II missions were recommended in Apr 1983 following a 2yr study by the NASA Solar System Exploration Committee:

1 Venus Radar Mapper, replacing the more expensive VOIR, for launch in 1988.

2 Mars Geoscience/Climatology Orbiter, to orbit Mars for 1 Martian yr (2 Earth yrs) and study the surface and atmosphere, for launch in 1990. Estimated total cost $350 million.

3 Comet Rendezvous/Asteroid Flyby. A Shuttle launch in Jul 1990 is proposed, passing the main-belt asteroids Namaqua and Lucia on the way to a several-yr rendezvous with Comet Kopf from May 1994.

4 Titan Probe/Radar Mapper. A candidate for international co-operation, this would be launched in 1988-92 and could possibly include a Saturn orbiter.

Mercury

History Project Mercury, the first US manned spaceflight programme, was initiated in Nov 1958 and achieved its aim of demonstrating with a 1-man vehicle that men could be sent into space and return safely to Earth. The 6 flights started with Alan Shepard's 15min sub-orbital "lob" on 5 May 1961 and ended on 16 May 1963 with Gordon Cooper's 22-orbit, 34¼hr MA-9 mission. The first research and development spacecraft was successfully launched by Atlas D on 9 Sep 1959, and various test firings of boosters, spacecraft and escape systems followed. The main flights were named and numbered according to the launcher employed. Mercury-Atlas 1, a ballistic-trajectory test of an unmanned spacecraft in Jul 1960, was unsuccessful because of rocket failure. Mercury-Redstone 1 in Nov 1960 also failed because the rocket engines cut out after it had lifted just 2·54cm above the pad; the flight was completed successfully under the same designation a month later. MR-2, L 31 Jan 1961 from C Canaveral, carried Ham the chimpanzee. Despite various malfunctions which led to him enduring 17g during lift-off and splashing down 212km from the target in a leaking spacecraft after 6min of weightlessness, he was none the worse for his experience. The flight gave the astronauts confidence in the system. (Ham died in Jan 1983 in a N Carolina zoo, aged 26.) MA-2 to 4, further unmanned tests, were completed by Sep 1961. The spacecraft's bell shape was dictated primarily by heating conditions during re-entry, and its small size by the limited US launch capability at that time: Redstone's thrust was only 347kN (78,000lb), barely enough to launch the astronaut and essential equipment. Mercury was orginally designed so that control could be exercised entirely from the ground in case the astronauts became incapacitated. But the astronauts were soon demonstrating that there were times when only they could save the mission from failure. Total cost of Project Mercury was $392·1 million.

MR-3 On 5 May 1961 at 9·34am Alan Shepard, 37, became the first US man in space, flying in *Freedom 7* for 15min 22sec. He was weightless one-third of that time, rose to an altitude of 187·5km, attained a maximum speed of 8,336·2km/hr and landed 486km downrange from the Cape. He experienced 6g during acceleration and 11g on re-entry. Recovery operations went perfectly, the spacecraft was undamaged, and Shepard was exuberant.

Miss Sam was one of several Rhesus monkeys who showed that spaceflight was safe for US astronauts. She tested Mercury's escape tower with a 14km flight on a Little Joe rocket. ***(NASA)***

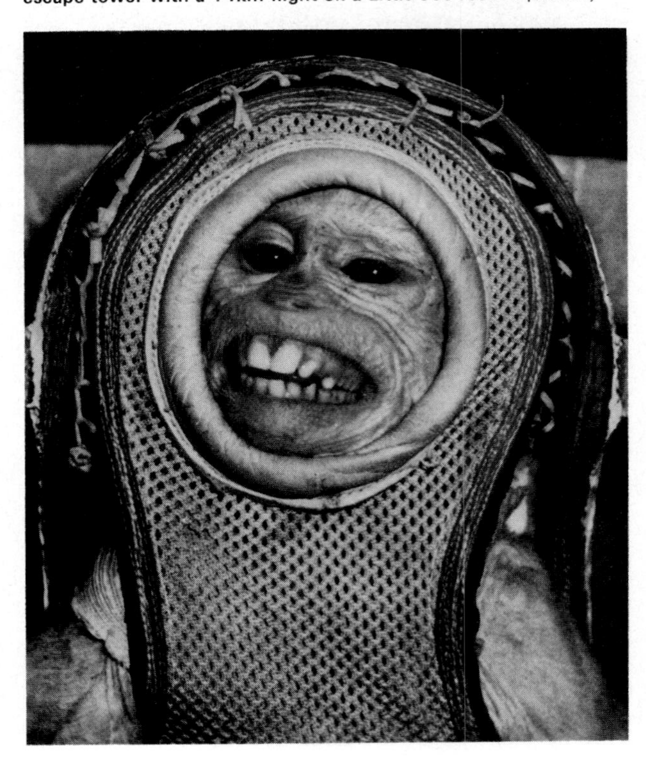

MR-4 On 21 Jul 1961 Virgil Grissom admitted that he was a "bit scared" at lift-off in *Liberty Bell 7* on the 2nd suborbital mission. He was weightless for 5min 18sec of the 15min 37sec flight, reached a height of 190·4km and endured 11·1g on re-entry. The new explosive hatch cover, incorporated for quick rescue of an injured astronaut, blew off after splashdown. As *Liberty Bell* shipped water and sank, Grissom swam clear and was picked up by helicopter.

MA-5 On 29 Nov 1961 Enos the chimp went into a nearly perfect orbit of 159 × 237km but had to be brought back after 2 instead of 3 orbits due to ECS troubles. Recovery was perfect after a 3hr 16min flight, with Enos surviving a max 7·8g during re-entry, as well as 79 undeserved electric shocks after a malfunction of control levers which he had been trained to operate to avoid shocks. He continued to operate it all the same, and it was decided that had an astronaut been on board he would have been able to correct the ECS problem.

MA-6 On 20 Feb 1962, after delays and postponements starting on Jan 23, John Glenn, 41, became the first US man in orbit, 11 months after Gagarin. His apogee was 261km, perigee 161km, and velocity 28,234·3km/hr. He was weightless for 4hr 48min of the 4hr 55min flight. A telemetry fault started a major alarm by indicating that the heatshield was no longer locked in position. A difference of opinion on the ground was resolved with the decision to order Glenn not to jettison the retropack before re-entry; it was hoped that this would help retain the heatshield in position. As the retropack burned and broke away, Glenn commented: "That's a real fireball outside". But all was well and *Friendship 7* splashed into the Atlantic 64·5km short of the predicted area. Glenn's only injury was knuckle abrasions when he punched the detonator to blow off the hatch after *Friendship 7*, with Glenn still inside, had been hoisted on to the deck of the recovery destroyer. Recovery forces totalled 24 ships, 126 aircraft and 26,000 personnel.

MA-7 On 24 May 1962 Scott Carpenter in *Aurora 7* became the 2nd US man in orbit. By the end of 2 orbits he had used more than half his fuel, and re-entry mistakes contributed to *Aurora 7* overshooting the Atlantic splashdown target by 420km. It was 50min after splashdown before it was established that Carpenter had struggled out of the spacecraft and was safe in his liferaft. He had been weightless for 4hr 39min out of 4hr 56min, had reached a velocity of 28,241·6km/hr and had endured 7·8g.

Mercury-Redstone 3 takes the first US astronaut into space. *(NASA)*

Atlas rocket, topped by *Friendship 7* and escape tower, being prepared to launch John Glenn. *(NASA)*

John H. Glenn, then 40, became 1st US astronaut in orbit in 1962, resigned from NASA 2yr later, became a Senator in 1974, and made an unsuccessful bid for the Presidency in 1984. *(NASA)*

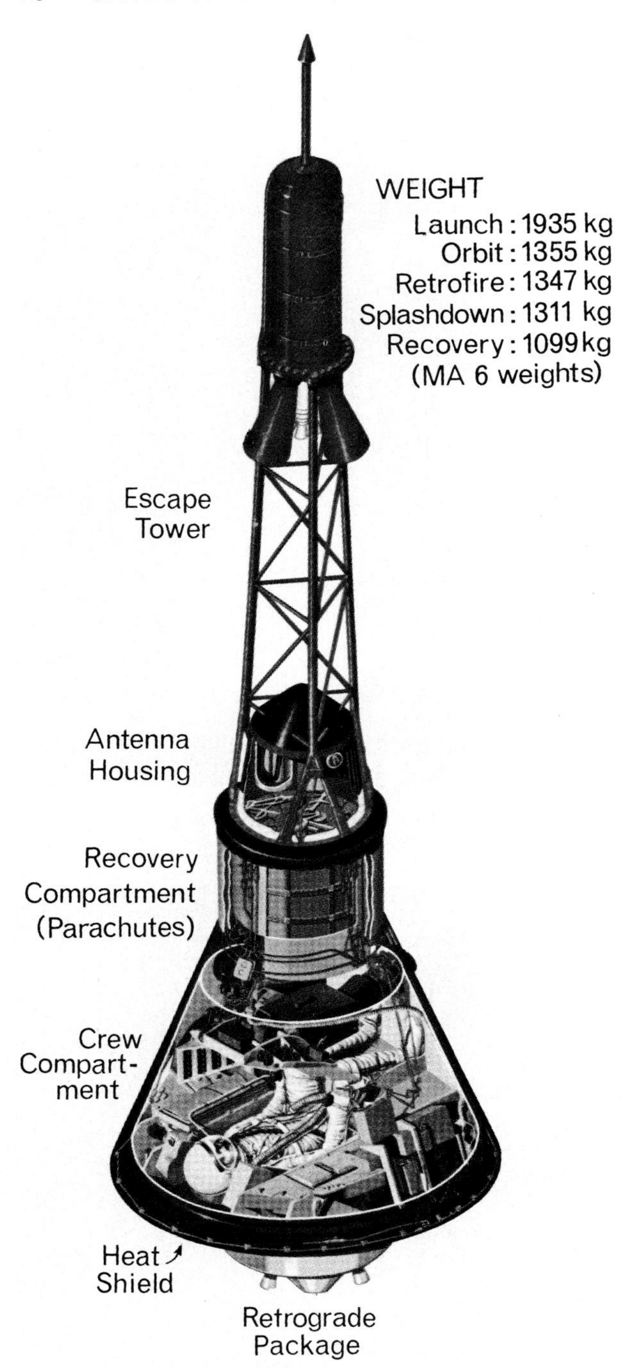

MERCURY SPACECRAFT

MA-8 Launched on 3 Oct 1962 with Walter Schirra in *Sigma 7*, this was the "textbook flight". The 6-orbit flight - twice as long as previous missions - went well from beginning to end. Schirra carried out disorientation tests and practised conserving his fuel so effectively that a final, 1-day MA-9 mission at last became possible. He splashed down only 7·24km from the recovery ship in full view of the crew and waiting newsmen. He had been weightless for 8hr 56min of the 9hr 13min flight.

MA-9 On 15 May 1963 Gordon Cooper in *Faith 7* started a 22-orbit flight which lasted 34hr 20min. Extra equipment, fuel, oxygen, water, etc, for the longest Mercury mission had raised the in-orbit weight to 1,373kg. Cooper also carried out experiments aimed at developing a guidance and navigation system for Apollo. On the 19th orbit a warning light suggested that the craft was decelerating and beginning to re-enter; by the 21st his automatic stabilisation and control system had short-circuited and the carbon dioxide level was rising in the spacecraft. Despite his dry comment, "Things are beginning to stack up a little," Cooper manually conducted his re-entry so efficiently that he splashed down only 6·44km from the recovery carrier.

Mesa

Developed by Boeing from Sweden's Viking and Applications Explorers, Mesa is a small satellite bus being offered at about $20 million for payloads such as Sarsat. It will fit between Ariane's 3rd stage and the main payload, or into the Shuttle.

Nimbus

History The US Navy came to regard Nimbus as indispensable to shipping operations in the Arctic and Antarctic. Satellite pictures showing the location and movement of ice masses enabled USN ships to move in these areas for several extra months, and were ultimately expected to make operations possible right through the 6-month polar night. (Russia is using special spring and autumn Cosmos launches for the same purpose.) The Nimbus series were originally conceived (as its Latin name, meaning "raincloud", implies) as meteorological satellites to provide atmospheric data for improved weather forcasting. But as increasingly sophisticated sensing devices were added to successive spacecraft, the series grew into a major programme studying Earth sciences. From it, too, sprang ERTS (now renamed Landsat), the results of which proved to be sensational and are described separately. By mid-1973 data provided by Nimbus 1-5 covered oceanography (the geography of the oceans), hydrology (the study of water in the atmosphere, on land surfaces, and in the soil and underlying rocks), geology (the history of Earth, especially as recorded in rocks), geomorphology (the study of the Earth, its distribution of land and water and evolution of land forms), geography (description of land, sea and air, and distribution of life, including man and his industries), cartography (chart and map-making), and agriculture (data on moisture and vegetation patterns over various land surfaces).

By 1975 all 7 of the originally planned Nimbus series, designed and built by General Electric, had been launched at an estimated total cost of $340 million. They were designated A-F before launch, and given numbers only if successful. All except B, lost as a result of a launch vehicle failure, exceeded their mission objectives. Equally successful was the 8th, launched in 1978 as NASA's first satellite to be devoted mainly to studying atmospheric pollution; that added $79 million, plus $4·6 million for the Delta launch, to total costs.

Spacecraft description The basic Nimbus spacecraft is butterfly-shaped when deployed in orbit, 3m long, and with a span across the solar panels of 3·4m. The panels, each 2·4 × 0·9m, provide more than 200W of power and are supplemented by 2 SNAP-19 nuclear generators. The body of the spacecraft consists of 2 main components separated by struts; the larger carries the sensors and other equipment; the smaller, between the solar panels, houses the stabilisation and control system. Working on the principle that the Earth is warm and space is cold, the infra-red sensors keep the satellite's cameras pointed towards Earth at all times. The system is controlled by a computer which maintains the correct attitude to within 1° in each axis by means of cold-gas jets and inertia wheels. Pitch and roll stability is sensed by horizon scanners, while a gyroscope controls stability in the yaw axis.

Nimbus 1 L 28 Aug 1963 by Thor-Agena B from Vandenberg. Wt 376kg. Orbit 422 × 932km. Incl 98°. This was fully operational for 1 month until failure of the solar-array power system. A short 2nd burn of the Agena rocket resulted in an eccentric orbit instead of the intended circular 1,110km orbit. Nimbus 1 returned 27,000 cloud-cover photos, providing the first high-resolution TV and infra-red weather photos and proving to meteorologists that such pictures could be received at small, inexpensive portable stations as the satellite passed overhead. Hurricane Cleo's "portrait" was taken during Nimbus 1's first day in orbit. Subsequently many other hurricanes and Pacific typhoons were tracked. Inaccuracies on relief maps were corrected, and the Antarctic ice front more accurately defined. Orbital life 15yr.

Nimbus 2 L 15 May 1960 by TAT-Agena B from Vandenberg. Wt 413kg. Orbit 1,100 × 1,181km. Incl 100°. A nearly perfect orbit. Designed for 6 months (2,500 orbits) of operations, Nimbus 2 actually provided 33 months before finally ending on orbit 13,029 on 17 Jan 1969. Following the success of Nimbus 1, over 300 Automatic Picture Transmission (APT) stations in 43 countries were able to receive its high-resolution infra-red pictures. Temperature patterns of lakes and ocean currents were obtained for shipping and fishing industries, and thermal pollution could be identified. An additional Medium Resolution Infra-red Radiometer, which measured electromagnetic radiation emitted and reflected from Earth in 5 wavebands from visible to infra-red, permitted detailed study of the effect of water vapour, CO_2 and ozone on the Earth's heat balance. Orbital life 800yr.

Nimbus 3 L 14 Apr 1969 by Thor-Agena D from Vandenberg. Wt 575kg. Orbit 1,070 × 1,131km. Incl 99°. The 3rd success in the series following the launch failure of Nimbus B on 18 May 1968, Nimbus 3 carried 7 meteorological experiments, plus SNAP-19, a nuclear power unit. Vertical temperature measurements of the atmosphere by Satellite Infra-red Spectrometer (SIRS) were acclaimed as one of the most significant developments in the history of meteorology. Previously, because data over the oceans had been scanty, only 20% of the world had detailed weather information. SIRS made it possible to obtain temperature data over the entire Earth with an accuracy of 2°F above 6,100m, and within 4°F below that level. A 10·4kg electronic collar fitted to a wild elk in the National Elk Refuge in Wyoming was interrogated twice daily to study the migratory habits of large animals. The Nimbus 3 rocket also orbited Secor 13, a US Army geodetic satellite weighing 20·4kg. Life of Nimbus 3 is 800yr, Secor 2,000yr.

Nimbus 4 L 8 Apr 1970 by Thor-Agena D from Vandenberg. Wt 675kg. Orbit 1,093 × 1,107km. Incl 107°. In its circular, near-polar orbit Nimbus 4 was still making worldwide weather observations twice daily (once in daylight and once in darkness) more than 2yr later. Of its 9 experiments, 4 were new and 5 were improved versions of experiments on earlier flights. New experiments included Interrogation Recording and Location System (IRLS); examples of targets tracked included weather balloons floating around the world, floating ocean buoys, a wild animal, and Miss Sheila Scott on a solo flight over the N Pole. One task was to measure the thickness of island ice floes floating in the Arctic by interrogating buoys placed in the water. One such, T3, was 11 × 6km long and 30m thick. They melt in summer in brackish swamps covered by fog and are impossible to reach. Details of their behaviour are providing information on the "cradle" of much of the weather affecting the US and Europe. Other Nimbus 4 tasks included the analysis of water quality and sewage pollution of the Great Miami River near Cincinnati and of lakes Erie and Ontario. Low oxygen content indicates raw effluent, the oxygen decrease being caused

Nimbus 4 weather satellite. *(NASA)*

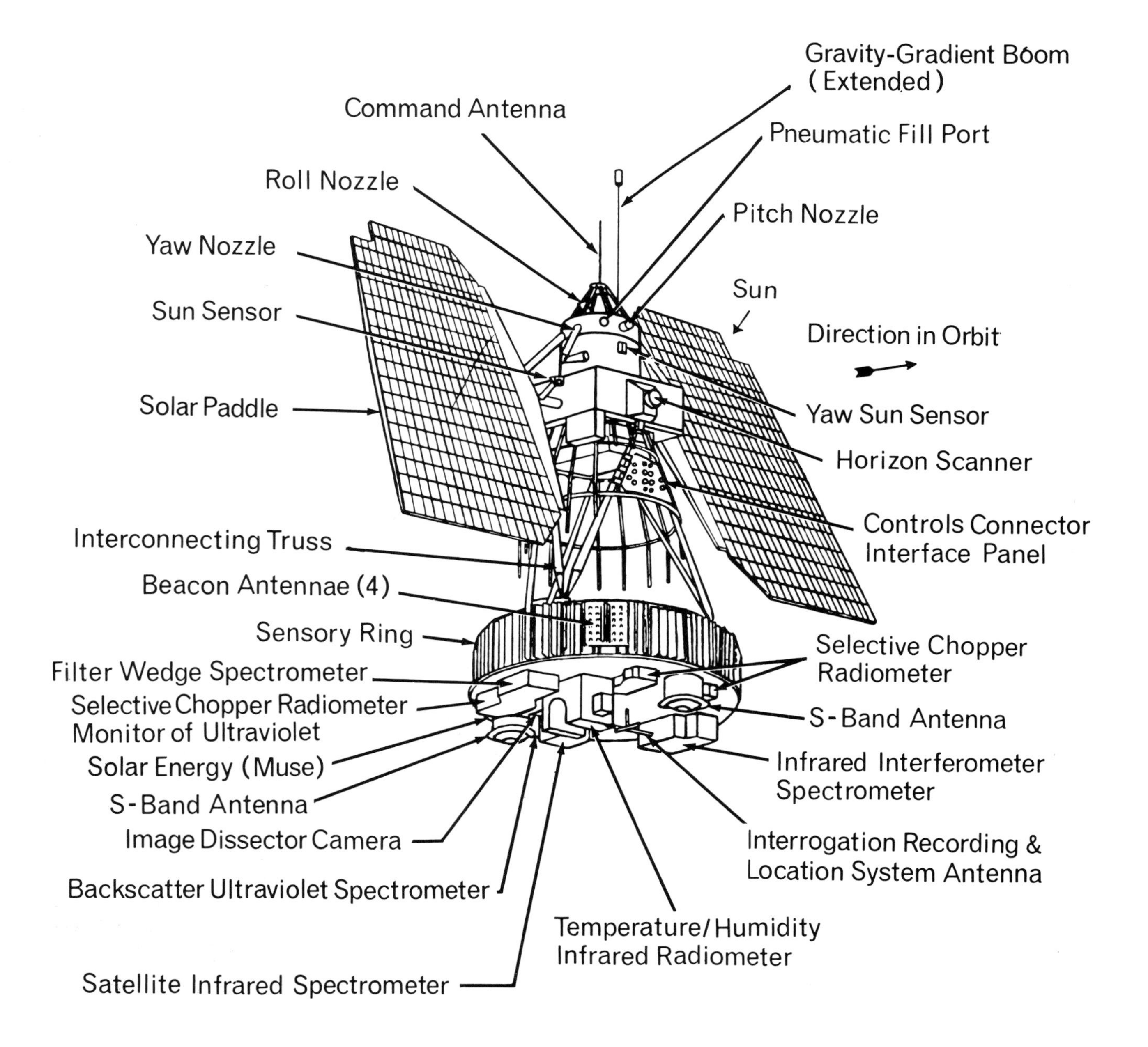

by bacterial decomposition. Other IRLS platforms were also placed in Mt Kilauea, Hawaii, probably the world's most active volcano, to measure the relationship between temperature rises and eruptions, and in a bear's den in Montana to monitor its environment during hibernation. Each platform will only respond when the satellite interrogates it with an individual 16-bit digital "telephone number"; the address of the bear den in Montana was 0111100001111001. The Nimbus 4 rocket also placed TOPO-1, a US Army Secor-type satellite, wt 21·7kg, into an identical orbit for use in space-ground tactical exercises. Orbital life: Nimbus 4 1,200yr, TOPO-1 2,000yr.

Nimbus 5 L 11 Dec 1972 by Delta from Vandenberg. Wt 768kg. Orbit 1,089 × 1,102km. Incl 99°. Carrying 6 new instruments, Nimbus 5 was designed to take the first vertical temperature and water readings of Earth's atmosphere through cloud, a major step forward since many parts of the world are under cloud for more than 50% of the time. False-colour pictures enabled investigators to say that rain was falling in certain areas "at a rate of five-hundredths of an inch per hour". Measurements were also made of water vapour which evaporated from the oceans, condensed and fell back to the surface. Sensors were also carried to map the Gulf Stream off the US E coast and the Humboldt Current off S America's W coast. Plotting the daily position of the Gulf Stream enables southbound ships to avoid it, while northbound ships, by riding in the stream, get the benefit of several extra knots.Planned life was 1yr, but it was still operational at the end of 1982 after 300 million operations. Orbital life 1,600yr.

Nimbus 6 L 12 Jun 1975 by Thor-Delta from Pt Arguello. Wt 827kg. Orbit 1,092 × 1,104km. Incl 99·9°. 1,600yr life. In addition to maintaining the regular weather, rainfall and icepack watch, Nimbus 6 was allotted the task of investigating the dangers involved in recovering the huge oil and gas deposits already located in the Arctic in the Prudhoe Bay and Beaufort Sea areas. A series of data-collection platforms were dropped to enable Nimbus 6 to track icepack movements, the amount of summer melting, etc. This information was used to select locations for oilrigs, and to decide whether pipelines from them would have to be laid under the ocean floor, on the sea bottom or over the icepack. Nimbus 6 ultimately acquired data from a total of 130 buoys, icebergs and other platforms around the world. In 1978 it was used to track the 6,000km journey of a lone Japanese explorer travelling by dogsled from N Canada to the N Pole, and in 1980 to follow a British seaman's attempt to cross the Pacific in a rowing boat.

Nimbus 7 L 24 Oct 1978 by Delta from Vandenberg. Wt 907kg. Orbit 943 × 953km. Incl 99°. Circling Earth 14 times per day, Nimbus 7 was the first satellite designed to monitor the atmosphere for man-made and natural pollutants. It carries 8 instruments - 7 from the US, 1 from Britain - to help scientists from 8 participating countries to decide whether the concentration of ozone in the Earth's upper atmosphere is changing; whether the Earth is warming up or cooling down; and the extent of pollution in the world's oceans. By sensing the colours of the oceans the Coastal Zone Colour Scan instrument enabled oceanographers to map chlorophyll concentrations, sediment distribution and salinity over large areas of coastal or ocean water. In its first 2yr Nimbus 7 transmitted over 20,000 2min scenes, making it possible among other things to identify and track oil spills over 11 major ocean areas traversed by tankers. A 2-month experiment in Mar-May 1981 combining the Ozone Mapping Spectrometer and research aircraft flights produced promising results in the search for ways of helping high-flying airliners avoid areas of high ozone concentration (which causes eye and throat irritation) and clear-air turbulence.

NOAA

History The National Oceanic and Atmospheric Administration manages and operates the US weather satellites, currently called NOAA and GOES, once they have been launched by NASA. The daily cloud-cover pictures, with much other weather forecasting information, is now the daily diet of millions of TV watchers around the world. By 1982 over 30 American weather satellites had been launched; the first series, launched during 1960-65, were called Tiros; 9 more, from 1966-69, were designated Tiros Operational Satellite (TOS) or ESSA, for the Environmental Science Services Administration, which managed the programme in that period. In 1970 a 2nd generation, called Improved TOS (ITOS), was introduced, with letter designations before launch and a number when in orbit. The prototype Synchronous Meteorological Satellites (SMS) were added in 1974 to help weather forecasters by providing continuous observation of short-term weather features. They were the forerunners of the Geostationary Operational Environmental Satellite System (GOES), which now provides continuous day and night weather observation of N and S America and its oceans; Russia, ESA and Japan are all working towards their own geostationary weather satellites to complement the US system for global use. The NOAA spacecraft now distribute their unprocessed data to Earth stations in more than 120 nations as they pass overhead, and at least 1,000 schools, private individuals and even Boy Scout troops pick up NOAA imagery, often on home-made receivers. Britain's Meteorological Office, France's Space Centre and Canada's Communications and Research Centre participate in the NOAA system.

Spacecraft description Tiros satellites had a "hat-box" shape: 18-sided polygons, 1·07m dia, and 0·55m high. Solar cells covered the sides and top, with apertures for 2 TV cameras on opposite sides. Each camera could take 16 pictures per orbit at 128sec intervals, though the intervals could be decreased to 32sec. 2 tape recorders could store up to 48 pictures when ground stations were out of range. The weight of 119kg for Tiros 1 had risen to 138kg for Tiros 9 and 10. ESSA satellites, similar but more advanced, had 2 Automatic Picture Transmission (APT) cameras able to photograph a 3,000km-wide area with 3km resolution at picture centre. Pictures were taken and transmitted every 352sec, allowing a typical APT station to receive 8-10 per day. ITOS/NOAA satellites remained box-shaped, 1m × 1m × 1·24m, wt 340kg, with 3 solar panels, and momentum flywheel providing spacecraft stabilisation. Tiros N/NOAA satellites, with lift-off wt of 1,421kg and 736kg in-orbit wt, are 3·7m long, with 1·88m dia and 2·37 × 4·91m solar array.

Tiros The Television and Infra-red Observation Satellite Programme began as a joint NASA/Defence Department project to develop a meteorological satellite. As soon as Tiros 1, L 1 Apr 1960, began orbiting at 692 × 740km, it was clear that the US had successfully established both a meteorological survey and a military reconnaissance satellite. During the 78 days that its batteries lasted, it sent back 22,952 cloud-cover photographs. As the first of them came in, a meteorologist at a ground station observed that the programme had gone "from rags to riches overnight". It is believed that its photographs included some of the Soviet Union and China so detailed that aircraft runways and missile sites could be readily identified. By the time 10 had been launched, the more advanced Nimbus and ESSA satellites were taking over. The first 8 Tiros, all operating in similar orbits, sent back several hundred thousand photographs, together with information about the flow of heat the Earth was

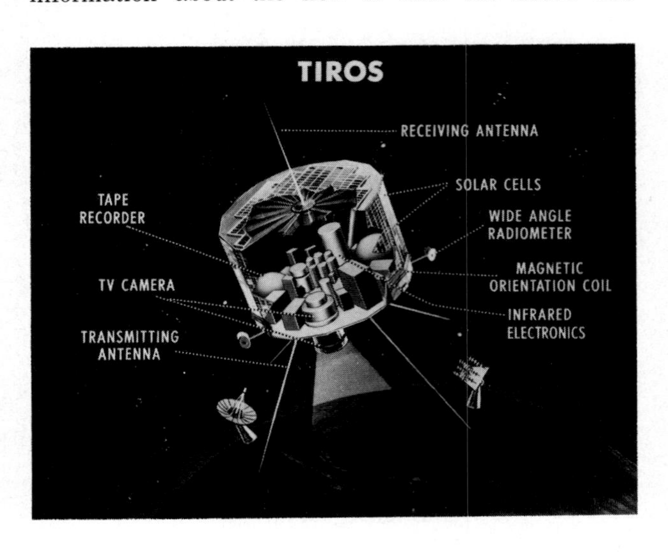

reflecting back into space; this was vital meteorological information unobtainable until then. Tiros 9, L 22 Jan 1965, was the first attempt to reach polar orbit from Cape Canaveral. A series of 3 Delta dogleg manoeuvres duly placed it in the planned 82° Sun-synchronous orbit. But a 2nd-stage overburn meant that the orbit, instead of being 644km circular, was 700 × 2,578km. In the event, the higher apogee provided more Earth cover than planned. On 13 Feb the first photomosaic of the entire world's cloud cover was provided by 450 excellent pictures. By the time Tiros 10, L 2 Jul 1965, was shut down on 3 Jul 1967 more than 500,000 cloud-cover pictures had been returned.

TOS This system, based upon Tiros technology, began operating with the launch of ESSA-1 on 3 Feb 1966 (orbit 702 × 845km) and ended with ESSA-9 (L 26 Feb 1969, orbit 1,427 × 1,508km). By that time 400 receiving stations were in operation around the world, and weather services of 45 countries, as well as 26 universities, up to 30 US TV stations, and an unknown number of private citizens who had built their own receivers, were receiving and using their weather photographs each day. In 1969 a picture from ESSA-7 made history by revealing that the snow cover over America's Midwest, in Minnesota and the Dakotas, was 3 times thicker than normal. Measurements showed that it was equivalent to 15-25cm of water covering thousands of square miles. A disaster area was declared before it happened, and when the floods came much had been done to control the situation.

ITOS/NOAA The Improved Tiros Operational System more than doubled TOS capability because its cameras were able to work at night as well as in the daytime. They were designated ITOS before launch and once in orbit given a new series of NOAA designations because that agency had then taken over ESSA. Thus NOAA-1 was launched 11 Dec 1970 from Vandenberg into a 1,429 × 1,472km orbit with 101° incl. But ITOS-B, L 21 Oct 1971, was never given a NOAA designation because it failed to achieve a satisfactory orbit. NOAA-3 and 4 were launched in Nov 1973 and 1974, and NOAA-5 (wt 340kg) in Jul 1976, all into similar orbits.

Tiros N/NOAA A 4th generation with still further improved capabilities began with the launch of Tiros N (the NASA prototype) on 13 Oct 1978 (wt 734kg; orbit 850 × 866km; incl 102°); it was joined by the complementary satellite NOAA-6 on 27 Jun 1979. Developed like the earlier generations by RCA, these were the start of a series based on the Block 5D bus developed for the USAF DMSP. Launched into a near-polar, Sun-synchronous orbit by Atlas F from Vandenberg, each can view virtually all of Earth's surface at least twice every 24hr. It was claimed that their new sensing devices would enable forecasters to predict climatic trends accurately 1 week, and ultimately, 2 weeks, in advance. Their sensors measure temperature and humidity in the Earth's atmosphere, surface temperatures, surface and cloud cover, water-ice-moisture boundaries, and proton and electron flux near the Earth. Sea surface temperatures are measured to an accuracy of 2·7°F. They can receive, process and re-transmit data from free-floating balloons, buoys and remote automatic stations distributed around the globe, and can also track moving stations. NOAA-B, L 29 May 1980, failed to achieve the correct orbit, but by the time NOAA-7 was in orbit (L 23 Jun 1981) it was possible to claim that fishermen in California, Oregon, Washington and Alaska were improving their catches of salmon, albacore and herring by using the sea surface temperature charts. One 60-vessel towing and transportation company announced fuel savings of 20-40% as a result of choosing routes with the help of NOAA's stream and loop current information. NOAA-7 cost $15 million plus $7·5 million launch costs. The series has a 2yr operational life and 350yr orbital life.

NOAA-8 L 28 Mar 1983 by Atlas from Vandenberg. Wt 1,712kg. 1st of the advanced RCA Tiros N spacecraft for use as part of the international Cospas-Sarsat system as well as for monitoring global weather, this spacecraft was at first in an unstable, 833km circular orbit with 98·3° incl. Advice on stabilisation was given by the USAF, which had had similar problems with its DMSPs. NOAA-8's high-resolution instruments measure both surface and vertical temperatures; a UK stratospheric sounding unit measures upper-atmospheric temperatures, and a French system gathers data from balloons, buoys and remote weather stations.

NOAA-9, 10 and 11 Will also carry search and rescue instruments. RCA Astro Electronics has a $64·1 million contract for delivery of NOAA-9 and 10 in Mar 1986 and Jan 1987, with the possiblity of a further $16·9 million for NOAA-11 in 1988.

The NOAA-10 weather satellite is also used as part of the Cospas/Sarsat international search and rescue system, relaying distress calls from both ships and aircraft. *(RCA)*

SMS/GOES Weather-watching from geostationary orbits was developed by NASA with 2 prototype Synchronous Meteorological Satellites and then taken over by NOAA with the Geostationary Operational Environmental Satellites. The main instrument is the Visible/Infra-red Spin Scan Radiometer, which detects reflected sunlight, or radiated heat during the night, to differentiate between water, land and clouds. By GOES-4 more advanced versions of the telescope added data on atmospheric temperatures and water vapour content. Ford Aerospace built SMS-1 and 2 and GOES-1 to 3; Hughes Aircraft provided the main instruments for the whole series and built GOES-4 to 6. As the series progressed data were being collected from more than 1,500 remote platforms on land, at sea, and carried by balloons and aircraft; solar activity was also being monitored, with warning given of solar flares. Hughes' 1977 contract to build, test and deliver the last 3 GOES was worth $77 million. In 1981 cost of GOES-5 was said to be about $20 million, plus $16 million for launch.

SMS-1 L 17 May 1974 by Delta from C Canaveral. Wt 627kg. Stationed initially over the E Atlantic, and then at 75° W over Bogota, this provided the first day-and-night stormwatch, with full-disc pictures of the W hemisphere every 30min. As part of the Global Atmospheric Research Programme (GARP) it provided the first continuous coverage of a major hurricane, designated Carmen, in Sep 1974. SMS-2, L 6 Feb 1975, was placed at 115°W, E of Hawaii, so that the two together could cover the W hemisphere. One of its tasks was to keep watch on California's forest areas, including the famous redwoods, to give warning within 90min of fire outbreaks.

GOES-1 L 16 Oct 1975 by Delta from C Canaveral. Wt 293kg. First positioned over the Indian Ocean at 60°E, with ESA's ground station in Spain processing data; then moved to 90°W.

GOES-2 L 16 Jun 1977. Placed at 105°W to assist developing countries.

GOES-3 L 16 Jun 1978. Replaced GOES-1 before its move to 60°E.

GOES-4 L 9 Sep 1980. Stationed at 135°W.

GOES-5 L 22 May 1981 at 85°W.
(When GOES-4's imaging system failed in Nov 1982 GOES-1, then at 116°W, was reactivated and its images re-transmitted by GOES-4.)

GOES-6 L 28 Apr 1983. Placed at 135° in mid-May.

GOES Next NOAA has given this designation to an FY 1985 budget request to develop a further advanced series which can image weather patterns and take atmospheric temperature soundings simultaneously.

NOSS The proposed National Oceanic Satellite System is designed for joint civil and military use. Costing $800 million and drawing on experience from Nimbus, GOES, Seasat and NOAA, NOSS will be placed in orbit and recovered by Shuttle (see also Military - US).

OAO

History A programme of 4 Orbiting Astronomical Observatories, begun in 1959 and completed with the launch of the 4th in Aug 1972 at a total cost of $364·4 million. Although only 2 of the launches were successful, their achievements ranged from the first observation of ultra-violet (UV) emissions from the planet Uranus to new insights into the structure and composition of Earth's upper atmosphere. OAO-2's success in studying young, hot stars which emit most of their energy or light in the UV - the blue portion of the spectrum not visible to the human eye or ground observatories because of Earth's atmosphere - proved that with such unmanned spacecraft astonomers could conduct sustained viewing of the universe. It enabled the 82cm UV telescope installed on OAO-3 (renamed Copernicus once in orbit to celebrate the 500th anniversary of "the father of modern astronomy") to be pointed within 3 hundredths of an arc-second. In 1975 Copernicus began the study of 3 Sun-like stars located 11 light-years from Earth to see if it was possible that intelligent civilisations on their planets were trying to communicate with us. Astronomers continued this work with IUE (see International - ESA), and that in turn should be followed by a similar effort with the Space Telescope.

OAO-1 L 8 Apr 1966 by Atlas-Agena D from C Canaveral. Wt 1,776kg. Orbit 792 × 805km. Incl 35°. The battery failed after only 3 days in orbit, but OAO-1 provided engineering data showing that the concept of astronomical investigation from space was feasible, and led to improvements in later craft.

OAO-2 L 7 Dec 1968 by Atlas-Centaur from C Canaveral. Wt 2,016kg. Orbit 770 × 780km. Incl 34·99°. When it was shut down on 13 Feb 1973 because it had developed an electrical fault it had operated for over 4yr instead of 1. During 22,000 orbits its 11 telescopes had viewed 1,930 celestial objects and made 22,560 observations. These included the detection of a huge hydrogen cloud around Comet Tago-Sato-Kisaka, the first evidence that such clouds exist; observations of stars with magnetic fields over 10,000 times stronger than that of the Sun; the first UV observations above the atmosphere of a supernova, the momentary outburst of a star to a brightness millions of times greater than that of the Sun; and observations of UV emissions from Uranus.

OAO-B L 30 Nov 1970 by Atlas-Centaur from C Canaveral. It failed to achieve orbit when a protective shroud could not be jettisoned, and fell back to Earth.

OAO-3 (Copernicus) L 21 Aug 1972 by Atlas-Centaur from C Canaveral. Wt 2,220kg. orbit 748 × 740km. Incl 35°. It was still transmitting on command 8yr later. By then, Princeton University's 80cm UV telescope, with a 47·6kg mirror made from thin fused silica ribs, had detected large concentrations (more than 10%) of molecular hydrogen in the denser interstellar dust clouds, and surprisingly large amounts of deuterium (heavy hydrogen) in interstellar dust clouds. (Deuterium is a basic element for fusion in the formation of stars, and current theories had suggested that most of it should already have been used up. The "Big Bang" theory may therefore have to be revised.) These observations are made by collecting UV light from a star and directing it to a spectrometer, which then sends digital readings to Earth. The 2nd experiment, consisting of 3 smaller X-ray telescopes developed at University College, London, and provided by the Science Research Council, successfully followed up the work of earlier satellites (notably Explorer 42, L Dec 1970) which established that X-rays were present in the universe in much larger quantities than were generated by the Sun. The object was to chart about 200 other X-ray sources already identified, as well as finding new sources.

OAO-2 discovered hydrogen clouds. *(Grumman)*

Studies of the black hole Scorpius V-861 (discovered by HEAO) showed that the super-giant star was apparently losing matter to its invisible companion. The X-ray telescope was also used to confirm the identity of another black hole candidate, Cygnus X-1, within a binary system. Another striking finding from observations of 55 unique objects was the fact that the rotation rate of the Cygnus X-3 binary system increased perceptibly over a period of only one month. Clusters of galaxies in Perseus, Coma and Virgo, as well as the supernova remnants in Puppis, were also studied. Copernicus started work on trying to establish whether other civilisations were trying to contact us with UV laser beams. US scientists, by studying a bright star first when it was high in the sky and then through Earth's atmosphere as it sank below the horizon, also used Copernicus to estimate how far Earth's ozone layer is being broken up by chlorine resulting from freon, the gas used in millions of aerosol spray cans. Orbital life 500yr.

OGO

Details of vast clouds of hydrogen, the most abundant element in the universe, were provided by the 6-satellite Orbiting Geophysical Observatory series. They also contributed, after difficulties with the first 3, to the design of standardised 3-axis-stabilised observatories which could be used repeatedly. OGO-1 was launched 5 Sep 1964 by Atlas-Agena from C Canaveral. Wt 487kg. Orbit 35,743 × 114,040km. Incl 57°. OGOs 2-6 were launched during 1965-69. Each was launched into a different elliptical orbit and into a different sector of the cislunar space quadrant. The last (L 5 Jun 1969; wt 620kg) transmitted until 23 Jun 1969. The 130 experiments they carried sent back 1·5 million hr of data and added a number of notable firsts to space history. They included: first observation of protons responsible for a ring of current surrounding the Earth at a distance of several Earth radii during magnetic storms; first satellite global survey of Earth's magnetic field; first observation of daylight aurorae; first worldwide map of airglow distribution; and new insights into the bow shock, the interaction between the leading hemisphere of the Earth's magnetic field and the solar wind. In Apr 1970 OGO-5 measured a huge hydrogen gas envelope, 12 million km across and 10 times larger than the Sun, surrounding Comet Bennett. Not visible from Earth, the hydrogen cloud was measured by a French device while the comet was 104 million km from Earth and OGO-5 was operating in a spin-scan mode 22,500 × 107,800km above the Earth. The existence of large amounts of hydrogen around comets was first discovered in Jan 1970 by OAO-2. OGO orbital life 10-16yr.

Orion

Orion Satellite Corporation has reservations for Shuttle launches of Orions 1-4 in 1986-7 and is offering Ku-band transponder capacity for sale or lease to large corporate users. Intelsat is complaining that such systems would threaten its global viability, but the US Commerce Department has been advised that the service is in the US national interest.

OSO

History 8 out of 9 Orbiting Solar Observatories succeeded in keeping the Sun under continuous observation from Mar 1962 to Sep 1978, covering 1½ of the 11yr solar cycle. With observations added by Skylab's Apollo Telescope Mount in 1973, astronomers were well placed to study the 1980-82 peak of solar activity with SMM, so it was all the more unfortunate when that spacecraft suffered a major failure. One of 4 inter-related series (the others were OAO, OGO and HEAO), the spin-stabilised,

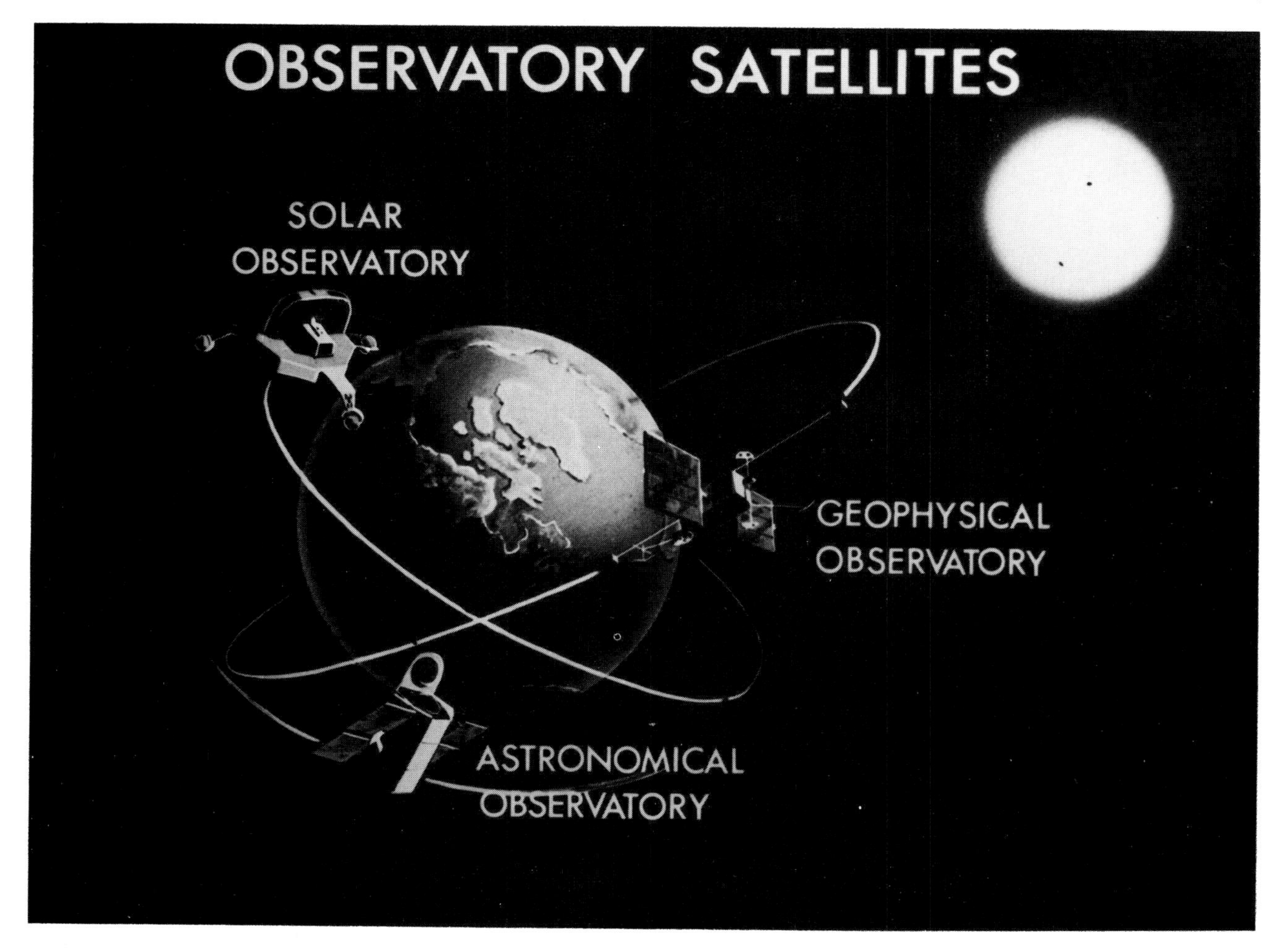

Sun-oriented OSOs were NASA's first standardised observatory spacecraft. They studied the Sun and its atmosphere in the X-ray, ultra-violet and infra-red wavelengths, as well as celestial objects. Discoveries included the fact that solar flares - the sudden releases of energy and material from the Sun - have temperatures of over 30 million °C; and a single flare, which may last minutes or hours, can release as much energy as the whole Earth uses in 100,000yr. Solar radiation takes 8½min to reach Earth, and it has been found that several minutes before that there is an increase in soft X-ray emissions, suggesting that it may be possible to predict major flares. "Holes" in the Sun's corona, with temperatures much lower than average, were another discovery. The coronal holes are somewhat similar to the solar polar caps, which have temperatures of about 1 million °C (1·8 million °F) compared with 2 million °C (3·6 million °F) in other parts of the solar corona.

Spacecraft description Launches were by Thor-Delta from C Canaveral into 550-600 × 510-550km orbits with 33° incl. Wt 208kg until OSO-7, which at 571kg was similar in appearance but much more sophisticated. It had a lower, rotating "wheel" section with 152·4cm dia and 71·6cm ht. The wheel spun at 6rpm, automatically controlled by a pneumatic system. The upper "sail" section was 234cm high and 209cm wide. Non-spinning, it provided power by means of solar cells, and served as a platform for the US and French Sun-pointing spectrometers. A gyroscope in the sail acted as a "Sun-position memory" so that the Sun-pointing experiments re-acquired the Sun as quickly as possible when the vehicle emerged from Earth's shadow. Gas jets and magnetic torque coils kept the spin axis perpendicular to the Sun.

OSO-1 L 7 Mar 1962. Transmitted for nearly 2yr and provided 77 days of "near perfect" solar observations. Observed more than 140 solar flares. 30yr life.

OSO-2 L 3 Feb 1965 after an original attempt on 14 Apr 1964 had killed 3 technicians and injured 9 when the 3rd stage exploded during a test. Operated 10 months and brought total data hours to 6,000.

OSO-C What was to have been OSO-3 was lost on 25 Aug 1965 when the Thor-Delta 3rd stage fired prematurely.

OSO-3 to 6 L Mar and Oct 1967, Jan and Aug 1969. Continued observations, the latter 2 carrying British and Italian equipment.

OSO-7 L 29 Sep 1971. Despite incorrect orbit of 323 × 571km, which led to decay on 9 Jul 1974, provided 1st photographs of rapidly moving structures in the Sun's white light corona, and studied coronal holes and cool regions above the Sun's rotational poles. Other achievements included measuring the X-ray light curve of binary stars in the Milky Way, and the binary system SMC X-1 in the Small Magellanic Cloud.

OSO-8 L 21 Jun 1975. Provided data until 26 Sep 1978. Wt 1,064kg. Carried US and French UV telescopes to investigate transfer of solar energy between layers of the Sun's atmosphere.

Pageos

Part of the Geodetic Satellite Programme to create a world survey network with an accuracy of 10m, Pageos was L 23 Jun 1966 by TAT from Vandenberg. Wt 111kg. Orbit 5,432 × 3,016km. Incl 85°. Pageos (Passive Geodetic Earth-Orbiting Satellite) was an aluminium-coated Mylar 41m-dia balloon of the Echo 1 type, folded and packed inside a spherical canister which was separated into halves by an explosive device after orbital insertion. The balloon then automatically inflated. Intended to provide an orbiting point source of light which could be photographed for 5yr to determine the size and shape of Earth to a degree never before possible, it was brightly visible and a source of unfailing interest to professional and amateur astronomers alike for 10yr. Reflecting the Sun, it was as bright as the star Polaris and was observed simultaneously from 41 portable camera stations around the world which were used to construct a 3-dimensional geodetic reference system. The resulting triangulation network made it possible to obtain the distance between 2 surface points 5,000km apart to an accuracy of 10m. Orbital life was 50yr but on 20 Jan 1976 it mysteriously broke up into 14 pieces. Speculation about the cause included a build-up of static electricity or some deliberate interference by either America or Russia.

Pioneer

History For many years the most exciting of the unmanned projects, living up to its name with discoveries about the Sun, Jupiter, Saturn and Venus. Since surpassed by Viking and Voyager. Pioneer 1, launched in 1958, was the first NASA spacecraft, having been handed over by the USAF. The series was complementary to the Mariner flights. In 20 years 13 out of 17 launches added immensely to man's knowledge of Earth's cosmic surroundings. Designated with letters before launch, missions were numbered in the series only if they could be successfully used. Pioneers 1-4, intended as lunar probes, failed to get there but made many other discoveries. Three more Pioneer launches by Atlas-Able on 26 Nov 1959, 25 Sep 1960 and 15 Dec 1960, and what should have been Pioneer 10, launched by Delta on 27 Aug 1969, were failures and do not appear below. The interplanetary Pioneers began with 5, which with its predecessors was used to study solar energy and to provide up to 15 days' warning of solar flares so that astronauts could be protected during Moon flights. Originally designed for 6 months' operation, Pioneers 6-9, were used to study space from widely separated points during an entire solar cycle of 11yr and were all still predicting solar storms at the end of 1981.

A major achievement of this series was to discover Earth's long magnetic "tail", measuring about 5·6 million km, on the side away from the Sun. In 1971 the alignment of Pioneers 6 and 8 at points more than 161 million km apart enabled the solar wind's density to be measured more accurately than ever before. In Sep 1972 NASA's Ames Research Centre in California succeeded in locating and reviving Pioneer 7, which had turned its transmitters off, although it was on the far side of the Sun and 312 million km from Earth. Then came the triumphant flights to Jupiter in 1972-4. Pioneer 10 was the first spacecraft placed on a trajectory to escape from the solar system into interstellar space; the first to fly beyond Mars; the first to enter the Asteroid Belt; and the first to fly to Jupiter. It is hoped it will be the first to sense the interstellar gas beyond the Sun's atmosphere.

In achieving all this Pioneer 10 established that the much feared Asteroid Belt contained less material than at one time believed, and presented little hazard to either unmanned or manned spacecraft.

Pioneer 10 description

Lift-off weight (inc launcher)	146,673kg
Lift-off height	40·4m
Spacecraft weight overall	260kg
Spacecraft payload	27·8kg
Spacecraft radius	6·4m
Spacecraft antenna dia	2·7m
Launch vehicle	Atlas-Centaur-TE-M-264-4

Pioneer spacecraft were designed to carry a variety of instruments tailored to the study of either individual or a series of planets. Nos 10 and 11 were the first craft designed to travel into the outer solar system, to operate there for 7yr, and to go as far from the Sun as 2,400 million km. Since the launch energy needed to reach such distances is far higher than for shorter missions, the spacecraft must be very light. The 258km wt includes 30kg of scientific instruments and 27kg of propellant for attitude changes and midcourse corrections. 6 thrusters each provide 0·2-0·6kg thrust. They could adjust the place and time of arrival at Jupiter by changing velocity, or merely adjusted the attitude. This was done by pulse thrusts, timed by a signal from a star sensor which saw Canopus once per rotation, or by one of 2 Sun sensors which saw the Sun once per rotation. Velocity changes totalling 670km/hr could be made during the mission.

The spacecraft were spin-stabilised, giving the instruments a full-circle scan 5 times a minute. Because solar radiation at Jupiter is too weak to operate an efficient solar-powered system, 4 nuclear units provided electric power; they were placed on 2·7m booms so that their radiation would not affect the scientific experiments (13 on Pioneer 10, 14 on Pioneer 11). Controllers used 222 different commands to operate the spacecraft; during the 4 days it took to pass Jupiter commands took 45min to reach it. The heart of the communications system is the fixed, 2·7m-dia dish antenna, which focuses the radio signals in a narrow beam. The onboard experiments are still returning data on the solar atmosphere from Earth beyond Jupiter. Pioneers 10 and 11 both carry a 15 × 23cm plaque showing the origin of the spacecraft in the solar system, and drawings of a man and woman related to the spacecraft's size in case they should one day be seen by another intelligent species.

Pioneer 11's historic first encounter with Saturn, followed by the two Pioneers which at last penetrated the mysteries of Venus, are described below. The series is to continue, though what was originally planned as Pioneer Jupiter is now to be found under the new title of Project Galileo.

Pioneer 1 L 11 Oct 1958 by Thor-Able from C Canaveral. Wt 38kg. Failed to reach the Moon but looped 113,854km into space. In 43hr 17min of flight it discovered the extent of Earth's radiation bands.

Pioneer 2 L 8 Nov 1958 by Thor-Able from C Canaveral. This 39·5kg probe, intended to reach the Moon, failed as a result of unsuccessful 3rd-stage ignition.

Pioneer 3 L 6 Dec 1958 by Juno 2 from C Canaveral. The 5·9kg probe again failed to reach the Moon, but did attain a height of 102,333km and discovered Earth's 2nd radiation belt.

Pioneer 4 L 3 Mar 1959 by Juno 2 from C Canaveral. 5·9kg lunar probe, passed within 59,983km of Moon and then into solar orbit 0·9871 × 1·142 AU.

Jupiter and Red Spot: Pioneer 10 confirmed that the Sun's biggest planet was mostly liquid hydrogen. *(NASA)*

Pioneer 5 L 11 Mar 1960 by Thor-Able from C Canaveral. 43kg probe, sent into solar orbit 0·8061 × 0·995 AU. Returned solar flare and wind data until 26 Jun 1960 from a distance of 37 million km.

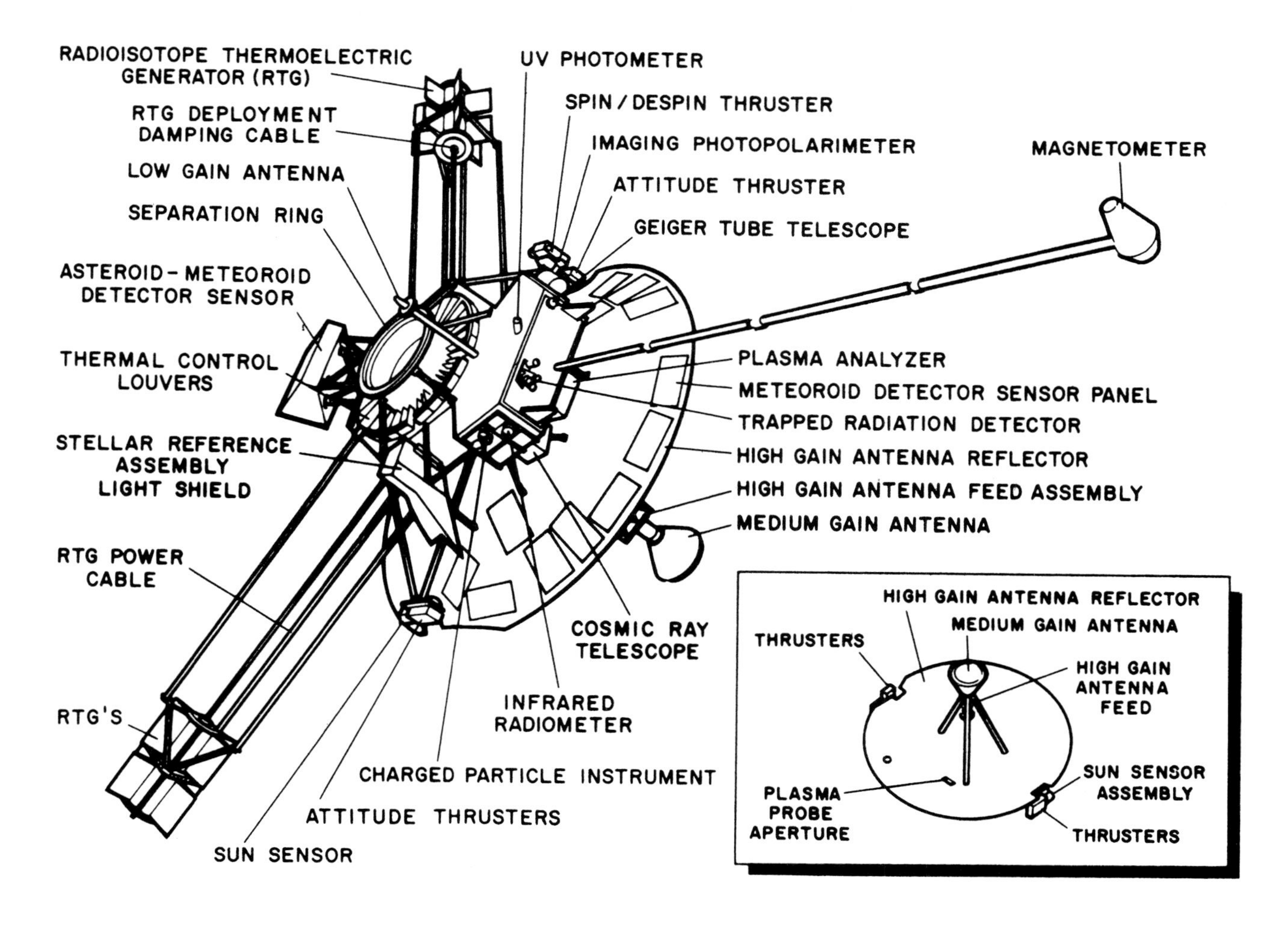

PIONEER/JUPITER SPACECRAFT

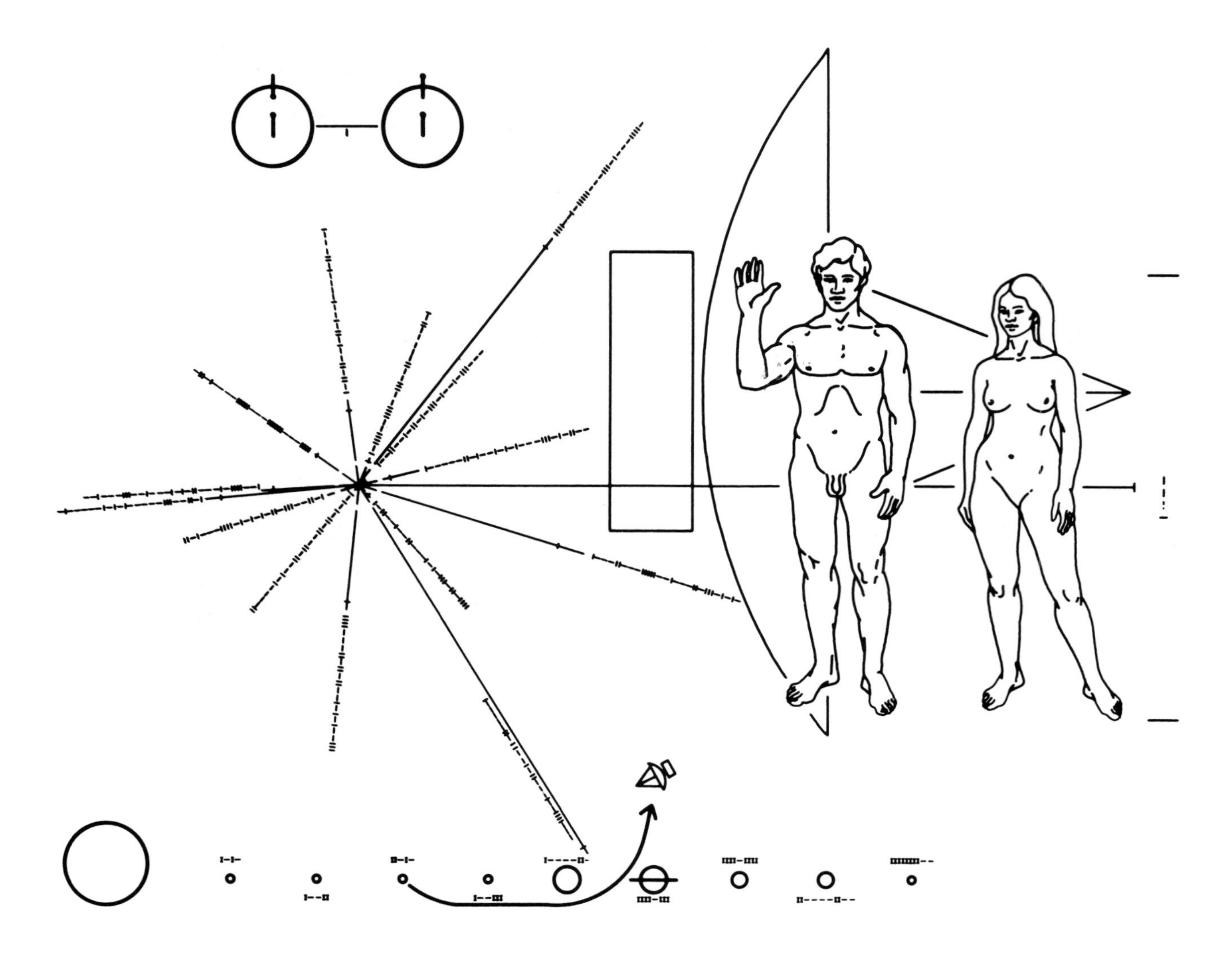

Pioneer plaque. The key is the hydrogen atom (top left), the most common in the universe. 14 lines from Sun indicate cosmic sources of radio energy. Human figures - man's hand raised in goodwill gesture - are compared with spacecraft to give scale. Bottom, plan of solar system, showing Pioneer coming from Earth.

Pioneer 6 L 16 Dec 1965 by TAD from C Canaveral. 63kg cylinder, 1m dia and 0·9m high, launched towards Sun into 0·814 × 0·985 AU solar orbit. It sent back first detailed description of the tenuous solar atmosphere. With Pioneer 7 gathered continuous data on events on a strip of solar surface extending nearly halfway round the Sun. It also measured the tail of Comet Kohoutek. By Dec 1980, having circled the Sun 17½ times and covered 14·5 billion km, it had been operational for 15yr - a record for an interplanetary spacecraft - compared with its planned 6-month lifetime. With Pioneers 7, 8 and 9 it was then forming a network of solar weather stations, often millions of km apart.

Pioneer 7 L 17 Aug 1966 by TAD from C Canaveral. 63kg cylinder launched away from Sun into 1·010 × 1·125 AU solar orbit (see Pioneer 6). In 1977 it observed the Earth's magnetic tail 19·3 million km out, 3 times deeper into space than the tail had ever before been detected.

Pioneer 8 L 13 Dec 1967 by TAD from C Canaveral. 65·3kg cylinder placed in 1·0 × 1·1 AU solar orbit to join previous 2 Pioneers in obtaining data on solar wind, magnetic field and cosmic rays. Also defined tail of Earth's magnetosphere. Launched piggyback with Pioneer 8 was NASA's first Test and Training Satellite, used to exercise the Apollo communications network.

Pioneer 9 L 8 Nov 1968 with 2nd Test and Training Satellite by TAD from C Canaveral. Wt 66·6kg. Placed in 0·75 × 1·0 AU solar orbit. This was the 4th of what was to be a series of 5 solar probes, but Pioneer E, intended to be Pioneer 10 and carrying 3rd piggyback Test and Training Satellite, failed to orbit on 27 Aug 1969.

Pioneer 10 L 3 Mar 1972 by Atlas-Centaur from C Canaveral. Wt 270kg. Initial speed of 51,800km/hr, faster than any previous man-made object, placed it on a trajectory which took it past Jupiter's cloud tops at a distance of 130,300km on 4 Dec 1973. Initial speed was achieved by adding for the first time to an Atlas-Centaur booster (lift-off thrust 186,590kg) a solid-fuelled TE-M-364-4 3rd stage developing 6,800kg thrust. (This is an uprated version of the retromotor used for the Surveyor Moon lander.)

Pioneer 10 swept past the Moon's orbit in just over 11hr, compared with the 89hr taken by Apollo 17. In Jul the Asteroid Belt, believed to be 270 million km wide, was entered amid speculation as to whether Pioneer 10 would be damaged or destroyed by a 48,000km/hr collision with an asteroid fragment. It was estimated that a particle only 0·05cm in size could penetrate vital spacecraft areas. But its closest pass to any known body was 8·8 million km from the 1km-dia Palomar-Leyden. Pioneer 10 emerged unscathed 7 months later, having established that the Asteroid Belt could also be penetrated by men without difficulty. On 7 Aug it was teamed with Pioneer 9 to send details of one of the most violent solar storms ever known; one estimate was that during it the Sun produced as much energy as the whole Earth would use in 70 million yr. When Pioneer 11 was launched on 6 Apr 1973 Pioneer 10 was within 190 million km of Jupiter. In Aug, when the Earth passed between Pioneer 10 and the Sun, the spacecraft's attitude had to be adjusted slightly to ensure that the alignment sensors would not lose count of the spin rate and confuse the automatic orientation system. From 26 Nov 1973, when command-and-return time had increased to 92min, a total of about 300 pictures of the approach and flyby were obtained. Early ones were mainly for calibration, but about 40 yielded much information after lengthy computer processing. In addition to the Great Red Spot, pictures of the moons

Ganymede, Callisto and Europa were obtained. The biggest disappointment was the failure to obtain good pictures of the orange moon, Io, but nonetheless much was learned, as detailed below. 26 days before closest encounter Pioneer 10 began threading its way through the orbits of the Jovian moons.

On 8 Nov it entered the Jovian environment by passing the orbital path of Hades, the outermost moon, at 23·6 million km. 3 days later the path of Andrastea, the tiny, 16km-dia, innermost of Jupiter's 4 outer satellites, was passed at 20·5 million km. Accelerating steadily under the pull of Jovian gravity, Pioneer 10 travelled another 9 million km in the next 11 days to pass the 3 inner moons, Demeter, Hera and Hestia, between 11·5 million and 11·4 million km from the planet. By now the velocity was 450km/sec and on 26 Nov it encountered Jupiter's bow shockwave, crossing the magnetopause 24hr before the earliest predicted encounter, re-entering the magnetic field on 1 Dec, and crossing the magnetopause for the 2nd time slightly more than 3·2 million km from Jupiter. The 4 Galilean moons still lay ahead: on 2 Dec the orbital path of Callisto, as big as Mercury and circling Jupiter at 1·8 million km, was crossed. 16hr later, Ganymede, the largest moon, was passed 798,000km from the planet. On 3 Dec, 6hr after that, Europa, similar in size to Earth's Moon and 583,260km out, was passed. Finally Io, 356,000km from Jupiter and the most interesting of the Jovian moons because it is the most reflective object in the whole solar system, was passed. The big disappointment was that although pictures of Callisto, Ganymede and Europa were obtained, it was not possible to photograph Io, which brightens for about 15min when it reappears from behind Jupiter. It is believed that this is because Io has an atmosphere of methane or molecular nitrogen which freezes on the surface during occultation and evaporates to gas when heated by the Sun.

Closest approach came at 02·25 GMT on 4 Dec, and 16min later Pioneer 10 passed behind Io and then behind Jupiter itself. The spacecraft reached a speed of 132,000 km/hr as it passed within 130,300km of the cloud tops, crossing the Jovian equator at an angle of 14°. 15,000 commands were sent to Pioneer 10 during the flyby; 6 of its 11 instruments were operated continuously. They were the UV photometer, the magnetometer, 2 of the 4 high-energy particle detectors, the asteroid/meteoroid telescope and the meteoroid detector. The solar wind instrument and the 2 remaining high-energy particle detectors were calibrated several times each day and operated in accordance with the changing approach conditions. The infra-red radiometer was sent several hundred commands each day for observations of the inner satellites, though the bulk of the commands directed the imaging photo-polarimeter to obtain views of the satellites and of Jupiter itself. The immense volume of data returned took nearly a year to correlate.

Though scarred by the intense Jovian radiation belts - some of the instruments, particularly the asteroid-meteoroid detector, were degraded - the spacecraft survived so well that it was decided that Pioneer 11 could make a closer approach. Still working 2½yr later in Feb 1976, when 1,384 million km from Earth, Pioneer 10 crossed the orbit of Saturn and sent back data showing

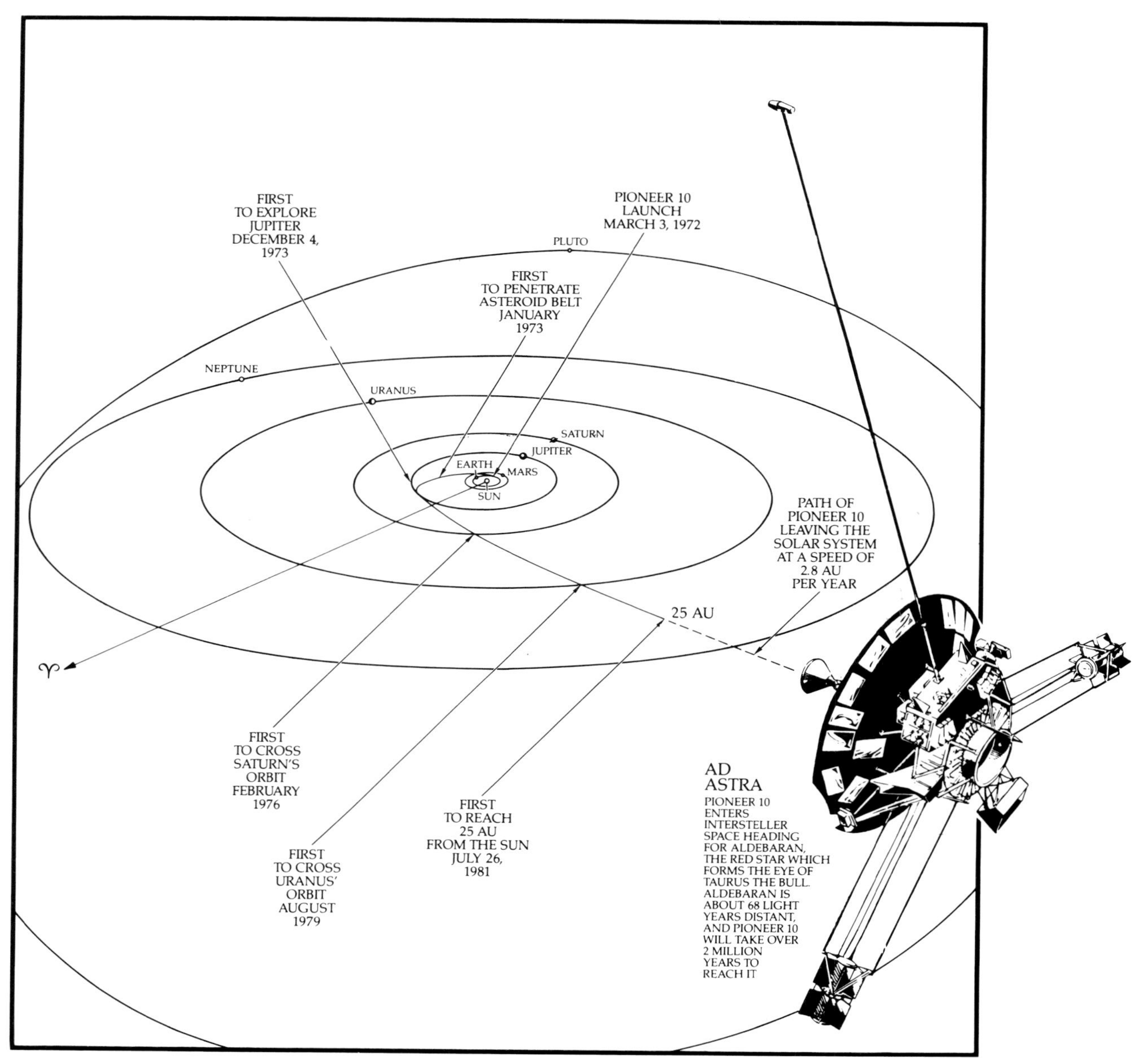

THE FLIGHT OF PIONEER 10

that Jupiter's enormous magnetic tail, almost 800 million km long, spanned the whole distance between the 2 planets.

10yr after launch, when 4·6 billion km from the Sun and with its data taking 3hr 42min to reach the Ames Research Centre, California, Pioneer 10 was studying the extent of the heliosphere. On 13 Jun 1983 Pioneer 10 crossed the orbit of Neptune; further from the Sun than any known planet, it was considered to have left the solar system. But since it should be possible to track it beyond 9·25 billion km, it is now being used to seek either a possible dark star companion to the Sun or a 10th planet. Cruising at about 48,000km/hr, it is heading towards Aldebaran (the "eye" of the Bull constellation). But it will take 2 million yr to reach that point, and by then Aldebaran will have moved on. Not enough is yet known about the motions of the stars to permit calculation of its ultimate destination, but Pioneer 10 and its plaque should encounter another solar system about once every 1 million yr.

Pioneer 11 L 5 Apr 1973 by Atlas-Centaur from C Canaveral. Identical to Pioneer 10 apart from the addition of a 2nd magnetometer to measure Jupiter's high magnetic fields. Following Pioneer 10's success it was possible to re-target Pioneer 11 so that it would fly 3 times closer to Jupiter and then fly on to Saturn. On 19 Apr 1974, at a distance of 675 million km, the spacecraft's thrusters were fired for 42min 36sec, adding 230km/hr to its velocity of 46,200km/hr. The manoeuvre used 7·7kg or 28% of the original propellant load. Now targeted to pass under Jupiter's S pole, Pioneer 11 sent back 130 pictures during its final approach. Its imaging system and UV photometer studied Callisto, Ganymede and Europa. Changes in cloud structure near the Red Spot were observed since Pioneer 10's flyby, and Callisto's S pole ice cap was revealed.

Pioneer 11's data confirmed that Jupiter was primarily a liquid planet consisting mostly of hydrogen, and that it radiated more heat than it absorbed from the Sun. At 17.22 GMT on 3 Dec 1974, travelling faster than any previous man-made object at 171,000km/hr, Pioneer 11 passed 42,940km below the Jovian S pole. Mission controllers at NASA's Ames Research Centre had an anxious 1hr wait. But then Pioneer 11 emerged unscathed from the radiation belts behind Jupiter, and it was flung over the N pole to double back across the solar system towards the planned Saturn encounter. Since it was travelling in the opposite direction to Jupiter's rotation scientists were given a look at the magnetic field, radiation belt and surface during a complete revolution of the planet. Pioneer 11 confirmed the conclusion, drawn from its predecessor's data, that Jupiter's magnetic fields consist of 2 very different regions: a weak outer field with the solar wind impinging on it, and a much stronger inner field. And it found that the cloud tops in Jupiter's polar regions (which cannot be seen from Earth) are much lower than at the equator and covered by a thicker transparent atmosphere, with some "blue sky" similar to Earth's.

On 20 Nov 1975 Pioneer 11 made the first observations of Saturn from a spacecraft in a prelude to photo-polarimeter scanning planned for Jul 1976. On 18 Dec 1975, by which time Pioneer 11 had covered a 3rd of the 640 million km between Jupiter and Saturn and was travelling at 64,000km/hr, controllers successfully fired the thrusters to increase velocity by 108km/hr. Finally, after a 6yr journey of over 3·2 billion km, Pioneer 11 achieved man's first visit to the ringed planet, spending 10 days taking photographs and making measurements. Closest approach, on 1 Sep 1979, took it within 20,800km of Saturn's cloud tops at a speed of 114,500km/hr. As the spacecraft swung under the plane of the rings Saturn's gravity changed its course by nearly 90° towards the edge of the solar system, in the opposite direction to Pioneer 10. Major discoveries included the 11th moon (1979S1), with a 400km dia, and the fact that Saturn has a magnetic field which is 1,000 times stronger than Earth's but 20 times weaker than Jupiter's. There are radiation belts comparable in intensity to Earth's but 10 times larger. Low temperature measurements suggest that there is unlikely to be life on Titan. Two new rings were identified: the F-ring, clearly visible in a picture taken 943,000km away; and the G-ring, lying between the orbits of Rhea and Titan and about 500,000km above Saturn's cloud tops. Many more new ring details showed up, included the French Division between the middle and inner visible rings (B and C). Measurements indicated that the rings have a low density and are largely made up of ice.

Although Pioneer 11 suffered 2 meteoroid hits above the rings and 3 more below them, no damage was caused by high-velocity ring particles. Spacecraft can thus be operated safely in the vicinity of the visible rings, making it possible to target with confidence the more advanced Voyagers already on their way.

Pioneer Venus Orbiter L 20 May 1978 by Atlas-Centaur from C Canaveral. Within 2yr this spacecraft had radar-mapped 93% of the Venusian surface and sent back 1,000 pictures, with measurements, of the clouds. It is expected to continue working until 1985. The 2·5m-dia cylindrical craft had a launch wt of 583kg and 368kg in-orbit wt. The 12 scientific instruments weighed 45kg; the magnetometer sensors were mounted on a 4·7m boom to avoid magnetic interference from the craft, and communications are handled by a high-gain dish antenna mounted on a 3m mast. On 4 Dec 1978 the spacecraft was placed in a 378 × 64,645km Venusian orbit, adjusted by 1 Jan 1979 to 149 × 65,983km at 103° incl to Venus's equator, compared with the planned 150 × 67,000km orbit at 75° incl. The 24hr orbit permits orbital events to be referenced to Earth time, and the 1 million-bit memory enables periapsis (low point) data to be returned when two 64m antennas can see the spacecraft at the same time.

In May 1980 NASA was able to issue a series of vivid colour pictures of the surface of Venus, buried as it is beneath its dense atmosphere. There are 2 continent-sized highland areas, named Ishtar and Aphrodite, the first in the northern hemisphere and about the size of Australia. It is comparable in height to the Tibetan plateau but twice as large. The eastern end consists of a mountain massif rising to 11,800m above "sea level" (if there were any sea), higher than Earth's Mount Everest. This was provisionally named Maxwell Montes. Aphrodite Terra, near the equator and larger than Ishtar, is about the size of the northern half of Africa. At its eastern end is a dramatic rift valley measuring 2·9km below sea level at its deepest, 280km wide and 2,250km long. Two shield-shaped mountains, probably volcanoes, were found rising to about 4,000m at 30°N. Some 60% of Venus's surface was found to be relatively flat, rolling plain, with large, relatively shallow craters of 400-600km dia and 200-700m deep. They were compared with those on Mars and the Moon, and could be explained by "surface rebound," like those on Jupiter's moons. Discoveries relating to the "greenhouse effect" are discussed below.

Pioneer Venus Multiprobe L 8 Aug 1978 by Atlas-Centaur from C Canaveral. With 904kg total wt, this consisted of the Multiprobe Bus (309kg), a Sounder Probe (316kg) ejected by springs 24 days before arrival, and 3 identical 93kg North, Day and Night Probes, spun off by increasing to 48 the 15rpm bus rotation 20 days and 12·8 million km before arrival. These 4 atmospheric-entry spacecraft were highly successful in conducting a detailed examination of Venus's atmosphere and weather in conjunction with the Pioneer Venus Orbiter, which had arrived 5 days earlier. It is believed that these results will contribute to our understanding of the forces that drive Earth's weather. On 9 Dec the bus burned up in the atmosphere at 33°S 43°W. The Sounder Probe landed by parachute at 0°S 43°W and transmitted from the surface for 68min. The North Probe landed at 75°N 20°E, the Day Probe at 26°S 45°W, and the Night Probe at 27°S 45°E. (The co-ordinates are with respect to Venus's disc seen from Earth at time of encounter.) Entry speeds were around 41,600km/hr, with forces peaking at 315g at 78km above the surface for the Sounder, and at 200-560g, depending on atmosphere entry angle, for the smaller probes. To withstand these forces, plus Venus's 480°C (900°F) heat, corrosive atmosphere and 100 times Earth-atmosphere pressure, the probes were constructed as sealed, spherical titanium pressure vessels fitted with conical carbon-phenolic heatshields. Providing access for the sampling and observing instruments was the most difficult problem: the Sounder Probe had 15 sealed windows or other pressure vessel penetrations, the smaller probes 8; there were one diamond and 14 sapphire windows. During their 57min descent the Probes transmitted their data directly to Earth.

Investigators consider that the spacecraft have "virtually proved" the theory that Venus's high surface temperature is due to the "greenhouse effect," namely that sunlight passes easily through the atmosphere but has difficulty escaping because of its 96% carbon dioxide content. (The relevance to Earth of this finding lies in the fact that 80 years of burning fossil fuels has increased atmospheric carbon dioxide by 15% and could double it in the next 50 years. The resulting increase in atmospheric temperatures would have a catastrophic effect on agriculture and ocean levels.)

Comparison of the cloud and wind patterns with the 1974 Mariner flyby, plus Orbiter observations of 2 Venusian 243-day years, showed that wind patterns change dramatically over such periods. The high-altitude smog layer which completely envelops the clouds, lying about 30km above them, appears and disappears over several-year periods. The major remaining question is why, on a planet with almost no axial rotation, do the upper-level winds circle the planet every 4 days at the tremendous speed of 360km/hr, covering it completely and blowing at every latitude from equator to pole? The 4 probes showed that these winds are coupled to lower-altitude winds which also have very high speeds. The 360km/hr cloud-level winds are found at an altitude of 65km; speeds then range down to 192km/hr at 50km and 80km/hr at 20km. But over half of Venus's surface, with its very dense atmosphere, is almost stagnant; wind speeds of only 3-18km/hr were measured from the surface up to 10km.

In May 1982 Dr Thomas Donahue announced that evidence obtained from the Sounder Probe indicated that at one time Venus had enough water for oceans 30% or more the size of Earth's.

A low-cost project using much equipment developed for previous missions, Pioneer Venus was managed by NASA's Ames Research Centre. The spacecraft were built by Hughes.

Future Projects New versions of the spin-stabilised Pioneers are being proposed for the role of geochemical orbiter over the Moon and Mars, with other Mars-orbiter applications to include surface penetrators and upper-atmosphere and hydrology studies. A near-Earth asteroid mission is also being studied. Venus Radar Mapper, for Shuttle launch in 1988, has replaced the proposed Venus Orbiting Imaging Radar (VOIR), cancelled in 1982 as a result of budget cuts. Costing up to $250 million, this will cost half as much as VOIR. Combining sub-systems from Voyager, Galileo and Viking, it will be placed in a 250 × 1,900km Venusian orbit to spend a minimum of 243 days mapping the planet. The synthetic-aperture radar will provide 1km resolution and cover 90% of the surface.

Rainbow

Rainbow Satellites Inc is planning to operate 2 16-transponder Rainbow satellites, plus 1 ground spare, for video transmissions. Working in the Ku band, they are due for launch in Aug and Nov 1986 into the 79° and 132°W slots.

Ranger

History The first of 3 projects (the others being Lunar Orbiter and Surveyor) aimed at obtaining sufficient knowledge and pictures of the lunar surface to make the manned Apollo landings possible. The first 6 missions were failures, beginning with Ranger 1, launched from C Canaveral on 23 Aug 1961, right through to Ranger 6, launched on 30 Jan 1964. Plans to eject and hard-land a 42·6kg capsule to measure seismic activity on the Moon's surface as a mission "bonus" were abandoned, as failures continued right through the manned Project Mercury flights and the need for close-up pictures of the lunar surface became desperate if Project Apollo was not to be held up. For Rangers 6-9 a 170kg conical structure containing 2 wide-angle and 4 narrow-angle TV cameras, plus video combiner, transmitters, etc, was mounted on the spacecraft to send back about 14min of TV pictures from 2,261km above the Moon until the spacecraft was destroyed by impact at 9,330km/hr. Ranger 6 looked like being successful until it was found that the TV system had been destroyed by being inadvertently turned on early in the flight. Rangers 7, 8 and 9, however, justified the persistence of NASA's Jet Propulsion Laboratory at Pasadena, California. Ranger 7, launched on 28 Jul 1964 on a 68hr flight, returned 4,308 excellent pictures – man's first close-up lunar views – before impacting in the Sea of Clouds. Ranger 8, launched on 17 Feb 1965 on a 64hr flight, returned 7,137 high-quality photographs covering 2,331,000 sq km in the Sea of Tranquillity. Finally Ranger 9, launched on 21 Mar 1965 on a 64hr flight covering 417,054km, relayed 5,814 excellent pictures before impacting within 4·8km of the target point in Crater Alphonsus. 200 of the pictures were shown live in the first TV spectacular from the Moon. The total of over 17,000 pictures, some showing craters as small as 25cm across, revealed the lunarscape several thousand times better than the best Earth telescope had ever done, and finally justified Ranger's cost of $260 milllion. But while they provided the evidence needed to complete the design of soft-landing spacecraft, it was left for Project Surveyor to establish whether the Moon's surface was firm enough to support manned landings.

Spacecraft description Rangers 6-9: wt 366kg; ht 3·1m, solar panels 4·5m. Launchers: Atlas-Agena D. Conical camera structure mounted on hexagonal spacecraft bus. Battery of 6 TV cameras (2 wide-angle, 4 narrow-angle) covered by polished aluminium shroud with 33cm opening near top. High-gain antenna and 2 solar panels hinged to base. Midcourse motor provided 22·6kg thrust for up to 98sec. Attitude control, maintained by nitrogen gas jets, Sun and Earth sensors and 3 gyros programmed by computer and sequencer, enabled spacecraft to aim cameras at Moon while pointing high-gain antenna to Earth for transmission.

RCA Satcom

History Designed and built by RCA Astro-Electronics of Princeton, NJ, and owned and operated by RCA American Communications Inc. By 1983 there were 6 RCA Satcoms providing cable TV, voice channels and high-speed data transmission throughout the US and Puerto Rico. More than 4,000 Earth stations had direct access to the system. RCA-Satcom 1 (L 13 Dec 1975, in-orbit wt 463kg), Satcom 2 (L 26 Mar 1976) and Satcom 3R (L 20 Nov 1981 to replace Satcom 3, L 7 Dec 1979 and lost when boost motor failed) had approx 8yr operational lives. By Satcom 4 (L 15 Jan 1982 by Delta 3910/PAM to 83° W, in-orbit wt 550kg) design life had been increased to 10yr. Launches were by Delta from C Canaveral. Satcom 4 was designed for launch by Shuttle or Delta 3910/PAM-D, PAM taking the place of the 3rd stage and considered part of the payload. 3-axis-stabilised, Satcom 4 has 11·2m span, main body 1·6 × 1·27 × 1·29m, and offers 24 channels, each designed to carry 1,000 voice circuits, 1 FM/colour TV transmission, or 64 million bits/sec of computer data. In Nov 1981 leases for 7 of Satcom 4's 24 channels were auctioned to cable TV companies for a total of $90·1 million. Satcom 5 L 27 Oct 1982 by Delta/Thiokol from C Canaveral into synchronous orbit at 143 W. Wt 589kg. Ownership transferred to Alascom Inc, supplying communications within Alaska and with other US states.

Satcom 4 and Satcom 1R, L 11 Apr 1983 for commercial and government use, are claimed to be the world's 1st completely solid-state satellites. Satcom 2R (video), L 8 Sep 1983 to 72°W, and Satcom 6, due for launch in May 1986, operate in C band. Satcom Ku-1 to 3 have been authorised for launch in May 1985, Jan 1986 and Aug 1987 into 77°W, 87°W and 126°W slots.

Advanced versions of RCA Satcom are being used both commercially and by the US Government for high-speed data transmission. *(RCA)*

SBS

History An all-digital commercial communications system, the first of its kind, Satellite Business Systems is based on 3 geosynchronous satellites. Formed in 1975 by a partnership among subsidiaries of Comsat General Corp, IBM Corp and Aetna Life and Casualty, and with funds totalling $373 million, its aim is to be able to send a column of business data the size of Tolstoy's *War and Peace* anywhere in the world in one second. With a network control centre near its HQ in McLean, Va, and an ability to offer customers slots measured in milliseconds, distance no object, SBS had 13 customers for its Communications Network Service by the time the first satellite was launched on 15 Nov 1980; they included Boeing, Westinghouse, General Motors and Wells Fargo. In addition to telephone, computer and electronic mail services, SBS will make teleconferences practical because the video signal becomes just another digital message sent via the satellite. Participants in teleconferences can gather in meeting rooms thousands of miles apart and see each other through a video window, possibly a 1·2m wall screen. Documents and charts can be handed back and forth and studied in a scanner built into the table. Comsat hopes that by the mid-1980s it will be able to use 4 SBS satellites for a subscription TV service direct to homes; each satellite would cover a different time zone. SBS lost $21 million in 1981 and losses were expected to continue through 1984, despite estimated 1983 revenues exceeding $140 million.

Spacecraft description SBS-1 to 3, built by Hughes Aircraft Co, are a new series designated HS-376 and the first domestic system to utilise Ku-band (12/14GHz) frequencies, enabling small, inexpensive Earth stations to be used. Fitted with a folding antenna reflector and deployable solar panel, they can be stowed upright in the Shuttle bay, thus reducing the launch charge, which is based on length of bay used. With 550kg on-station wt, they are spin-stabilised and incorporate a high-speed all-digital 10-transponder system capable of relaying up to 480 million data bits per second, equivalent to more than 10 million words. With solar panel and antenna deployed they are 6·60m long and 2·16m in dia.

SBS-1 L 15 Nov 1980 by Delta/PAM from C Canaveral to geostationary orbit at 106°W.

SBS-2 L 24 Sep 1981 by Delta/PAM to 97°W.

SBS-3 L 11 Nov 1982 by STS-5 to 94°W. 1st commercially launched Shuttle satellite. SBS-4 is due for Shuttle launch in Aug 1984.

SBS-5 and 6, ordered from Hughes in Nov 1983 under a $100 million contract which includes PAM-D upper stages, will have double the communications capacity of SBS-1 to 4, and an operational life extended 3yr to 10yr. Spot beams will add Alaska and Hawaii to the 48 states already covered. Greater capacity is provided by 4 110MHz wide-band transponders, complementing the 10 43MHz narrow-band units of the 4 earlier craft. Modifications and increased power will add 65kg to in-orbit weight. SBS-5 will be launched to 124° W by Shuttle in Aug 1986, with SBS-6 serving as a ground-based spare.

Seasat

History Intended as a proof-of-concept mission providing worldwide ocean surveillance for 1-3yr, Seasat 1 (L 27 Jun 1978 by Atlas-Agena from Vandenberg into 776 × 800km polar orbit; wt 2,300kg; incl 108°; orbital life 200yr) abruptly lost power and ceased operating on its 1,502nd orbit, on 9 Oct 1978. The cause was thought to be an electrical short in the solar array. However, 70 of the 106 operational days produced data which took 4yr to process and yielded much new information about the seabed in little known areas of the southern oceans. Its 5 microwave sensors were intended to measure sea temperatures and wave heights, convertible into wind speed and direction with an accuracy of 2m/sec; the information was to enable ships to avoid bad weather, save fishing fleets time by indicating where particular species were shoaling, and detect even small icebergs and navigable openings in icefields. Savings of over $2 billion for the US alone were predicted by the year 2000, against the satellite's cost of $36·5 million. Early radar images of Alaskan ice packs, etc, promised fulfilment of these predictions. But it was Nov 1982 before JPL scientists were able to announce new bathymetric (water depth) charts, showing for instance that the Louisville Ridge was a nearly continuous chain of

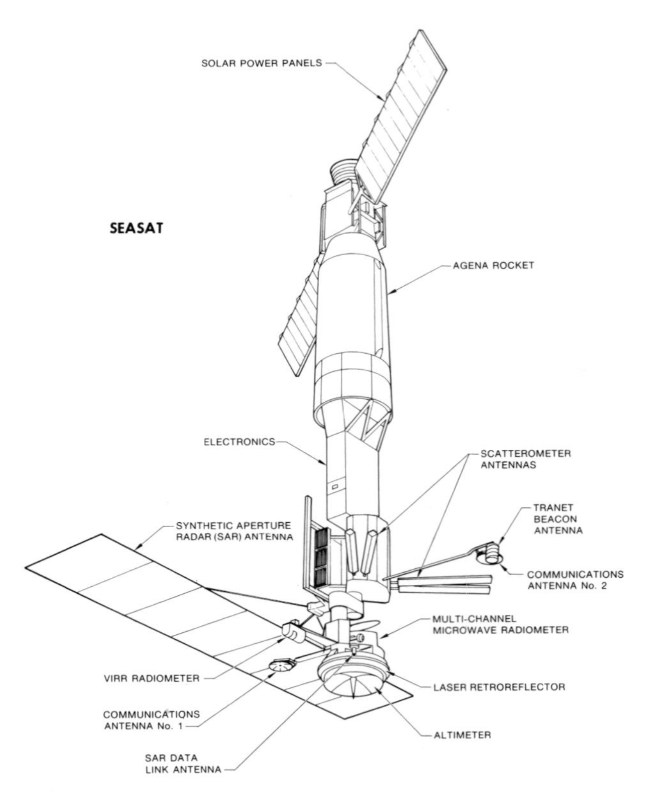

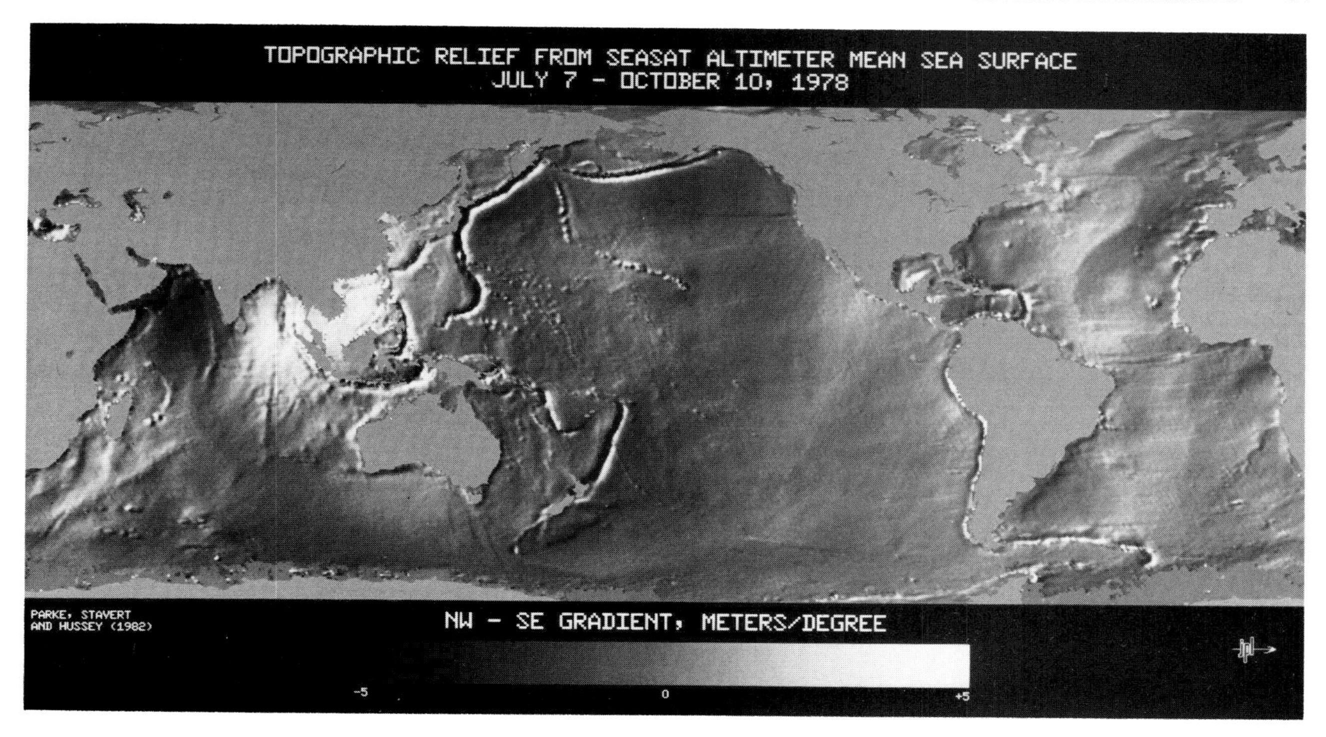

mountains running beneath the sea SE of the Tonga-Kermadic Trench. Sea floor mapping is possible because the sea surface broadly conforms to the sea floor: thus a mountain on the sea floor causes a peak on the ocean surface detectable by a satellite altimeter. A successor, Topex, is planned for launch in 1989.

SERT

History For over 20yr NASA's Lewis Research Centre has been developing a new type of space propulsion based on an ion-electric rocket engine under the Space Electric Rocket Test (SERT) Programme. The principle is to use electricity (generated either by nuclear power or solar panels) to ionise a propellant (mercury in the case of SERT) into ions and free electrons, and then to create electric fields with charged plates which would drive the ions rearward at very high speed. Small amounts of thrust are produced economically over a long period and could be used to extend the life of many spacecraft, enabling those near Earth to keep station with minute positional changes. Another use would be the steering of deep-space probes.

SERT-1 L 20 Jul 1964 by Scout. Suborbital test lasting 30min and providing 0·03 newtons (0·0006lb) thrust.

SERT-2 L 4 Feb 1970 by TAT Agena from Vandenberg. Wt 3,100kg. Orbit 618 × 625km polar at 99° incl to keep spacecraft continuously in sunlight. Considered unsuccessful at the time because the 2 engines failed due to corrosion well before the 6-month target. But Lewis maintained contact for 11yr and recently announced that all power-processing components had remained fully functional and that the efficiency and reliability of the system had been proved. A more powerful version, designed for main propulsion clusters of future spacecraft, was due for test in 1983.

Skylab

History The Skylab space station, launched on 14 May 1973 and subsequently visited by 3 Apollo astronaut crews who lived and worked in it for periods of 28, 59 and 84 days, was a remarkable example of human ingenuity overcoming what at first seemed inevitable technical disaster. By the time the 3rd crew splashed down on 8 Feb 1974 America had demonstrated man's ability not only to live and work in space almost indefinitely (with all its implications for flights to the planets) but also that faulty equipment could be replaced and repairs improvised to an extent never thought possible when spaceflight began.

Nearly 26,000 scientists and other workers were employed on preparations for the mission, which, in addition to tackling steadily increasing flights of long duration, was aimed at studying solar activity from outside Earth's dusty atmosphere, and looking inward towards Earth to study its resources, problems of pollution, and the possibility of giving early warning of natural disasters such as floods and volcanic eruptions.

Skylab, as photographed from the Skylab 4 CM. The parasol sunshade is just visible below the later, twin-pole sunshade. *(NASA)*

The Skylab Workshop was launched by Saturn 5. Although the Skylab cluster was placed in the correct 440 × 427km orbit with 50° incl, it was soon apparent that all was not well. The launch vibrations had torn away the combined meteoroid/thermal shield, which in turn ripped away one of the pair of solar array wings and caused the other to be jammed in a partially opened position by debris. With no thermal protection, and limited power available from the Apollo Telescope Mount (ATM) batteries and the partially opened wing, internal temperatures soared. At the time NASA refused to confirm suspicions that exact details of the damage to Skylab had been provided by the US military Spacetrack system. More than a year later an article in America's *Air Force Magazine* disclosed that the long-range tracking radars used by Spacetrack to reconstruct the shape and size of every object in space had in fact been used to establish that the solar panel had failed to deploy.

SKYLAB CLUSTER ELEMENTS

Apollo Telescope Mount (ATM)
Solar observation unit
Length : 3.35 m
Dia : 2.13 m
Weight : 11,180 kg

Total cluster weight : 90,265 kg *
Total cluster length : 36.12 m
Total work volume : 361.4 m³
*Excludes Payload Shroud : 11,795 kg

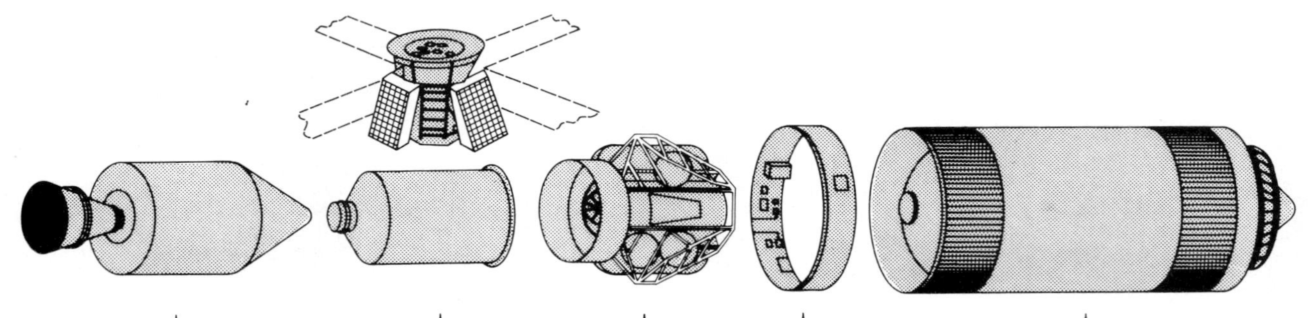

Command Service Module (CSM)
Crew ascent and descent vehicle
Length : 10.45 m
Dia : 3.96 m
Weight : 13,970 kg

Multiple Docking Adapter (MDA)
Docking interface
ATM/EREP controls and displays
Length : 5.30 m
Dia : 3.05 m
Weight : 6,260 kg

Airlock Module(AM)
Environmental control
Utility centre for cluster
EVA port
Length : 5.37 m
Dia : 3.05 m (STS)
1.68 m (TNL)
Weight : 22,225 kg

Instrument unit (IU)
Launch vehicle control centre
Length : 0.91 m
Dia : 6.58 m
Weight : 2,065 kg

Orbital workshop (OWS)
Primary living and working quarters
Length : 14.66 m
Dia : 6.58 m
Weight : 35,380 kg

Left: Skylab ready for launch by Saturn 5. Right: Much smaller Saturn 1B perched on "milkstool" to launch first Skylab crew. ***(NASA)***

Launch of the first crew, Charles (Pete) Conrad, overall commander of the Skylab astronauts, and Dr Joseph Kerwin and Paul Weitz, should have taken place 24hr after Skylab 1 but was immediately postponed for 5 days. At first it seemed very unlikely that it would ever be possible to inhabit Skylab. Various options were studied, including the possibility of a simple "fly-around" mission to examine and report on the damage. Internal temperatures at first went as high as 190°F but were stabilised at lower levels as Mission Control learned to orient the cluster at the most favourable angle to the Sun. Then, emergency teams and the astronauts themselves, carrying out watertank rehearsals at the Marshall Spaceflight Centre, Huntsville, finally agreed on two possible ways of replacing Skylab's thermal shield. First, they could deploy a parasol-type sunshade by pushing it from inside Skylab through one of the scientific airlocks and then opening it like an umbrella. Second, a twin-pole sunshade could be pushed back over the Workship during an EVA. With a variety of special tools and other equipment, they added 180kg to the planned weight of the CSM. It was also decided to make a stand-up EVA before docking, with 2 astronauts in the open Command Module hatch trying to pull out the jammed solar panel with a pole and sending back TV pictures of the damage. These operations were ultimately successful, as related in the following mission descriptions.

First view of the Skylab damage. Top left, broken wires indicate one solar wing torn away. Bottom right, 2nd solar wing jammed partially open; compare with picture (p 91) showing wing fully deployed. *(NASA)*

Note NASA's official numbers were Skylab 1 for the unmanned space station launch and Skylabs 2, 3 and 4 for the subsequent manned flights. Newsmen referred to the manned missions as Skylabs 1, 2 and 3 to avoid confusing the public. The official NASA designations are used here.

Skylab 2 Conrad, Kerwin and Weitz were finally launched on 25 May. It was Conrad's 4th flight, the first for the others. Launch was by Saturn 1B (total lift-off wt 689,680kg, CSM wt 6,062kg). Because Pad 39B had been built for Saturn 5 Moon launches, the much smaller Saturn 1B had to be launched from a 40m steel pedestal (nicknamed "the milkstool" by Cape workers). 7hr later Conrad was describing how one of Skylab's 9m-long solar panels had been torn completely off while the other was jammed, partially opened, by the torn and buckled heatshield. After much conferring the hatch was opened, and with Conrad at the controls, manoeuvring round Skylab and in imminent danger of colliding with it, Weitz tried in vain with toggle-cutters fixed to a long pole to release the panel. As the CM swept into darkness Weitz scrambled hurriedly inside. Then came a major emergency: after 4hr and 6 attempts Conrad failed to dock with Skylab. An emergency flight home seemed inevitable. But spacesuits were donned for the 3rd time in a day, the docking tunnel opened, and the probe dismantled. A nut floated off into space, which was to cause anxiety later, but the last possible docking attempt by the indomitable Conrad succeeded. With doubts which later proved groundless as to whether the faulty probe would ever permit undocking, the crew rested. Then Conrad remained in the CM while Weitz and Kerwin, wearing masks and testing for possible dangerous gas, opened up the 5

hatches to gain access in turn to the Multiple Docking Adapter, the Airlock Module and Orbital Workshop. The MDA, with undamaged thermal shield, was clean and cold; in the Workshop they could feel "radiating heat". At Mission Control we were able to see, via the TV camera set up in the CM window, the orange tip of the parasol emerging from the airlock as Conrad and Weitz, taking frequent rests as they worked in the blistering heat, slowly forced it through the 20cm scientific airlock on its 7·6m handle. With difficulty mosts of its wrinkles and folds were eliminated, and the handle pulled in again section by section so that the sunshade, a 6·7 × 7·3m rectangle, rested snugly just above the Workshop. Protected at last from the direct rays of the Sun, the Workshop began to cool off, though more slowly than hoped. On the 4th day the sleeping cubicles were still too hot to use, but with the crew sleeping in the MDA and the ATM deployed, Skylab, though short of power, was operational.

On the 6th day a power failure, with the loss of 2 out of the 18 remaining batteries, followed the first full day of scientific experiments. Finally, after 3 days of discussions and rehearsals, Conrad and Kerwin transformed a deteriorating situation by means of a hazardous spacewalk. Conrad's efforts with the toggle-cutters on a 7·6m pole to cut away the jagged heatshield and free the jammed solar wing were unsuccessful. So, without the planned handrail to support him, he made his way along the outside of the Workshop and lay across the 1·2m-wide and 9·4m-long beam of the solar wing, guiding the toggle-cutters at close quarters while Kerwin, remaining in the relative safety of the Workshop's "roof", operated them by means of a lanyard. Then Conrad fastened a 9·1m rope to the beam and, as rehearsed, stood up with it to obtain leverage over his shoulders. Suddenly the beam swung out and clicked into place. Conrad and Kerwin "literally took off" as it happened. But within a few moments electric power began flooding into the 8 Workshop batteries, which had been useless since lift-off. Thereafter Skylab had plenty of power for its whole mission despite the loss of the other solar panel.

By the end of the 28-day mission, triumphantly completed, the released solar wing was producing more than 5,500W. Adequate electrical power enabled the astronauts to complete more than 80% of their planned activities. Solar observations totalled 81hr compared with the planned 101hr, the major accomplishment being the monitoring of a solar flare on 15 Jun. 11 out of 14 planned EREP (Earth-resources) passes were completed, and 7,460 out of 9,000 planned photos obtained. Data were obtained over 31 US states, 6 foreign countries and over the Pacific, Atlantic and other oceans. On a final 1½hr spacewalk to recover film from the ATM Conrad even succeeded in getting one of the 2 failed ATM batteries to resume operating by thumping it with a wooden hammer. From then until the end of the 3rd mission 23 of Skylab's 24 batteries continued to give good service.

Skylab 3 The 2nd manned mission, launched on 28 Jul (total lift-off wt 593,560kg, CM 6,085kg), ran into trouble almost at once. An emergency flight home was considered, and preparations for a rescue mission (using the Skylab 4 spacecraft) hurriedly mounted. The flightplan had been extended just before launch from 56 to 59 days to provide a shorter sea voyage after splashdown because Dr Kerwin had become very seasick at the end of Skylab 2. Alan Bean, commander of Skylab 3, was making his 2nd flight; Owen Garriott, a civilian with a doctorate in engineering, and pilot Jack Lousma were both making first flights.

Docking was achieved with no repetition of the Skylab 2 problems. 2 days were spent switching on lights, power systems and air-conditioning, and trying to bring down the Workshop temperature of 80°F to more comfortable levels. Unstowing of the CM stores, which included 6 pocket mice (killed on the 3rd day by a short-circuit), minnows and 2 spiders named Anita and Arabella, was delayed when all 3 men developed nausea which persisted for about a week and made it necessary to postpone a major spacewalk to deploy the twin-pole sunshade brought up by the previous crew as a back-up to the parasol. The major crisis came on the 5th day, when a leak developed in a 2nd CM thruster. An emergency flight home within 24hr before too much manoeuvring gas was lost was seriously considered, with the preparation of a rescue mission (which would require 35 days before it could be launched) as the main alternative. The latter option was selected, since it meant the astronauts, still not fully adapted to weightlessness, would have time to settle down and carry out most of their programme.

After 3 postponements of the EVA, on 7 Aug Garriott and Lousma spent a record 6hr 31min outside. It took them nearly 4 difficult hours to get the new sunshade in position over the top of the parasol, which was to be retained in case the replacement should be unsuccessful. The new sunshade worked well, however, and Workshop temperatures quickly dropped to a comfortable average of 75°F. An EVA inspection of the CM thrusters failed to find the cause of the leaks, but as an extra precaution ground engineers worked out procedures which would permit the CM to provide a safe passage home despite its 2 faulty RCS clusters.

In the big Workshop dome the crew successfully tested astronaut manoeuvring units (AMUs) for use during spacewalks on future flights. Garriott, who had had no previous training, quickly mastered their use. Both Anita and Arabella showed that they could spin webs while weightless (though Anita died before the end of the flight, and Arabella was found to be dead after splashdown). During 2 more EVAs film was changed and recovered, and when the CM was finally prepared for re-entry it was packed with 75,000 pictures of the Sun taken during over 300hr of solar observations (which included 6 solar flares); film and photographs and tape data obtained on 39 instead of the planned 26 EREP passes; the results of welding and materials-processing experiments: and much other material. The astronauts had travelled 38·6 million km.

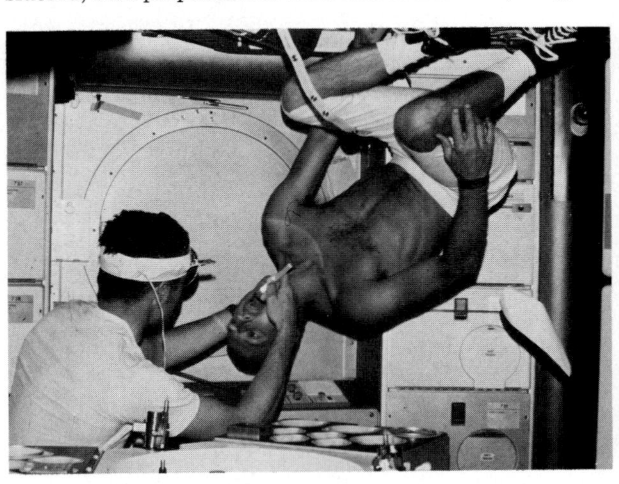

Say "Ah!" Medical checks on Skylab 2, which set a duration record of 28 days, were vital before deciding whether it was safe to embark on the later flights of 59 and 84 days. *(NASA)*

Skylab 3: Jack Lousma, seen during a strenuous 6.5hr EVA, deployed the sunshade which finally provided good working conditions inside. The Earth is reflected in his facepiece. *(NASA)*

Dr Owen Garriott needs no chair to "sit" at Skylab's console. The 3 crew members each averaged 10hr a day on scientific experiments. *(NASA)*

Skylab 4 The launch finally took place on 16 Nov 1973, 6 days late because of fatigue cracks in the Saturn 1B's 8 fins. Gerald Carr, Dr Edward Gibson and William Pogue - all making first flights - were aiming at an 84-day mission. Lift-off wt was 595,370kg, CM wt 6,104kg. A prime object had become observation of Comet Kohoutek, discovered the previous Mar and due to pass within 21 million km of the Sun on 28 Dec. Since its orbit brings it round the Sun and past the Earth only about once in 80,000 years, it was expected to be the biggest astronomical event since Halley's Comet in 1910. It was decided that 2 EVAs should be made, on Christmas Day and 29 Dec, for the crew to mount cameras on the ATM truss to obtain pictures of the comet's changing composition as it became heated by the Sun. The CM was packed with spare parts for Skylab, a portable treadmill to help the crew keep their leg muscles in good condition during 3 months of weightlessness, and extra food; this included 23·1kg of chocolate bars, plus 11·3kg of survival food to enable them to exist for another 10 days if they had to be rescued towards the end of the mission.

On the 8th day, recovering from sickness and irritability, Pogue and Gibson broke the Skylab 2 record with a spacewalk lasting 6hr 34min 35sec. Tasks included the repair of a jammed antenna, involving the tricky job of removing 6 screws with gloved hands, and installation of a new electrical switching box.

Crew morale improved as they settled down to a steady routine of scientific work, including observations of Comet Kohoutek. They were able to disprove ground reports that it had broken up, though it rapidly became clear that it was not going to provide the historic astronomical spectacle which had been prophesied, and which had led to the whole Skylab 4 flightplan being rewritten.

On 18 Dec, when Soyuz 13 was launched on its 8-day flight, history was made because astronauts and cosmonauts were in space together for the first time. But no direct communications were possible between the 2 spacecraft, nor were any sightings reported by either crew. On Christmas Day Carr and Pogue spacewalked for 7hr, photographing Kohoutek before it passed behind the Sun at a distance of 21 million km. Because of its proximity to the Sun they were unable to see Kohoutek, but aimed the cameras on directions from Mission Control.

On 28 Dec Carr and Pogue carried out another EVA to observe and photograph Kohoutek moving away from the Sun and Earth on its 80,000-year orbit. On 30 Dec they discussed their observations via TV with Dr Lubos Kohoutek, the Czech astronomer who discovered the comet.

Carr and Gibson carried out the 4th and last EVA on 3 Feb. During it they retrieved the final film of the Sun and Comet Kohoutek, and recovered materials and collection devices placed outside the Workshop for analysis of the effects of the space environment. They also placed an extra micrometeorite experiment outside for possible collection by a future visiting US or Soviet spacecraft. This brought the total EVA time for the mission to 22hr 19min.

On 8 Feb 1974 the crew prepared a "time capsule" - a bag containing 5 food and drink items, unused film, camera filters, clothing, flightplan pages and small electronic devices - which could be retrieved to permit study of long-term effects if Skylab was later revisited.

When the crew undocked they were on their 1,213th revolution, Skylab on its 3,898th. During a final, nostalgic fly-around Gibson reported that both sunshades had faded a lot and, as photographs later proved, one had developed a split. Before they splashed down Skylab was a dead space station, having been depressurised by Mission Control. In its 454km orbit (raised by an RCS burn a few days earlier) its life before re-entry was estimated at between 6 and 10 years. As a result of 1½hr exercise per day (treble the amount taken by the Skylab 2 crew), all 3 astronauts were in far better physical condition that either of the earlier crews. They had brought home with them about 75,000 solar pictures, 17,000 of Earth and 2,500 of Comet Kohoutek, plus the results of many scientific experiments. They had become the joint holders of the record of man's longest spaceflight, and had travelled about 54,804,000km. The 3 crews had orbited the Earth 2,475 times and travelled about 112,804,000km.

Summary and conclusions For America the successful completion of the Skylab missions marked the end of the first era of manned spaceflight. For those who doubted whether the estimated cost of about $2,400 million was justified, it is worth recording that although the Earth-resources equipment was considered the least effective, just one potential copper deposit located by it in Nevada is estimated to have an ultimate value of billions of dollars, far exceeding the cost of the whole US space programme so far. As one astronaut said: "Skylab worked better broken than anybody had hoped for if it was perfect." Man's adaptability was effectively demonstrated by the fact that the last crew, although they had flown much the longest mission, returned to Earth in the best physical condition. Discussion of the defence aspects of the flight is discouraged, but visual observations made by the astronauts of activities at Soviet missile centres and along the Soviet/Chinese border also established that, however good automatic camera and sensors may be, on-board astronauts will be an essential part of future military space reconnaissance systems.

Re-entry crisis Before Skylab's launch the Russians had begun to demonstrate their ability to place their somewhat smaller Salyut stations on a safe re-entry trajectory into the Pacific at the end of their lives. The author's questions at pre-launch press conferences about the danger of damage from surviving Skylab debris - its final weight was at least 77 tonnes - were brushed aside by NASA, which did not then consider it a problem. The agency also maintained that Skylab would have an orbital life of at least 11yr. In fact the Royal Aircraft Establishment's estimate of 6yr proved exactly right. As Skylab's orbit, taking it over Europe and many other densely populated areas, began to decay, concern grew about the bomb-like effects which might result from many tons falling on towns. The furore was increased by Russia's Jan 1978 loss of control of Cosmos 954, which finally fell with its nuclear reactor in northern Canada. Efforts to resume control of Skylab and prolong its life so that it would remain in orbit long enough for the Space Shuttle to reach it were only partially successful. Rescue plans were based on the Teleoperator Retrieval System, designed to be carried into orbit by the Shuttle, docked with Skylab by remote control, and then fired to take the station into higher orbit for later re-use, or at worst to place it on a safe re-entry trajectory. This scheme was finally abandoned because of delays to the first Shuttle launch and the rising cost of the TRS. Finally, with its last orbits carefully monitored and followed worldwide, Skylab re-entered over the Indian Ocean on 11 Jul 1979. Disintegration began at an altitude of 19km, lower than expected and resulting in a much smaller distribution of debris. The affected area, in Western Australia, proved to be about 74km wide and 4,400km long, compared with the expected 185 and 7,400km. No one was hurt, and NASA appealed to residents of this thinly populated area to search for pieces of debris and return them. Those who wanted to keep the pieces as souvenirs were assured that they would get them back after examination, together with a plaque expressing NASA's gratitude. More than 20 such plaques have been issued, and recovered debris included 2 large oxygen tanks, a pair of titanium spheres which held nitrogen, and an 80kg piece of aluminium from the door of the film vault in the Skylab workshop.

SMM

The Solar Maximum Mission satellite, L 14 Feb 1980 by Delta 3910 from C Canaveral (566 × 569km at 28·5° incl; wt 2,315kg) in Apr 1984, became the world's 1st satellite to be retrieved, repaired and redeployed in orbit. Carrying 7 instruments to observe solar flares over a wide band of wavelengths in the ultra-violet, X-ray and gamma-ray regions of the spectrum, it was intended to cover the maximum period of the sunspot cycle and act as a major key to understanding the violent nature of the Sun and its effects on Earth. During the 1980 period of maxium activity SMM observed over 1,500 flares; 10, including a 4min burst with a flare core temperature probably exceeding 100 million degrees F, were observed in detail by all 6 flare instruments. The 7th instrument measured the total optical energy which reached Earth in the form of heat, probably affecting the climate and length of the growing season. A fuse failure in Dec 1980 meant that the attitude-control system could no longer point the craft precisely at the Sun, but by spinning up SMM so that it rotated every 6min Goddard engineers enabled 3 of the instruments to continue acquiring good data. From Feb 1980 to Aug 1981 they detected a persistent decrease of 0·01% in the total amount of energy reaching Earth, which was thought to be a factor behind the harsh 1981-82 winter.

The partial failure led to the original plan for the Shuttle

The Solar Maximum satellite made history as the first to be repaired and redeployed in orbit. The dish antenna at bottom, used to transmit via TDRS, was not deployed until after the repair.

to use the SMM (which had been equipped with a grapple fixture) for its 1st retrieval operation becoming a much more ambitious in-orbit repair mission. Two major repairs were necessary: replacement of the attitude-control module, and the much more complicated task of replacing failed microprocessors in the coronagraph-polarimeter experiment. The repair was first considered for STS-4 in 1982, but SMM was at twice the height planned for what was to be the last Shuttle test flight, and no EVAs had been rehearsed. The rehearsal carried out on Mission 10, including Bruce McCandless's 1st MMU flight, and the successful repair on Mission 11 are covered in detail in the Shuttle missions section. Restored to health - subject to a 30-day checkout by Goddard - SMM was replaced in orbit with its dish antenna deployed for the 1st time, enabling it to transmit via TDRS-1. In addition to resuming its solar observations until about 1990, SMM may now be used to photograph Halley's Comet in 1986, when it will be too close to the Sun for Earth-based observations.

SMM originally cost $79 million, exclusive of launch

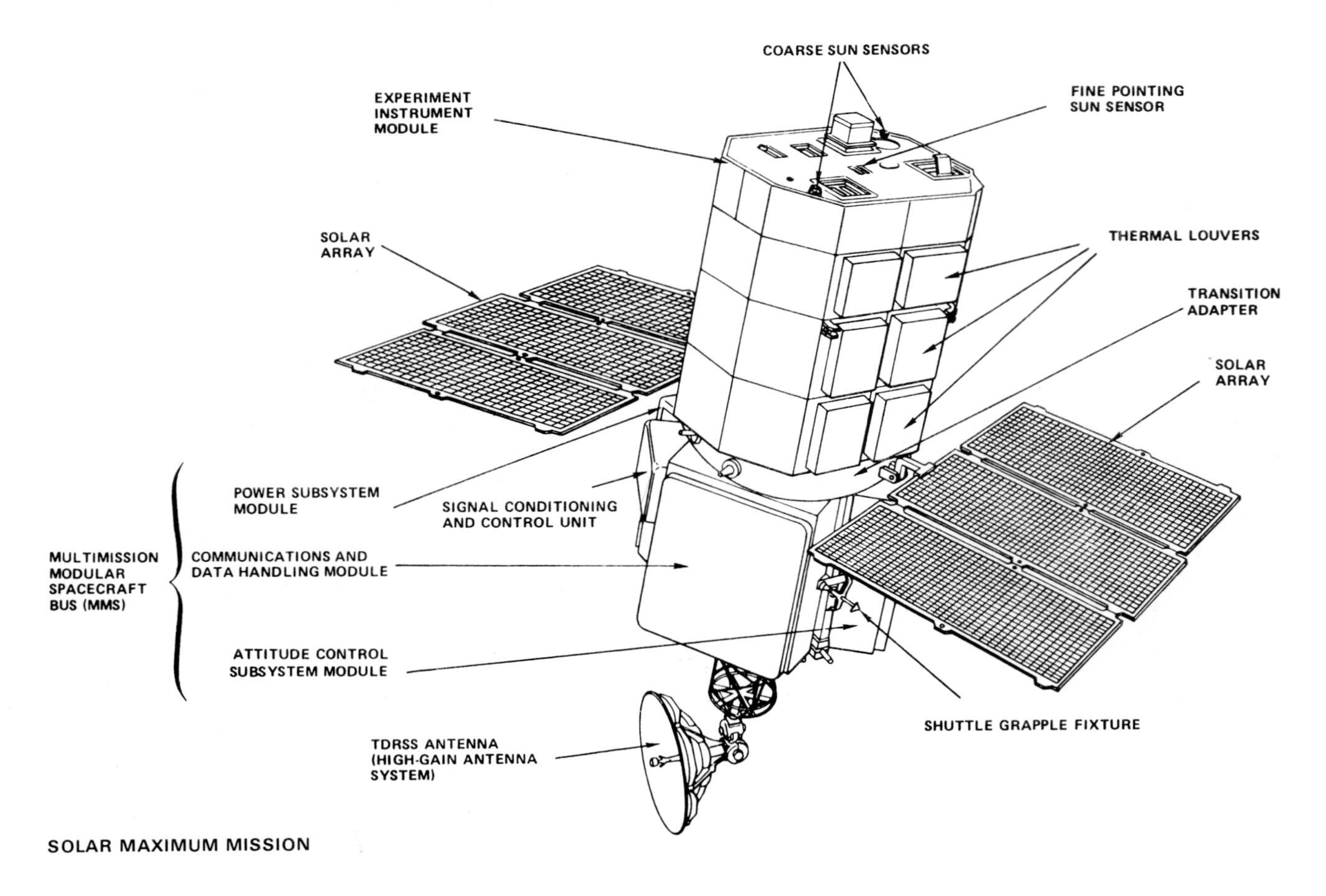

SOLAR MAXIMUM MISSION

SMM, repairs completed, seen at rear of *Challenger's* cargo bay in April 1984. Astronaut James van Hoften, in foreground, had enough EVA time left to try out the MMU. *(NASA)*

and tracking; replacement now would have cost about $240 million, compared with repair costs estimated at $50 million. Following this success, NASA revived plans to repair Landsat 4 and recover the lost Westar 6 and Palapa B2 satellites.

Solar Optical Telescope

Due for its first one-week mission on a Spacelab flight in 1990, this will be a 1·25m-dia telescope capable of making several hundred images on film during observations of the Sun through various optical filters in the near-infra-red, visible and ultra-violet regions of the spectrum. It will expand on the Skylab ATM results of 1973, and will be flown on a series of Spacelab missions.

Spacelab

Although Spacelab has been provided by the European Space Agency and is described in the International section, its missions will be operated by NASA, though usually with other countries participating.

Space Shuttle

The US Space Shuttle will dominate the development of spaceflight in the West for the rest of this century. Its progress and prospects are covered in the following sections:

Status

With 4 test flights and 7 operational flights triumphantly completed by Apr 1984 (though latterly with worryingly narrow safety margins), the Space Shuttle had successfully demonstrated its ability to rescue and repair damaged spacecraft in orbit. It should also be ferrying a new breed of rigger-astronauts into orbit to assemble space stations for permanent occupation before the end of this decade.

Columbia, the first winged and wheeled spacecraft, was first launched on 12 Apr 1981. Its landing 2 days 6hr 20min later on the baked clay of the Californian desert evoked worldwide admiration - and immense relief among the US space fraternity. Ten years after its conception - 8 of them spent on active development - it had at last ceased to be "trouble-plagued" and had shown that it could, as predicted by ex-President Nixon, "transform the space frontier of the 1970s into familiar territory, easily accessible for human endeavour in the 1980s and 1990s."

But the original idea that the Space Shuttle - officially known as the Space Transportation System (STS) - would become the sole US launcher had had to be abandoned. By the end of 1983 a new unmanned launcher, possibly based on the MX ICBM's propulsion system, was being planned.

The originally planned total of 570 Shuttle missions for 1980-91 (of which 106 were to be for the Defence Department) was reduced first to 487 by 1996 and then to 312 by 1994 - a target which now seems barely possible.

Columbia* inaugurated reusable Space Shuttle flights on 12 April 1981. *(NASA)

By May 1984 111 Shuttle flights carrying 325 payloads up to Sep 1989 were listed in the monthly manifest. There was however a good deal of space available on many of these flights, and following STS-8 NASA began sending its successful astronauts on sales tours. They repeated an offer first made in 1982 that major foreign users could nominate their own payload specialists to accompany their satellites; 100hr of training spread over 6 months and costing about $150,000 would be required.

Canada, Germany and Japan already have long-term plans to select astronauts and fly them on the Shuttle. France too is eager to match her Soyuz/Salyut experience with a flight aboard the Shuttle. Australia was the first country to accept the NASA offer, with Brazil and Britain also showing interest. In Britain's case, the chance to send up her own astronauts in 1985 and 1986 tipped the scale against political pressures to use Ariane rather than the Shuttle to launch her forthcoming military communications satellites. Cost comparisons are discussed under "User costs".

By the end of 1983 it was no longer intended that the Shuttle should become the sole US launcher system. As the Launcher section shows, the failure of the Shuttle system to become as cheap as expected has led to commercial organisations providing a whole series of alternatives, while a new, MX-based, unmanned launcher may also be developed.

With only 4 launches achieved in 1983, the 12 planned for 1984 looked increasingly unattainable; indeed, 2 of them had to be cancelled even before the year began. All the same, 12 were manifested for 1985 and 17 for 1986 in the hope that a regular 24 a year could be reached in 1987. But even that rate gives a total of only 252 by 1994.

Until Oct 1985 all launches will be from Cape Canaveral's Kennedy Space Centre. Then the first polar launches should be carried out at the USAF's Vandenberg site. The inaugural Vandenberg mission was originally scheduled for 1984 until it was found that the prevailing winds were too strong to permit assembly of the Shuttle cluster on the launchpad. A protective building has since been designed. Vandenberg launches will total only 8 by the end of 1988. The hope is for "a moderate growth" to about 10 a year, of which less than half will be military.

NASA will launch DoD missions from Kennedy, with the USAF reciprocating by launching NASA flights from Vandenberg. Current estimates show that the 4 Orbiters now in service or under construction could carry out 38 flights a year from the 2 sites if a 2-shifts-a-day, 5-day week were worked, and 53 a year with a 7-day week.

Inevitably, with NASA constitutionally committed to an "open," unclassified programme and the USAF fearful about the security of its military satellites and space systems, there are signs of friction between the two. As part of extensive security modifications which the USAF is funding at the Johnson, Goddard and Kennedy centres, DoD has successfully insisted upon having its own classified control room at the Johnson Centre at Houston. At Kennedy "security modifications that will satisfy specific DoD mission requirements" are well under way.

It is hardly surprising that proponents of NASA's internationally admired open programme are anxious about the extent of the military penetration. Kennedy was originally created on Merritt Island, across the Banana River from the USAF's Cape Canaveral launchpads, so that NASA could free itself of the restrictions surrounding joint use of the military facilities.

It is already too late to apply the obvious solution – construction of a 5th Orbiter for the exclusive use of DoD – and in any case funding for this continues to be put off from year to year. By 1983 NASA Administrator James Beggs was expressing confidence that a 4-Orbiter fleet would provide plenty of capability, including Space Station development, into the 1990s. But other managers were talking of the need for at least 5 and possibly 7. They were also increasingly concerned (following *Challenger*'s pre-launch problems) about the effects of losing an Orbiter. Against this, White House advisers were said to believe that "even a 3-Orbiter fleet would be viable".

One reason why Shuttle costs have failed to come down is that they were based on the assumption that most Western nations would employ STS to orbit their national spacecraft instead of developing independent launch facilities of their own. But with the European Space Agency's Ariane rocket becoming operational about the

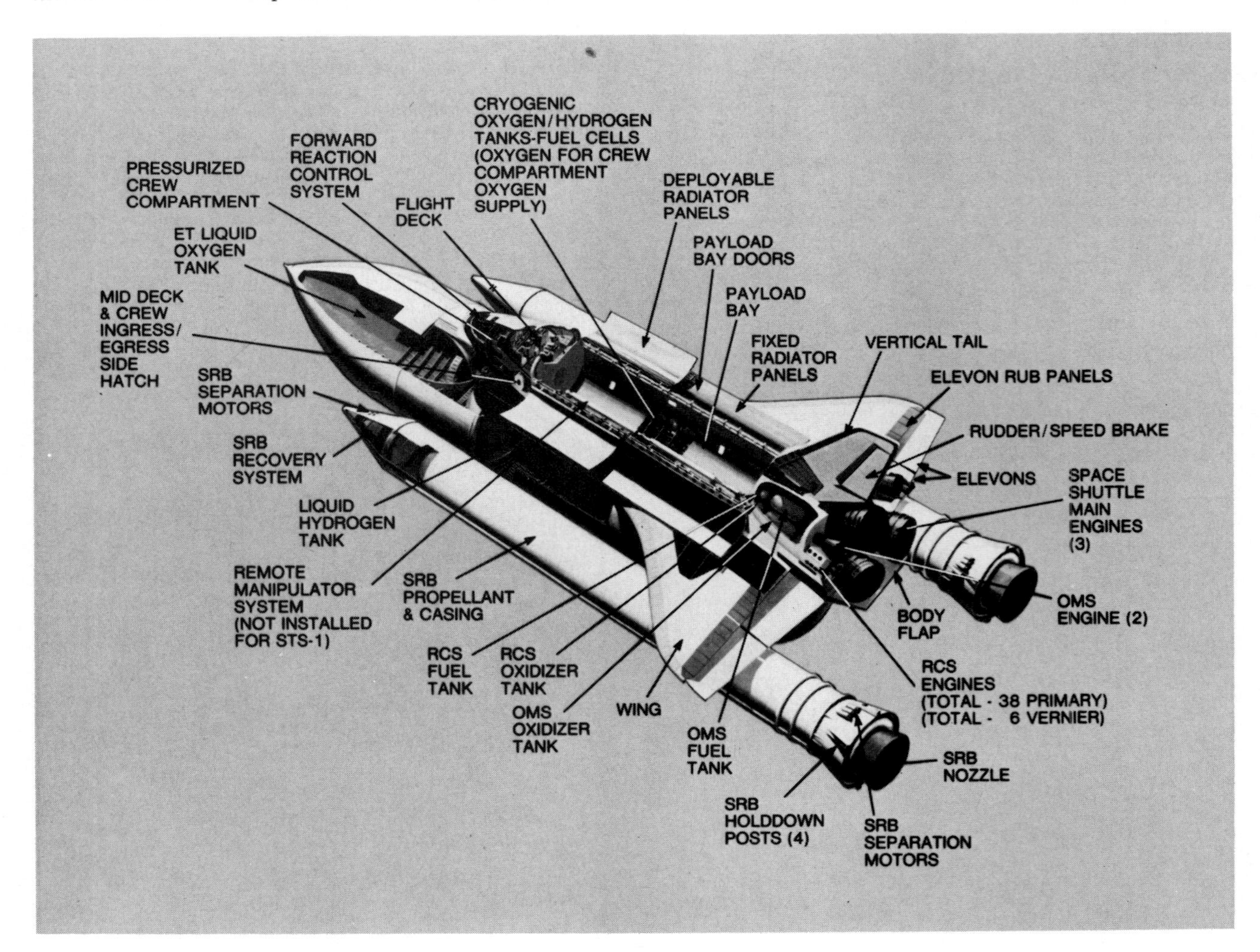

Flight director Gary Coen, at JSC's Mission Control, Houston, monitors *Challenger*'s landing at Edwards at the end of STS-6. Astronauts Bridges and O'Connor, at right, served as spacecraft communicators as part of their training for their own future flights.

same time as the Shuttle, fierce international competition has developed. Reliability has become as vital a factor as cost for customers, particularly Third World nations, which urgently need to establish their own satellite systems before the geostationary slots are all filled. Recent policy statements to the effect that DoD will if necessary take over Shuttle launches reserved years ahead by commercial customers - coming on top of NASA budget cuts at the expense of collaborative projects with Europe - have strengthened the belief that the West needs alternatives to the Shuttle. This conviction supported Ariane through its own delays and technical setbacks, and stimulated the continued commercial use of "conventional" launchers like Delta.

Future development

Originally conceived as being capable of 30 days in orbit, the Shuttle is currently limited by fuel-cell life to 9-10 days. In May 1983 Rockwell announced a 2-stage plan to extend this to 15-18 days by 1985 and 45 days by 1988-89. The 1st stage could be achieved for $200 million by the addition of a "cryogenic wafer" at the rear of the payload bay. The 2nd stage, costing $200-250 million, would require an orbiting solar array and battery system with which the Shuttle could dock. This would require relocation of the airlock and mid-deck modifications, among other things, and gained some DoD support as a cheaper alternative to space station development. NASA countered by pointing out that 35% of the Shuttle's cargo-carrying capacity was at present wasted; immense savings could be achieved by filling it with spare parts and fuel for storage in space. Increasing the payload from 65 to 82% could be easily achieved and would save $5·1 billion by the year 2000. In the meantime, however, NASA faces the need to spend about $1 billion in the next 10yr on improving the Space Shuttle Main Engine (SSME).

This concept of a possible shuttle development shows an External Tank replacing the Orbiter as an unmanned, heavy cargo carrier or large space station module. *(Martin Marietta)*

User costs

The rival US and European systems are both subsidised, and consequently set the price level which commercially produced launchers must match. Under a Reagan Administration directive Shuttle flights will continue to be subsidised until 1988, while Europe's Arianespace must make a profit after taking over operations from Ariane's 10th launch. The US policy is intended not only to enable Shuttle to be competitive with Ariane, but also to allow it to be sold for less than conventional US launches. The level of subsidy seems to be very high: an Aerospace Corp technical analysis in May 1983 claimed that synchronous orbits actually cost twice as much by Shuttle as by expendable launcher.

In 1978 both Shuttle and Ariane were said to cost around $20 million for a "dedicated launch" (use of the whole payload capacity). NASA's User Handbook, issued in that year, included a price list and launch conditions for prospective national and commercial customers. Starting price then was $19 million for a dedicated flight, with payment of $100,000 non-returnable "earnest money" being required before negotiations started for a flight 3yr later. By Feb 1978 advance payments for 175 payloads had been made. It was pointed out that costs could be much reduced by sharing payloads with other customers, while a small, self-contained package requiring no Shuttle power and weighing less than 91kg could be orbited for as little as $3,000. Both NASA and ESA sought to build up their order books by offering launches to foreign customers at less than cost. By the time the competition had settled down in 1982 NASA was offering to launch 2 Brazilian satellites by Shuttle/PAM-D for $34 million each, with Arianespace hoping to win the contract with a bid of $29 million each and an earlier 1st launch. The actual cost per STS launch was about $70 million by then, and the day when the Shuttle could be operated profitably was still many years off. Then NASA announced that from 1 Oct 1982 to 30 Sep 1988 the standard charge for a dedicated launch would be $71 million in 1982 dollars. This meant about $26 million for a Delta-class payload to synchronous transfer orbit, including the upper stage and "nominal optional services"; an Atlas-Centaur-type payload would be about $41 million. NASA expected Shuttle prices "to remain competitive" with those of expendable vehicles, including Ariane. There was no suggestion that they might be significantly lower, and by the end of 1983 the 1984 price per dedicated launch was being given as $86 million.

While doubts about the cost-effectiveness of men in space have largely faded away, the fact that 20-25% of Shuttle costs are due to the need to protect the crew inevitably stimulates suggestions that much greater commercial use could be made of the system if it incorporated an unmanned Orbiter with the minimum of heat-shielding and no life-support equipment.

Development costs

These will be written off, with no effort made to recover them in charges to users. In Mar 1982 they were estimated at $5·15 billion in 1971 dollars. In 1982 dollars that had risen to $9·912 billion, equal to $6·654 billion in 1971. The 1971 estimate of $250 million for an Orbiter rose to about $2 billion in 1983 dollars, though this included for the first time engines and equipment, Remote Manipulator System, galley and closed-circuit television. Rockwell International, which was responsible for Apollo, is prime contractor for the Shuttle. The major Shuttle contracts, with their value in 1982 dollars, are as follows:

	$ million
Rockwell (Shuttle Orbiter)	3,560
Grumman (Rockwell subcontractor responsible for wings)	45
McDonnell Douglas (Rockwell subcontractor responsible for OMS/RCS pods)	85
McDonnell Douglas (Support)	52
Rocketdyne (Shuttle Main Engines)	1,546
Thiokol (Solid Rocket Booster manufacture)	206
USBI (Solid Rocket Booster assembly and retrieval)	89
Martin Marietta (External Tank)	529

General description

When first conceived in 1969 the whole system was to be recoverable. But a proposed manned launch vehicle which was to fly back to base was abandoned because of its high development cost in favour of the present system, with Solid Rocket Boosters recovered from the sea and the External Tank being irrecoverably jettisoned. This compromise, it was hoped, would still enable the system to reduce launch costs by five-sixths compared with current expendable boosters – a hope which is still far from realisation.

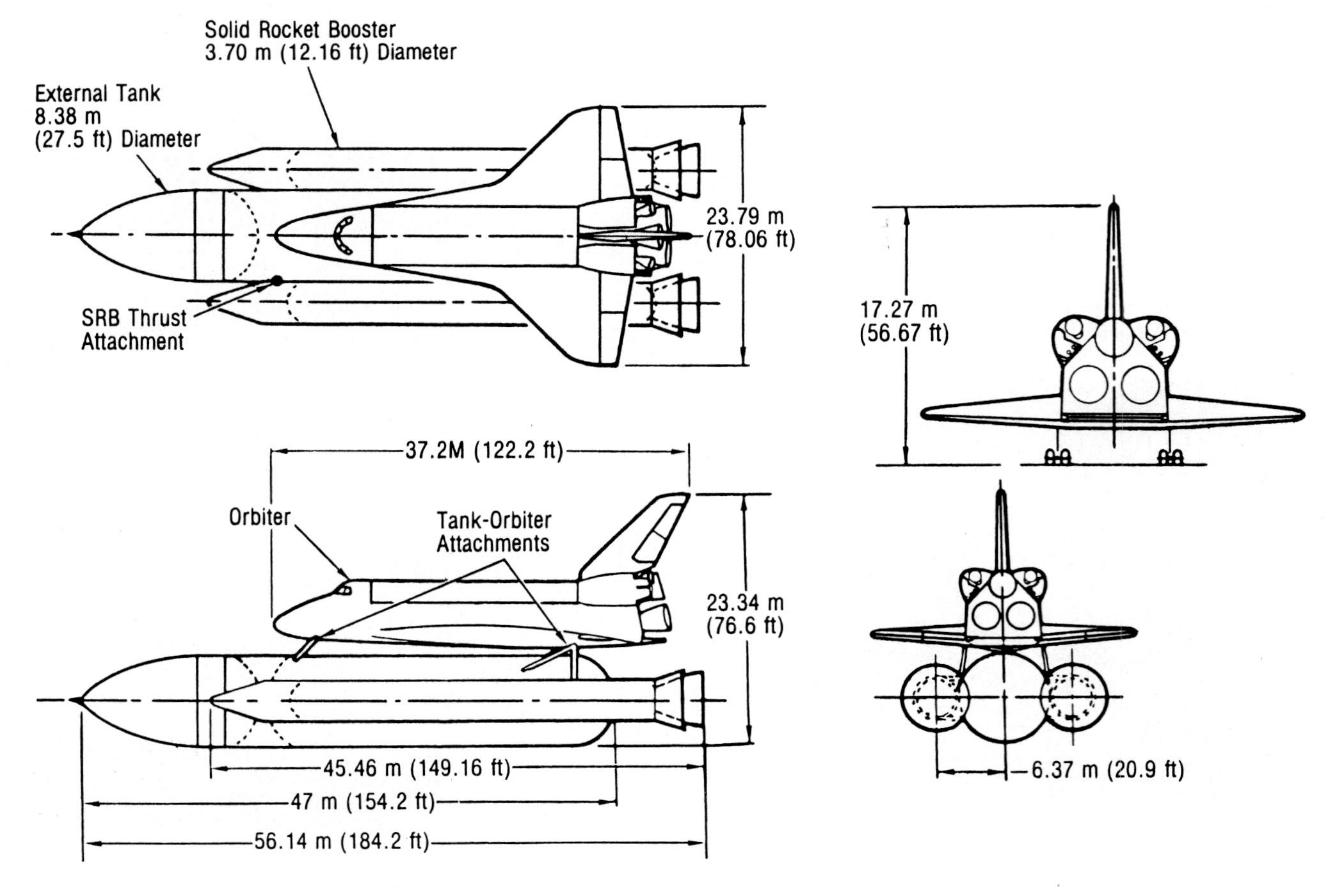

External dimensions

Wing span	23·79m (78ft 0·68in)
Length overall	56·14m (184ft 2·4in)
Length of External Tank	47m (154ft 2·4in)
Length of Boosters	45·46m (149ft 2·0m)
Height overall	23·35m (76ft 7·2m)

Weights

Shuttle complete (OV-102)	2,010,625kg (4,432,667lb)
Orbiter (empty)	74,844kg (165,000lb)
External Tank (full)	756,441kg (1,667,667lb)
Boosters (2), each	589,670kg (1,300,000lb)

Thrust

Total, at lift-off	28,590kN (6,425,000lb)
Orbiter, main engines (3), each	1,670kN (375,000lb)
Boosters (2), each	11,790kN (2,650,000lb)

Performance

Payload:

In 185km (115-mile) orbit, due east
29,485kg (65,000lb)

In 500km (310-mile) orbit, 55° inclination
11,340kg (25,000lb)

In 185km (115-mile) polar orbit
14,515kg (32,000lb)

Orbiter

This is usually described as being about the size of a DC-9 airliner. Long-term planning has been based on the production of 5, each capable of at least 100 flights.

Before being named, what was intended to be the ground test Orbiter Vehicle was designated OV-099, and the 5 flight-standard vehicles OV-101 to 105. Just before the Approach and Landing Tests (ALT) began in 1977 NASA wanted to name OV-101 *Constitution*, but nearly 100,000 "Star Trek" fans wrote to President Ford asking him to name it *Enterprise* after the spaceship in the TV series. Following a 45min meeting between James Fletcher, then NASA Administrator, and President Ford, NASA gave way to popular demand. This caused some internal embarrassment, since it was already suspected that OV-101 would never go into orbit. So much had been learned during its construction that it was decided shortly afterwards to upgrade OV-099 as the second vehicle to fly, replacing it with *Enterprise* as the ground-test vehicle after the ALT programme. The first 4 operational Orbiters were then named after famous pioneer sailing ships: *Columbia* (OV-102) after one of the first Navy frigates to circumnavigate the world (this was also the name of the Apollo 11 Command Module); *Challenger* (OV-099) after the ship that explored the Atlantic and Pacific in 1872-76 (also the name of the Apollo 17 Lunar Module); *Discovery* (OV-103) was named after 2 ships, the vessel that discovered Hudson's Bay in 1610-11, and the one in which Captain Cook discovered the Hawaiian Islands; and *Atlantis* (OV-104), after the ketch which did oceanographic research during 1930-66.

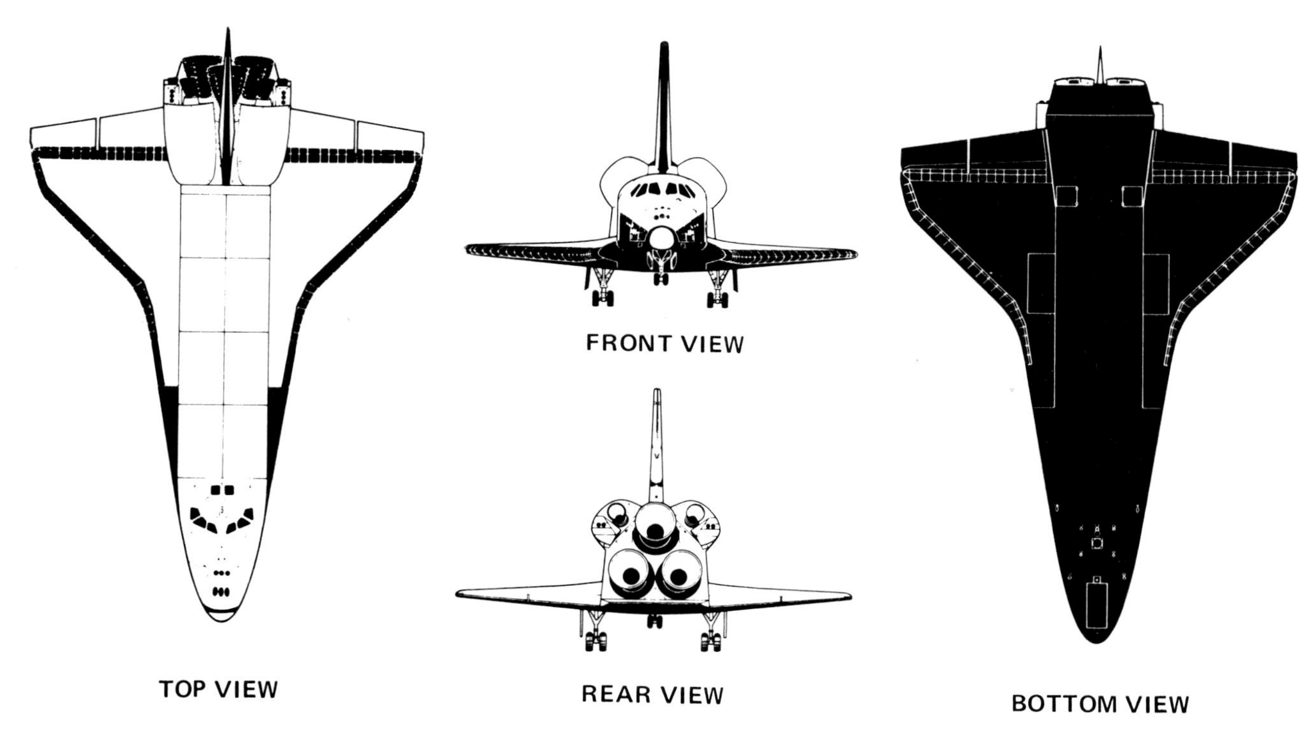

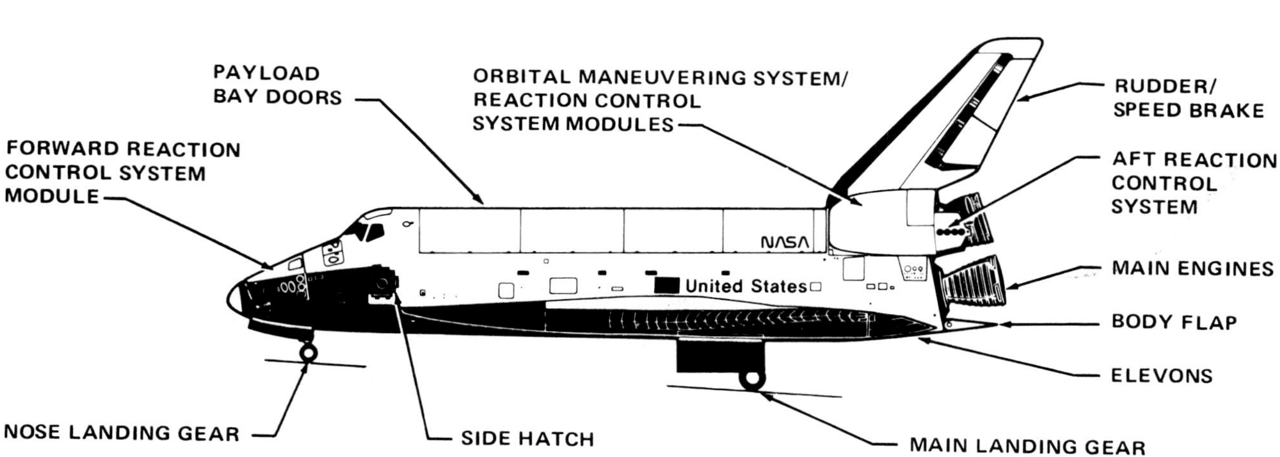

Columbia carried out the first Shuttle flight, STS-1, in Apr 1981. *Challenger* made its debut on STS-6 in Apr 1983. *Discovery*, delivered to KSC in Oct 1983, is scheduled to fly first on Mission 12 in 1984. The Jun 1984 manifest revealed that after 6 flights from there it is to be permanently stationed at Vandenberg for the polar missions. *Atlantis*, due for delivery in Dec 1984, is scheduled to start flying on Mission 29 in Nov 1985. *Columbia* had been due to have its ejection seats removed, together with other major modifications, after STS-5 in late 1982. But the DoD insisted that 2 Orbiters should be available in case of damage or loss, and this was delayed until after STS-9/Spacelab 1. As that mission approached, Shuttle managers decided to remove *Columbia* from flight status for 2yr and use it as a source of spare parts for *Challenger* and *Discovery* as they took alternate flights. This decision was hurriedly reversed by Administrator Beggs, who described it as bad political and engineering policy. He was clearly anxious about the decision's effect on the funding of a 5th Orbiter, but the difficulty was then to find missions for *Columbia* in the revised manifest.

NASA continues to have difficulty in getting a full go-ahead for OV-105; this is not surprising, since even within NASA there are varied views about the necessity for it. The result is that it could not be completed before 1988, and at a cost of over \$1·5 billion. Partial funding of long-lead-time parts totalling \$400 million has so far been approved.

Shuttle managers who once talked about the need for a 6th Orbiter by 1990 at present speak in low voices. Early 1982 saw an offer by Space Transportation Co, a subsidiary of a large investment banking firm, to invest \$1 billion with a down payment of \$300 million in buying and operating OV-105. In return, the company would take over from NASA all marketing of the launch services to commercial and foreign users. This arrangement was at first viewed favourably by the US Government, and at the end of 1983 SpaceTran was still expressing interest. But Administrator Beggs said the proposal "was moribund, if not dead".

While he was a senator, former astronaut Harrison Schmitt was advocating a fleet of 8 Orbiters, 4 for civilian use and 4 for the military. It seems likely however that such proposals will be overtaken by developments like a quick-reaction mini-shuttle for military use, accompanied perhaps by the proposed unmanned SRB-X cargo vehicle.

Heat-shielding such a large spacecraft against re-entry temperatures of up to 2,300°F has proved less of a problem than feared. The Orbiter's 31,000 thermal tiles, each a different size, shape and thickness and consisting of 99·7% pure silica fibre derived from high-quality sand, were designed to resist cracking as a result of the vibrations of launch, contraction in the cold of space, and expansion in the heat of re-entry. Difficult and time-consuming to apply, some fell off on the early flights. But they did their job well, and flight experience led to rapid development of improved methods of bonding, and the production of cheaper, lighter and more effective "thermal blankets". These were being used to replace the white, low-temperature tiles, leading to a 26% reduction in tile numbers.

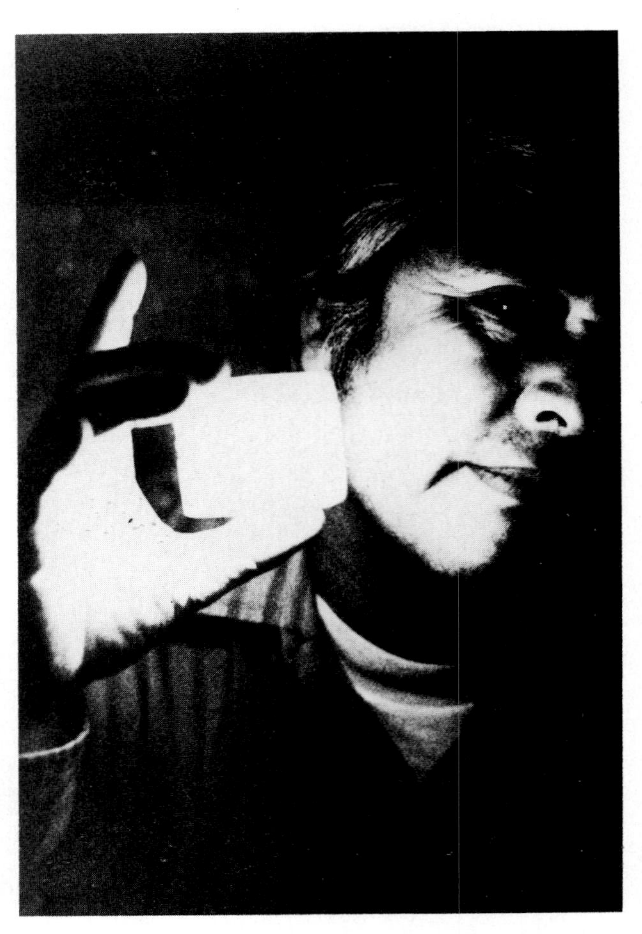

White-hot from a 2,300°F oven, an Orbiter silica insulation tile can be held in the bare hand without injury. The demonstration illustrates the speed with which heat is cast off. *(Lockheed)*

Columbia*'s second landing, at Edwards in November 1981, was carried out by astronauts Engle and Truly. *(NASA)

External Tank The External Tank, which dominates the cluster and contains 703,000kg of propellant for the Orbiter's 3 liquid-fuelled main rocket engines, is the only part of the stack not to be recovered. But since it almost reaches orbital speed before it is jettisoned, and its empty weight of about 33,000kg is greater than the Shuttle's main payload, it is hoped that later in the programme it can be sent into orbit with the Shuttle for re-use as a major building block in the construction of space platforms. (Few are aware, even inside NASA, that much less fuel would be used if the ET were taken into orbit now. It is jettisoned just before orbital speed to ensure a controlled re-entry and descent. Jettisoning it at the most economic moment would leave this large object tumbling in low orbit, to re-enter within days - sooner or later over a populated area.) First results of a continuing effort to reduce ET weight from the original 34,640kg appeared on STS-3, when a 272kg reduction was achieved by leaving off the white paint; the brown of the spray-on foam insulation then contrasted vividly with the stark white of the SRBs. A further 2,720kg was saved under a $42·9 million Martin Marietta contract by reducing the thickness of the aluminium skin panels and changing the liquid oxygen anti-geyser system. 1st use of these modifications was on STS-5, each reduction increasing Shuttle payload by about the same amount. On STS-6, *Challenger*'s 1st flight, ET weight was 30,311kg, over 4,530kg lighter than on STS-1. Martin Marietta was also keen to take an ET into orbit for inspection by an astronaut with the MMU, to see how much more could be saved by reducing the insulation. 10 tanks to the original design were produced, 7 for flight use. Development of the lighter ET led to delivery delays in the early stages. Then as the number of Shuttle flights in the 1981-5 period was repeatedly cut, a major storage problem developed. With Martin Marietta due to deliver another 42 by 1985, plans for storing them on barges had to be considered, since KSC had space for only 3 at a time. The pre-flight cost of an ET rose from $1·8 million in 1971 to $10·1 million 10yr later. A Sep 1983 contract for 26 ETs plus material for another 21 (following the existing contract for 15 ETs) was worth $505 million.

The "beanie cap hood" for the ET was a response to one of the many problems that delayed the first flight. It was found during simulated Shuttle countdowns that vented oxygen tends to form ice on the sides of the ET. This shakes off during lift-off, hitting and damaging the Shuttle's thermal protection tiles. The hood, designed to be lowered over the ET during countdowns to siphon off the vented oxygen, was fully used for the first time on STS-2. It diminished but did not cure the problem.

Plans to take the ET into orbit for use as a space station building block went out of favour for a couple of years on the grounds that the cost of adapting it would equal or exceed that of sending up a tailor-made unit. But NASA has recently revived the idea, arguing that delivering the ET into orbit would increase total payload, and make its residual fuel available for storage in a space station. Other proposals include use of the ET's aft end as a cargo carrier. In this mode it could carry an Orbit Transfer Vehicle, to be extracted by the RMS and attached to a much larger satellite in the cargo bay for onward routing. These ideas seem likely to be given impetus by the disclosure that on STS-9/Spacelab 1 and subsequent missions in such high inclinations under- or overspeeds during launch could result in surviving chunks of the ET descending upon areas as widely spaced as Scotland, East Germany, the Soviet Union and the Middle East.

Solid Rocket Boosters These produce their thrust instantly upon ignition, burn out in about 2min 12sec, and descend on parachutes for recovery at sea. The plan is to reuse them 20 times and to order 5 pairs for each Orbiter from Thiokol and McDonnell Douglas. After recovery each is broken down into 14 major components for refurbishing, with the sections being reused separately. First to be reused, on STS-4, were the rate gyro panels and instrument SEP from STS-1, with the frustums, forward skirts and 2 APUs following on STS-6. Not until the other 2 APUs were reused on STS-9 were all STS-1 sections finally recycled. STS-2 components were due for reuse on STS-7, 8, 9 and 11. Plans to replace some metal segments of the booster motor case with composite filament materials to reduce SRB weight and increase Shuttle payload by a further 2,720kg were announced in Mar 1982. First use of this new version is scheduled for late 1985.

Concept for using External Tanks as modules for a space facility. But problems associated with safe removal of residual fuel have yet to be overcome.

STS-9 launch, November 1983. Note "beanie cap" hood, hurriedly developed to fit over ET peak to siphon off vented oxygen and diminish ice formation on ET's outside. *(NASA)*

Re-using the SRBs: this SRB forward skirt, already used on STS-1 and STS-6, was prepared for a 3rd flight on Mission 41-B. It is seen being attached to a frustum previously used on STS-5.

SRBs cost about $50 million a pair, and pressure upon NASA to economise has recently led to a proposal to use the SRBs as sounding rockets on each mission, since each could carry 90kg of scientific instruments without loss of performance. Following the loss of the SRBs on STS-4 as a result of excessive water impact speeds, larger, 41·4m main parachutes, compared with the current 35m chutes, have been tested.

Extravehicular Mobility Unit Spacewalks, still referred to by the term "extravehicular activity" used in Apollo days, will play a major part in future Shuttle activities, and the EMU includes spacesuits, Manned Manoeuvring Units, emergency life support and rescue equipment. Crew members must be able to leave the Orbiter for much routine work, as well as in emergencies. These tasks include inspection of the Orbiter or its payload; photography; installation, removal and transfer of film cassettes on sensors; operation of outside equipment; cleaning optical surfaces; repair or calibration of modular equipment; and the freeing of the payload bay doors or RMS arm if they get stuck. At first crew members will continue to use 30·5m tethers during spacewalks. On later missions they will use the free-flying Manned Manoeuvring Unit to service or recover spacecraft hundreds of metres away from the Orbiter. In emergency they will be able to use the Personal Rescue System if an Orbiter becomes stranded in orbit.

The Shuttle spacesuit, cheaper and more flexible than those used on the Moon, starts with a zip-on one-piece liquid cooling and ventilation garment made of Spandex. Weighing 3kg and with an expected life of 15yr, it includes urine collection facilities for up to 950mlit and a drink bag containing 0·6lit of water. A "Snoopy cap" with headphones and microphone fits over the head and chin, and also provides caution and warning tones. Over this goes the 2-piece spacesuit itself, consisting of hard upper torso, made in 5 sizes with hard waist ring; lower torso in various sizes with hard waist ring; the helmet, available

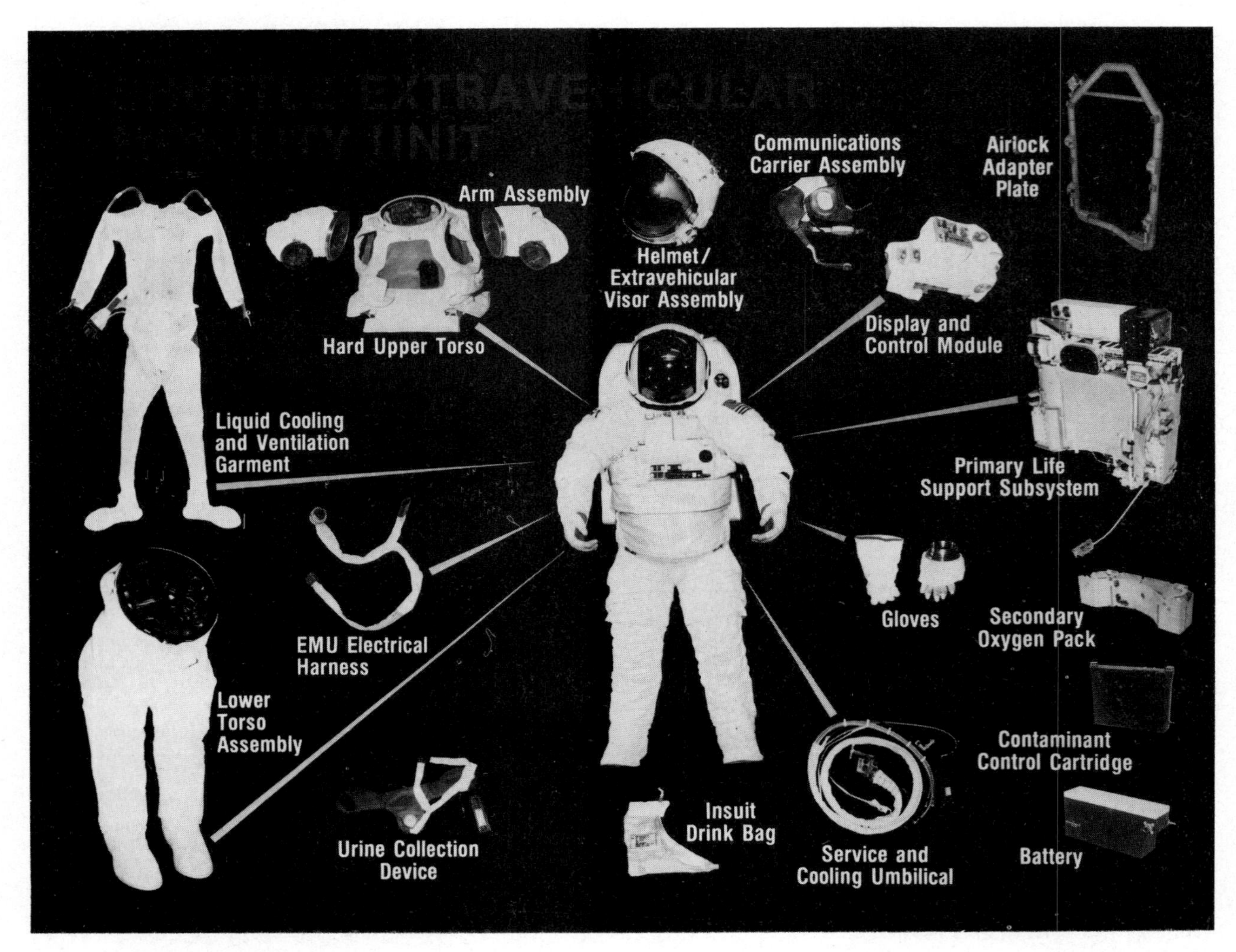

Shuttle Extravehicular Mobility Unit.

in only one size; and the visor assembly, which protects against micrometeroids and the Sun's ultra-violet and infra-red radiation and fits over the helmet. The spacesuit and associated garments weigh a total of 30kg. There are no zippers on the spacesuit, the components connecting with hard snap-rings. The suit is composed of several bonded layers, beginning with a polyurethane-on-nylon pressure bladder, followed by many Kevlar layers with folded and tucked joints (for mobility), and ending with a Kevlar, Teflon and Dacron anti-abrasion layer. The hard upper torso has an aluminium shell. The materials are designed to prevent fungus or bacteria growth: unlike the individually made Apollo suits, they will be used by scores of different astronauts, and must be cleaned and dried after each use. Only 2 suits are carried on each mission.

The Portable Life-Support System (PLSS), attached to the back of the upper torso, provides enough oxygen and water for 7hr inside the suit, including 6hr for EVA and a 30min reserve. There is also a secondary pack to supply oxygen and maintain suit pressure for 30min if the primary system fails. Astronauts can plug themselves in to the Orbiter airlock for fresh supplies of oxygen and water. The PLSS adds another 72·6kg of wt. A display and control module on the front of the upper torso tells the crewman how much EVA time he has left, and warns of any excessive consumption or of anything wrong with the water pressure and temperature in the cooling garment.

The Manned Manoeuvring Unit (MMU), which enables an astronaut to become a human spacecraft, snaps on to the back of the PLSS. Normally only one MMU will be carried. Two hand controllers operating 24 nitrogen thruster jets provide 6 degrees of freedom; the left-hand controls direction, the right-hand rotation. There is enough nitrogen for several Orbiter fly-arounds, 2 round-trip Orbiter-to-Orbiter rescue flights, or numerous payload servicing trips. Recharging the nitrogen is easily done by one man in the payload bay; the batteries are also easily replaced but take 16hr to recharge in the airlock. The 102kg MMU has attach points and power outlets for lights, cameras and tools, and thus serves as a portable space work station.

Bruce McCandles making the historic 1st untethered MMU flight during Mission 41-B, a rehearsal for the Solar Max recovery mission. Wearing the chest-mounted docking device, he successfully docked with the target pin seen at bottom in payload bay. *(NASA)*

Because only 2 spacesuits are carried, the suitless crew members will each be provided with a Personal Rescue System, an 86cm-dia inflatable sphere made of the same material as the Shuttle suit. Sitting tailor-fashion inside and clutching a portable oxygen system, the occupant may not be very comfortable, but he will have a communications system and viewing port, and enough pressurising gas for a 1hr rescue period. If an Orbiter becomes disabled, suited crew members can carry the others across to a rescue Orbiter; alternative methods include use of the RMS, or passing the balls along a line as in a ship-to-ship rescue.

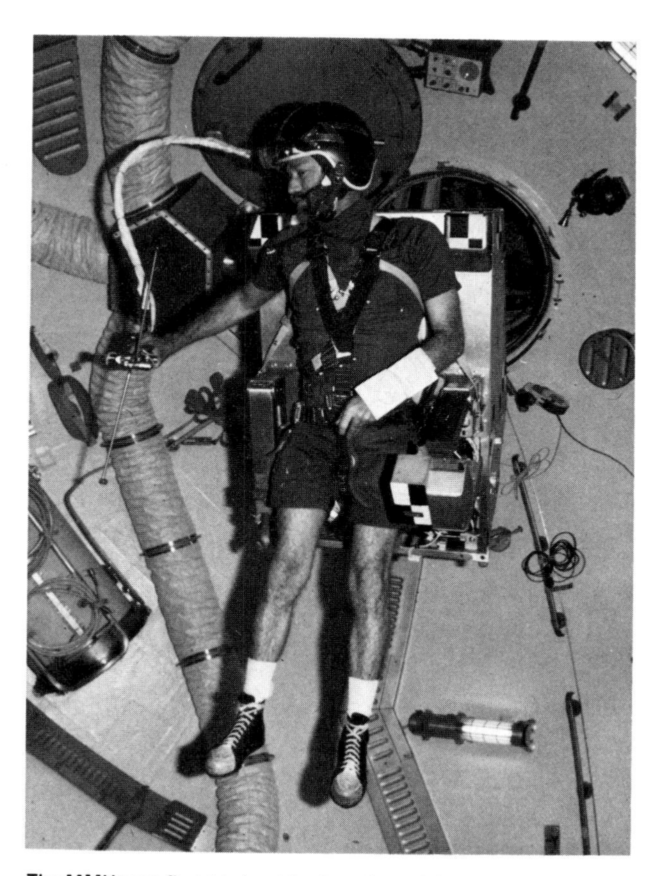

The MMU was first tried out in the safety of Skylab's spacious dome in 1974. Gerald Carr, seen using it here, is joint holder of the US 84-day longest flight record. *(NASA)*

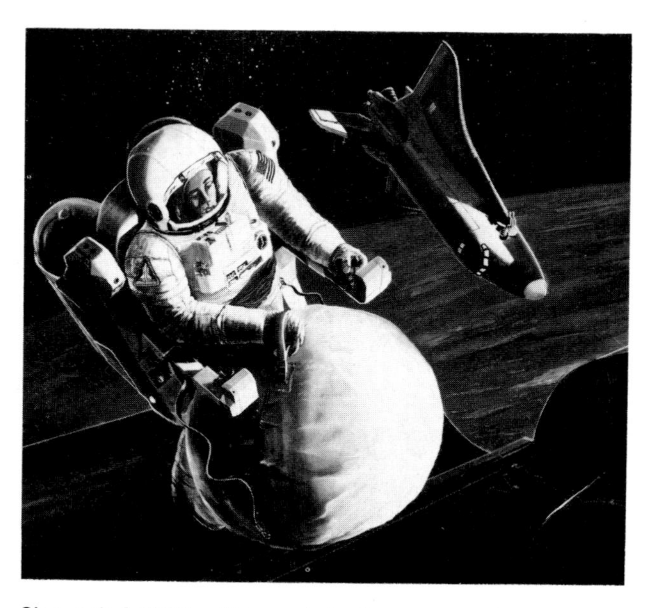

Since only 2 MMUs will be carried, other crew members will be provided with rescue balls inside which they can be transferred to a rescue Shuttle. *(Martin Marietta)*

Remote Manipulator System Called Canadarm by Spar Aerospace Ltd of Canada, which designed and built it, the RMS is 38cm in dia and 15·3m long and weighs 408kg. In space it can manoeuvre a load 4·6m dia by 18m long and weighing up to 30,000kg. Designed as an extension of the human arm, it has shoulder, elbow and wrist joints driven by DC electric motors, and can be controlled in 5 different ways, ranging from manual spacecraft-type hand controls for the crew, using direct observation and TV cameras on the elbow and wrist, to complete computer control. Precision control is essential to avoid risk of collision between the loaded arm and the Orbiter. Stowed on the left side of the payload bay and latched into 3 cradle pedestals, it can be restowed by EVA if it jams while in use, or in the last resort jettisoned by pyrotechnics. In highly successful tests on STS-3 Gordon Fullerton had little difficulty in locking the "end effector", or hand, on a 160kg scientific package measuring 62 × 107cm and swinging it out above *Columbia* with no apparent danger of collision with the doors or side. A larger, desk-sized package could not be lifted because the TV camera on the wrist had failed, but all the same 100% success was claimed for the tests. Maximum tip velocity of the unloaded arm is 2·0ft/sec; loaded it is 0·2ft/sec. Each RMS is designed for a 10yr, 100-mission life. The first, with its associated ground equipment, was provided by Canada at a cost of C$100 million; 3 additional systems were ordered by NASA at a cost of C$75 million for delivery in May 1982, Nov 1983 and Nov 1984.

Launch aborts There are 3 possible abort alternatives during the launch period. Return to Launch Site (RTLS) would be used in the event of a main engine failure in the first 2min. The remaining engines, plus RCS thrusters, would be used to achieve a pitch-around manoeuvre enabling the Orbiter to jettison the ET 45km from the coastline and glide back to the KSC runway. Abort Once Around (AOA) would be used from 2min after SRB separation, again in the event of a main engine failure. The procedure relies on 2 OMS firings after ET jettison to place the Orbiter in a sub-orbital coast and "free return" orbit for re-entry and glide back to the runway. Abort To Orbit (ATO) is available in the event of a main engine failure after passing the AOA point. Again the procedure relies upon 2 OMS firings, one to insert the Orbiter into orbit, the other to circularise the path. An alternative mission might then be possible. When asked about procedures for other possible failures, the favourite astronaut response is: "Start worrying." No mission would survive an SRB failure.

Flight procedures Mission Control at Houston can exercise full automatic command from launch to orbit, and from re-entry to landing. Team members there are assigned to each flight about 9 weeks before launch. In addition, 3 payload operations centres have been set up: at the Jet Propulsion Laboratory, Pasadena, for planetary missions; at Goddard Space Flight Centre for Earth-orbital free-flyers; and at Johnson Space Centre (where Mission Control is also based) for Spacelab and other payloads. The crew can however override the automatic controls and take over manually. They have full responsibility for control while in orbit, though Mission Control can initiate automatic sequences for all orbital activities except docking; this will be manually controlled by the commander and pilot. At lift-off all 5 engines – 2 SRBs and 3 Space Shuttle Main Engines (SSMEs) – burn simultaneously until the SRBs are jettisoned at about 2min 12sec. The 3 SSMEs continue burning for a total of about $8\frac{1}{2}$min before being shut down for the rest of the flight. They can be throttled over a range of 65-109% of their rated power level. Designed for $7\frac{1}{2}$hr of operation, they should have a lifespan of 55 flights. The twin OMS (Orbital Manoeuvring System) engines, sited in pods on each side of the fin and above the main engines, each give 26·69kN (6,000lb) thrust and can if needed assist in achieving the chosen orbital inclination. They are needed for major orbital changes, rendezvous and deorbit. Their fuel tanks are interconnected with the Reaction Control System (RCS), used for attitude manoeuvres (pitch, yaw and roll) and small velocity changes along the Orbiter axis (translation manoeuvres) when the Orbiter is above 21,336m (70,000ft). The forward RCS has 14 primary engines each providing 3,870N (870lb) thrust and 2 vernier engines each providing 106N (24lb) thrust; the aft RCS has 12 primary and 2 vernier engines in each pod.

"Canadarm," the 1st Remote Manipulator System, being assembled in Toronto. *(Spar Aerospace)*

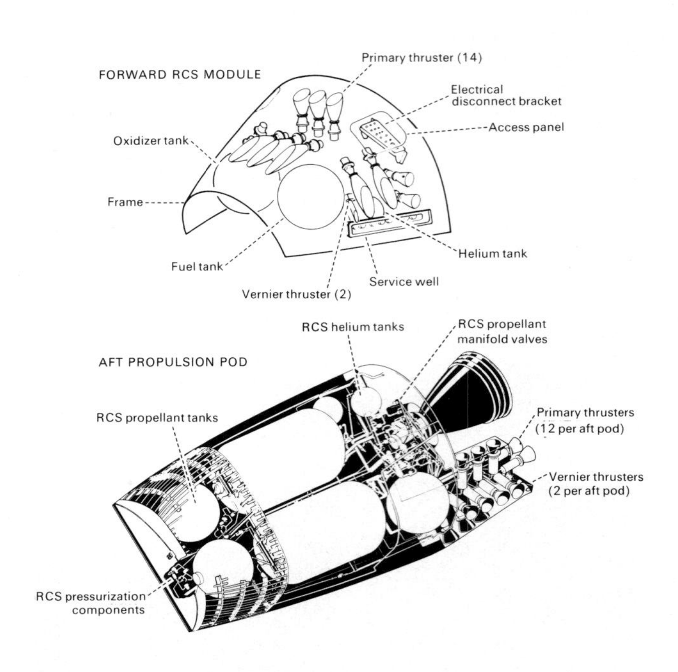

Orbiter Reaction Control Subsystem. ***(NASA)***

The primary engines, which provide the normal control of the Orbiter, are reusable for at least 100 missions and capable of carrying out 50,000 starts. Once on station in orbit, commander and pilot leave their forward-facing, side-by-side seats (with the commander, as required by aviation tradition, in the left-hand seat) and attach themselves to rearward-facing control stations. There the commander has a hand controller and instruments for manoeuvring, while the pilot has controls for opening the payload bay and operating the payload handling system. In addition to windows, the pilot will have a TV monitor giving him pictures from small TV cameras mounted on the joints of the Remote Manipulator System. The mission specialist seated behind the pilot during launch swivels his seat to a display panel and will be responsible for ensuring that the proper electrical, cooling and other needs are provided for whatever experiments or equipment are being carried. The payload specialists, if any, will activate the experiments and monitor data-collection.

By STS-8, with Kennedy, Edwards and Northrup all equipped for night landings, the need for support for possible emergency landings at several foreign airfields around the world had been much reduced. Night landings at Kennedy, to get Orbiters back to the launch site at times when thunderstorms and clouds were less likely, were also under consideration.

Payload delivery The Shuttle's primary mission is to deliver payloads into low Earth orbit up to its own maximum altitude of 1,110km, though the highest planned so far is 580km for Landsat retrieval in 1986. Shuttle's minimum orbital height is 217km. Since many satellites are intended for geosynchronous, elliptical and other orbits much higher than 1,110km they will have to be ejected with an upper stage attached. The development of such stages continues to be a subject of debate. From the Kennedy Space Centre at C Canaveral the assistance of the Earth's 1,665km/hr easterly rotation permits a maximum payload of 29,500kg to be carried on a due east launch. From the USAF's Vandenberg site, where Earth's rotation neither helps nor hinders the Shuttle, a maximum of 18,144kg can be carried into polar orbit. On a more westerly orbit from Vandenberg the possible payload drops to 14,500kg. For interplanetary and military payloads exceeding Shuttle's current capability, strap-on rockets and other ways of augmenting the thrust have been considered. Until the end of 1981, 3 different upper stages were being produced. Boeing was developing a 2-stage Inertial Upper Stage (IUS) for joint DoD and NASA use, but as expense mounted the NASA version was cancelled. McDonnell Douglas was developing for NASA a "wide-body" Centaur to launch planetary missions, but that too was cancelled on cost grounds. McDonnell Douglas was working a Spinning Solid Upper Stage (PAM/SSUS) in 2 sizes and capable of injecting payloads of 1,111kg and 1,996kg into geosynchronous orbit. In early 1982 a joint NASA/DoD committee was set up to decide on the more powerful upper stage likely to be needed about 1987 for sending heavier military payloads into synchronous orbit and for possible space platform development.

The procedure under which the Orbiter moves to a safe distance while the combined satellite/upper-stage systems are checked is described under Shuttle mission STS-5. In such cases, with upper-stage firing dependent upon a timing mechanism, it would not be safe for the Orbiter to attempt inspection or retrieval in the event of a malfunction.

Larger, liquid-fuelled Orbital Transfer Vehicle seen leaving Shuttle with comsat for synchronous orbit. After delivering the satellite it would return to the Shuttle for re-use. ***(Boeing)***

Indonesia's Palapa B satellite, spinning at 50rpm, lifts out of its protective cradle in the payload bay in perfect line with *Challenger*'s fin. In front can be seen Germany's SPAS-01 and the OSTA-2 experiment package, with the robot arm in the stowed position on the right. ***(NASA)***

Orbital Manoeuvring Vehicle reusable space tug. 1.2m thick and 4.5m in dia, OMV is proposed by Boeing as a means of delivering and retrieving satellites between Shuttle and altitudes of 3,700km. Seen here after leaving payload bay. ***(Boeing)***

Frisbee-type launch for the proposed Hughes 393 domestic communications satellite. With double the power and capacity of current comsats, and occupying only a quarter of the payload bay, it is deployed by a spring mechanism at 2rpm. ***(Hughes)***

Processing and turnround In Sep 1983 it was announced that Lockheed Space Operations, a consortium including Grumman, Morton Thiokol Chemical Corp and Pan American World Services, had won the contest for the NASA/USAF Space Shuttle processing contract, effective from 1 Oct 1983. A 3yr contract with an option on a further 3yr, it was estimated to be worth over $2 billion over 6yr. Renewal options covering 15yr could take this total to $6 billion. Lockheed's share as consortium leader is 75%; Grumman, responsible for launch processing, has 13%; Morton Thiokol's task of processing the ETs and SRBs (including their recovery after launch) is worth 11%; and Pan American, responsible for establishing "airline" turnround procedures, work control and maintenance planning, has 1%.

NASA's policy is to hand over responsibility for the whole STS processing operation at the Cape and Vandenberg, thus freeing itself for development work on the projected space station. It is also hoped that the single processing contract will improve safety and mission effectiveness while reducing both costs and turnround time. Turnround time was down to 69 days between STS-6 and 7. Processing had taken 649 days for STS-1, compared with 198, 114, 80 and 120 days for the next 4 missions.

Spacelab This is described in the International section Provided by ESA as the European contribution to the Space Shuttle system, Spacelab hardware is due to be used in a variety of configurations on 41 of the 91 Shuttle flights so far manifested; 5 of them will include the Manned Module (see US manned flights: Shuttle). Charges will range from $82 million for a Spacelab 1-type mission to $42 million for an Igloo/2-Pallet flight, $18 million for 1 Pallet, and $5 million for a single experiment rack in the Manned Module.

Double exposure shows Spacelab installed in *Columbia*'s payload bay ready for the STS-9 launch. Picture by Klaus Wilkens of Technicolour Govt Services Inc.

Life aboard the Shuttle

Environment Earthlike in composition and pressure (80% nitrogen/20% oxygen at 1,033g/sq cm, 14·7lb/sq ft), the air is cleaner than Earth's and, for hay fever sufferers, enviably pollen-free. Temperatures can be regulated over 16°-32°C, and the fact that launch and re-entry forces should not exceed about 3g means that conditions are well within tolerable limits for ordinary healthy persons.

Clothing Comfortable, relaxed clothing can be worn throughout the flight. From the 1st operational flight, STS-5, bulky pressure suits were no longer needed, reducing suiting-up time by 40min and total garment weight by 18kg. Fire-retardant cotton jumpsuits are now worn, with an unpressurised launch and re-entry helmet connected to a small portable air supply as a protection against smoke and toxic fumes in a fire or unexpected release of hazardous substances (as in ASTP re-entry). Personal items like these, tools, instruments and food packages are colour-coded on the Velcro patch used to secure the items. On STS-5 red and yellow were used for commander and pilot, green and blue for the MSs.

Astronaut Dr Judy Resnik revealed in a light-hearted lecture that no plans had been made to provide night clothing for Shuttle flights, since the men had always "slept in their knickers or less". But on the insistence of the women astronauts T-shirts will be made available for use as pyjama tops. NASA had also "swallowed the weight penalty" associated with allowing each woman one container of wrinkle cream and a tube of lipstick. One problem associated with the selection of women astronauts was with the suction-cup shoes used to anchor crew members to the deck. The women found that when using them they were not high enough to see out of the Orbiter's aft flight deck windows. Stools had to be made with suction cups on top. There is no prospect in the foreseeable future of a woman commanding a space flight, since all those selected so far are mission specialists and command goes to a pilot.

Food and sanitation Most foods can be eaten with ordinary spoons and forks in zero gravity so long as there are no sudden accelerations, decelerations or spinning; food, it was found, is held on the utensil by its own surface

Once in orbit astronauts can enjoy relaxed clothing. Bob Crippen, seen here experiencing weightlessness for the 1st time on STS-1, is now the most experienced Shuttle astronaut. *(NASA)*

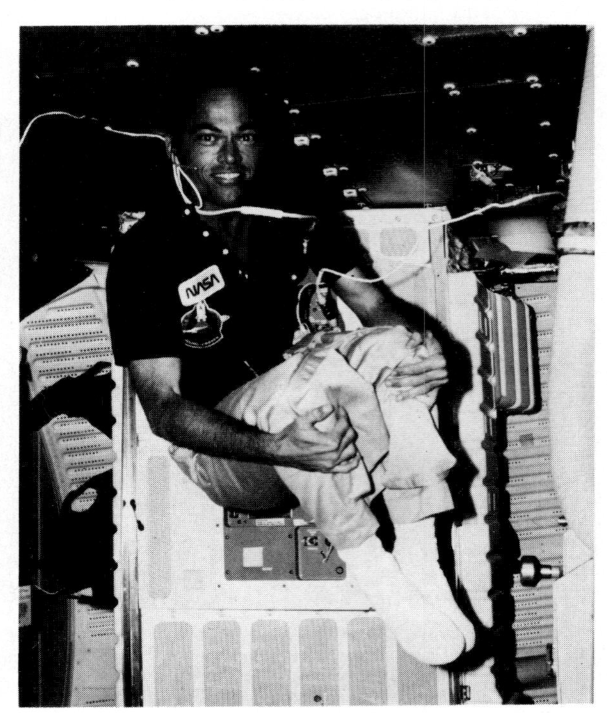

tension. Dining in space can be almost as pleasant a communal activity as on Earth, although no alcohol is permitted aboard. Shuttle menus, providing 3,000 calories daily, include 70 food items and 20 beverages. A different menu is possible every day for 6 days, with 3 1hr meal periods and "help yourself" snacks available from the pantry in between. Sufficient food for 7 crew members for 30 days can be carried. Most Shuttle foods are preserved by dehydration to save weight and storage space, and there are no freezers or refrigerators on board. The galley in the crew working area has hot and cold water dispensers and an oven. Preparations start about 30min before a meal. On a typical day breakfast might consist of an orange drink, peaches, scrambled eggs, sausage, cocoa and a sweet roll; lunch would be mushroom soup, ham and cheese sandwich, stewed tomatoes, banana and cookies; and dinner would comprise shrimp cocktail, beefsteak, broccoli au gratin, strawberries, pudding and cocoa. (Astronauts appear to be partial to cocoa, but tea and coffee are also available.) Housekeeping chores are shared among the crew - about 15min per head per day - and cleaning utensils and food trays with disinfected wet wipes is vital. The population of some microbes could increase rapidly in a confined, weightless spacecraft, possibly spreading illness to everyone on board, and crumbs could get into delicate equipment and the lungs of the astronauts. There are no washing machines, so trousers are changed weekly, socks, shirts and underwear every 2 days, and sealed in airtight plastic bags after being worn.

Sleeping Apollo-type sleeping bags, hung on the crew quarters' walls, were used on early test flights and are appropriate for short routine flights. For longer operational flights 4 rigid "sleep stations" - 3 horizontal and 1 vertical - are provided. Each has a sleep pallet, sleep restraint, personal stowage, a light, ventilation inlet and outlet, and overhead light shields. Each crew member has a 2·3kg personal hygiene kit containing shaving kit, brush and comb, etc; pockets, loops and Velcro (of which astronauts say there is never enough) are included to prevent the articles drifting away while in use. Each crew member is also given 7 cotton 30·5cm sq washcloths and 3 46·6 × 68·8cm cotton towels for a week's flight.

Waste management The Shuttle is the first spacecraft to have a conventional toilet able to accomodate both men and women. Located in a corner of the living accommodation, it has a curtain for privacy, with handwashing facilities immediately outside. Airflow substitutes for gravity, and the urinal can be used either sitting or standing. The commode has a "slinger motor" which should carry wastes to the bottom of the toilet for

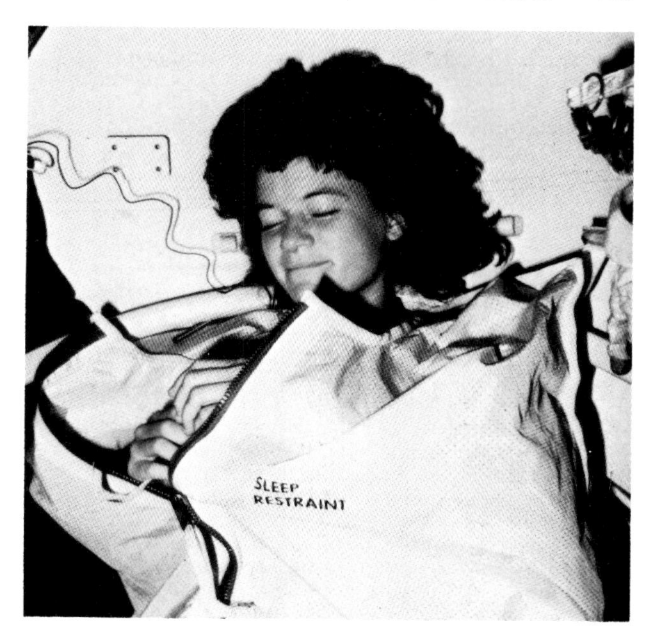

Sally Ride, the 1st US woman astronaut, chose to use the sleep restraint rather than a rigid bunk during her 6-day flight on STS-7. *(NASA)*

processing and vacuum-drying. But up to Mission 10 it had never remained fully operational. Astronauts usually use conventional shaving cream and a safety razor, cleaning off the face with a disposable towel. Dry electric shaving could result in whiskers floating about the cabin, where they might be inhaled or drawn in to clog equipment.

Shuttle missions

Test flights *Enterprise* (Orbiter 101) was rolled out from Rockwell's final assembly facility at Palmdale, Los Angeles, on 17 Sep 1977. A total of 13 Approach and Landing Tests (ALT) were carried out, with prime crew Fred Haise and Gordon Fullerton and second crew Joe Engle and Richard Truly making alternate flights. Starting on 18 Feb, the 9-month series began with 5

Deke Slayton, who managed the Orbiter Flight Test Programme, monitored the 1st piggyback flight to KSC by flying alongside in a T-38 chase aircraft. *(NASA)*

"captive" flights. *Enterprise*, unmanned and inactive, was carried on the back of a former American Airlines Boeing 747-100, modified by NASA to act as a Shuttle Carrier Aircraft for ferrying Orbiters to and from the various Shuttle facilities. Then, starting on 18 Jun, 3 more captive flights were flown, but with the astronaut crews taking turns to man the flight deck and check the performance of its systems. On 13 Aug 1977 the first of 5 free flights took place. *Enterprise*, weighing 68,040kg, was launched from the 747 at a height of about 7,600m after the carrier jet had gone into a shallow 7° dive; the flight lasted 5min 15sec following separation at 250-270kt equivalent airspeed (EAS). The 4th and 5th flights were made without the tailcone, carried earlier to reduce the effects on the 747's tail of turbulence around the rear of the Orbiter. (The cone is also routinely used on ferry flights.) The final ALT, lasting 1min 55sec on 26 Oct, touched down at 200 instead of 189kt EAS, resulting in a heavy bounce and skip. A repeat flight was considered, but it was decided that Haise had overcorrected. *Enterprise* was declared a "crisp and accurate" flying machine, the ALT programme was completed 3 months early, and at that stage the 7yr Shuttle development plan was still on schedule. At that time, despite financial problems and turbopump fires during tests of the main propulsion engines, it was still hoped that first flight would not slip beyond Jun 1979.

Enterprise was then used for 4 ferry flight tests, modified for vertical ground vibration tests at Marshall Space Flight Centre, and ferried to Kennedy and used for mating tests with the ET and SRBs. It then went back to Palmdale for ground tests, and will eventually be used at Vandenberg for practice and verification checks.

STS-1 After a 48hr postponement at T-9min on 10 Apr 1981 because the 5th computer was 40msec out of synchronisation with the other 4, *Columbia* made a majestic ascent from Pad 39A at 3·98sec past 0700 EST on Sunday 12 Apr. It was the 20th anniversary of Yuri Gagarin's first spaceflight. The Orbiter's 3 main engines fired at T-3·46sec, followed by the SRBs at T+2·64sec, when the hold-down bolts were triggered. Total cluster weight was 2,023,750kg; total lift-off thrust was 3,129,250kg. The commander, John Young, admittedly self-selected since he was Chief of the Astronaut Office, was making his 5th flight. Robert Crippen, 43, originally chosen as an MOL astronaut 14yr earlier, was at last making his first. Launch Control needed 150 personnel on this occasion, compared with 500 for the Apollo launches; later Shuttle missions will not need more than 75. Crippen was to describe later how, during the first few seconds, debris and ice from the ET were striking *Columbia*'s windows. That was the result of a pressure pulse of unexpected force as the 2 SRBs ingnited and reached their combined thrust of over 2,270,000kg in only three-hundredths of a second. Rebounding from the launchpad, the pulse shook the Orbiter with a force of several g, deflected the controls on the wings and bent several fuel tank support struts. It was also responsible for damage to tiles on the OMS pods, to be revealed 1hr 43min later when the payload bay doors were opened. With SRB thrust higher than expected, capcom astronaut Brandenstein called: "Columbia you are looking a little hot; all your calls will be a little early." It led to a 5° steeper climb, and a trajectory 8km above nominal. That could have made abort options difficult, but overall the effects were beneficial. As the cluster cleared the pad, twin gold flames from the SRBs grew to a dazzling 600m long. *Columbia*'s onboard computers put the cluster into a breathtaking 120° roll and then pitched it on to its back. That gave Young and Crippen a visual horizon and evoked the commander's favourite adjective: "Flying this thing is just outstanding." SRB separation came at 2min 11sec, 38·6km downrange at an altitude of 45·8km, and on NASA's long-range cameras spectators saw the rockets thrusting themselves clear of *Columbia* and the ET. Since this first flight was mainly a test of the system's reusability there was much interest in the first SRB recovery. In each case one of the 3 main parachutes failed and was lost, but Capt James Bond, commander of the recovery ships *Liberty* and *Freedom* stationed 225km off Florida, spotted the black smoke trail marking the boosters' descent. They splashed down 25km from the ships, which reached them within an hour. They had fallen nearly 5km apart, with their accompanying parachutes and frustums (the cones housing the main parachutes) in a line about 300m apart. A remotely controlled plug, looking rather like a large torpedo with fins and engines on the tail, was successfully inserted into one SRB with the help of control sticks and a TV screen, and the sea water pumped out so that the casing floated horizontally. A back-up retrieval boom had to be used for the second SRB, but that too was successfully recovered. 12hr after *Columbia* went into orbit the SRBs began their journey under tow at 8kt back to Port Canaveral.

With *Columbia* not yet quite in orbit, main engine cut-off came at T+8min 38sec, followed by ET separation at T+9·02 and an evasive manoeuvre by *Columbia* to avoid the possibility of a collision. From its altitude of 116km the ET continued on a sub-orbital trajectory resulting in impact in the Indian Ocean off the W coast of Australia. Orbit insertion was achieved at T+10·38 by an 86sec burn of the two 6,000lb constant-thrust OMS engines, resulting in an initial 244·4 × 109·2km orbit. A 2nd burn of 75sec at T+43min gave a 246km circular orbit, with the final orbit of 274km being achieved after 6hr with 29sec and 33sec burns. Before that, however, came the first anxious opening of the 18m-long, 4·5m-high payload bay doors to check whether the expansion and contraction caused by exposure to temperatures ranging from –170° to +350°F would affect their ability to close again. Various emergency procedures, including finally an EVA, had been worked out, since re-entry was not possible until the doors were closed. The crew seemed less concerned than Mission Control about the missing tiles revealed by the TV cameras when the doors were opened. Having closed and reopened the doors and deployed the space radiators, the crew were given a "go" for continued flight. They doffed their suits and had a meal break. With no galley available on the mid-deck – its installation had been delayed by economies – Young and Crippen were dependent upon a "carry-on food warmer". RCS tests and fuel purges were then carried out in a general atmosphere of high good humour, and after 12hr in orbit the first rest period was taken. They took their two sleep periods in their flight deck couches instead of in the "sleep restraints" to be

STS-1 lift-off, 7am, 12 April 1981. Start of a new age in spaceflight, with astronauts John Young and Bob Crippen alone on the flight deck. ***(NASA)***

STS-1: the 1st Shuttle landing at Edwards, perfectly executed, was as much admired and commented upon worldwide as the 1st lift-off. *(NASA)*

provided for operational missions. The first was somewhat marred for the crew because cabin temperature fell to 37°F and they found it necessary to don extra socks and shirts to keep warm. Mission Control was later able to advise on procedures for pumping hot water into the flight deck temperature-control system. Two other minor problems were the failure of a tape recorder which resulted in a loss of Shuttle performance data, and trouble with what Young called "the commode" – the first proper spaceborne toilet and an eagerly awaited luxury for veteran spacemen like him. "It started out poor and became completely inoperable," said Young. Apparently the slinger fan, meant to direct human waste into the disposal system, worked in reverse. (Subsequently there were suggestions that the crew had not operated it correctly; their reaction to that was never published.) The second day in orbit was spent mostly rehearsing the de-orbit and re-entry procedures, and during the second sleep period *Columbia* was flown in a nosedown tail-first attitude. This provided opportunities for secret cameras at Malabar, near Melbourne, Fla, and in Hawaii to scan the vulnerable underside for missing tiles. Spaceborne sensors were probably also used when *Columbia* was upside down in relation to Earth. But all that was revealed about these operations was the fact that DoD had been co-operating with NASA, and that the missing tiles were not a problem. A master alarm, warning that one of the auxiliary power units needed for the landing was malfunctioning, disturbed the crew's rest but was quickly put right. During orbit 36, 53hr into the mission and with *Columbia* flying tail-first, a 2·5min burn of the OMS engines started the re-entry procedures over the Indian Ocean. The crew then manually manoeuvred *Columbia* to a nose-first attitude by means of the RCS jets. Entry interface (the first significant traces of the atmosphere) was at 400,000ft (122,000m) at a speed of Mach 25, which had to be lost in the following 30min. Angle of attack and bank angle were used to control the trajectory and to balance the loads on the thermal protection system. *Columbia* emerged from the 16min communications blackout to cross the US coast at 45,600m and Mach 6·7, about 8,200km/hr, and Crippen

STS-1: 1st door opening revealed missing tiles on OMS pod, but it was soon decided they posed no landing hazard. *(NASA)*

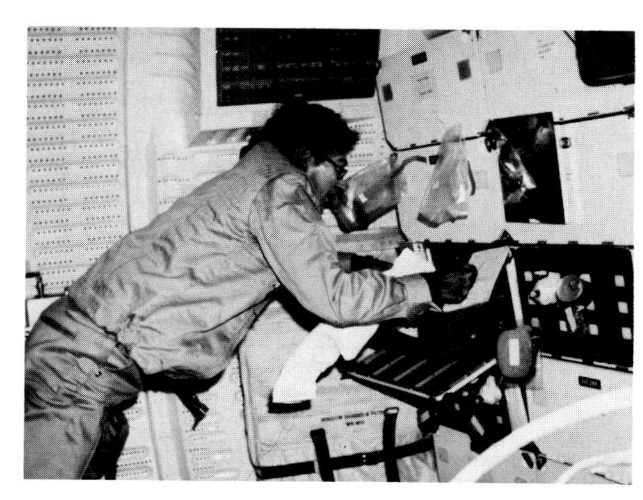

STS-1: John Young combines shaving with breakfast and a run-through of overnight teleprinter messages. *(NASA)*

STS-1: Young and Crippen receive congratulations as they emerge 63min after touchdown. ***(NASA)***

STS-2: remotely operated cameras capture the drama of *Columbia*'s 2nd launch. ***(NASA)***

exulted: "What a way to come to California." Airborne TV cameras picked up the approaching spacecraft for the final minutes as it continued to lose speed and in a long turn acquired the final heading at 11,000m and Mach 0·8. At the end of a 54hr 21min 57sec flight covering 1,733,450km *Columbia* touched down smoothly 854m beyond the planned point at a speed of 333-341km/hr. When the crew emerged 63min later John Young was able to claim that 135 of the 170 flight objectives had been accomplished and that only 0·7kg of the 29·4kg of manuals carried had proved necessary. Pointing out that *Columbia* had weighed 89,014kg on landing, he said that an extra 4·5 tonnes could have been returned from space. Additional payload could also be made available by removing, for instance, many of the "unnecessary" tiles on the top of the Orbiter. The fast, long landing was a bonus, too, showing that drag was less than expected. But he took a conservative view of suggestions that the success had been so complete that the flight-test programme could be speeded up, notably by landing back at the Cape before STS-5, and asserted that the pre-planned programme should be strictly adhered to.

STS-2 With Joe Engle, 49, as commander and Richard Truly as pilot (having celebrated his 44th birthday at the pre-launch breakfast), *Columbia* finally became the first spacecraft to make a second flight, on 12 Nov 1981. But there had been 4 delays, building up to 7 months instead of the planned 5 between the flights. Flight director Neil Hutchinson pleaded for patience at a critical press conference, pointing out: "This is the first time we have done this a second time." The most serious delay, costing one month, resulted from a spill of nitrogen tetroxide on the Orbiter's nose during fuel loading. The highly toxic fuel soaked through the tiles, destroying the adhesive and soaking into the thermal blankets. A total of 370 more tiles either fell off or had to be replaced; 26 thermal blankets beneath the tiles also had to be replaced; and there were fears, later proved unfounded, that the forward RCS had been damaged. A first attempt at a 73hr countdown (spread over 102·5hr) reached T-31sec on 4 Nov - the point at which ground-based computers were due to hand over to *Columbia*'s 4 computers for auto-sequence start - before being abandoned. Launch Control tried but failed to override the computers, which stopped the countdown because of indications of contaminated oil in the auxiliary power units (APUs), which, among other things, start the rocket engines. It was believed that this contamination, probably due to overheating during a series of holds, would have cleared itself if the launch had continued. Engle and Truly left *Columbia* after being inside for 5hr to face another 8-day wait - at a cost to NASA of nearly $2 million. Countdown was picked up again at T-35hr on 10 Nov but minor technical problems persisted, resulting in the launch being 2hr late. The flight was finally cut to a "minimum mission" of 54hr 13min 10sec - 7min shorter than STS-1 - compared with the planned 124hr (originally extended from 101hr "to ease the crew activity load"). NASA's claim afterwards that this shortened mission had been "90-95% successful" was thus difficult to accept.

For this second launch VIPs had been moved back, outside a 6·4km radius, because environmental checks had found that on STS-1 hydrochloric acid and aluminium oxide dust from the SRBs had fallen on their site; this fallout could have caused red spots and a burning rash. Newsmen however stayed put on their adjacent site, 5·6km from the launchpad. Few took official advice to buy $20 covers for their (mostly hired) cars, and in the event the wind carried the "acid rain" safely south and out to sea.

Between countdowns the toilet was once again checked to ensure it was working (newsmen having been told that its failure during the first flight was due to faulty operation by the crew), and the flight deck windows repolished. Total lift-off weight was 2,029,528kg, 5,737·5kg heavier than STS-1, mainly due to the OSTA experimental package and the RMS.

We watched *Columbia* roll 118° 7sec after lift-off, taking it from a 120° heading on the pad to 58° over the Atlantic and a 38° orbital inclination. At +23sec computer-controlled elevon and rudder movements provided data on minor structural vibrations during powered flight. Following SRB separation at 2min 10sec all 6 parachutes opened. The chutes and 2 frustums were quickly recovered 270km E of C Canaveral. But 6m swells and high winds delayed recovery of the SRBs, which had to be left bobbing in corrosive seawater for 4 days (they were due to be reused on STS-6).

During lift-off the crew again observed debris from the ET striking the Orbiter, though there was apparently less of it than on STS-1. (Later examination revealed 334 small nicks on tiles, which proved to be repairable without removal.) At main engine cut-off (MECO), at 8min 42sec and about 111km altitude, the speed was 25,668ft/sec, only 1ft/sec below nominal, but troubles were already developing. First there was a fault once more in one of the 3 APUs, and then one of the 3 fuel cells began overloading with water. The crew were so busy troubleshooting that it was 4½hr before they had time to remove their flight suits. At T+5hr it was decided that fuel cell No 1 must be drained and shut down for good, and in the meantime OMS burns 3 and 4, due to raise the orbit to 257km, were postponed for one orbit. Truly fell 25min behind with opening the payload bay doors, and the first colour TV pass was lost because of difficulties with one of the cameras. But finally the doors were open, and reassuring views showed no missing tiles on the OMS pods. With a one-third power loss, there was some thought of not activating the OSTA payload at all. At the same time, astronauts on the ground started work in the simulators, flying re-entries with degraded fuel cells to develop possible emergency procedures. But it was finally decided to start the experiments, and to the scientists' relief OSTA-1's power use had no significant effect on other Orbiter operations. During the first night's sleep, however, minimum lighting and power were used to ensure there was enough to run OSTA.

On the second day, while Truly was busy starting to exercise the RMS, Mission Control first announced that the "minimum mission" rule would be applied, then that they hoped to continue with the full mission from day to day, and finally that it would definitely be a minimum mission. For the RMS activities, astronaut Dr Sally Ride, 30, became the first woman capcom. An astrophysicist, she had worked extensively on development of the RMS and was soon engaged in highly technical exchanges as the crew unlatched the arm. On TV the arm could be seen slowly extending, and it was soon stretched like a clock hand across the Earth's disc, which was visibly revolving above the inverted *Columbia* as the Orbiter's sensors and atmospheric probes did their work. As Truly reported success, Ride commented: "You guys do good work". Truly, acknowledging the first woman capcom, replied: "You sound mighty good too." 4hr of RMS work was achieved, compared with the 15hr scheduled in the flight-plan. While recradling the arm Truly had trouble with the back-up drive system, though the prime system worked well. Fears that the arm might get stuck and have to be stowed manually by an EVA, or even "amputated" and jettisoned, proved groundless. The worst that happened was that the 2 TV cameras attached to the wrist and elbow joints failed towards the end of the test.

At T+31hr Engle successfully replaced the failed No 1 cathode ray tube in front of his seat on the flight deck with a unit from the aft crew station; this 1½hr job had not been practised during training.

The fact that the OSTA scientists had planned and simulated a 54hr mission made the transition from a full mission relatively simple, and the major data loss resulted from the inability to observe ground targets at specific times. But the synthetic-aperture radar used all 8hr, 3,400ft, of film, of which 3,200ft contained "viable" pictures obtained over every continent lying under the mission's ground track. 108min of ocean colour data was obtained, compared with the 120min planned for the full mission. Primary targets were located on the US E and W coasts, in the Sea of Japan, and in an area from S Europe to the W African coast.

9 months later JPL announced "astounding revelations" following study of the imaging radar pictures. The radar had penetrated the dry sand of the Sahara Desert to reveal ancient rivers and landscapes beneath. The information should help archaeologists in their search for evidence of prehistoric inhabitants, as well as providing the expected assistance in locating oil and water. Another success announced a year later was confirmation of the Multispectral Infra-red Radiometer's finding of an ore-rich area in a remote Mexican desert.

Re-entry preparations, complicated by reduced power reserves and doubts about landing site weather, resulted in a heavy workload for both crew and Mission Control. The earlier return involved dumping 635kg of propellant by simultaneously firing 2 RCS yaw jets on each side, thus moving the centre of gravity aft without perturbing the attitude. Re-entry began with a 2min 55sec retrograde burn of the OMS engines, which slowed *Columbia* by 313·4ft/sec and consumed 590kg of propellant. The need to obtain experience (pressed for by DoD) of crosswind landings led to several changed decisions. Before blackout Engle was instructed to land manually on Runway 15 in the 10-20kt crosswind. But following John Young's report that gusts were reaching 21kt on his practice landing approach, a surprised crew were instructed when they emerged 16min later from blackout to make a last-minute switch to Runway 23 and to do an autoland; this was another test which needed to be done.

STS-2: 1st deployment of RMS by astronauts Engle and Truly. End effector, resembling human hand, can be seen near rear of payload bay; TV camera is on elbow. ***(NASA)***

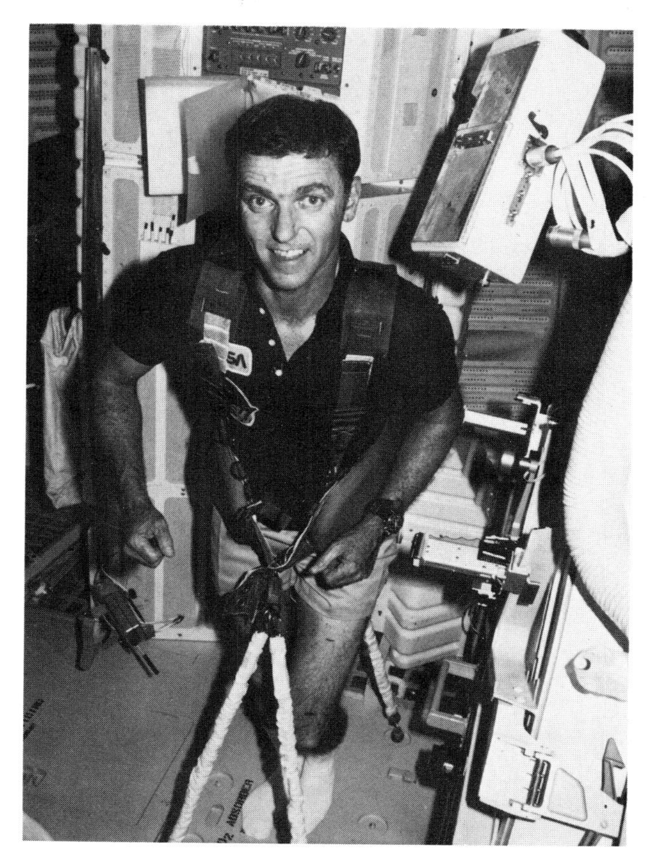

STS-2: commander Joe Engle, who qualified as an astronaut on the X-15, enjoys an exercise session on the treadmill. Richard Truly took the photo. ***(NASA)***

STS-2: Truly, on the mid-deck, is apparently not too alarmed by the length of the teleprinter instructions updating the 2-day 6hr flight plan. ***(NASA)***

During and after blackout, from which *Columbia* emerged at Mach 10·5 and 165,000ft, 19 groups of manoeuvres were manually commanded and automatically flown, including a speedbrake sweep to determine why *Columbia* landed long on STS-1. Some were only 20sec apart, needing close co-ordination between Engle and Truly. There was some DoD concern when some of these manoeuvres - which may prove necessary when Orbiters return to Vandenberg - confirmed the STS-1 finding that they caused greater temperatures on the OMS pods than predicted: 700° instead of 500-600°F. Because of the manoeuvring *Columbia* was 40km S of its nominal track at 200,000ft, and NASA's C-141 flying laboratory - itself 16km off track - was again unable to obtain infra-red pictures of the re-entry from its height of 41,000ft. However, astronaut Kathryn Sullivan, using a hand-held camera in a T-38 chase plane, got good views of *Columbia*'s underside in order to check damage before and after landing. Touchdown was at 185kt, about 1,000ft short of the aim point as a result of the 20kt headwind, but once again right along the centreline. Perhaps the biggest advance achieved on this second flight was the fact that no tiles were lost and only about 12 damaged, including some which had mysteriously lost their upper surface. Landing weight was 92,534kg. Out of 258 priority flight-test objectives, 233 had been accomplished. 9 days later - one day late - *Columbia* arrived back at KSC and was redesignated STS-3.

STS-2: astronaut Kathryn Sullivan took this view of *Columbia*'s belly from a T-38 chase aircraft just before the 2nd landing. The object was to check whether any damage occurred during re-entry or after touchdown. *(NASA)*

STS-2 landing: note the use of airbrakes on the fin to avoid overloading the wheel brakes. *(NASA)*

STS-2: Engle and Truly, wearing the ejection/escape suits used for ascent and entry on the 1st 4 flights, leave *Columbia*. Their mission had been halved by fuel cell problems. *(NASA)*

STS-3 With launch achieved for the first time on the planned day, 22 Mar 1982, and only 1hr late, this mission was prolonged by a day beyond the planned week's duration because of bad weather at the landing sites. Commander Jack Lousma, 46, making his 2nd flight, and MOL astronaut Gordon Fullerton, 45, presided over a smooth countdown towards a 10.00 EST lift-off. A failed heater on a nitrogen gas purge line caused the slight delay. A pre-launch drama was provided by the flooding by rainstorms of the Edwards Air Force Base landing site. Trainloads of equipment were hurriedly despatched 1,200km south to the gypsum-sand landing strips at Northrup, White Sands, New Mexico, where preparations for a possible abort landing were completed in only $2\frac{1}{2}$ days. *Columbia*, with its 10,000kg payload, followed a much steeper, more efficient ascent trajectory which brought higher levels of noise and vibration to the press stands than on previous launches, accompanied by thoughts of the possibility of a return-to-launch-site abort 26min later. *Columbia* disappeared into broken white clouds before its roll was completed, depriving spectators of a sight of the new look created by the unpainted ET, now brown instead of white and lighter by 270kg. Target lift-off weight was 2,031,578kg, 2,049kg heavier than STS-2. Warnings that APU-3 was overheating were reported by the crew and confirmed by Mission Control, which instructed that it should be shut down at 7min 30sec. This led to SSME No 3 being shut down 20-30sec before MECO but did not affect total performance; the remaining 2 APUS and engines provided adequate power. The SRBs, jettisoned at 2min 6sec, were recovered quickly with the loss of only 1 of the 6 parachutes and towed triumphantly into Port Canaveral only 36hr after launch. MECO came at the predicted 8min 34sec, velocity of 15,463mph and

Astronauts Gordon Fullerton (left) and Jack Lousma, STS-3 pilot and commander, take the elevator to *Columbia* for the simulated countdown which completes each mission's "dry countdown demonstration test". *(NASA)*

STS-3 lift-off: for this and subsequent flights the ET was brown instead of white, the decision to leave it unpainted saving 270kg in weight. By STS-6 the ET was 5 tonnes lighter and the SRBs each 2 tonnes lighter. *(NASA)*

105·5km altitude. 2 OMS burns, 10min and 42min into the mission, completed the ascent phase, resulting in a 242 × 243km orbit only 0·55km from that planned.

Just as he did on Skylab 3, Lousma suffered from nausea and vomiting soon after becoming weightless, and Fullerton was also affected for the first 3 days. Following wake-up on the second day Lousma noticed tile damage on the nose, and inspection with the RMS revealed about 25 white tiles missing or damaged on both sides of the nose, and about 12 black tiles missing or damaged on the base heatshield and body flap upper surface. During launch Lousma had reported "a real blizzard" as debris - probably a mixture of ice from the ET and dislodged tiles - rattled on his windscreen; subsequently pieces of tile were found around the launchpad and on the nearby beach. (The occurrence led to a decision to remove, densify and replace about 1,000 tiles. Of these, 200 were damaged during launch and landings, and 38 were "debonded" from the forward fuselage and upper body flap during launch.)

The most serious equipment problem involved failures in the S-band receiver-transmitter communications system, leading to consideration of the need to shorten the flight. As usual, however, ways of living with the problem were found.

Lousma, who slept on the flight deck, with Fullerton below in the crew quarters, complained that his sleep was interrupted over China and Iran by mysterious noises in his headphones, possibly caused by radar tracking of *Columbia* by the Soviet Union. They also had trouble with the toilet, and discomfort because the thermostat kept the atmosphere either too warm or too cold.

Most important result of the mission was the fact that *Columbia* could survive extremes of heat and cold more severe than would be encountered in a normal flight. The OMS engines were fired without difficulty after 80hr in the shade created by placing *Columbia* in a nose-to-Sun attitude. The payload bay doors refused to latch after prolonged exposure to intense cold away from the Sun, but closed normally after the Orbiter was rolled around so that the top could heat up. Another major achievement was the first removal and return of a payload by the RMS, carried out without any difficulties. Because of crew sickness, the busy 3rd day was switched with the lighter 4th day. Then Fullerton grappled and lifted the 160kg Plasma Diagnostics Package, swung it out into space, and subsequently had no difficulty in dropping it back into the Y-shaped guides to the retention mechanism. Because of the failure of the wrist-mounted TV camera, a more ambitious sequence of manoeuvres with the 363kg Induced Environment Contamination Monitor (IECM) had to be postponed to the next mission. However, results of experiments made known some weeks later showed cause for major concern about the way the Shuttle tends to contaminate its own environment. The IECM sensors showed unexpectedly large amounts of debris - vented propellant, vehicle outgassing, water dumps and dust particles from the payload bay and other sources - surrounding *Columbia* in its passage through space. First reaction was that the USAF infra-red sensors would probably be ineffective on STS-4, and that ultimately sensors intended to track Soviet spacecraft would tend to lock on to the debris. There was concern too about its effect on future astronomy payloads, like that planned to observe Halley's Comet. The Orbiter also actually glowed in the dark, probably because oxygen atoms combined to form luminescent molecular oxygen. The Plasma Diagnostics Package, monitoring the firing of *Columbia*'s electron beam generator at 4 other satellites (DE-1, ISEE-1 and Isis 1 and 2) to determine whether the beam's radio waves could be acquired, found that very strong, low-frequency waves at 100Hz to 100kHz were produced. The RMS proved to be unaffected by temperature extremes. Fullerton, unaware of the findings to come much later, commented: "Any surprises have all been pleasant."

The crew reused the electrophoresis equipment first carried on ASTP for tests involving separation of human blood and kidney cells in zero g by the use of small negative electric charges. The results were then frozen and returned to Earth for study, but were lost in the weekend after the landing because a freezer at the Johnson Space Centre either failed or was left turned off. Lousma, who had been involved in the famous spider experiments during Skylab, conducted a student investigation of insect flight in zero g. He reported that moths soon discovered they did not need wings to fly when weightless, while bees continued to tumble confusedly around.

Only 39min before the scheduled re-entry firing on 29 Mar the crew were told: "You are waved off for 24hr." This was on John Young's recommendation, as winds rose to

STS-3: a heavily clouded Earth is the background in this TV view of the RMS carrying the Plasma Diagnostics Package, used to study the extent to which *Columbia's* engines and discharges pollute its own immediate environment. *(NASA)*

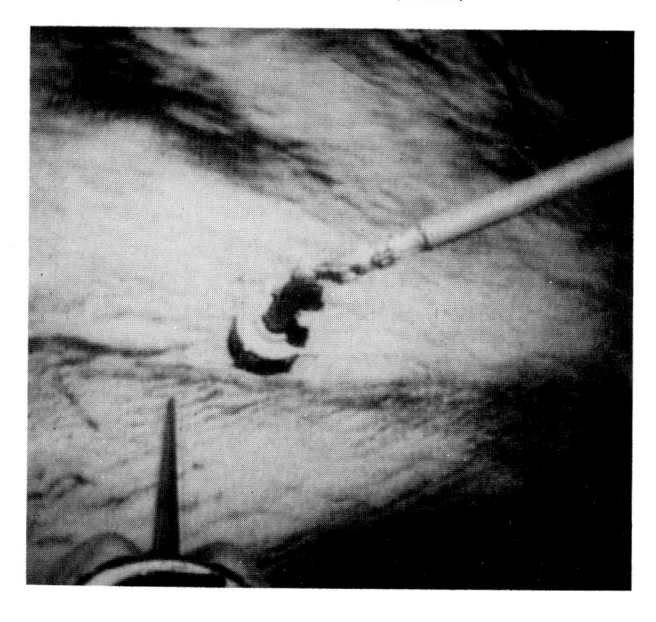

STS-3 landing: bad weather at Edwards led to a landing "wave-off" only 39min before re-entry. *Columbia* touched down instead – to the dismay of the engineers – on the gypsum Northrup Strip at White Sands Missile Range. *(NASA)*

98km/hr and dust drifted on the runway. Hurried preparations were made for a landing on the concrete runway at Kennedy next day if the bad weather continued. This proposal had the backing of engineers worried about the effects of a Northrup landing on vulnerable systems and structures. However, conditions at Northrup improved and *Columbia* finally landed there on the 129th orbit after a flight lasting 8 days 0hr 5min 3sec. Re-entry was initiated by a 2min 29sec OMS firing, which produced a 296km/hr reduction in velocity. Once again the planned crosswind landing was deleted because gusts were reaching 48km/hr, but the crew flew another sequence of test manoeuvres during a modified re-entry profile involving 444km of cross-range tracking, and all 3 APUs worked perfectly. Spectators were concerned that the undercarriage came down less than 5sec before touchdown, and then at the way *Columbia*'s nose pitched up sharply during deceleration. Lousma, who had taken over manual control at 36m and touched down 460m short of the optimum point, explained later that the nose pitch-up (which some observers feared might lead to *Columbia* pitching right over on its back) occurred because he had thought, wrongly, that the nose was dropping too fast, and had pulled back too hard on the hand controller. "Not my best landing," he said. "I wish I could have done better." *Columbia* rolled to a halt without braking 4,500m later, with Lousma testing nosewheel steering en route by doing an S-turn along the centreline.

The reserves remaining at the end of the mission would have permitted at least another 2 days in orbit if this had been necessary. And for the first time NASA's C-141 observatory aircraft had succeeded in obtaining an infra-red image of *Columbia*'s belly during re-entry. The picture, taken when the Orbiter was at 56,500m and a speed of Mach 15·6, showed hot spots in the centre of the wing, on the body flap and on the nosewheel doors, but the wing leading edges were cooler than expected. This time it took only 1 week to get *Columbia* back to the Cape so that work could start on preparing it for STS-4. Among the repairs needed was replacement of 10 manoeuvring thrusters, 3 in the forward RCS and 7 in the aft pods, because of contamination by gypsum dust during landing. The first were taken from *Challenger*, the latter from *Discovery*,

STS-4 The 1st NASA mission to carry a classified Defence Department payload, and the 1st to achieve an on-time lift-off, at precisely 11.00 EDT on 27 Jun 1982. Commander Thomas "Ken" Mattingly was making his 2nd flight, and Henry Hartsfield, a MOL astronaut, was making his 1st. The countdown had been extended to 90hr plus 24hr of hold time because of the need to open and close the payload bay doors in the vertical position for servicing of the DoD's 3,630kg Cryogenic Infra-red Radiance Telescope (Cirris). Equipped with an ultra-violet sensor and space sextant, Cirris is designed to identify Soviet missiles and aircraft against Earth's background.

STS-3: Fullerton and Lousma are congratulated by Flight Operations Director George Abbey (centre). Lousma, who said "It was not my best landing," left NASA soon after. *(NASA)*

The pre-launch period proved to be a testing time, with 26·67cm of rain falling on the pad in a month, culminating in 6·4cm of rain and ¼in hailstones the day before. About 200 out of 400 indentations and nicks in the tiles were filled with ammonia-stabilised slurry before launch. Total lift-off weight was 2,034,208kg, about 2,268kg more than STS-3, and it was probably the saturated state of the cluster which led to a lower than expected ascent performance. *Columbia*'s trajectory was 2,430m lower than planned, and after separation the SRBs hit the sea at 550km/hr instead of 100km/hr and plunged 600 fathoms to the sea bottom, representing a loss of $50 million. The 2 drogue and 6 main parachutes – 4 of which, in the 1st example of recycling, were being reused from STS-1 – were also lost, due to the premature firing of explosive bolts. Although the main engines burned for 2-3sec longer, using 907kg-worth of the ET's 5,440kg payload margin, the ability to make an emergency landing at Dakar, Senegal (which had replaced Rota, Spain, on this occasion) if 1 OMS engine failed was not reached until T+3.10 instead of 2.40. An orbit of 241km after 2 OMS burns was finally selected in preference to the planned 305km. On the 5th day, however, an OMS burn followed by RCS burns finally resulted in a record Shuttle altitude of 324km, placing the Orbiter in the most suitable position for cross-range manoeuvring during re-entry.

Thermal tests, which called for 66hr tail-to-Sun, 33hr underside-to-Sun and 5hr payload bay-to-Sun, had to be changed because of the hail damage. Even so, on the 4th day, after a long period of underside-to-Sun, the port door overlapped the aft bulkhead and would not close. On the 5th day, however, after temperatures had been evened out by 10hr in the "barbecue mode," door closing and re-opening were successfully accomplished.

So that not even Houston's Mission Control should see details of Cirris and its operation, the crew had been instructed not to send any TV of the payload bay area unless satisfied that no views of it would appear. Cirris activity was to be videotaped on board with the minimum of conversation between the crew and the Paycom (payload communicator) at the USAF Satellite Control Facility at Sunnyvale, Calif. However, there was trouble almost at once when the telescope's protective lens cover refused to open. Since only designation numbers and letters ("Payload C") could be used, there was some confusion between the crew and Sunnyvale about switch positions. More detailed information was passed by coded telex messages, and it was not until after the mission that we learned that the crew had tried to free the cover by tapping Cirris with the RMS. When that failed Houston considered authorising Mattingly to do the 1st Shuttle EVA by crawling into the payload bay and attempting to pull the cover free. But this would have taken up a whole day of what was still a test flight, involving many changes of altitude to boil off the water trapped in the Shuttle tiles. Presumably useful data were obtained from other instruments on the DoD payload, however, since Paycom, congratulating Mattingly before re-entry, added: "We think this mission has been a great success in opening a new era, and we're looking forward to many missions in the future." Mattingly replied: "You guys do good work, and we know you've had some disappointments, but hopefully some success too." (The USAF later decided to refly Cirris in 1984.)

Non-classified activities included much use of the RMS, which this time picked up the Induced Environment Contamination Monitor and manoeuvred it for several hours to sample contamination around *Columbia*, particularly the effects of the RCS jets. Perhaps the mission's major failure in the non-military field was that much of this information was lost because the RCS jet plume monitor was not correctly configured (though this was not the fault of the crew). Knowledge of the plumes' effects was needed, among other things, to help plan Shuttle activities in the vicinity of other spacecraft.

On their 1st full day in orbit the crew devoted 6hr 50min to the first experiment by a commercial enterprise. This was the Continuous Flow Electrophoresis System, designed by McDonnell Douglas to process organic material in a continuous stream for Johnson & Johnson's pharmaceutical division. They completed the work, involving human liver and kidney cells, 40min early. This activity had been given priority because the experiment

STS-4: Apollo 16 veteran Tom Mattingly commanded the final test flight.

The electrophoresis system (see Notes), first tried out on STS-4, produces purer and more effective medicines. Charles Walker, seen with the unit, was later selected to operate it continuously on Mission 41-D. ***(McDonnell Douglas)***

was directly applicable to the production of important medicines that are more difficult to make under Earth conditions. On the 3rd day egg and rat albumens were processed for comparison with parallel samples produced on Earth. Samples of blood and urine were taken from both crew members before and during flight in connection with student experiments concerning cholesterol distribution and the possibility of reduced chromium levels in microgravity. The first Getaway Special, carrying 9 Utah University student experiments which had cost a total of $10,000, at first failed to operate. But on the 4th day the crew succeeded in starting it by using a jumper cable to bypass a broken wire, earning themselves a congratulatory message from the students: "One small switch for NASA; a giant turn-on for us."

Amid all these activities, Mattingly, who has specialised in spacesuit development, found time to don the Extravehicular Mobility Unit spacesuit in the airlock, and proved that this was easy to do; this was a preparation for the planned EVA activity on STS-5. On their 7th day *Columbia* passed within 12·8km of a Soviet rocket stage from a 1975 launch. Norad was monitoring the rocket, so there was no collision danger. The crew did not see the object; the flight director said the encounter was so brief that they would have had to be watching exactly the right place at the right time, "and not blink". Only 16 Orbiter problems were logged during the mission, compared with 61, 50 and 47 on the first 3 flights.

Re-entry began 7 days 10hr into the mission with a 338·7km/hr retrograde OMS burn of 2min - longer than earlier missions because of the greater altitude. A steeper angle of attack was selected to provide about 50°F higher temperatures; their effects at the thermal tile/aluminium skin bond line were subsequently assessed. 9 flight-test manoeuvres were executed between the 4 hypersonic S-turns that are a part of all Shuttle re-entry profiles. At Mach 13 *Columbia* unexpectedly developed a 550-650km/hr descent rate compared with the predicted 330km/hr, resulting in a 32km downrange error; this was corrected automatically by the navigation system. The landing, the first on a concrete runway, was smooth but slightly short. Landing gear was extended 20sec before touchdown, Mattingly having resumed manual control at 762m. He held *Columbia* just above the approach overrun, bleeding airspeed down to 360km/hr until the right mainwheel touched down 7·6m to the left of the centreline, 290m from the runway threshhold. "Moderate braking"

STS-4: President Reagan and his wife Nancy greet Mattingly and Henry Hartsfield at Edwards after their successful 7-day flight. *(NASA)*

was used to stop *Columbia* in about 1min. Yet again the important first crosswind landing was not attempted, although the required 18-28km/hr crosswind was available on the dry lakebed. And again the C-141 observatory aircraft failed to obtain infra-red imagery of *Columbia*'s peak heating, due to a mechanical failure of the telescope. But damage was minimal: only 1 tile was lost, from the base of an OMS pod, and only 37 out of 300 slightly damaged tiles needed replacement. It was Independence Day - America's 206th birthday - and for the 1st time the President was at Edwards to watch the landing and greet the crew. The landing had been delayed for 1 orbit so that the arrival time of 09.30 PDT would be more convenient for him. He then watched the departure for the Cape of *Challenger* on the 747 carrier, and addressed a crowd of 10,000 against the background of *Enterprise*. He did not fulfil hopes that he would announce funding for a space station, but he did say that the US "must look aggressively to the future by demonstrating the potential of the Shuttle and establishing a more permanent presence in space."

STS-5 With a maximum launch window of 40min, and a preferred window of only 6min for the deployment 8hr later of the first commercial satellite to be Shuttle-launched, *Columbia*'s 5th countdown and launch, on 11 Nov 1982, were so smooth that the planned 07.19 EST lift-off occurred only 0·0678sec late. With commander Vance Brand, 51, making his 2nd flight, and Robert Overmyer, 46 (5th of the 7 MOL pilots), as pilot, the many firsts of the inaugural Shuttle operational flight included the 1st 2 mission specialists, Drs William Lenoir, 43, and Joe Allen, 45. Average age of the 1st 4-strong crew was a mature 46. With the ejection seats made inoperable so that in the event of a catastrophe there was no question of half the crew being left behind to die, Lenoir sat on a jump seat behind the commander and pilot for launch. Allen had a foldaway seat in the windowless mid-deck, where he sat with his back to the airlock. (They changed places for re-entry.)

Lift-off weight was 2,033,750kg, including the 14,557kg payload, half of which was to be left in orbit. SRB separation was about 1,220m lower than the targeted 47,943m, headwinds causing the slightly depressed trajectory. This time the SRBs came down as planned about 225km off the coast and 12km from the recovery ships. For the 1st time everything - 2 drogue and 6 main chutes, 2 frustums and 2 booster casings - was recovered.

A new call - "Negative return," passed to the crew at T+3.55, 85km altitude and 62km downrange - meant that they could no longer return to the launchsite and were committed to space flight. At the Cape the cloud of aluminium oxide and gaseous hydrogen chloride left by the launch was for the 1st time blown over the VAB and the press site. There were no apparent ill-effects on the spectators, though "a very small fish kill" in a lagoon area just beyond the pad was later reported.

At 2hr 5min, following 2 OMS burns, a nominal parking orbit of 296 × 298km at 28·45° incl had been achieved. The only serious malfunction was the failure of a cathode ray tube. While completing the 3rd orbit *Columbia* passed 80km below Salyut 7. The 2 crews, probably in darkness

STS-5: test firing of *Columbia's* OMS engines. The white plumes, visible for only a split second at ignition, may be ice being blown from the nozzles. *(Aerojet General)*

over the Indian Ocean, were unable to communicate even if they had wished to do so. But the record-breaking Salyut 7 crew may have been able to exercise their sensors by monitoring *Columbia*'s passage.

SBS-3, 1st of the 2 satellites carried in the payload bay, was successfully deployed 8hr into the mission. TV showed the 4·2m stack (the compressed SBS-3 plus PAM-D upper stage) being spun up to 52rpm on its turntable, and then, when its clamp was released by explosive bolts, smoothly lifting clear of *Columbia*'s sides and fin. 3min before release the Orbiter's computer had started the PAM timer, which 45min later began an 83sec burn to place SBS-3 in a synchronous transfer orbit. 30min before that *Columbia* had performed a 17·5km/hr orbital manoeuvre to increase separation to 29km and rolled so that the belly faced SBS-3 - a precaution to ensure that its windows were not damaged by the burn. Not everyone agreed that this was necessary, and it meant that the PAM firing could not be seen by either the crew or the TV cameras. The total weight of this 1st payload ejected from the bay was 3,268kg, comprising the PAM (190kg plus 1,963kg of solid propellant for thrust to the transfer orbit), apogee motor for injection into synchronous orbit (493kg), and SBS-3 itself (622kg, including 148kg of hydrazine fuel for 8-9yr of stationkeeping).

The following day, on the 22nd revolution, a similar sequence resulted in Canada's Telesat/Anik C3 being even more accurately deployed: it was ejected within 151m of the target point, compared with 1,372m for SBS-3. In-orbit weight of Anik C3 was 632kg. NASA's responsibility for the satellites ended with the PAM firings; both were successfully transferred to their synchronous orbits after an initial telemetry problem with Anik unconnected with the launch operations.

The remaining in-flight activities were less successful. Both Overmyer and Lenoir suffered from spacesickness, necessitating postponement from the 4th to the 5th day of the planned $3\frac{1}{2}$hr EVA in the payload bay. During the exercise manual methods for emergency closing of the doors, and the use of new tools for assembling space stations, were to be tried out. On the 5th day the EVA had to be cancelled altogether after Allen and Lenoir had spent some hours in the airlock vainly trying to overcome different malfunctions in their spacesuits. (NASA later deducted $131,250 from Hamilton Standard's bonus as a penalty. The total spacesuit contract is worth $350 million, which includes $11·6 million of cost overruns.)

Major changes in the landing plans had been made before lift-off: the autoland had been abandoned because of unresolved variations in pre-flight simulations, and another hard-surface landing at Edwards had been necessitated by flooding of the lakebed runways. Following a 2min 23sec retrograde burn, and with Joe Allen taking his turn in the jump seat, Brand flew a fully manual landing from Mach 0·97 to touchdown. At Edwards it was only 6min after sunrise, and *Columbia*, rolling to a stop along the centreline, glowed pink in the early morning sun. Maximum braking at 260km/hr had been planned but because of wet runways this was delayed to 222km/hr. Although the roll-out appeared to be the smoothest so far, it was found that the inboard wheel on the left main landing gear had locked during the last 15m, shredding the tyre. During re-entry Allen had had time to take the 1st pictures of the event, and these showed the whole flight deck bathed in a red glow. He said it was like flying down a neon tube, with the glow moving disconcertingly closer during Brand's manoeuvres.

During and after the mission much concern was expressed at the continued 50% incidence of sickness among Shuttle crews. USAF managers in particular need to be confident that astronauts will be able to deploy spy satellites from the Shuttle as soon as it enters orbit. NASA replied that the sickness was no worse than that encountered by many sailors, and the astronauts themselves, worried that it might lessen their chances of future flights, protested that there had been too much public discussion of their malaise. A few weeks after the mission NASA announced a 5yr, $51 million programme of research into the causes and prevention of spacesickness, and astronaut Drs Thagard and Thornton were added as 5th crew members on STS-7 and 8 to start the investigation.

STS-6 The 1st flight of *Challenger*, planned as a 2-day mission to deploy TDRS-1 but later extended to 5 days to include the EVA not achieved on STS-5, was finally launched on 5 Apr 1983 with a 93hr countdown after 7 postponements due to fuel leaks, engine cracks and storm contamination of the satellite. Mission commander Paul Weitz, 50, had had to wait 10yr for his 2nd flight, while pilot Karol Bobko, 45 and an MOL astronaut, had waited 17yr for his 1st. Donald Peterson, 47, who occupied the centre seat to assist in ascent emergencies, and Dr Story Musgrave, 47, who sat to the right of Peterson, were the mission specialists. The "Geritol crew," as Mission Control dubbed them, had an average age of 48, making them as a team the oldest likely to fly. A 3rd spacesuit

Astronaut Dr Joe Allen, dressed in the Shuttle's "constant wear" garment, takes part in biomedical tests. Electrodes on head and face monitor his responses to zero-g. *(NASA)*

STS-5: perfect deployment of Canada's Anik C3 comsat gives good view of PAM-D upper stage nozzle, which may have failed twice after the Mission 41-B deployments.

STS-5: pre-set camera exposure enables all 4 crew members to celebrate Shuttle's 1st commercial successes. Commander Brand, with notice, is surrounded by (left to right) Lenoir, Overmyer and Allen. *(NASA)*

placed aboard following the previous failures became Musgrave's prime suit even before lift-off when his own proved faulty in final pre-flight checks – a situation described by Shuttle chief Gen Abrahamson as "less than desirable". Total lift-off weight was 2,036,469kg, including the 20,412kg payload. *Challenger* contributed 116,530kg. Hydrogen leaks through engine cracks might have caused a disastrous launch explosion had they not been detected during the flight-readiness firing. Despite being throttled to the 104%, 488,800lb-thrust level, main engine performance was slightly lower than predicted, resulting in a slightly depressed trajectory and flightpath angle 1° lower than planned; neither was considered a safety problem. The SRBs, each 1,814kg lighter but including several tonnes of re-used components, splashed down only 2·4km apart and 302km E of the launch site, and were recovered in calm seas. Parts had been previously used on STS-1 and were to be re-used on STS-11 and STS-13. The 30,311kg ET, 4,691kg lighter than that used on STS-1, splashed in the Indian Ocean at the predicted impact point. Its descent was tracked by Space Command with a DPS early-warning spacecraft to obtain infra-red burn data and to monitor the spread of the re-entering debris, both for the benefit of NASA and as a missile-tracking exercise.

Soon after opening the payload bay doors the STS-6 crew reported that some of the flexible insulating blanket which had replaced 600 white tiles on the OMS pods had curled off. Mission Control announced that it did not pose a re-entry problem, but the damage was inspected later with interest during the EVA. The 285km, 28·5°-incl orbit was finally achieved with the 2nd OMS burn 43min into the mission. Apart from occasional references to debris-clogged filters (the debris was carefully collected and packaged for return to Earth and detailed examination), the first 10hr was busily devoted to preparations to deploy the 18½-tonne TDRS/IUS. *Challenger*'s payload bay was manoeuvred towards Earth, since the TDRS systems could not tolerate direct sunlight. Musgrave deployed the tilt table to 29° at T+08.25.00 and then to 59° at T+09.48.00. The USAF Satellite Control Facility at Sunnyvale transmitted commands to the IUS and brought on-line all 5 gyros, 2 of which were earlier thought to have failed. The Payload Operations Centre at White Sands was successful in checking out the TDRS. Just before raising the tilt table for the second time Musgrave transferred the IUS from Orbiter to internal power, and the TDRS from Orbiter power to the IUS batteries.

STS-6 crew: left to right, MS Donald Peterson, commander Paul Weitz, MS Story Musgrave and pilot Karol Bobko. ***(NASA)***

Lifting sling hoists *Challenger* vertically inside VAB's Bay 3 for mating with ET and SRBs in preparation for its 1st flight, STS-6. ***(NASA)***

With the commander ready for any emergencies and flying the Orbiter with rotational and translational hand controllers at the starboard aft station, Bobko in the forward cockpit, and Peterson in the centre position observing, Musgrave had a 2min window at 10.01.58 Mission Elapsed Time (MET) in which to deploy the payload by manually selecting a number of pyrotechnic and other switches. Later TV showed the TDRS/IUS floodlit and moving slowly out of the payload bay above the flight deck, followed by a 2·2ft/sec RCS burn by Weitz to dip *Challenger*'s nose away from the payload. For Earth observers the need for a TDRS system was emphasised many times on this mission when they were unable to see crucial activities live because *Challenger* was out of alignment with a ground station.

The IUS 1st stage was successful in placing the vehicle into transfer orbit 55min later, but the 2nd burn at 16.16 MET left it tumbling at 30rpm in a 35,317 × 21,786km orbit. For several hours it was thought the whole vehicle was lost, but increasingly desperate transmissions from all 3 control centres unexpectedly succeeded in regaining some measure of control. The tumbling was checked, separation achieved, and the 17·3m solar panels extended just in time to save the batteries. It was later decided that over a period of weeks it would be possible, using the 24 1lb thrusters and 385kg of the 590kg total of hydrazine propellant, to manoeuvre TDRS into its correct orbit. Since enough propellant had been loaded to cover commercial activities since eliminated from the programme, TDRS-1 should still achieve its full 10yr life. The TDRS rescue effort was greatly helped by the fact that a Norad camera in New Mexico was successfully monitoring the IUS at the time of its failure at an altitude of 32,300km. This also helped explain the later discovery that the 2 roll-control thrusters were inoperable, with possible pitch thruster damage as well, since it appeared that the IUS and TDRS had scraped or bumped during separation. Although at the time NASA remained optimistic about the final outcome, the cause of the IUS problem remained undetermined, leaving doubts about the launch of STS-8 and 9, both of which depended upon a successful TDRS-1 deployment.

On the 4th day Musgrave and Peterson successfully climbed into their spacesuits and endured 3·5hr of pre-breathing 100% oxygen in the airlock. TV confirmed their reports that the only difficulty they encountered in 4hr 10min of EVA was in reading the suits' digital displays in bright sunlight. Musgrave, designated EV1, collected and

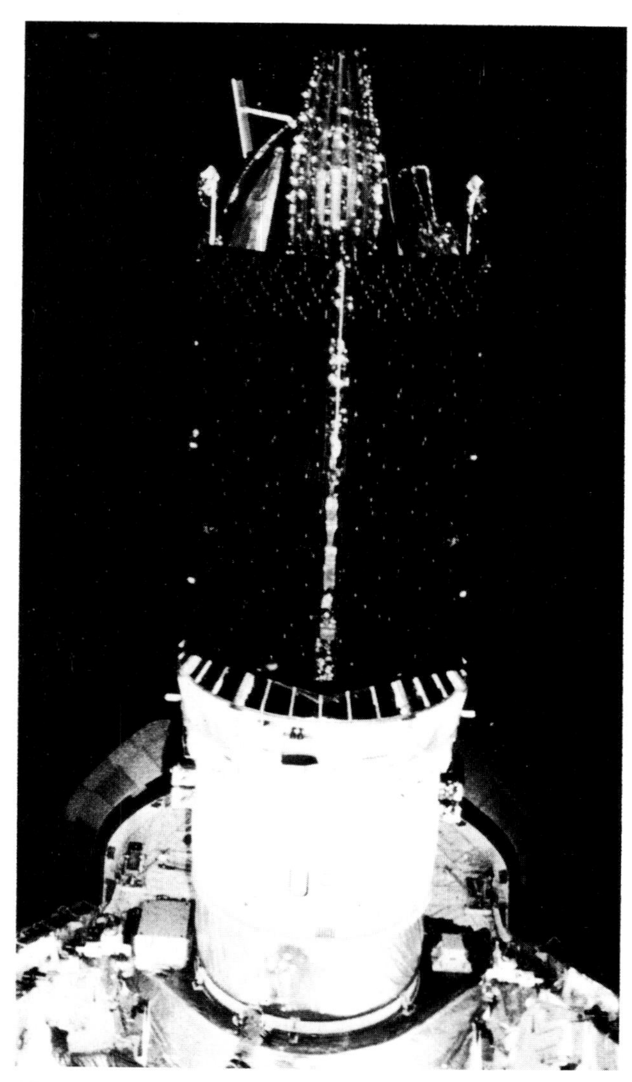

STS-6: TDRS-1 being raised for deployment from payload bay, which it almost filled. Subsequent partial IUS failure was the start of a series of upper-stage problems which seriously delayed DoD and other projects. *(NASA)*

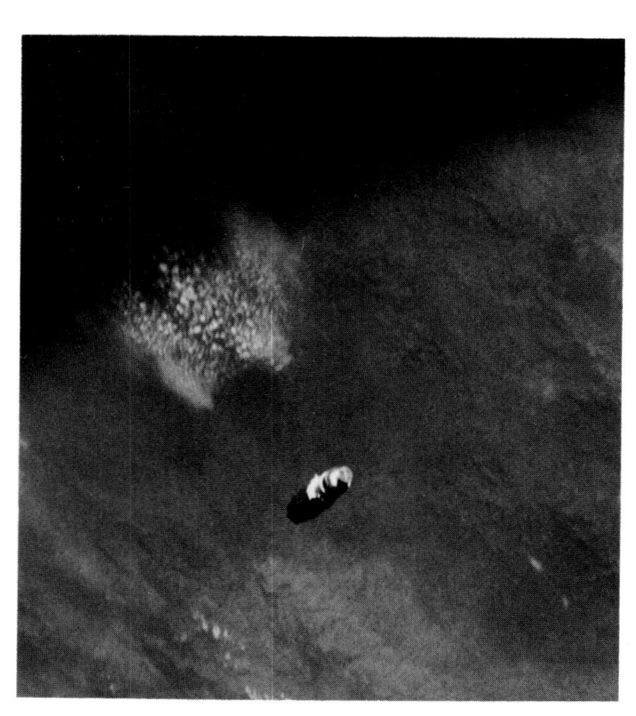

Perfect deployment of TDRS-1 from *Challenger* against background of Kenya's Rift Valley gave no hint of troubles to follow. *(NASA)*

STS-6: astronaut Story Musgrave performing the 1st Shuttle EVA, rescheduled from previous mission following suit failures. He is attached to slidewire, centre. Right, TDRS's support equipment is visible. *(NASA)*

STS-6 EVA: with Donald Peterson, Musgrave evaluates the handrail system. New blanket insulation is clearly visible on OMS pods against Mexican background. *(NASA)*

donned a "mini work station" stowed on the right forward side of the payload bay. Having attached themselves to tethers, the 2 astronauts monitored each other's progress as they moved down each side of the bay, trying out various points designated as work stations, practising the use of tools, demonstrating suit mobility, etc. They showed their ability to winch the payload bay doors shut if this should ever become necessary and to lower the satellite deployment tables manually if they should stick. Checking on whether moving a large mass in zero g might cause problems, Musgrave "translated" along the bay's hinge line without apparent difficulty. His heartbeat during these operations was around 60/min; Peterson's reached 130-140, tripping a high oxygen usage suit alarm.

For *Challenger*'s re-entry Weitz and Bobko had the first use of the Kaiser Electronics head-up display (HUD) on the final glideslope; on this occasion they used it as an aid rather than a primary instrument. 8 flight-test manoeuvring sequences, involving 30 jet firings, were added to the normal hypersonic S-turns, but a Mach 9 test failed to occur, perhaps because of a software problem. *Challenger*'s weight at the 2min 27sec de-orbit burn was 89,552kg; the Orbiter touched down on the centreline of Edwards' Runway 22 and stopped in 2,188m.

STS-7 Although the planned first landing back at KSC was thwarted by bad weather, STS-7, launched within milliseconds of the planned time on 18 Jun 1983, proved that the Shuttle could retrieve and repair satellites. Commander Robert Crippen, 45, became the 1st man to make 2 Shuttle flights. Frederick Hauck, 42, was pilot, and John Fabian, 44, the lead mission specialist (MS1). Dr Sally Ride, 32, became both the 1st US woman and the youngest astronaut to fly. Dr Norman Thagard, 39, and designated MS3, had been added to study spacesickness. Lift-off weight was 2,034,667kg, 1,926kg lighter than on *Challenger*'s 1st flight. Main engine cut-off was on target at 7,824m/sec and 105·5km altitude. 2 OMS burns placed *Challenger* in a 296km circular orbit for the 1st satellite deployment.

Sally Ride's refusal to acknowledge that there was anything different about this flight because it included a woman was not shared by Hugh Harris, the "Voice of the Cape". As the SRBs' flame joined that of the SSMEs he cried: "Lift-off, lift-off of STS-7 and America's 1st woman astronaut." Later Ride relaxed sufficiently to enthuse: "Have you ever been to Disneyland? That was definitely an E ticket," which reporters soon identified as the equivalent of first-class. Meanwhile the SRBs fell close to the recovery vessels 251km offshore, and were quickly recovered with all the chutes. As on previous flights, however, the chutes had suffered burn damage.

In *Challenger*'s payload bay were Telesat Canada's Anik 3 and Indonesia's Palapa B satellites. Also carried was Germany's SPAS-01, described as a "space platform" which could operate either inside or outside the bay. Anik 3 was deployed at 09.29 MET and Palapa B 24hr later, after being spun up to 50rpm on their turntables and then pushed out by springs at 2·8ft/sec. The release of their clamps produced a "solid bang" throughout *Challenger*. On TV they could be seen rising perfectly in line with the edge of the Orbiter's fin. The PAM motors fired 45min after deployment for 85sec to place the satellites in transfer orbit. As before, *Challenger* was rolled to ensure that its windows were not contaminated by the firing. It was estimated that Anik was deployed within 450m of the target point and within 0·085° of the desired pointing vector. With such precise positioning it was estimated that commercial sponsors could expect an extra 2yr of operational lifetime, worth perhaps $50 million.

Highlights of the mission, as expected, were the proximity operations ("proxops") with SPAS-01 and its subsequent retrieval. This entirely German concept, designed and developed by MBB with a team of only 20 at a cost of $23 million, was unberthed by Fabian with the RMS on the 4th day. He released it over the payload bay, and with Crippen and Hauck monitoring closely through the overhead windows in case evasive action should be necessary, recaptured it after 60sec. Fabian said he could

STS-7: Canada's Anik C2 satellite, separating from *Challenger*, is framed by the darkness of space, Earth's horizon and part of the overhead flight deck. *(NASA)*

detect only 0·6cm oscillations in the RMS as he approached SPAS. Care had to be taken because of excessive temperatures in the SPAS data-handling system, but despite that observers on the ground were able to share with the crew a 9hr "orbital ballet" as the two vehicles manoeuvred together. TV cameras on both the RMS and SPAS itself provided views of the two spacecraft. With SPAS at times 300m away, we saw at last what the Shuttle looked like against Earth's background. Crippen manoeuvred *Challenger* without difficulty for a series of rendezvous radar tests, closing in smoothly for SPAS recapture. Ride described later how she performed one capture with the arm in single-joint mode, in which there is no computer assistance and toggle switches must be flipped to operate only one arm joint at a time. It went beautifully, she said; the secret was to go in quickly, "not take extra time to pretty it up". Thagard made a capture without difficulty while SPAS was rotating at 0·1°/sec, and with the satellite attached to the RMS Fabian and Ride rehearsed the manoeuvres which will be necessary in 1986 with the Space Telescope.

For 30hr from the 3rd day cabin pressure was reduced and maintained at 10·2lb/sq in in a successful test aimed at eliminating the need for 3·5hr pre-breathing periods in the airlock for crewmen embarking on EVAs. An in-flight problem was not revealed until after landing: Hauck said that on the 3rd day he had noticed an "impact crater" on a window to the right of the pilot's seat, caused by either man-made debris or a micrometeorite. Only the outer of 3

STS-7: Dr Sally Ride, about to become 1st US woman in space, addressing the press at the Shuttle Landing Facility after flying in from Houston. Left to right, commander Bob Crippen, MS Dr Norman Thagard and John Fabian, and pilot Fred Hauck. *(NASA)*

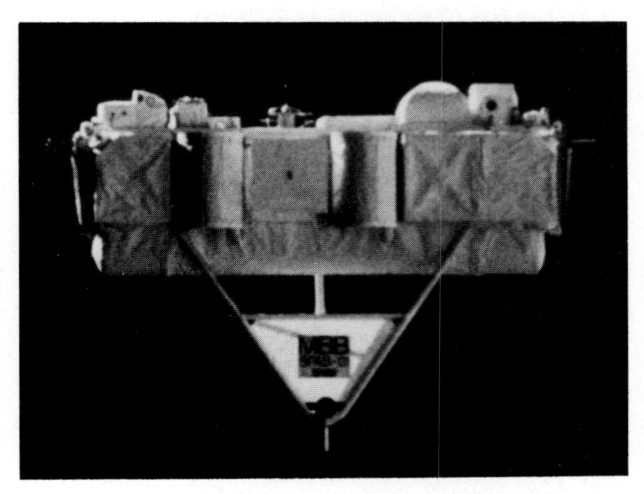

STS-7: Germany's SPAS-01 flying free above *Challenger* after deployment by Fabian. After release and capture practice with the robot arm, Ride said the secret was to go in quickly. *(NASA/MBB)*

panes was affected, but the crater, later measured at 0·0178in thick, was deep enough to necessitate replacing the window. Chances of such a strike were said to be 1 in 270 space days; a *Columbia* window had had to be replaced after STS-5. On the 5th day the toilet's slinger failed once again, but the crew were able to continue using it.

Because of bad weather at the Cape *Challenger* was "waved off" on orbits 96 and 97, and trouble with one of the 3 APUs and loss of pressure in a hydraulic pump accumulator meant that earlier plans to stay in orbit for up to 2 extra days while the Cape weather cleared had to be abandoned. After flying 1,367km cross-range (within 20km of the limit) N from its 98th-orbit groundtrack, *Challenger* was brought down at Edwards after a mission lasting 6 days 2hr 24min. Ride assisted with the re-entry checklist in the centre cockpit seat. Pre-entry burn weight was 96,205kg, 7·5 tonnes heavier than STS-6, due to SPAS-01 and materials-processing hardware. Some damage found on black tiles on the nose and lower sides was attributed to ice falling from the ET on lift-off. While *Challenger* was being towed away from the usual perfect centreline landing there was brake trouble, necessitating a wheel change. Praise for Sally Ride's performance as the 1st US woman came from her bosses, but she herself refused to regard herself as anything but just another member of the crew, merely commenting that the mission had been "the most fun" she was ever likely to have.

The electrophoresis experiments - which had demonstrated on STS-6 the ability to separate 700 times more material with 4 times the purity than a comparable effort in 1g conditions - were again highly successful. They led to a decision to send the 1st industrial payload specialist, Charles Walker, on 41-D to operate a prototype production unit continuously throughout the 5-day mission. He was expected to bring back enough vaccine to start clinical tests on human subjects. Most publicised among a number of student experiments carried was a colony of 150 carpenter ants, the idea of disadvantaged children at two schools at New Camden, NJ, and funded by RCA, the area's biggest employer. Known as "Orbit 81" because that was when the experiment was first expected to fly, it kept the children interested in science for 4yr as they planned their investigation of the ants' ability to continue their colony in weightless conditions. Film and still cameras were arranged to take regular pictures throughout the flight, but after the landing it was found that they had all died, apparently due to the heat, before lift-off.

STS-7: first view of Shuttle against Earth, taken by 70mm camera mounted on SPAS-01. RMS deployed to form a figure seven; empty satellite containers at rear of payload bay. *(NASA/MBB)*

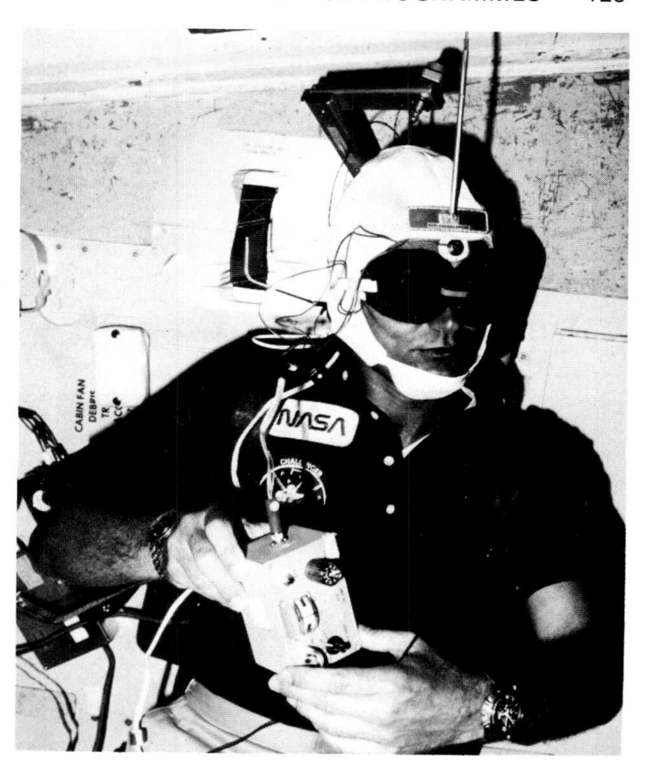

STS-7: Dr Norman Thagard, using himself as the subject, evaluates physiological reactions of astronauts in mid-deck experiments. *(NASA)*

STS-8 Although it included the 1st night launch and night landing, this 6-day flight was at first sight the least eventful to date. Launched 17min late, due to a rainstorm, at 02.32 EST on 30 Aug 1983. Commander was Richard Truly, 45, making his 2nd and last flight. Pilot Daniel Brandenstein (40), MS1 Dale Gardner (34), MS2 Dr Guion Bluford (40 and 1st US black astronaut) and MS3 Dr William Thornton (54 and oldest astronaut so far) were all making 1st flights.

Lift-off weight was 2,037,400kg, 2,477kg heavier than STS-7, but only 100% thrust was required from the SSMEs because of improved thrust from the SRBs, reduced main engine stresses, and improved guidance control. The crew said the Orbiter appeared to be "engulfed in fire" and they had the impression they were "inside a bonfire" to such an extent that Gardner feared, mistakenly, that an engine was unstable. (Not until mid-October, when the STS-9 stack was rolled back and dismantled, causing yet another delay to that flight, did we learn that STS-8 had been, according to one of the crew, within 4sec of disaster. There had been excessive erosion of the carbon-fibre lining of one of the SRB nozzles. The official NASA version was that it would have burned through in 14sec, causing the hot gases to break through sideways and sending the cluster into a spin.)

Deployment of main payload Insat 1B (weight with PAM 3,376kg), which necessitated the night launch, was successfully achieved at 296km altitude during orbit 17. It was spun to 40rpm on the turntable and ejected within 1,500m of its orbital target with no more than 0·09° error. 45min later the PAM fired to start Insat's journey to synchronous orbit, and though initially the solar panels refused to open, this was later achieved by rolling Insat towards the Sun to thaw out the hinges. Although nothing was seen by the crew, videotape recordings suggested that Insat may have been struck by debris from the Orbiter at a distance of about 15m. Insat, STS-8's only reimbursable payload apart from the 260,000 stamped envelopes for sale to collectors, brought NASA a return of only $8·3 million in launch fees. But there was compensation in the completion of a very full and successful programme of experiments.

Bluford and Gardner began electrophoresis tests early in the flight, for the first time using living cells. Commercial secrecy has obscured some of the earlier work, but this time we learned that pancreas cells from living dogs were used. The aim is to achieve precise separation of beta cells, which produce insulin in the body. This cannot be done efficiently in Earth conditions, and there is hope that space-processed cells may restore diabetics to normal health with one implantation. Other tests involved human kidney and rat pituitary cells.

Earth geological observations were claimed to be the most detailed since Skylab. The night launch gave *Challenger* the benefit of long periods of daylight over the Southern hemisphere for the first time. Active volcanoes in the Pacific, features of S Africa and central Australia, and ocean currents were all described and photographed in great detail.

The 3,384kg, dumbell-shaped Payload Flight Test Article (PFTA), which had replaced the scheduled TDRS-2, was unberthed by Gardner on the 3rd day and used for extensive tests spread over 3 days. Manoeuvres with the aluminium, lead-ballasted structure were designed to gain experience for deployment of the LDEF and the SMM repair on 41-C. As the 4·6×3·96m dumbell was held by the RMS above the payload bay, the RCS jets were fired to check the effect on *Challenger*. On TV the PFTA could be seen to shiver slightly, a dynamic motion that took about 20sec to damp out. Another RMS test showed that the robot arm could be hooked over the Orbiter's side and used to obtain views of the underside, wings and control surfaces if damage was suspected.

STS-8: 1st night launch, 30 August 1983, lights up Florida sky. With Florida weather more reliable during darkness, night launches and landings are likely to become common. *(NASA)*

STS-8: Dr Guion Bluford, 1st US black astronaut, uses the mid-deck treadmill in a medical test. *(NASA)*

Thornton had a busy time studying spacesickness (now officially "space adaptation syndrome"). He said that "the usual range" of such symptoms occurred, and that he had learned more in his first 90min of flight than in all the previous years he had studied the problem.

Much work was done on the use of communications links via TDRS-1, and TV transmitted this way was noticeably better. Though the crew had to be woken one night to throw switches when the link via TDRS could not be maintained, confidence was expressed that "with appropriate adjustments" it would work well for STS-9/Spacelab 1.

The re-entry for the first night landing, described as "a hypersonic descent profile," was based on the STS-6 pattern, which had subjected the tiles to less damage. 9 manoeuvres involving 30 thruster firings were conducted between the 4 hypersonic S-turns needed for correct alignment with the runway. An attempt to maintain communications during the usual 17min blackout period by using antennas on top of *Challenger* to penetrate the ionised gas surrounding the Orbiter and transmit via TDRS-1 failed because of a White Sands computer malfunction. But this system should ultimately put an end to the blackout period, which until now has always been a feature of manned re-entries.

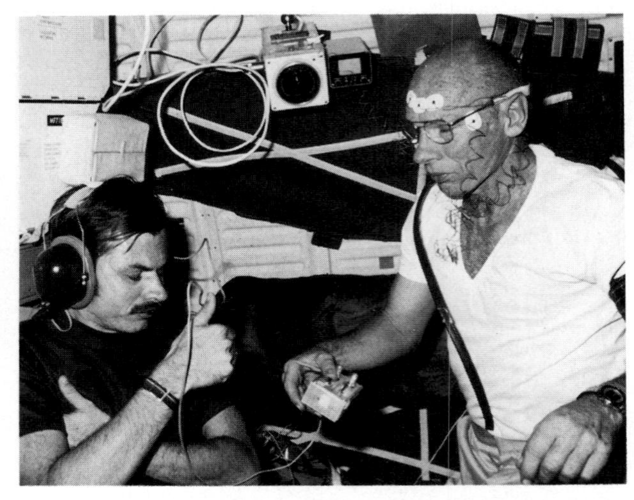

STS-8: Dr William Thornton (right), at 54 the oldest astronaut so far, conducting an audiometry test on Dale Gardner. *(NASA)*

STS-8: 1st night landing at Edwards was made possible by xenon lights each producing 800 million candlepower, which made runway visible 80km away. *(NASA)*

Truly reported good head-up display performance on the final approach, and sighted Runway 22 - floodlit by xenon lights each producing 800 million candlepower - while *Challenger* was still 80km away at an altitude of 23,000m and a speed of Mach 2. He touched down in the centre of the target area 850m along the runway. Gen Abrahamson said that the landing gave confidence that ultimately most Shuttle recoveries would be at night, since the Cape weather was usually better then.

STS-9/Spacelab 1 This 10-day flight, with sufficient reserves for 2 more days in orbit, began on 28 Nov 1983 after another 3 months of delays occasioned by TDRS difficulties and the SRB nozzle problem of STS-8. Commander was John Young, 53, making more history with a 6th flight. Pilot Brewster Shaw, 38, was making his 1st. MS1 was Dr Owen Garriott, 53, who spent 59 days aboard Skylab 3. MS2 was Dr Robert Parker, 46, and the 1st payload specialists were America's Dr Byron Lichtenberg, 35, and Dr Ulf Merbold, 42, of Germany. It was the largest crew ever flown in 1 spacecraft, and Merbold, representing ESA on this joint NASA/ESA flight, was the 1st non-American to fly in a US spacecraft. (He fled from his home town of Greiz, E Germany, at the age of 19; Sigmund Jähn, who had represented E Germany on Soyuz 31 5yr earlier, came from Voigtland, less than 50km away.) Total lift-off weight of *Columbia* on her 6th flight was 204,686kg, including 15,233kg for Spacelab 1 and 1 Pallet. With the SSMEs operating at 104% thrust for 6min 4sec, she reached her 57° orbit (taking her over Europe for the 1st time) in 49min 29sec. Parker was on the flight deck with the pilots; the others were on the mid-deck, 2 seated with their backs to the closed hatch leading to the 5·75m tunnel to Spacelab (carried in the rear of the payload bay for centre-of-gravity reasons), the other to the left.

The SRBs splashed down 201km E of Jacksonville, arrangements having been made to examine the nozzle linings as soon as the recovery ships got them back to Port Canaveral. The ET, which could have fallen in Scotland if lift-off thrust had been below nominal, came down as planned in the Indian Ocean.

All 6 crew members had to join in before the sticking mid-deck tunnel hatch could be opened at 3.30 MET. Remote-controlled TV provided perfect pictures via TDRS, enabling Mission Control and newspeople to watch, 12min later, as Garriott, Lichtenberg and Merbold joyfully floated into Spacelab. Round-the-clock operation was achieved by Young, Parker and Merbold, the Red Team, working 12hr day shifts in GMT and Shaw, Garriott and Lichtenberg, the Blue Team, working the "night shift". They had adjusted their circadian rhythms by staggering their meal and sleeping times before flight, with the result that the Red Team lunched while the Blue Team had the traditional breakfast before boarding. A new call, "Marshall Operations," referred to the Marshall-managed Payload Operations Control Room, established below Houston's Mission Control. Here the international scientists could monitor and control the flow of data from their experiments, and consult and instruct the crew in their operation. This they did with enthusiasm, especially in the early experiments involving blowing hot and cold air into the ears of Parker and Merbold, which were to prove that in 1914 a Nobel Prize had been wrongly awarded for the theory that air blown into the ears at different temperatures could cause a sensation of turning, even if the subject was not moving. It was announced at the daily science briefings that the ear, nose and throat specialists must face the fact that all their textbooks must be rewritten. Parker and Merbold, faced with demands to switch on other experiments when wearing the elaborate harnesses and headgear required for these tests, protested sharply at times. "I think you might be quiet until we get one or other done," said Parker on one occasion. Lichtenberg had to suspend the "drop and shock" tests, combining small electric shocks applied to the calf of one leg with unexpected movements, because he found them disturbing; Garriott seemed less affected. An investigation of mass discrimination involved 24 golfball-sized spheres of equal diameter but with lead centres of varying weights; the astronauts were required to evaluate them in pairs before, during and after the mission, entering on a card which they thought to be heavier. Five of the crew

STS-9 mating: Orbiter *Columbia*, hoisted high in the VAB, is about to be lowered for attachment to ET and SRBs in preparation for the much delayed Spacelab 1 mission. *(NASA)*

STS-9/Spacelab 1: a cake on the pre-launch breakfast table is decorated with the crew patch. From left, Ulf Merbold, 1st European aboard the Shuttle; MS Dr Robert Parker, commander John Young, making his 6th flight; pilot Brewster Shaw; PS Byron Lichtenberg; and MS Dr Owen Garriott. *(NASA)*

STS-9 launch, 28 November 1983, carried a 6-man crew, the largest ever on one spacecraft. ***(NASA)***

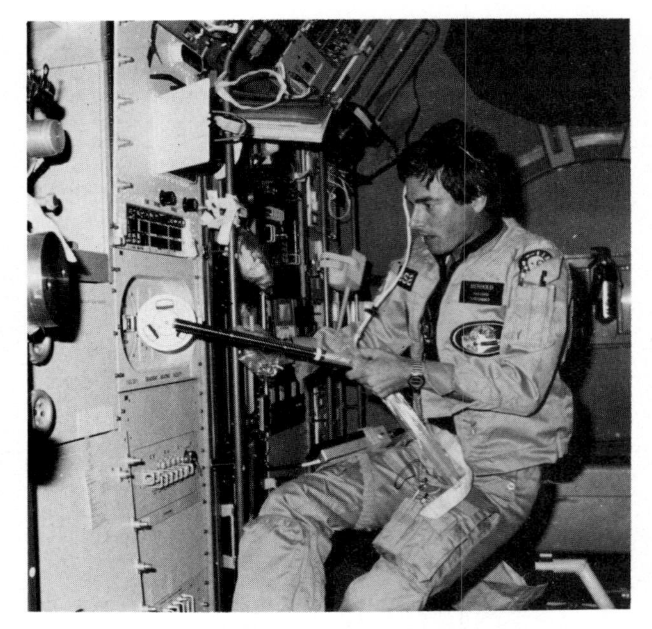

Inside Spacelab, Dr Merbold uses the gradient heating facility for a materials-processing experiment. ***(NASA)***

members played the game but characteristically John Young refused, to the annoyance of the Scottish psychologist who devised the experiment.

Much satisfaction was expressed at the daily science briefings at the amount of data being returned, though the investigators seemed shocked when the author asked whether future missions might therefore concentrate more on productive activities like materials processing and crystal-growing, and less on life sciences. Studies of the solar constant meant keeping the payload bay, with its Pallet-mounted instruments measuring energy output in the band from ultra-violet to infra-red, exposed to the Sun for hours at a time. One 7hr period was ended abruptly when Young climbed from his bunk in alarm at what he described as "rattles, bangs and groans" throughout *Columbia*. Mission Control thought they were unimportant thermal effects of the uneven heating, but scientists with other Pallet-mounted instruments who feared their data would be ruined by excessive heating fully approved of Young's insistence on rolling *Columbia* around. A theory that this led to a fracture in an OMS fuel pipe, which froze and started a fire when it thawed during the landing, was unproven as we went to press.

A live TV hookup with the US President, the German Chancellor and the crew demonstrated how little is known nowadays of what is really going on. During a TV test, Garriott, Parker and Shaw lined up in Spacelab covering their eyes, ears and mouth in a "see, hear and speak no evil" demonstration. It was a protest against a White House "script" for the hookup, which had decreed that only Young, Lichtenberg and Merbold were to appear, and even telling them what they should say and when. All this, and no doubt many other things such as Garriott's not-so-spontaneous radio-ham talk with King Hussein of Jordan, were discussed over the teleprinter. Small wonder that when re-entry plans had to be revised there was no teleprinter paper left, and the numbers had to be read up as in earlier days.

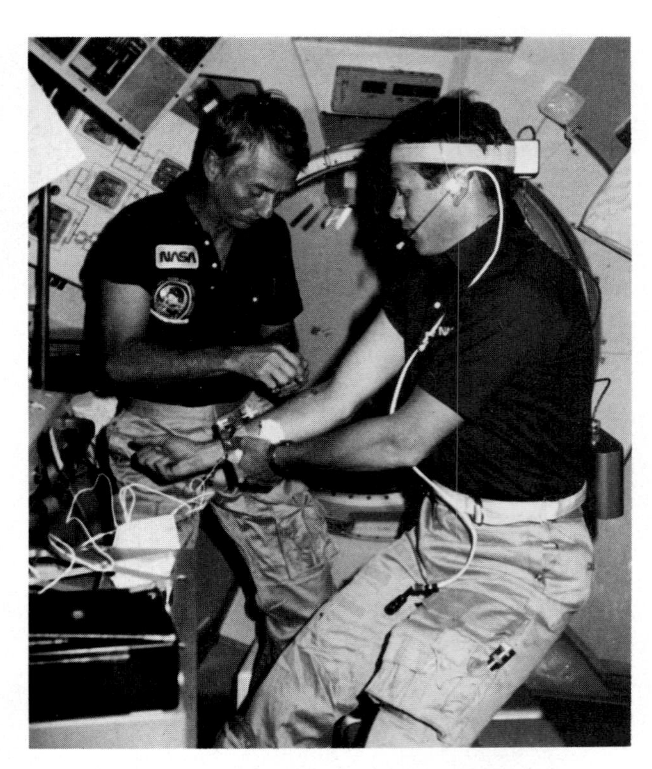

Half the Spacelab experiment time was devoted to life science. Here Garriott draws blood from Lichtenberg before connecting him to portable oscilloscope at left. Young refused to give blood in case it impaired his judgment as commander. ***(NASA)***

At 103,735kg *Columbia* was 6,804kg heavier than for any previous landing. The plan to re-enter on orbit 162 was abandoned when John Young, twice firing nose thrusters to put *Columbia* in the right attitude for door closing, experienced "jarring which hit the vehicle really hard," with a computer failing on each occasion. One was set up again but the other resisted all Young's efforts to restore it. The mission had to be extended another 5 orbits while the computers were checked and realigned, and during this time one of the 3 inertial measuring units also failed. *Columbia* made what appeared to be its usual perfect landing 5 orbits later after 10 days 7hr 47min of flight time, with Young doing an impressive nosewheel steering test along the centreline. But at nosewheel touchdown the restored computer dropped out again and the TV showed what appeared to be smoke coming from the tail. "That's a picture of the APU exhaust . . . . that's not a fire," Mission Control said reassuringly. It was misplaced confidence, and the transcript shows one of the pilots saying immediately afterwards: "Hey, we've got a thermal APU." In what appeared to be a repeat of an incident on the previous landing, a fuel pipe had ruptured. The smoke had come from a smouldering fire which reached 400°F, causing the explosion of 2 of the 3 APUs. Once again the Shuttle had been close to disaster. The crew, apparently unaware of how close they had been to an emergency evacuation, emerged normally to walk around *Columbia* and inspect what seemed to be more than the usual amount of scorching around the OMS pods and fuselage. The mission and payload specialists disappeared into 2 trailers for a week of medical tests on readaptation to 1g, while Young and Shaw made the usual brief statements before returning to Houston. Young, who had been working for nearly 24hr, was generally regarded as having made his last flight, after reported pre-flight comments that at his age eyes and ears begin to let you down. Jokingly, he commented that when the 1st computer failed his knees knocked, and the 2nd failure turned him to jelly - but Shaw already had the books out looking up the emergency procedures. Despite its extended mission, *Columbia* could have managed another 2 days in orbit. Within the hour the scientists representing 100 international groups were rejoicing in a 98% successful mission, though NASA announced that never again would a flight attempt to cover so many different disciplines.

The increasing influence of the military on the conduct of Shuttle missions was demonstrated after this flight by the fact that, in spite of its international character, scientists were not allowed to view more than 1,000 Earth scenes taken by Germany's metric camera until they had been vetted by NASA's DoD liaison office. It was not clear whether this scrutiny was applied to the 80 frames "donated" to the crew for their personal Earth photography (a reward for Parker's skill in saving this experiment by repairing the camera). But John Young did take a picture of Tyuratam, to the embarrassment of his employers, and this was published in *Aviation Week* and subsequently made available to the public.

Mission 41-B With a precise lift-off on 3 Feb 1984, what was still in the computers as STS-11 became Flight 10 or Mission 41-B in the revised designation system. The last STS mission of 1984 suffered mixed fortunes: the breathtaking success of the 1st use of the Manned Manoeuvring Unit for untethered flight on days 5 and 7 brought much relief after the disappointment of the loss of both the Westar 6 and Indonesian Palapa B2 satellites. Thus it was that commander Vance Brand, 52 and making his 3rd flight, found ill luck continuing to dog his missions. His 4-man crew, all making 1st flights, comprised Robert "Hoot" Gibson, married to astronaut Rhea Seddon; 46yr-old Bruce McCandless (MS3 and EV1), an astronaut for nearly 18yr and involved in MMU development for most of that time; 41yr-old Robert Stewart (MS2 and EV2), who made history as the 1st Army man in space and acted as 3rd pilot for launch and landing; and 31yr-old Ronald McNair (MS1), who had had a mere 5yr wait for his 1st flight and was a specialist in laser physics and molecular spectroscopy.

With its 6,662kg payload - most of that accounted for by the twin commercial satellites - *Challenger* climbed into an initial 305km orbit. This phase was uneventful except that on each of the SRBs 1 of the 3 recovery parachutes failed to open, resulting in a 120km/hr sea impact and some extra damage. Less than 8hr later Westar 6 was spun up on its turntable and ejected amid much audible satisfaction; McNair commented that "it made quite a thump - really got your attention" when it left. Some hours later, however, it began to be rumoured that the satellite had been lost. Confirmation that it had indeed been lost and that deployment of Palapa 2B - a similar Hughes spacecraft, to be sent to geostationary orbit by an identical PAM-D upper stage - would be postponed came at the end of Day 1. It was Day 3 before NASA was able to announce that NORAD had found 2 large objects and a number of small ones in a 1,218 × 307km orbit; one of the large objects was a healthy Westar 6 in a useless orbit, and the other almost certainly the PAM-D. The upper stage was suspected of failing to complete its burn because of a nozzle problem of the kind which had led to the IUS failure. It fell to the Indonesian Government to decide whether Palapa B2 should be deployed or brought back home. Since it was going to cost $10 million in any case for the launch, and the satellite was fully insured, the Indonesians opted to go ahead. This was a relief to NASA, since bringing back such a heavy payload, though technically possible, might well have led to cancellation of the proposed 1st Kennedy landing. With 16 successful launchings behind PAM and no other failures, NASA was content to rely upon the law of averages. It was a law that had to be repealed when, on Day 5, Palapa 2B ended in an orbit similar to that of Westar 6 and just as useless. Told of the disaster by teleprinter, the astronauts said the news "blew their minds," even though they were assured that they had played their part impeccably. Meanwhile NASA faced the fact that, with another 5 PAM satellites in the 1984 Shuttle manifest, its whole year's programme was in jeopardy. Worse still, with the IUS problem still unsolved, the geostationary orbit was out of reach until PAM could be given a clean bill of health. Western Union and Indonesia had the consolation of making claims on their 5·5% insurance policies for $105 and $75 million respectively for the satellites, launch costs and loss of business, while Lloyd's of London predicted that in future satellite insurance would cost up to 12%.

Mission 41-B crew: seated, commander Vance Brand and pilot "Hoot" Gibson; behind, left to right MS Robert Stewart, Ron McNair and Bruce McCandless. *(NASA)*

In the meantime, the crew had suffered another setback. A 2m balloon intended to be used as a rendezvous target as part of the SMM repair rehearsals exploded when deployed from its canister in the payload bay because lanyards intended to trigger the opening mechanism failed to work correctly. A planned rendezvous from a distance of 278km had to be abandoned, but its 90kg lump of ballast - the cause of some concern about collision hazards - was enough of a target to permit the performance of 3 vital rendezvous instruments to be checked.

After the setbacks the untethered MMU flights on Days 5 and 7 were a triumphant success, providing spectacular TV as McCandless and Stewart took turns converting themselves into human satellites. After cabin pressure had been reduced first to 12·5lb/sq in and then to 10·2lb/sq in, McCandless and Stewart entered the airlock, donned their Extravehicular Mobility Units (EMUs), prebreathed pure oxygen for 40min and egressed into the payload bay. Their Martin Marietta Manned Manoeuvring Units (MMUs), each costing $10 million and the only 2 in existence, were hanging on either side of the bay in Flight Support Stations where they can be recharged for 6hr use. The astronauts used the port-side MMU on Day 5 and the starboard MMU on Day 7.

Mission 41-B: Bruce McCandless in the MMU, making space history with the 1st untethered flight. A few metres from *Challenger*'s cabin, he is moving by using his hand controllers. *(NASA)*

Mission 41-B: McCandless, with the MMU on his back and docking device on his chest, about to dock with pin (bottom centre) on SPAS-01 satellite, which remained in payload bay following an RMS failure. *(NASA)*

Mission 41-B: McCandless tries out a powered screwdriver in another rehearsal for the Solar Max repair mission. *(NASA)*

McCandless began by checking out the port-side MMU, weighing 154kg and resembling an armchair without a seat. The arms end with hand controllers for the two nitrogen thruster systems, each with 12 7·6N (1·7lb) thrusters. The right hand controls rotation and attitude, the left hand translation (forward movement). A right-hand button instructs the backpack computer to maintain the astronaut in a given position, freeing his hands for work.

It took nearly an orbit before McCandless had backed into and fastened his MMU. Brand reported: "Bruce is in free flight," and as TV from the RMS showed him tentatively manoeuvring in the forward payload bay, McCandless commented: "That may have been one small step for Neil, but it was a heck of a big leap for me." We had a close-up view of him through his helmet as he flew up to the flight deck windows and peered through. He then backed out, first to 50m and then to 100m, making a lonely figure against the blackness of space and the blue and white of Earth. His only complaint was about "shudders and rattles and shakes" when forward movement was commanded while the system was in attitude-hold mode. Otherwise both he and Stewart, who later also went out to 100m, seemed satisfied with the MMU's performance. The 1st, 5hr 55min, EVA ended with McCandless standing on a platform fixed to the end of the robot arm as it was manoeuvred by McNair in a simulation of the SMM repair activities.

The second EVA was preceded by a disappointment when Gibson, carrying out RMS checks, found a malfunction in its wrist yaw system, which meant that it could not be used to lift Germany's SPAS satellite above the bay for use as a rotating docking target. However, the crewmen managed successfully to demonstrate the procedure that would be used to dock with the SMM satellite on 41-C. With a docking unit mounted at chest level on the MMU, they flew high above the bay, then came down both feet and head-first to line up with the 2 docking units on SPAS. They then engaged the pins on the satellite with surprising ease, even though to the onlookers it seemed like trying to thread a needle in impossible circumstances.

We now had an unexpected demonstration of the Shuttle's ability to overtake and recapture a crewman even if his thrusters stuck when propelling him at their full velocity of 20m/sec (70km/hr). Later in the EVA, as Stewart was parking the MMU, a movable foot restraint floated away from McCandless. There were a few moments of confusion, with McCandless, now on a tether, and Stewart, still partly in the MMU, offering to go after it. After a quick check with Mission Control Brand went through the emergency procedures for rescuing a stranded astronaut. While he manoeuvred *Challenger*, McCandless scrambled along the starboard lip of the bay to the full extent of his tether, giving advice to Brand at the same time. In a few seconds the foot restraint, which seemed to have floated far out of reach, was overhead again so that, amid cheers, McCandless could reach up and grab it. Shortly after Mission Control invited the crew to sit and look around for a while – apparently a pre-arranged signal that President Reagan was about to call up from California. McCandless hovered above the bay and Stewart hung down outside the starboard side while the usual pleasantries were exchanged. Pilot Gibson admitted that when his wife, Rhea Seddon, made her flight around August he would be more nervous than when making his own. The President asked about the mid-deck experiment with 6 white rats, 3 of which had been injected with substances to simulate the symptoms of rheumatoid arthritis. He was assured that the astronauts were pleased with the results so far. The EVA was then completed, with the astronauts standing on work platforms to practice pumping fuel from one point to another. This technique is officially applicable to Landsats, but is in fact more urgently needed for the very expensive low-flying spy satellites which will enter service once the Vandenberg launch facility is available.

All this brought an air of confidence back to the mission, and official doubts about the wisdom of attempting the first Kennedy landing faded away as the weather forecast improved. On orbit 127 *Challenger* was programmed to make a 2min 49sec retrograde burn NW of Australia and committed to the KSC landing at a weight of 91,446kg an hour later. 10 flight-test manoeuvres were called for, and the 1,046km of cross-range needed to line the Orbiter up with the 4,571m-long, 91m-wide runway required numerous attitude changes. While the runway itself was being patrolled to ensure that no alligators and bobcats chose to cross at an inconvenient moment, John Young was flying above it, worrying about the fog and gloomily describing it as "a good old marginal situation". Mission Control decided correctly that the fog would lift for the landing, due 13min after dawn. When *Challenger* appeared directly over Kennedy, a brilliant speck in the

Mission 41-B: "cherry-picker" in space. Bruce McCandless is moved around on the end of the robot arm as he stands on a foot restraint or work platform connected to it. *(NASA)*

rising sun and over 18,000m high, the author certainly shared Brand's own conviction, expressed later, that he was far too high to have any hope of touching down on the runway. But Brand's instruments assured him that all was well, and in a graceful sweep out over the Atlantic *Challenger* came down, south-to-north and on the centreline. The 3,282m roll-out left a comfortable 700m of unused runway plus the 300m of overrun. The first Shuttle birdstrike left some feathers on the windshield. As soon as it was safe the Orbiter was jacked up to check the brakes, and once again "significant damage" was found in the right main gear. But the need to remove the left OMS pod and change the RMS, in addition to the usual tile damage, was not expected to prevent a record turnround of about 55 days, permitting the SMM repair mission to go ahead during the early-April launch window.

Mission 41-C The Solar Maximum Satellite repair mission, planned and anticipated for years, was successfully carried out after an early setback that led to the SMM coming within 5min of being completely lost. Launched 4 Apr 1984, 41-C was commanded by Robert Crippen, 47, making his 3rd flight; 1st-timers were Francis ("Dick") Scobee, 45, pilot; Terry Hart, 38, MS1; Dr George ("Pinky") Nelson, 34, MS3 and EV1; and Dr James ("Ox") van Hoften, MS2 and EV .

Challenger was for the 1st time placed on a direct-insertion trajectory - initially 462 × 220km - designed to conserve onboard propellants by avoiding the need for an OMS burn. But uneven burning of the SRBs led to extra use of ET fuel, with the computers coming within 1sec of commanding an OMS burn which would have ruled out any attempt at SMM recovery. The Long Duration Exposure Facility (LDEF) carrying 57 experiments, including 6 for the USN and USAF, was placed in orbit by the RMS on the 2nd day without difficulty despite its record 21,322kg wt. The world's 1st retrievable satellite, it is to be recovered in 10 months' time. Mission Control was surprised and visibly dismayed on Day 3 when, after a slow rendezvous with Solar Max in its 490km orbit, Nelson flew across to it in his MMU but failed to dock after 3 attempts. A small object against the 4m-high satellite, his face within inches of it, he was heard reporting: "It's pitching into me a little," and that although he had achieved a good dock, the trunnion pin jaws had failed to fire. An attempt to stabilise Solar Max by grasping a solar wing only added yaw. Nelson reported that he was running low on fuel, and Crippen ordered him to return to *Challenger*. "I'm not flying right either, Crip," he complained. Crippen offered to bring *Challenger* to the rescue, but Nelson realised that he was being disoriented by the Orbiter's manoeuvres. Crippen and Hart then used the RMS in 4 unsuccessful attempts to grapple Solar Max, now wobbling about all 3 axes. With forward RCS propellant much depleted, Crippen ruled out another MMU effort and clearly had little hope that further RMS attempts would be possible. Goddard, vainly trying to regain control of Solar Max, now afflicted by steadily weakening batteries, concluded that its gyros were saturated and giving false readings. Miraculously, within 5min of final extinction, the satellite emerged from eclipse with its panels pointed sunwards and its batter ies charging. While *Challenger*'s crew took a day off, the batteries were recharged, the computers reloaded and Solar Max restabilised using its magnetic torque bars. Mission Control devised a slow, economic re-rendezvous, NASA gloom was replaced with renewed confidence, and from then on everything seemed easy. On Day 5 *Challenger* slid beneath Solar Max and Terry Hart reached up and grabbed it with the RMS at the first attempt. The spacecraft was lowered into its cradle, tilted so that its 8m span cleared *Challenger*'s fin, and turned ready for repair. On Day 6 van Hoften and Nelson, secured to the RMS platform and manoeuvred into position by Hart, used an electric wrench and replaced the attitude-control box within an hour. A baffle was placed over a vent port which had been allowing space plasma to enter and affect the X-ray polychromator detectors; and finally scissors and a Japanese-made screwdriver were used to remove and replace the failed electronics box on the coronagraph/polarimeter. Nelson used a tape

Mission 41-C: RMS placing Shuttle's heaviest payload yet, the LDEF, in orbit. Carrying 57 experiments, it was due to be retrieved 10 months later. *(NASA)*

Mission 41-C: astronaut George Nelson, left between solar panels, vainly attempting to dock with the 4yr-old Solar Max satellite. *(NASA)*

measure to check Solar Max's trunnion pins while van Hoften carried out 50min of engineering tests in the MMU. A 7hr 7min EVA had completely restored Solar Max to health and next morning the RMS was used to replace it in orbit. The triumph of landing on Friday 13 Apr was unmarred by yet another "wave-off" from Kennedy, and *Challenger* touched down an orbit later at Edwards. But unpredictable weather at Kennedy led to Mission Control's Operations Directorate recommending that Edwards should once again become the primary landing site for at least the next 11 missions, pending improvements in weather forecasting and a study of alternate landing sites at Orlando, Tampa and Jacksonville.

Mission 41-C: Nelson, right, on RMS work platform and James van Hoften complete two major repairs on Solar Max, now securely held at rear of payload bay. *(NASA)*

Space Station

History President Reagan directed NASA to begin development of a permanent Space Station, crewed by 6-8 astronauts, in his Jan 1984 State of the Union Address. Echoing a predecessor's 1961 Moon landing pronouncement, he proposed that it should be in orbit "within a decade" and should constitute a bold and imaginative programme to maintain US leadership in space well into the 21st century. It is to start as an 8yr, $8 billion project needing perhaps 7 Shuttle flights for initial assembly. NASA Administrator James Beggs was asked to discuss participation with European nations, Canada and Japan. The author was told that NASA would not regard anything less than 15-20% foreign participation, equal to at least $1 billion, as worthwhile. Decisions would be required by the end of 1984 so that participating nations could join in on Phase B, 2yr of engineering definition studies.

The Presidential decision overrode continuing opposition from DoD. Worried about the need to spread its space defences as widely as possible to make them less vulnerable, the Defence Department had said in mid-1983 that it saw no need for a space station. DoD had been supported in end-1983 reports by the National Research Council, which also saw no scientific need for a station during the next 20yr. Behind all this is the fact that defence needs call for a polar-orbiting station, whereas NASA's scientific and commercial activities would be more economically and conveniently conducted in low (and vulnerable) equatorial orbit.

Space station and space platform studies had been under way for some years before the Reagan decision,

Space Station: this Rockwell concept includes a large unpressurised rack for storage, facilities for assembling large structures, and a "safe haven" docking hub. Also shown is an OTV delivering a comsat to synchronous orbit. *(NASA)*

with each NASA centre doing its own research. JSC's Space Station Office (since designated NASA's lead development centre) was advocating an evolutionary facility with "modular add-ons". Tasks would include servicing of satellites in low orbits and refuelling of Orbital Transfer Vehicles for the placing and supervising of geostationary satellites. JSC also advocated design of the station as a stepping stone towards a lunar base to be used for mining raw materials, military surveillance and communications, and perhaps also for hazardous industrial activities. In May 1982 NASA headquarters set up a

Space Station Task Force to co-ordinate all the studies, with English-born John Hodge (who had played a pioneering role in Mercury, Gemini and Apollo) as director. His 3yr evaluation of "needs, attributes and architectural options" included Phase A study contracts totalling $6 million to Boeing, General Dynamics, Grumman, Lockheed, McDonnell Douglas, Martin Marietta, Rockwell and TRW.

NASA emissaries seeking international support in Europe in 1983 said that both polar and low-Earth orbit stations would be needed by the year 2000, by which time returns from the stations should be worth $2·4 billion a year. NASA, having learned that more colourful long-term targets were politically unpopular on cost grounds, was surprised when President Reagan and his science adviser criticised its approach as bureaucratic and not visonary enough: lunar bases and manned expeditions to Mars were back in favour. It was also being pointed out that, with Shuttle missions currently operating at only 65% load factor, the rest of the capacity could be used economically to carry fuel and equipment to a space station warehouse.

Subject to Congressional approval, funding was expected to total $30 million in 1984, $150 million in 1985 and $400 million in 1986 before rising steeply to $1 billion a year.

Objectives The Reagan directive aims to encourage private investment in space ventures by assisting industry with research facilities and "seeded" money. The communications industry, it is pointed out, has already placed 80 satellites in orbit and does $3 billion of business a year, directly and indirectly employing 1 million people (mostly Americans); another 300 such satellites will be needed by the year 2000. Potentially profitable activities already identified by NASA and industry include: the refining of biological materials to produce improved treatments for cancers, diabetes and certain kidney diseases; development of ultra-pure semiconductor crystals for use in super-fast computers and devices of interest to defence electronic industries; production of super-light, high-strength materials for high-performance aircraft and high-strength, super-insulating materials for homes, cars and aircraft; and electronic mail and computer-to-computer communications. The joint development of new medications by Johnson & Johnson and McDonnell Douglas, cited as the most striking example of this kind of activity, was followed in Feb 1984 by an announcement that 3M of St Paul, Minnesota, is to join NASA in experiments on organic crystalline materials and thin film, which have applications in electronics, imaging, energy conversion and biology.

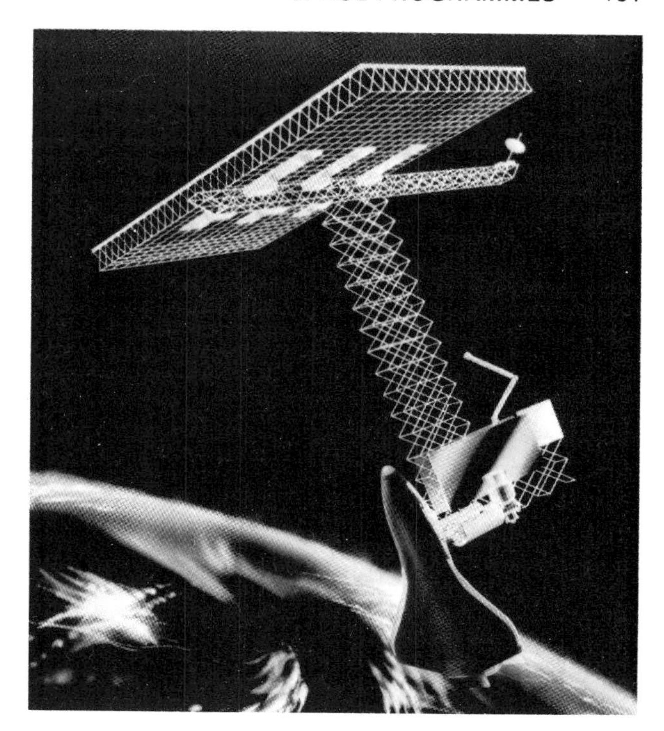

Johnson Space Centre's own concept for a space station. JSC has now been made the lead centre for the Station. ***(NASA)***

Starting his European tour to seek foreign support in London in Mar 1984, James Beggs told the author that the Space Station could also be a "way station" to the Moon, Mars and the asteroids. But a lunar base has a low priority, since scientists have still not examined all the lunar samples brought back during Apollo. Asked whether this would change rapidly if the Soviets established a lunar base, Beggs said: "We don't march to

Space Station: TRW concept. ***(NASA)***

A McDonnell Douglas proposal for an 8-man, 5-module building-block station carrying 3 months' supplies. This company built Skylab and is expected to play a major part in Station development. *(McDonnell Douglas)*

An electrically driven automatic assembly machine, moving along one edge, could instal columns and joints to construct a platform 3 times the size of a football pitch in 2 days. 5,000 columns could be delivered on 1 Shuttle flight. *(Lockheed)*

the same drum beat as the Soviets any more – nor do they march to ours." They were in fact going in opposite directions, having gone to a "rudimentary space station" before tackling a reusable shuttle. He gave the impression that a visit to an asteroid had the highest priority, followed by Mars. If the US military become interested in manned space station activities, Beggs believes, they will probably use NASA's work to build one in higher orbit later.

Elements NASA is likely to act as prime contractor for the Space Station, issuing perhaps 4 Phase C contracts to groups of manufacturers (1 of which might be European). Initially it will be entirely Shuttle-launched and supported, ruling out the use of expendable launchers. First of 5 modules, possibly launched at monthly intervals starting towards the end of 1991, will be a power unit. It will comprise a large solar array and storage batteries since the preferred nuclear power option is as yet politically unacceptable; the module will include attitude-control and reboost systems. Next will come a habitable module with multiple docking adaptor to act as the station's hub. Only at this stage is EVA likely to be necessary, in connection with solar array deployment; later modules are likely to be docked with the Shuttle robot arm. The modules will be cylinders 6·7m long with 4·2m dia; total station weight will be about 36,000kg. A NASA presentation in Washington in Jul 1983 envisaged that 2 of the 4 habitable modules would be used by NASA for life science and materials-processing research, the others by McDonnell Douglas and Johnson & Johnson for biological production and by Microgravity Research Associates for crystal growth. Studies showed that each module would require 3-4hr work each day by crew members. During the 1990s a 2yr trial by 2 crewmen would validate closed-cycle life-support systems, ultimately permitting ground supply requirements to be much reduced. Another 10 modules would be needed by 1996, with station weight reaching 94,000kg by 2000.

Included in the $8 billion funding are 2 free-flying platforms. One, called the Astro Platform, will accompany the station in its 300km, 28° orbit, carrying advanced astrophysics sensors, materials-processing equipment and other research activities which are affected by the disturbance caused by crew movements; the platform would be serviced as necessary by astronauts. The second platform, Shuttle-serviced in a 98° Sun-synchronous polar orbit, would be used for Earth-atmosphere and Earth-resources studies; this project is more likely to attract DoD investment.

By Feb 1984 NASA had set up 7 inter-centre teams to study the required space station systems: Attitude Control and Stabilisation, Data Management, Auxiliary Propulsion, Environmental Control and Life Support, Space Operations Mechanisms, Thermal Management, and Electric Power.

An unmanned Orbital Manoeuvring Vehicle (OMV), also known as the Transfer Orbit Stage (TOS), will be needed by 1990 to deliver and retrieve payloads up to altitudes of 3,700km, beyond the reach of the Shuttle. Paid for from STS funds, this is expected to cost $375 million to develop.

This Shuttle-derived Space Station cargo launch vehicle consists of a shortened ET with a single Shuttle main engine, and a large canister containing upper stage and payload on top. The SRBs on each side are also shortened. ***(Boeing)***

Space Telescope

History Planned for launch by Shuttle in 1986, and renamed in Oct 1983 the Edwin P. Hubble Space Telescope after the creator of the Big Bang theory, this much postponed project is described by NASA as "the most important scientific instrument ever flown". Observations of objects such as quasars, Seyfert galaxies and the remnants of large supernova events should almost certainly lead to new information about the origins of the universe. With an expected 9,070kg weight, it is the largest payload under development for Shuttle launch. Operating in a 593km orbit inclined at 28·5°, this 2·4m-dia reflecting telescope (accommodating 5 instruments at its focal plane) should enable astronomers to spend about 20yr observing 350 times the volume of space available to ground-based observatories. Objects 50 times fainter and 7 times further away than are visible from Earth-based telescopes should be detectable. In the most distant views astronomers will be observing the universe as it was around 14 billion years ago; it should also be possible to study the planets of nearer stars.

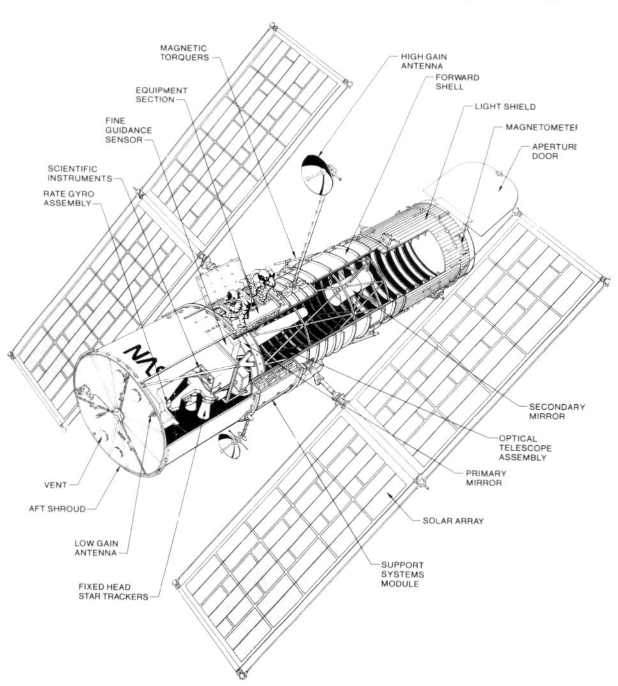

The Space Telescope is expected to expand the observable universe 350 times, enabling astronomers to study objects 14 billion light years away. ***(Lockheed)***

The Space Telescope is a cylinder 14·3m long and 4·7m in diameter. Two retractable solar arrays with a total area of 33 sq m will still supply over 4,000W of electricity after two years in orbit. It will be operated remotely from the specially built, $24 million Space Telescope Science Institute at Johns Hopkins University, Baltimore, Md. Every 30 months it will be visited by Space Shuttle so that spacewalking astronauts can service it and replace instruments if necessary. Every 5yr the solar arrays will be retracted and the whole Telescope brought back to earth for refurbishment and relaunch about a year later.

ESA originally agreed to contribute about 15% of the cost when that was estimated to be $595 million, in return for 15% observation time for European astronomers. Europe's contribution covers provision of the faint-object camera and the solar arrays. But technical troubles and a further 18 months' delay announced in 1983 will bring the cost to $1 billion by launch date. Lockheed is the main NASA contractor for the Telescope's structure and support systems; the Perkin Elmer Corp is building the optical telescope assembly.

Star

5 Shuttle missions are planned under the Shuttle Telescopes for Astronomical Research programme, 3 of them directed at Halley's Comet. First of these, now designated OSS-4, is due to be flown in Nov 1985 to catch Halley at its closest approach to Earth. The next, in Mar 1986, will observe Halley at its closest to the Sun. The 3rd, in May 1986, will observe the comet after its closest solar approach.

Surveyor

History The 3rd of the 3 unmanned lunar exploration projects carried out in parallel with the 3 manned projects aimed at placing men on the Moon before 1970. Following the 3 successful Rangers, Apollo planners needed more detailed photographs of the lunar surface and data about the composition of the lunar soil and its ability to support the 6,800kg weight of the Apollo Lunar Module. The Hughes-built Surveyor's launch weight of 998kg was reduced by fuel consumption to 283kg at touchdown. Other objectives were to develop the technology for soft lunar landings and survey potential manned landing sites. Because Surveyors 1, 3, 5 and 6 successfully landed at sites spaced across the lunar equator and achieved all Apollo objectives, Surveyor 7 was landed on the rim of Crater Tycho to conduct digging, trenching and bearing tests. Launches were by Atlas-Centaur from C Canaveral.

Surveyor 1 L 1 Jun 1966. After a 63hr 36min flight, made the world's first fully controlled lunar soft landing, in the Ocean of Storms. In the following 6 weeks it sent back 11,150 pictures, from horizon views of mountains to close-ups of its own mirrors.

Surveyor 2 L 20 Sep 1966, crashed SE of Crater Copernicus when one of the vernier engines failed to fire.

Surveyor 3 L 17 Apr 1967, landed safely despite a heavy bounce in the Ocean of Storms 612km E of Surveyor 1. In addition to returning 6,315 photos, it used a scoop to make the first excavation and bearing test on an extraterrestrial body. (In Nov 1969 Apollo 12 landed almost alongside Surveyor 3 and Conrad and Bean brought back to Earth the TV camera, scoop and other parts for studies of the effects of 31 months of lunar exposure.)

Surveyor 4 L 14 Jul 1967, lost radio contact 2½min before touchdown and crashed in Sinus Medii (Central Bay).

Surveyor 5 L 8 Sep 1967. Major technical problems were successfully solved by tests and manoeuvres during the flight. 18,000 photos were obtained from the southern part of the Sea of Tranquillity, and the first on-site chemical soil analysis was carried out.

Surveyor 6 L 7 Nov 1967, landed in Sinus Medii in the centre of the Moon's front face. In addition to sending over 30,000 pictures, it performed the first take-off from the lunar surface. Its 3 vernier engines, fired for 2½sec, lifted it to 3m and landed it again 2·4m away to prove that the surface was firm enough for a manned landing.

Surveyor 7 L 7 Jan 1968, was then sent to Crater Tycho; chemical analyses suggested that the debris there had once been in a molten state. 21,000 photos were obtained. By successfully detecting a 1W laser beam transmitted from Earth Surveyor 7 proved that laser communications in space were feasible. Apollo astronauts' requests to visit this site were turned down because of the extra fuel needed to reach and return from such a remote area.

Spacecraft description A 3-legged vehicle 3·05m high, with a triangular aluminium frame providing mounting surfaces and attachment points for landing gear, main retro-rocket engine of up to 4,536kg thrust, and 3 vernier engines of up to 47kg thrust. A central mast supported the high-gain antenna and single solar panel of 0·75 × 1m. Aluminium honeycomb footpads were attached to each leg of the tripod landing gear. The TV camera was pointed at a mirror which could swivel 360°, and at Earth command be focused from 1·2m to infinity. Narrow or wide-angle views and 200 or 600-line photos could be returned. Surveyors did not go into lunar orbit before landing; in a direct approach, automatically controlled by radar, the main retro-rocket was fired at an altitude of about 96km for 40sec. At 40km altitude, with speed down to 402km/hr, the verniers took over to slow the craft to a soft landing at 12km/hr.

TDRSS

History The Tracking and Data Relay Satellite System (TDRSS) is a key factor in Space Shuttle activities, for which it will ultimately provide continuous data and voice communications instead of the current 20% coverage via Earth stations around the world. But the programme has proved difficult and expensive. First use was scheduled for Spacelab 1, for which 2 were needed to provide a continuous flow of scientific data in real time. Other important uses should include relay of Landsat Thematic Mapper images and, later, communications with unmanned free-flying platforms. But repeated TDRSS launch delays led to postponements of Spacelab 1 and protests to NASA by ESA.

The project became excessively complicated when NASA entered into its first large-scale joint venture with the commercial sector in the form of a lease-type contract with Western Union. Funding problems, including interest charges totalling $600 million, led to delays. Then the Defence Department, which also wants to use the system for data relay, caused additional postponements when its modifications and additions made the satellites too heavy for Centaur launches. This meant that the programme had to wait until the Shuttle became available. TDRS-2 was still further delayed when DoD decided it should have additional protection against possible Soviet radio interference or interception. With DoD modifications adding over $300 million, the original award of $796 million for 3 satellites, rising to $841·8 million for 6, grew to $2·156 billion.

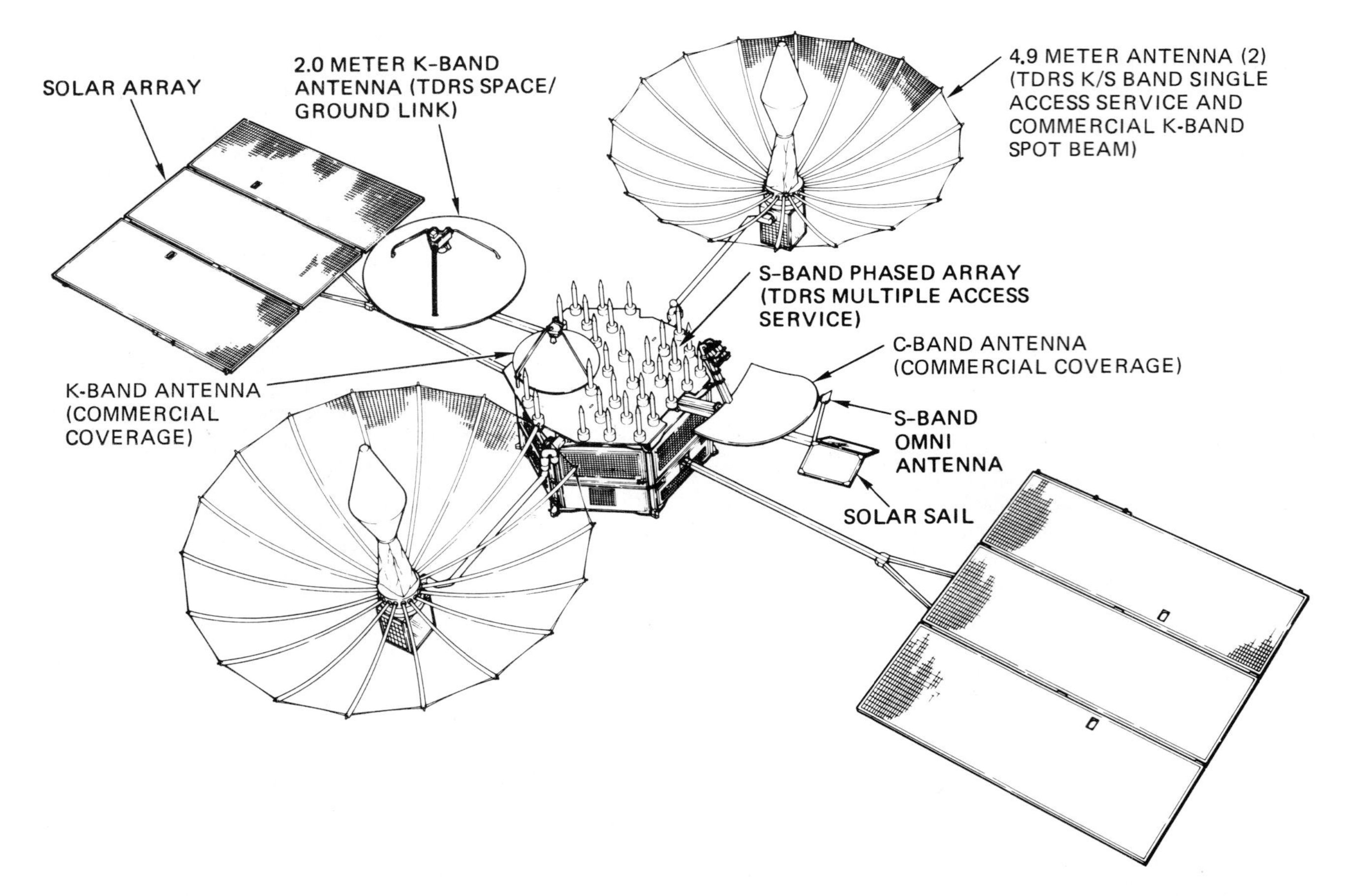

America's Tracking and Data Relay Satellites, though much delayed, will ultimately mean that many foreign-based Earth stations will no longer be needed. ***(TRW)***

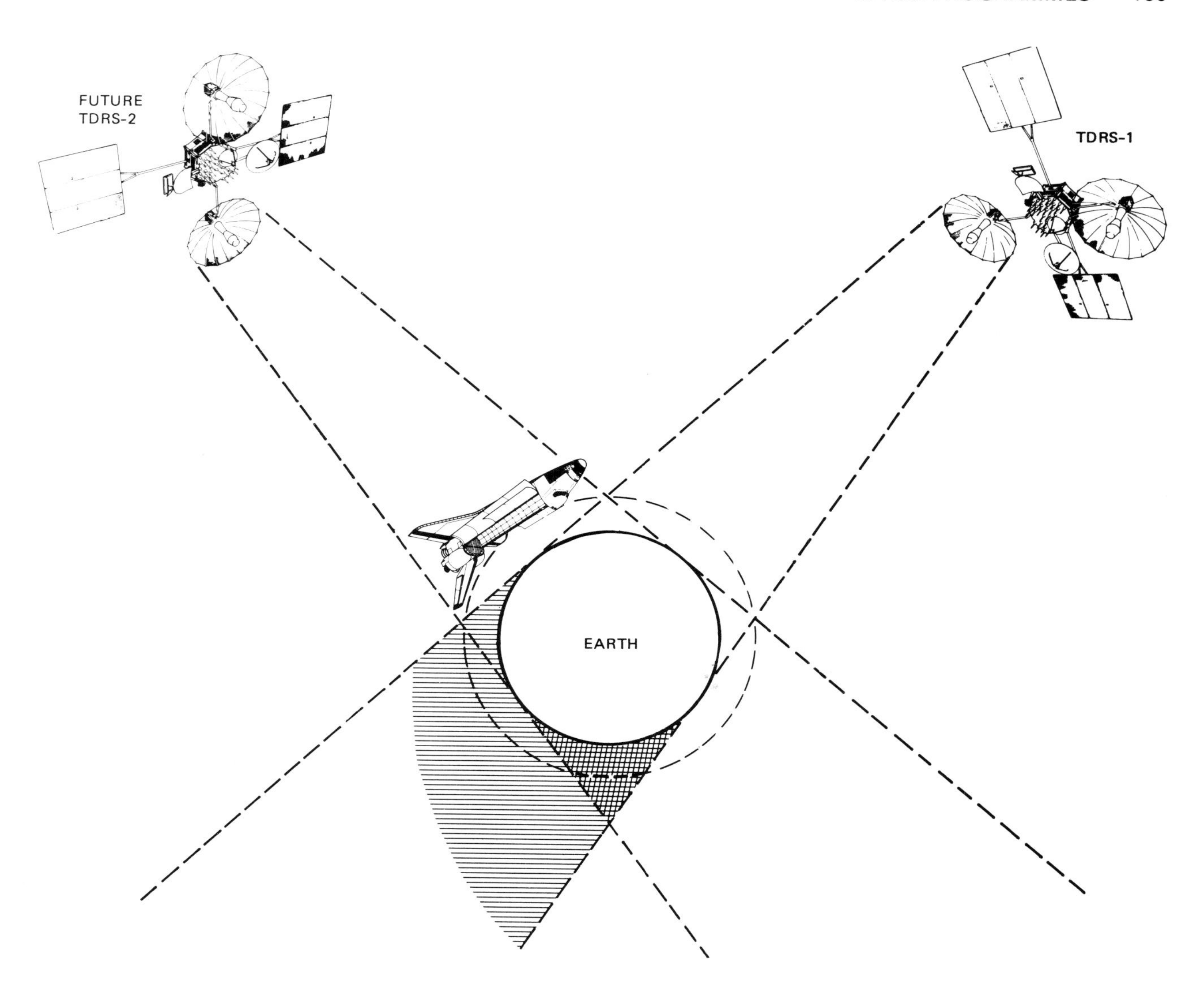

TDRS 2-satellite system will give the Shuttle and low-orbit satellites 85% coverage; only the small double-shaded zone will be uncovered. *(TRW)*

Prime spacecraft contractor was TRW until Jun 1980, when the contract was transferred to a partnership consisting of Spacecom (set up by Western Union Space Communications Inc), Continental Telephone Space Communications Inc and Fairchild Industries. It was then announced that TRW would build 6 satellites: 2 for lease by NASA, 1 for Spacecom for commercial Westar users, the 4th as a joint on-orbit spare for NASA and Spacecom, and 2 as flight-ready spares. The USAF also dropped its plan to instal costly additional capability to prevent "Soviet radio penetration" of TDRS-2.

Two months before launch of TDRS-1 was due NASA and Spacecom agreed on cancellation of the commercial service, making it a 100% NASA system, though still owned by Spacecom. Though increasing NASA's costs by $216 million, this decision extends TDRSS life in the relay role to 15yr, and will enable numerous ground stations to be closed. NASA will now retain TDRS-4 as a ground back-up to the 2 operational and 1 back-up craft in orbit. TDRS-1 was to be placed at 41°W, east of Belem, Brazil, and TDRS-2 at 171°W over the Pacific, with TDRS-3 going to 79°W in 1984 as an in-orbit spare. Launch weight of TDRS/IUS from the Shuttle payload bay is 16,783kg, TDRS in-orbit weight 2,120kg. TDRS-1's launch problems are described in full under Shuttle mission STS-6: a 2nd-stage malfunction of the IUS and a collision between them at separation damaged a number of RCS nozzles and left the satellite in a 21,786 × 35,317km orbit while NASA, TRW and Spacecom experts tried to calculate how to get it into the correct orbit in time for Spacelab 1. After 30 thruster burns consuming 370kg of hydrazine fuel but still leaving a residue of 227kg compared with the 90kg needed for the planned 10yr life, TDRS-1 was triumphantly placed in its correct orbit on 29 Jun 1983 and finally moved to its permanent 41°W location by 17 Oct.

Telstar 3

American Telephone and Telegraph Co is replacing its Comstars, leased from Comsat General. Hughes was given a $137·7 million contract in 1980 for 3 domestic communications satellites. With launches from C Canaveral into slots at 96°, 76° and 88·5°W, they have 24 transponders operating at 4/6GHz. Spin-stabilised, they have 650kg in-orbit weight and 10yr life. Telstar 3A L 29 Jul 1983; 3B and 3C due in Aug 1984 and May 1985.

Topex

Successor to Seasat, Topex is a $270 million project being planned by NASA's JPL for launch in 1989. With lift-off weight of about 1,500kg and based on an existing design, it will spend at least 5yr mapping ocean currents, wave heights and seabed contours. It should mean safer navigation in coastal areas - for supertankers in the English Channel, for instance - and provide vital information for oil rig designers by building up data on maximum wave heights. With an accuracy of 0·14m, these figures should be twice as precise as Seasat's. Ocean floor details and depths will also help in the making of decisions about dumping of toxic waste. Subject to agreement on an Ariane 4 launch, France's CNES may join the project, which would then be known as Topex/Poseidon. In Apr 1984 NASA awarded 8-month, $1 million satellite definition contracts to Fairchild, RCA and Rockwell; joint and US-only projects are being studied.

UARS

Two Upper Atmosphere Research Satellites are to be launched by NASA in the late 1980s to study the Earth's atmosphere between 10 and 120km altitude. 26 investigations are to be carried out by teams from the US, Britain and France in response to growing anxiety about the effects on stratospheric ozone of emissions from high-flying aircraft, and by chlorofluorocarbons from aerosol spray cans, refrigeration, air-conditioning and foam-blowing operations. Each country is to pay its own costs.

USAT

United States Satellite System Inc is planning 3 Ku-band satellites for launch by a NASA vehicle or Ariane. Equipped with regional and spot-beam transponders, 2 will be launched in Feb and Aug 1984 to 85° and 120°W.

Vanguard

History The Vanguard rocket, a US Navy development of sounding rocket technology built on Viking and Aerobee, was selected in 1955 as the most suitable launcher for America's first satellite. This was an unfortunate choice, as it turned out, since Russia had launched Sputnik 1 and the US Army had launched Explorer 1 before the first Vanguard satellite went into orbit. The choice had been between sounding rockets and military rockets - in this case the US Army Redstone - for the first satellite launch. Russia choise military rockets in the same year and got there first. In America, against a background of bitter service rivalry with the US Air Force advocating the use of its Atlas rocket, the deciding factor was the decision that development of military rockets, known to be lagging behind Russia's, must not be held up.

The Vanguard 1st stage, using liquid oxygen and kerosene, developed 12,250kg thrust. The 2nd stage, burning white fuming nitric acid and unsymmetrical dimethyl hydrazine, provided 3,402kg thrust. The 3rd stage, with solid propellant, yielded 1,406kg. The complete vehicle was 22m long, with 1·1m dia, and weighed 10,250kg. The first full-scale Vanguard Test Vehicle, designated TV2, was successfully launched with dummy upper stages from C Canaveral on 23 Oct 1957, sending a 1,814kg payload on a 175km-high 491km trajectory. But TV3 toppled over on the launch pad and exploded. 2 months later the back-up vehicle veered off course and broke up at an altitude of 5·8km. By then Sputnik 1 was in orbit. TV4 finally launched Vanguard 1 (L 17 Mar 1958; wt 1·47kg), a remarkably long-lived and successful satellite which transmitted temperatures and geodetic measurements until Mar 1964. But TV5 failed and so did the first 3 operational Vanguard Satellite Launch Vehicles, designated SLV1, 2 and 3. There were 2 more failures between the successful Vanguard 2 (L 17 Feb 1959; wt 9·8kg) and Vanguard 3 (L 18 Sep 1959; wt 23kg), the latter marking the end of the programme. Dr Wernher von Braun, rejoicing in the triumph of Redstone, later said Vanguard was a "goat" through no fault of its own. For all its failures, Vanguard's 2nd and 3rd stages were later bequeathed to Thor and Atlas, and its 3rd stage to Scouts, while the swivelling motors on its 1st stage worked perfectly on Saturn.

Venus Radar Mapper

This $300 million JPL project replaces the Venus Orbiting Imaging Radar (VOIR), which would have cost twice as much. Using available hardware and a left-over Voyager 3·7m-dia antenna, it is scheduled for Shuttle/Centaur launch in Apr 1988 to orbit Venus 8 times a day at 250 × 8-10,000km. It would take about 243 days to map the Venusian surface through its cloud cover with a synthetic-aperture radar system. Costs have been cut by operating from an elliptical instead of circular orbit, and using the one antenna to map the surface during the close approach, then turning it round for data transmission from the elliptical orbit.

Viking

History Whether there is microbial life on Mars is still being debated as the scientists continue to study the findings of the two Viking Landers 8yr after the event. A US Government report published in Jan 1981 says: "Although scientists found what could be considered evidence of life, they could not find evidence of death.

Vanguard 3 collapses in sea of flame during US Navy test firing, December 1957. ***(NASA)***

Since we know only of finite life spans, the lack of 'bodies' led some scientists to suggest that the indications of life were in fact some new unknown exotic chemistry unrelated to life processes." The Viking 1 and 2 Landers, which touched down with remarkable success on 20 Jul and 3 Sep 1976 after separating from their parent Orbiters, provided man's first surface pictures of the Red Planet: they showed a rocky desert which was indeed red, set against a sky of pink or red.

The Landers sent back a total of over 4,500 pictures, the Orbiters 51,500, a yield which surpassed all expectations in quantity and quality. The last Viking 1 Orbiter pictures were not processed until Mar 1982. A Viking 1 Lander picture of Jul 1981 showed how soil piles created 2yr before by the robot arm had been noticeably but not completely eroded by wind. The low wind velocities were a surprise: neither Lander recorded gusts of more than 120km/hr, and mean velocities were much lower. Nevertheless, a dozen small dust storms and 2 global dust storms were observed, the latter obscuring the Sun from the Landers and most of the planet from the Orbiters. The Orbiters mapped 97% of the surface at 300m resolution and 2% at 25m or better. Their thermal mappers and atmospheric water detectors showed that the N Pole cap was water ice; the S Pole cap probably retains some carbon dioxide throughout the summer. Subsurface water in the form of permafrost covers much if not all of the planet. The N and S hemispheres have very different climates because the global dust storms originate in the south in summer. Surface changes occur extremely slowly, only 2 small events having been observed at the landing sites in 4yr.

These missions continued the Martian exploration started by Project Mariner. Originally scheduled for launch in 1973, Viking was postponed for financial reasons to the much less favourable 1975 launch window. A political flavour was added when it was decided to attempt the first landing, after a journey lasting nearly a year, on 4 Jul 1976, the 200th anniversary of US independence. But the pre-launch history of the project was most inauspicious: trouble with both the Titan launcher and battery failures and helium leaks in the spacecraft led to Vikings 1 and 2 having to be switched. Viking 1 finally lifted off 9 days late on 20 Aug 1975, Viking 2 19 days late on 9 Sep 1975. By then only 12 days remained of the 40-day launch window. The first team of 750 scientists, engineers and technicians, working with traditional NASA determination and flexibility, overcame technical problems and setbacks during the launch and the 740 million km chase of Mars around the Sun. Then they had to search for new landing sites (and abandon the 4 Jul landing) when the first orbital pictures showed the original selections at Chryse and Cydonia to be much too rough. These areas were chosen because Chryse is a valley NE of the giant 4,800km-long Martian Grand Canyon discovered by Mariner 9. About 5km lower than the mean surface, this area may once have been a drainage basin for a large portion of equatorial Mars. Cydonia is at the southernmost edge of the N Pole "hood" – a hazy veil shrouding each polar region during the winter and, at 5·4km below mean surface, even lower than Chryse. It was hoped (though the scientists were to be disappointed) that these would be smooth, calm areas, with enough moisture to carry organic life.

Viking 1 Orbiter took this picture of Martian atmospheric phenomena on Revolution 802. Cloud plumes 150km long may have been cooler air breaking through a warm layer in the Kepler Crater area. *(NASA)*

Spacecraft description Design was based on the successful Mariner 9, with the addition of the Lander dictating a larger structure. The total lift-off weight of 3,520kg was one factor which led to the 11-month journey to Mars, instead of 5 months for the Mariner missions.

Orbiter With a total weight of 2,325kg (propellant 1,422kg) it is octagonal, 2·4m across, 3·3m high and 9·7m wide with solar panels extended. These panels have an area of 15 sq m and provide power for the radio transmitter and, during flight, to the Lander. Rechargeable batteries provide power when the spacecraft is not facing the Sun, and during correction manoeuvres and Mars occulation. The Orbiter carries 3 instruments and 4 experiments. 2 narrow-angle TV cameras were first used to check the landing site, then provided high-resolution imaging of the Martian surface. A water detector mapped the Martian atmosphere, while an infra-red thermal mapper searched the surface for signs of warmth. Other instruments provided data on the planet's size, gravity, density and other physical characteristics. The communication system was used as a relay between the Lander and Earth via an antenna on the outer edge of a solar panel. Data (a total of 1,280 million bits) could be stored aboard the Orbiter on 2 8-track digital tape recorders and transferred to Earth at the rate of 16,000 bits/sec.

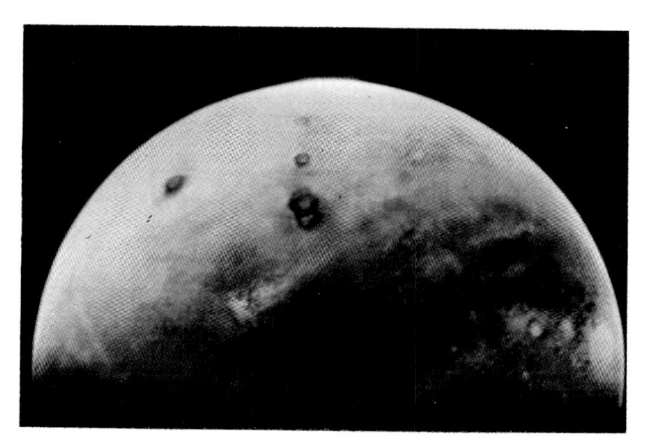

Mars from 560,000km as seen by Viking 1. *(NASA)*

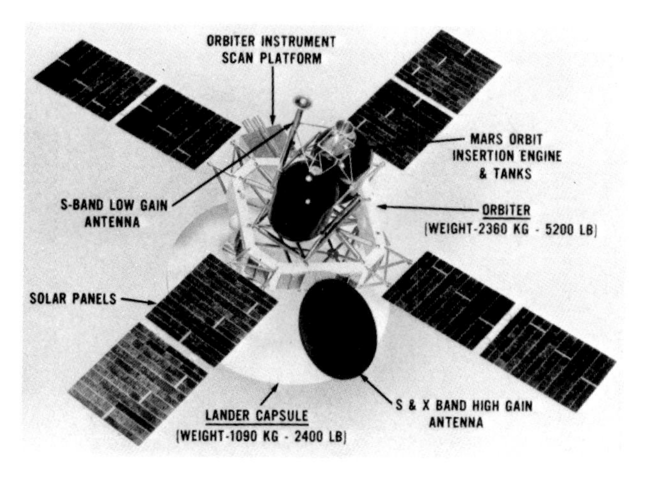

Viking spacecraft: pre-launch sterilisation ensured that the chances of contaminating Mars with Earth Micro-organisms were less than 1 in a million. *(NASA)*

Lander With total weight of 1,067kg (fuel 491kg), it is approx 3m across and 2m tall. Before launch it was encased in a bioshield to ensure that it did not contaminate Mars with Earth organisms. Inside was an aeroshell to act as heatshield during Martian entry. The main body is a hollow, 6-sided aluminium box with alternate 109cm and 56cm sides. 3 landing legs were provided because a 3-legged structure will always rest on all 3 legs, while a 4-legged structure also usually rests on 3 legs; they are attached to the shorter sides. Mounted on the outside of the body are 2 TV cameras with 360° scan, meteorology boom, surface-sampler boom, seismometer, power generators, antennas, 3 main descent engines (each containing 18 tiny nozzles to spread exhaust gases and minimise contamination on touchdown), inertial reference unit, various control boxes and the soil inlets for the organic, inorganic and biology instruments. Inside the body is an environmentally controlled compartment for the biology instrument, gas chromatograph/mass spectrometer, computer, tape recorder, data-storage memory, batteries, radios, data-acquisition and processing unit, and command control and support units. Basic power was provided by 2 SNAP nuclear generators. The spacecraft cost $930 million, plus $132 million for the 2 Titan-Centaurs. Martin Marietta was prime contractor to NASA's Langley Research Centre, with TRW having designed and built the biology instrument which searched for life.

Viking 1 L 20 Aug 1975 by Titan 3E-Centaur from C Canaveral. On 19 Jun 1976, after a 310-day flight, Viking 1's main engine was fired for 38min, consuming 1,800kg of the total fuel load of 2,260kg and slowing the spacecraft by 4,296km/hr to the required orbital speed of 16,500km/hr. This placed it in a 42·6hr elliptical orbit significantly higher than that scheduled, and a trim manoeuvre was required after one orbit to put it in the originally planned 24·66hr orbit inclined at 38° to the Martian equator, with a 1,500km periapsis and 32,597km apoapsis. The Orbiter was soon transmitting spectacular pictures of the proposed Chryse landing area, revealed to be much rougher and more cratered than expected, with islands and channels cut by massive flows of water which had long since disappeared. The 4 Jul landing was postponed for a fortnight so that, with a combination of radar data from Earth and high-resolution photographs from orbit, a smooth area in the western reaches of Chryse Planitia (Gold Plain) could be found. On 20 Jul the Lander separation command was transmitted, initiating a 3hr 21min automatic separation and entry sequence during which pyrotechnic devices and springs pushed the Lander away from the Orbiter. The Lander deorbit sequence started with a 23min propulsion burn using 8 of the 12 hydrazine RCS engines; the remaining 4, each producing 3·6kg of thrust, were used for attitude control. During a 3hr coast period the Lander was rolled 180° to sterilise the base cover of the aeroshell with solar ultraviolet radiation. At 244,000m the Lander dipped into the upper levels of the Martian atmosphere. At 6,000m the base of the aeroshell was jettisoned and a 16m-dia parachute deployed. The initial entry speed of 16,500km/hr had been reduced to about 1,090km/hr, and the 1min parachute descent further reduced this to 220km/hr. By that time the altitude was 1,430m. Finally, 3 hydrazine engines slowed the Lander to a vertical touchdown speed of about 8·7km/hr. More than 19min after the actual touchdown, signals confirming a successful landing reached Earth. It was 7yr to the day after Armstrong and Aldrin had landed on the Moon. Already dramatic cliff falls into valleys and meteorite impacts into watery mud had been revealed by orbital photographs. Now man's first view of the Martian surface, transmission of which began 25sec after touchdown, showed a red jumble of boulders and rocks interspersed with what was probably sand. Later panoramic pictures showed a creamy pink sky. Scientists were relieved that only about 1·5% of the atmosphere was argon gas, instead of the 30% suggested by Russia's Mars 6 lander, since this gas could have had adverse effects on the equipment used to analyse soil samples. The only major disappointment was that the seismometer package, which should have dropped onto the surface to detect Marsquakes, remained stuck in its protective cage because the locking pin failed to eject. The search for Martian life by testing soil samples proved to be dramatic. The locking pin on the sampler arm, stuck at first, was finally shaken out 5 days after touchdown by means of a technique worked out in the Viking simulator. 8 days after landing the first soil sample was scooped up and transferred to the internal laboratories. Confirmation was provided by a clear picture of the straight trench which had been dug on the Martian surface. On 31 Jul the amount of oxygen released in the gas-exchange experiment was announced to be 15 times greater than expected. This could have been caused by biological activity, but the biology team cautiously decided it could equally have been caused by an inorganic chemical reaction in the soil which was not understood. The temptation to announce that there was life on Mars was resisted. The Orbiter's pictures of the surface commanded less attention but had their own fascination. Early morning ground fog was shown in low-lying areas such as craters and channels; pre-dawn pictures, when compared with those of the same area taken 30min later, showed white patches where slight warming of the sub-zero surface had caused small amounts of water vapour to condense in the colder air just above the surface. Pictures of the Cydonia region showed strange geometric markings, like an aerial view of ploughed fields.

Viking Orbiter 1 passed within 480km to get this view of Phobos. Project scientists now believe that Phobos and Deimos were captured from the asteroid belt. *(NASA)*

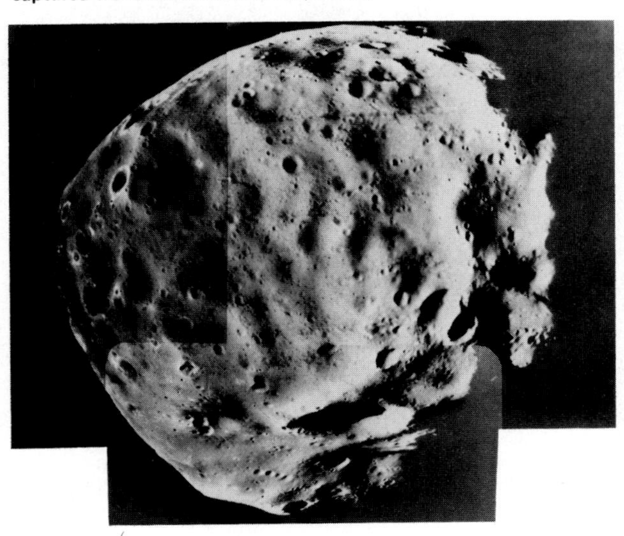

Mars' Crater Yuty, photographed by Viking 1 near the landing site from 1,877km, is 18km in dia, with central peak, and was probably caused by a meteorite collision. Layers of broken rock have been thrown out by the impact. *(NASA)*

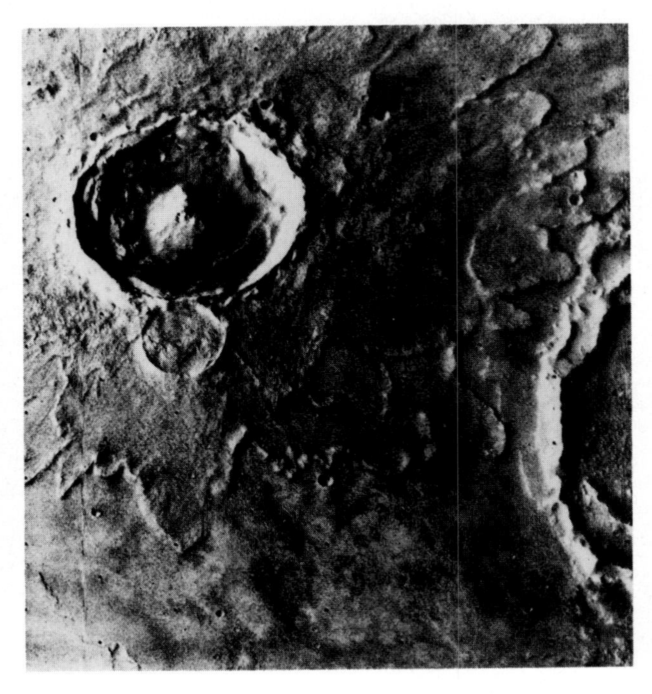

During its final phase Orbiter 1 took 30 pictures per day of the region SW of Olympus Mons and of the 3 volcanoes on Tharsis Ridge; the latter were of especial interest because of the large riverlike channels in their vicinity. On 7 Aug 1980, with attitude-control gas running out, the orbit was raised from 357 × 33,943km to 320 × 56,000km in order to avoid impact, and therefore contamination, before the year 2019. It was then switched off after 2 full Mars years (4 Earth years) of observations.

Suggestions in 1981 that the Viking 1 Lander should be abandoned as an economy measure, even though it should continue reporting from the Martian surface until the end of the century, brought an unexpected reaction. Stan Kent, a 25-year-old British-born space scientist, organised a "Save a Starving Robot" fund and soon collected donations totalling $60,000. After wrestling with administrative problems NASA accepted the money - the cost of keeping Viking going for 2 months - and was promised further support from the same source.

A system was devised for reconditioning each of Lander 1's 4 batteries, and it was then expected to return imaging, meteorology, radio science and engineering data on 8-day cycles until 5 Dec 1994. However, with only 1 full-time and 2 part-time staff involved, an inaccurate transmission sent in Nov 1982 resulted in contact being lost. Efforts to restore transmissions continued until May 1983, when Lander 1 had finally to be abandoned. In Jan 1982 it had been named the Thomas Mutch Memorial Station in tribute to the leader of the Viking imaging team, who had been lost in a Himalayan climbing accident.

Viking 2 L 9 Sep 1975 by Titan 3E-Centaur from C Canaveral. Wt 3,399kg. Viking scientists, fully occupied with the Viking 1 Orbiter and Lander, had to start running down those activities as Viking 2 approached Mars. On 7 Aug 1976 this was successfully placed in an orbit with 1,500km periapsis, 27·6hr period and 55° incl. With the Lander still attached, the Orbiter was soon transmitting pictures of huge striped patterns in northern regions which resembled the sand dunes created by wind in the Sahara Desert. In the light of Viking 1 experience a fresh search was begun for a suitable landing area. Sites in the Cydonia and Alba Patera regions were rejected because they had too many craters, ridges and channels. 12 days after Viking 2's arrival a new site, Utopia Planitia 7,420km NE of Viking 1, was chosen. Final confirmation on 3 Sep that Lander 2 was safely down, with one footpad perched on a rock and tilted at an angle of 8·2°, was not obtained until 8hr later. The shock of separation from the Orbiter caused a temporary electrical fault which deprived mission controllers of their primary communications link through the Orbiter. However, the Lander's onboard computer guided it down to a point later established as 225·86°W, 47·97°N. Because the radar was apparently misled by a rock or highly reflective surface, an unscheduled extra firing of the retrorockets 0·4sec before touchdown cracked the surface crust and scattered dust. It also gave an extremely soft touchdown. This time the seismometer was successfully freed to start its watch for Marsquakes. First pictures showed a bleak landscape covered with pitted rocks and small sand drifts. Scientists were somewhat disappointed that the area did not differ as much as expected from the Lander 1 site. The weather at the new site was found to be less complex than at Chryse Planitia. Winds followed a daily pattern, with average velocity of 15km/hr and maximum gust speed of 26km/hr. With the night 2·5hr shorter at the Utopia site, temperatures were slightly higher: minimum, at 04·08 Martian time, was -81°C (-114°F); maximum, at 15.50, was -30·5°C (-23°F). Surface pressure was 0·6 millibars higher than at Chryse, suggesting that the Utopia site was 1·3km lower. 9 days after touchdown the first soil sample was scooped up and transferred to the 3 biology experiments. But as weeks passed and the samples were analysed and reanalysed, Lander 1's first dramatic results were not repeated and no positive evidence of even the most primitive life forms was found. By the end of Oct, with Lander 1 having been virtually shut down for 50 Sols (Martian days) so that work could be concentrated on Lander 2, a small rock had been carefully pushed aside by the surface sampler arm and soil which had been

Viking 2 Lander looked across an orange-red plain strewn with rocks up to 1m across. Cylindrical mast is the Lander's low-gain antenna, which receives commands from Earth. *(NASA)*

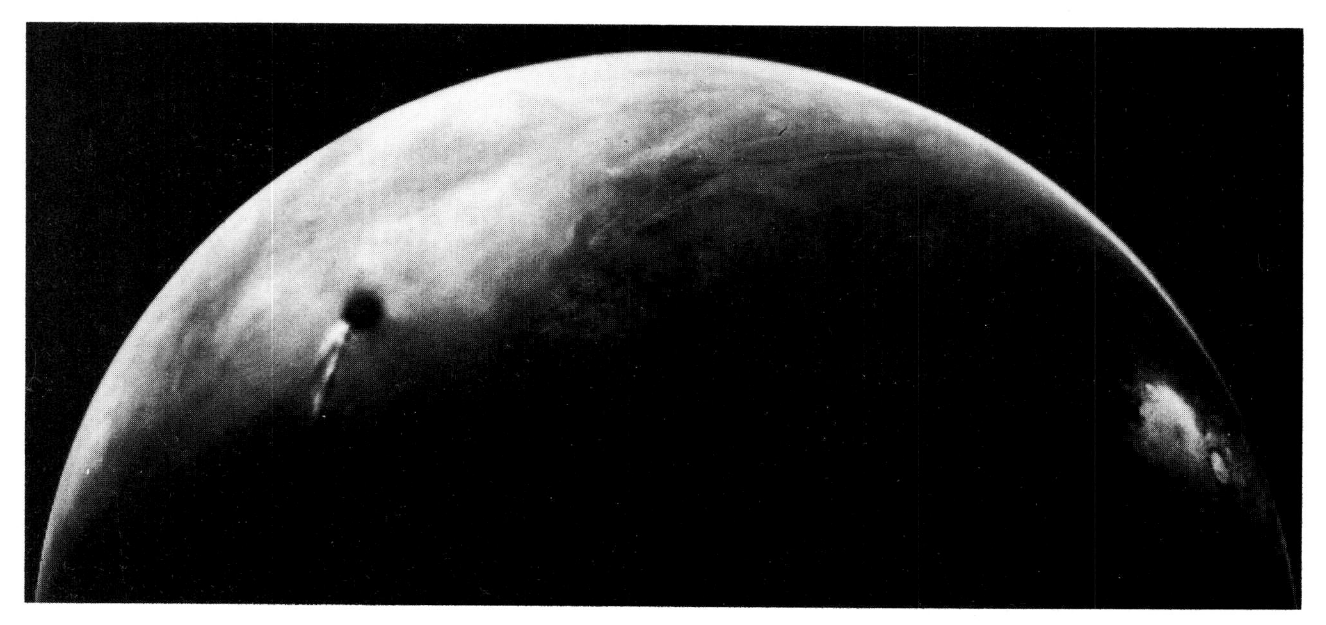

Giant Martian crater Ascreaus Mons was seen by Orbiter 2 with water-ice cloud plumes on its western flank. Centre, a great rift canyon is visible; right, a large frosty crater basin called Argyre. *(NASA)*

protected beneath it transferred and tested. When this was completed, just before the end of the primary mission on 10 Nov 1976, it was admitted that if there was life on Mars the Vikings were unlikely to detect it.

One of Lander 2's most exciting discoveries was water frost on the surface at the end of the Martian northern winter; it appeared again 1 Martian year (2 Earth years) later. The biology experiments of both Landers were ended in May 1977 when their supplies were exhausted, though both craft continued to send back pictures and data until Lander 2 was turned off on 11 Apr 1980. Orbiter 2, with its attitude-control gas depleted, was placed in a 302 × 33,176km orbit and turned off on 24 Jul 1980.

Future missions With the question of whether there is any sort of life on Mars still open - especially as one interpretation of the Viking evidence suggests that Mars may still be developing its atmosphere - there are already plans for further missions. One is for a Viking Rover able to operate for 2yr and travel up to 500km. That would go some way to answering critics who say that if the two Vikings had been visiting Earth and had descended in the Atlantic and the Sahara Desert they would have sent back very inaccurate reports about this teeming planet. Other proposals include a Martian glider, a ball which could be blown about the surface, and a sample-return mission as carried out by the Soviet Union on the Moon. But none of them is likely to get financial support at present, and the possibility of sending men to Mars has receded far into the 21st century.

Voyager

History Conceived in the 1960s as a "Grand Tour" of the outer planets - using their gravitational fields to accelerate one spacecraft in turn from Jupiter to Saturn, Uranus and Neptune, and another from Jupiter to Saturn and Pluto - the project was finally pared down by financial cuts to an extension of the Mariner programme, which had so successfully explored the inner planets. Work on a "reduced programme," which by mid-1981 had in fact been wonderfully successful in exploring the Jupiter and Saturn systems, with high hopes of at least reaching Uranus in 1986 (and perhaps even Neptune in 1989), was begun by NASA in Jan 1972. Between that time and the launches in 1977 the programme underwent many changes, with proposals for Jupiter orbiters and landers finally being postponed for later missions like Galileo. There were so many changes to the spacecraft that instead of being designated Mariners 11 and 12 they were christened Voyager following a competition to find a

Searching for life on Mars: left, Lander 2 pushing aside rock to obtain soil sample. Scientists believed that any life forms would seek shelter beneath rocks. Right, trench after soil removal. No conclusive evidence was found. *(NASA)*

new name. The spacecraft were designed, built and operated by the Jet Propulsion Laboratory at Pasadena, California. Difficulties before launch and during the flight cast doubts on the eventual success of this mission. After a 4-day road journey from JPL to Cape Canaveral mysterious faults were discovered in the attitude and articulation control and flight-data subsystems of Voyager 2, due to be launched first because its trajectory included the Uranus option. The two spacecraft had to be interchanged as a result, with consequent swapping of equipment.

The launches - actually achieved comfortably within the 20 Aug-23 Sep window, though with 16 days instead of the desired 11-day gap between them - give no indication of the effort and pessimism that characterised that period. The doubts are now long forgotten, however, and there are no superlatives left to describe the quality of the data and the nearly 70,000 images of the Jovian and Saturnian systems received by the time that both spacecraft had completed their close encounters in mid-1981. And the astronomers can now enter in their diaries with some confidence the expected dates - Jan 1986 and Aug 1989 - of Voyager 2's encounters with Uranus and Neptune. By then the cost of this project, $320 million (not including launch vehicles and flight support activities), will seem modest indeed.

Spacecraft description The two Voyagers are identical apart from the higher-output Radioisotope Thermoelectric Generator (RTG) required for Voyager 2's Uranus option. Their design had to ensure operation at distances from Earth and Sun greater than on any previous mission, and a working life of at least 4 but preferably 12 years. The combined planetary vehicle and propulsion module had a 2,016kg launch wt, of which Voyager itself accounted for 792kg. Each Voyager uses 10 instruments and telescope-equipped TV cameras to study the outer planets and their satellites, rings and environment. The 3·7m antenna dish, always kept pointing back to Earth, is the largest to be flown on a planetary mission. Because the amount of sunlight reaching the outer planets is only a fraction of that reaching Earth, the Voyagers must use nuclear power instead of solar energy. Each is equipped with 3 RTGs, each weighing 39kg, installed inside a beryllium case on a deployable boom. To avoid their radiation the most sensitive science instruments are located at the end of another boom 2·3m long and mounted 180° from the RTG boom. A data-storage capacity of 536 million bits enables each craft to store the equivalent of 100 full-resolution photos. The spherical propellant tank at the centre of the basic 10-sided aluminium framework supplies fuel for the 12 attitude-control and 4 trajectory-correction thrusters. Like the Mariners from which they were developed, the Voyagers are stabilised on 3 axes, using the Sun and a star (Canopus) as celestial reference points.

Voyager 2 began sending back detailed pictures of Saturn's complex ring system from 3.4 million km. Note the shadow cast by the rings on Saturn itself. ***(NASA)***

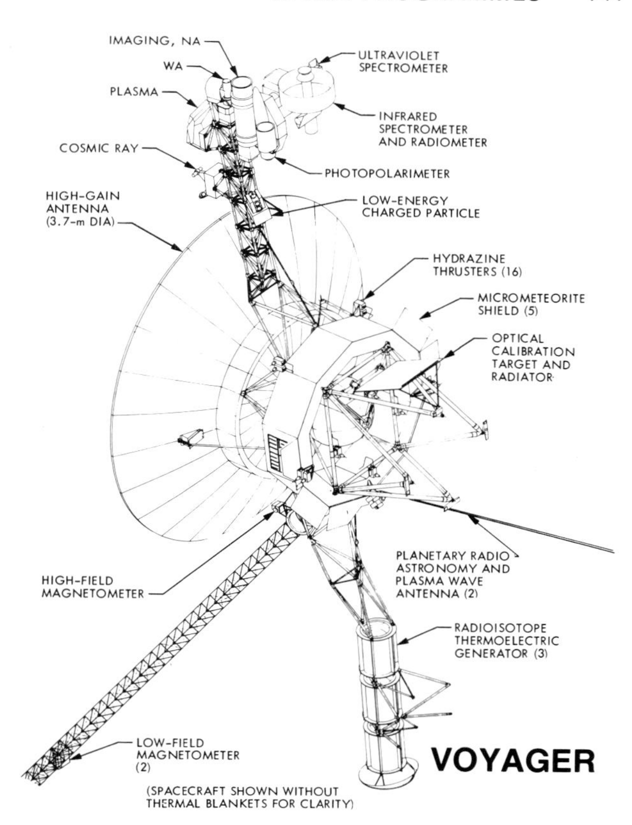

Both craft carry a 12in copper record called "Sounds of Earth" and containing greetings in 60 languages, sounds of birds, whales and animals, samples of music, and other information so that if they are intercepted millions of years hence by other civilisations the finders will have some idea of what life was like during Earth's 20th century. The record's aluminium jackets give information on how to play it, using the cartridge and needle thoughtfully provided!

Voyager 1 This craft should have been launched, by the last Titan-Centaur from C Canaveral, on a shorter trajectory and 12 days later than Voyager 2. But it was held back an extra 4 days until 5 Sep 1977, when doubts as to whether Voyager 2's science boom had been fully deployed were resolved. The first 3 months of Voyager 1's flight were anxious ones: during calibration of the scan platform the azimuth gears slowed and stuck, jammed by soft debris retained in the unit during assembly. Careful manoeuvring of the platform crushed out the debris, however, and the platform was freed. After that things looked up: in a TV test a fortnight after launch the crescent Earth and Moon, 11·5 million km away, were successfully shown in the same picture for the first time. On 15 Dec, with both spacecraft 124·7 million km from Earth and 17 million km from each other, Voyager 1 overtook 2 on the 18-month journey to Jupiter. By the end of Oct 1978 both were safely through the 360 million km-wide Asteroid Belt, making them the 3rd and 4th spacecraft to survive such a journey.

From 4 Jan 1979 a 26-day test observation, starting from 61 million km, gave the 106-strong team of astronomers and scientists a long-distance look at the whole Jovian system so that they could prepare for the near-encounter operations 2 months later. During a subsequent 4-day period the narrow-angle TV camera took a picture of Jupiter every 96sec, and 3,500 images were processed into a colour film record of its rapid rotation. A time-lapse film of the Great Red Spot was also made. Despite a radiation environment probably 100 times as high as the lethal level for human beings, only two problems arose with Voyager 1's sensitive electronics during the close encounter with the Jovian system. High-accuracy photometry, which would have provided the best data on cloud heights and composition, was lost when circuitry failed, and a brief loss of synchronisation between 2 computers smeared some of the Io and Ganymede images. The computers were resynchronised when radiation intensity diminished.

Jupiter's Great Red Spot and turbulent region to its west, seen by Voyager 1 from 5 million km. Middle right is one of several white ovals which are visible from Earth. ***(NASA)***

Voyager 1's trajectory curved past Jupiter just south of the equator, 278,000km above the cloudtops at closest approach on 5 Mar 1979. The images sent back to Earth at that time, some of the nearly 19,000 finally obtained during the 98-day observation period, were taking 37min 43sec to travel the 679,675,000km to the Deep Space Network Station at Canberra, Australia. The clarity of the pictures was unaffected by the fact that Jupiter's gravity was accelerating the spacecraft to 135,000km/hr and placing it on course for an encounter with Saturn 20 months later. Closest encounters with the Galilean satellites occurred after passing Jupiter, when the Voyager flew within 19,000km of Io, 112,000km of Ganymede, 124,000km of Callisto and 732,000km of Europa. Active volcanism on Io was first identified on a picture taken on 8 Mar.

First evidence of a ring around Jupiter came on 4 Mar in one of a series of "black space" images obtained in search of rings or additional moons. A mosaic of these pictures dramatically revealed it, no more than 30km thick and 128,000km from the centre of the planet. It was found in good time for Voyager 2, following 4 months later, to be targeted to inspect more closely both that and the 8 active volcanoes found on Io by Voyager 1. But Voyager 2 proved to be near enough for preliminary observations as early as 15 Apr, so 2 days before that Voyager 1's Jupiter activities were ended so that its companion craft could be brought into use.

Voyager 1's 20-month journey from Jupiter to Saturn, marked by occasional alarms when contact was lost and then recovered by the engineers, gave the scientists time to study the 33,000 Jovian pictures and, using them, to bring the number of known Jovian moons to 16. Closest approach to Saturn itself was 124,000km from the visible cloud tops on 12 Nov 1980, when the data and images were taking 1hr 25min to travel the 1·5 billion km back to Earth. The same day Voyager 1 passed Titan at 4,000km, the closest approach to any body in the Voyager mission. That day the craft was also finding time to study 4 other major moons: Tethys from 415,320km, Mimas (88,820km), Enceladus (202,521km) and Dione (161,131km). Next day Rhea was examined from 72,000km. The Voyager 1 trajectory also permitted occultations of the Earth and Sun by Titan, Saturn and the rings: as the craft disappeared behind each body its radio signals passed through the atmospheres of Saturn and Titan and through the rings, providing accurate knowledge at last about the atmospheres and ionospheres of the planet and moon and the size and density of the ring particles. Voyager 1's 117-day surveillance of the Saturnian system ended on 15 Dec 1980, having yielded more than 17,000 photographs. After nearly 2yr of analysis it was announced in Sep 1982 that the rings were formed from original planet-building material rather than from debris of a moon.

The final task of the Voyagers is to find the exact location of the heliopause, the outer limit of the solar wind. This is the presumed boundary in our part of the Milky Way at which the influence of the Sun gives way to that of other stars of the galaxy. Voyager 1 is likely to reach the boundary in about 1990, though before that the craft may signal that it has detected low-energy cosmic rays penetrating the outer reaches of the solar system from nearby supernovae remnants.

Voyager 2 So many technical problems accompanied Voyager 2 on its flight to Jupiter that few of the scientists concerned could have had much confidence that it would perform so brilliantly on both the Jupiter and Saturn

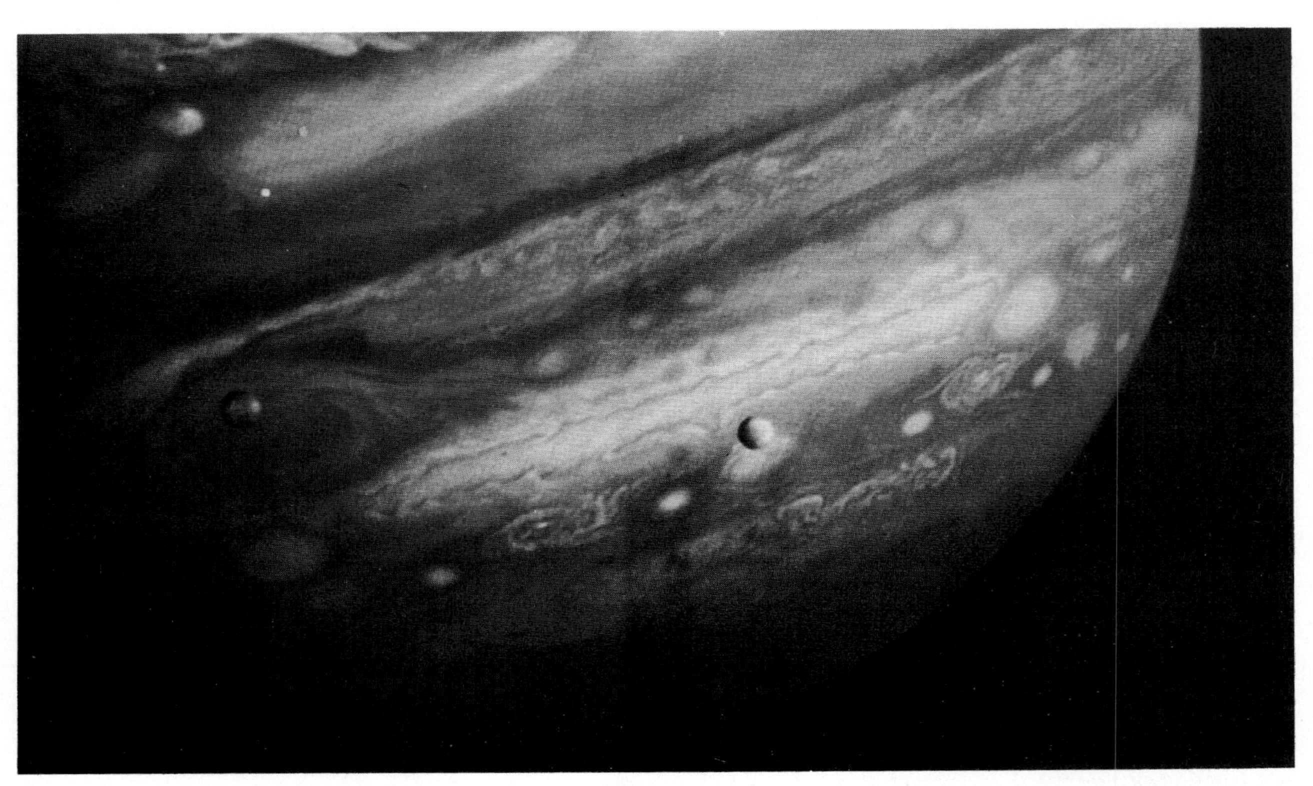

Voyager 1 photo of Jupiter and its satellites Io (left) and Europa (right). Io is 350,000km above the Great Red Spot; Europa 600,000km above Jupiter's clouds. Jupiter was about 20 million km from Voyager when this picture was taken. ***(NASA)***

encounters. An unexplained pitch and yaw 18hr after the Titan-Centaur launch on 20 Aug 1977 led to unfounded fears that it had collided with the discarded propulsion module. There was concern also that the science boom had failed to deploy fully, leading to 5 additional springs being installed on Voyager 1 as it waited its turn on the launchpad. Pasadena's engineers worked around or found cures for circuitry failures, and decided to fly the spacecraft "upside down," locked on to the star Deneb instead of Canopus because the solar wind had been steadily altering its attitude. For 6 months from 5 Apr 1978, when the primary radio receiver failed, a series of communications problems seriously endangered the whole mission. Patient work with the computers uncovered the fact that Voyager 2 tended to change its listening frequency, and identified the new frequency. By Sep it had been established that ground commands could if necessary be transmitted "in reverse". But the single remaining receiver worked well throughout the Jupiter and Saturn encounters, and there are high hopes that it will continue to work for at least another 7 years for the Uranus encounter.

By encountering the Galilean satellites during approach to Jupiter rather than on the outward passage as Voyager 1 had done, Voyager 2 was able to obtain high-resolution pictures and measurements of their opposite sides. (Like the Moon, they always present the same face to the parent planet.) Closest approach to Callisto was 215,000km, to Ganymede 62,000km, to Europa 206,000km, and to Amalthea 559,000km. Closest approach to Jupiter itself, 650,000km from the cloud tops deep in Jupiter's southern hemisphere, was on 9 Jul 1979. Its signals at that moment took 51min 49sec to travel the 931 million km to Earth, 14min longer than Voyager 1's. Voyager 2 added another 15,000 pictures of Jupiter and its 5 major satellites to those taken by Voyager 1, including additional dramatic views of 7 of the 8 erupting volcanoes; the largest, it seemed, had "shut down". It became clear that in the 5½yr since Pioneer 10 flew past Jupiter in Dec 1973 there had been some changes in the Jovian magnetosphere, probably caused by the sulphur, oxygen and sodium being ejected by Io's volcanoes. From 1·4 million km Voyager 2 was also able to add a colour picture of the Jovian ring.

On 25 Aug 1981 Voyager 2 passed within 101,300km of Saturn's surface, 23,000km closer than Voyager 1 10 months earlier. The encounter yielded another 18,000 pictures, justifying one project scientist's claim of "200% success" for the mission.

This Voyager 2 view of Europa, smallest of Jupiter's 4 Galilean satellites, reveals a fractured icy surface, with the cracks probably filled with dark material from below. There are few impact craters, suggesting active processes. ***(NASA)***

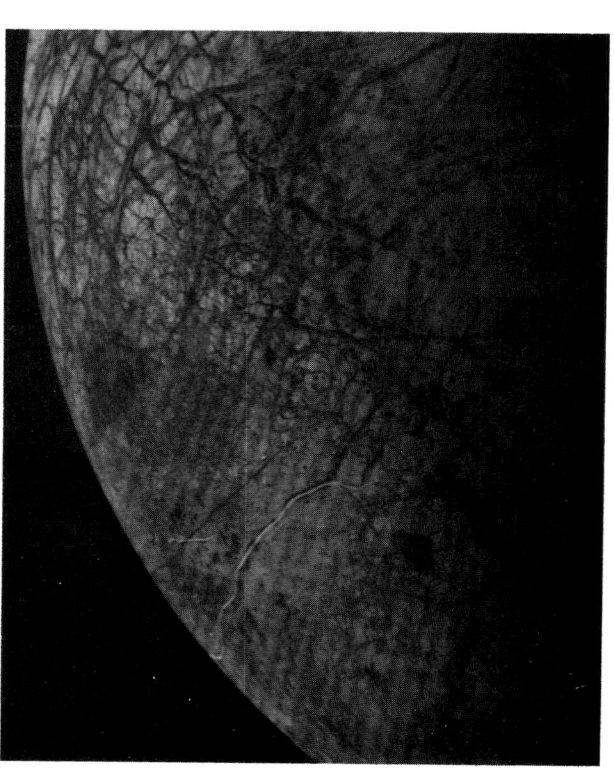

Following the Voyager 1 revelations of the 69,000km ring plane, areas of concentrated study by Voyager 2 included the B-ring and its spokes; the braided F-ring and its satellites; the eccentric rings (C-ring and one in the Cassini Division); and the Encke Division. Photopolarimeter observations of ring occultation by the star Delta Scorpii were highly successful.

Voyager 2 discovered a new "kinked" ring inside the Encke Division, and its similarity to the F-ring made scientists rethink their theory, suggested by Voyager 1 results, that its wobbly character was caused by the "shepherding" effect of satellites 1980S26 and 1980S27. They changed their minds because the kinked ring has no nearby satellites, while in the case of the F-ring further studies showed no ring perturbations due to the satellites which could have caused the braiding effect. Complex activity of the B-ring and its spokes was revealed by a film made from a series of images; some spokes seemed to form, others to shear out, and yet others to fade away. The A-ring was shown to be reddish in colour, while the Cassini Division has a bluish colour similar to that of the C-ring, indicating a similar composition. Colour differences in the C-ring also indicate that there is imperfect separation of the ring material.

The origin of electrical discharge signals first recorded by Voyager 1 was shown by Voyager 2 to lie in two patches of Saturn's northern and southern hemispheres. Data from the infra-red interferometer spectrometer showed that Saturn's temperature ranges over 80-92°K, varying with latitude and height. Saturn's jetstreams were also identified in much greater detail: 3 definite easterly flows could be seen at northern latitudes, and the jetstreams extend to higher latitudes than on Jupiter. Saturn's zonal profile does not appear to correspond with the light and dark bands in the clouds. An anticyclonic (anti-clockwise) feature resembling a reversed "6" could be seen at 72°N; 249km across, it has a centre convective region of about 60km across. There is a similar feature at about 50°N.

Voyager 2 took higher-resolution photographs of 5 of the Saturnian moons: Enceladus, Tethys, Hyperion, Iapetus and Phoebe. While the Voyager 1 photos of Enceladus suggested that it was intriguingly smooth and uncratered, the closer look revealed truncated and collapsed craters, with grooved terrain in some areas and an absence of craters in others. These variations are believed to be the result of frictional heat produced by the expansion and contraction of the planet's crust. Tethys, seen from 999,200km, proved to have the largest crater found in the Saturnian system: about 402km dia, with a central peak and surrounded by a ring of mountains. The crater is so large that Mimas could fit inside it. Hyperion was shown to be an irregular disc about 300km across and 200km thick. With its axis at an angle to the orbital plane, it may have been hit by some object which jostled it into that position; this theory is supported by its cratering. One side of Iapetus was shown to be very dark, with a surface reflectivity of only 4-5%, equivalent to the darkest surfaces on Earth. Radio data indicated that earlier theories that it consisted of 60% water ice and 40% rock were wrong; its composition is more likely to be 55% water ice, 35% rock and 10% methane. The methane would contain billions of tons of carbon, possibly accounting for Iapetus's dark area.

Because Voyager 1 passed too far from Phoebe to attempt photography, Voyager 2's images, though taken from just over 2 million km, were eagerly awaited. They were due to be sent back just after the 95min Saturn/Earth occultation, the "Uranus aim point" required to bend the spacecraft's trajectory towards Uranus. But Voyager 2 re-emerged from behind Saturn on 26 Aug with its cameras pointed into space and its scan platform stuck. After two anxious days project scientists succeeded in moving the platform sufficiently to start the outbound imagery of Saturn and later rolled the entire spacecraft to bring Phoebe into view, hoping that analysis would show a surface likeness with Iapetus. Before that, high-resolution photos of 7 of Saturn's newly discovered satellites had been taken: 1980S26/7, the "shepherds" of the F-ring; 1980S6, in Dione's orbit; 1980S1 and S3, which share an orbit and are probably one satellite split in two;

and 1980S25 and S13, recently discovered by Earth observations to be orbiting Saturn 60° behind and ahead of Tethys. All 7 were observed to be irregularly shaped and heavily battered by impacts with cosmic debris. They range in size from 10km to 320km across.

Project scientists have now learned to live with the sticking scan platform and still hope that at the end of Jan 1986, by which time the spacecraft will have travelled an additional 2·8 billion km on top of the 2 billion km from Earth to Saturn, all their objectives will be achieved by rotating the spacecraft instead of the scan platform. Its nuclear power should keep Voyager 2 operating right through the decade for a final encounter with Neptune in Aug 1989. At the end of 1982 67kg of its original 104kg of attitude-control and trajectory-correction propellant still remained as it flew on at a speed of 19·9km/sec in relation to the Sun. At that point the greatest hazard to this almost incredible mission lay not in space but in the budget cuts which might deprive NASA of the $85 million needed to complete it.

Mission summary So far as Jupiter is concerned, the discovery of active volcanism on the satellite Io was the greatest surprise, even though Io was always regarded as the most mysterious of the moons within the solar system. Material associated with Jupiter's Great Red Spot was found to move in an anti-clockwise direction, with a rotation period of 6 days. The two Voyagers have provided the scientists with data on 160 days of Jovian weather, showing that the 300yr-old Red Spot drifts westwards 1/2° per day; why it is red is still uncertain. There are also smaller spots, some of them white ovals. Only about 40 years old, they appear to interact with the Red Spot and each other. Cloud-top lightning, similar to "superbolts" in Earth's high atmosphere, was detected. Auroral emissions, similar to Earth's "Northern Lights," were seen in Jupiter's polar regions. Ultra-violet auroral emissions, not observed during the Pioneer 10 encounter in 1973, are probably associated with the material from Io that falls into Jupiter's atmosphere; the planet's hydrogen atmosphere appears to contain about 11% helium. Plumes from Io's volcanoes, with material being ejected at 3,600km/hr (compared with 180km/hr from Mount Etna on Earth), extend 250km above the surface. The violence of Io's volcanoes is probably caused by "tidal pumping": its orbit is perturbed by two other large moons nearby, Europa and Ganymede, and then it is pulled back again by Jupiter itself, causing bulging as great as 100m (compared with tidal bulges of 1m on Earth). Europa's interesting linear features remain a mystery; this moon too may be internally heated by tidal forces, though to a much lesser extent than Io. Ganymede, now established as the solar system's largest moon, is cratered and grooved, with much ice, and has appparently been under tension from global tectonic processes. Callisto is heavily cratered, with enormous impact basins which have then been partly erased by the flow of the ice-laden crust. While all these satellites are spherical, little Amalthea, at 265km across about 10 times bigger than Mars' Phobos, is elliptical. The outer edge of Jupiter's rings is 128,000km from the centre of the planet and no more than 30km thick; the material appears to be gradually falling into the planet.

The fascinating complexity of Saturn's ring system was the highlight of Voyager 1's discoveries there. While the major surface features of Jupiter and Saturn were found to be similar, Saturn's belt-zone structure extends to much higher latitudes; Saturn's wind speeds, at about 1,600km/hr at the equator, are substantially higher too. Radio emissions, primarily from the N Pole region, indicate that Saturn and its magnetosphere rotate in 10hr 39min 26sec, 2sec longer than formerly believed. The 7 main ring systems were found to consist of hundreds of small ringlets. From the planet outward they are designated D, C, B, A, F, G and E (the inconsistency being due to the fact that the first rings discovered were lettered inwards). The newly discovered F-ring was found to have at least 3 "braided" or intertwining rings, believed then to be "shepherded" by the gravitational pull of two small satellites, S13 and S14. The tenuous E-ring, like the D, consists mainly of particles less than 0·0002in in diameter. The A, B and C-rings consist of hundreds of small ringlets, a few elliptical in shape. There was much excitement at the discovery that the B-ring had "spokes" of very fine material radiating outwards, and that the A-ring was also apparently shepherded by a small satellite, S15, orbiting just outside it. The C-ring's components are about 1m in size. As for the satellites, Enceladus (500km dia), composed mostly of water ice, may be the source of particles for the E-ring. Like Enceladus, Mimas (390km), Tethys (1,050km), Dione (1,120km) and Rhea (1,530km) are all spherical and apparently consist mostly of water ice. Mimas is notable for a large, breast-shaped crater covering about a quarter of its surface. All these moons except Enceladus are cratered, and Tethys is scarred by a 60km-wide, 750km-long valley. Iapetus (1,440km), probably mostly water ice, is "peculiar," with one bright and one dark hemisphere. Titan (5,120km), now the solar system's second largest moon, is an extremely important and interesting body, with an atmosphere similar to the one that would have evolved on Earth had it been the same distance from the Sun. Like Ganymede, Titan appears to be composed of about equal proportions of rock and ice, its surface hidden by a thick atmosphere consisting mostly of nitrogen. With a surface temperature of -294°F and pressure of 1 1/2 atmospheres, it probably has rivers of methane, playing the same role as water on Earth.

Westar

History Owned by Western Union and using Hughes-built satellites, this was the first US domestic communications system. It relays voice, data, video and facsimile to the continental US, Hawaii, Alaska, Puerto Rico and the Virgin Islands. Following expiry of Westars 1 and 2 this system is expected to settle down to long-term use of 4 slots.

Westars 1 and 2 L 13 Apr and 10 Oct 1974 by Delta from C Canaveral. Wt 572kg, approx 291kg in orbit. Placed in stationary orbit at 99°W over San Antonio and 123·5°W near San Francisco. Spin-stabilised cylinders 1·8m in dia and 1·6m high, each has a spoon-shaped mesh antenna which is despun in orbit so that it points continuously to Earth, and more than 20,000 solar cells providing 300W. Each handles 12 colour TV channels or 14,400 1-way phone circuits through 5 Earth stations located near New York, Atlanta, Chicago, Dallas and Los Angeles. 7yr design life, 9yr actual life. Westar 1 retired Apr 1983.

Westar 3 L 10 Aug 1979 by Delta from C Canaveral after being stored for 5yr. Placed at 91°W over Baton Rouge, La. Expected life 8yr+.

Westar 4 L 26 Feb 1982 by Delta/PAM from C Canaveral. Wt 1,110kg, 585kg in orbit. Double the size and capacity of earlier Westars, this series has 24 transponder channels and a cylindrical solar array producing over 800W. Design life 10yr and positioned at 99°, it supplements - and with Westar 5 will take over from - Westars 1 and 2.

Westar 5 L 8 Jun 1982 by Delta/PAM. Successfully reached 123°W despite problems with both 1st stage and PAM. All its 24 transponders were sold before launch.

USSR

USSR SPACEFLIGHTS 1957-1983

	57	58	59	60	61	62	63	64	65	66	67	68	69	70	71	72	73	74	75	76	77	78	79	80	81	82	83	Total
1. Sputnik	2	1	—	3	4	—	—	—	—	—	—	—	—	—	—	—	—	—	—	—	—	—	—	—	—	—	—	10
2. Luna (Lunik)	—	—	3	—	—	—	2	—	4	5	—	1	1	2	2	1	1	2	—	1	—	—	—	—	—	—	—	25
3. Vostok, Voskhod	—	—	—	—	2	2	2	1	1	—	—	—	—	—	—	—	—	—	—	—	—	—	—	—	—	—	—	8
4. Cosmos	—	—	—	—	—	12	12	27	52	34	61	64	55	72	81	72	85	74	85	101	86	96	79	88	94	97	94	1,521
5. Venera	—	—	—	—	—	3	—	—	2	—	1	—	2	1	—	1	—	—	2	—	—	2	—	—	2	—	2	18
6. Mars	—	—	—	—	—	3	—	—	—	—	—	—	—	—	2	—	4	—	—	—	—	—	—	—	—	—	—	9
7. Polyot	—	—	—	—	—	—	1	1	—	—	—	—	—	—	—	—	—	—	—	—	—	—	—	—	—	—	—	2
8. Electron	—	—	—	—	—	—	—	4	—	—	—	—	—	—	—	—	—	—	—	—	—	—	—	—	—	—	—	4
9. Zond	—	—	—	—	—	—	—	2	1	—	—	3	1	1	—	—	—	—	—	—	—	—	—	—	—	—	—	8
10. Molniya	—	—	—	—	—	—	—	—	2	2	3	3	2	5	3	6	8	7	10	7	6	6	5	4	8	5	7	99
11. Proton	—	—	—	—	—	—	—	—	2	1	—	1	—	—	—	—	—	—	—	—	—	—	—	—	—	—	—	4
12. Soyuz (Union)	—	—	—	—	—	—	—	—	—	—	1	2	5	1	2	—	2	3	4	3	3	5	4	6	3	3	2	49
13. Meteor	—	—	—	—	—	—	—	—	—	—	—	—	2	4	4	3	2	5	4	3	4	—	3	2	2	2	1	41
14. Intercosmos	—	—	—	—	—	—	—	—	—	—	—	—	2	2	1	3	2	2	2	2	1	2	2	—	2	—	—	23
15. No designation	—	—	—	—	—	—	—	—	—	2	—	—	—	—	—	—	—	—	—	—	—	—	—	—	—	—	—	2
16. Salyut	—	—	—	—	—	—	—	—	—	—	—	—	—	—	1	—	1	2	—	1	1	—	—	—	—	1	—	7
17. Oreal	—	—	—	—	—	—	—	—	—	—	—	—	—	—	1	—	1	—	—	—	—	—	—	—	1	—	—	3
18. Prognoz	—	—	—	—	—	—	—	—	—	—	—	—	—	—	—	2	1	—	1	1	1	1	—	1	—	—	1	9
19. Launches for other countries	—	—	—	—	—	—	—	—	—	—	—	—	—	—	—	1	—	—	2	—	1	—	1	—	1	1	—	7
20. Raduga	—	—	—	—	—	—	—	—	—	—	—	—	—	—	—	—	—	—	1	1	1	1	1	2	3	1	2	13
21. Ekran	—	—	—	—	—	—	—	—	—	—	—	—	—	—	—	—	—	—	—	1	1	—	2	2	1	2	2	11
22. Progress	—	—	—	—	—	—	—	—	—	—	—	—	—	—	—	—	—	—	—	—	—	4	3	4	1	4	2	18
23. Radio	—	—	—	—	—	—	—	—	—	—	—	—	—	—	—	—	—	—	—	—	—	2	—	—	6	—	—	8
24. Gorizont	—	—	—	—	—	—	—	—	—	—	—	—	—	—	—	—	—	—	—	—	—	1	2	1	—	1	2	7
25. Iskra	—	—	—	—	—	—	—	—	—	—	—	—	—	—	—	—	—	—	—	—	—	—	—	—	1	2	—	3
26. Astron	—	—	—	—	—	—	—	—	—	—	—	—	—	—	—	—	—	—	—	—	—	—	—	—	—	—	1	1
Total	2	1	3	3	6	20	17	35	64	44	66	74	70	88	97	89	107	95	111	121	105	120	102	110	125	119	116	1,910

This table, which includes some launches not announced by Soviet sources, was compiled by NASA.

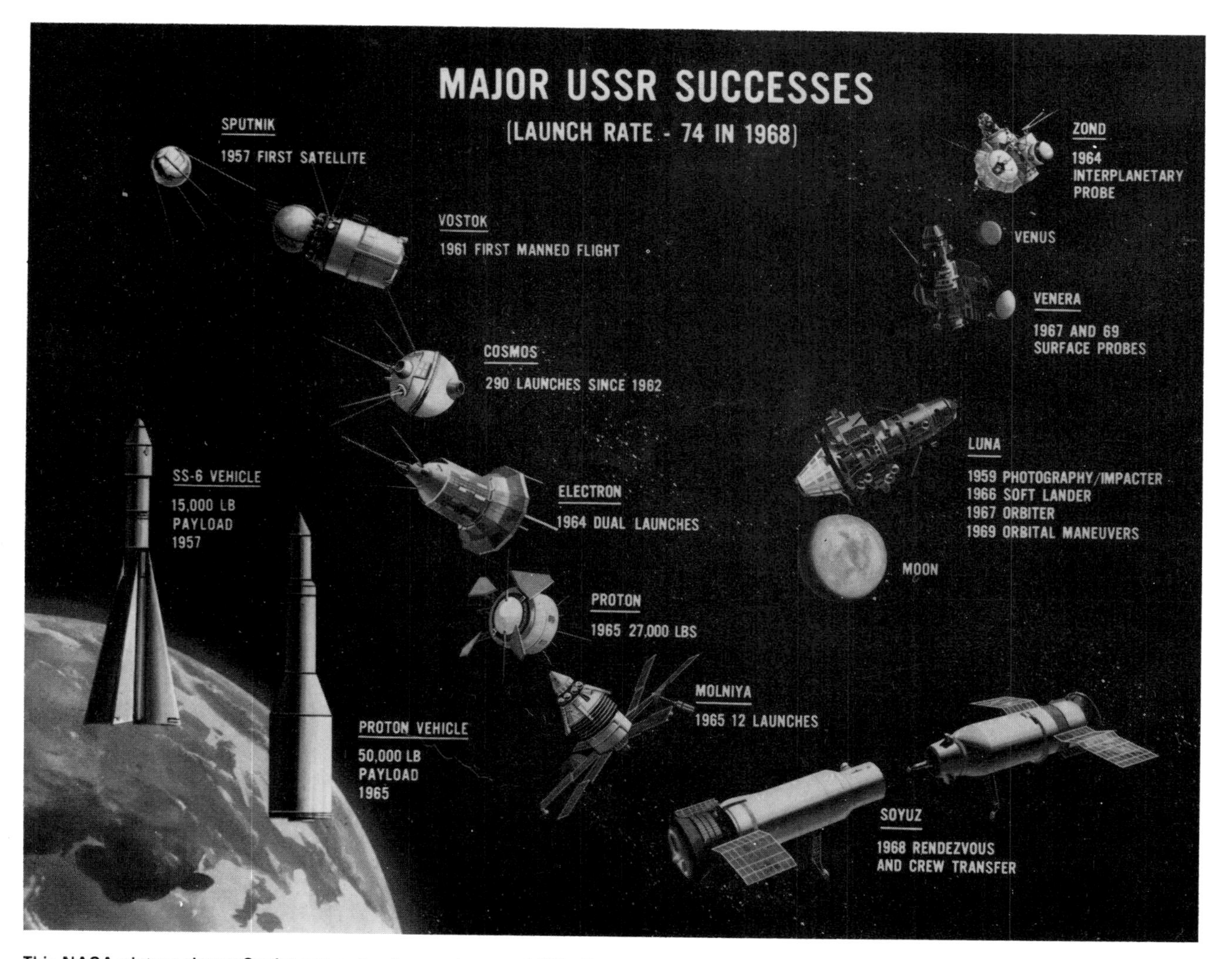

This NASA picture shows Soviet space developments up to 1968. The Salyut space stations have followed since, and the annual launch rate is now well over 100.

Astron

Designed to observe any designated galaxy, star or quasar, with the Taurus Constellation as its main target, Astron, L 23 Mar 1983 by D1e from Tyuratam (wt ? 3,900kg, orbit 1,950 × 201,000km, incl 51°, period 98hr) carries the largest ultra-violet telescope so far. Mounted on a Venera-type body, the 5m-long telescope, with an 800mm-dia main mirror and 400kg weight, was jointly developed by the Crimea Astrophysical Observatory and France's Astronomical Laboratory at Marseilles. In Jan 1984 Astron registered the disappearance of a powerful source of radiation in the constellation Hercules. 6yr orbital life.

Cosmos

LAUNCH RATE

Year	Numbers	Total
1962:	1–12 =	12
1963:	13–24 =	12
1964:	25–51 =	27
1965:	52–103 =	52
1966:	104–137 =	34
1967:	138–198 =	61
1968:	199–262 =	64
1969:	263–317 =	55
1970:	318–389 =	72
1971:	390–470 =	81
1972:	471–542 =	72
1973:	543–627 =	84
1974:	628–701 =	74
1975:	702–786 =	85
1976:	787–887 =	101
1977:	888–973 =	86
1978:	974–1069 =	96
1979:	1070–1148 =	79
1980:	1149–1236 =	88
1981:	1237–1330 =	94
1982:	1331–1427 =	97
1983:	1428–1521 =	94

History On 28 Sep 1983 the Soviet Union launched the 1,500th Cosmos, an ocean surveillance satellite. Cosmos 1000 was launched on 1 Mar 1978. As Moscow pointed out at that time, it took more than 10yr to launch the first 500, and only just over half that time for the second 500. Since the series began on 16 Mar 1962 the average has been over 70 per year and is steadily rising; over the last 5yr the average has been 90, the record standing at 101 in 1968. And it should be noted that since many launchers carry more than 1 satellite (there are usually 2 octuple launches annually) the number of payloads usually tops 100.

Cosmos of unspecified purpose. Note solar cells on both octagonal body and deployable arrays.

The main value of the general name is that it can be used to cover the astonishing number of Soviet military launches. Although it has never been admitted, well over half the Cosmos series are either wholly or partly for military use; occasionally, as in 1976, it can be as high as 90 out of 101. The practice of announcing launches under the Cosmos title, accompanied by serial numbers, orbital parameters (which will in any case be picked up immediately by the US and UK tracking systems) and a vague description of its scientific mission, was adopted when the series began in 1962. This directory attempts to disentangle the civil from the military Cosmos programmes; those classified (somewhat arbitrarily) as military are described in the Military section. The 94 Cosmos launches in 1983 can be studied in detail in the complete launch tables for those years at the start of this Directory.

The Cosmos numbers are also used as a convenient way of concealing the inevitable failures during the development of a new series. For instance, interplanetary probes are given designations in the Venera, Mars or other series only when they are on course and working well; otherwise they are merely allotted the next Cosmos number and a routine announcement is made about the launch. Nowadays, however, launch systems are so reliable that this is seldom necessary.

The Cosmos series is launched from all three Soviet sites, but mainly from Plesetsk. From this site are launched both military and civil applications satellites, including navsats, ferrets, octuple comsats, store/dump comsats, Molniya and Meteor. Statsionars, which need to be launched as far south as possible, start from Tyuratam, along with retrograde Meteors, (which cannot be safely launched from Plesetsk), Rorsats and Eorsats. Spy satellites are launched from both Plesetsk and Tyuratam, probably to achieve an efficient matching of launch vehicles to available pads, which may explain the occasional Molniya launch from Tyuratam. Kapustin Yar is used for C-vehicle launches to relatively low inclinations.

The non-military Cosmos satellites continue to range over a wide area of research, and many have achieved remarkable results. Many of these are published, though sometimes not for months or even years afterwards. The fact that they make little impression in the West is largely due to the shortcomings of our media.

Soviet scientists pointed out when the 700th Cosmos was launched that the tremendous role which space was beginning to play in the Soviet economy was becoming increasingly obvious. Most Cosmos satellites were standardised both in design and in the composition of the onboard systems. This made "flow line production of these space-workers possible," and together with the use of standard rocket carriers had enabled large savings to be made in the cost of space exploration.

According to the Soviet *Encyclopaedia of Spaceflight* (1969), the research programme "includes study of the concentration of charged particles, corpuscular fluxes, radio-wave propagation, distribution of the Earth's magnetic field, solar radiation, meteoric matter, cloud formations in the Earth's atmosphere, solution of technological problems of spaceflight (including docking, atmospheric entry, effect of space factors, means of attitude control, life support, radiation protection, etc), and flight testing of many structural elements and spaceborne systems." The review issued after the 700th launch broke the series down more simply into: 1) research into physical phenomena in space and the upper Earth's atmosphere; 2) tackling technical problems in connection with spaceflights, development work on structural members and onboard systems of future spacecraft; and 3) experiments of an applied character in the interests of terrestrial sciences and the national economy.

In addition to studying the ionosphere, etc, the first group also had the task of probing the radiological situation in space before manned craft were launched. Development work carried out by the 2nd group included automatic link-ups, "very important for cosmonautics of both today and of the nearest future". The 2nd group also included biosatellites, such as Cosmos 690, to study the

Soviet picture of an early Cosmos.

effects of spaceflight on living organisms. Clearly Group 2 is connected with the building up of large space stations by automatically joining a series of Salyut-type vehicles and then manning them permanently. Group 3, said briefly to include meteorological, radio and TV satellites, no doubt also includes the still unmentioned military craft.

A further official review of the series when the 900th was launched in Mar 1977 also stressed the "unified space vehicle" which had made series production possible and substantially reduced costs. "One of the aims is to check on radiation safety for manned flights. Detailed measurements have been recorded, enabling a thorough study to be made of spaceship routes, and radiation charts to be drawn up for certain altitudes. A great deal has been done to study the Sun, the nature of cosmic rays, and the Earth's magnetic field." Cosmos had given Soviet scientists their first opportunity to take telescopes outside the Earth's atmosphere; radio-astronomical observatories with infra-red emission receivers had opened up immense possibilities for the measurement of physical parameters of the atmosphere and Earth surface temperatures. The meteorological forecasting system now in operation had been made possible by Cosmos craft. On them too had been conducted the first studies of superconductive systems and molecular generators in space. Lunokhod assemblies and automatic docking had been tested by them, opening up the prospect of multi-purpose orbital stations.

Little was added when Cosmos 1000 was launched on 1 Mar 1978; that was said to be a navigational satellite which would locate shipping precisely by use of the Doppler effect. It was however pointed out that "Cosmos launchings follow no regular pattern, varying according to research requirements; but the general level of activity has increased as the programme has advanced."

The series as a whole – ranging from small uninstrumented spheres to large vehicles like Cosmos 110, which carried two dogs with sufficient supplies to enable them to be recovered after 22 days, and the 15,000kg Cosmos 1267, which is apparently a future space station module – is too long to give in full. Some interesting examples follow, however. It is also interesting to note that, so far as I can trace, no public mention was made of one significant event in Soviet space history: the launch of Russia's first geosynchronous satellite, Cosmos 637, on 26 Mar 1974. Perhaps this was because America had by then been using synchronous satellites for 12yr.

Cosmos 1 L 16 Mar 1962 by B1 from Kapustin Yar. Wt ? 200kg. Orbit 217 × 980km. Incl 49°. At first classified as Sputnik 11; used radio methods to study structure of the ionosphere. Dec after 70 days.

Cosmos 2 L 6 Apr 1962 by B1 from Kapustin Yar. Wt ? 400kg. Orbit 212 × 1,560km. Incl 49°. At first classified as Sputnik 12. Returned data on radiation belts and cosmic rays and re-entered after 499 days.

Cosmos 3 L 24 Apr 1962 by B1 from Kapustin Yar. Wt ? 400kg. Orbit 228 × 719km. Incl 49°. Returned radiation belt and cosmic ray data. Dec after 176 days.

Cosmos 4 L 26 Apr 1962 by A1 from Tyuratam. Wt ? 4,000kg. Orbit 298 × 330km. Incl 65°. The 1st spacecraft to be recovered; it was about 5m long and 2m in dia, and re-entered after 3 days. The 1st military satellite, since its task was to measure radiation before and after US nuclear tests.

Cosmos 7 L 28 Jul 1962 by A1 from Tyuratam. Wt ? 4,000kg. Orbit 209 × 368km. Incl 65°. Its job was to watch for solar flares during the manned Vostok 3 and 4 flights. Re-entered or dec after 4 days.

Cosmos 97 L 26 Nov 1965 by B1 from Kapustin Yar. Wt ? 400kg. Orbit 220 × 2,098km. Incl 49°. 1st experiment with measuring masers; tested a molecular quantum generator, which makes it possible to communicate with and control other spacecraft, and to send information great distances. Also checked aspects of the theory of relativity. Dec after 492 days.

Cosmos 110 L 22 Feb 1966 by A2 from Tyuratam. Wt 5,500kg. Orbit 186 × 904km. Incl 52°. Biological satellite carrying dogs Veterok and Ugolyok, who were successfully recovered after 330 orbits in 22 days.

Cosmos 122 L 25 Jun 1966 by A1/2 from Tyuratam. Wt 2,500kg. Orbit 625 × 625km. Incl 65°. Meteorological satellite. Launch witnessed by Gen de Gaulle. Expected life 50yr.

Cosmos 133 L 28 Nov 1966 by A2 from Tyuratam. Wt 6,500kg. Orbit 181 × 232km. Incl 52°. First flight of Soyuz spacecraft. Recovered 30 Nov.

Cosmos 144, 156, 184, 206 L between Feb 1967 and Mar 1968 into circular 628km orbits at 81° incl with lifetime of 50-60yr. Part of Meteor system. C 144 and 156 carried equipment for TV and infra-red photography, providing pictures of cloud layers and of snow and ice-fields for about 8% of Earth's surface. Also measured radiation streams reflected and emitted by Earth and its atmosphere over about 20% of Earth's surface on each orbit. The area scanned by one was reviewed by the 2nd 6hr later. C 206 was about 20min behind 184, so that forecasters could check data received from the first.

Cosmos 166, 215 L 16 Jun 1967 and 19 Apr 1968 by B1 from Kapustin Yar. Wt ? 400kg. Orbit 260 × 577km. Incl 48°. Studied solar radiation. The 2nd had 8 mirror telescopes, an X-ray telescope and 2 photometers to observe the radiation of hot stars in various wavebands – a first step, according to Pravda, towards placing a big telescope beyond the confines of Earth's atmosphere. Dec after 130 and 72 days respectively.

Cosmos 149, an early meteorological satellite, was in orbit for 17 days.

Cosmos 186, 188 L 27 and 30 Oct 1967 by A2 from Tyuratam. Wt 6,000kg. Orbit 209 × 276km. Incl 51°. Soyuz test vehicles. Carried out world's 1st automatic docking, and Russia's 1st docking of any kind. C 186, the active craft, automatically manoeuvred to rendezvous and dock with C 188. After 3½hr they were commanded to undock and manoeuvred into different orbits. The mission, performed in typical Soyuz orbits, was a rehearsal for manned flight. Both craft were recovered during the 4th day.

Cosmos 212, 213 Cosmos 212 and 213 achieved a 2nd automatic docking on 15 Apr 1968, 5 months before the joint Soyuz 2 and 3 flights, which did not succeed in docking.

Cosmos 243 L 23 Sept 1968 by A1/2 from Tyuratam. Wt 6,000kg. Orbit 209 × 319km. Incl 71°. Regarded by Soviet scientists as landmark in the series; 1st satellite to study heat emissions from Earth and its atmosphere. It enabled an Antarctic ice map to be made, showing temperature distributions around the world. Registering heat radiation in this way enabled scientists to determine moisture content of the atmosphere, and to discover focal points of intensive precipitation concealed by thick clouds. According to Soviet scientists, oceans are huge accumulators of solar energy, which they emit as evaporation heat; this heat "feeds" the cyclones which make the Earth's weather. Water surface temperatures in the Pacific from the Bering Sea to the Antarctic were "mapped in tens of minutes". C 243 was probably recovered after 11 days.

Cosmos 261 L 20 Dec 1968 by B1 from Plesetsk. Wt ? 400kg. Orbit 217 × 669km. Incl 71°. This paved the way for the Intercosmos programme. Bulgaria, Czechoslovakia, GDR, Hungary, Poland and Romania collaborated in experiments exploring air density in the upper atmosphere, and the nature of the polar auroras. Dec 12 Feb 1969.

Cosmos 262 L 26 Dec 1968 by B1 from Kapustin Yar. Wt ? 400kg. Orbit 262 × 965km. Incl 48·5°. 1st satellite to study vacuum ultra-violet (VUV) and soft X-ray radiation (SX) from the stars, Sun and Earth's upper atmosphere. Carried 3 16-channel photometers. Results announced Oct 1969. Dec after 4 months' operation, 18 Jul 1969.

Cosmos 336-343 L 25 Apr 1970 by C1 from Plesetsk. Wt 40kg. Orbits 1,313 × 1,554km. Incl 74°. 1st octuple launch. Each believed to be spheroid about 1m long, 0·8m dia.

Cosmos 359 L 22 Aug 1970 by A2e from Tyuratam. Wt 1,180kg. Orbit 208 × 889km. Incl 51°. Almost certainly intended to be Venera 8 but failed to achieve escape velocity. (Venera 7 was launched 5 days earlier.)

Cosmos 381 L 2 Dec 1970 by C1 from Plesetsk. Wt ?. Orbit 971 × 1,013km. Incl 74°. Provided information on physical characteristics of ionosphere layers covering almost entire global surface. Expected life 1,200yr.

Cosmos 419 L 10 May 1971 by Proton from Tyuratam. Orbit 145 × 159km. Incl 51°. Attempted Mars probe; failed to leave Earth orbit. Dec after 2 days.

Cosmos 433 L 8 Aug 1971 by F1r from Tyuratam. Orbit 159 × 259km. Incl 49°. FOBS test; possibly recovered just short of 1 orbit.

Cosmos 482 L 31 Mar 1972 by A2e from Tyuratam. Wt ?. Orbit 204 × 9,800km. Incl 52°. Intended to be Venera 9 to accompany Venera 8, launched 4 days earlier. Escape stage fired only partially. Life 6yr.

Cosmos 496 L 26 Jun 1972 by A2 from Tyuratam. Wt ?. Orbit 195 × 343km. Incl 51°. Probably test of equipment for manned spaceflight, possibly redesigned Salyut/Soyuz hatch. Life 10 days.

Cosmos 520 L 19 Sep 1972 by A2e from Plesetsk. Wt ? 2,000kg. Orbit 650 × 39,300km. Incl 63°. 1st operational missile early-warning satellite. 30yr orbital life.

Cosmos 637 L 26 Mar 1974 by D1e from Tyuratam. Wt ?. Orbit 35,407 × 35,760km. Incl 25°. Russia's first ever synchronous satellite, and a rehearsal for the new Statsionar series. Orbital life 1 million yr.

Cosmos 929 L 17 Jul 1977 by D1 from Tyuratam. Wt ? 20,000kg. Orbit 214 × 278km. Incl 52°. Test of 14m-long module with manoeuvring engine aimed at increasing size of space stations. Later test flight, C 1267 L 25 Apr 1981, docked with Salyut 6, when it was referred to as the "Star" module. C 929 re-entered 2 Feb 1978.

Cosmos 932 L 20 Jul 1977 by A2 from Tyuratam. Wt ? 4,000kg. Orbit 150 × 358km. Incl 65°. Identified S African nuclear device apparently about to be exploded in atmosphere; test was stopped by international pressure. Recovered 13th day.

Cosmos 1000 L 31 Mar 1978 by C1 from Plesetsk. Wt ? 700kg. Orbit 964 × 1,010km. Incl 83°. 1st satellite acknowledged as part of navsat system.

Cosmos 1001 L 4 Apr 1978 by A2 from Tyuratam. Wt ? 6,500kg. Orbit 200 × 228km. Incl 52°. 1st orbital test of new Soyuz T. Several manoeuvres, with recovery on 11th day.

Cosmos 1383 L 1 Jul 1982 by C1 from Plesetsk. Wt ? 700kg. Orbit 964 × 1,010km. Incl 83°. Test for Sarsat search and rescue system.

Cosmos 381: equipment included a Mayak radio transmitter to probe and analyse the ionosphere. *(Novosti)*

Cosmos 1500 L 28 Sep 1983. Wt ?. Orbit 635 × 667km. Incl 82°. Ocean surveillance, said to be 1st equipped with side-looking, all-weather radar. Emergency use of its pictures was made in Dec 1983 to map escape routes for "dozens of ships" trapped in ice in the E Arctic. Supplied to 53 research stations in the Soviet Union, its pictures show the boundaries of strong ocean currents and the ocean-atmosphere interaction.

Cosmos 1508 L 11 Nov 1983. Wt ? 550kg. Orbit 400 × 1,966km. Incl 82·9°. Logged by RAE as 2,500th successful launch. 30yr orbital life.

Cosmos 1514 L 14 Dec 1983. Wt ? 5,700kg. Orbit 215 × 260km. Incl 82°. International Biosatellite carrying monkeys and rats. Rec after 5 days.

Cosmos 1517 L 27 Dec 1983 by Proton from Kapustin Yar. Wt ? 1,000kg. Orbit 180 × 221km. Incl 50·66°. Subscale spaceplane. Rec in Black Sea after 1 orbit.

Cosmos 1519-21 L 29 Dec 1983 by Proton. Wts ?. Orbits approx 19,005 × 19,140km. Incl 64·7°. Triple Glonass (Navstar-type) launch.

Navsat Cosmos These started with C 192, L 23 Nov 1967 into orbit just below 800km at 74° incl. They are now launched at a rate of about 8 per year. Starting with C 514, L 16 Aug 1972, the inclination was changed to 83°, with the altitude remaining the same. Orbital life is about 1,200yr. Currently there are 2 separate systems, with satellites spaced at 30° and 45° intervals in orbit to ensure availability at all times. First publicly announced navsat was C 1000, launched in 1978. It is a cylinder with a dome at each end (this is probably the standard small Cosmos), all enclosed in a drum-shaped solar panel; length and dia are approx 2m, wt 700kg. C 381 had a smaller, approx 1·3m drum, and may have represented the standard for

Cosmos test vehicle for Soviet Shuttle, probably Cosmos 1445, being recovered from the Indian Ocean on 16 March 1983. *(RAAF)*

earlier navsats. The system is called Cicada ("cricket"), presumably because the signals on 150 and 400MHz frequencies resemble an insect chorus.

Geodetic Cosmos These fly in high, near-polar and usually circular orbits. Their existence is known by inference from Soviet statements and by studying orbits to find vehicles which could perform such a role. The study of the Earth's shape, geodesy has military applications, accurate mapping of the Earth's surface being needed for strategic missile targeting. Both the US and USSR had to solve this problem, since the relative position of the continents was not known accurately enough to ensure that the missiles could be guided to within kill distance of their targets. These launches have been fewer recently, the last known being C 1067, L 26 Dec 1978 (orbit 1,159 × 1,213km, incl 83°), and possibly C 1132, L 30 Sept 1981 (orbit 1,492 × 1,503km, incl 82°). Geodetic measurements can however be taken with the position-fixing equipment on navigation satellites, so some navsats may have a dual purpose (see US - Navsat).

DRS

The Soviets have given notice that DRS direct-broadcast satellites will be launched in 1983-85 to 95°E, 16°W and 160°W.

Elektron

Elektron 1 and 2 L 30 Jan 1964 by A1 from Tyuratam. Wt 350 and 445kg. Orbits 394 × 7,126km, and 441 × 67,988km. Incl 61°. Russia's 1st dual launch, the satellites being separated from the launch vehicle while the last stage was still firing. They studied the inner and outer zones of the Van Allen radiation belt and Earth's magnetic field, gathering information for radiation protection of manned spacecraft. Orbital life 200yr and 30yr.

Elektron 3 and 4 L 11 Jul 1964 by A1/2 from Tyuratam. Similar wt, orbits and missions. Orbital life 200yr and 23yr.

Gamma

Gamma 1, delayed from 1984 to mid-1985, is a Progress modified as a 1,500kg gamma-ray observatory. France is contributing to the payload.

Iskra

These amateur radio satellites were produced for the use of Comecon countries including the USSR, Bulgaria, Hungary, Vietnam, GDR, Cuba, Mongolia, Poland, Romania and Czechoslovakia. Designed by students of the Ordzhonikidze Aviation Institute in Moscow, they carry a retransmitter, memory device, command radio channel, and radio-telemetric system for relaying scientific information and data about spacecraft functioning. Iskra 1 L 10 Jul 1981 by Meteor 31; orbit 638 × 663km, incl 98°; dec 9 Jul 1982. Iskra 2 L 17 May 1982 by Salyut 7/Soyuz T-5; wt 28kg, ? 0·6m dia; orbit 335 × 345km, incl 51·6°; dec Jul 1982. Iskra 3 L 18 Nov 1982 by Salyut 7/Soyuz T-5.

Luna 1 being prepared for launch. First spacecraft to reach "2nd cosmic velocity," and intended to impact the Moon, it passed less than 600km away. *(Novosti)*

Luna

History As this book goes to press Soviet scientists are due to resume their lunar exploration programme, extending back 25yr, with the first Lunar polar orbiter. They announced at the 9th Lunar Science Conference at Houston in 1978 that this would be flown in 1983, and that they were also planning soil-recovery operations on the Moon's far side. This long programme has encountered many setbacks, especially in recent years, but 3 out of 6 soil-return missions - Lunas 16, 21 and 24 - have been successful. Lunas 17 and 21 landed the Lunokhod rovers.

Soviet scientists must have been planning lunar flights and exploration several years before their 1st satellite, Sputnik 1, was orbited in Oct 1957. Luna 1, 15 months later, was only Russia's 4th space launch. Details given below, and also under Lunokhod, show the Russian technique of tackling and solving each problem with a group of spacecraft and then pausing for a year or two of research and development before going on to the next phase. Thus the flyby technique was mastered in 3 flights in 1959, culminating in the transmission by Luna 3 of the first historic pictures of the Moon's far side. The next phase, preceded by a failure (Sputnik 25 on 4 Jan 1963), began the development of soft-landing techniques. Though none appears to have been completely successful, much was learned from Lunas 4-8. Cosmos 60, on 12 Mar 1965, and Cosmos 111, 1 Mar 1966, were almost certainly 2 of only 5 complete failures in this long programme. The group of 5 Lunas launched in 1966 marked the biggest leap forward, demonstrating both soft-landing and orbiting techniques. It was only during Lunokhod 1's first activities on the lunar surface that we were told that Luna 12, 4 years earlier, had space-tested the electric motor for the robot's wheels. There were follow up trials of the gears on Luna 14, which in Jul 1969 crashed on the lunar surface during an attempt to make an automatic recovery of lunar samples and get them back to Earth a few hours before the Apollo 11 crew. This was the first time that Russia had tried a soft-landing from parking orbit; previous attempts had always been by direct flight. The more sophisticated parking-orbit technique, which gives time for orbital corrections and makes pinpoint touchdown possible, has always been preferred by the Americans and was finally employed by Russia from Luna 16 onwards. (Luna 16 was apparently preceded by 2 launch failures, Cosmos 300 and 305, in Sep and Oct 1969.) When Lunokhod exploration began, Academican Boris Petrov claimed that robots could carry out such missions for less than one-tenth the cost of a similar manned flight. Such a claim is difficult to sustain against such a background of Luna failures. Luna 22, which was successfully manoeuvred in a whole variety of lunar orbits over a period of 15 months, provided the data and experience for Lunokhod and soil-recovery operations on the Moon's polar regions and far side under the command of at least one orbiting satellite.

Luna 1 L 2 Jan 1959 by A1 from Tyuratam. Wt 361kg. Intended to impact on the Moon, this was the world's 1st spacecraft to reach "second cosmic velocity", or 40,234km/hr. Launch believed to be by 3-stage vehicle with total thrust of 263,000kg. The spherical craft, equipped with instruments for measuring radiation, etc, and its separated 3rd stage both passed within 5,955km of the Moon and went into solar orbit. Also named Mechta, it was only Russia's 4th space launch.

Luna 2 L 12 Sep 1959 by A2 from Tyuratam. Wt 390kg. The 1st spacecraft to reach another celelestial body. Impacted E of Sea of Serenity, carrying USSR pennants.

Luna 3 L 4 Oct 1959 by A1 from Tyuratam. Wt 278·5kg. 1st spacecraft to pass behind the Moon and send back pictures of far side. Placed on an elliptical Earth orbit with apogee of 480,000km so that without midcourse corrections lunar gravity pulled it around the Moon at a distance of about 6,200km. Equipped with a TV, processing and transmission system, it took an unannounced number of far side pictures which were sent back as Luna 3 moved back towards Earth. 3 were published, including a composite full view of the far side, and 2 large seas were named Mare Moscovrae (Moscow Sea) and Mare Desiderii (Dream Sea). Dec after 11 orbits totalling 177 days.

Luna 4 L 2 Apr 1963 by A2e from Tyuratam. Wt 1,422kg. First of 5 spacecraft aimed at solving problems of soft-landing instrument containers. Contact lost as it missed Moon by 8,529km, passing into 89,782 × 692,300km Earth orbit.

Luna 5 L 9 May 1965 by A2e from Tyuratam. Wt 1,476kg. First soft-landing attempt. Flight lasted 82hr. Retro-rocket, due to be fired at about 64km from surface, malfunctioned and spacecraft impacted in Sea of Clouds 5min earlier than planned.

Luna 6 L 8 Jun 1965 by A2e from Tyuratam. Wt 1,442kg. During midcourse correction manoeuvre engine failed to switch off. Spacecraft missed the Moon by nearly 161,000km and passed into solar orbit.

Luna 7 L 4 Oct 1965 by A2e from Tyuratam. Wt 1,506kg. After 86hr flight retro-rockets fired early; crashed in Ocean of Storms.

Luna 8 L 3 Dec 1965 by A2e from Tyuratam. Wt 1,552kg. After 83hr flight retro-rockets fired late; crashed in Ocean of Storms.

Luna 9 L 31 Jan 1966 by A2e from Tyuratam. Wt 1,583kg. 1st successful soft landing, followed by 1st TV transmission from surface. After 79hr flight the spacecraft ejected an egg-shaped 0·6m-dia instrument capsule weighing 100kg as a probe touched the surface and switched off the retro-rocket. The capsule, weighted so that it rolled into an upright position, was stabilised on Earth command by 4 spring-ejected "petals". 3 panoramas of the lunar landscape on the eastern edge of the Ocean of Storms, with different Sun angles, were transmitted over a 3-day period.

Luna 10 L 31 Mar 1966 by A2e from Tyuratam. Wt 1,600kg. 1st lunar satellite. Wt was 245kg when it was fired into 349 × 1,017km lunar orbit with 71° incl. In addition to broadcasting the "Internationale" several times it studied lunar surface radiation, magnetic field intensity, etc. Communications were maintained for 2 months and 460 orbits, providing opportunities for monitoring the strength and variation of lunar gravitation.

Luna 11 L 24 Aug 1966 by A2e from Tyuratam. Wt 1,640kg. 2nd lunar satellite. Placed in 159 × 1,200km lunar orbit with 27° incl. Possibly carried TV system which failed to operate. Data were received during 277 orbits until 1 Oct 1966.

Luna 12 L 22 Oct 1966 by A2e from Tyuratam. Wt 1,640kg. 3rd lunar satellite. Placed in 100 × 1,739km orbit. A TV system transmitted large-scale pictures of the Sea of Rains and Crater Aristarchus areas, showing craters as small as 15m across. Tested electric motor for Lunokhod's wheels. Communications continued for 602 orbits until 19 Jan 1967.

Luna 13 L 21 Dec 1966 by A2e from Tyuratam. Wt ?1,583kg. 2nd successful soft landing. Capsule was again bounced on to the Ocean of Storms. In addition to sending back panoramic views, 2 Meccano-like 1·5m-long arms were extended to measure soil density and surface radioactivity. Communications lasted for 6 days from landing.

Luna 14 L 7 Apr 1968 by A2e from Tyuratam. Wt ? 1,615kg. Placed in 160 × 870km orbit with 42° incl. Studied Moon's gravitational field and "stability of radio signals sent to spacecraft at different locations in respect to the Moon". Made further tests of geared electric motor for Lunokhod's wheels.

Luna 15 L 13 Jul 1969 by D1e from Tyuratam. Wt ? 5,600kg. A bold but unsuccessful attempt to obtain lunar samples and return them to Earth a few hours before America's first men on the Moon (Apollo 11) could do so. Launched 3½ days before Apollo 11, Luna 15 was placed in lunar orbit and remained there during the Apollo flight and first manned landing. 2 orbital changes were made, however, and American scientists, concerned about possible conflict of radio frequencies, asked Frank Borman, who had recently visited Moscow, to seek assurances from Russia that there would be no interference. Such assurances were quickly given. 2hr before Apollo 11 was due to lift off from the Sea of Tranquillity Luna 15, on its 52nd revolution, began descent manoeuvres, apparently aimed at the Sea of Crises. But Russia's first attempt at an automatic landing from lunar orbit went wrong and the spacecraft crashed at the end of a 4min descent. It was more than a year later that the Luna 16 flight revealed what had been intended.

Luna 12 orbiter: *1* **gas containers;** *2* **TV;** *3* **thermal radiator;** *4* **radiometer;** *5* **instrument compartment;** *6* **chemical battery;** *7* **orientation system;** *8* **antenna;** *9* **electronic unit;** *10* **control jets;** *11* **retro-engine.** ***(Novosti)***

Luna 16 L 12 Sep 1970 by D1e from Tyuratam. Wt ? 5,600kg (lander ? 1,880kg). 1st recovery of lunar soil by automatic spacecraft. Luna 16 was placed in a 110km orbit with 70° incl, corrected prior to landing to 15 × 110km with 71° incl. On Sep 20 Russia's Space Communications Centre gave the command to fire the descent engine. At 20m above the surface this was switched off and final touchdown on the Sea of Fertility, 1·5km from target point, was controlled by 2 vernier engines. Luna 16 consisted of landing and ascent stages; as in the case of America's Lunar Module the landing stage served as a launch platform for the ascent stage. On Earth command an automatic drilling rig was deployed; with a 0·9m reach, it was capable of penetrating the surface just over 30cm. About 100gm of material was lifted by the drill into the loading hatch of a spherical capsule at the top of the ascent stage, which was then hermetically sealed. After 26½hr on the surface the ascent stage was launched on a ballistic trajectory back to Earth. It consisted of the lunar soil container, an instrument compartment and a parachute compartment containing braking and main parachutes and 2 gas-filled balloons, presumably in case of descent on water. No midcourse corrections were made on the return flight. 3hr before re-entry the instrument compartment was jettisoned. The sample capsule's transmitters enabled it to be located and recovered in Kazakhstan on 24 Sep.

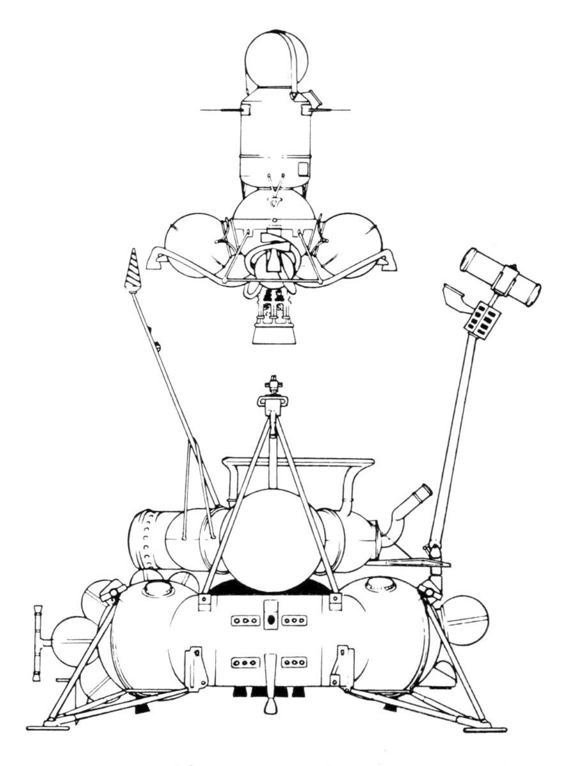

Luna 16, illustrating the separated Earth-return vehicle after transfer of lunar material from the drill unit into the recoverable capsule. ***(D.R. Woods)***

Luna 16: touchdown in Kazakhstan with 100gm of lunar soil. ***(Novosti)***

Luna 17 L 10 Nov 1970 by D1e from Tyuratam. Wt ? 5,600kg (Lunokhod 756kg). Carrying the first Moon robot, Lunokhod 1, Luna 17 was placed in an initial circular 84km orbit with 141° incl; the following day the perilune was lowered to 19km. On 17 Nov a successful soft landing was made in a shallow crater in the NW Sea of Rains (Mare Imbrium). After checks by TV cameras that there were no boulder obstructions, one of 2 alternative ramps was lowered by commands from the Deep Space Communications Centre and Lunokhod 1 rolled down on to the surface. It proved to be the Soviet Union's greatest technical space success. Full details are given under Lunokhod.

Luna 18 L 2 Sep 1971 by D1e from Tyuratam. Wt ? 5,600kg. Intended as a soft-lander, most probably with an ascent stage for a 2nd soil recovery, this was placed in a 100km circular lunar orbit with 35° incl. After 54 orbits taking 4½ days (twice as long as Luna 16 and 17), an attempt was made to land in highland terrain in the Sea of Fertility. Communications ceased shortly after Earth command had started the descent engine, probably as a result of impact.

Luna 19 L 28 Sep 1971 by D1e from Tyuratam. Wt ? 5,600kg. 5th Soviet lunar orbiter, initially placed in circular 140km orbit with 40° incl. Communications continued for over a year, at the end of which Luna 19 had made more than 4,000 lunar orbits. A systematic study of the Moon's gravitational field and the effect of its mascons was made, coupled with TV pictures of the surface. At least 10 powerful solar flares were observed. Lunar radiation was compared with similar measurements made by Mars 2 and 3 in Martian orbit and by Venus 7 and 8 and Prognoz 1 and 2.

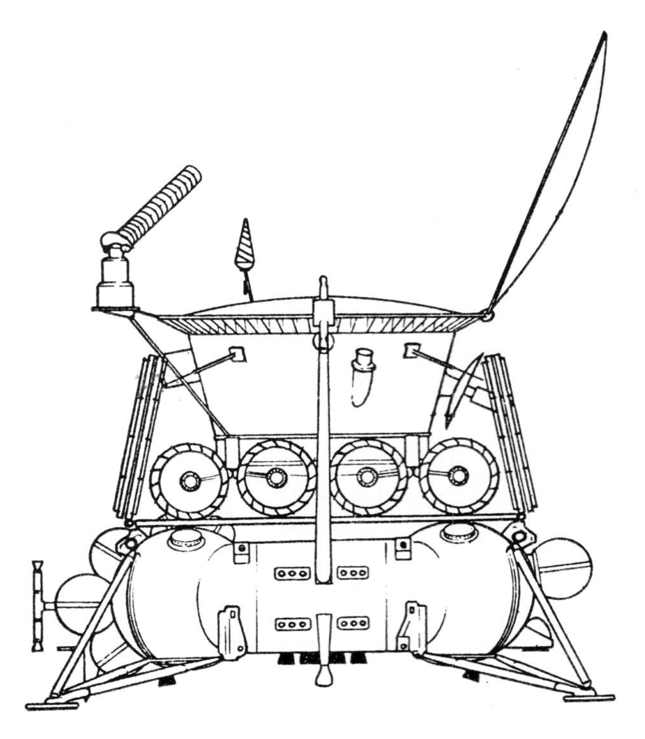

Luna 17, comprising the 2nd-generation orbiter/lander, with Lunokhod 1 vehicle mounted on it. Lunokhod ramps are shown folded. *(D.R. Woods)*

Luna 20 L 14 Feb 1972 by D1e from Tyuratam. Wt ? 5,600kg. 2nd successful soil recovery. Placed in initial circular 100km lunar orbit at 65° incl. This was lowered 1 day later to 21 × 100km. After site selection photography the descent engine was fired for 4min 27sec, and 7½ days after launch a safe landing was made on a mountainous isthmus pockmarked with large craters S of the Sea of Crises and on the Sea of Fertility's extreme NE; it was 120km N of Luna 16's sampling point. A "photo-telemetric device" relayed to Earth pictures of the surface, and from these a site with "a grey cloudy structure" was chosen for sampling. A rotary-percussion drill, of improved design as a result of Luna 16, drilled into the rock. It sank quickly to a depth of 100-150mm; then, because of the rock's hardness, drilling had to be done in stages so that the drill did not overheat. The samples were then lifted into the return capsule on Luna 20's ascent stage, which was fired back to Earth on a ballistic trajectory 1 day after the touchdown. It was recovered with some difficulty on an island in the River Karakingir, 40km NW of Dzhezkazgan in Kazakhstan after landing in a blizzard on 25 Feb. The ash was light-grey compared with the "black slate colour with a metallic glitter" of the Luna 16 samples. Scientists expected these samples to be about 1,000 million yr older than the Luna 16 samples, which were estimated at 3 to 5,000 million yr old.

Luna 21 L 8 Jan 1973 by D1e from Tyuratam. Wt ? 5,600kg. Carried Lunokhod 2 to the Moon. Descent was made on the 41st orbit from a height of 16km. Lunokhod 2, wt 840kg (84kg heavier than Lunokhod 1), was placed on the E edge of the Sea of Serenity. Touchdown was only 180km from the Apollo 17 landing point (see Lunokhod 2).

Luna 22 L 29 May 1974 by D1e from Tyuratam. Wt ? 5,600kg. Placed in 220km circular lunar orbit at 19·35° incl with 2hr 10min revolution period on 2 Jun 1974. For 4 days from 9 Jun the orbit was lowered to 25 × 244km to obtain TV panoramas of high quality and good resolution. Simultaneously, altimeter readings were taken of the relief of the area, and chemical rock composition determined by gamma radiation. Luna 22 was then returned to a 181 × 299km orbit to continue studying the gravitational field. It was still operational when 5 months later it was joined in orbit by Luna 23. But there is no evidence that Luna 22 was used to help control the Luna 23 operations on the surface. Completion of its programme after 15 months of operation and over 4,000 lunar revolutions was announced in Oct 1975. Final manoeuvres included lowering the orbit to within 30km of the surface for TV pictures. It was finally abandoned in a lunar orbit of 100 × 1,286km with 21° incl.

Luna 23 L 28 Oct 1974 by D1e from Tyuratam. Wt ? 5,600kg. Placed in a lunar orbit of 94 × 104km with 130° incl on 2 Nov. By 5 Nov the orbit had been reduced to 17 × 105km, and a landing was attempted the following day on the southern part of the Sea of Crises. Moscow Radio reported that touchdown was in a sector with unfavourable relief, and the device for taking lunar rock samples to a depth of 2·5m was damaged. No drilling or soil collection was possible, and a revised and reduced programme had to be abandoned after 3 days. It was the 5th attempt at soil recovery, with only 2 successes.

Luna 24 L 9 Aug 1976 by D1e from Tyuratam. Wt ? 5,600kg. Russia's 3rd successful sample recovery operation. With a much bigger soil-carrier 2m in length, Luna 24 achieved a larger sample than the 100gm and 150gm now believed by US scientists to have been obtained by Lunas 10 and 16. The initial circular lunar orbit of 115km at 120°, achieved on 14 Aug, was lowered to 12 × 120km on 16-17 Aug. Touchdown, after about 53 lunar orbits, was in the SE of the Sea of Crises (12·75°N 62·2°E). Luna 23 had attempted to land in the same area. The site was 17km from a large crater 10km in dia and 2km deep, and the hope was to obtain ejecta samples. Core samples were drilled out from a depth of 2m, and the return rocket was launched from the lunar surface 23hr after touchdown. After recovery 200km SE of Surgut, W Siberia, the sample was taken to the Moscow Institute of Geochemistry and Analytical Chemistry, where it was said to be "silvery in colour with a brown tint". The material was lighter, with small grains up to 8mm across, and included more large grains than the Luna 16 sample. US scientists were once again given samples totalling 3gm, probably because US techniques for high-quality age-dating and chemical analysis are superior to Soviet methods. Divided into 6 soil samples and one rock fragment, they were allotted to 30 investigators. 4 samples totalling 0·91gm were studied in British laboratories, which found that the most common rock type at the Luna 24 site was a low-titanium, high-iron basalt. This had not previously been recognised among samples from other Luna and Apollo sites.

Lunokhod

General description The 756kg Lunokhod had a circular 2·1m-dia instrument compartment mounted on an 8-wheel chassis. The spoked, wire-mesh wheels, each 51cm in dia, were in 4 pairs. Each had an independent electric motor; if any one seized up or got stuck, a powder charge could be fired to break the drive shaft, enabling the wheel to become a passive roller. Movement would still have been possible with only 2 wheels on each side operational. Sensors provided automatic braking, over-riding Earth commands if slope angles became so steep that there was a danger of overturning. Disc brakes held the craft at rest on slopes and during the lunar nights. There were 2 forward speeds and possibly 2 in reverse.

The instrument compartment, made of magnesium alloy, warmed up easily and released heat slowly, thus enabling it to survive the lunar nights, equivalent to 14 Earth nights. The compartment was topped with a lid which was opened when the instruments were in use and kept tightly shut at other times. During lunar nights the instruments, cameras and other systems were kept warm and operational despite outside temperatures of -150°C by the circulation of heated gas. One TV system transmitted a frame every 3–20sec, enabling the operators to monitor the Lunokhod's progress. The 2nd obtained panoramic pictures of the locality, the horizon, Sun and Earth. For this it used 4 identical telephoto cameras, 3 looking to the sides and rear and the 4th, mounted alongside the low-rate camera, looking through forward portholes. The fore and aft cameras had a combined range of 30° horizontally and 360° vertically, the side cameras 180° horizontally and 30° vertically.

Operating technique The Lunokhods were driven by a 5-man team - commander, driver, navigator, systems engineer and radio operator - working in the Deep Space Communications Centre, believed to be near Moscow. The technique was said to be so difficult that it beat many highly experienced drivers and pilots. This was mainly because of the time delays resulting from sending commands and waiting to observe the response. They had to remember that the robot was already several metres ahead of the slow-scan TV picture they were watching, and if there were rocks or boulders to be avoided it would have moved on still further before the signals to take avoiding action had reached it. However, the automatic stop system worked so well that, whatever happened to Lunokhod 2 in the end, Lunokhod 1 never once overturned during its 11-month life.

Lunokhod 1 Landed on the Sea of Rains (Mare Imbrium) by Luna 17 on 17 Nov 1970. Expected life was 90 days but it continued to operate for 11 months. During that time it travelled 10,540m, photographing an area of more than 80,000 sq m. Over 200 panoramic pictures and 20,000 separate photographs were returned; physical and mechanical soil analyses were carried out at 500 locations and chemical analyses at 25. When at last Lunokhod's equipment froze during the 11th lunar night following exhaustion of its isotopic fuel, the vehicle had been placed so that its French laser reflector could continue to be used indefinitely for Earth-Moon measurements. The first task at the beginning of each lunar day was to open the lid to enable the solar cells on the lid's inside to start recharging the depleted batteries. When being parked at the end of the previous lunar day it had been carefully turned towards the E to obtain the maximum intensity from the Sun's rays as soon as the lid was opened. Soon afterwards it became necessary to keep the upper part of the hull cool, and a mirror-like surface was used to reflect heat into space.

Initially the wheels were the main source of anxiety: it was thought that they would build up charges of static electricity because of the vacuum conditions on the lunar surface. It was feared that without air it would be impossible to run off the electricity, with the result that particles of soil would cling to the wheels and ultimately clog them and make the robot inoperable. The fear was unjustified: in practice the soil fell off like volcanic sand. Orientation proved another problem at first: on the 3rd lunar day scientists became excited by a big rock, estimated to be 150m away. They asked the crew to move up close to the boulder, but when they did so it was nowhere to be seen. It turned out to be only a small stone, which on the screen appeared to be a huge rock. Attempts to steer Lunokhod closer to it resulted in its being passed and left further behind. It was soon found that TV pictures of the wheel ruts provided a useful distance scale for navigators and drivers, as well as information for scientists about the physical and mechanical properties of the lunar soil. By the end of the 2nd lunar day the robot was climbing slopes with gradients of 23°, weaving, turning and zig-zagging to avoid steep craters and large boulders, and being sent with confidence through shallow craters 150m across. Probably its most successful instrument was RIFMA (Roentgen Isotropic Fluorescent Method of Analysis), able to analyse the chemical composition of lunar rocks by sending out a stream of X-rays. Under the action of the X-rays the atoms of the rocks rearranged their electronic shells and sent back X-radiation of their own. Measurements of this response showed how much aluminium, silicon, magnesium, potassium, calcium, iron and titanium was contained in the rocks On 12 Dec 1970 RIFMA reported a solar flare which, according to Pravda, would have created hazardous radioactivity for men on the Moon.

Lunokhod 2 Landed on the eastern edge of the Sea of Serenity in the Le Monnier crater on 16 Jan 1973, Lunokhod 2 appears to have ceased operating abruptly early in its 5th lunar day, which began on May 8. But though it operated for only half as long as its predecessor, it was much more active. Its 9th, pedometer, wheel recorded that it had covered 37km, $3^1/_2$ times the distance achieved by Lunokhod 1, and 86 panoramic pictures and 80,000 TV pictures of the lunar surface were transmitted. The Le Monnier landing site was chosen to provide information about a transitional "sea/continental" area, and because there is a 16km tectonic fault nearby. Improved equipment and additional instruments added nearly 100kg to Lunokhod 2's weight. Control and communications systems had been improved, and one of the panoramic TV cameras, located in the vehicle's centre, had been raised to improve picture quality and reduce the time taken to relay signals back to Earth, thus in turn giving the crew better control.

Equipment carried included a corner reflector provided by French engineers, to continue the Franco-Soviet collaboration on laser reflectors begun with Luna 17. Measurements made from Earth and circumlunar orbit had already established that the visible side of the Moon averages 2km lower than the middle radius, and that the far side is $9^1/_2$km higher. Additional measurements by means of the French corner reflector were aimed at mapping lunar contours with more precision, checking the theory that the Moon's continents, like those on Earth, are subject to drift.

During the 3rd session there was a near-collision with the Luna 21 landing stage; the 2 vehicles were less than 4m apart when Lunokhod 2 stopped. Its speed was more than double that of its predecessor, and operating it over the rising, terraced ground towards the Taurus Mountains about 6km away was described as "very taxing" for the crew.

First tests by the RIFMA instrument suggested that the rock composition was much the same as had been found on the Sea of Rains. The landing area was however generally more rugged, with more craters and boulders. During the first lunar night the temperature at the end of the measuring probe fell to a record -183°C, though the wheel temperatures never fell below -128°C.

During its 2nd lunar day, occasionally sinking to the hubs of its wheels in loose rock, Lunokhod 2 moved steadily towards the Taurus Mountains, taking magnetic measurements on the way. The robot zig-zagged up the mountain slopes, wheel-slip sometimes registering 80%. From a 400m peak views were obtained of the opposite shore of the Le Monnier bay, and of the main mountain peaks 55–60km away. It also sent back an unexpected view of Earth, appearing as a "thin sickle" in the local sky.

During the 3rd day and throughout the 4th Lunokhod 2 was worked along a narrow tectonal abyss 30–50m deep and 300–400m wide. Moving cautiously along the westward edge, and then in the reverse direction along the eastern slope, it established that the feature was solid basalt rock which had been split by tectonic forces. On May 9, at the start of the 5th day, it was reported that

Lunokhod 2 photographed at 1973 Paris Air Show. Cone, left, is omni-directional antenna; top right, narrow-beam directional antenna. The TV cameras returned 80,000 pictures. *(Graham Turnill)*

movement away from the abyss in an NE direction had begun. There were no further reports until, on Jun 3, it was announced that the robot's research programme had been completed. No indication was given as to whether there had been a mechanical failure, or whether it had been lost or damaged by falling into a crater or fissure.

Perhaps Lunokhod 2's most interesting discovery was the fact that while the Moon almost certainly has admirable conditions for astronomical observations during the lunar night, conditions during the day are not nearly so good as had always been imagined. An astrophotometer (an electron telescope without lens) registered glow in the lunar sky, both in the visible and ultra-violet bands of the spectrum, to establish whether the lunar "atmosphere" contained cosmic dust which could affect observations from the surface. The after-sunset glow proved much more intense than expected: 10-15 times brighter than on Earth, suggesting that the lunar sky is full of dust particles which scatter solar light. This high luminosity would make daytime astronomy ineffective.

Conclusions Academician Blagonravov said that performance had surpassed all expectations, and that Lunokhod could in principle make a round-the-Moon journey - though this would presumably involve the use of a lunar satellite for communications and control on the far side. Soviet scientists pointed out that Lunokhod was specifically designed for the Moon; planetary robots would be different, walking, crawling or hopping, according to conditions. In the case of Mars, with its atmosphere saturated with carbon dioxide, frictional surfaces would probably be unsuitable because they would wear off much too fast, and the great distances from Earth would mean revision of the remote-control methods employed for Lunokhod, which generated great difficulties. Radio signals from Earth to Moon take 1·3sec; from Earth to Mars they take 20min. A "Marsokhod" would therefore have to be self-controlled by means of an onboard computer.

Lunokhod 2 rear view. *(Aviation Week)*

In May 1978 Soviet scientists said that they had decided against launching any more Lunokhods in the near future because they cost so much to build and operate. They had decided instead to concentrate on a lunar polar orbiter in the early 1980s, and a Luna far-side sample return.

Lutch

Lutch, L Mar 1982 to geostationary orbit at 53° E, is evaluating super-high frequencies in the 11, 20 and 30GHz ranges for future communications networks. Intercosmos members participating include Bulgaria, Hungary, GDR, Poland and Czechoslovakia. Details of the experiments were given at the 1983 IAF Congress.

Mars

History As with the Moon, Russian scientists were well advanced with plans for an interplanetary expedition long before the launch of Sputnik 1. First official Martian launches were Sputniks 22 and 24 in late 1962, preceded by Sputniks 19-21 aimed at Venus, and followed by Sputnik 25 aimed at the Moon. It was Soviet practice from the start to make 3 launches on each occasion; in 1962 none was successful, although much was learned from the only one successfully placed on course. Then in May 1971 the 2 planets were so favourably placed - factors such as the relative inclinations of the orbital planes providing good radio communications are important, as well as distance - that it was possible to put spacecraft on suitable trajectories with a velocity only a few hundred m/sec above the minimum escape speed. This time 2 out of 3 spacecraft reached the planet. The flights of Mars 2 and 3 occupied 192 and 188 days respectively. Such favourable conditions will not occur again until 1986, which explains why Russia prepares triple launches and the Americans, usually, double launches. It was no coincidence that Mars 2 and 3 were in Martian orbit at the same time as America's Mariner 9.

The Soviet technique, at least in these early landing attempts, is to separate the landing craft for a direct approach before the parent craft is placed in Martian orbit. The US preference is first to place the whole vehicle in parking orbit, thus providing time for adjustments before the descent attempt is made. In the case of Mars 2 and 3 it meant that the landings had to go ahead despite the dust storm raging on the surface. There was no possibility of holding the mission in Martian orbit, though presumably it was possible to adjust the targeted landing site during the final, pre-separation manoeuvre. It should also be noted that a Martian landing presents technical and design problems which are the exact opposite of those encountered on Venus. Atmospheric pressure on the Venusian surface is about 100 times greater than on Earth; on the Martian surface pressure is about 100 times less. Thus, while the dense Venusian atmosphere will quickly decelerate an approaching spacecraft, on Mars, if the approach angle is not exactly right, it will either pass through the thin atmosphere and into space without slowing down, or enter so fast that the parachute will be ineffective. No doubt these factors contributed to the bitter disappointment of 1973-4 after Russia had successfully despatched a fleet of 4 spacecraft. This occasion again emphasised the shortcomings of the "slingshot" approach. One lander missed the planet completely, the other crashed just before touchdown; the

Mars 3: lander is mushroom-shaped capsule at top. Piped section inside right solar panel is radiator for temperature control. Antenna for French stereo experiment is mounted on right solar panel. ***(Graham Turnill)***

orbiters too were unsuccessful. It seems likely that the decision not to exploit the 1975 launch window was made because Soviet techniques were being reviewed in the light of the Mariner 9 success and the more sophisticated methods being tried by America's Viking programme.

Experience from Mars 2 and 3 and Lunokhods 1 and 2 is being designed into "planetokhods" for robot exploration of the Martian surface. These vehicles would have to be operated mostly by onboard sensors rather than by Earth commands because of the time required for signals to be transmitted and obeyed.

Since the 1973 failures and the US success with Viking Russia has suspended Martian missions, possibly having decided to wait for the 1986 launch window, when the new heavy launcher should be available.

Spacecraft description Photographs released shortly after Mars 2 and 3 had been placed in Martian orbit showed that they were about 3·6m in height and that more than half the launch weight consisted of fuel. The landing craft, protected by a heatshield shaped like an inverted saucer, consisted of a sphere resting in a toroid, with its engine nozzle projecting through it. The toroid contained the parachute system; the sphere was described as the "automatic Mars station". The heatshield provided maximum aerodynamic braking as the spacecraft entered the thin Martian atmosphere, and it was discarded shortly before the drogue and main parachutes were explosively deployed. The parachutes were jettisoned at about 30m to reduce final landing weight just before a solid-fuel braking engine was fired by a radar altimeter. Separation of the lander, followed by the placing of the orbiter in an elliptical path around Mars, was carried out by an onboard navigation/computer complex, updated by Earth commands. The Mars 3 lander carried mass spectrometers for chemical analysis of the atmosphere, equipment to analyse the chemical and physical content of the surface soil, anemometer, thermometer, pressure gauge and at least 1 TV camera. Data were to have been relayed via the orbiters but only 20sec of signals were obtained. Each orbiter carried 2 cameras, 1 wide-angled, 1 with a telescopic 4° narrow-angle lens. The photographs were automatically developed and the image transmitted directly to Earth by the 200cm S-band aerial. The system was however apparently less successful than the US Mariner arrangement, which breaks down the pictures into coded dots which are then reconstructed on Earth by computer. Other orbiter instruments provided ultra-violet, infra-red and optical examination of Mars and its atmosphere. By using infra-red sensors to measure the thickness of the carbon dioxide atmosphere, variations in surface height could be found and a relief map built up. Surface colouring could also be defined. Mars 3, 6 and 7 carried French equipment to collect solar emission data for comparison with similar measurements on Earth. With both US and Soviet spacecraft in Martian orbit at the same time, a teleprinter link was established between NASA's Jet Propulsion Laboratory at Pasadena, California, and the Soviet Co-ordinating and Computing Centre for the exchange of information. Soviet information supplied over this link was not allowed to be published even though, apparently, it duplicated Russian official statements.

Sputnik 22 L 24 Oct 1962 by A2e from Tyuratam. Wt 893·5kg. Placed in an initial Earth orbit of 180 × 450km with 64° incl, the spacecraft and final rocket stage blew up when being accelerated to escape velocity. The debris passed within range of the West's BMEWS radars, which indicated a possible ICBM attack. The incident was especially alarming at that moment, occurring as it did during the Cuban missile crisis. BMEWS computers, which assess trajectory and impact points, proved the alarm to be false within a few seconds.

Mars 1 L 1 Nov 1962 by A2e from Tyuratam. Wt 893·5kg. World's first Mars probe, successfully launched from Earth orbit similar to that of Sputnik 22. Mars 1 was 3·3m long; with solar panels and radiators deployed following trans-Martian insertion, width was 4m. It consisted of 2 hermetically sealed compartments, orbital and planetary. The orbital section contained the mid-course rocket engine, solar panels, and high and low-gain antennas; the planetary section contained a TV system and other instruments for studying the Martian surface during the period of closest approach. The Soviet Deep Space Tracking Station at Yevpatoria in the Crimea held 61 communication sessions, first at 2-day then at 5-day intervals until 21 Mar 1963. Contact was then lost because the craft's antenna could no longer be pointed towards Earth, following a fault in the orientation system. At that point the craft had travelled 106 million km from Earth. It should have passed Mars 3 months later at a distance of between 998 and 10,783km.

Sputnik 24 L 4 Nov 1962 by A2e from Tyuratam. Orbit 196 × 590km. Incl 64°. This too disintegrated during an attempt to place it on course for Mars from Earth parking orbit; 5 major pieces of debris were tracked by BMEWS.

Cosmos 419 L 10 May 1971 by D1e from Tyuratam. Wt ? 10,000kg. 1st use of Proton launcher for a planetary mission. 1st of 3 craft intended for Mars; placed in 145 × 159km Earth orbit but failed to separate from 4th stage and decayed in 2 days.

Mars 2 L 19 May 1971 by D1e from Tyuratam. Wt 4,650kg. Reached Mars on 27 Nov after 192-day flight during which 3 mid-course corrections were made. Following the final manoeuvre a landing capsule separated from the orbiter and made a 1st, unsuccessful attempt to soft-land. Few details were given, except that it had delivered a pennant with the USSR hammer-and-sickle emblem to the Martian surface. It was presumably destroyed on impact. After the lander was released the orbiter's retro-rockets were fired and it became the 2nd artificial Martian satellite, with an 18hr, 1,380 × 25,000km orbit at 48° incl. Within a few days Mars 2 had established that at least nine-tenths of the Martian atmosphere consisted of carbon dioxide; there was almost no nitrogen, and water vapour was very scarce.

Mars 3 L 28 May 1971 by D1e from Tyuratam. Wt 4,650kg. Reached Mars 5 days after its predecessor, on 2 Dec. About 4½hr before atmospheric entry the Martian lander was automatically separated from the parent craft and placed on a shallow approach angle for a 3min descent. Aerodynamic braking on the saucer-shaped heatshield began at a speed of 21,600km/hr; the speed was still supersonic when the drogue parachute was explosively deployed. After main parachute deployment the heatshield was jettisoned and radio antennas deployed. The soft-landing engine was switched on at about 30m and the parachute ejected to one side to prevent its canopy covering the landing capsule. 90sec after touchdown a timing device switched on the video-transmitter.

Mars 3: close-up view. *(Novosti)*

For 20sec "a small part of a panoramic view" was transmitted by its TV camera, but in any case surface details were probably obscured by the dust storm then in progress. It was probably a combination of a heavy landing and the dust storm that overwhelmed the capsule. It had touched down in a pale-coloured region in the southern hemisphere, between Electris and Phaetontis (45°S, 158°W).

After releasing the lander the orbiter stage of Mars 3 fired its retro-rockets and went into an 11-day, 1,500 × 200,000km orbit from which it sent back data for about 3 months. US observers believe that, because it seldom approached Mars closely, its total observing time was about 2hr, and because it could not be reprogrammed, all of its film was automatically exposed towards a planet completely hidden by the great dust storm.

Summary of results Like America's Mariner 9, Mars 2 and 3 were able to study the 3-month dust storm which began in Oct 1971. Soviet scientists concluded that the winds were not constant but most likely occurred only in the initial phase, and that the particles, mostly of silicate composition, took months to fall "in the absence of supporting vertical currents in the atmosphere". The dust clouds were at least 8-10km high, and during the storm the surface temperature dropped by 20-30°C while the atmosphere got warmer as it absorbed solar radiation. When the storm ended surface temperatures, taken from the southern hemisphere, where summer was ending, across the Equator to the northern hemisphere, ranged from 13°C to -93°C; at the N Pole cap they dropped to -110°C. The Martian "seas", or dark areas, were warmer than the "continents", or light areas, and in some cases the seas took longer to cool, suggesting a rock surface with lower heat conductivity. Atmospheric pressure varied with altitude but at average surface level was about 1/200th of Earth's. Mars 3 measured heights of 3km and depressions of 1km below average level. During the storm the dust clouds rose as high as 10km. Water-vapour content of the Martian atmosphere did not exceed 5 microns of precipitated water - 2,000 times less than the Earth figure. Immediately above the surface the atmosphere was mainly carbon dioxide; at 100km altitude the CO_2 was broken up by ultra-violet solar radiation into a carbon monoxide molecule and an oxygen atom. Traces of oxygen were recorded at altitudes of up to 700-800km, where its concentration was 100 atoms per cu cm. The severe climatic conditions did not rule out the possibility of some form of life, though at best it was likely to be no more than micro-organisms or simple plants.

Mars 4 and 5 L 21 and 25 Jul 1973 by D1e from Tyuratam. Wt 4,650kg. The start of Russia's most ambitious onslaught on Mars, with 2 pairs of spacecraft (see below) being launched within 3 weeks. This was the 1st time a complete series had been successfully launched, but although all 4 reached the vicinity of Mars and returned some data the results were disappointing. Mars 4's retro-rockets failed fo fire and it passed the planet on 10 Feb 1974 at a distance of 2,200km, returning one swath of pictures and some radio occultation data as it flew by. Mars 5, which like Mars 4 had travelled about 460 million km since launch, was successfully placed in a 1,760 × 32,500km orbit with 35° incl and 25hr period of revolution 12 Feb 1974. However, it continued to operate for only a few days; during that time it returned a few photographs of excellent quality showing a small portion of the southern hemisphere in the region where a landing was intended.

Mars 6 and 7 L 5 and 9 Aug 1973 by D1e from Tyuratam. Wt 4,650kg. The 2nd pair was launched only a few hours before the 2-yearly launch window ended. The object of the 4-craft fleet was stated to be "comprehensive exploration of the planet". Both contained French instruments for investigating radio emissions from the Sun in the 1m waveband, as well as solar plasma and cosmic rays. Mars 6 was said to differ from the previous pair and would carry out part of its scientific exploration with the use of equipment on Mars 4. Since it was also stated that Mars 6 and 7 were "analogous", presumably 7 was intended to work with 5. 6 reached the planet on 12 Mar 1974 and the final trajectory correction was performed automatically "by the onboard system of astro-navigation". The lander module separated, its descent engine fired as planned and, following aerodynamic deceleration, the parachute deployed. For 148sec of parachute descent man's first direct measurements of the Martian atmosphere were transmitted back to Earth via the parent craft. Its telemetry ceased abruptly when the landing rockets were fired, and it presumably crashed at approx 24°S 25°W. But during its descent through the thin atmosphere measurements of the electric current drawn

Mars 4/5 orbiter after release of lander. TV cameras are exposed at top; optical sensors for astro-orientation system are at bottom right. *(D.R. Woods)*

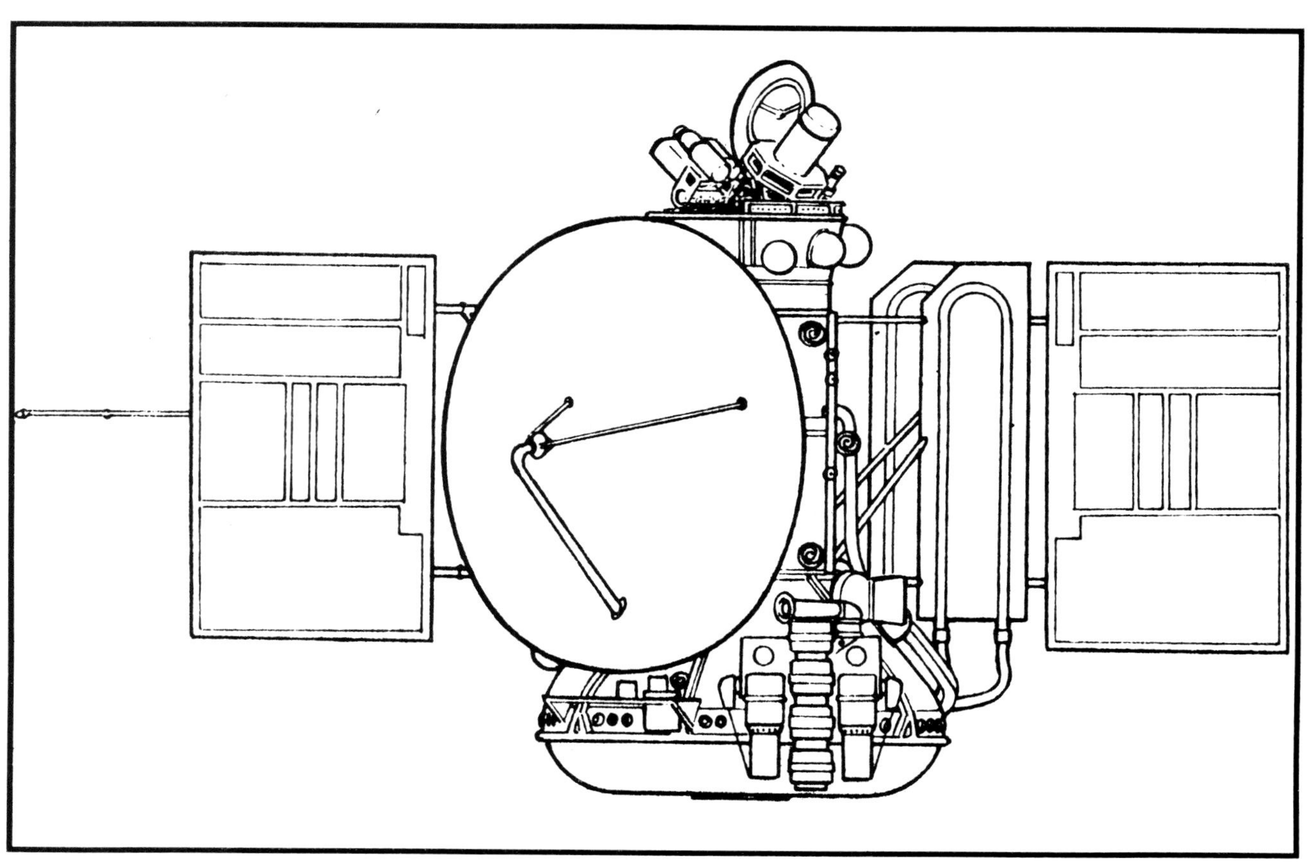

Mars landing sequence: *1* **separation of lander;** *2* **lander's engine fires;** *3* **aerodynamic braking;** *4* **parachute descent;** *5* **lander's retro-jets fire;** *6* **deployment on surface.** *(Tass)*

by the mass spectrometer pump had been made. The pump current increased in a completely unexpected manner, and Soviet scientists interpreted this as indicating the presence of argon, a rare gas. Since large quantities of this gas would affect both the landing sequence and the operation of the atmospheric gas spectrometer of the US Viking, NASA scientists made changes as a direct result of Mars 6. Mars 7 was an almost complete failure. It reached the planet 3 days before Mars 6; the official report merely said that "the descent module was separated from the station because of a hitch in the operation of one of the onboard systems, and passed by the planet 1,300km from its surface". However, Soviet scientists were able to claim that the flights had yielded new information about the Martian surface and atmosphere, including the fact that the atmosphere over some areas contained several times more water vapour than had been thought.

Future projects After a 13yr gap the Soviets are planning a return to Mars in 1986. They hope to bring a Martian orbiter within a few thousand metres of Phobos and then fly in formation with the Martian satellite to obtain high-resolution photographic and other data. This may establish whether Phobos is in fact a captured asteroid.

Meteor

History Soviet weather satellites, with a history going back 18yr, are now operated in 2 series in different orbits; by the end of 1983 31 Meteor 1s and 10 Meteor 2s had been launched, all by A1. They provide a daily weather review for more than two-thirds of the globe. Cosmos 122 (L 25 Jun 1966 from Tyuratam; wt ? 2,000kg; orbit 589 × 643km, incl 65°; orbital life 50yr) was the first to be identified by Russia as meteorological. Its launch was witnessed by President de Gaulle of France, the first Westerner to visit a Soviet launch site. By 1978, according to Moscow Radio, 1 million photographs had been received from Meteors, and the width of territory surveyed from the two heights now in use ranged from 2,200 to 2,600km. Advance warning of tropical storms had saved many lives. Irrigation planning had been improved as a result of data on snow cover in the Tien Shan and Himalaya Mountains, and sailing time of Soviet ships was estimated to have been cut by 10% because Meteors could provide the best courses to avoid storms, winds and ice conditions. This maritime contribution alone was officially estimated to be worth "over a million roubles in profit". The 12th International Botany Congress, held in Leningrad in Jul 1975, was also told that TV pictures from Meteors had enabled shepherds in Soviet Central Asia to begin using new pastureland in a vast area between the Caspian and the Pamirs. This too was worth "millions of roubles" because it helped both to conserve plant resources and to use them rationally.

Meteors are also monitoring clear-air turbulence – strong currents in the upper layers of the atmosphere – so that the crews of intercontinental airliners can be warned of potential hazards. 1980 saw the introduction of hail warnings, of particular value in Soviet Georgia, with its 600,000 hectares of vineyards. By mid-1983 Soviet scientists were working on an earthquake forecasting system; rock fractures shortly before a tremor send out electromagnetic impulses which can be recorded. At present, however, it is still very difficult to identify these indications early enough for palliative steps to be taken.

Spacecraft description Weighing about 2 tonnes each, both types are believed to be 5m long and nearly 2m in dia. The main cylindrical body is similar to those of Ekran and Gorizont, and carries meteorological sensors mounted on an Earth-pointing platform. A pair of solar panels - about 10m in span and rotated by a drive mechanism so that they remain Sun-oriented - charge the electrical storage batteries. The instruments and a fold-out reflector aerial for communications are kept pointing towards Earth by a 3-axis attitude-control system. In 1hr the 2 TV cameras aboard each satellite cover 30,000 sq km, automatically adjusting their exposure times to match lighting conditions on the daylight side, and storing and then transmitting the pictures. On Earth's dark side heat emissions and cloud formations are measured by infra-red sensors. Radiometers measure the radiation balance of the atmosphere; clouds, ice and snowfields reflect about 80% of solar radiation, and measurements of incoming energy are regarded as essential for reliable forecasts. 2 diagrammatic maps are compiled every 24hr.

Meteor 1 Launches, beginning with 01 on 26 Mar 1969 from Plesetsk (orbit 644 × 713km, incl 81°), began at the rate of 3–4 per year so that 2–3 were always operational and providing a continuous survey of atmospheric conditions from pole to pole in a band up to 1,500km wide. Information must be dumped in passes lasting only a few minutes to the 3 reception centres of the USSR Hydrometeorological Service in Moscow, Novosibirsk (Siberia) and Khabarovsk (Pacific Coast); processing the data takes 1½hr. From Meteor 1 (10), L 30 Dec 1971, all 81° orbits have been at about 890km altitude. That was the first Soviet metsat to carry automatic picture transmission compatible with Western receivers. It also had a small ion plasma engine for orbital adjustments. From Meteor 1 (28), L 29 Jun 1977, orbits have been retrograde, Sun-synchronous and circular, at 630km latitude and 98° incl. These launches must be carried out from Tyuratam, since northward launches from Plesetesk would pass over the N Pole and the US. Operational life has increased, so that only about 1 per year of each type is now required.

Meteor 1(31) Priroda L 10 Jul 1981 by A1 from Tyuratam into 630km Sun-synchronous orbit with 98° incl. The Soviets' 1st Earth-resources satellite to provide continuous imaging from space. Main sensors are a 32-channel scanning radiometer and 33-channel microwave radiometer for surface observations; some of the equipment was provided under the "Bulgaria 1300" programme. At the end of 1983 the spacecraft was being used to determine snow depths on all Soviet agricultural regions to assist in planning the spring work season and forecasting the 1984 harvest. It is claimed that its warnings of hurricanes and typhoons have already saved hundreds of thousands of dollars, with direct reception on large ships enabling them to avoid storms and ice cover.

Meteor 2 Introduced experimentally at first with Meteor 2 (01) (L 11 Jul 1975), these satellites operate in the 890km, 81° orbit, have a longer life, and apparently use scanning radiometers rather than the TV cameras fitted to earlier Meteors. (Scanning radiometers also equip US Tiros satellites.) Various combinations of orbital plane have been tried, usually with 2-3 satellites giving passes at 6–8hr intervals; the intermixing of north and south-bound tracks provides pictures at about 4hr intervals. Soviet sources say that the Meteor 2s were used to "perfect equipment for obtaining meteorological information" and data on the streams of radiation penetrating near-Earth space, and that they are capable of sending back pictures of cloud cover. Their mission is "to watch high-speed natural phenomena; the path of cyclones, tidal action, dust storms and tsunami waves."

Molniya

History By the end of 1983 98 Molniyas (the name means "Lightning") had been launched in 3 series. They form a major part of the Soviet Union's communications system, supplying TV, telephone and telegraph links to nearly 100 Orbita ground stations in the USSR and other communist countries, including Cuba and Mongolia. They are regularly used to transmit facsimiles of newspaper pages to remote areas. According to Moscow Radio in Feb 1972, the 24hr TV service to the Far North and Far East territories had "saved the national economy

Meteor 2 second-generation weather satellite.

hundreds of millions of roubles" by eliminating the need for a network of ground relay stations. Since launches started in 1965 they have continued at the rate of 1 every 110 days despite the addition of the Statsionar geostationary series from 1975. While the US moved steadily from low-altitude, tape-storage satellites and passive, balloon-type reflectors to synchronous satellites like Intelsat, Russia followed a different path with the Molniya 1, 2 and 3 series. This was mainly because of the severe payload penalties resulting from launching to equatorial orbit from Tyuratam's relatively northerly latitude of 47°.

The Molniya 1s, much heavier than the corresponding US comsats, have about 10 times the power output of the original Early Bird. They fly in an orbit inclined 63–65° to the equator, with a low point of about 500km over the S hemisphere but spending most of their time north of the equator as they climb to nearly 40,000km and then descend once more. With a 12hr period, these orbits enable each satellite to provide 18hr of communications per day with one pass over Central Asia and N America. The satellites' high power permits the use of ground terminals to distribute Moscow's TV programmes to remote areas, and they also carry telephone and computer data traffic. 3 satellites at 120° intervals can provide 24hr of communications.

Starting with the Molniya 1(11) in 1969, the number above the horizon at any one time was increased from 2 to 3 by reducing the spacing to 90°. Following the introduction of the Molniya 2 series in 1971 and the Molniya 3s in 1974, they orbited in groups of 3 - one of each type - using different radio frequencies to avoid inter-satellite interference. As traffic increased, the spacing between Molniya 1s was decreased to 45° from 1976, and Statsionars began to take over from Molniya 2s, the last of which was launched in 1977. The first 12 Molniya 1 launches were from Tyuratam, but since 1970 all Molniyas have been launched from Plesetsk, with the single exception of Molniya 1(52) from Tyuratam on 23 Dec 1981. Standard inclination was 65° until 1973 and 63° since. Nowadays most Orbita stations have 3 Molniya 3s and 5 Molniya 1s in view at any time, in addition to the Statsionars - though not all of the Molniyas may be operational.

Under a US-USSR agreement of 30 Sep 1971 Molniya 2s were used as "hotline stations," with a complementary service being provided by US Intelsats. This move was intended to eliminate the danger of accidental war by making direct communications possible between the superpower leaders in times of crisis. This system replaced a ground cable route. The Molniya Earth station in the US may now be linked to the Statsionar system.

Molniya 1 The 1st, L 23 Apr 1965 by A2e (orbit 2,303 × 38,177km, wt 1,750kg), provided "many months" of TV and phone links between Moscow and Vladivostok, and finally decayed on 16 Aug 1979. Molniya 1(2), L 14 Oct 1965 (orbit 342 × 39,950km), provided experimental TV, phone and telegraph communications before decaying. Molniya 1(3), L 25 Apr 1966 from Tyuratam, was used for an exchange of colour TV with France, and for Earth pictures showing cloud patterns on a global scale. Colour photos of Earth started in 1967. Molniyas also provide communication links during manned Soyuz flights. Performance has been steadily upgraded to improve radiation resistance, decrease noise levels and increase power output. Some of this series also carry a steerable camera to add wide-area weather views to the Meteor data. The 20th Molniya 1, L 4 Apr 1972 from Plesetsk, also carried France's SRET-1 (environmental research satellite) into orbit under a 1970 collaborative agreement. SRET-1 was released 2sec later than Molniya. SRET-2 was given a similar piggyback launch by Molniya 1(30) on 5 Jun 1975. Cylinders with a conical end, Molniyas are 3·4m long and 1·6m in dia, with 6 paddle-wheel solar wings providing 500-700W of power. 2 0·9m dish antennas extend from the base, which also contains an orbital correction engine. 3 transceivers have an estimated life of over 40,000hr; TV transmissions are carried out at 625 lines per frame. Molniya 1(55) L 21 Jul 1982 by A2e. Orbit 603 × 39,750km. Incl 63·1°. Molniya 1 average orbital life is 15-20yr.

Molniya 1S L 29 Jul 1974 by D1e from Tyuratam. Orbit 35,787 × 35,790km. Incl 0·4°. No reliable details of weight and size available. The official announcement made no reference to the fact that this was Russia's first operational synchronous satellite. It stated that 1S was launched "under the programme of further perfecting communication systems with the use of satellites". The period of revolution in its circular orbit was given as 23hr 59min. It proved to be a test vehicle for the Statsionar series. Orbital life 1 million yr.

Molniya 2 1st launch of this more advanced series using higher frequencies was 24 Nov 1971. Wt ? 2,000kg. Orbit 466 × 39,887km. Dec 10 May 1976. Last was Molniya 2(19), L 11 Feb 1977, with 20yr life. An important difference was that the 2-section fold-out solar panels were replaced by 3-section units, bringing power output up to about 1kW. The orbit-correction motor, probably also used on current Molniyas, gives 200kg thrust and 65sec of operation.

Molniya 3 1st launch 21 Nov 1974 by A2e. Wt 2,000kg. Orbit 628 × 40,685km. Length about 4·2m, dia 1·6m. Molniya 3(19) launched 27 Aug 1982. They have a 12yr life.

Molniya 2 communications satellite shown at 1973 Paris Air Show. ***(Aviation Week)***

Polyot

Polyots 1 and 2 L 1 Nov 1963 and 12 Apr 1964, they were described as "the first manoeuvrable Earth satellites". With estimated weight of 600kg, Polyot 1, placed in an initial orbit of 339 × 592km, was successfully manoeuvred several times, ending in a 343 × 1,437km orbit with a 25yr life. Polyot 2, which had an 18-day life, made several similar manoeuvres and also changed its inclination from 58° to 60°. America's first orbital manoeuvres were made by Gemini 3 on 23 March 1965.

Prognoz

History For over 12yr this series has maintained a continuous watch on solar radiation, the physical processes taking place on the Sun, and the interference they can cause with space communications. Sphere-shaped, with 4 cruciform solar panels and approx 900kg weight, they are placed in highly eccentric 500 × 200,000km orbits

Prognoz 2, displaying instrument array for studying solar radiation. ***(Graham Turnill)***

at 65° incl by A2e launches from Tyuratam. This provides comparative measurements from both near-Earth space and the upper atmosphere and from outside the magnetosphere. Equipment has been provided by France, Sweden, Czechoslovakia, Hungary and Poland, with parallel observations being made from ground stations. Prognoz ("Forecast") 1, L 14 Apr 1972, was at its apogee while Prognoz 2, L 29 Jun 1972, was at perigee. Prognoz 3 L 15 Feb 1973. Prognoz 4 L 22 Dec 1975. Prognoz 5 L 25 Nov 1976. Prognoz 6 L 22 Sep 1977. Prognoz 7 L 30 Oct 1978. Prognoz 8, L 25 Dec 1980, tested instruments to be used by Vega to study VLF plasma waves. Prognoz 9 L 1983; radio-astronomy observatory.

Progress

Since the first was launched on 20 Jan 1978 Russia has sent up her cargo ferries with the regularity of a train service. They have become central to Salyut space station operations, their arrival eagerly awaited by the long-duration crews spending months at a time in orbit. Their use as "space dustbins" is almost as valuable as the cargoes they deliver. Before they were available Soyuz crews had to eject through an airlock packages of waste which took a long time to drift clear and tended to pollute their environment. Now all their waste material is packed aboard the departing Progress and burned up when it is placed on a destructive re-entry course after its engines have been used to lift the Salyut and its docked Soyuz into a higher orbit.

With a total weight of 7,020kg, Progress carries 2,300kg of cargo, 30% of its lift-off weight. Like Soyuz it is 2·2m wide, but at 8m slightly longer because an extra instrument section has been added. The high proportion of cargo is made possible because no emergency rescue system is needed at launch, nor is the heavy heatshield necessary for re-entry. With no solar panels, it has an independent flight time of 8 days, though as part of an orbital station it can remain docked for several months. Payload consists of about 1 tonne of fuel and 1,300kg of dry cargo; 4 propellant tanks and several other special tanks containing compressed air and nitrogen are carried in what was originally the Soyuz descent module. As on Soyuz, Progress uses 14 10kg and 8 1kg thrusters. Its automatic control systems differ substantially, however. 3 external lights and 2 external TV cameras are carried, and rendezvous data are sent both to ground mission control and the Salyut crew, who can if necessary assist in the docking operation.

Soyuz 20 was used unmanned to rehearse the system with a 91-day mission to Salyut 4 starting in November

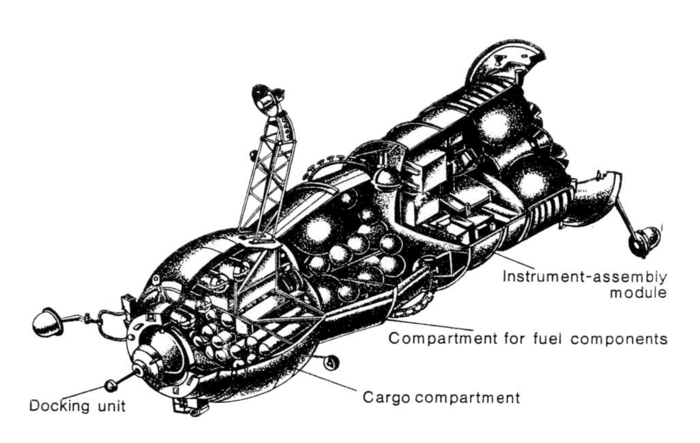

Progress cargo spacecraft.

1975, 3 years before Progress became operational. Launch and re-entry dates up to the time of going to press were as follows:

Progress	1	20 Jan-7 Feb 1978
	2	7 Jul-4 Aug 1978
	3	8 Aug-24 Aug 1978
	4	4 Oct-26 Oct 1978
	5	12 Mar-5 Apr 1979
	6	13 May-9 Jun 1979
	7	28 Jun-18 Jul 1979
	8	27 Mar-26 Apr 1980
	9	27 Apr-22 May 1980
	10	29 Jun-19 Jul 1980
	11	29 Sep-11 Dec 1980
	12	24 Jan-19 Mar 1981
Progress flights to Salyut 7		
	13	23 May-6 Jun 1982
	14	10 Jul-11 Aug 1982
	15	18 Sep-16 Oct 1982
	16	31 Oct-13 Dec 1982
	17	17 Aug-18 Sep 1983
	18	20 Oct-16 Nov 1983
	19	21 Feb-2 Apr 1984
	20	15 Apr-7 May 1984
	21	8 May-26 May 1984
	22	28 May-

Proton

History A series of 4 satellites to study the nature of cosmic rays of high and super-high energy, and their interaction with the nuclei of atoms. Intercosmos 6, launched nearly 4yr later, was described as a logical development of the Proton stations.

Proton 1-3 L 16 Jul and 2 Nov 1965 and 6 Jul 1966 by D1 from Tyuratam. Weighing 12,200kg, they were of record size at that time. Orbits were 190 × 630km, and they had a 3-month life.

Proton 4 L 16 Nov 1968 by D1 from Tyuratam. Weight set a new record of 17,000kg. Orbit of 255 × 493km gave an 8-month life. Experiments included studies of the energy spectrum and chemical composition of primary cosmic particles, and of the intensity and energy spectrum of gamma rays and electrons of galactic origin.

Radio

Small radio relay satellites for amateurs built by Soviet students. Radio 1 and 2 L 26 Oct 1978 with Cosmos 1045 by F1f from Plesetsk. Wt 40kg each. Orbits 1,688 × 1,709km. Incl 82·55°. Could be used by hams around the world for only 1yr due to excessive vibration at launch. Radio 3-8 L 17 Dec 1981 by C1 from Plesetsk. Orbits 1,564-1,659 × 1,660-1,780km. Incl 82·96°. Orbital life 15,000yr.

Salyut

Status Although 1983 proved to be a year of setbacks so far as Salyut 7 operations were concerned, 12yr of space station operations had by then proved that it was quite practicable to establish permanent bases in Earth orbit, manned near-permanently by rotating crews. Both the early Salyut failures and the more recent reverses could be written off as valuable experience. Successfully using unmanned Progress cargo ferries to supply long-stay crews, Soviet scientists had proved that it was possible to create advanced manufacturing bases in near-Earth orbit and for the men and women living there to endure flights to the nearer planets at least. A US Office of Technology report published in Dec 1983 awoke to the fact that Soviet cosmonauts had accumulated 2·5 times as much experience as their US rivals, and that "the Soviets are more knowledgeable than the US in space biology and medicine; and in a number of technical areas, notably in automated docking systems, they routinely use techniques that the US has never demonstrated. ... The Soviet space station programme is the cornerstone of an official policy which looks not only towards a permanent Soviet human presence in low Earth orbit, but also towards permanent Soviet settlement of their people on the Moon and Mars. The Soviets take quite seriously the possibility that large numbers of their citizens will one day live in space."

Another step in this direction appears imminent. Gen Shatalov spoke in Jan 1984 of "heavy transport ships" which would launch space complexes consisting of a number of separate free-flying units. Some laboratories and workshops would be docked to the basic station only for repairs and transfer of equipment and supplies. This at last gave a clear indication of the purpose of the Cosmos 1267 and 1443 experiments carried out with Salyuts 6 and 7; Soviet scientists have observed for some years that cosmonaut movements in manned stations interfere with materials-processing activities in zero g.

History Salyut development began in the 1960s. 1st launch, in 1971, was followed by 2 failures, Salyut 2 and Cosmos 557. With the launch of Salyut 3 it seemed that different types were to be used for parallel military and civil operations. Crew composition and the lower orbits suggested that the failed Salyut 2, followed by the successful Salyuts 3 and 5, were basically military. The failed (and unannounced) Cosmos 557, followed by the successful Salyuts 4 and 6, were basically civil. But the improved Salyut 7 is apparently an all-purpose vehicle.

The fact that Salyut 3 remained operational for 7 months, compared with the planned 3 months, encouraged the Soviets to try for much longer with Salyut 4, which ultimately survived for over 2yr. It seems doubtful that it was intended to be occupied by 3 rotating crews, though this was certainly the plan for both 4 and 5. Success came at last with the splendid Soyuz 18B flight, when the crew spent 63 days in Salyut 5. Much of that success was due to running repairs by the Soyuz 17 and 18B crews - possibly inspired by the example of the US Skylab crews - to the solar telescope and cosmic ray equipment, as well as to more fundamental things like air conditioning. But there was also a whole series of failures of both Soyuz and Salyut equipment between the aborted Soyuz 18A launch and the Soyuz 25 docking attempt, amounting to 8 failures out of 13 space station missions. Then, apart from the "random" Soyuz 25 and 33 failures, came consistent success with Salyut 6. The steady development of the use of the Progress cargo ferry - 19 up to the beginning of 1984 - was another notable achievement.

Salyut 4 mock-up on display at 1975 Paris Air Show. *(Graham Turnill)*

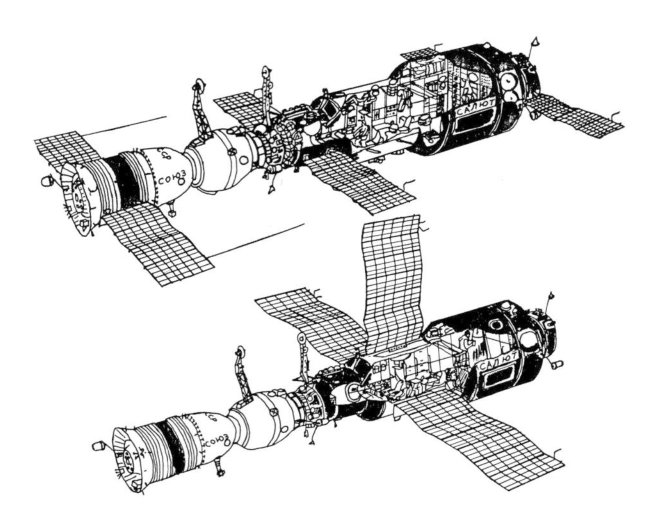

Upper: Soyuz 11 docked with Salyut 1, showing interior arrangement. Lower: Soyuz 17 docked with Salyut 4; major revisions include absence of solar panels on Soyuz, and switch to 3 large, steerable solar panels on Salyut. ***(C.P. Vick)***

Soviet hints that Salyut 6, with its 2 docking ports, would also be used to resume spacewalking activities materialised only partially. Flight Control still seems reluctant to authorise EVAs: the Soyuz 32 crew had less difficulty in clearing the jammed radio-telescope antenna than they did in overcoming Flight Control's reluctance that they should undertake the task. More recently, there was the case of the T-9 crew, who lived with the folded solar panels delivered by Cosmos 1443 for 4 months before they got outside to deploy them. But apart from EVAs, full advantage was taken of the Salyut 6 success: not only were manned crews sent up to the station for a period of 4yr instead of the expected 1yr, but there was a final evaluation of the behaviour of larged mated bodies lasting more than 1yr. After the last men had left, the 15,100kg Cosmos 1267 was docked with the 19,000kg space station until final controlled re-entry. (A US report that C 1267 had a number of portholes, and was a first test of a Soviet "battle station," was regarded with extreme scepticism by British observers.)

Salyut 7 was obviously delayed by several years, and one cosmonaut told the author that it was being completely redesigned. There certainly seemed to be a need for a 3rd docking port so that rotating Soyuz craft did not have to be switched from one end to the other to clear the refuelling port for Progress cargo ferries; that would however have called for right-angle dockings, which have not yet been tried. It was therefore a surprise when Salyut 7 was at last launched in early 1982 to find that it was only an "improved" version of its predecessor. Did the Soviets now have such confidence in their quick-launch capability that rotating crews would be left without a Soyuz?

The Salyuts have always had one major advantage over America's original Skylab, which finally crashed back to Earth amid so much anxiety. From the start, Salyuts have been equipped with orbital manoeuvring engines which permit re-entry of this large object to be commanded safely over the Pacific, with surviving debris falling harmlessly into the sea. It may however well be that the prime reason for the engine has been the necessity to keep Salyut in a very low Earth orbit to compensate for the limited lifting power of Soviet rockets and, until the T series, for the restricted manoeuvring capability of the Soyuz spacecraft. In its orbit of about 220 × 270km, periodically raised to about 350km, Salyut loses 142m of altitude every 16 revolutions, i.e. every 24hr. For a long period Salyut 6's main engine was unusable, but then it was shown that the orbit of the whole Soyuz/Salyut/Progress complex could be safely raised by using the Progress's engines before it was undocked and destroyed. Soon, spare fuel on Soyuz spacecraft was also being used in this way before they returned to Earth, resulting in a very satisfying economy.

The cost of the 5yr Salyut 6 mission has been estimated at around $9 billion, with the overall Soyuz/Salyut programme so far costing about the same as America's Mercury-to-Apollo Moon-landing programme.

Salyut description Current Salyuts have 3 habitable compartments with 100 cu m of interior space, compared with Skylab's 357 cu m. From Salyut 6 a major redesign of the original engine system replaced the single nozzle with twin nozzles on each side of the 2nd docking port added at the rear. More than 20 portholes are provided for experiments, visual observations and photography. The work is conducted from 7 control posts.

SOYUZ-SALYUT-PROGRESS COMPLEX

Weights (kg)	
Total	25,700-32,900
Salyut in orbit	18,900
Soyuz	6,850
Progress	7,020
Orbital parameters	
Altitude	300-400km
Inclination	51·6°
Dimensions (m)	
Maximum length	29
Salyut length	15
Maximum diameter	4·15
Span with extended solar battery panels	17
Crew	
Host crew	2-3
Guest crew	2-3
Active service life	about 5yr
Weight of on-board research equipment	up to 2,000kg
Power supply	4kW

Forward Transfer Compartment The front end, 3m long and 2m in dia, is basically an access tunnel with an inner bulkhead enabling it to be sealed off for spacewalks from the EVA hatch cut in the side; 2 spacesuits stored here are for the shared use of all visiting crews. There are 7 portholes, some carrying astro-orientation instruments, and activities are controlled from Posts 5 and 6. Outside equipment includes antennas for approach and docking, with lights for manual docking.

Working or Operations Compartment This, the main living and working area, consists of 2 cylindrical shells connected by a conical section. The 1st and smaller section is 2·9m in dia and 3·5m long, the conical section is 1·2m long, and the 2nd section is 4·1m in dia and 2·7m long. Instruments and equipment are arranged in standard racks around the sides of the interior. Post 1, Salyut's central control position in the lower part of the 1st section, has 2 seats; Post 2, in the same section but closer to the cone, is for astro-orientation and navigation. Between them the crew eat and rest; a table has food-heating devices and provides hot and cold water. It is also used for repairs. On one side is the system which regenerates water from the atmosphere, controlled by Post 7, and on the other the onboard computer (which according to US sources has no back-up). To the outside of the 1st section are attached 3 solar panels, 2 horizontal and 1 vertical, with a total span of 17m. They can be swivelled through 340° so that they always face the Sun, and produce 4kW of power. (On Salyuts 1-3 there were 2 smaller pairs of horizontal panels, with the Soyuz providing supplementary power from a similar pair. From Salyut 4 onwards the Soyuz began to fly without solar panels since they were able to recharge internal chemical batteries from Salyut's power supplies. But this limited Soyuz flying time to 48hr in the event of failing to dock, which happened so often that the solar panels were later restored.) The main manoeuvring engine and the 4 clusters of attitude-control thrusters, spaced at 90° intervals around the large-dia section, are controlled from Post 1. Post 3, located in the rear of the large-dia section, controls the scientific equipment. In this area, on opposite sides, are the crew's bunks and food stowage. At the aft end of this section is a toilet and 2 airlocks for ejecting waste. Beside it is a vacuum cleaner, dust filters, and storage for water and other consumables. The shower is at the forward end of this section. Post 4, in the lower centre of the large-dia section, is used for medical experiments and photography, with the MKF-6M Zeiss camera in one of 2 portholes. Much cosmonaut time is spent nearby on the treadmill, bicycle ergometer and lower-body vacuum suit.

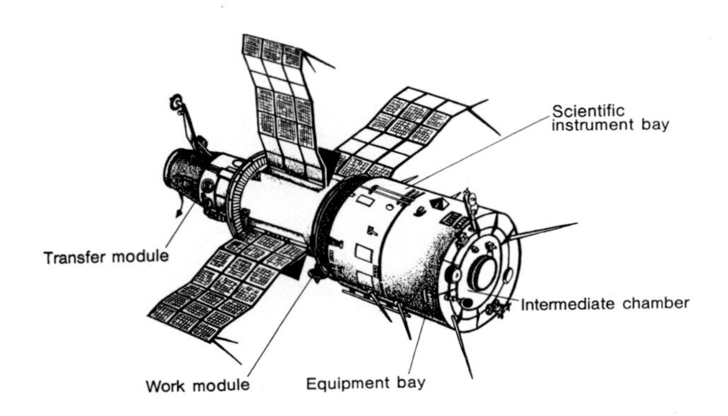

Salyut 4. Mast-mounted dish antennas projecting from transfer module and equipment bay are associated with the docking system.

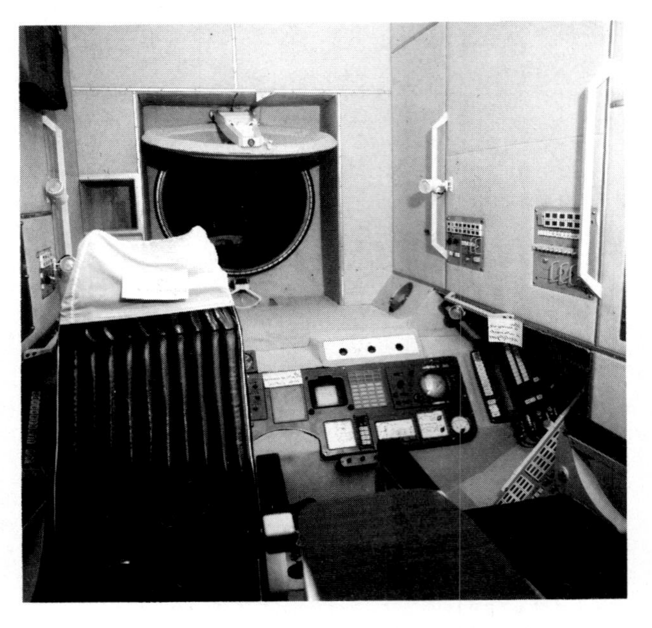

Salyut 4 control centre, as shown at Paris 1975. Console with rotating globe at left showed spacecraft position with respect to Earth. Note shade for covering window when opposite sun. *(Graham Turnill)*

Rear Transfer Compartment On the early Salyuts this was an unpressurised service module 2·17m long and 2·2m in dia. On Salyut 6 it was widened to the same 4·15m dia as the working compartment to take the second docking port. Here the short-term Intercosmos visitors are received when they dock, as are the unmanned Progress ferries, which plug into the fuel resupply lines. With the main engine and thrusters using the same fuel and oxidiser, refuelling is virtually automatic. However, the 1kW compressor needed for this operation uses so much solar-generated electricity that refuelling has to be spread over 6 shifts so that enough power remains for other vital systems.

Scientific equipment More than 2 tonnes of equipment is used for materials processing, biology, astronomy and Earth-resources surveys. The Working Compartment is dominated by the 650kg BST-1M submillimetre telescope and the MKF-6M multispectral camera. The telescope, which measures radiation in the infra-red, ultra-violet and sub-millimetre wavebands, has been used to study the galaxies, individual stars, a full lunar eclipse and the Earth's atmosphere. The Zeiss multispectral camera, operating in 6 bands of the electromagnetic spectrum, can cover 10 million sq km of the Earth's surface with each film cassette. Its images are correlated with photographs of the same area taken simultaneously by an An-30 aircraft at an altitude of 6-8,000m while specialists on the ground take measurements of soil characteristics, humidity, crop ripeness, etc. The 2 furnaces - Splav ("Alloy") for smelting metals, and Kristall ("Crystal") for crystal-growing - were delivered by Progress 1 and 2. Splav weighs 23kg and has 3 heat chambers; inside, molybdenum reflectors concentrate the heat on the sample and ensure that furnace wall temperatures do not exceed 40°C. Heat is ejected via a small airlock, which also enables melting to be carried out in a vacuum. The Kristall kiln produces semi-conductor materials by the zone-melting technique and has also been used to fuse optical glass.

Cosmonauts' living conditions From the first mission in Salyut 1, much attention has been devoted to making living and working conditions as pleasant as possible on long-duration missions. A major aim has been to increase cosmonaut productivity by improved design and increased automation. This was the 24hr "time budget" for the 2-man Salyut 6 missions:

Sleep	9hr
Meals	2
Exercise	2
Earth communications	1½
Recreation	2
Working time	7½

That meant working days spread from 8am to 11pm. With two men much of that working time was taken up with routine station maintenance and operation, and it soon became clear that not until there was a third crew member able to devote his whole working time to scientific activities would space station productivity really rise. In Salyut 2 a psychological trick was tried, the illusion of a floor and ceiling being created by painting one interior surface dark (the "floor") and the opposite one light ("ceiling"). It was so successful that by Salyut 6 a whole team of psychologists had worked on the interior, using soft pastel colours to produce "a homely atmosphere," with a "cosy corner" for leisure in addition to the kitchen and dining compartment. Even more important to the cosmonauts was noise reduction. After Grechko had complained about the high noise level during his Salyut 4 stay much work was done on reducing the background noise of ventilators, fans, motors and other systems. Many were muffled, and the life-support system motors were moved to the side of the Working Compartment. Basic foods were prepared to last for 18 months, long enough for a manned Mars mission. But in Earth orbit, with increasingly frequent visits from short-stay crews and Progress ferries, long-stay crews had the benefit of regular supplies of fresh bread and vegetables, resulting in a noticeable improvement in their general health and ability to resist spacesickness. The daily calorie intake was increased to 3,000, with a choice of 60 different meals and 5 different types of bread. The daily menus looked like this:

Breakfast Tinned ham, white bread, cottage cheese with blackcurrant jam; cake, coffee and vitamin pills.
Dinner Vegetable soup with cheese and biscuits; tinned chicken; plums with nuts and vitamin pills.
Supper Tinned steak, black bread; cocoa with milk, and fruit juice.

Although most foods are in bite-sized mouthfuls to avoid the dangers caused by floating crumbs, the cheese, soups and other liquids are in tubes. Some of the food can be warmed on a heating unit, while most of the solid, bite-sized foods must have water added, followed by kneading to a pasty consistency. Soviet doctors say that a cosmonaut must consume daily 700g of food (dry weight), 800g of oxygen and 2·5lit of water. Another 2lit of water is needed for hygiene. Water recovery and recycling, including collecting from the air the moisture exuded by the cosmonauts' bodies, had reached an advanced stage by Salyut 6. It was then estimated that the life-support requirements of a 3-man crew on a 1yr mission would amount to 11 tonnes, or about half the weight of the whole space station. Longer flights would require more advanced recycling processes for human wastes. A more immediate improvement in water handling on Salyut 6 was the installation of a 180kg water tank which could be pumped full by Progress. This replaced a number of 5kg rubber containers in which the water was kept potable by the addition of a small quantity of ionic silver.

Salyut 1 L 19 Apr 1971 into an initial orbit of 200 × 222km, it remained in near-Earth orbits for nearly 6 months until a final re-entry manoeuvre caused it to burn up on Oct 11 after about 2,800 revolutions. Following the death of the Soyuz 11 crew on 29 July, temperature and atmospheric pressure were maintained at life-supporting levels. It seemed likely that at least one more Soyuz docking was intended, until the inquiry into the Soyuz failure showed that major modifications would be needed. Salyut 1's engine was then retro-fired. The docking of Soyuz 10 and 11 are described under Soyuz.

Salyut 2 L 3 Apr 1973. Orbit 215 × 260km. Incl 51°. This launch was the start of a further series of setbacks to Soviet spaceflight. The object was said to be "to check the improved design of onboard systems and equipment and to conduct scientific and technical experiments in space". A week later, after 130 orbits, the station was said to be stable and the orbit had been raised to 261 × 296km. No indication was ever given as to whether it had been intended for manned occupation, except that on Cosmonauts' Day, 10 Apr, Gen Shatalov said in an interview that Salyut 2 was similar "in design and purpose" to Skylab, due to be launched the following month. Shortly after, Western observers reported that it had become unstable. On 14 Apr it apparently broke up into 25 fragments, and it finally decayed on 28 May.

Cosmos 557 L 11 May 1973. Orbit 214 × 243km. Almost certainly a back-up for Salyut 2. Launched, possibly in haste, 3 days before America's Skylab. (A rocket, thought to have been carrying yet another Salyut, had blown up during a launch attempt between Salyut 2 and Cosmos 557.) This too was a failure and re-entered 11 days later.

Salyut 3 L 25 Jun 1974. Initial orbit 270 × 219km. This launch took place 2 days before President Nixon arrived in Moscow. He had already left when, 8 days later, Soyuz 14 was launched to dock with it. Improvements and changes included 3 solar arrays, automatically rotating through 180° so that they could always face the Sun, instead of the 4 arrays on Salyut 1. Its use for military reconnaissance is discussed under Soyuz 14. A completely successful mission was at least partially based on the parallel use of a ground-based Salyut in which problems were examined and solutions checked before being passed to the crew in orbit. Before Popovich and Artyukhin left after their 15-day stay they prepared Salyut 3 for continued automatic use and for a possible further manned visit.

The failure of Soyuz 15 to dock with Salyut 3 is described under the Soyuz entry. The planned 90-day programme was stated to have been "completely fulfilled" when a "re-entry module" was ejected on revolution 1,451 from a 253 × 275km orbit and recovered in Russia. There had been American speculation that one of 2 spherical airlocks on Salyut 3, officially said to be for ejecting body wastes during manned missions, could be used for returning ground reconnaissance film. (The code words used during Soyuz 14 may have been associated with this.) The flight ended on 24 Jan 1975 when Salyut 3's retro-rocket was fired by ground command. It had functioned for 7 months, more than twice as long as originally planned.

Salyut 4 L 26 Dec 1974 by D1 from Tyuratam. Wt 18,900kg. Initial orbit 219 × 270km. Incl 51·6°. Salyut 3 was still in orbit when its successor was launched. The purpose of Salyut 4 was stated to be further tests of Salyut design, systems and equipment. 11 days later its orbit was raised to 343 × 355km. The Soyuz 17 crew docked, apparently without difficulty, 17 days later, the longest gap between launch and boarding of a space station. Their activities are described under Soyuz 17; when they left Salyut 4 on 9 Feb 1975 it was in a 334 × 361km orbit and remained under automatic control. It was to have been occupied a 2nd time 2 months later by the Soyuz 18A crew, whose launch was aborted. It therefore had to continue under automatic control until the back-up Soyuz 18B crew boarded it for their record 63-day stay from 23 May to 26 Jul. Their final orbital change left the space station at 349 × 369km, with about a year's life before natural decay. In Oct 1975 its equipment was being used to study the constellation Monoceros. 4 months after the departure of the 2nd crew the unmanned Soyuz 20 was automatically docked, remaining there for 91 days before being brought back to Earth. Feoktistov said that it demonstrated not only a re-supply capability but also the possibility of using an unmanned Soyuz to collect sick crew members. On 3 Feb 1977, only 1½ days before natural decay would have resulted in re-entry, Salyut 4 was manoeuvred into a re-entry trajectory over the Pacific. It had completed more than 12,000 revolutions and 2 crews had occupied it for a total of 93 days.

Salyut 5 L 22 Jun 1976 by D1 from Tyuratam. Wt? 19,000kg. Orbit 219 × 260km. Incl 51·6°. The 2-man Soyuz 21 crew successfully docked and boarded 14 days after launch. Details of their 63-day flight and unexpected return can be found under Soyuz 21. Indications were that, like Salyut 3, this was mainly for military rather than civil purposes.

Following the failure of the Soyuz 23 docking attempt Salyut's 5's orbit steadily decayed and re-entry was predicted for Mar 1977. However, trajectory corrections were made on 14 and 18 Jan, raising the orbit to 256 × 275km. Then, on 7 Feb, came the successful boarding and 17-day stay by the Soyuz 24 crew. One day after their departure the automatic return module was separated and brought back to Earth "with materials from research and experiments," probably film cassettes from external cameras and material loaded into the module by the crew by means of an internal airlock. On 8 Aug 1977 Moscow's Flight Control Centre announced that after 6,630 revolutions Salyut 5's braking engines had placed it in a descent trajectory to the Pacific Ocean. More than 300 experiments had been carried out and 65 million sq km of Earth's surface had been photographed, including the basins of the Indian and Atlantic oceans and a large part of the Soviet Union. An infra-red extra-atmospheric spectrum of the Sun, and of outer space near the Sun, had been obtained, and research carried out to assess the effect of man's industrial activity on Earth's atmosphere. The Reaction device had been used to master the soldering and melting of metals.

Salyut 6 What appeared to be a routine launch on 29 Sep 1977 into an initial 219 × 275km orbit at 51·6° incl actually represented a "giant leap" towards a permanently manned space station and proved to be a mission equal in long-term importance to America's Moon landing. Over a period of 3yr 8 months 30 spacecraft successfully docked with Salyut 6: 16 of them carried a total of 33 crewmen, Soyuz 34 and T-1 docked unmanned, and the other 12 were unmanned Progress cargo ferries. Five long-term missions of 96, 140, 175, 185 and 75 days, plus 11 short-stay visits (by crews including 9 cosmonauts from Intercosmos countries), meant that Salyut 6 was occupied for a total of 676 days, far surpassing the 171 days' occupation of the US Skylab in 1973-4. More than 200 experiments were carried out in Salyut 6's various furnaces, and 48 million sq km of the Earth's surface was photographed. The Progress transports delivered more than 20 tonnes, enabling crews to replace more than a quarter of the station's equipment and to add to the range of its scientific apparatus. Details of the missions are given under Soyuz and Soyuz T.

No mention was made of a 2nd docking port when, 10 days after launch, the new space station's career started inauspiciously with the failure of Soyuz 25 to achieve a docking. The rumoured existence of a 2nd port was confirmed when Soyuz 26 docked without difficulty on 19 Dec at the aft, instrument, end. Konstantin Feoktistov explained that the 2 docking modules were identical, and the only difference in procedure was that Salyut turned to the approaching spacecraft according to whether the nose or stern port was to be used. Other innovations in Salyut 6 were thermo-regulating and attitude-control systems, an operational water-regeneration system, and a teleprinter. TV cameras, developed from those used on the Apollo-Soyuz mission, consisted of 2 black-and-white (B/W) and 1 colour inside and 3 B/W outside. A small polythene cabin into which hot water was sprayed under pressure provided a shower for the cosmonauts. An automatic orientation and control system, Delta, relieved the cosmonauts of many routine tasks.

Following the Soyuz 33 failure in Apr 1979 there was a 1yr gap in manned launchings, and it seemed that Salyut 6 must be time-expired. But the cosmonauts were so successful in replacing and repairing worn-out components that a new lease of life began in Apr 1980. This phase continued through 1981, when 4 missions were flown to it, beginning with Progress 12 in Jan with supplies for the last long-stay crew in Soyuz T-4. After that Salyut flight director Alexei Yeliseyev said that there

Salyut 6/Soyuz 27: right to left, cosmonauts Romanenko, Grechko, Gubarev and Remek preparing to open the first "post office" in space. *(Tass)*

Salyut 6 training: cosmonauts Romanenko and Grechko rehearsing EVA activity in an underwater facility at the Yuri Gagarin Cosmonaut Training Centre in preparation for their 96-day mission. *(Tass)*

would be a break in manned flights while a new station was designed. There did however seem to be second thoughts about that later on, possibly because there was a political preference for sending the promised and much publicised Franco-Soviet mission to Salyut 6 rather than Salyut 7. Then came the final docking, on 19 Jun 1981. Salyut 6 was in a 335 × 377km orbit with 51·6° incl, and the docking craft was the 15,100kg Cosmos 1267. There had been much Western speculation about the purpose of this large vehicle ever since its launch on 25 Apr, and it carried out many manoeuvres before the docking. It was then described as "a large Soviet spaceship," more than twice the size of Soyuz and bringing the total weight of the complex to 34 tonnes. The aim was said to be the perfecting of methods of assembling large permanent complexes. The RAE *Table of Satellites* described Cosmos 1267 as a cylinder with 3 solar panels, about 14m long and with 2·4m dia. Its engines were used on 30 Jun to raise the orbit to 470 × 521km, with 83° incl. Soviet scientists apparently felt a need for long tests of the behaviour and manoeuvrability of two such large bodies before they were occupied. US suspicions that Cosmos 1267 was an experimental "battle station" seemed to be based on evidence that it had a number of portholes and torpedo-like tubes. A US Defence Secretary responsible for research insisted that it might be "the forerunner of a weapons platform". The final assessment, revealed months later, was that it carried "miniature attack vehicles designed as a space station defence system". After manoeuvres on 28 Jul 1982 Cosmos 1267's retro-rockets were fired the following day to steer the 2 vehicles - apparently still joined - to a controlled re-entry over the Pacific. Moscow Flight Control said Salyut 6 had been an important stage on the road to creating permanently operating manned orbital complexes.

Salyut 7 L 19 Apr 1982 into a 212 × 260km orbit with 51·6° incl amid Western expectations that it would have 3 docking ports and be the start of a 6-12-man station, Salyut 7 was officially described as a base for studies of the Earth's surface and atmosphere; astrophysical, medical and biological studies and technical experiments; and tests and adjustments of improved systems. It was 28 days before the first crew joined it, leading to speculation that something might be wrong. Then Valeri Ryumin (now described as "flight director") revealed that Salyut 7 was the same size as its predecessor but had a number of

Artist's impression of an advanced Soviet space station made up of Salyut and, possibly, Star modules.

improvements. The most important apparently affects the two docking ports, which are bigger and stronger and capable of taking larger craft. The solar panels and battery produce more power, and a much improved computer enables the crews to spend more time on scientific work. (Ivanchenkov, the 2nd flight engineer to visit Salyut 7, later said that "without the participation of ground control" this Delta system "is able to calculate all the 6 elements of space orbit on the basis of information which it receives, and to fulfil any regime required of it by the crew in aligning the station to some particular part of the stellar sky.") Crew comforts include a refrigerator and the new Rodnik ("Spring") hot and cold water system, which according to Berezovoi is as reliable a water tap in a city flat. Salyut 7 is "full of light, comfortable and cosy, with wall panelling painted in bright colours which are a pleasure to the eye. It is easy to keep the washable panels spotlessly clean." Instead of pre-planned daily menus, cosmonauts can choose their own dishes, as in a restaurant, merely advising Flight Control periodically of what needs replacing when the next Progress is sent up. Dzhanibekov, Salyut 7's first commander, also says that there have been big improvements in the provision of light and that much work has been done to reduce noise (a significant source of complaint during earlier missions). This refers to a new, quieter ventilation system which also absorbs dust. Other changes include 2 portholes transparent to ultra-violet light, and the replacement of the submillimetre telescope used on Salyut 6 with a group of telescopes for the study of X-ray sources.

The record 211-day stay of the Soyuz T-5 crew, with visits from T-6, T-7 and Progress 13-15, is described under Soyuz T. Some observers were surprised that the Soyuz T-5/7 crew returned to Earth on 10 Dec 1982 without being replaced first with an overlapping crew. Flight Control had more ambitious plans, however, and on 2 Mar 1983 came the launch of Cosmos 1443, which docked on 10 Mar after a slow, 8-day approach. Full details of its nature and equipment, apart from the fact that it had a total weight of 20 tonnes, were slow to emerge, and it was a disappointment when Soyuz T-8 was unable to dock on 22 Apr and had to return to Earth. With the successful docking on 28 Jun of T-9, commanded by the veteran Lyakhov, it became clear that C 1443 had the dual role of space station module and space tug. It included a large re-entry vehicle capable of returning 500kg to Earth for water recovery. It also carried a large propellant load, transferable between Salyut and C 1443 as required and giving the whole station much more manoeuvrability than was possible with Progress. The Earth-return facility, apparently rehearsed on C 939 and C 1267, was intended for the bulk transport back to Earth of materials produced in zero g. C 1443 delivered 3 tonnes of cargo to Salyut and was said to be 13m long and 4m in dia; 3kW of power was provided by about 40 sq m of solar arrays. Later we learned that its cargo included the additional solar panels deployed 8 months later by the T-9 crew. Why deployment was delayed so long, and why C 1443 had to be jettisoned, were unexplained mysteries as we went to press (see Soyuz T-9). Alexandrov gave the total weight of Salyut 7/T-9/C 1443 as 47 tonnes and length as 35m.

Solar power station

Soviet scientists said in 1980 that they were considering a plan to place a solar power station near Mercury, where solar activity is hundreds of times greater than in near-Earth space. They said that methods of using rays in the super high-frequency range to transmit energy over great distances were already known. To cover the 19 million km between Earth and Mercury it would be necessary to set up a relay system to "bounce" the energy back in stages.

Soyuz

History For over 15yr Soyuz ("Union") launches have averaged 3 per year. The original Soyuz series ended with the 40th flight, in May 1982, with the improved "T" series, totalling 11 at the time of writing, overlapping from Dec 1979. Despite many failures (which have surprisingly recurred during the T series) Soyuz has achieved the long-term aims stated at its inception: the development of sustained flight, orbital manoeuvring, docking, scientific and technical investigations in Earth orbit, and ultimately the assembling of manned orbiting stations. The original design was by Chief Design Engineer Sergei Korolev and his assistant, Voskresensky, shortly before their deaths (see Voskhod). The spherical and cylindrical shapes of the Soyuz modules appear to be logical developments of their earlier vehicles. Soyuz is equipped for Earth-orbit missions up to altitudes of about 1,300km, and can be used either automatically or manually for both manoeuvring and docking. The system was successfully tested with an automatic docking of the unmanned Cosmos 186 and 188 satellites on 30 Oct 1967, the 1st Soviet docking of any sort to be achieved. This was successfully repeated with Cosmos 212 and 213 in Apr 1968 in preparation for the Soyuz 4/5 mission the following Jan.

But until recently the regular failures which marred the series have tended to blur the effect of its successes. The first trial flight, Soyuz 1 in Apr 1967, ended in cosmonaut Komarov's death; the Soyuz 10 crew, having docked with Salyut 1, were unable to enter it; the Soyuz 11 crew were killed by depressurisation during re-entry, with the result that future crews had to be reduced from 3 to 2 to make room for the unwisely discarded spacesuits. Soyuz 15, 23 and 25 all failed to dock with Salyuts, the last of these robbing Russia of what was obviously intended to be a spectacular to celebrate the 60th anniversary of the October 1917 Revolution. What should have been Soyuz 18 ended in spaceflight's first abort between lift-off and orbit. Against these failures could be set the development of long-duration spaceflight, culminating in the 61 days spent in Salyut 4 by the back-up Soyuz 18B crew and the 96-day mission by the Soyuz 26 crew in Salyut 6 which set a new world record by surpassing the 84 days of the final US Skylab. By then the first space welding experiments had been accomplished, and the first crew transfers from one vehicle to another in a simulated space rescue had been carried out.

Soyuz on display at Paris Air Show 1973. *(Aviation Week)*

Two docked Soyuz orbital modules. *(Graham Turnill)*

Even more important was the adaptation of Soyuz as the unmanned Progress cargo ferry. A 3-day flight in Sep 1975 by an unmanned Soyuz, designated Cosmos 772, was thought to be a test of the improved 3-man version but was more probably a preparation for the unmanned Soyuz 20 cargo test flight to Salyut 4. Since then, in 18 flights spread over 5yr, Progress has been highly successful and has undergone constant improvement, as shown in the detailed accounts of the wonderfully productive long-duration missions. Soyuz 31 gave a rare glimpse of the military research included in these missions, and as confidence grew with success the cosmonauts were allowed to become more human. It was only after Ryumin had spent almost as much time in space as would be needed for a trip to Mars that we heard about his worries over the failure of his first mission. And not until Soyuz 38, the 7th international flight, did we learn that long missions result in flat feet.

Spacecraft description

Overall length	10·36m
Diameter	2·29 to 2·97m
Docking probe	2·74m
Solar wingspan	10·06m
Total weight	6,000kg
Total volume	9 cu m

There are 3 compartments: the Re-entry or Landing Module in the centre, flanked on one side by the Orbital Compartment, used as a workshop and for resting while in orbit, and on the other side by the Instrument Compartment. The Orbital and Instrument compartments are jettisoned just before re-entry.

Re-entry Module This is shaped like a car headlight, with an exterior coating as protection against re-entry heat and interior insulation to provide both heat and sound protection. During re-entry interior temperatures do not exceed 25-30°C. The shape ensures aerodynamic lift, with deceleration normally producing no more than 3-4g compared with the 8-10g endured by cosmonauts during the ballistic re-entries of Vostok and Voskhod. The cosmonauts take their seats 2hr before launch. The panel in front of the commander contains instruments and switches for the spacecraft systems, a TV screen, and an optical orientation viewfinder set up on a special porthole next to the panel. Orientation controls are on the commander's right and manoeuvring controls on his left. There are portholes to starboard and port for visual and photographic observations. There are 4 TV cameras (2 mounted externally and 2 inside) which provide 625-line transmission and 25 pictures per second. The atmosphere regeneration system contains alkali metals which absorb carbon dioxide and simultaneously release oxygen, maintaining a 1kg/sq cm conventional nitrogen/oxygen atmosphere (compared with Apollo's all-oxygen atmosphere). Heat-exchangers condense excessive moisture and direct it to collectors. Water is carried in containers, since Soyuz does not use fuel cells providing water as a by-product, as in Apollo. Multi-channel telemetry systems store information in onboard memory units and transmit it to Earth during regular radio sessions. The hatch in the upper part, or nose, is used for entry before launch and for transfer after launch to the orbital compartment. The re-entry procedure is started by a retrograde firing lasting about 146sec. The single main parachute, for which a back-up is available and which is preceded by a drogue, is deployed at 8,000m; solid-fuel retro-rockets, fired at about 0·91m above the ground, ensure that landing velocity does not exceed 3m/sec. A direction-finder transmitter sends out one signal during descent and another after landing to aid search parties.

Orbital Module Mounted on the nose of the completed craft, with an estimated volume of 6·26 cu m, this provides sufficient room for the cosmonauts to stand up, and an area for work, rest and sleep. It has 4 portholes for observation and filming, controls and communications systems, a portable TV camera, and still and cine cameras. A "sideboard" contains food, scientific equipment, medical kit and washstand. Communications and rendezvous radar antennas are mounted on the exterior. It is used as an airlock by closing and sealing the hatch communicating with the Re-entry Module; the Orbital Module is then depressurised and the external hatch can be opened for egress into space. For entry into a Salyut space station a hatch in the nose is used. In the case of an "active" spacecraft, a 2·74m docking probe is added to the nose; on a "passive" craft an adapter cone is fitted to receive the probe.

TOP VIEW OF SOYUZ SPACECRAFT

Earth communication antenna
Onboard orientation lights green
Apollo VHF antenna
Solar panels
white
Docking target
Sun sensor
Command radio link/trajectory measurement antennas
ø 2,72 m
8,37 m
Radio/television antenna
7,48 m
Flashing light beacons
VHF antenna
white
Onboard orientation lights red
Radio telemetry antenna

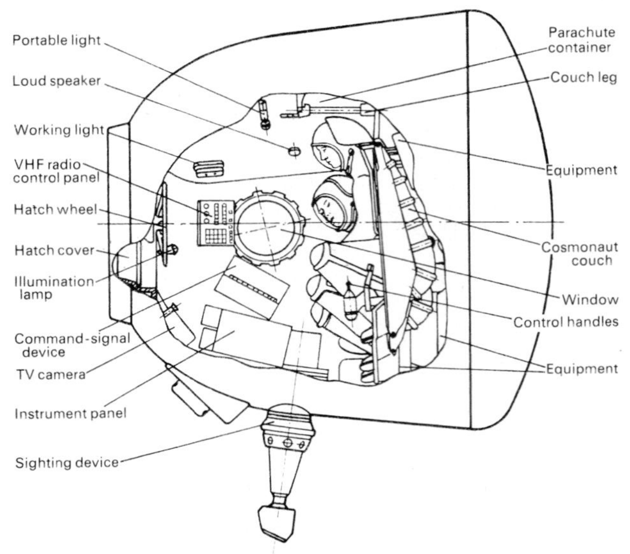

Soyuz re-entry configuration.

This Soyuz mock-up demonstrates how cramped the accommodation was for 3 cosmonauts, and why Salyut crews had to be reduced to 2 until the improved Soyuz T was available. ***(Graham Turnill)***

Instrument Compartment Cylindrical and about 2·97m in dia, this is at the rear and, like the Apollo Service Module, cannot be entered by the cosmonauts. It has a hermetically sealed instrument section housing the thermo-regulation system, electric supply system, orientation and movement control systems and computer, and long-range radio communications and radio-telemetry. In the non-sealed section are 2 liquid-propelled rocket motors (the main and standby engines), each providing a thrust of 400kg. These are used for orbital manoeuvres up to a height of 1,300km and for braking to start re-entry. A separate, low-thrust engine system provides attitude control. Mounted at the rear are 2 wing-like solar panels, each 3·66m long; when deployed in orbit they have a span of 10·06m and 14 cu m of solar cell area. A single whip antenna extends forwards from the leading edge, near the tip of each panel.

Escape tower For lift-off and launch an escape tower is mounted on top of the Orbital Compartment. Capable of lifting the Orbital and Re-entry compartments clear of the rocket in the event of an emergency either on the launchpad or immediately after lift-off, it contains 3 separate tiers of rocket motors for boost, trajectory-bending and vernier control. 20yr after being designed, it was to prove its worth on Soyuz T-10A.

Soyuz 1 This was Russia's first manned flight for 2yr. Vladimir Komarov, the first cosmonaut to be given a 2nd flight, was launched on 23 Apr 1967 into a 201 × 224km orbit with 51·7° incl. The fact that he was alone in the 3-man Soyuz may have indicated some lack of confidence in the new craft. In any case, there appear to have been problems, and re-entry was ordered on the 18th orbit. At a height of 6·5km the main parachute harness twisted, the spacecraft crashed to the ground and Komarov was killed. He was the first man to have been killed during a mission in 6yr of spaceflight. Earlier in the programme centrifuge tests had shown that Komarov had developed an irregular heartbeat and he had been suspended from the flight programme. Only after strenuous appeals had he been allowed to continue with command of Voskhod 1. There is no evidence, however, that he had any health problems during either of his flights. (Donald Slayton, one of the original 7 Mercury astronauts, developed a similar heart condition; as a result he was the only Mercury man not allowed to fly on that programme.)

Soyuz 2 and 3 Soyuz 2, used as an unmanned target vehicle, was launched on 25 Oct 1968 into a 185 × 224km orbit with 51° incl. It was followed on 26 Oct by Soyuz 3, flown into a 205 × 225km orbit by 47yr-old Georgi Beregovoi. The Soyuz 2 launch was not announced until after Soyuz 3 was in orbit. Soyuz 2 is believed to have been equipped with a docking collar, and Soyuz 3 made an automatic approach and rendezvous to within 180m. The spacecraft then separated to 565km and Beregovoi carried out a 2nd, manually controlled approach; no distance was given. On 28 Oct Soyuz 2 made an automatic re-entry, thus successfully testing the parachute system. Beregovoi continued his flight, making regular TV reports, photographing Earth's cloud and snow cover, and studying typhoons and cyclones. The re-entry went so well that the search party was able to see and photograph the final descent at the end of a 64-orbit flight lasting 94hr 51min.

Soyuz 4 and 5 World's 1st manned docking, followed by the spacewalk-transfer of 2 cosmonauts from one craft to another, was achieved on this mission. Soyuz 4, piloted by Vladimir Shatalov, was launched on 14 Jan 1969. Its initial orbit of 173 × 225km at 51° incl was corrected on the 4th orbit to 207 × 237km. This was in preparation for the launch of Soyuz 5 on 15 Jan into a 200 × 230km orbit. On board were 3 cosmonauts, Boris Volynov (commander), Yevgeni Khrunov (research engineer), and Alexei Yeliseyev (flight engineer). On 16 Jan, while Soyuz 4 was completing its 34th orbit and Soyuz 5 its 18th, the 2 vehicles were automatically brought within 100m. Shatalov then took over manual control and docked Soyuz 4 with 5. Khrunov and Yeliseyev donned spacesuits, egressed and used external handrails during their 37min spacewalk into Soyuz 4.

The 2 craft were undocked after 4hr and the following day Soyuz 4, on its 48th orbit after 71hr 14min of flight, successfully re-entered and landed. Volynov, alone in Soyuz 5, successfully re-entered and landed on 18 Jan after 50 orbits and 72hr 46min. In effect, the mission was the first rehearsal of a rescue in space.

Soyuz 6, 7 and 8 These spacecraft were launched on successive days and each remained in orbit for 5 days. The group flight, the 1st time 3 manned vehicles had been in orbit simultaneously, therefore covered a total of 7 days. Soyuz 6, launched on 11 Oct 1969 into an initial orbit of 185 × 222km, carried Georgi Shonin (commander) and Valeri Kubasov (flight engineer). Inclination of all 3 orbits was 51·7. Soyuz 7, launched on 12 Oct, carried Anatoli Filipchenko (commander), Viktor Gorbatko (research engineer) and Vladislav Volkov (flight engineer) into an initial orbit of 206 × 225km. Soyuz 8, on 13 Oct, carried Vladimir Shatalov (overall commander) and Alexei Yeliseyev (flight engineer) into an initial orbit of 204·5 ×

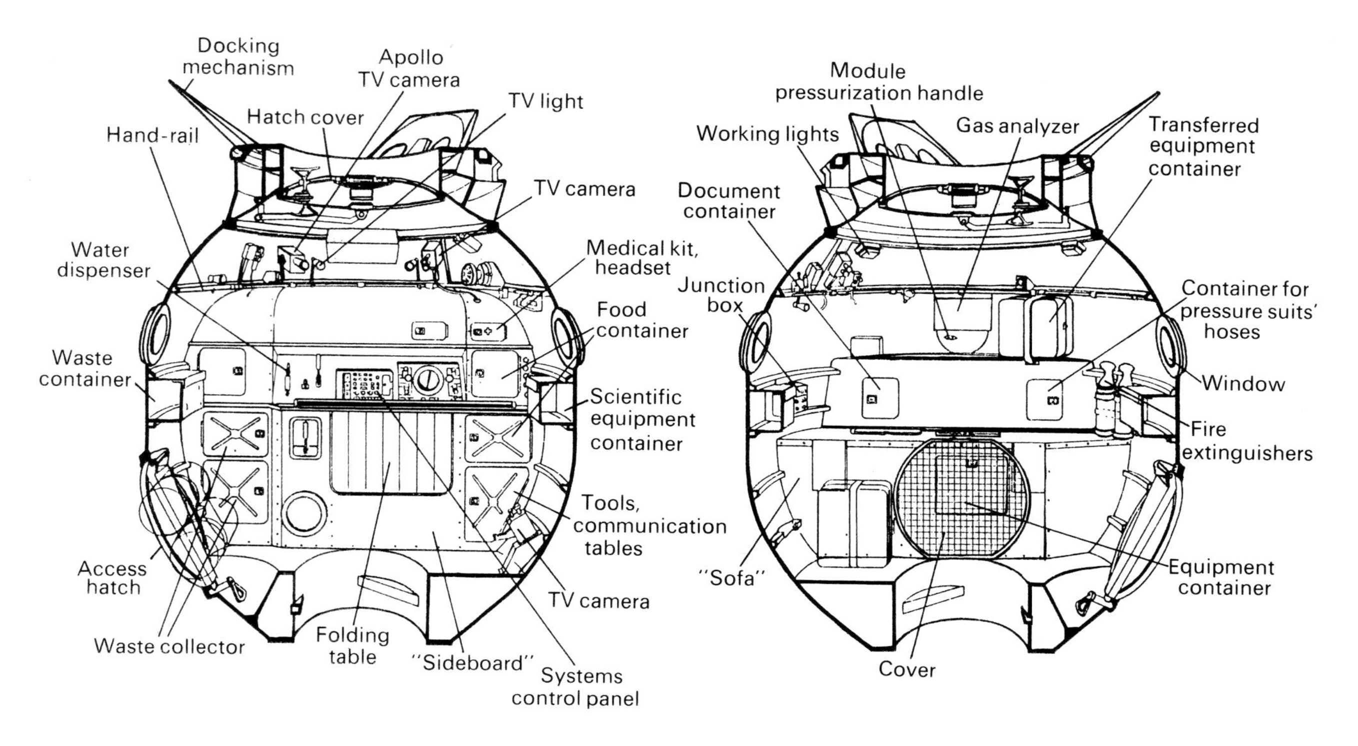

Soyuz orbital module arrangement.

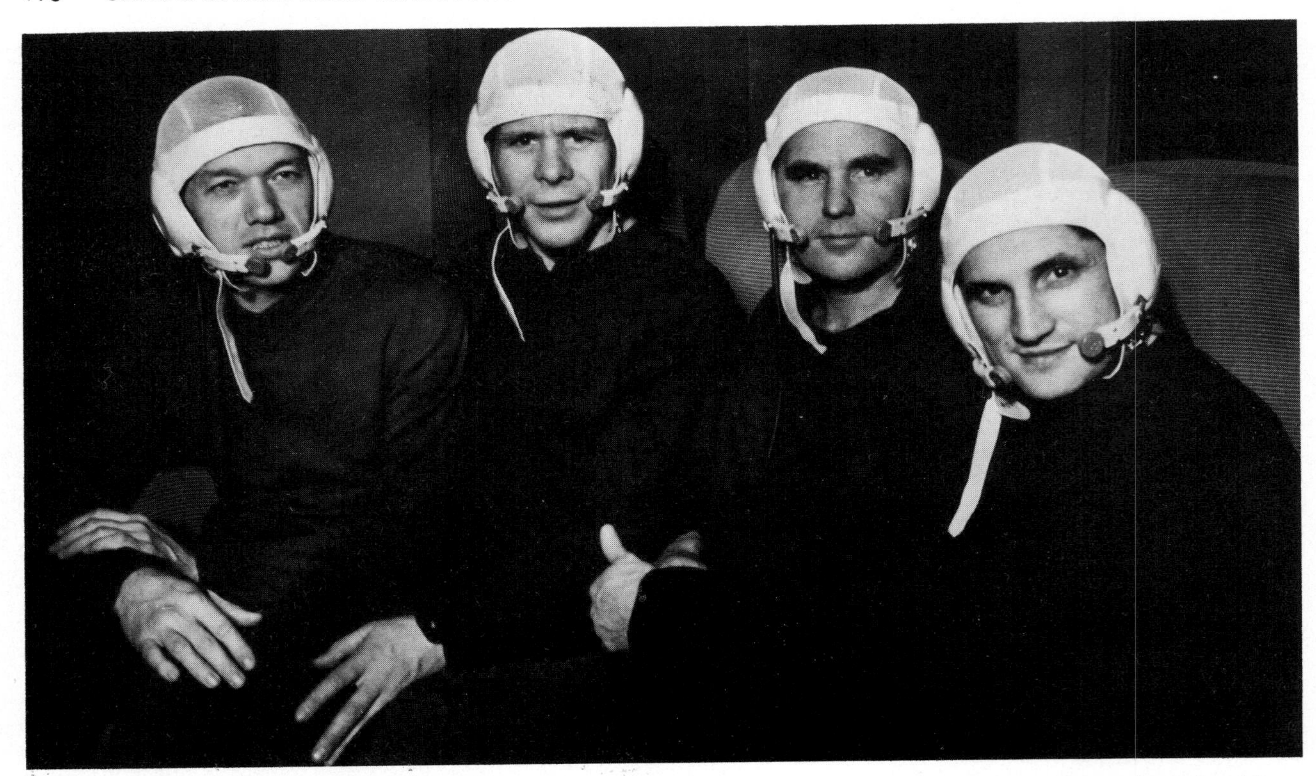

Alexei Yeliseyev, Yevgeni Khrunov, Vladimir Shatalov and Boris Volynov took part in the world's 1st manned docking, on Soyuz 4 and 5.

Anatoli Filipchenko, Vladislav Volkov and Viktor Gorbatko after their return from Soyuz 7. *(Novosti)*

222km. Each crew member of Soyuz 8 was making his 2nd flight. Despite much speculation during the flight that 7 and 8 would dock, on 15 Oct they merely conducted a manual rendezvous to within 488m, observed by Soyuz 6 from a distance of several kilometres. This position was maintained for about 24hr. On their 77th orbit Shonin and Kubasov in Soyuz 6 sealed themselves in their re-entry module, depressurised the orbital workshop and, using a remote-control panel, tried out the Vulkan experimental welding equipment in the workshop. Soyuz 6 landed after completing 80 orbits and 118hr 42min. Soyuz 7 landed 1 day later on 17 Oct after 80 orbits and 118hr 41min, and Soyuz 8 on 18 Oct after 80 orbits and 118hr 41min.

Soyuz 9 This established a new long-duration record of nearly 18 days (424hr), surpassing the record established 5yr earlier by Gemini 7. It was the first manned launch at night. The 2-man crew was Andrian Nikolayev, commander, who had waited 8yr for his 2nd flight, and civilian Vitali Sevastyanov as flight engineer. The spacecraft was launched on 1 Jun 1970 into a 207 × 222km orbit, increased by manual firings on the 5th and 17th orbits to 248 × 267km. The work included the study and photography of ocean phenomena, including coastal currents and surface temperatures, and was aimed at developing fish-finding techniques for use by Soviet trawlers. Final descent was shown on Soviet TV for the first time.

Soviet doctors' concern about the long-term effects of spaceflight appeared justified by the fact that both cosmonauts had difficulty in re-adjusting to Earth conditions. For 5 to 8 days they said they felt as if they were in a centrifuge, being subject to 2g. They found difficulty in walking and lying in bed was uncomfortable for 4 or 5 days. During the flight Nikolayev's daughter Elena was brought to Flight Control to talk to him and celebrate her 6th birthday. She was born a year after Nikolayev married Valentina Tereshkova, the first woman in space, following their pioneering flights.

Soyuz 8 lift-off, 13 October 1969. Last of 3 Soyuz launched on successive days, this carried the overall mission commander, Vladimir Shatalov. *(Novosti)*

Soyuz 9 being prepared for the Soviets' 1st long-duration flight, which lasted 18 days in June 1970. *(Novosti)*

Soyuz 10 perched on the edge of the huge flame trench at the Tyuratam launchpad. *(Novosti)*

Soyuz 10 L 23 Apr 1971, 3 days after Salyut 1, into a 209 × 248km orbit. Soyuz 10 had a 3-man crew: Vladimir Shatalov (commander) and Alexei Yeliseyev (flight engineer), both making their 3rd flights, and Nikolai Rukavishnikov (test engineer). When the automatic procedures had brought Soyuz 10 to within 180m of Salyut the crew took over manual control and Shatalov completed the final approach and docking during the 12th orbit of Soyuz 10 and the 86th orbit of Salyut. The 2 vehicles remained docked for 5½hr. The crew did not enter Salyut, and Western speculation was that either they had been unable to open the docking tunnel or that the flight had been cut short because Rukavishnikov became ill. Soyuz 10 remained in orbit for 16hr after undocking. Then, after 30 orbits, Soyuz 10 re-entered and landed in darkness (02.40 local time) 120km NW of Karaganda on 25 Apr.

Soyuz 11 The 3-man crew made history by spending nearly 24 days in space – 23 of them working in Salyut 1 – but they died during re-entry following the failure of an air valve on the spacecraft. Soyuz 11, crewed by Georgi Dobrovolsky, 43 (commander), Vladislav Volkov, 36 (flight engineer and making his 2nd flight), and Victor Patsayev (test engineer), who celebrated his 38th birthday aboard, was launched on 6 Jun 1971. On entering orbit Soyuz 11 was 3,000km behind Salyut 1. The following day, on the 16th orbit, a 2nd manoeuvre initiated rendezvous and automatic approach was carried out from 6km to 100m. The crew then took over, completing the docking manually with the 2 craft in a 185 × 217km orbit.

The preparation of Salyut 1 as a manned orbital station took 2 days, during which the cosmonauts complained about having too much work. On 29 June the cosmonauts transferred their research materials and logs to Soyuz 11 and prepared to return to Earth. Undocking took place at 21.28, the crew reporting that separation was normal, with all systems functioning. At 01.35 on 30 Jun (23 days 17hr 40min and 380 orbits since launch) a normal firing of the braking engine was carried out. From that moment all communications with the crew ceased. But the automatic

Soyuz 11: test engineer Viktor Patsayev, commander Georgi Dobrovolsky and flight engineer Vladislav Volkov died during re-entry after a successful 24-day flight. *(Novosti)*

re-entry procedure of aerodynamic braking, parachute deployment and soft-landing engines continued smoothly, with an on-target touchdown. A helicopter recovery crew, landing simultaneously with the spacecraft, opened the hatch and found the crew lifeless in their seats. On 2 Jul, following a Moscow funeral attended by Brezhnev, Kosygin and other Soviet leaders, their cremated remains joined those of Komarov in the Kremlin wall. A Soviet Government commission announced later that a rapid depressurisation in the re-entry vehicle as a result of a loss of sealing had led to the cosmonauts' deaths. Nearly 3 years later, after NASA had insisted on a full report on the accident before the joint ASTP flight could take place, it was learned that it was not a hatch seal (as had been assumed in the West) but a faulty valve which had caused the fatal decompression.

Soyuz 12 Announced as a 2-day flight immediately after launch on 27 Sep 1973 into a 194 × 249km orbit. Only 2 crew were carried: Lt-Col Vasili Lazarev (commander), a physician as well as a test pilot, and Oleg Makarov (flight engineer); both were making their 1st flights. Objects were stated to be the testing of structural modifications, improved flight systems and controls, and Earth spectrography to obtain data "for the solution of economic problems". Following the previous disaster spacesuits were carried instead of a 3rd crew member, though it seems they were worn only for re-entry. 1st-day manoeuvres changed the orbit to 326 × 345km. Soft-landing 400km SW of Karaganda, Soyuz 12 was said to have functioned faultlessly.

Soyuz 13 The immediate result of this launch on 18 Dec 1973 was that for the first time Soviet cosmonauts and American astronauts were in space simultaneously. Maj Pyotr Klimuk (commander) and Valentin Lebedev (flight engineer), both making first flights, joined in Earth orbit the 3rd crew of 3 Skylab astronauts, then in the 32nd day of their 84-day mission. Differing frequencies did not permit direct exchanges between the 2 crews. Soyuz 13's initial orbit was not given, but after manoeuvres on the 5th orbit it was stated to be 225 × 272km, incl 51·6°. The orbital compartment contained an Oasis 2 "greenhouse," an experimental biological system designed to grow nutritive protein during prolonged spaceflights. Re-entry and landing on 26 Dec was admitted by Shatalov to have been an anxious time because of a heavy snowstorm in the landing area, located 200km SW of Karaganda.

Soyuz 14 A night launch at 21.51 Moscow time on 3 Jul 1974, 8 days after Salyut 3. The commander, Col Pavel Popovich, was making his 2nd flight after an interval of 12 years; the flight engineer, Lt-Col Yuri Artyukhin, was making his first flight despite having been a cosmonaut for 11 years. The launch, followed by a trajectory correction, placed Soyuz 14 (wt approx 6,570kg), in a 255 × 277km orbit at 51·5° incl, and 3,500km behind Salyut 3. Docking, which took place about 26hr after launch, was Russia's first since the Soyuz 11 tragedy. Code words used by the cosmonauts suggested that military reconnaissance was among their activities; special targets were reported by America (presumably having been seen by

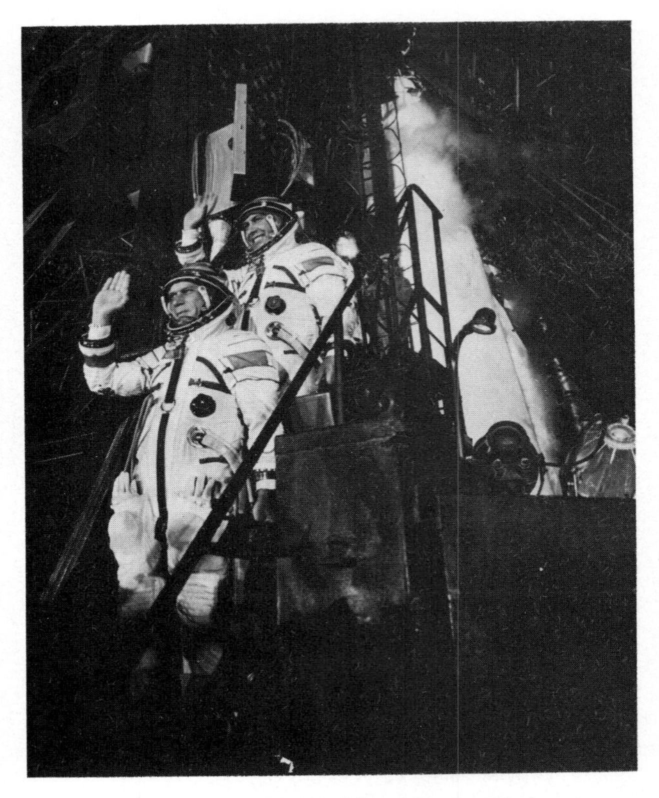

Soyuz 14: Pavel Popovich and Yuri Artyukhin about to board. They used code words during the 15-day mission, leading to speculation that their activities included military reconnaissance. *(Novosti)*

spy satellites) as having been laid out near the Tyuratam launch site for the cosmonauts to observe and photograph in order to assess the reconnaissance potential of Salyut. The crew landed on the 16th day, 140km SE of Dzhezkazgan, Kazakhstan, and only 2,000m from the target point.

Soyuz 15 Another disappointing failure. Instead of making a 2nd visit to Salyut 3 and probably staying in it for about a month, the Soyuz 15 crew were unable to achieve what should have been Russia's 5th manned docking, and after about 36 revolutions had to make an emergency return to Earth at night. Launch was also at night (22.58 Moscow time), on 26 Aug 1974. The commander, Lt-Col Gennadi Sarafanov, 32, and flight engineer, Col Lev Demin, 48 and the first grandfather in space, were both making their 1st flights.

Over a year later Gen Shatalov revealed that the failure was associated with a fully automatic docking system being tested for later use on Soyuz tanker spacecraft (later

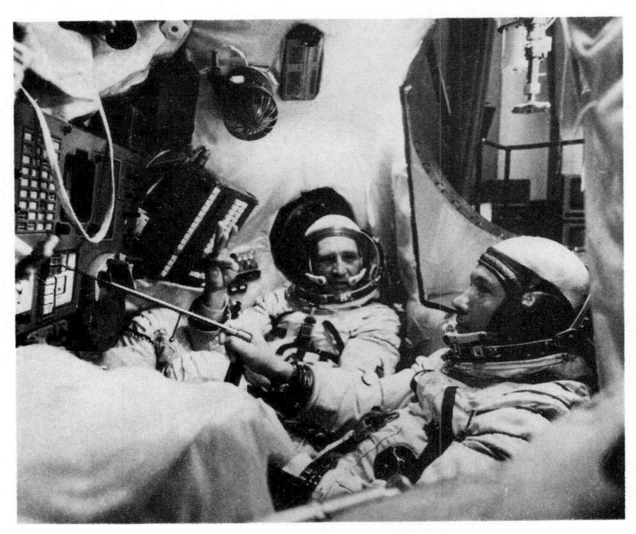

Soyuz 15 crew in simulator. Demin demonstrates "reach extender" to Sarafanov on left.

revealed as the Progress series) designed to replenish space station consumables and thus give them a much longer orbital life. Twice the system went wrong, throwing Soyuz 15 out of control when the distance between it and Salyut 3 was only 30-50m. Following what must have been a major emergency, Sarafanov and Demin returned to Earth.

Soyuz 16 L 2 Dec 1974, this 6-day flight was a near-perfect rehearsal for ASTP. Flown by the ASTP 2nd prime crew of Anatoli Filipchenko (commander) and Nikolai Rukavishnikov (flight engineer), Soyuz 16 entered an orbit of 177 × 223km with 51·8° incl after a correction on orbit 5. Touchdown was 30km NE of Arkalyk in Kazakhstan. Only 5min after the landing Soviet ASTP chief Dr Bushuyev phoned his US counterpart in Houston to tell him that the flight had been successfully concluded. The crew had not made a single mistake.

Soyuz 16: artist's concept made for NASA from a 35mm film of the Soviet ASTP rehearsal. Mechanism to test Soviet ASTP androgynous docking system is attached to the orbital module, left. *(NASA)*

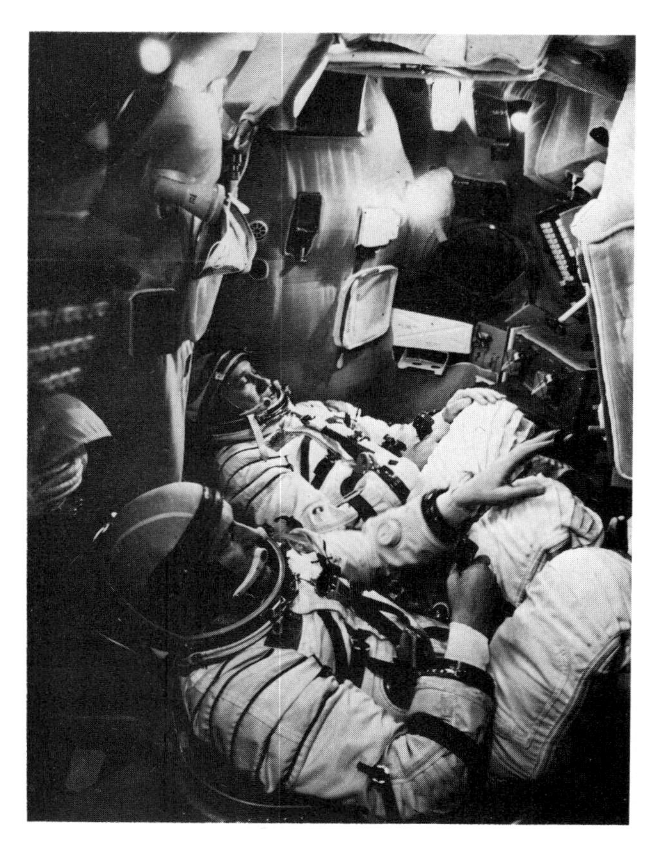

Nikolai Rukavishnikov and Anatoli Filipchenko training for Soyuz 16, the Apollo-Soyuz rehearsal. *(Novosti)*

Soyuz 17 Another night launch, at 00.43 Moscow time (21.43 GMT) on 11 Jan 1975, appears to have gone well from the start. It followed Salyut 4 by 16 days. After correction the orbit was 293 × 354km with 51·6° incl. Both crew members, Lt-Col Alexei Gubarev (commander) and Georgi Grechko (flight engineer), were 43 and making 1st flights. The latter helped to design the original Luna soft-landers, and this flight was possibly in part a reward for that achievement. Docking with Salyut 4 was achieved apparently without difficulty after about 30hr. With Salyut 4 in a circular 350km orbit, approach was made automatically up to 100m. The crew, without spacesuits, opened the hatchway, floated through the transfer tunnel, switched on the Salyut's lights, energised the radio transmitters and began inspecting their equipment. On opening the Salyut hatch they had found a notice left by Soviet engineers where a normal home would have the doormat: "Wipe your feet". During the first week they worked between 15 and 20hr a day, and as a result became fatigued and had some difficulty in sleeping. By the 8th day Grechko had recorded over 4·8km of weightless movement around the 2 compartments. Like Skylab, Salyut 4 was equipped with a teleprinter, used to send up instructions in connection with the taxing programme of experiments. 1hr a night was allowed for leisure before bed, and there were 4 small meal breaks during the day.

A major experiment not tackled even by America's Skylab crew concerned the recovery of evaporated water produced by the body; up to 1lit per man per day could be recovered by Salyut 4's regeneration unit, it was claimed. The crew used this water for washing and also tasted it, reporting that "it was the same as running water". Skylab, it was pointed out, had carried 3 tonnes of drinking water; the Salyut 4 experiment would make it possible to reduce stocks of water to a minimum on future flights. On some days over 6hr were devoted to studying the Sun with the orbital solar telescope in the Salyut's funnel-shaped housing (in which ground reconnaissance equipment is also located). 2 X-ray telescopes and an infra-red telescope were also used to study the Crab Nebula and the planets Mars, Saturn and Jupiter. Looking inwards, Gubarev and Grechko measured the transparency of the Earth's upper atmosphere in experiments to check the water vapour and ozone which protect the surface from the Sun's ultra-violet radiation. Several cameras were also used to study and photograph the surface; particular attention was paid to selected areas of Soviet territory in southern Europe, and parts of Central Asia, Kazakhstan and the Far East.

Before returning the crew used a remote-control system to spray new reflecting layers on to the lenses of the solar telescope to counteract the dust and condensed vapours which had gathered on them. This was caused, it was explained, by micrometeoroids which knock particles out of the space station skin and create a fine dust around it.

Although the crew had kept fit during over 500 orbits by conscientious use of the Salyut's ergometric bicycle and running track, they were advised to wear gravity suits for re-entry. Touchdown was 110km NE of Tselinograd, Kazakhstan, and the crew were promptly located and recovered by the search and rescue team. The mission had lasted 29 days 14hr 40min. Gubarev was found to have lost 2·7kg and Grechko 4·5kg. They recovered this weight about a week after landing, but both were said to be still tiring easily more than a fortnight after returning to Earth.

Soyuz 18A What was supposed to be Soyuz 18 (and now classified in Russia as the "Soyuz anomaly"), sending a 2nd crew to Salyut 4 for a stay of up to 60 days, ended dramatically after about 9min. According to a Soviet announcement issued 1½ days after the launch on 5 Apr 1975, "on the 3rd-stage stretch the parameters of the carrier rocket's movement deviated from the pre-set values, and an automatic device produced the command to discontinue the flight under the programme, and detach the spaceship for return to earth." It was the first time on either a Soviet or US manned flight that an abort had been necessary between lift-off and orbital insertion. But the 2-man crew, Col Vasili Lazarev and Oleg

This view of a Soyuz being taken by rail to the launchpad provides a close-up view of the escape tower on the nose. It saved the lives of the Soyuz 18A crew by lifting the re-entry module (beneath the nozzles) clear of the blazing rocket. *(Novosti)*

Makarov (who flew together on the 2-day Soyuz 12 mission), landed safely near the Siberian town of Gorno-Altaisk in the foothills of the rugged Altai mountain range, N of a point where the Mongolian, Soviet and Chinese frontiers meet. The cosmonauts were later said to be "feeling well' but it seems certain that they had a very uncomfortable time after landing. Subsequent US reports said that the upper stage of the SL-4 rocket burned for only about 4sec before cutting off. During the emergency the cosmonauts, who, unlike Apollo crews, have no control over the automatic abort sequences, "indicated substantial concern" about whether the system was working properly; they also sought repeated reassurance that they were not going to land in China. For short periods during the descent the spacecraft was pulling up to 14-15g. The delay in announcing the launch and abort arose probably because it took about a day to locate and rescue the crew. Later, to allay American concern about ASTP, a scientific attaché at the US Embassy was told that the Soviet booster was an early version, "less diligently checked" than the rocket to be used on the joint mission.

Soyuz 18B This flight was successful from the start. 63 days in space (61 of them in Salyut 4) was a record for Soviet spaceflight and 2nd only to the 84 days of the final US Skylab. Shortly after the launch, at 14.58 GMT on 24 May 1975, Moscow was able to announce that "continued experiments with Salyut 4" were intended. Commander was Lt-Col Pyotr Klimuk, flight engineer Vitali Sevastyanov, and both were making 2nd flights. According to the Novosti press agency, Sevastyanov was the first journalist in space: not only did his wife work at Novosti, but he himself was a commentator for a popular science series, "Man, Earth, Universe" on Soviet TV. Gen Shatalov, director of cosmonaut training, aroused curiosity by announcing that the aims included the "search for new possibilities for using both an individual spacecraft and a group of spacecraft to carry out research and applied scientific tasks in near-Earth orbit." Soyuz 18's initial orbit was 193 × 247km with the usual 51·6° incl. Docking was achieved on 26 May, apparently without difficulty, with Salyut 4 then on its 2,379th orbit, at 344 × 356km. Sevastyanov reported that "not one gramme of fuel" was wasted in absorbing the shock. As they entered Salyut 4 they found a notice, "Welcome to our communal house," left by previous crew Gubarev and Grechko and intended for Lazarev and Makarov. Having checked the space station's life-support systems and found them in order, the crew settled down to studies of the Sun, planets and stars in various bands of the electromagnetic radiation spectrum; geological and morphological studies of the Earth; investigations of physical processes in the Earth's atmosphere; and biological research on themselves.

Commander Pyotr Klimuk and flight engineer Vitali Sevastyanov boarding Soyuz 18B. *(Novosti)*

About 2 tonnes of scientific equipment had to be checked and reactivated, a task taking 48hr. "A few problems" were reported, but nothing serious. By the 4th day both cosmonauts had adapted to weightlessness, and their health remained good throughout the flight. Afterwards Klimuk said that the body "remembers" weightlessness and "knows ways" of adapting itself. The running track introduced on Salyut 3, together with the special suits worn by the Soyuz 14 and 17 crews to put loads on the muscles and so simulate gravity, had been so successful that both the loads and the exercise periods were increased on this flight. The suits were taken off only at night. Studies of ripening crops, etc, on Earth provided a contrast with the earlier winter photographs taken by Gubarev and Grechko. Using the Oasis plant-growing system, the 1st crew had succeeded in getting only 4 out of 28 peas to sprout in 29 days. But although 13 of the 16 peas planted by the 2nd crew sprouted within 4 days, the young plants died 4 weeks later; Sevastyanov thought that this was because they had been scorched by the lamps switched on for a live TV transmission from Salyut 4. A

10-day period in June, during which Salyut 4 was passing over Russia in daylight, was devoted to Earth-resources work: in 60 orbits 2,000 photographs were obtained. Sevastyanov celebrated his 40th birthday on 8 Jul, their 45th day in space. A celebration dinner included spring onions which had been grown on board. When Leonov and Kubasov were launched in Soyuz 19 on 15 Jul into an orbit 125km below that of Salyut 4, Klimuk and Sevastyanov had just completed their 52nd day in space. Both were in excellent health and spirits, and their only problem seemed to be that the space station was overheating slightly; extra fans were needed to keep the temperature down. As the 2 Soviet crews passed the next day Leonov congratulated the "old-timers" in space in the 1st of 2 such exchanges. Major scientific achievements claimed for the mission included 600 pictures from the solar telescope; recordings of X-radiation from 10 sources in deep space; photography of 8·5 million sq km of Soviet territory; and 2 scientific firsts: spectrographic investigation of the polar lights and of noctilucent clouds. Data showing that the star Cygnus X-1 was a black hole had also been obtained.

While Apollo was landing on 24 Jul at the end of the joint Apollo/Soyuz 19 flight Klimuk and Sevastyanov began preparations for their own landing in Soyuz 18. The Salyut/Soyuz orbit was changed to 349 × 369km and about 50kg of film and other research material packed into the descent module. They also prepared Salyut 4, which by then had been in orbit for over 7 months and had completed 3,352 revolutions, for continued automatic operation under ground control. Undocking occurred 3hr 22min before landing, which was at 14.18 GMT on 26 Jul. They touched down well within the target area, 56km E of Arkalyk, in the same area as the Soyuz 19 landing 5 days earlier. During the flight Klimuk lost 3·8kg in weight and Sevastyanov 1·9kg. The latter recovered his lost weight in his first day back on Earth. Opening a space exhibition in Belgrade a few weeks later, Sevastyanov said that one result of their flight was that the whereabouts of even the smallest ore deposits under 22 million sq km of Soviet territory (more than one-third of the country) was now known.

Soyuz 19 The first joint Soviet/US flight, launched on 15 Jul 1975 with Alexei Leonov and Valeri Kubasov as the Soviet crew, is described in the International section under Apollo-Soyuz.

Soyuz 20 Launched unmanned on 17 Nov 1975 in an apparent rehearsal of both a space station re-supply mission and a 90-day manned flight. Soyuz 20 docked automatically with Salyut 4 on Nov 19 "with the aid of onboard radio devices and computer installations." It was the 1st time an unmanned spacecraft had been given a Soyuz designation, and the 1st time Russia had achieved 3 dockings with a Salyut. The docked orbit was 343 × 367km. This mission picked up the tanker experiments started on Soyuz 15, when the automatic docking system failed. Soyuz 20, with its load of tortoises and other livestock, gladiolus bulbs, etc, returned to Earth on 16 Feb 1976 after undocking and making 2·6 revolutions before re-entering. 4yr later Moscow botanists announced that they had named a new variety of gladiolus after Soyuz 20. Following their exposure to radiation and weightlessness they had flowered in the depth of winter 23 days earlier than normal in the USSR Academy of Sciences' botanical gardens, displaying an orange-red instead of the normal red colour. Subsequent generations of the gladioli lost the early-flowering characteristic but retained the distinctive orange-red colouring.

Soyuz 21 With Col Boris Volynov as commander (making his 2nd flight) and Lt-Col Vitali Zholobov as flight engineer, Soyuz 21 was launched on 6 Jul 1976. What became Russia's 2nd longest flight to that date was launched 14 days after Salyut 5 into an orbit of 193 × 253km; Salyut 5's orbit at that time was 220 × 275km. After 1 day, during which its orbit was corrected to 254 × 280km, Soyuz 21 docked with Salyut 5. The operation took only 10min, with minimum use of fuel, and was probably Russia's smoothest docking up to that point. The combined weight of Salyut 5/Soyuz 21 was given as 25 tonnes and the length as just over 26m. Gen Shatalov warned that no dramatic spacewalks or spectaculars should be expected, and when the Salyut 3 practice of using code words to conceal the nature of crew exchanges with flight controllers was again followed, it became clear that this mission was mainly military. There were however detailed descriptions of other activities. By 10 Jul the crew were studying an aquarium which included "a very pregnant guppy"; the melting and hardening of

Apollo-Soyuz: the world saw lots of pictures of Soyuz 19 taken from Apollo during the joint flight in 1975, but not until after the landing did pictures of Apollo taken from Soyuz become available. *(NASA)*

metals in weightlessness, using ingots of bismuth, lead, tin and cadmium; and, by means of a hand spectrograph rather like an amateur cine camera, aerosol and industrial pollution of Earth's upper atmosphere. The instrument could be used for spectrography of layers of the atmosphere against the background of the rising and setting Sun, and against various types of land or water surface. Another experiment, operated by Zholobov, demonstrated how liquid in a sphere connected to another empty container above it could be transferred upwards without the use of pumps or other equipment, "under the effects of surface tension". Zholobov said the method could be used for spacecraft refuelling.

The crew's main task was however almost certainly to observe Operation Sevier (North) in Siberia, where massive land, sea and air exercises were being carried out to coincide with their flight and give them an opportunity to assess the potential of a manned spacecraft for monitoring and participating in such operations. And when Cosmos 844 (L 22 Jul 1976, orbit 172 × 353km) failed due to a manoeuvring engine malfunction the crew were believed in the West to have performed some of its data-gathering duties until a replacement was available. On 30 Jul the cosmonauts were reported to be taking pictures of areas likely to contain mineral deposits, studying seismic activity, "appreciating the danger of mud streams in mountains" and examining areas on which hydro-engineering structures were to be built. By 6 Aug, having completed a month in orbit, they had also carried out studies of the solar corona. 3 days later they extracted from the crystalliser one of the crystals they had grown in space over a period of 26 days, placed it in a container for return to Earth, and started a 2nd experiment with a dye added to indicate the diffusion of the coloration into the crystals grown. There was a note of anxiety in a report on their medical condition at this point by Dr Igor Pestov, who said the doctors were "sometimes worried" because, due to non-stop work, the cosmonauts did not always manage to get proper rest. They had each lost about 1·5kg in weight and had higher than normal blood pressure. There was nothing exceptional or dangerous about this but steps were being taken to "normalise" these reactions. During the 5th week 2 days were devoted to medical studies of the cosmonauts' hearts, body mass, etc, following which they were said to be "feeling fine". First details were also given of an instrument aboard Salyut 5 designed to test the cosmonauts' reactions. It consisted of a small unit with 3 coloured lamps and 3 key switches. The cosmonauts had to answer lamp signals by pressing the appropriate keys, and their reactions were said to be "somewhat slower" in space. The 6th week included exercises in manual control of the space station, and more medical checks. 14 Aug was "the most festive and joyous day of their flight," when schoolchildren from Vietnam, "Congo," Cuba, Poland, Russia, the US, Czechoslovakia, Sri Lanka and Chile travelled from an international camp on the Crimean coast to the Yevpatoria control centre. They talked to the cosmonauts and were given "exhaustive replies to all their questions". By 18 Aug, Salyut 5 had completed 900 orbits, of which 700 had been with the crew on board. At that point they were studying the Sun with an infra-red telescope and working with enthusiasm. In an interview Sevastyanov said on 18 Aug he did not know whether Volynov and Zholobov would break the 63-day Soviet record set by Klimuk and himself on Soyuz 18: "The time of the flight is of secondary importance, though it is not to be ignored." On 24 Aug Volynov and Zholobov were said to be starting their 7th week in "good health and spirits" and to be carrying out infra-red studies of the Earth's land and ocean surfaces. Then, on the same day, another report said that the flight was nearing completion. The crew were putting Salyut 5 into automatic regime and preparing Soyuz 21 for return with their film, samples and materials. Usually at least 2 days are devoted to preparations for return, but this time touchdown took place only 10hr after the announcement. They were said to have achieved a smooth landing 200km SW of Kokchetav in Kazakhstan, but the fact that it was made in darkness, at 21.33 Moscow time on 24 Aug, seemed to support Western theories that there had been an element of emergency about the return. US reports several weeks later stated that an emergency return became necessary because of an acrid odour flowing from the Salyut 5 environmental control system. Zholobov and Volynov endured it for some time but were unable to locate the cause before it became unbearable. Thus for the 6th time in 9 manned missions it had not been possible to complete the flightplan. The cosmonauts were said to be "walking a little shakily" and suffering from the usual post-flight effects, but they quickly recovered.

Soyuz 22 L 15 Sep 1976, and at first expected to carry a 2nd crew for Salyut 5, Soyuz 22 was in an inclination (64·75°) that had not been used for a Soviet manned flight since Vostok, so ruling out a docking. Commander Col Valeri Bykovsky, 42, was making his 2nd flight, 13yr after his first. Flight engineer Vladimir Aksyonov, aged 41, was getting his first trip only 3yr after becoming an astronaut. Orbital parameters, omitted from the official announcements, were initially 200 × 281km, circularised on revolution 16 at 251 × 257km. The docking mechanism had apparently been replaced with a multispectral camera developed and built by the GDR company Karl Zeiss Jena. It was said to be "the first occasion on which foreign instruments had been installed in a Soviet manned spaceship" (ASTP was obviously excepted). During their 8-day flight individual sections of Soviet and GDR territory were photographed as part of a co-operative study of geological and geographical features in the interests of the 2 national economies.

The cosmonauts were said to be "enjoying a large measure of independence in controlling the craft and making various investigations." This, coupled with an unusual orbit taking it over the main areas of activity in a massive NATO land, sea and air exercise stretching from Norway to Turkey, was one of the many factors which suggested that this was a first which Russia was unlikely to claim: the first manned spycraft. Other factors were the seniority of Bykovsky, said to have graduated at the Zhukovsky Military Air Engineering Academy in the 13-year interval since his first flight, and the unexpected choice of Aksyonov (also with a specialised air force background, with no rank given) rather than any of the other cosmonauts with greater seniority waiting for 1st or 2nd flights. Their experience and classified knowledge might well have enabled them to select and point their new cameras at NATO activities of reconnaissance interest much more quickly than ground control could have achieved. During the flight it was revealed that it was the start of increased manned spaceflight co-operation between the USSR and Soviet-bloc countries. Selection of non-Soviet cosmonauts would begin in 1977, with flights for them in the 1978-83 period. All missions would be commanded by a Soviet cosmonaut, with a foreign country providing the flight engineer. (When the flights began, all foreigners were in fact described as "cosmonaut researchers," with no flight control responsibilities.) The landing, on 23 Sep (10.42 MT), was 150km NW of Tselinograd, Kazakhstan. Both cosmonauts were in good health; they had taken 2,400 detailed photographs, in 6 different bands of the spectrum, of Soviet and GDR territory.

Soyuz 23 L 14 Oct 1976 with the announced aim of docking with Salyut 5, this was apparently intended to lead to a 60-90-day stay. Soviet scientists were presumably satisfied that the fault in Salyut 5's air-conditioning system had either been corrected automatically or could be put right by the new crew. But the flight proved to be the 7th Soviet failure in 11 attempts to complete space station missions, and the 2nd successive Soyuz failure. Commander Lt-Col Vyacheslav Zudov and flight engineer Lt-Col Valeri Rozhdestvensky were both making 1st flights after 10yr training. The latter, a former diving unit commander, somewhat prophetically said in the prelaunch press conference: "I think my pre-cosmic profession may yet come in handy." It did, since the emergency return, 48hr 06min after lift-off, ended in the first Soviet splashdown. After being placed in a 243 × 275km orbit with 51·6° incl Soyuz 23 began automatic rendezvous with Salyut 5 during its 17th revolution, 25hr after launch. The space station's orbit was 253 × 268km. Soviet sources said later that the Soyuz rendezvous approach electronics malfunctioned and the spacecraft never got to the point, about 100m from the Salyut, at which the crew would have been able to take over manual control for final docking. Since this version of Soyuz was flown without solar panels to make it lighter and more manoeuvrable, and the internal battery power limited flight independent of a Salyut to 2½ days, what should have been Russia's 9th manned docking had to be cancelled and preparations begun immediately for return to Earth. Retro-rockets were fired 44min before landing, and Gen Shatalov warned the crew to stay in their seats after touchdown because of wind conditions. He congratulated them on working calmly and confidently and said their reports were very clear and good. Possibly because high winds dragged the main parachute in the final 7km of descent, the spacecraft came down in darkness and a snowstorm in the 32km-wide Lake Tengiz, 195km SW of Tselinograd, thus fulfilling Rozhdestvensky's prediction. Aircraft located Soyuz 23 and helicopters dropped frogmen with flotation devices. The official announcement said "evacuation was under difficult conditions at night in heavy snowfalls". A major problem was that the lake was salt water and therefore only frozen along the shoreline. The ice hazard was made worse by both high winds and fog, and the search and rescue teams were forced to wait for daylight before completing the rescue. But while officials worried about the cosmonauts' ability to survive the night's bitter frost, they were relatively warm and safe, even if uncomfortable, inside the spacecraft. It was the rescue teams who needed medical assistance as the night wore on. However, there were no serious casualties and the cosmonauts were finally flown to Arkalyk, 140km W of the lake, where they were reported to be in good condition.

Soyuz 24 L 7 Feb 1977 "to continue the experiments started by Soyuz 21 with Salyut 5". Commander Viktor Gorbatko was making his 2nd flight, with flight engineer Lt-Col Yuri Glazkov making his 1st. They had been the standby crew for Soyuz 23. It had taken 4 months to launch their back-up mission. From a corrected orbit of 218 × 281km they successfully docked with Salyut 5 the next day and took a sleep period before boarding.

Apparently to discourage speculation that a record-breaking flight was intended, Gen Shatalov warned at the start that the flight would be "routine," and on 16 Feb it was announced that the crew were halfway through their programme. It is likely that their first task was to renew elements of the environmental system, following the hurried departure of the Soyuz 21 crew nearly 6 months earlier. On the 3rd day they were still "recommissioning" Salyut 6, and had also started medical and crystal-growing experiments. Salyut 3 veteran Yuri Artyukhin, now head of communications at Flight Control, said that water reserves were being supplemented by regeneration from moisture gathered in the space station. On 12 Feb Gorbatko and Glazkov successfully completed repairs to a computer "and replaced some units of other systems." A 300mm solar telescope was used for studying the Sun. Twice a day physical exercise sessions totalling at least 2½hr were taken to "a nice musical accompaniment" with 3 different rhythms. The cosmonauts wore the latest loaded suits, with shock-absorbers pressing them on to a

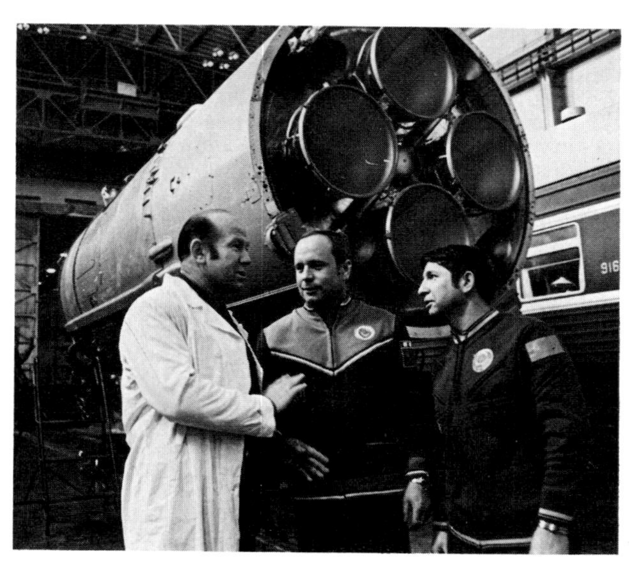
Alexei Leonov, deputy head of cosmonaut training, briefing Viktor Gorbatko and Yuri Glazkov for the Soyuz 24 mission. *(Novosti)*

moving track with a force of 50kg. On 22 Feb, after completing spectral photography of the Earth's surface and atmosphere, Gorbatko and Glazkov started to return the space station to automatic control. Material obtained during the research programme was transferred to the "transport craft," presumably a reference to the automatic return module brought back to Earth the day after the crew had landed 37km NE of Arkalyk on 25 Feb.

Soyuz 25 What was confidently expected, even in the West, to be the start of a new, adventurous era in space station development, involving overlapping crews as Soyuz exploited its potential as a genuine space ferry, proved to be a repeat of the Soyuz 23 failure. Launch was on 9 Oct 1977, with Lt-Col Vladimir Kovalyonok, 35, as commander and Valeri Ryumin, 38, as flight engineer both making their first flights. By the 3rd revolution Soyuz 25 was in a 205 × 228km orbit with a 51·6° incl. Moscow announced that it would "link up with orbital station Salyut 6, launched on 29 Sep." On their 2nd morning and the 17th revolution, with the Soyuz in a 339 × 352km orbit matching that of Salyut 6, the crew were recorded by the Kettering (England) space group calling out the usual rendezvous phrases. But on revolution 18 the expected post-docking exchanges on equalising pressure, etc, were missing. In America, NORAD showed Salyut 6 and the Soyuz close to each other on revolutions 20-23, then tracked them as one for a period during revolution 23. Then came the Moscow announcement about the rendezvous: "From a distance of 120m the vehicles performed a docking manoeuvre. Owing to deviations from the planned procedure for docking the link-up was called off. The crew has begun making preparations for a return to Earth." In fact the Soyuz 25 crew waited until the following morning, when they were ideally placed for re-entry, before firing the retro-rockets on the 32nd revolution. A successful descent was made after a 48hr 46min flight, with a landing 185km NW of Tselinograd. A subsequent absence of praise for the cosmonauts suggested there might have been some human failure during the final manual attempt at docking. The subsequent Soyuz 26 EVA inspection indicated that Moscow was worried that the Salyut docking port had been damaged in what might have been a collision rather than a docking.

Soyuz 26 This launch on 10 Dec 1977, 9 weeks after the humiliating Soyuz 25 failure, at last brought the patient Soyuz space scientists their hard-earned breakthrough. Lasting into its 97th day, the flight broke Skylab's 4yr-old long-duration record of 84 days – with little likelihood that the Americans could do better for a decade. With Yuri Romanenko, promoted Lt-Col since being named as ASTP backup, as commander and Georgi Grechko, with 29 days in Salyut 4 behind him, as flight engineer, the mission's achievements had such long-term importance that they are dealt with here at length.

The crew docked with Salyut 6 on their 17th orbit without difficulty, despite the fact that the link-up was with the Salyut's second docking port. Salyut 6's orbit was then 337 × 363km. The crew entered the station 3hr after docking. Improved living conditions for their 14-week stay included a Skylab-type shower and the Salyut 4 experimental water recycling system built in as an operational feature.

First highlight, on 20 Dec, was a 1hr 20min EVA, an exercise never undertaken by Russia unless strictly necessary. Using a new type of semi-rigid suit with self-contained breathing equipment in backpacks, Romanenko and Grechko depressurised the Salyut transfer compartment and Grechko egressed through a hatch. With Romanenko controlling the EVA from inside, Grechko, carrying the colour TV camera, transmitted to Moscow Flight Control pictures of the forward docking unit and inspected joints, sensors, guide pins, fasteners and seals to see if the Soyuz 25 impact had done any damage. Secured with a safety line, Grechko also used special tools before reporting that there were "no scratches, traces or dents," that the docking equipment's lamps, electric sockets and latches were unbroken and that the receiving cone was also unharmed. He was outside for 20min. Romanenko, longing to see the view, then made an unauthorised exit, something not revealed until long after the mission. How long he remained outside we still do not know. But no doubt as a conscientious commander he waited until Grechko was

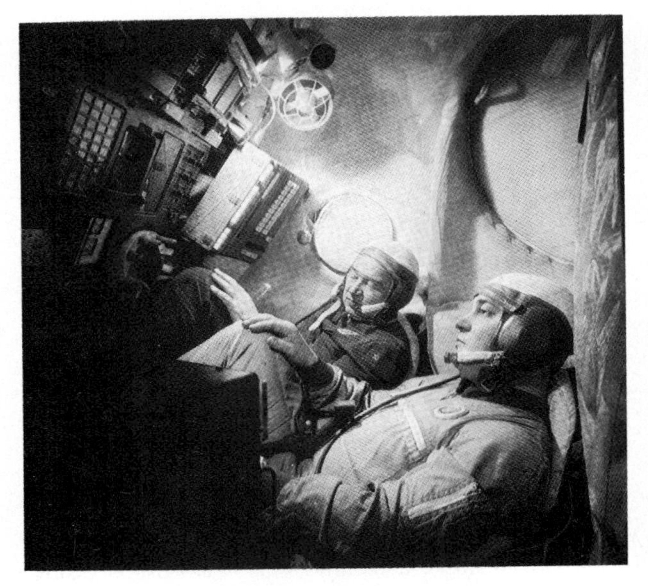

Soyuz 26: commander Romanenko, right, and flight engineer Grechko, left, broke Skylab's long-duration record. They were also the first to return in a fresh spacecraft - Soyuz 27 brought up by a short-stay crew.

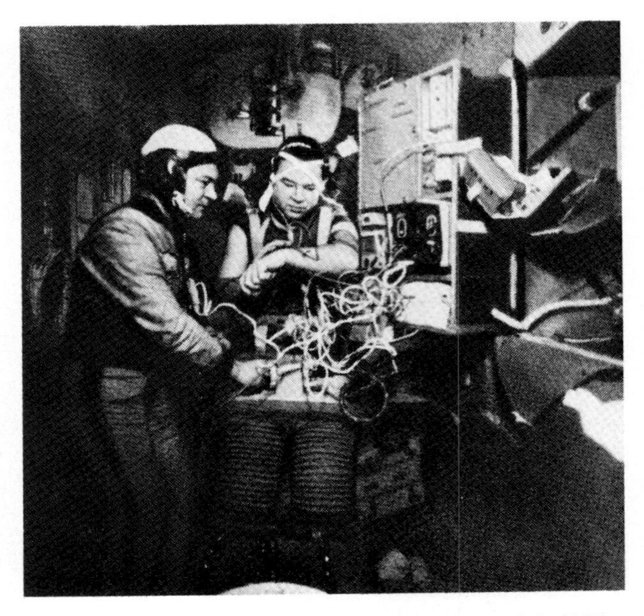

Soyuz 26: Romanenko and Grechko preparing to use their Chibis vacuum suits in preparation for return to Earth after 96 days in orbit.

safely inside again before venturing out. Russia's first spacewalk since the Soyuz 4/5 transfer in 1969 did not pass without a major alarm. Not until 2 months later did the Russians disclose that when they had returned to the airlock and closed the outer hatch the instruments indicated that a valve was stuck and that repressurisation of the transfer compartment and return to the main Salyut station was impossible. And entry to Soyuz 26 for return to Earth was of course also impossible without passing through the Salyut. Repressurisation was tried, however, and proved that it was a cable fault giving a false alarm. It was then feared that the fault was due to condensation, with a danger of short-circuiting or corrosion. Romanenko and Grechko were ordered to check for water in the electric circuits, but found none. The crew then settled down to a routine of observations and experiments, celebrating the New Year as they swung in and out of it 15 times with a decorated fir tree which they had taken up in Soyuz 26.

The EVA having established that the primary docking port was undamaged, Soyuz 27 was launched on 10 Jan 1978 and next day docked without problems with Salyut 6, then in a 334 × 367km orbit. In a historic first double link-up, which also created the first 4-man space station, Dzhanibekov, commander of Soyuz 27, and flight engineer Makarov joined Romanenko and Grechko for a 5-day stay. During the docking the Soyuz 26 crew sealed themselves in their own spacecraft to ensure survival if Soyuz 27 made violent contact. When it was successfully completed they returned to the Salyut and TV viewers around the world saw the emotional bearhugs as 3hr later Romanenko and Grechko pulled the new crew through the forward hatch to join them. They celebrated with cherry juice drunk through tubes. The triple space complex was about 30m long and 32 tonnes in weight, with 7 habitable compartments. To check its rigidity the cosmonauts were instructed "to jump up and down to make the station shake" while instruments monitored the vibrations caused. The tests proved that the long "space sausage" was perfectly safe, and cosmonaut Titov pointed out that the ability to overlap rotating crews would save the cosmonauts much tedious work in closing down and opening up space stations when there was a gap between visits. But the most strenuous part of this visit resulted from the need to exchange the individual couches, spacesuits and other personal equipment in Soyuz 26 and 27, manoeuvring them through 6 docking hatches in each direction. This was to enable the new crew to return home in Soyuz 26, thus freeing the aft port for a later refuelling mission. (Soyuz 26 had had to dock at the "wrong" port because of the Soyuz 25 failure.) This work took up part of 14 Jan, intended as a rest day. During preparations for the return of Dzhanibekov and Makarov, Grechko reported

Soyuz 26: Romanenko in his EVA suit about to leave Salyut 6 to inspect the docking port for possible damage caused by the Soyuz 25 impact.

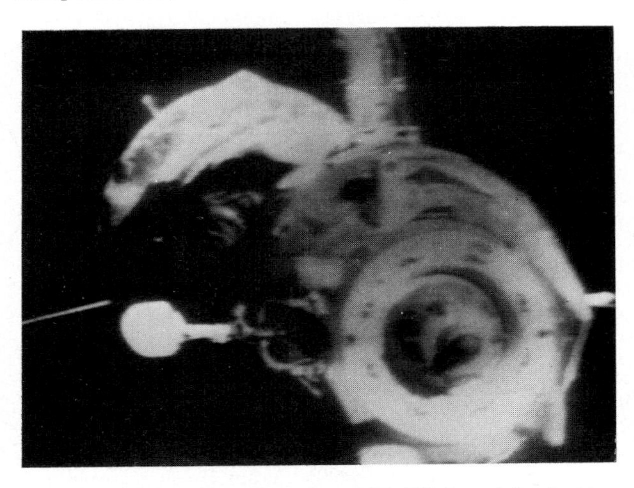

Soyuz 26: Mission Control monitored this TV view of the docking with Salyut 6 - always an anxious moment for the Soviets in view of their frequent docking failures. *(Tass)*

that their visit had been of great help, both physically and psychologically, on their long mission. For the next 6 days the cosmonauts spent much of their time on the Raduga ("Rainbow") experiment, following up the work of Soyuz 22 with a GDR multi-zonal Zeiss camera by photographing vast areas of the USSR, particularly the Central Asian republics, Kazakhstan, the Altay and Volga areas and the central black-soil zone for Earth-resources purposes. Then came the next mission highlight, the launch on 20 Jan of Progress 1, an unmanned, one-way cargo ferry bringing fuel and other supplies to sustain the long-duration mission. Launched into a 194 × 262km orbit with 88·8° incl, Progress 1 was moved to 246 × 334km at 90·2° incl the next day. Because ground commands could only be given when the Progress was within range of ground stations, 2 days were spent on the automatic docking. This time Romanenko and Grechko remained in the Salyut, monitoring the approach and acting as a stand-by system. They had the ability to take over or cancel the docking, or to turn on their own engines to escape if the approaching freighter got out of control. In fact the automatic docking - a repeat of the Soyuz 20/Salyut 4 rehearsal - was apparently achieved without difficulty on Progress 1's 33rd revolution. The first in-orbit refuelling operation proved however to be long and difficult, though it was said to have been flawlessly completed 12 days later on 3 Feb.

On 23 Jan Romanenko and Grechko entered Progress 1 and made safety checks. Next day they transferred dry cargo such as carbon dioxide absorbers, filters, clean towels, water containers, food, films and research equipment into Salyut 6 and began preparations to transfer fuel. 2 days were taken to link the pipelines and check their tightness by test pumping. 2 more days were needed to transfer the propellant in a process complicated by the lack of gravity. The gas was first pumped into high-pressure balloons, which could then be used to force it into the Salyut tanks. The oxidiser refuelling, "the last and most labour-intensive operation," was reported as under way on 30 Jan. The previous day, a rest day, had been the crew's 50th in space.

On 31 Jan Salyut 6's air pressure was increased to 810mm, using the new supplies from Progress 1; it was said to have dropped as a result of losses during the spacewalk and "operations for waste disposal." By 1 Feb the crew had transferred ½ ton of freight and were refilling Progress with used life-support equipment, filters, and empty food and water containers - turning it, in fact, into the first space dustbin. It took the crew 12 shifts to unload Progress, and on 5 Feb its engines were fired twice to raise the orbit of the whole Progress/Salyut 6/Soyuz 27 complex. Novosti's space correspondent commented that this seemingly routine technical experiment heralded the use of space tugs to assemble space stations, to move sections from one orbit to another, or to take individual spacecraft to space-based repair shops. Because only 3 instead of the expected 5 trajectory corrections had been needed before Progress 1 docked, the ferry had plenty of fuel left. The manoeuvre was an excellent way of consuming this fuel and conserving the Salyut's before Progress was "cast off" on 7 Feb and placed on a re-entry trajectory causing it to burn up over the Pacific. Before it did so, however, the back-up automatic approach system was tested and the whole docking system thoroughly checked by making 3 more link-ups and separations.

Much time was needed to assemble and install in a special airlock the Splav electrical smelting furnace delivered by Progress 1. With one end exposed to space, this had 3 heating areas capable of producing temperatures of up to 1,100°C, with computer control ensuring 5° accuracy in use. It was 15 Feb, 26 days after Progress 1's arrival, before it was ready for use. For the rest of the mission Splav was used with semi-conductor materials like aluminium-antimony combinations and for impregnating porous molybdenum with gallium. Samples of what space factories should soon be able to produce in quantity were brought back to Earth for study.

The crew's attention was briefly diverted from Splav by an extraordinary auroral display on 15 Feb. Ground stations separately registered a major magnetic storm. Romanenko and Grechko described how, on the night side of Earth in the N hemisphere, they spent 10min travelling "through thousands of searchlights in a wavy line" stretching upwards above America. Near Earth the lights were green, becoming red as the spacecraft passed through them at a height of 450km. Pictures they took were finally published in the West in Apr 1980. Kovalyonok spoke of seeing a "rich colour palette" and dubbed the spectacle "colour music".

The final highlight in this historic mission came with the launch on 2 Mar of Soyuz 28 with an international crew. Commander Alexei Gubarev was accompanied by Vladimir Remek, a Czech military pilot and the first man from outside Russia and America to get into space. The following day, 3hr after docking, there was an emotional meeting with Romanenko and Grechko; the latter had spent 29 days with Gubarev in Salyut 4. On 4 Mar the visitors were able to join Romanenko and Grechko in celebrating with more cherry juice toasts the breaking of the 84-day record established by the Skylab 4 crew; astronauts Carr, Gibson and Pogue telegraphed their congratulations.

The main experiment during their week together was the placing by Remek in the Splav furnace of capsules containing silver chlorides and copper and lead chlorides prepared by the Czechoslovak Academy of Sciences. They took the results home with them on 10 Mar, when Salyut 6's orbit was 338 × 357km.

By 12 Mar Romanenko and Grechko had begun their "descent programme," using their new Chibis vacuum suits to create negative pressure on the lower part of the body, increasing their time on the running track, and doing other exercises to prepare their leg muscles and blood vessels for the return to Earth. Next day they began closing down Salyut 6 for a period of unmanned operation, and on 16 Mar they sealed themselves in Soyuz 27's re-entry module. They finally touched down at the end of a 96-day 10hr mission 265km W of Tselinograd. As a safety precaution high-tension lines were switched off in the area as they descended. TV pictures showed Romanenko and Grechko standing outside the spacecraft as they were greeted with hugs and kisses. A week later Moscow reported that they were still trying to "swim" out of bed in the mornings. Among the marks of recognition announced was the erection of bronze busts at the birthplaces of Grechko, then holder of the world record total of 126 days in space, and of Gubarev, his companion aboard Salyut 4 and for part of the Salyut 6 flight.

A month later, at a Moscow news conference called to give details of future Salyut 6 flights, it was stated that Romanenko and Grechko had both lost 5kg in weight. They had found standing erect rather a strain, but Grechko had no repetition of the chest pains he had experienced after his Soyuz 17/Salyut 4 flight. They had spent 10-12hr per day in walk-around lower-body negative-pressure suits which created additional work for the cardiovascular system by pulling blood to the lower extremities. They had also spent 1-3hr per day exercising. Both lost 1·5-2cm from leg circumference as a result of inactivity of the leg muscles in zero g, and had reduced heart volume. The leg and heart data were comparable with those of the Skylab crews, who did not use lower-body negative-pressure devices.

Grechko said that small gravity loads - presumably resulting from thruster firings or the cosmonauts' own movements inside the space station - had reduced the effectiveness of some of the materials-processing experiments. And it was 3yr later before we learned that Romanenko had suffered from toothache during the flight and had had to be given a course of treatment in consultation with doctors at Flight Control. For this they used the Salyut first-aid box, which cosmonauts are forbidden to open without first asking permission. As a precaution, additional medical supplies were sent up in Progress 1.

Soyuz 27 L 10 Jan 1978, with commander Lt-Col Vladimir Dzhanibekov making his 1st flight and flight engineer Makarov making his 2nd, this completed the 1st triple-vehicle space station, described above. Docking with Salyut 6's forward port was routinely completed on the 17th revolution, finally proving that it had not been damaged by Soyuz 25. Salyut 6 was then in a 338 × 343km orbit. Solar corona and zodiacal light observation and Earth photography were included in their 5 days of joint work with the Soyuz 26 crew. But though they delivered messages, presents and letters from Earth, and took back

Launch of Soyuz 27, which would go on to make history with the first double docking at a space station. *(Tass)*

Soyuz 28: Col Alexei Gubarev (lower) and Capt Vladimir Remek (Czech). It was the 1st Intercosmos manned flight. *(Novosti)*

Soyuz 27: Vladimir Dzhanibekov, left, and Oleg Makarov. *(Novosti)*

scientific experiments, they apologised after landing 310km W of Tselinograd on 16 Jan for forgetting to bring their fellow cosmonauts' mail back as well. Their 94-revolution flight lasted 6 days 5min.

Soyuz 28 L 2 Mar 1978, this was the 1st international spaceflight, with commander Alexei Gubarev making his 2nd flight, accompanied by cosmonaut-researcher Vladimir Remek, a Czech military pilot who became the first non-Soviet or American to get into space. The orbit, initially corrected to 269 × 309km, took them smoothly on the 18th revolution to docking with Salyut 6 (orbit on 8 Mar was given as 338 × 357km). Taking part in the Intercosmos research programme, "to check the capacity of an international crew to work successfully in space," the crew were given only light tasks, including the growing of super-pure crystals. More details can be found under Soyuz 26 above. After 6 days aboard Salyut and a 124-revolution flight of 7 days 22hr 17min they landed 310km W of Tselinograd.

Soyuz 29 At 20.17 GMT on 23 Sep 1978 mission commander Col Vladimir Kovalyonok and flight engineer Alexander Ivanchenkov became the world's 1st spacemen to spend 100 continuous days in orbit. Launched on 15 Jun, with re-entry on 2 Nov, this 139-day mission also saw Russia pass the total of 1,000 man-days in space, compared with America's 937 at the end of Skylab. The long-term significance of this eventful flight was at the time missed or largely ignored in the West. It gave Kovalyonok a second chance following the failure of his 1st command, Soyuz 25, to make the 1st docking with the much improved Salyut 6. Despite that failure he had been promoted, while Ryumin, his original flight engineer, had been replaced. (However, if Ryumin did indeed spend some time out of favour, he went on to high achievements on Soyuz 32 and 35 and subsequent promotion to flight director of Salyut 7.) Ivanchenkov had had a 5yr wait for his 1st flight since being named in 1973 as an ASTP back-up crew member.

Soyuz 29 docked at the forward port, with Salyut 6 in a 339 × 353km orbit. The crew began work by replacing a defective ventilator on the Splav materials-processing furnace in the airlock, and activating the water recycling system. By their 12th day they had completed a 3-day Splav experiment which was related to future plans for large structures in space. During this period Salyut 6 was gravitationally stabilised, all thruster firings for the period being eliminated. Past experience had shown that thruster firings, and even the movements of the cosmonauts, would generate some gravitational forces and affect attempts to create new alloys in zero-g conditions.

28 Jun saw the arrival of Soyuz 30, the first of 2 manned Soyuz and 3 unmanned Progress ferries which kept Kovalyonok and Ivanchenkov continuously busy throughout this arduous mission. The 2nd international flight, Soyuz 30 took the 1st Pole, Maj Hermaszewski, into space. He and commander Pyotr Klimuk joined the Soyuz 29 crew in producing a cadmium-tellurium-mercury semiconductor in Splav, and returned to Earth with that and the results of other experiments after 8 days.

By the time Progress 2, the next unmanned ferry, had been launched on 7 Jul and docked after 25hr 33min and 17 revolutions, Soviet scientists were talking confidently of using such resupply missions to keep Salyut 6 operational for up to 5yr. Transfer of fuel - this time only about 600kg - for the main propulsion and attitude-control engines was on this occasion conducted almost entirely under ground control. In addition to 200lit of water and other cargo, a new kiln (Kristall) for processing optical glass was delivered. This allowed glass samples to be suspended in 0g while melting took place, ruling out the influence of sidewall contact with the glass. Relieved of much of the fuel transfer work which had occupied the Soyuz 26 crew, Kovalyonok and Ivanchenkov were soon devoting their time to glass fusing, attempting to produce a monocrystal of gallium arsenide. While they were doing so Moscow scientists were examining the 47gm of semiconducting alloy produced in Splav and taken back by Soyuz 30.

On 29 Jul came yet another highlight. In a 2hr 5min EVA during which they wore the same EVA suits and portable life-support systems used by the Soyuz 26 crew the previous December, they retrieved experiments placed outside before Salyut 6 was launched. These included a micrometeorite sensor, cassettes carrying bio-polymers, plates registering the amount of radiation impinging on Salyut 6, and samples of rubber, plastic and other materials for use in future spacecraft. They also attached a new radiation detector. Over the S Pacific the cosmonauts had to use hand-held lights while working in the dark. During the daylight periods at the start and finish of what seems to have been a problem-free spacewalk they carried a portable colour TV camera with which they sent back excellent pictures of themselves and of the Earth 346km below. Having finished their work ahead of schedule, they rejected Flight Control's advice to re-enter the Salyut. "We would just like to take our time," said Kovalyonok. "This is the first time in 45 days that we have got out into the street to take a walk." The Progress 2 ferry was retained until after the spacewalk to make maximum use of its supplies to restore the atmosphere lost during depressurisation. Before it was loaded with waste and undocked on 2 Aug, the crew carried out further tests of the oscillations caused by activities inside the 3-vehicle combination. Progress 2 was then commanded to a destructive re-entry over the Pacific on 4 Aug.

Progress 3 was launched only 4 days later and docked on 10 Aug at the end of revolution 32 under ground control, with Kovalyonok and Ivanchenkov merely observing from inside the Salyut. This time there was no refuelling, but 280kg of food and 450kg of oxygen were included in supplies intended to build up the space station's long-term reserves. Soviet estimates were that Progress 3 alone carried enough supplies for 2 men for 31-45 days, indicating consumption of 20-30kg of food, water and oxygen per day by a 2-man Salyut crew. The resupply flights, it was disclosed, had enabled some improvements to be made in crew diet. There were canned meats that could be heated and other foods in aluminium tubes, amounting to 10 varieties in all and increasing the daily calorie intake from 2,900 to 3,200. Letters, newspapers and presents, including a guitar for Ivanchenkov, were unloaded first. (It seems that the guitar was received without enthusiasm by Kovalyonok: I could find no further reference to it in the progress reports on the remainder of the mission!) So many improvements had been made in cargo-delivery techniques that Progress 3 delayed normal work far less than its predecessors. The day after its arrival the crew once again switched off all engines for a semi-conductor experiment in Splav. Much activity with the E German-built MKF-6 multispectral camera was also reported. Each image covered 220 × 165km, with each cassette carrying enough film to cover 16 million sq km. Moscow announced that space photography had helped to discover underground water reserves of up to 4 billion cu m in the desert area of Mangyshlak Peninsula, on the E coast of the Caspian.

Progress 3 remained docked for 12 days for unloading of cargo and reloading with waste. On 17 Aug its fuel reserves were used to raise the whole complex to 343 × 358km; on 21 Aug it was undocked and 3 days later placed on a burn-up trajectory over the Pacific. Two days after that, with the Soyuz 29 crew on their 76th day, Soyuz 31 was launched, carrying the 3rd international crew. Commander was the veteran Valeri Bykovsky, with Lt-Col Sigmund Jahn of the GDR Air Force as cosmonaut-researcher. After 7 busy days, with Jahn concentrating on photography of the Earth and oceans, the Soyuz 31 crew returned on 3 Sept in Soyuz 29, which had spent 83 days docked with Salyut 6. The long-stay crew were thus left with a fresh spacecraft. Experimental results taken back by Bykovsky and Jahn included not only the photographs but zero-g crystal samples and the results of tests of man's ability to distinguish subtle nuances of sound in orbit and to react speedily in such conditions to commands given by either men or machines. The latter experiments clearly have a military significance.

Yet another first was achieved on this remarkable mission 4 days after Bykovsky and Jahn left. In a bold manoeuvre which took 2 days of preparation Kovalyonok and Ivanchenkov entered Soyuz 31, undocked and backed off to a distance of 100-200m. Salyut 6 then did a half-somersault, at first drifting round naturally in its gravitationally stabilised mode and then being powered with the minimum use of fuel to complete the 180° turn. Salyut 31 moved in again and redocked with the transfer compartment. One report said this took 40min, another 90min. The object was of course to clear the instrument compartment docking port for yet another Progress visit. Then came a fortnight of routine work, operating the Kristall furnace to produce lead telluride crystals (a semi-conductor material) and observing the Earth's atmosphere. There were more highlights when on 20 Sep the crew broke the previous record of 96 days in space and on 23 Sep they became the first men to complete 100 days in space. Soviet doctors, expressing satisfaction with their health, said that they seldom used Salyut's first-aid kit but preferred to buck themselves up with "a tonic drink made from eleuthera bark". Occasionally the crew complained about their food. According to one of their doctors, although they were supplied with "luxury tinned foods" after a while they were asking for "anything fresh," particularly meat, bacon fat, bread, onions, garlic, potatoes and gherkins. On 17 Sep, for only the 2nd time in flight, they took a shower, moving within range of the TV camera so that they could be observed wearing a transparent plastic mantle, mask and goggles. It seems likely that the doctors were concerned about the possibility of ill-effects from inhaling the weightless water-drops. On 28 Sep Ivanchenkov celebrated his 38th birthday. Progress 4, for which they had so successfully prepared, was launched on 4 Oct. Unloading was spread over 10 days and because it involved a good deal of physical work the crew were instructed to vary it with observations, for the 5th week, of the timber massifs of Siberia and the Far East. They used a manual spectrometer which enabled them to distinguish between deciduous and evergreen trees, and to assess timber stocks and fire damage. A trained eye was needed, since from space the forests look brown rather than green. Progress 4's cargo of over 2,000kg included 300kg of food, 46kg of air and 176kg of drinking water. There was clothing and footwear, since the cosmonauts had worn out their shoes on the running track. There were fresh music cassettes to help relieve the boredom experienced by the crew when taking exercise. Having been reloaded with waste material, Progress 4 was used to push Salyut 6 higher and then cast off on 24 Oct; it was steered to a safe re-entry over the Pacific 2 days later.

Kovalyonok and Ivanchenkov then began intensive

preparations for return to Earth. Not only did they have to mothball Salyut 6 for what proved to be nearly 4 months of unmanned operation, but they also had to prepare themselves physically with long sessions in their Chibis vacuum suits for the resumption of life in 1g conditions. Their record flight ended with a landing in Soyuz 31 at 14.05 Moscow time on 2 Nov 180km SE of Dzhezkazgan in Kazakhstan. By then Salyut 6 had been in orbit 13 months and had made over 6,300 revolutions. The results of this mission, during which they travelled 90 million km, included 18,000 Earth pictures; over 50 experiments with 60 meltings had been conducted in the Splav and Kristall furnaces; and the crew had filled hundreds of pages with drawings and descriptions of their observations of phenomena in the atmosphere and on the Earth's surface. Specific observations included 7 forest fires in Australia, and the discovery of regular glacial movements in Latin American mountains. Perhaps the most reassuring aspect of the flight was the speed – within about 10 days – with which Kovalyonok and Ivanchenkov readapted to Earth conditions with no apparent physical ill-effects.

Soyuz 30 L 27 Jun 1978, the 2nd international flight and the 3rd for commander Pyotr Klimuk took Maj Miroslaw Hermaszewski of Poland into space. Docking was apparently routine, 22hr 27min into the flight and in a 264 × 310km orbit. Principal activities were materials-processing experiments in which they joined with the Soyuz 29 crew in combining cadmium, tellurium and mercury to obtain a new semi-conductor material, and Earth-resources photography of Poland. The results were brought back when the Soyuz 30 crew undocked on their 8th day and landed 300km W of Tselinograd after a 7-day 22hr flight.

Soyuz 31 The 3rd international flight, L 26 Aug 1978 with commander Col Valeri Bykovsky making his 3rd flight and Lt-Col Sigmund Jahn of the GDR as cosmonaut-researcher, was also intended to provide the Soyuz 29 crew, completing 3 months in orbit, with a fresh spacecraft for their return. Soyuz 31 was the 13th spacecraft to be launched to Salyut 6, including the 3 Progress resupply craft. Docking was achieved after 14 orbits at 271 × 326km. Working with Kovalyonok and Ivanchenkov, the Soyuz 31 crew carried out experiments aimed at developing new metallic and non-metallic materials in zero g, and brought back the results of many Salyut 6 experiments. They included detailed Earth photos, the results of tests of new photographic techniques inside the space station, examples of rare crystal growth in zero g, and the results of the sound discrimination and reaction tests mentioned under Soyuz 29. They returned in Soyuz 29 on 3 Sep after 7 days 20hr and 125 orbits, touching down at Dzhezkazgan, Kazakhstan.

Soyuz 32 This mission, extending Russia's experience of long-duration flight to almost 6 months, was a significant step towards manned flight to the planets. L 25 Feb 1979, it was commanded by Lt-Col Vladimir Lyakhov, making his first flight 12yr after selection. Flight engineer was Valeri Ryumin, apparently unshaken after his alarming experiences on Soyuz 25 17 months earlier. (His wife Natalya did however later describe the period after the failed mission as "the hardest in both his life and mine," with the cosmonaut unable to sleep as he "went over and over that miserable docking operation, trying to find what went wrong and not being able to.") On this occasion they docked without trouble. The first week was devoted to checking out the improved sanitary and hygienic arrangements, the running track and cycle machine, the Penguin weighted suits, the Chibis vacuum suits and other facilities gradually introduced during previous flights and which had finally made the 139-day stay of the Soyuz 29 crew possible. Another vital task was a photographic inspection of the portholes to make sure that there had been no dangerous erosion as a result of micrometeorite hits. Traces of a meteorite were found on one window but there was no serious damage. Sample materials recovered from the outside by Ivanchenkov and Romanenko had revealed over 200 impacts, indicating that far more dust and particles were striking Salyut than expected, though the likelihood of encountering a dangerously large rock was only a fraction of one per cent. Because only half as much fuel as expected had been used for the docking, Soyuz 32's engine was used on 1 Mar to raise the Salyut's orbit by 35km to 308 × 338km. The new orbit took Salyut 6 over the same areas on Earth every 2 days, as required for a number of scientific experiments. The crew used a soldering iron – another first in space – to repair the videotape recorder, and started growing cucumbers in the pots used by the previous crew to grow onions. The cucumbers were "more for the soul than for science," and dill, parsley and onions were also grown and eaten. Progress 5, L 12 Mar, docked 2 days later on its 32nd orbit. Docked at the Salyut's instrument bay, it was loaded with fuel and 300 replacement components to enable the crew to complete their comprehensive overhaul of the 18-month-old space station. Items replaced included the solar power batteries, the worn-out teletype and the walkie-talkie system used by the cosmonauts to communicate when 30m apart at opposite ends of the station. The most significant advance provided by this cargo ferry was a black-and-white TV set, which allowed the crew to receive visual advice and instructions and to view TV programmes (mostly videotaped) for recreation. Flight Control was rearranged so that one TV camera looked at the controller and another at his working desk, allowing him to show the crew technical documents, drawings and sketches. It was first tested on 24 Mar with the transmission of a page of *Pravda*. It also made family exchanges possible, with each side able to see as well as hear the other. An improved component for Kristall, the materials-processing furnace already used for 40 operations, was also delivered and installed.

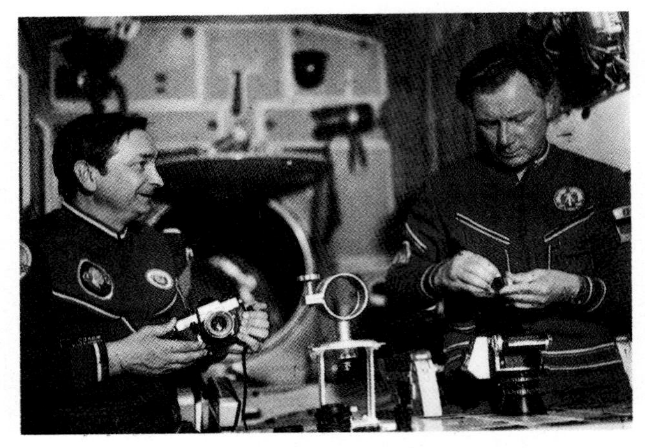

Soyuz 31: Soviet cosmonaut Bykovsky, left, with Sigmund Jahn (GDR) loading film during their 7 days in Salyut 6, during which they experimented with new photographic techniques.

Then came the most ambitious piece of space engineering yet tackled by the Russians, namely the isolation of one of the three Salyut fuel tanks. The membrane dividing liquid fuel and gaseous nitrogen had been damaged and it was feared that the two were becoming mixed. On 16 Mar Flight Control placed the whole Soyuz 32/Salyut 6/Progress 5 complex in a slow turn around its transverse axis, so using centrifugal force to separate the fuel and nitrogen in the defective tank. The crew were then able to transfer most of the fuel to another tank and the residue to an empty container on Progress 5. After that a valve was opened and the tank vented into space. Over the following 7 days the tank continued to be "vacuumed" and blown through. Finally it was filled with compressed nitrogen and isolated. Having ensured the safety of the ship, the cosmonauts were then able to move on to improving their personal comfort by replacing the elastic shell of their shower unit with the new one provided. They had every reason for the high spirits reported by Flight Control doctors.

All the same, it was almost a month before Lyakhov and Ryumin were ready to start work on technological experiments. By then they had settled down to a routine of 5 working days, starting at 8am and ending at 11pm, and 2 "rest days," which in fact were usually taken up with housekeeping and the daily routine of 2-2½hr of exercise. Every day too they had to reply to 25 medical questions, and every 8-10 days there was a very detailed medical examination. Lyakhov, who admitted to having been a smoker for 28 of his 38 years, said he was now "managing very well" without it. Their first experiment involved smelting semi-conductor material provided by France in the improved Kristall furnace. They also planted seeds in

This Soviet postage stamp commemorating Lyakhov and Ryumin's 175-day record also revealed first details of the spacewalk, during which Ryumin used pliers to free the jammed umbrella-shaped antenna so that the 2nd docking port could be re-used *(Graham Turnill)*

a small centrifuge because previous experiments had shown that while germination was normal, later development was disrupted by zero-g conditions. Dwarf trees were also planted. After its engines had been used for two more orbital corrections, boosting Salyut 6/Soyuz 32 into a 284 × 357km orbit, Progress 5, loaded with discarded components and waste materials, was undocked on 3 Apr, 21 days after its arrival, and placed on course for a destructive re-entry on 5 Apr. A miniature gamma-ray telescope designated Yelena, also delivered by Progress 5, was used for a long series of experiments, recording gamma rays and electrons possibly originating in black holes.

The crew had a week after the departure of Progress 5 in which to prepare for a week-long visit by a Bulgarian cosmonaut under the command of Rukavishnikov in Soyuz 33. When that mission subsequently failed much replanning was needed. While Flight Control worried about the reliability of Soyuz 32's engine, and whether it might have a fault similar to that of Soyuz 33, Lyakhov and Ryumin, with their 50th day in orbit approaching, busied themselves producing a monocrystal semiconductor material, indium antimonide, in the Kristall furnace. In a broadcast they pointed out that such in-orbit production was both practical and economic, since Russia's electronic industry needed only a few kilogrammes per year of certain types of semi-conductor, and production costs on Earth were extremely high. 10 samples of materials provided by France had been delivered by Progress 5, and by 27 Apr this series of tests had been completed. Semi-conductors, magnetic materials and metal compounds were included, and one experiment involved observations of diffusion during the melting and cooling of tin-lead and aluminium-copper alloys. According to a report in *Aviation Week*, a number of coded transmissions were sent both by telemetry and in conversation between the cosmonauts and Flight Control, suggesting that military observations were included in their duties.

When May Day arrived, 9 weeks into the mission, this crew became the first to enjoy the nearest thing possible to a holiday in space. For 5 days they had no scientific work to do and were able to watch reports of the Red Square celebrations and have TV conversations with their families. They made full use of their 64 sound and 50 videotape cassettes, the latter containing films and cartoons. But their 70 different foodstuffs were not on this occasion supplemented with the fresh fruit and vegetables brought up by the visiting international crews. Nor did the holiday include any exemption from the need to take at least 2hr exercise per day. In a lecture to the International Astronautical Federation in 1981 Ryumin made clear how much he had disliked these compulsory sessions on the treadmill and other machines.

Following the break the crew alternated metallurgical experiments with use of the BST-IM sub-millimetre telescope – the 1st time it had been used for 7 months – to measure changes in the Earth's atmosphere, identify fish shoals and detect forest fires. Work with the sub-millimetre telescope was so complicated, they reported, that they really needed the help of a 3rd person.

Progress 6, with another two months' fuel and other supplies, was launched on 13 May and safely docked under the control of the Soyuz 32 crew 2 days later. Its contents included tulip buds - which obstinately refused to flower. About 100 other items included letters, newspapers and presents, packed last so that they could be extracted first; materials for the Splav and Kristall furnaces; a replacement component for the computer; new lighting equipment for use in TV broadcasts; and electric light bulbs. Progress 6 also delivered materials for the planned Soviet-Bulgarian experiments which Soyuz 33 had been forced to take back to Earth. The Soyuz 32 crew carried out the most important of these, an attempt with the Splav furnace to obtain foam metals of low specific weight and high mechanical properties.

News that Soyuz 34, with a Hungarian cosmonaut aboard, had been postponed because of the Soyuz 33 failure came first from the Hungarians. But it was a surprise when Soyuz 34 was launched unmanned on 8 Jun, with Progress 6 still docked and no free port available on Salyut 6. However, Progress 6 was jettisoned on 9 Jan and after 4 main engine orbital manoeuvres Soyuz 34 successfully docked at the rear port. It was now revealed that Soyuz 32 - which had been in orbit 109 days, longer than any other Soyuz - was to be sent home unmanned, leaving the new vehicle for Lyakhov and Ryumin.

With a crew aboard, a Soyuz could carry only 50kg of cargo back to Earth; without a crew it was possible to increase this to 180kg. Flight Control felt confident enough of the craft's reliability to order that the load include 29 ampoules containing the results of the smelting experiments, 50 film canisters, biological experiments, and some defective equipment for examination. The latter would normally have been jettisoned in a Progress. Following undocking on 13 Jun, extra tests on Soyuz 32 meant that 3 orbits instead of the normal 2 were flown before it was placed on a re-entry course for a landing 295km NW of Dzhezkazgan.

With doubts about the Soyuz main engine not completely resolved, it was decided to postpone all further Intercosmos missions to Salyut 6 until the next long-duration flight. On 14 Jun Lyakhov and Ryumin completed a busy week by sealing themselves in Soyuz 34 and using the now well tried routine of backing off, commanding Salyut 6 to rotate 180° and redocking at the forward port. Their routine of experiments and observations then continued for another fortnight, the highlight being the spotting of an agitation in the sea on the Equator and about 300km east of the African coast. They described it excitedly as two waves colliding in an otherwise calm sea. Cosmonaut Grechko, at Flight Control, said that he had once seen a similar phenomenon.

Progress 7 was launched on 28 Jun and docked on 30 Jun. Its engines were fired on 3 and 4 Jul for 114sec and 75sec to raise Salyut 6's orbit to 399 × 411km, the highest ever for a Salyut. On 15 Jul, as they slept, Lyakhov and Ryumin passed the previous long-duration record of 139 days 14hr 48min - and the most exciting month of their long mission still lay ahead.

Not until Progress 7 had been jettisoned on 18 July after a stay of 18 days were we told that its cargo had included a radio telescope designated KRT-10 (the acronym stands for Cosmic Radio Telescope). It took the crew several days to lay out the mast along Salyut's axis, to unfurl the 10m parabolic mirror antenna, and to connect the multichannel, high-sensitivity receiving apparatus to the timing and recording instruments. We were not told exactly how it was "moved out into open space" - presumably through the docking port - but the operation was watched by means of TV on board Progress 7 as that craft drifted slowly away. Viewers at Flight Control saw the outline of Salyut 6 suddenly disappear as the antenna's dish suddenly opened. The Earth weight of such an instrument was said to be several tons. For use in space it needed to be only one-tenth that, and later the weight was said to be "about 300kg". Once calibrated by the cosmonauts and operated in conjunction with the 70m dish at the Crimea Long-Range Space Communications Centre at ranges varying from 400km to 10,000km, the instrument could achieve a sensitivity equal to that of a single telescope with a receiver the diameter of the Earth.

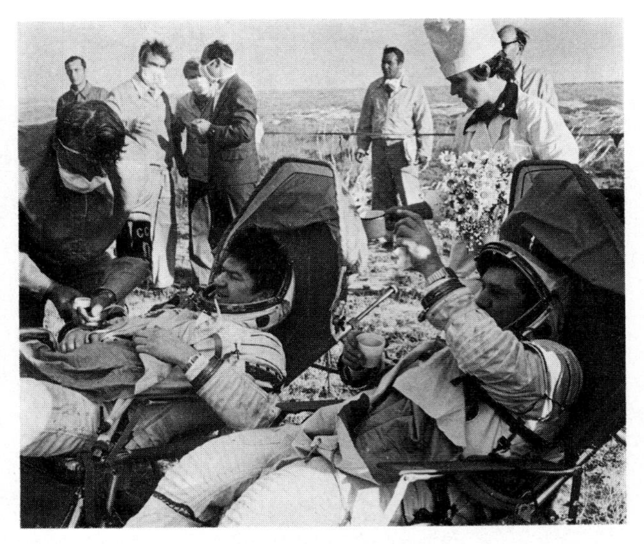

Ryumin and Lyakhov in reclining chairs after recovery. For a few days they had some difficulty in speaking. *(Novosti)*

For the following three weeks KRT-10 was used every other day, probably because its power consumption was so high that a further day was needed to recharge the batteries. Observation targets comprised the Milky Way, the Sun, Pulsar 0329 and deep-space sources of radio emissions. Studies of the Earth related to meteorological phenomena, soil humidity and moisture, and wave heights and sea salinity.

On 5 Aug it was stated that the mission was "approaching its end". As it turned out, however, an anxious fortnight still lay ahead. KRT-10 was due to be jettisoned on 9 Aug, but during the attempt the wire mesh of the antenna became entangled, possibly with Salyut's own antennas or engine nozzles. Over the next 6 days various attempts verging on the desperate were made to dislodge it; they included sharp forward manoeuvres of the whole Salyut/Soyuz complex and rocking the station around its axis. None worked and the cosmonauts, due before returning to go outside and retrieve samples next to the hatch, were eager to make an EVA to solve the problem. Flight Control was reluctant: it was the end of their longest mission and might be hazardous for a fresh crew, let alone two men with 170 days of weightlessness behind them.

Faced with the alternative of bringing the Soyuz 32 crew home with KRT-10 still entangled, blocking the refuelling port and ending Salyut 6's honourable career, Flight Control finally agreed to an EVA. Several days were spent preparing for it - rehearsals were carried out on Earth on a Salyut mock-up - and the crew's return was almost certainly delayed as a result. On 15 Aug Lyakhov and Ryumin donned the station's two EVA suits, used by two previous crews. They sealed themselves in the forward transfer compartment and depressurised it. Ryumin opened the hatch, which stuck at first, and floated out on a 20m line. Lyakhov remained just outside the hatch. A subsequent Soviet postage stamp gave us our first picture of what it all looked like. Because of the delay in opening the hatch the crew had to wait outside for 36min while they passed through the Earth's shadow. Ryumin said the stars "looked like huge diamond pins on black velvet; I felt I only had to stretch out my hand to hold one in my palm." When daylight returned he made his way along the Salyut's handrails and reached the entangled antenna. Standing on a projection from the space station, he cut the antenna free with four snips of a pair of pliers. A final kick from his boot sent the antenna tumbling into the distance. After that, in an EVA that took 1hr 23min, they brought inside for return to Earth instruments which had been registering micrometeorite hits for 2 years, as well as samples of optical and heat insulation materials attached to the outside.

After that Ryumin's 40th birthday next day was a particularly happy one. On 19 Aug, Russia's Aviation Day, they entered Soyuz 34. Cosmonaut Leonov, in charge of the recovery, cleared the airways and ordered high-tension power lines in the area to be switched off. After a flight lasting 175 days 36min they touched down 170km

Soyuz 33: Georgi Ivanov (Bulgaria), left, and Nikolai Rukavishnikov preparing for the 4th international flight. But they had to make an emergency return after failing to dock, and so far there has been no 2nd chance for Bulgaria.

SE of Dzhezkazgan. The hatch was opened and they were helped out into special reclining chairs while an inflatable tent was prepared for preliminary medical examinations. Lyakhov's weight had dropped by 5·5kg but Ryumin's was unchanged. Within 3 days they had recovered from the nausea and dizziness usually associated with a return to normal gravity, and had also overcome a rather alarming new effect - a slight difficulty in speaking. It was pointed out that in 18 years 92 cosmonauts had been into space, and almost half had made repeat journeys. Between them they had logged a total of 6 years in space.

Soyuz 33 The 4th of the international flights, this was a hazardous failure which led to major rearrangements of later flights. Marked by the issue of a special international spaceflight postage stamp, launch of the 6,800kg Soyuz 33 on 10 Apr 1979 took place in the worst conditions ever recorded, with winds gusting to over 60km/hr. "Regard it as a rehearsal for weightlessness," commander Nikolai Rukavishnikov, making his 3rd flight, told his Bulgarian companion, cosmonaut-researcher Georgi Ivanov. A 7-day stay in Salyut 6 with the record-breaking Soyuz 32 crew had been planned. Progress 5, which had re-entered 5 days before, had not only refuelled Salyut but had delivered equipment for experiments which included a Bulgarian-designed spectroscopic Earth survey (this was ultimately used by the Soyuz 35 crew). Ivanov was also said to have "a special surprise" in store for the resident crew, "something never before seen in space". At the time of writing the surprise remained undisclosed, possibly awaiting a subsequent, more successful Bulgarian mission. On the 18th orbit, after 5 course corrections, Soyuz 33 had reached a 273 × 330km orbit and was only 3km from Salyut 6, when Rukavishnikov reported that the main approach engine was malfunctioning. The engine had shut down 3sec into a 6sec burn, accompanied by abnormally high vibrations. It was decided to abandon the docking and prepare for a return 13 orbits later by using the reserve engine. A ballistic re-entry meant that the crew underwent 9g, and Rukavishnikov later described it as a difficult descent. "We were practically sitting inside a flame," he said. "I felt as if the craft had been thrust into the flame of a blowlamp." The landing, after a 23hr 59min flight, was made in darkness 320km SE of Dzhezkazgan. But there were no further problems and by the time the recovery helicopter reached the crew they were standing outside the spacecraft. Both men were warmly praised and much decorated for their handling of the emergency.

Soyuz 33: Maj Georgi Ivanov (Bulgaria) and Nikolai Rukavishnikov after their failure to dock with Salyut 6. This was the only failure in the 9 Intercosmos flights. *(Novosti)*

Soyuz 34 returns to earth with Soyuz 32's Lyakhov and Ryumin. It had been sent up unmanned as a safety precaution to provide them with a fresh craft at the end of their 175-day flight. *(Novosti)*

Soyuz 34 L 6 Jun 1979 unmanned to check modifications to the main engine following the failure 2 months earlier. When Soyuz 34 was launched its Salyut 6 docking port was still occupied by Progress 6, which had been sent up on 13 May. However, on the morning of 8 Jun the Progress was undocked and Soyuz 34 successfully took its place later that day. Progress 6 was placed on course for a destructive re-entry next day. But for the previous failure, Soyuz 34 would no doubt have carried another 7-day crew whose main task would have been to deliver a replacement spacecraft for Lyakhov and Ryumin, whose Soyuz 32 had been in orbit for 109 days, far more than its previously proved operational capability of 60 days. Once Lyakhov and Ryumin had checked out their new craft they packed Soyuz 32 with film cassettes, processed metals and other completed experiments. The spacecraft was then undocked on 13 Jun and successfully returned to Earth unmanned.

Soyuz 35 On this mission Valeri Ryumin came within 3 days of being the first man to spend a whole year of his life in space. L 9 Apr 1980, it took Ryumin back to Salyut 6 less than 8 months after he had returned to Earth aboard Soyuz 34. The 185-day flight broke his own previous record by 10 days. It was a bold decision to send him back for another flight so soon, and an important step towards gaining experience of the effects of Mars-duration flights on cosmonauts. Ryumin's inclusion seems not to have been entirely planned, since he replaced flight engineer Valentin Lebedev, who had been waiting 7 years for his 2nd flight since Soyuz 13. But Lebedev injured a knee during training on a trampoline and needed an operation. Despite his experience, Ryumin was not given command: that had already gone to Leonid Popov, making his 1st flight.

Soyuz 35 completed the 8th Soviet manned docking, linking up with Salyut 6 25hr 28min after launch; 2½hr and 2 orbits later the crew entered the space station. Ryumin had the strange experience of opening a "welcome aboard" letter he himself had written for the next crew before leaving the previous August. The first 4 days were mainly concerned with reactivating Salyut 6's

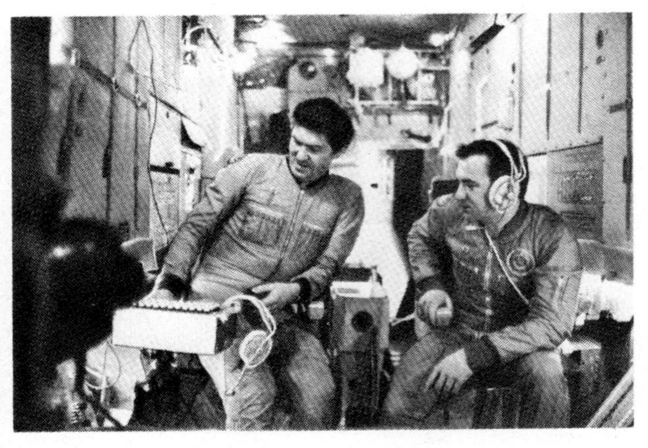

Soyuz 35: Valeri Ryumin (left) and Leonid Popov preparing for their 185-day flight, starting in April 1980. It brought Ryumin's total space time to 362 days.

hot water and other systems and setting up a miniature "greenhouse" (Malakhit). Carried up aboard Soyuz 35, it was intended mainly for the growing of orchids, which it was thought would not only yield useful data but "also increase the cosmonauts' sense of comfort and well-being". Its main purpose, to solve the riddle of why flowers refused to bloom in zero g and to find ways of obtaining oxygen and nutrients from plants during long flights, was not achieved. Much of the first 2 weeks was spent on maintenance and repairs, for Salyut 6 had then been in orbit for over 30 months and had passed 14,500 revolutions. It was not until the 5th day that they began to unload Progress 8, which had preceded them by 11 days and been used to trim the orbit for their docking. The supplies included an 80kg replacement battery for the steering unit. Before Progress 8 was undocked on 25 Apr its engine was fired for 81sec to raise the Salyut apogee by 20km. Then, only 1 day after 8's destructive re-entry, Progress 9 was launched to deliver additional replacement components ordered by Popov and Ryumin. Among its supplies was a device for moulding polyurethane: one of the tasks of this crew was to produce small Mischa Bear symbols to be taken back to Earth by visiting crews for presentation to favoured guests during the forthcoming Moscow Olympics. A more practical advance was the pumping of 180kg of water into a Salyut tank to replace the previous 5kg water bottles.

The end of April to mid-June was a busy period featuring much precisely timed work. Progress 9's load of equipment and consumables was cleared by 20 May, when it was undocked after the routine use of its engines to raise the Salyut orbit. Amid the regular processing experiments was a new one, the production of polyurethane foam to see if this light but strong material could be used for building large space structures. Two chemical components were mixed in a vessel to produce the foam, which was then "poured into a mould". No indication was given as to whether artificial gravity was used to help this process.

Seven days after Progress 9's departure Soyuz 36 arrived, bringing the Hungarian Bertalan Farkas under the command of Valeri Kubasov. Hungarian-supported experiments included studies of the formation of interferon in human cells under zero g, intended to lead to means of combating viral infections; cultures used were Interferon 1, derived from human cells and a compound to boost the production of the protein; and Interferon 2, "consisting of interferon in powder, ointment or emulsion form". An experiment which may have been less popular with the crew was a questionnaire in which the cosmonauts were required to assess their own psychological state by answering questions covering 9 areas. The Soyuz 36 crew departed on 3 Jun in Soyuz 35, leaving Popov and Ryumin with a fresh spacecraft. The next day Popov and Ryumin entered Soyuz 36, undocked and backed away 36m. Salyut 6 was then turned through 180° so that they could re-dock with the forward port, thus clearing the aft port for the imminent arrival of Soyuz T-2 and future Progress missions.

Three days later Soyuz T-2 docked without difficulty. Crew for this first manned flight of the new 3-man craft

Soyuz 35: Valeri Ryumin and Leonard Popov relax after their 185-day flight. *(Novosti)*

was Lt-Col Yuri Malyshev and Vladimir Aksyonov. Lasting 4 days, of which only 3 were spent aboard Salyut 6, this mission appears to have been devoted entirely to trying out the new vehicle, its systems and new lightweight spacesuits.

The resident crew settled down again to their materials-processing routine, varying it with tasks such as the observation of forest areas of Russia which were prone to fires. On 24 Jun it was reported that Salyut 6 had been in orbit for 1,000 days and had been manned for almost half of that time. On 1 Jul the crew carried out experiments with the Isparitel ("Evaporator") device, in which an electronic gun was used to apply metallic coatings in a vacuum; examples of such coatings are the aluminium film covering the mirrors of telescopes in a space station, porthole coverings and the thermo-regulating coating of the craft itself. This work was interrupted by the arrival of Progress 10. L on 29 Jun, it docked on 1 Jul; 30% of its cargo had been ordered by Popov and Ryumin. Unloading and refuelling were fitted in with the regular work, and after the Progress had been refilled with refuse in its final role as a celestial dustbin it was as usual used to adjust Salyut's orbit to 228 × 355km before being jettisoned and destroyed on 19 Jul. This placed Salyut 6 in the correct position to receive Soyuz 37, which docked on 24 Jul. That crew departed nearly 8 days later in Soyuz 36, once again leaving the long-stay crew with a fresh spacecraft. On 1 Aug they repeated the now familiar routine of switching Soyuz 37 to the other end of Salyut 6 in order to free the service bay.

During August some of the discussions between Flight Control and Popov and Ryumin sounded like radio gardening programmes on radio. Biologists were called in to discuss why the many different plants cultivated during Salyut 6's 3-year life had died as they reached the flowering stage, despite variations in their conditions and the addition of a wide variety of nutrients. Certain crops, including radishes, carrots, cucumbers, peas and wheat, were however being grown with some success. It had been established that the seeds had to be planted in the right direction, towards the light source, though the periods of light and darkness when they were placed beside a porthole were quite different from those on Earth. On each 90min orbit there was 60min of daylight and 30min of darkness, and it was thought that these frequent sharp changes might be having a bad effect. Finally, on 2 Sep, it was reported that the crew had achieved another notable first: for the first time a plant - an arabidopsis - had been persuaded to flower in conditions of weightlessness.

When in mid-August Ryumin became the first man to complete 300 days in space, Soviet doctors were at last beginning to feel confident that early health problems caused by weightlessness could be overcome. Both cosmonauts were maintaining a high work rate, and for the first time showed no loss of weight. On the contrary, Popov had gained 2kg and Ryumin 1kg. But the maintenance of good health demanded much time spent on exercises: each man had to walk 4½km and run another 4km on the treadmill, or do an equivalent amount of exercise on the bicycle. During these periods the Chibis suits were worn. These are ribbed and made of rubberised plastic, and when air is evacuated from within the resulting vacuum facilitates the blood flow to the legs.

By early Sep, with 5 months behind them, the crew had performed 50 smelting operations using the Splav and Kristall electric furnaces, as well as photographing 100 million sq km of the Earth's surface. In a striking example of the progress made in materials-processing techniques, the infra-red viewers sent up to monitor their body temperatures included a semi-conductor produced during earlier experiments aboard Salyut 6 and then presumably sent back to Earth for use in the instrument. This was KRT, obtained by fusing cadmium telluride and mercury telluride. Leonid Kurbatov of the USSR Academy of Sciences said that KRT had been obtained aboard Salyut 6 "in its almost ideal crystal structure" for the first time, and was a "triumphant advance in infra-red technology". On Earth these two ingredients are prone to stratification when fused, making it expensive and time-consuming to produce the perfect KRT crystals needed for the production of semi-conductors. It is also expensive to do it in space: the Soyuz 35 crew carried out their smelting experiments in a capsule furnace at a temperature of about 1,000°C and a pressure of 100 atmospheres. Droplets of mercury were applied to the outside of the furnace, and their evaporation balanced the pressure inside. The results "were definitely better than on Earth, and this method of production has economic possibilities". Moreover, all the cyrstals produced in space could be used, while only part of the Earth-produced alloy was usable.

Having been prepared by the cosmonauts, crystallisation trials were usually carried out during sleep periods because even careful movement by the crew imparted some gravitational forces to the space station. Later, as knowledge accumulated and Soviet scientists began to develop designs and operating methods for future space smelting and crystal-producing factories, the Vrashcheniye ("Rotation") experiments were embarked upon. They called for the Salyut to be rotated like a centrifuge to permit studies of the effects of directed micro-gravitation on the crystallisation of solid solutions of metals. On 8 Sep the Rezonans experiment was carried out, with one cosmonaut running or walking on the running track and deliberately trying to shake the space station. It was at this point that the crew reported heavy micrometeorite damage to the screen vacuum insulation protecting the solar battery panels from heat radiation.

On 19 Sep yet another visiting crew arrived in Soyuz 38. Commander Yuri Romanenko, a previous long-stay resident of Salyut 6, was accompanied by Cuban cosmonaut-researcher Arnaldo Mendez. When they left after nearly 8 busy days it was in their own spacecraft. With the main crew nearing the end of their mission a new spacecraft was not needed, and the departure of Soyuz 38 left the rear docking port clear for a final supply craft. Progress 11 was launched on 29 Sep, the 3rd anniversary of the launch of Salyut 6, which by then had made over 17,000 orbits and travelled 700 million km. It docked on 1 Oct; Popov and Ryumin completed unloading by 4 Oct and began preparing to return to Earth. 4 Progress ferries had brought a total of 10 tonnes of supplies to Salyut 6 during their stay. Unloading and repair work had occupied about a quarter of their time, with Earth observations topping the list of their scientific activities. Using a 6-zone MKF-6M camera, a KATE-140 topographic camera, a Bulgarian spectrograph, SPEKTR-15, and several conventional cameras, they had covered more than 100 million sq km of the Earth's surface and oceans. In addition, mission director Dr Yeliseyev said that nearly 100 samples of new materials, including semi-conductors, and over 150 samples of surfacing had been obtained in the Splav and Kristall equipment, with some space-grown crystals already being used in experimental instruments.

Leaving Progress 11 still docked, Popov and Ryumin finally landed in Soyuz 37 180km SE of Dzhezkazgan on 11 Oct at the end of a 184-day 19hr 2min mission. The initial medical examination produced a surprising result: for the 1st time spacemen had not lost weight, with Popov gaining 3·2kg and Ryumin 4·7kg. This was ascribed to the more regular and well ordered conditions aboard Salyut 6, their excellent appetites and their high level of physical activity. This included up to 5km each each per day on the running track. No doubt the fresh vegetables and fruit delivered by 3 Progress ferries and 3 manned Soyuz visits during their 6 months in orbit had also contributed.

Soyuz 36 The 5th international flight, L 26 May 1980, took the 1st Hungarian, Capt Bertalan Farkas, on an 8-day mission to Salyut 6. It also provided Valeri Kubasov, making his 3rd flight, with his 1st command. He had been flight engineer on Soyuz 6 in 1969 and on Soyuz 19 (ASTP), and it is rare for a Soviet flight engineer to get a command. The 10 experiments which they carried out with Popov and Ryumin included materials processing with gallium arsenide crystals and chromium; studies of the misalignment of instruments caused by the 300°C difference between the sunlit and shadowed sides of the space station; solar photography, and monitoring of the cosmonauts' psychological state. They returned in Soyuz 35 on 3 Jun, landing 140km SE of Dzhezkazgan.

Soyuz 37 The 6th Intercosmos flight, L 23 Jul 1980. Commander Viktor Gorbatko, making his 3rd flight, was accompanied by Vietnamese cosmonaut-researcher Pham Tuan. Docking with Salyut 6 took place 25hr 29min after launch. A joint message from the Soviet Union and Vietnam pointed out that the Olympic flame was burning in Moscow for the 22nd Games. During their flight, lasting nearly 8 days, this crew worked with Popov and Ryumin on a number of experiments relating to Earth resources and the environment, and photographed areas of Vietnam in connection with studies of tidal flooding and erosion on the coastline. Also studied were silting in river mouths and its effect on fish feeding habits, and hydrological features of the Mekong and Red River basins. Azola water ferns, fast-growing nitrogen-rich plants from SE Asia which are used as a fertiliser in growing rice, were taken to Salyut 6 to see if they would be useful on long flights. Pham Tuan himself was the subject of a breathing experiment. Return was in Soyuz 36 on 31 Jul, landing 180km SE of Dzehezkazgan.

Soyuz 36: left to right, Bertalan Farkas (Hungary), Popov, Ruymin and Kubasov in Salyut 6. They carried out materials processing and solar photography.

Soyuz 37: Lt-Col Pham Tuan (Vietnam) and Viktor Gorbatko are congratulated after their return. Tuan studied the effects of wartime defoliants on Vietnam's forests. ***(Novosti)***

Soyuz 38 The 7th Intercosmos flight, L 18 Sep 1980, carried the first Cuban, who was also the first black man, to go into space. He was Lt-Col Arnaldo Tamayo Mendez; Commander was Yuri Romanenko, who had spent 96 days in Salyut 6 on the Soyuz 26 mission. Docking went smoothly after 25hr 38min, and the crew was soon aboard receiving congratulations from Leonid Brezhnev and Fidel Castro. The 2 leaders described the flight as "a new contribution to the strengthening of Soviet-Cuban friendship" which would facilitate the solution of economic problems in the USSR and Cuba. This was a reference to the fact that the 15 experiments allotted to this crew included a study of Cuba's natural resources from space. For the "Support" experiment Mendez was wearing Cuban-made shoes designed to retain the pre-weightless

Soyuz 38: Romanenko (right) with Cuban cosmonaut Arnaldo Mendez. 15 experiments on their flight included studies of Cuba's natural resources.

shape of his feet: we learned for the first time that after long missions cosmonauts had a tendency to develop flat feet when they returned to Earth. On the 3rd day Mendez reported to the doctors that he was "feeling OK", though not as well as he had on the previous days. He was heard clearly conducting the re-entry preparations, but there appeared to be no mention of the crew's health on their return in the original spacecraft on 26 Sep, so it seems possible that Mendez was a victim of spacesickness. Like the launch, the landing was carried out in darkness, 175km SE of Dzhezkazgan. Gen Shatalov gave the crew advance warning that there might be a "slight delay" in the arrival of the rescue helicopters.

Soyuz 39 The 8th Intercosmos flight, L 22 Mar 1981, was crewed by Col Vladimir Dzhanibekov, making his 2nd flight, and cosmonaut-researcher Jugderdemidiyn Gurragcha of Mongolia. The latter's fluent Russian made possible an ambitious programme in which the cosmonauts worked with ground stations and aircraft to assess Mongolia's ore, oil and gas resources, tectonic faults and highland pastures. For this they used a visual polarising analyser of Soviet-Mongol design, having first checked how far the optical characteristics of the portholes had been affected by prolonged space exposure. They also participated in TV exchanges of holographic images (see Soyuz T-4 for further details). They landed 170km SE of Dzhezkazgan on 30 Mar after a 7 day 20hr 43min flight.

Soyuz 39: Vladimir Dzhanibekov, left, with Jugderdemidiyn Gurragcha (Mongolia) during rehearsals. They studied Mongolia's oil resources and tectonic faults.

Soyuz 40: Dumitru Prunariu (Romania) and Leonid Popov. Theirs was the last mission in the original Soyuz series. *(Novosti)*

Soyuz 40 L 15 May 1981, this was the 9th and last Intercosmos flight, completing the promised programme 2yr early. The commander, Leonid Popov, added another 7 days to his former world record total of 185 days. With him as cosmonaut-researcher was Sr Lt Dumitru Prunariu, 28, of the Romanian Air Force. Although 4 engine firings proved necessary, a routine docking early in the 17th orbit enabled them to join the Soyuz T-4 crew, Kovalyonok and Savinykh, aboard Salyut 6. A new experiment provided by Romanian scientists, who have specialised in studies of the Earth's upper atmosphere and changes in its magnetic field, included dielectric detectors "to catch super-heavy nuclei and establish the existence of a state of matter which had so far only been predicted theoretically". When they landed, 225km SE of Dzhezkazgan after a flight lasting 7 days 20min 38sec, it was announced that the 2-man Soyuz had been used for the last time.

Soyuz T

History Although Soyuz T was described by cosmonaut Aksyonov as a "ship of the next generation," Gen Shatalov said when it finally carried its first 3-man crew that it had not been given a new name because "in form, volume and weight it remains the same old Soyuz craft". All the same, it had taken 9yr following the Soyuz 11 disaster in 1971 before Soviet scientists felt confident enough to return to 3-man flights in the modernised version in late 1980. Cosmos 1001 and 1074 in Apr 1978 and Jan 1979 were probably unannounced tests, and the cautious approach to the new craft was evidenced by the fact that T-1, taken through a complete 100-day cycle of launch, docking with Salyut 6 and return, was unmanned. There was no official confirmation that it was a 3-man craft until T-3, though in the meantime there had been frequent references to the desirability of having 3-man crews in Salyut space stations because 2 men were kept so busy with operational activities that insufficient time was left for scientific work. When T-3 was launched we were told that total lift-off weight was 300 tonnes; spacecraft, crew and equipment alone totalled 7 tonnes. The 3-stage T-3 launcher stood 50m high, with the spacecraft and 3rd stage alone standing 15m high.

No doubt the docking failures of Soyuz 15, 23, 25 and 33 with Salyuts 3, 5 and 6, necessitating emergency returns, had added to the delay in returning to 3-man crews. The most obvious change in Soyuz T was the return to the use of solar panels to provide power for separate flight for longer than 2 days if there were docking problems. Other changes, as demonstrated by T-1, included an autonomous onboard computing complex capable of flying the complete mission without human assistance. At the same time, provision of a digital computer with visual display meant that cosmonauts would also have much more ability to fly their own craft instead of relying on Flight Control to do it for them. Soyuz T was also given a combined engine installation compatible with that of Salyut 6, enabling the main engine and 4 attitude-control thrusters to use the same fuel supply, and giving higher thrust and greater manoeuvrability. Another effective economy measure was to jettison the orbital module before instead of after the re-entry burn, reducing the fuel required for that manoeuvre by 10%. The craft had a stronger, lighter frame and a new parachute system for the final descent. A new rescue system to cope with launch failures, "involving a new engine running on solid fuel and automatic equipment," had also been added, no doubt as a result of the aborted Soyuz 18 mission in 1975. Further evidence of the importance being given to the cosmonauts' ability to monitor and control their own survival was the fitting of jettisonable outer panes to the descent module portholes. This enabled crews to get rid of the outer windows, sooted-up by re-entry heating, in order to obtain a better view of the approaching Earth through the inner panes. Finally, the many minor improvements included colour TV in place of the previous black-and-white system, and almost certainly Apollo-type couches and layout. However, when T-6 was displayed at the 1983 Paris Show it was apparent that there is still barely enough room for 3 men crowded together in the launch and re-entry module.

Soyuz T is designed for use unmanned or with crews of either 2 or 3. In the latter configuration the commander sits in the centre of the descent module, with the flight engineer on his left and the cosmonaut-researcher on his right. Welcoming the return of 3-person crews, Makarov looked forward to 6-strong teams on Salyuts, with round-the-clock shift work and every member "having his own favourite corner, as if at home".

Soyuz T-1 L 16 Dec 1979 unmanned. Before docking with Salyut 6 on 19 Dec, 3 days of extensive manoeuvres were carried out, starting behind the space station and much further away than is typical of Soyuz and Progress rendezvous manoeuvres. T-1 was finally pulled ahead of and above Salyut 6, then dropped in front to dock with the forward port. Undocking on 23 Mar was followed by 2 further days of autonomous manoeuvres before a night landing on 25 Mar, completing a mission of 100 days 9hr.

Soyuz T-2 L 5 Jun 1980, only 2 days after the Soyuz 36 crew had returned from Salyut 6, this 1st manned mission docked 24hr 39min after launch. Commander was Lt-Col Yuri Malyshev, making his 1st flight, with Vladimir Aksyonov as flight engineer. Aksyonov, who had previously flown on Soyuz 22, had specialised in new spacecraft systems for many years. Pre-docking manoeuvres, both manual and computer-controlled, were carried out to test the new systems, and Malyshev used manual control for the final 180m and the actual docking. Mission director Dr Yeliseyev later said that this was not intended: for some reason the planned automatic approach sequence was unfamiliar to the crew and Flight Control personnel. Afterwards it was found that the proposed automatic mode would have worked if it had been allowed to continue. Mention of manoeuvres to test the solar batteries provided confirmation that Soyuz T had returned to the use of solar panels so that flights would no longer be restricted to 2 days, with a docking failure necessitating an immediate return. Rescue facilities and the new life-support systems were also checked during the 4-day flight. The new spacesuit, weighing 8kg compared with Gagarin's 20kg suit, is made of more advanced plastics, with new hinges for elbow, wrist and knee joints, and with better vision from the pressure helmet. So easy is the new suit to work in that a cosmonaut wearing one can pick a single match out of a box.

After undocking the crew were able to fly around Salyut 6 for photography and inspection purposes – the 1st time a Soyuz vehicle had had enough fuel for such manoeuvres. (US spacecraft had been able to make flyarounds since the Gemini programme in the mid-1960s.) Finally, a new

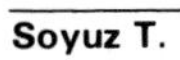
Soyuz T.

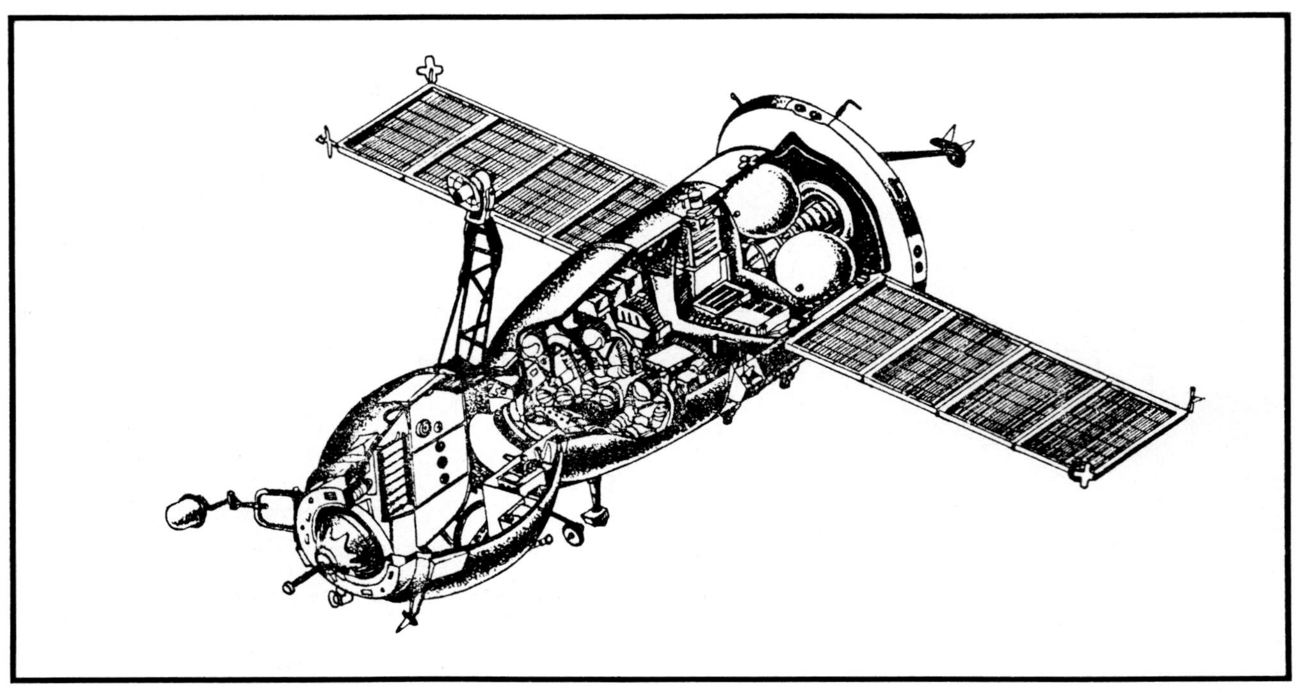

Soyuz T-2: Vladimir Aksyonov, specialist in spacecraft systems, demonstrates the new 8kg spacesuit before the launch. ***(Novosti)***

parachute system was successfully used during the landing on 9 Jun, with touchdown 200km SE of Dzhezkazgan.

Soyuz T-3 L 27 Nov 1980, this mission marked the resumption at last of flights by 3-man crews, the minimum always thought necessary by both Russia and America for space station operations. Commander Leonid Kizim, making his 1st flight, was a new name in the West even though he had been a cosmonaut for 15yr. Flight engineer Oleg Makarov, officially stated to have made 2 previous flights, had also been aboard the aborted Soyuz 18A in 1975. Having also worked under Korolev preparing the pioneering flights of Gagarin and Titov, Makarov is one of the most experienced of all the cosmonauts. Gennadi Strekalov, making his 1st flight and designated researcher-engineer, is a designer and had apparently played a prominent part in developing Soyuz T. Docking, on the 17th orbit after 25hr 36min, took place "exactly on schedule". It is presumed to have been completely automatic, since it was cited as evidence that the new onboard computing complex, in effect an automatic pilot, was working perfectly. Leonov, deputy head of the Gagarin Training Centre, said that the crew's job was to try to breathe new life into Salyut 6, and they apparently succeeded. They took with them and installed a new hydraulic unit incorporating 4 pumps for Salyut 6's temperature control system. This was a difficult task, since it was necessary to open the system, exposing the fluid in zero g. Other repairs included replacing a programming and timing device for the control system; installing a new transducer in a compressor in Salyut's in-orbit refuelling system; and replacing air-to-ground telemetry components. While they carried out this work with specially developed tools, they were in radio contact with Popov and Ryumin so that they could have the benefit of their advice and intimate knowledge of the systems. As usual the scarred portholes were carefully checked, though no way of replacing them had been found.

Within 3 days the Splav electric furnace was also in use again, producing a semiconductor compound of cadmium-mercury-tellurium; and a new experiment - holographic photography in space - was tried for the 1st time. A joint Soviet-Cuban idea 1st flown but not tried out during Soyuz 38, this involved the 1st use of a spaceborne laser. Because vibrations blur the image, in ground-based holographic photography the laser and optics are mounted on a base sometimes weighing several tons. The Salyut 6 equipment, based on a gas (helium-neon) laser, weighed less than 5kg and was found to be vibration-free when tested on a moving train. The T-3 crew used it to film the dissolution of salt crystals in water under weightless conditions, and later brought the film back to Earth. It was planned to use the process, if successful, to help solve the problem of the porthole windows, and to measure the outflow velocity from the Salyut's attitude-control nozzles.

Attention was also given to Salyut 6's 3 "greenhouses," in which orchids, which faded and lost their flowers in 2-3 days, had been replaced with arabidopsis. These plants had produced flowers - the 1st space blooms - but when returned to Earth were found to be sterile even though the species does not normally need pollination. The crew's success in extending the life of the space station meant that later crews would be able to study the progress of the peas, flax and wheat also growing in the greenhouse; this work is intended to demonstrate that interplanetary spacemen could produce their own Vitamin C during their journey. Their final task was to pack all waste material into Progress 11 and supervise its undocking on 9 Dec, ready for de-orbiting 2 days later. Altogether the crew was kept so busy that a proposed rest day and 3 TV programmes were cancelled, and they did not complete their planned exercises. The medical leader pointed out that "such a routine would not be suitable for a prolonged expedition". Following landing on 10 Dec 130km E of Dzhezkazgan, Kizim and Strekalov "suffered a certain amount of tension"; the more experienced Makarov was less affected.

Soyuz T-4 L 12 Mar 1981 as the start of the final series of manned missions in the 3½yr-old Salyut 6, this flight took commander Vladimir Kovalyonok back for a 2nd, 74-day stay, following his 140 days aboard in 1978. Flight engineer Viktor Savinykh, a specialist in mapmaking and aerial photography who was making his 1st flight, became the 50th Soviet cosmonaut and the 100th person in space. Once again the 3rd seat was unoccupied, extra film and instruments being carried instead. The engines of Progress 12, docked with Salyut 6 since 26 Jan 1981, were used to adjust the station's orbit for the T-4 docking on the usual 17th orbit. The crew's 1st major task was to unload the scientific equipment which had been awaiting use aboard Progress 12 for nearly 3 months. Savinykh photographed details of the Salyut's ageing structure, and in another series of holographic experiments 3-dimensional images of objects outside the space station were transmitted. There was daily attention to the 3 greenhouses - Oasis, Vazon and Svetoblok - as efforts to grow peas, onions, orchids and arabidopsis continued.

Progress 12 was jettisoned on 19 Mar in readiness for the arrival of Soyuz 39 with cosmonauts Dzhanibekov and Gurragcha aboard 4 days later. A busy 7 days followed, with the crews carrying out materials-processing work and using, among other things, the Soviet/Mongolian visual-polarising analyser to observe and photograph Mongolian territory. While the visitors filmed the dissolution of a common salt crystal in weightless conditions, Kovalyonok and Savinykh used a helium-neon laser to take holographic photographs of micrometeorite hits on a porthole. 30 experiments were completed during the Soyuz 39 visit, and the crew returned with 3 containers loaded with hologram film, specimens of processed materials, tadpoles hatched on board, and greenhouse specimens. The long-stay crew's continued housekeeping chores included cleaning of the interior with wet cloths. The T-4 engine was fired periodically to raise the orbit, and the Splav furnace was used to obtain such semiconductor materials as gallium arsenide, gallium antimonide, gallium-bismuth and germanium-silicon, as well as crystals of lead-zinc and bismuth-antimony.

By the 30th day they were making meteorological observations of the so-called "Bermuda Triangle" in conjunction with the research ship *Akademik Mstislav Keldysh*. This was followed by use of Isparitel, which replaced Splav in the airlock. Experimental spraying of copper and then silver on titanium specimens at varying temperatures was aimed at developing in-flight repair techniques for spacecraft exteriors.

On 15 May Soyuz 40 brought the final Intercosmos crew for another 7-day stay, and as soon as they had gone Kovalyonok and Savinykh prepared for their own departure. They installed a new type of docking system which was later to receive the unmanned Cosmos 1267, launched on 25 Apr. After they landed on 26 May 125km E of Dzhezkazgan it was stated that there would be no more manned flights for several months. 5 days after the crew's departure an object was seen to separate from Salyut 6. It was later learned that this was T-4's orbital module, left behind for the first time in a new procedure which reduces the amount of fuel needed for re-entry.

Soyuz T-5 L 13 May 1982, 28 days after Salyut 7, T-5 carried a 2-man crew for the first long-duration flight in the new space station. It was to extend the record to 211 days. The commander was Anatoli Berezovoi, making his 1st flight, with flight engineer Valentin Lebedev making his 2nd. Lebedev had missed the record 185-day flight of Soyuz 35, having injured a leg and been replaced by Ryumin. Docking took place on 14 May, half-way through the 17th revolution and in a 343 × 360km orbit with 51·6° incl. Hatch opening and entry came 2 orbits later. As they were settling in for a long stay the crew began slowly, with 3 days allowed for unstowing equipment, activating the water-recovery system and the telex link with Flight Control, and transferring equipment brought with them in T-5. On 17 May they launched through the airlock Iskra 2, a small amateur radio satellite designed by Moscow students for the use of the 11 Comecon countries; its orbit was 342 × 357km with 51·6° incl. On 23 May Progress 13 was launched. Carrying more than 2 tonnes of cargo, it docked 5 days later. Its contents included 660kg of fuel, 290lit of water, a Kristall furnace, an improved Oasis plant growth unit, and an electro-photometer designed to obtain star data. The water was carried in spherical vessels "on the surface of the spacecraft" to leave more room inside. (There was no indication as to how the crew recovered the vessels after the docking.) Progress 13 also delivered the French instruments needed for the Soyuz T-6 mission, including the Echograph equipment and Posture, a moving platform on which the cosmonauts would stand for studies of muscle reactions in weightlessness. The crew were reported to be feeling fine, and Berezovoi was said to be in charge of the cooking. By mid-Jun the crew were conducting a diverse research programme covering geology, agriculture, biology, forestry and oceanology. Cotton could be clearly seen in fields in Central Asia. The MK-6M and KATE-140 cameras were used to observe and photograph farmlands, to search for mineral deposits, and to detect ocean areas suitable for commercial fishing. Attempts to grow green vegetables continued, with the help of a water dispenser, a magnetic field and a biological centrifuge. After Progress 13 had been used for 2 trajectory corrections it was separated on 4 Jun and placed on a re-entry course 2 days later. The crew's pre-flight geology course, with visits to possible oil, gas and mineral-bearing areas, was then followed up with studies of promising oil and gasfields on the left bank of the Volga near Astrakhan, and of a new lead deposit in E Yakutia. Another orbital correction, possibly carried out with Soyuz T-5 on 19 Jun, was stated at the time to be in preparation for the launch of the Soviet-French expedition; this was the 1st time since Apollo-Soyuz that the Soviet Union had given advance details of a manned mission. The work of the first 5-man team in space is described under Soyuz T-6. After the visitors had left, Berezovoi and Lebedev, apparently as exhausted as the other 3, asked for a period of rest and relaxation.

On 30 Jul, their 78th day, Lebedev made a 2hr 33min spacewalk, with Berezovoi controlling and assisting in the open hatch and using a portable TV camera. Lebedev recovered and replaced an instrument registering micrometeorite strikes, as well as sample panels "with biopolymers, optical and various structure materials". The crewmen also spent some time testing new tools for assembly work outside Salyut 7; this included trying out joints with thermo-mechanical and threaded connections. Progress 14, which had docked on 10 Jul, was undocked 8 days before the launch of Soyuz T-7 with Svetlana Savitskaya, the world's (and Russia's) second spacewoman. Accompanied by Popov and Serebrov, she arrived on the T-5 crew's 100th day. Work during their stay was devoted mainly to tests of means of avoiding spacesickness; they also performed an electrophoresis experiment and metals and materials-processing work. On 29 Aug, 2 days after the departure of the T-7 crew in T-5 (leaving the new Soyuz for the long-stay crew), Berezovoi and Lebedev redocked T-7 at the rear port, freeing the forward port for the arrival of Progress 15 on 18 Sep.

In Oct came a hint that this might develop into a record mission. Progress 15 was used in engine firings on 28 and 29 Sep to raise the orbit to 364 × 384km before being separated on 14 Oct and directed to re-entry 2 days later. Internal activities included growing peas, oats, parsley, borage and onions, and the development of flax and arabidosis seeds; the complete cycle from planting to obtaining seeds was achieved with the latter. The "kitchen garden" was used for experiments with 13 species.

The cosmonauts took a shower once a month, and just over 2hr per day was spent on exercise, half of it on the running track. By the end of Oct it was stated that they were "naturally getting tired" and would be resting an extra 2hr per day.

A Flight Control news conference revealed that a substance with a new molecular weight had been obtained in the biochamber on the exterior of Salyut 7, and had presumably been returned to Earth by a visiting crew. A woman scientist said that this gave rise "to fresh thinking about the origination and evolution of life on Earth". One of the pumps supplying hot water failed, and an elaborate plan for the crew to study the speed and direction of an expelled rubbish container in relation to certain stars was frustrated by the Sun's glare from the solar panels. The crew observed that this demonstrated how even the best experiments, carefully thought out on Earth, could come unstuck in orbit.

Observations of astronomical targets like the Crab Nebula alternated with studies of such Earthly features as Asian glaciers. On 19 Oct the crew began transmitting video images of the Earth's surface, though it was stressed that the process employed did not produce high-resolution pictures; about 5,000 pictures were transmitted. With the help of automated cameras, over 20,000 images were taken during the mission, much attention inevitably being paid (in the light of another unsatisfactory Soviet grain crop) to wheat-growing areas.

The 185-day record was passed on 14 Nov, and the 200-day mark reached at the end of the month. The crew, it was reported, had received less than a quarter of the radiation recorded by the US astronauts aboard Skylab for 84 days. On 18 Nov they injected another amateur radio satellite, Iskra 3, into orbit via the airlock chamber. Later Berezovoi described how they had kept cheerful and overcome their "nostalgia for Earth" with tape recordings of falling rain, the rustle of foliage and birdsongs provided by their "psychological support group". They also watched amateur films of their families brought up by the short-stay crews. Then, while they started preparing to conclude their mission, we heard details of Korund, a 136kg installation claimed to be the first pilot-scale production plant in space. With a revolving drum system feeding capsules into an electric furnace generating 20–1,270°C, it could turn out highly uniform single crystals for use in the electronics industry that were immeasurably better than those produced in gravity.

Progress 16, which had arrived on 31 Oct, was used on 8 Dec to adjust the orbit for re-entry. With no replacement crew taking over, Berezovoi and Lebedev took an inventory of everything on board and recorded its location. Re-entry was on 10 Dec, with landing at 22.03 Moscow time 190km E of Dzhezkazgan. For the crew the hours that followed must have been the most unpleasant of the whole mission. Fog and low cloud suddenly reduced visibility to less than 100m, and the crew were warned to keep their suits on because the helicopters would not be able to reach them immediately. Landing heavily, T-7 rolled down a hillside. The 1st helicopter to reach the area crash-landed in a dry river bed in a snowstorm, damaging its undercarriage. The commander talked down a 2nd, carrying a medical and rescue team, but no others were able to land. 4 cross-country vehicles then set off for the landing site, making their way from 40-50km away by compass across the steppes and through the night. The rescue team did extraordinarily well to extract Berezovoi and Lebedev, in acute discomfort following their return to 1g, in only 40min. They then spent 5hr in the "warm salon" provided by a cross-country vehicle until dawn, when a 50min helicopter flight transferred them to Dzhezkazgan. There they were greeted by cosmonaut Leonov, deputy head of the Cosmonaut Training Centre. Then came a 50min flight to Tyuratam, where they were said to have arrived walking unsteadily and with assistance. They were allowed to walk about 500 steps during their first 48hr and recovery gradually followed. The night landing in undesirable conditions apparently came about because missing that opportunity might have meant another week in orbit, and the crew had begun to complain of irritability with each other, as well as requesting that their sleep periods be increased to 12hr.

It is possible that a new crew was not sent immediately to Salyut 7, or even "overlapped", because the Korund furnace had been set up to process semi-conductor monocrystals after T-7's departure completely free of the motions caused by crew movements, with the products being collected by the next crew when they arrived later.

It came as a surprise in Aug 1983 when Lebedev described in a *Pravda* interview how homesick and bored he had become in space, longing to return to his family and obtaining consolation only by looking out of the windows at Earth. He said it had been difficult to avoid irritability with his partner. But on the basis of their 211-day mission, Soviet experts said it was unlikely that interplanetary ships of the future would need artificial gravity.

Soyuz T-6 L 24 Jun 1982, this was the 1st Soviet-French manned flight. Crew was Frenchman Jean-Loup Chrétien, flying as test researcher after 2yr training, commander Vladimir Dzhanibekov, making his 3rd flight, and flight engineer Alexander Ivanchenkov, making his 2nd. Chrétien, who spoke Russian throughout the mission, was surprised to discover that as late as 4 days before his flight the components of the vehicle were still dispersed around the launch centre. Assembly was completed only 2 days before launch, and it took only 6min to raise the cluster to the vertical position. Describing the launch after his return, Chrétien said the 530sec of powered flight went as predicted, with launch loads of between 3·5 and 4g and noise levels at 30-50dB. The 1st and 2nd stages, used primarily to provide altitude, boosted T-6 to 160km and a velocity of 4km/sec; the 3rd stage increased the speed to 7·9km/sec and the altitude to over 200km. Launch was to have been completely automatic, but then a major computer failure forced the crew to take over as Soyuz T-6 was approaching Salyut 7 in a 248 × 277km orbit 25hr after launch. Chrétien's description in a French phone-in programme 11 days after landing gave a vivid insight into what had probably happened on previous docking failures at this stage: "... following the computer breakdown the vessel began to rotate about all 3 axes, and was thus like a stone rolling over. We had to act very quickly." Dzhanibekov took over manual control 900m from Salyut and successfully completed the docking just before the end of the 16th orbit, 14min early. When they entered the Salyut 2 orbits later, to be greeted by Berezovoi and Lebedev, it was the first time 5 men had been present in a space station. Gifts, souvenirs and letters brought up by the visiting crew included a small plastic replica of the Eiffel Tower, provided by Chrétien.

Experiments in space medicine, materials processing, biology and astronomy were crowded into their week's stay. Afterwards Chrétien joined his fellow crew members in complaining that their number and complexity was overwhelming, and that they had to cut down on sleep periods in order to keep up with them. This was due in part to the number of advanced experiments provided by 17 French laboratories. Most notable of these was Echograph, designed to study by means of ultra-sound probes the functions of the heart, and of blood vessels in the neck, thighs and abdomen. Chrétien was seen on TV having some initial difficulty in locating his heart with one of the sensors, zero g having caused it to move upward in his body. Using this system the crew were able simultaneously to see images of their hearts on a TV monitor and hear them in their earphones, permitting them to study the redistribution of blood in weightless conditions and the effect of varying loads on the heart muscle when it no longer had to overcome the force of gravity. The visiting crew also studied the effects of zero g on vision and colour sensitivity while Berezovoi and Lebedev were doing physical exercises. Other French experiments included calibration and materials-processing work in the electric furnace, and the use of a highly sensitive camera to study the faint glow of various galactic objects and of dust clouds in interplanetary space and in the upper layers of the atmosphere.

Soyuz T-6: Jean-Loup Chrétien (France), Vladimir Dzhanibekov and Alexander Ivanchenkov about to board their spacecraft. Their 7-day stay in Salyut 7 was highly productive. Chrétien's back-up is to repeat similar spacesickness tests in the US Shuttle.

Preparations for the return of the T-6 crew began on 2 Jul with the transfer to the Soyuz of videotapes of the Echograph work, tapes of posture experiments, film from night sky photography, and samples of processed materials. Direct TV showed the handshakes, embraces and even flash photos taken by the long-stay crew before the T-6 hatch was closed. Separation was on the 126th orbit, a 200sec retro burn was completed on the 127th, and 11min later the service module was jettisoned. Later Chrétien said that he was surprised at the rate of descent (6·5m/sec) with the parachute fully open. The descent was much more dramatic, he thought, than the launch. The rate was cut to 3·5m/sec just before touchdown by automatic firing of the soft-landing rockets.

The landing, after 7 days 21min, took place in a wheatfield 65km NE of Arkalyk and only 4km from the target point. Chrétien was awarded the title of Hero of the Soviet Union, the Order of Lenin and a Gold Star. Despite France's officially cool policy towards the Soviet Union following events in Afghanistan and Poland, scientists of the two countries seemed well satisfied with the flight and eager at the time to make use of the fully trained back-up French cosmonaut, Patrick Baudry, for a follow-up flight in the near future.

Soyuz T-7 L 19 Aug 1982, carrying the world's 2nd spacewoman, 34yr-old Svetlana Savitskaya. Commander was Col Leonid Popov, making his 2nd flight. Alexander Serebrov, making his 1st, was "Flight candidate of technical services," and Savitskaya was cosmonaut-researcher. Serebrov and Savitskaya were said to have had a "full course" of training, but it was noticeable that the usual date of joining the cosmonaut team was not given. Savitskaya, a pilot with 1,500hr of flight time on 20 aircraft types and a former world aerobatic champion, seemed an ideal choice if the decision to send the Soviet Union's 2nd woman had been made at short notice in order to steal the thunder of 1st American spacewoman Sally Ride. Docking with Salyut 7 on 21 Aug was apparently routine. It was the 100th day of the T-5 crew's flight, and Berezovoi greeted Savitskaya with flowers, an offer of an apron and an invitation to do the cooking. Savitskaya, plainly a firm believer in feminism, replied: "Housekeeping details are the responsibility of the host cosmonauts." The week's experiments concentrated on adaptation to zero g, particularly by Savitskaya. This work suggests that spacesickness continues to be as much of a problem for the Russians as it is for the Americans. Garments restricting blood flow, and a device restricting head and neck movements, were tried as ways of reducing stimulation of the vestibular system. Having transferred their couches from T-7, the visiting crew returned in T-5 on their 127th orbit, leaving the long-stay crew a fresh spacecraft. They landed 113km NE of Arkalyk on 27 Aug. Doctors said that Savitskaya withstood spaceflight as well as her male companions. They were the last crew to be received and decorated by Mr Brezhnev before his death.

Soyuz T-8 An unexpected docking failure with Salyut 7/Cosmos 1443 resulted in a return 2 days after launch on 20 Apr 1983. Commander was Vladimir Titov, 36, making his 1st flight, with flight engineer Gennadi Strekalov, 43, and cosmonaut-researcher Alexander Serebrov, 39, making 2nd flights. All appeared to be going well, with T-8 about 80km behind Salyut 7 some 6hr before the scheduled docking time on 21 Apr. The crew already knew, however, that the Soyuz radar antenna which measures distance and speed during the approach had failed to deploy. Titov described later how it was decided to attempt a manual docking even though the chances of success were considered small. The approach was made using only a viewfinder, but when only 150m away Titov decided they were going too fast and likely to crash. He made a quick manoeuvre to take T-8 below the Salyut, and they were then told to return to Earth. Landing was near Arkalyk on 22 Apr. It had been the 5th docking failure since Soyuz 15/Salyut 3.

Soyuz T-9 This eventful flight of nearly 155 days, L 27 Jun 1983, made commander Col Vladimir Lyakhov, 41, the 2nd man to spend over 300 days in space. Flight

Soyuz T-7: left to right, Alexander Serebrov, Leonid Popov and Svetlana Savitskaya, who just beat America's Sally Ride to become the 2nd spacewoman. As a world aerobatic champion, she needed little training.

Soyuz T-8: left to right, cosmonaut-researcher Alexander Serebrov, commander Col Vladimir Titov and flight engineer Gennadi Strekalov in the Gagarin Training Centre. They suffered yet another docking failure. ***(Tass)***

engineer Alexander Alexandrov, 40, was making his 1st flight. Docking was completed at the end of the 17th orbit, with Salyut 7/Cosmos 1443 in the 325 × 337km orbit to which the complex had been raised a week earlier. The reduction of the crew to 2 seemed to indicate that it was a remedial mission designed to settle the questions raised by the T-8 docking failure, subsequently attributed to a computer problem. Soviet sources also indicated that T-9 would not be a long-duration mission. Nevertheless, the crew quickly settled down to the well established round of materials-processing work and observations of Earth and outer space, and they were largely forgotten outside Russia until they came into the news in mid-Oct. On 27 Jul the crew reported "an unpleasant surprise" when a micrometeorite or man-made debris struck a window with a loud crack and formed a 3·8mm-dia crater. The damage did not threaten the mission, however, and Soviet scientists thought the object had come from a meteorite shower through which the Earth was passing. (Shuttle Orbiter *Challenger* suffered similar damage on STS-7.) By coincidence, the incident occurred when the crew were practising emergency evacuation procedures. It would take only 15min to retreat into the more heavily protected Soyuz, though the basic emergency return schedule takes 90min, including some moth-balling activity. Tass and Moscow Radio reports stressed how well the crew were working together, reflecting the impact made by Lebedev's comments on relationships during T-5.

Unloading of Cosmos 1443 took a long time, and the 1st of a series of mysteries about the progress of this mission came when, after advance references to the crew's preparations to send back C 1443's re-entry vehicle, the whole module was separated on 14 Aug. It spent 9 days in free flight before the re-entry vehicle was in turn separated and recovered 100km SE of Arkalyk on 23 Aug with 317kg of film, processed materials, spent hardware, air regenerator parts and a malfunctioning navigation system. With C 1443 still within possible re-docking distance, on 16 Aug the crew switched T-9 to the Salyut's front port to make way for the arrival of Progress 17 on 19 Aug. That craft remained docked for a month until separated on 17 Sep and directed to a controlled re-entry on 18 Sep. Suspicions that C 1443 had suffered some serious malfunction were confirmed when the big space station module was also placed on a re-entry course on 19 Sep. In an interview broadcast on 31 Aug Lyakhov and Alexandrov had said that the C 1443/Salyut 7/T-9 complex was "very difficult to control, but we managed it". Navigation, they explained, was only semi-automatic because a fully automatic system would be too costly in weight and power consumption. US sources reported the rupturing of a main oxidiser line on 9 Sep, resulting in two-thirds of the propellant venting into space and the crew having to don their spacesuits and retreat into the Soyuz.

Reports that Salyut 7 had only 16 of its 32 attitude-control thrusters usable and could not use its back-up main engine seemed to be confirmed by a Flight Control report on 3 Oct that substantial changes had been introduced into the crew's work schedule for their 4th month. They would now concentrate on processing semiconducting materials and biologically pure substances for medicine. This followed an interesting TV report on 23 Sep in which the crew had explained that they did not have ideal conditions for such experiments because their own movements, plus the working of ventilators and other equipment, created micro-vibrations. A device containing 3 seismographs had been developed to measure these micro-vibrations with an accuracy of 0·000001g.

The launchpad explosion of what was to have been Soyuz T-10, with a new return vehicle and possibly repair equipment, came on 27 Sep. When this and the alleged propellant loss were reported in the West the Soviets denied that there had been a leak but offered no explanation of Salyut 7's apparent "drifting" mode. Equally interesting was the failure to send up either a visiting crew or a fresh Soyuz return vehicle for over 100 days. Western observers began to speculate that the crew were virtually stranded, but then Progress 18 was launched on 20 Oct.

October ended and the T-9 crew were still in orbit, with neither visitors nor a replacement Soyuz. Each morning, we learned, they were woken by birdsongs radioed from mission control. They completed more than 100 experiments, producing industrial monocrystals, pure alloys, compounds and medicinal preparations with Salyut's furnaces and other equipment, and bringing the total for Salyut 7 to over 300. Then, to the surprise of Western

Moscow's Flight Control Centre monitors the docking of Soyuz T-9 with Salyut 7/Cosmos 1443. ***(Tass)***

observers, came 2 spacewalks of 2hr 50min and 2hr 55min on 1 Nov and 3 Nov. TV pictures revealed that they were carried out with confidence and good humour. Two telescopic solar panels, 1·5m wide and 5m long when fully extended, had been delivered in C 1443 and stored since. They were added to the edges of one or more of the existing 3 side-mounted solar arrays, and Flight Control made it clear that the necessary mountings and electrical plugs had been provided before Salyut 7 was launched. During the EVA the work was duplicated by cosmonauts Kizim and Solovyev in a water tank at Star City in case problems arose. The Soviets claimed it was the first assembly work ever done in "raw space". This and other repair work, it was learned after the mission ended, should have been done by the Soyuz T-10A crew, who had been specially trained for it.

At this point Soviet space officials broadcast denials that there had been any propellant leak and that the crew were in any danger. In any case, fresh supplies had arrived on 20 Oct with Progress 18, which was used to adjust the orbit before being jettisoned on 13 Nov. By the time preparations for return to Earth were begun, the crew had spent 1,000hr at the portholes, photographing 100 million sq km of territory, and to their experiments had been added production of "highly efficient molecular anti-flu vaccines from the shells of influenza viruses". Made on Earth, such vaccines cost 20 times as much as gold, we were told, and similar US experiments were not planned until 1987. Fog added to the difficulties of the night landing on 23 Nov. The crew made a good recovery and the final surprise came in a mid-Dec news conference, during which they admitted that there had in fact been a propellant leak. According to *Aviation Week*, their last few weeks had been spent in cold, damp and uncomfortable conditions due to a shortage of electrical power.

Soyuz T-10A News of manned spaceflight's 1st launchpad fire and explosion, necessitating use of the escape tower to save the lives of the crew, came first from US sources but was later admitted by the Soviets. The launch attempt was on 27 Sep 1983, with Vladimir Titov as commander and Gennadi Strekalov as flight engineer; these two cosmonauts had been specially trained for the EVAs described in T-9. The fire apparently started at the base of the SL-4 launcher at about T-90sec when a propellant line valve failed to close. The fire engulfed the launcher so quickly that the necessary wiring was burned through before the automatic abort could be initiated;

Soyuz T-9: Vladimir Lyakhov during 1 of 2 spacewalks to enlarge Salyut 7's solar panels. ***(Tass)***

another 10sec passed before Launch Control recognised the situation and commanded a radio abort. The service module remained on the booster as the escape tower pulled the orbiter and descent modules, the latter containing the crew, to an altitude of 950m. An explosive charge jettisoned the orbital module and the fast-opening back-up parachute landed the crew, uninjured despite a heavy touchdown, about 4km away. The 1st use of the escape tower proved remarkably successful, but pad damage – some reports suggested the destruction of 2 out of 3 manned launchpads in the area – made it impossible to launch a relief crew to Salyut 7 before the end of the year.

Soyuz T-10B L 8 Feb 1984, with commander Leonid Kizim, 42, making his 2nd flight, flight engineer Vladimir Solovyev, 37, and cosmonaut-researcher Dr Oleg Atkov, 34, a heart disease specialist, this at last realised Soviet scientists' long-held ambition to have 3 crew members on a long-duration Salyut mission, giving a better balanced workforce. As we went to press, it looked like becoming one of the longest and most successful missions, with 5 EVAs beating Skylab totals and setting a world record of

17hr 50min. This mission did not start too well, however: docking, carried out manually by Kizim, proved difficult; he complained that he could not see the Salyut docking target against a bright white cloud background, and was told sharply by Lyakhov and Ryumin in the control centre to concentrate. But the crew quickly settled down to what was described as "a vast programme" of Earth studies and astrophysical and technical experiments. It was said that the physician would concentrate on the cardio-vascular system, and the mechanisms of water and salt exchange in the body. In practice this meant a medical exam every other day for months ahead for the unfortunate cosmonauts. A new exercise system enabled them to adapt quickly and to increase their working day to 8·5hr.

Progress 19 arrived with supplies on 23 Feb, and was used to adjust the Salyut's orbit for Soyuz T-11's arrival before being jettisoned 2 days before that mission was launched. Following the departure of that crew, which included Indian spaceman Rakesh Sharma, in Soyuz T-10B, Kizim, Solovyev and Atkov boarded T-11 and redocked it at the other end of the station in readiness for the arrival of Progress 20 on 17 Apr. The steady routine of experiments included repairs to thin-filmed metal-coated surfaces. Alloys and fluoroplastic were vaporised and applied to metal surfaces with electron guns; 88 samples had been obtained, it was reported.

The crew had been in orbit for 3 months when the EVAs began, the 1st 4 taking place within 12 days. Kizim and Solovyev started on 23 May by hooking a ladder outside to create a work platform. During the next 3 they used pneumatic punches to make holes in the exterior lining. Then, "with the help of a special cutter resembling a big can opener, they cut out part of the skin and gained access to pipes leading to the power plant". By 4 May additional pipes and valves had been installed and tested for leaks. On 19 May they celebrated their 100th day with a 5th EVA, during which Kizim and Solovyev added a 2nd additional solar panel, 4·6 sq m in area and equipped with high-efficiency gallium-arsenide photocells.

Another 1st came on 24 May when Kizim was advised amid cheers that his wife had given birth to a baby girl while he was in space. Progress 21 and 22 came and went with supplies, and what was described as a 6-week programme of visual observation of the oceans was begun. That took them well past 150 days.

Soyuz T-10B: left to right, cosmonaut-researcher Dr Oleg Atkov, flight engineer Vladimir Solovyev and commander Col Leonid Kizim. As we went to press it seemed possible that they would break the 211-day space duration record. *(Tass)*

Soyuz T-11 (see Addenda): TV view of the docking with Salyut 7/Soyuz T-10B. *(Tass)*

Soyuz T-11: left to right, Gennadi Strekalov, Maj Rakesh Sharma (India) and mission commander Yuri Malyshev see their launch rocket being assembled. *(Tass)*

Space shuttles

History After years of speculation, by the end of 1983 it had become certain that the Soviets were developing both a heavy-lift shuttle, similar to the US Orbiter but with twice the payload capacity, and a 3-crew mini-shuttle or spaceplane.

Spaceplane 3 test flights in the Cosmos series have revealed that this is a Dyna-Soar-type lifting body with delta wing and fins, aircraft-style cabin and thermal tiles. Weighing about 1 tonne, it is probably a prototype designed to evaluate the aerodynamic and re-entry characteristics of a 10-20-tonne version capable of carrying up to 3 crew to Salyut and other low-Earth-orbit platforms. With quick-launch capability it could be used as a "space fighter" and reconnaissance vehicle, and it could also serve as a Soyut T replacement. It is however unlikely to meet the Soviet requirement for a vehicle able to carry heavy payloads from space stations to Earth.

Cosmos 1374 was launched 3 Jun 1982 by Proton from Kapustin Yar; orbit 191 × 230km, incl 49·6°. Recovered 109min later 560km S of the Cocos Islands in the Indian Ocean by 7-ship Soviet task force. Operations were monitored by Orion aircraft of the RAAF, whose photos the US made an unsuccessful attempt to classify; it is not known whether the descent was made by parachute or gliding. The identical Cosmos 1445, L 15 Mar 1983, was recovered 556km S of the Cocos. Retrofire of Cosmos 1517, L Dec 1983, was monitored by N Atlantic tracking ship instead of from the main Soviet tracking station in the Crimea. A Soviet announcement said that it performed a controlled descent in the atmosphere and splashed down in the Black Sea. The use of the smaller area of water showed increased confidence in the system and provided greater security from Western observation.

Heavy-lift shuttle Identified by US reconnaissance satellites at Ramenskoye flight-test centre near Moscow. Similar in design to the US Shuttle but with an 80°-sweep delta wing, it has an estimated lift-off weight of 1·5 million kg and lift-off thrust of 1·8-2·7 million kg, compared with Shuttle figures of 2·2 million kg and 3·1 million kg. The lower lift-off weight could give a 60,000kg payload to low orbit, double that of the US Shuttle. The ET is larger, at 65m long, and carries the 3 main engines, which in the US system are fitted to the Orbiter. This improves the Soviet orbiter's payload but means that the main engines are lost with the ET unless this can be recovered. Another difference is that the strap-on boosters are liquid rather than solid-fuelled. The Soviet orbiter and ET are carried (presumably separately) on converted Myasischev M-4 Bison bombers. One of these was reported in Apr 1983 to have run off the side of the runway at Ramenskoye, damaging the aircraft but not the orbiter.

Sputnik

History When Sputnik 1, the world's first artificial satellite, soared into orbit on 4 Oct 1957 it acted as the starter's pistol in the Soviet-American race to put men on the Moon. In America its launching brought bitter disappointment: earlier the same year a US Army plan supported by Dr Wernher von Braun, who wanted to use his Redstone rocket to put up a satellite in Sep, had been turned down. Had it been successful the US would have been 1 month ahead of Russia. As it was, Russia was to put up Sputnik 2 before America joined the contest with Explorer 1 on 31 Jan 1958.

The whole series of 10 Sputniks appears to have been devoted to developing the ability to place men in Earth orbit, using at least 6 dogs for the purpose. All were launched from Tyuratam over a period of 3½yr. Cosmos 1 was also Sputnik 11, and for a time the numbers of the 2 series overlapped. Sputniks 19, 20 and 21 were believed to be early attempts at launching Venus probes. Sputniks 22 and 24 were attempts to launch towards Mars, and Sputnik 25 (4 Jan 1963) was a Luna failure. From that point Russia allotted Cosmos numbers to any launch failures occurring in non-Cosmos programmes.

Launchers The standard Vostok vehicle (A), based on the original Soviet ICBM and providing lift-off thrust of 509,840kg, was used for Sputniks 1-3. For most of the subsequent launches an upper stage providing 90,260kg was added to the Vostok vehicle.

Sputnik 1 L 4 Oct 1957. Wt 83·6kg. Orbit 228 × 947km. Incl 65°. The world's 1st artificial satellite. It consisted of a polished metal sphere, dia 0·58m, with 4 whip-type aerials from 1·5m to 2·9m long. Instrumentation included radio telementry and devices for measuring the density and temperature of the atmosphere and concentrations of electrons in the ionosphere. It transmitted for 21 days; dec after 1,400 orbits in 96 days.

Sputnik 2 L 3 Nov 1957. Wt 508kg. Orbit 225 × 1,671km. Incl 65°. The world's 2nd satellite, it contained a spherical pressurised chamber for the dog Laika, the 1st living creature in space; she was being flown to obtain data on the effects of weightlessness on living organisms. Transmissions lasted for 7 days, after which Laika presumably died painlessly when her oxygen supply ran out. Spacecraft re-entered and dec after 2,370 orbits in 103 days.

Sputnik 3 L 15 May 1958. Wt 1,327kg. Orbit 217 × 1,880km. Incl 65°. Cone-shaped, 1·73m dia at base and 3·57m long, it carried instruments to study Earth's upper atmosphere, solar radiation, etc; dec after 690 days.

Sputnik 4 L 15 May 1960. Wt 4,540kg. Orbit 312 × 368km. Incl 65°. 1st Sputnik for 2yr, with Lunas 1-3 intervening. This was believed to be a test flight for a manned Vostok; recovery failed when the cabin went instead into a higher orbit and finally re-entered over 5yr later on 15 Oct 1965.

Sputnik 5 L 19 Aug 1960. Wt 4,600kg. Orbit 305 × 339km. Incl 80°. 2nd Vostok trial. 2 dogs, Belka and Strelka, were ejected and recovered by parachute after 18 orbits.

Sputnik 6 L 1 Dec 1960. Wt 4,563kg. Orbit 186×265km. Incl 65°. 3rd Vostok trial. Recovery failed and canine passenger was killed 1 day later.

Sputnik 7 L 4 Feb 1961. Wt 6,482kg. Orbit 223 × 327km. Incl 65°. Apparently a test flight in preparation for Sputnik 8, the 1st attempt at a Venus flyby. Dec 25 Feb 1961.

Sputnik 8 L 12 Feb 1961. Wt 6,474kg. Orbit 198 × 318km. Incl 65°. Launched Venera 1, Russia's 1st Venus probe, from Earth parking orbit. Sputnik 8 dec 25 Feb 1961.

Sputnik 9 L 9 Mar 1961. Wt 4,700kg. Orbit 183 × 250km. Incl 65°. 4th Vostok trial. Dog Chernushka successfully recovered on same day.

Sputnik 10 L 25 Mar 1961. Wt 4,695kg. Orbit 178 × 246km. Incl 65°. 5th Vostok trial. Dog Zvezdochka successfully recovered after 1 orbit. Yuri Gagarin, world's 1st man in space, went into orbit 18 days later.

Sputnik 1, man's first satellite.

Statsionar

History Although the first Statsionar was not placed in geostationary orbit until 22 Dec 1975, the Soviet Union has now launched more than 30 of these TV and communications satellites in 3 complementary series: Raduga ("Rainbow") for telephone, telegraph and domestic TV; Gorizont ("Horizon") for international relays such as the Olympic Games; and Ekran ("Screen") to relay central TV services from Moscow to Siberia and the Far North and even to ships in the Arctic. The reasons for Russia being so much later than America in using geostationary orbits probably included a shortage of Proton (D1e) rockets, which were being used for planetary and lunar exploration, and the fact that the Molniya series were giving (and still give) excellent service in their elliptical orbits. The Soviet method of achieving geostationary orbit is first to place the satellite in low Earth orbit at 51·6° incl. The final rocket stage is then fired at the first northbound equator crossing, about 80min after launch. This results in an elliptical orbit with apogee at stationary height and 47° incl. At the southbound equator crossing the apogee boost motor is fired to circularise the orbit and reduce the incl to near 0°. With the period now slightly higher or lower than required, the vehicle is allowed to drift to the desired longitude, at which point the orbital period is adjusted to match the Earth's rotation speed.

Russia registered the Statsionar project with the International Telecommunication Union on 3 Feb 1969, and gave further details on 1 Dec 1970. But it was over 3yr later, on 26 Mar 1974, before Cosmos 637, the 1st Soviet stationary experiment, went into orbit, followed by Molniya 1S on 29 Jul 1974. The aim is to have one operating and one back-up satellite at each Statsionar location; this is particularly important in the case of Ekran, since a satellite breakdown could mean complete loss of domestic TV services. Initial anxieties that the Statsionar system would interfere with Intelsat and Symphonie frequencies appear to have been allayed, for a time at least, by the development of new frequencies. Operational life of these satellites is about 2yr.

Raduga Starting with Raduga 1, L 22 Dec 1975 by D1e from Tyuratam, 13 had been placed in Statsionar locations 1-3 by 25 Aug 1983. Probably similar to Gorizont, about which more is known, Raduga carries continuous telephone and telegraph communications, plus black-and-white and colour TV from Moscow Central

Ekran direct-telecasting satellite, 1st orbited in 1976. Its 200W transmitter with narrow-beam antenna permitted cheaper, simpler Earth receivers to be used in Siberia and the Far North, covering 40% of Soviet territory.

TV to the Orbita network of ground and receiving stations. It has 3-axis stabilisation for precise orientation, solar batteries which automatically follow the Sun, and an orbit correction system.

Ekran Starting with Ekran 1, L 26 Oct 1976, wt ? 2,000kg, 11 had been launched by Sep 1982 into Statsionar-T positions at around 99° E, over the Indian Ocean. Each is about 5m long and 2m in dia, and the transmitting aerial at the base is a flat plate supporting 96 small helical elements. To avoid possible shadows from the transmitting plate the 2 large solar panels are mounted on a long boom. Villages and individual buildings can pick up its TV transmissions. In the Kirghizia Mountains 42 relay stations now provide TV for 96% of the population, with the Aksay and Susamyr valleys receiving services in mid-1983 for the first time.

Gorizont Starting with Gorizont 1 on 19 Dec 1978 (though this achieved an orbit of only 22,553 × 49,023km with 11·3° incl), 6 Gorizonts had been launched by 15 Mar 1982 into Statsionar 4 and 5 locations. Despite its non-stationary orbit Gorizont 1 proved to be of some use, and it was joined by the other 3 in time to support the Intersputnik and Intelsat systems in covering the Moscow Olympics. Designed for international rather than domestic use, the Gorizonts are cylinders about 5m long and about 2m in dia. At the Earth-pointing end is an array of aerial horns and reflectors, and the sensors used for attitude control. 2 solar panels, perpendicular to the body, provide power. On 9 Jan 1984 transmissions began of *Pravda, Izvestia* and other national newspapers from Moscow to Saratov via Gorizont. Moscow Radio said that this new Moskva system used "a receiving aerial which can be fitted directly on any local printing shop". It not only took the load off communications cables, but improved picture quality, and would be introduced in other towns.

Vega

Vegas 1 and 2 are to be launched from Tyuratam on 24 and 28 Dec 1984 to pass Venus and jettison descent probes in the 11-22 Jun 1985 period. They will then pass within 10,000km of the nucleus of Halley's Comet in the 6-12 Mar 1986 period. Lasting only 10min and taking the spacecraft through the coma, the cloudy region surrounding the nucleus, the encounter will take place at a speed of 78km/sec. Russia, France and Hungary are sharing the design and manufacture of the TV system, which aims at providing black-and-white and colour images of the comet's nucleus and central coma. A 3-channel spectrometer is being designed by Russia, Bulgaria and France to provide data on the near-nucleus portion of the coma and perhaps on the nucleus itself. France is providing an infra-red spectrometer to estimate the size, emissive power and temperature of the nucleus. Russia and W Germany are working on a spectrometer to estimate the chemical composition and mass of particles in the coma. Dust-particle counters are being built by Russia. Hungary, W Germany, Czechoslovakia and Poland are participating in other spectrometers and plasma-wave analysers, and Austria is providing magnetometers to measure the constant magnetic field. Vega 1 and 2 will each release a free-floating balloon of about 5·4m dia with a lander as they pass Venus. The plan is to track the balloons for 2 days by very-long-baseline interferometry from Earth stations to obtain data on Venusian atmospheric dynamics. Since Vega 1 will be the 1st of 5 Soviet, Japanese and European craft heading for Halley, the Soviets have promised to provide ESA with trajectory data to allow them to fine-tune Giotto's course 7 days after Vega 1's encounter on 6 Mar 1986.

Vega: France and West Germany are among the countries collaborating with the Soviets on this Venus/Halley's Comet project.

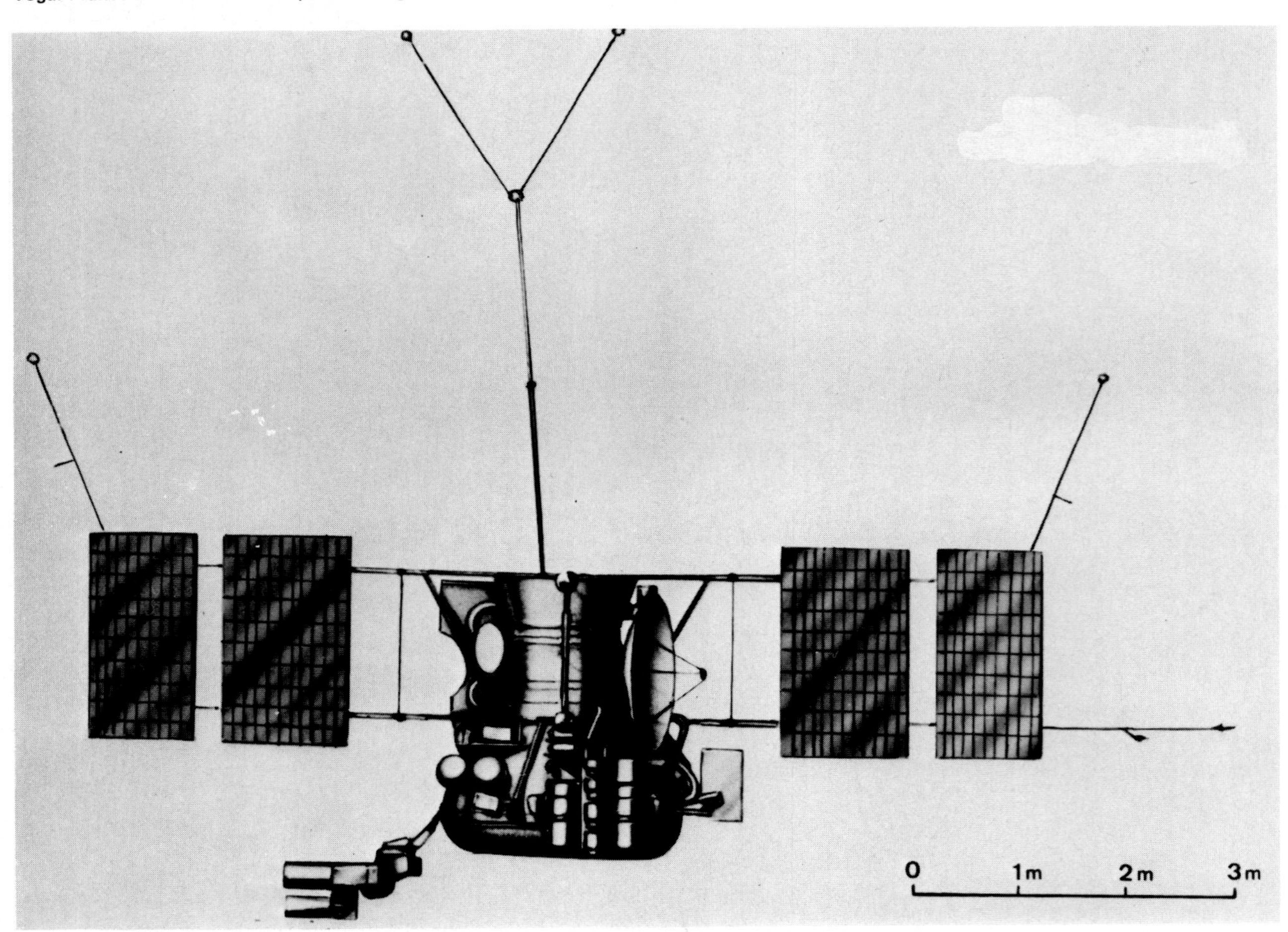

Venera

History After more than 20yr of persistent work, and many disappointments in the early years, Soviet scientists were rewarded in Oct 1975 with man's first pictures from the surface of another planet, sent back by Venera 9 and 10, and with both colour pictures and analysis of soil samples returned by Venera 13 and 14 in Mar 1982.

Venera 3 was the 1st spacecraft to reach a planet (1 Mar 1966) but it plunged into the hostile atmosphere without returning any planetary data. Russia was able to claim the first transmissions from a planet's surface following the successful descent of Venera 7 on 15 Dec 1970. While there may be some doubt whether this reached the true surface or fell on mountain peaks, with the result that it tumbled and abruptly ceased transmissions after 23min, there can be no doubt at all that Venera 8 transmitted from the surface for about 1hr on 22 Jul 1972 before its resistance to temperatures of up to 500°C and atmospheric pressures up to 100 times those on Earth was finally overcome.

The painstaking persistence of Soviet scientists in using the knowledge gained on each mission to improve both the techniques and spacecraft on each succeeding flight is evident in the missions detailed below. With its clouds reaching an altitude of up to 60km and surrounding the planet with what has been described as "a heavy oily smog," Venus presents much greater technical problems than Mars for exploring spacecraft. This may be one explanation for the extent of Soviet interest in Venus, which sometimes seems disproportionate to the likely rewards. By the time it was announced that Venera 8 was on course in Mar 1972 at least 16 launching attempts had been made. Venera designations are given to spacecraft only when they are safely on course; failures are merely given the next available Cosmos number. Russia prefers to operate her planetary probes in pairs so that transmissions can be compared. This was achieved with Venera 2 and 3 and with 5 and 6, but what should have been Venera 8, accompanying Venera 7 in Aug 1970, failed to achieve escape velocity and was written off as Cosmos 359. Similarly, what was intended to be Venera 9, accompanying 8 in Mar 1972 and launched 4 days later, had to be written off as Cosmos 482 when the escape stage fired only partially and left it stranded in an elliptical 205 × 9,800km orbit with a 6yr life. Venera 9 was intended to land on the planet's dark side to provide readings comparable with those sent by Venera 8 from the sunlit side.

However, as Soviet scientists developed the use of structural strengthening and ablative materials and found ways of adjusting the descent rates to achieve greater penetration before their spherical descent craft were destroyed, confidence grew. Following the successful Venera 8 launch, it was announced that a descent was intended – almost certainly the 1st time in Soviet space history that flightplan details were published in advance. The information gained on that flight was used in the 3yr gap that followed for a complete redesign of Venusian spacecraft. Coupled with a near-doubling of their size and weight, that yielded the successes of Venera 9 and 10. From Oct 1975 man could work from facts, not speculation, in his studies of Venus, its origin and development.

Soviet/US collaboration, despite the cool relations between the 2 governments over the Afghanistan situation, played a sizeable part in the success of the 1981-2 Venera 13/14 mission. US radar maps transmitted by Pioneer-Venus were passed to Soviet scientists to help them select landing points so that they could complement Pioneer's topographical work by analysing the surface chemistry. Reporting on the results of the mission to a planetary conference at Houston a few weeks after the landing, the Soviet delegation made it clear that they were hoping collaboration would continue on their 1984 mission.

Venera: full-size mock-up of the first spacecraft to return surface pictures of Venus. *(Graham Turnill)*

Spacecraft description Venera 1-8 consisted of a cylindrical main section containing mid-course propulsion engine, telemetry, attitude control, guidance, sensory and power systems, and solar panels. The spherical descent module was about one-third the usual payload of around 1,180kg. Total weight did not rise much as the series progressed, but increasing knowledge made it possible to use less of the weight on protection from pressure and temperature and more on scientific content. Centre of gravity was offset to assist self-orientation during descent, with separate hermetically sealed compartments for the instrumentation and dual-parachute descent system. Launch into Earth parking orbits was from Tyuratam by 43m-high Voskhod rockets providing lift-off thrust of 509,840kg. A 2-engine 2nd stage added 140,000kg thrust, and the 3rd stage accelerated the craft to escape velocity. In the case of Venera 8 it was announced that this burned for 243sec and accelerated the spacecraft to 11·5km/sec (41,433km/hr), "somewhat greater than second cosmic velocity".

Experience with Venera 4, 5 and 6 showed that the design of the main spacecraft was satisfactory, but the landing vehicle was redesigned to withstand external pressures of up to 180 atmospheres and 530°C, increasing its weight over Venera 5 and 6 by about 100kg. The parachute canopy was made of cloth able to withstand temperatures up to 530°C. Published weight figures vary slightly; the total given for Venera 7 was once again 1,180kg.

Since Venera 4-7 had all studied the planet's night side it was decided that Venera 8 should aim for the sunlit side to check the amount of light reaching the surface through the cloud layers. This, it was hoped, would indicate how far the solar rays penetrated the thick atmosphere and perhaps explain how the heating of the atmosphere to such high temperatures took place.

A major advance on Venera 8 was a dual antenna which enabled it to transmit 60min data when it succeeded in reaching the sunlit side on 22 Jul 1972. Additions like this were possible because Venera 7 experience showed that weight-consuming insulation and protection against the elements could be reduced. A refrigeration unit was added to cool the interior below freezing during the early part of the descent; the dual-parachute system, first tried on Venera 7, enhanced the effectiveness of this by enabling Venera 8 to drop rapidly through the upper atmosphere and then slow down as it approached the surface. At touchdown a separate antenna, tripod-mounted with large pads and able to withstand heavy winds, was thrown off, coming to rest a few metres away. This was designed to ensure transmission even if the main craft rolled or fell so that its primary antenna was not pointed to Earth. In the event, it was possible to use both antennas.

Venera 9 and 10, described as "2nd-generation" Venusian explorers, were completely re-engineered, providing the basis for the subsequent successful landers. Their braking discs and shock-absorbing rings gave them a cotton-reel appearance. Soviet engineers began by redistributing weights between the orbiter and lander. It was also decided that all transmissions back to Earth - on this occasion at a distance of 80 million km - should be passed via the orbiter, thus lightening and simplifying the lander's transmitter. As before, the lander started as a sphere. Protection against over 2,000°C and pressure equivalent of 300 tonnes on its front part during entry into the Venusian atmosphere was provided by 2 jettisonable semi-spheres. An astonishingly elaborate procedure - which nevertheless worked perfectly - was devised to ensure that the craft touched down safely and yet passed through the cloud layers quickly so that its brief surface life was not diminished. Details, given under Venera 9, apply equally to Venera 10. To help it survive the interior was chilled beforehand to -10°C. The sequence of retro-rockets, parachutes, parachute jettisoning, aerodynamic braking by means of a circular metal disc, and, in the final seconds, retro-rockets and a metallic ring to absorb the impact reduced the entry speed of 38,520km/hr to 24-27km/hr at touchdown.

Venera 1 L 12 Feb 1961 by Sputnik 8 from Tyuratam, wt 643·5kg. Radio contact lost at 7·56 million km but spacecraft passed 99,800km from planet. This was the 1st Soviet planetary flight.

Venera 2 L 12 Nov 1965 by A2e from Tyuratam, wt 963kg. Passed 23,950km from Venus 27 Feb 1966 but failed to return data.

Venera 3 L 16 Nov 1965 by A2e from Tyuratam, wt 960kg. After 105-day flight impacted on Venus 1 Mar 1966, becoming 1st spacecraft to reach a planet. It failed, however, to return any planetary data.

Venera 4 L 12 Jun 1967 by A2e from Tyuratam, wt 1,106kg, of which 383kg was the entry vehicle. After 128·4-day flight, presumed to have impacted on 18 Oct 1967, 1 day before US Mariner 5 flyby. Descent capsule transmitted data during 94min parachute descent. Highest temperature recorded was 320°C. Also sent measurements of pressure, density and chemical composition of the atmosphere before ceasing transmissions, possibly as a result of striking 1 of 3 mountain ranges recently detected by radar bounces. This flight enabled later spacecraft to be redesigned so that they could penetrate more deeply and survive longer.

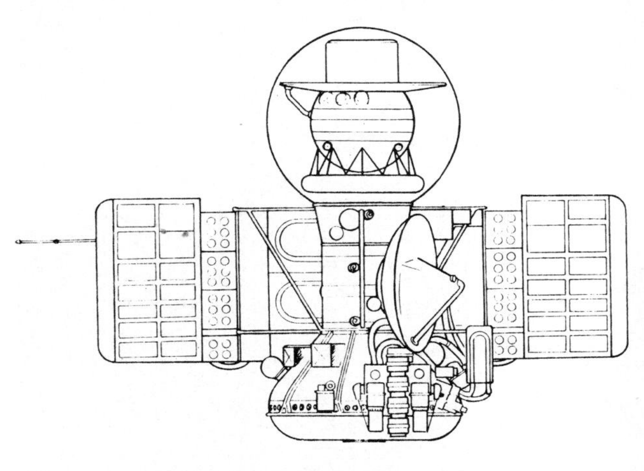

Venera 9 or 10 2nd-generation Venus satellite, showing the lander sphere on top and the orbiter relay station below. *(D. R. Woods)*

Venera 4 descent capsule with parachute attached. *(Graham Turnill)*

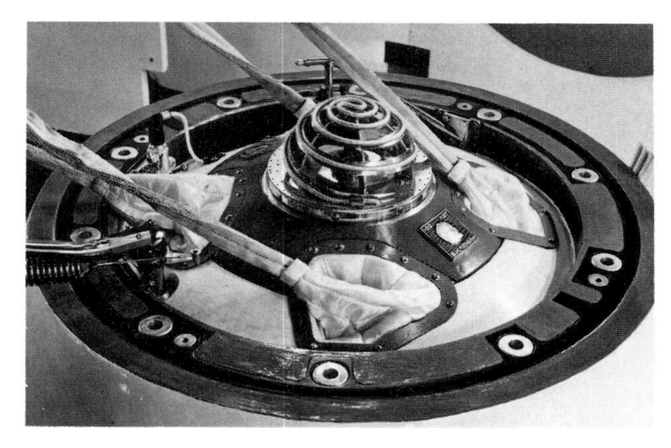

Close-up of Venera 4. *(Graham Turnill)*

Venera 5 L 5 Jan 1969 by A2e from Tyuratam, wt 1,130kg, of which 405kg was the entry vehicle. After 131-day flight the descent capsule separated at a distance of 37,002km from Venus. Initial entry velocity of 40,320km/hr was reduced by atmospheric braking to 756km/hr before deployment of main parachute one-third the size of that on Venera 4. The capsule entered the Venusian atmosphere on the planet's dark side on 16 Mar 1969 and transmitted data for 53min, travelling 36km into the atmosphere before being crushed.

Venera 6 L 10 Jan 1969 by A2e from Tyuratam, wt 1,130kg. After 127-day flight the descent capsule separated at a distance of 25,000km and, 24hr after Venera 5, entered the atmosphere on the planet's dark side. It transmitted data for 51min, descending 37·8km into the atmosphere before being crushed. Data from Venera 5 and 6 revealed that the concentration of nitrogen and inert gases in the atmosphere of Venus is 2-5%; oxygen does not exceed 0·4%; carbon dioxide represents 93-97%. Water-vapour content is very low. During descent temperature readings ranged from 28°C to 360°C. Pressure readings ranged from 0·5kg/sq cm to 28kg/sq cm.

Venera 7 L 17 Aug 1970 by A2e from Tyuratam, wt 1,180kg. After a 120-day flight entry into the planet's atmosphere began on 15 Dec. Distance from Earth at that time was 60·6 million km. After separation the descent craft's speed was reduced aerodynamically from about 41,400km/hr to 724km/hr. It was subjected to 350g, and temperature differences between the shock wave and vehicle reached 11,000°C. The parachute was opened 60km above the surface, when external pressure was about 0·7 atmospheres. The parachute canopy, of cloth designed to withstand temperatures up to 530°C, was bound by a Kapron cord. After a fast descent through the upper layers, the cord was made to fray and tear apart so that the parachute canopy opened fully, slowing the descent to allow fuller study of the lower layers. During this period only the gradually increasing temperatures were transmitted, the signals taking 3min 22sec to reach Earth and continuing for 23min after landing. During the flight the instruments had sent back data on the "powerful chromospheric flare" which began on 10 Dec 1970; it was possible to compare them with readings received simultaneously from Lunokhod 1 on the lunar surface and from satellites and ground observatories. When Venera 7 landed Soviet scientists were able to claim that for the first time information was being simultaneously received from 2 celestial bodies.

Venera 8 L 27 Mar 1972 from Tyuratam, wt 1,180kg. After 117-day flight and 300 million km Venera 8 reached the planet on 22 Jul, when the latter was 107·8 million km from Earth. It was manoeuvred to a touchdown in the narrow, crescent-shaped sunlit portion of the planet visible from Earth. The landing site, about 2,896km from the site of Venera 7, was selected to minimise the Earth distance. Earth was relatively low on the Venusian horizon, so that Soviet scientists could obtain readings from the sunlit portion; it was early morning at the landing site, with the Sun low on the local horizon. As the spacecraft entered the upper atmosphere the descent module separated while the service module went on to burn up in the atmosphere. The entry speed of about 41,696km/hr was reduced by aerodynamic braking and the parachute deployed at 900km/hr. During descent a refrigeration system comprising a compressor and heat-exchange unit was switched on to offset the 300°C temperatures found on earlier missions. Transmissions of temperatures, pressures, light levels and descent rates were interrupted for 6min during descent for calibration of onboard instruments; touchdown was at 09.29 GMT at 10° S, 335° Long. Surface light levels were considerably lower than Earth's. Surface temperatures were such that tin and lead would melt and iodine, mercury, bromine and sulphur evaporate.

Venera 9 L 8 Jun 1975 by Proton from Tyuratam, wt 3,376kg (lander 1,560kg). The 1st spacecraft to send to Earth a picture from the surface of another planet. Its launch and course were compared by Soviet experts to firing a rifle at a flying coin at a distance of about 1km. Having covered over 300 million km in 136 days it also became the 1st Venusian orbiter on 22 Oct with a closest approach of about 1,500km and a 2-day revolution period. Separation of the lander and orbiter took place 2 days beforehand. Since all the lander's signals were to be transmitted to Earth via the orbiter, the orbit had to be perfectly phased so that as the landing occurred the vehicles were approaching one another from opposite directions. At 09.58 GMT on 22 Oct the lander entered the Venusian atmosphere at 38,558km/hr. Aerodynamic braking reduced this to 899km/hr, then the elaborate new parachute system, using a total of 6 chutes, was automatically switched on by an overload detector. First the pilot parachute pulled out the deflection parachute, which pulled aside the upper part of the heat-protective sphere. That in turn brought out the braking chute, which cut the speed until it was possible for the 3 main chutes to be deployed. At 50km/hr the main chutes were jettisoned in their turn, with the final descent and landing once more being dependent on the aerodynamic landing shield. This elaborate system ensured a relatively fast descent through the cloud layers, in which earlier landers had tended to "hover", with the result that they had been overwhelmed by the intense heat. First signals were received 5min after the lander entered the atmosphere and continued for nearly 2hr 53min from the surface itself. The descent to the Beta region at 31° 41′ N, 293° 50′ Long had taken 75min.

The single but remarkable panoramic picture arrived 15min after touchdown. It made nonsense of centuries of theorising about what lay below Venus's dense cloud cover. It was expected that diffused, dim light would result in a poor picture. Instead, the scanning tele-photometers sent back a picture of startling sharpness. Far from landing in a sandy desert resulting from wind and heat erosion, Venera 9 was on a steep slope, possibly belonging to a volcano. The lander was lying 2,500m above Venusian mean level among a scattering of large, perhaps recently erupted rocks - possibly on top of one of them. In addition to the surface close-up picture, a 2nd panoramic view showed a section of horizon estimated at 200-300m and angular rocks of up to 1m; one, close to the lander, was about 30-40cm across.

Venera 10 L 14 Jun 1975 by Proton from Tyuratam. Identical with Venera 9, this spacecraft operated equally well. The lander was separated from the orbiter on 23 Oct. The descent procedure first used on Venera 9 was again successful, and this time took 75min. During it, measurements of the Venusian atmosphere and details of the physical and chemical contents of the cloud layers were transmitted to the orbiter 1,500km above and thence 80 million km to Earth. This landing, at 16° N, 291° Long on 25 Oct, was 2,200km from the first, much lower, and revealed a surface temperature of 465°C, atmospheric pressure 92 times that of Earth, and a wind velocity of 3·5m/sec. While the Venera 9 picture had showed a scattering of large rocks on what Boris Nepoklonov, Russia's chief Venusian topographer, described as "a typically young mountainscape," the Venera 10 picture showed "a landscape typical of old mountain formations". The rocks were not sharp, but resembled huge pancakes with sections of cooled lava or debris of weathered rock in between. Neither lander, when it hit the surface at a speed of 7-8m/sec, raised any dust, either of the lunar or Martian type. A Soviet planetologist said that while Venera 9 had landed on a high plateau, Venera 10 had landed in a stony desert. Venus should therefore be classed as a young, still evolving planet. In the month following the landings the Venera 9 and 10 orbiters made 15 and 13 revolutions around the planet, and Soviet scientists began to build up a large-scale composite picture of the planet's cloud blanket. Credit was given to the French-made ultra-violet spectrometer for measurements of the ratios of hydrogen and deuterium in the upper atmosphere.

Venera 11 and 12 L 9 Sep 1978 and 14 Sep 1978 by Proton from Tyuratam, wt ? 2,380kg (lander ? 1,560kg). The landers were apparently similar to the previous pair but, though successful in other ways, did not send back any pictures. During the 3½-month flight, when they were sometimes 1 million km apart, the 2 spacecraft, together with Prognoz 7, recorded 21 bursts of gamma radiation of varying intensity and dozens of weak X-ray bursts on the Sun in continuation of the Soviet-French programme. French instruments were also carried on Prognoz 7; use of the 3 vehicles permitted the origin of the bursts to be located with much greater accuracy. Venera 12, although launched 5 days later, arrived at Venus on 21 Dec, 4 days before its companion, after a 98-day, 240 million km journey. At separation, 2 days earlier, it was 66·4 million km from Earth, with data taking 8min 20sec to reach the flight control centre near Moscow. The transit module continued on a flightpath taking it within 40,000km of Venus so that it could relay the lander's data from "behind" the planet. Entering the atmosphere at 40,320km/hr, Venera 12 descended by parachute to 40km. From 62km chemical analyses of atmosphere and clouds were made, together with a spectral analysis of solar radiation in the atmosphere. "Rather frequent" electric charges - lightning - were also recorded. On the surface, at 7° S, 294° Long, transmissions continued for 110min; the amount of light was similar in intensity to the Earth level on an overcast day. On 23 Dec the Venera 11 lander was separated from the parent craft, which continued on a course taking it 35,000km from the surface. The lander touched down at 14° S, 299° Long, 800km from Venera 12, and operated for 95min in a temperature of 446°C and pressure of 88 atmospheres. Between them the 2 craft took 9 samples of the Venusian atmosphere at varying heights and confirmed that the basic components were carbonic acid and nitrogen; argon was also discovered. The expected sulphur was not found, though chlorine was. The 50km-deep clouds were found to be composed of droplets of concentrated sulphuric acid with a layer of gas beneath, and they let through 50% of radiation. The findings appeared to confirm theories that Venus's high surface temperature was the result of the greenhouse effect.

Venera 13 L 30 Oct 1981 by D1e from Tyuratam, wt ? 3,940kg. After a 4-month, 300 million km flight this provided the 1st soil analysis from the Venusian surface. During the flight galactic sources of gamma rays were studied with French instruments, and the interplanetary magnetic field with an Austrian magnetometer. On 27 Feb 1982 the descent module was separated from the parent craft, which passed within 36,000km of the Venusian surface, re-transmitting the lander's data and continuing into heliocentric orbit. The lander entered the atmosphere on 1 Mar at a speed of 40,320km/hr, descending by parachute to an altitude of 47km and continuing by means of aerodynamic braking. During the 62min descent measurements were made of the atmosphere's chemical and isotopic composition, electric discharges and cloud structure. Landing was at 7° 30′ S, 303° Long, on a plain called Phoebe, E of the volcanic Beta Shield region. 8 colour pictures, the 1st such views of Venus, were predominantly yellowish/orange, with an orange sky and clouds (the blue part of the spectrum is absorbed in the atmosphere), and showed an area of rust-coloured rocks, lava sheets and traces of chemical erosion rarely seen on Earth. Despite an 89-atmosphere pressure and 457°C temperature, the lander operated for 127min, 4 times longer than expected. The soil sampler successfully drilled into the surface and carried a sample to a sealed chamber in which temperature was maintained at 30°C and pressure at two-thousandths of that outside. Analysis showed a rock which Soviet scientists classified as leucite-basalt undergoing chemical weathering; it had a high content of potassium and magnesium, occurring on Earth only in volcanic areas like Hawaii and the Mediterranean.

Venera 14 L 4 Nov 1981 by D1e from Tyuratam, wt ? 3,940kg. After a flight similar to that of Venera 13 the lander was separated on 3 Mar 1982 and made a 62min descent, sending back details of atmosphere and clouds. Landing site was given as 13° 15′ S, 310° 9′ Long, a point about 1,000km SE of Venera 13; the site appeared to be on a 500m hill. Pictures showed an area covered with light-brown sandstone, lacking angular rocks but with some small potholes strewn with hillside waste. Venera 14 transmitted for 57min and the soil sample was taken in 465°C temperature and 94-atmosphere pressure. The rock proved to be a lava-like material similar to that found on volcanically active, mid-ocean ridges; it corresponded to "oceanic ... basalts widespread on Earth". Studies completed 2yr later showed that Venus once had a volume of water one-third that of Earth's, but the planet's proximity to the Sun led to its loss in the first 500 million years. Proof that Venus was seismically active was obtained with measurements of micro-shifts of the planet's crust. Shifts of just over 0·00001cm were detected at a point estimated to be 3,000km from Venera 14.

Venera 15 and 16 L 2 and 7 Jun 1983 by D1e from Tyuratam, wt ? 4,000kg. Possibly carrying terrain-imaging versions of the Soviet nuclear-powered radar ocean-surveillance craft. Having travelled 330 million km in 130 days, Venera 15 was placed in a 24hr, 1,000 × 65,000km Venusian polar orbit, the lowest approach being over the north pole on 10 Oct. The poles, it was pointed out, had not been accessible to Pioneer Venus, and on 16 Oct, in its 1st side-view radar session, Venera 15 obtained the 1st high-resolution pictures of the polar area. They showed impact craters, hills, major fractures, ledges and mountain ridges in imaging strips measuring 150km × 9,000km with 1-2km resolution. Venera 16 was placed in a similar 24hr orbit on 16 Oct; the Soviets said that it was planned to use Venera 15 for 150 days, followed by Venera 16 for a further 150 days.

By Jan 1984 67 communications sessions had been held with the orbiters, and a Soviet radio report from the Medvezhyi Gora long-distance space communications centre described Venus surface images building up on a TV screen as they were received by the 2,000-tonne dish aerial. Plains with small hills appeared first, followed by very high mountains similar to the Himalayas. Conditions "were similar to those in the firebox of a locomotive, with a 500° temperature and pressures in the order of 100 atmospheres." According to Alexei Bogomolov of the Soviet Academy of Sciences, conditions on Venus approximate to those on Earth several thousand million years ago. It was absolutely certain there was no life there.

Voskhod

History Controversy as to the origin and purpose of these 2 flights has never been completely resolved. One view is that the 1st resulted from pressure brought by the then Soviet Prime Minister, Mr Krushchev, on chief designer Sergei Korolev to perform a 3-man flight before America flew a 2-man Gemini spacecraft. This would certainly explain why, although Voskhod is officially described as "different from Vostok in both structure and equipment", no pictures have ever been released. It would also square with suggestions that it was so difficult to cram 3 cosmonauts into a stripped-down Vostok that it was necessary to fly them without spacesuits. However, these arguments cannot detract from the value and courage of Leonov's first spacewalk during Voskhod 2. The international sensation it caused - coupled no doubt with Leonov's demonstration of its relative safety and feasibility - led to a NASA meeting at Houston 11 days later (29 Mar 1965). There it was decided that on Gemini 4 the following Jun Edward White should carry out a similar spacewalk, although up to that stage the plan had merely been for the spacecraft hatch to be opened so that White could stand up, with head and shoulders protruding into space, without actually leaving the vehicle.

The Voskhod flights proved to be the last made under the leadership of Sergei Korolev, Russia's chief design engineer for rockets and spacecraft throughout the development of orbital flight. He died on 15 Jan 1966, 10 months after Voskhod 2, aged 60; only after his death was his identity publicly revealed. Korolev's health, like that of his deputy, Voskresenky, who died aged 52 after the Voskhod 1 mission, had been undermined by 6 years of imprisonment under the Stalin regime. Posthumous recognition came with the public burial of his ashes in the Kremlin wall. Cosmonauts present included Komarov, whose ashes joined those of Korolev less than a year later as a result of the Soyuz 1 accident.

Voskhod 1 The 1st 3-man flight and the 1st by a medical man. Vladimir Komarov (pilot), Konstantin Feoktistov (scientist) and Boris Yegorov (physician) were launched on 12 Oct 1964 and completed 16 orbits in 24hr 17min. The apogee was 409km, perigee 178km and incl 65°. For the 1st time the cosmonauts had no spacesuits; this was to save weight and space when fitting 3 men into the spacecraft. Dr Yegorov's seat was apparently set above and in front of the other 2. The usual telephone conversation between the spacecraft and the Soviet Prime Minister was notable because it proved to be Mr Krushchev's last public statement. While talking to them he observed that Mikoyan was "pulling the receiver out of my hand." Mr Krushchev was displaced the day after the landing.

Voskhod 2 Pavel Belyayev and Alexei Leonov were launched on 18 Mar 1965 into the highest orbit then attained: 172 × 495km, incl 65°. The flight lasted 26hr 2min and 17 orbits. Leonov later described how, over the USSR on the 2nd orbit, he donned his spacesuit with the help of Belyayev, entered the airlock and inflated his suit. When he emerged he pushed himself away from the spacecraft, and he could clearly distinguish the Black Sea with its very black water and the Caucasian coastline. He began rotating on his 4·88m tether 10 times a minute, but did not lose orientation and was able to maintain

Voskhod 2: a Leonov painting shows its telescopic EVA airlock. *(Theo Pirard)*

Alexei Leonov making history in 1965 with the 1st spacewalk. Now a general, he still plays a major role in Russia's manned space programme. *(Novosti)*

Voskhod 2: this view of the crowded interior explains why Leonov had difficulty in re-entering from his EVA. Note the globe on the tiny instrument panel: at least the cosmonauts always knew where they were. *(Novosti)*

communications with both Earth and spacecraft. His pulse rate was 150-160 when he left the spacecraft and peaked at 168. When he pulled too vigorously on the tether he had to put out his hands to avoid collision with the spacecraft. After 10min, during which his manoeuvres were watched by TV, Leonov was instructed to return to the airlock. Unofficial reports say he then ran into difficulties for the first time. His spacesuit had "ballooned", an effect predicted by British aerospace scientists, with the result that it took 8min of struggling before he was able to force his way in. Later in the flight the cosmonauts reported sighting an unidentified Earth satellite.

A major crisis occurred on the 16th orbit when the automatic re-entry system failed. For the first time it became necessary for a Soviet spacecraft to make a manual re-entry. An extra orbit was flown while Belyayev made preparations to fire the retro-rocket himself. When he did so he was at first uncertain about the spacecraft's attitude, but it proved to be correct. The landing was made in deep snow in a forest near Perm, about 2,000km N of the planned area. The cosmonauts had to wait 2½hr for the first helicopter, and then had to stay overnight at Perm before being flown back to base. Leonov suffered no ill effects from his spacewalk. Belyayev died of natural causes on 11 Jan 1970, aged 37.

Yuri Gagarin with the "Chief Designer," the only description of Sergei Korolev permitted until after his death at 60. His health was undermined by 6 years of imprisonment (during which he continued his work) under the Stalin regime.

Vostok

History Manned spaceflights began with Yuri Gagarin's 1-orbit flight in Vostok 1 on 12 Apr 1961. The name Vostok ("East") applies both to the spacecraft and the launch vehicle. Although the spacecraft were designed for automatic operation throughout the series, the cosmonauts' flightplans involved numerous astronomical and geophysical studies in addition to the development of manned spaceflight techniques.

Soviet preparations for the 1st manned flight started with Sputnik 2 in Nov 1957, when the dog Laika was placed in orbit and his behaviour monitored for 7 days. Recovery was not attempted and he died in space. Sputnik 4, launched on 15 May 1960, saw the launch of the 1st, unmanned Vostok prototype. The recovery failed and the re-entry section remained in space until it was burned up on re-entering 5yr later. Sputnik 5, the 2nd Vostok trial in Aug 1960, was successful, and 2 dogs, Belka and Strelka, were ejected and recovered in a parachute-borne container after 18 orbits. The 3rd trial Vostok (Sputnik 6) was successfully orbited in Dec 1960 but again the recovery system failed. The trials were repeated with 2 more Vostoks (Sputniks 9 and 19) in Mar 1961, and this time dog passengers were successfully recovered on both occasions. 18 days after the 2nd of these dogs, Zvezdochka, had flown, the first man went into space.

Vostok 1 Gagarin was launched from Tyuratam (though official records still give the site as Baikonur, 370km to the SE) at 09.07 Moscow time on 12 Apr 1961 and landed 1hr 48min later at the village of Smelovka, Saratskaya. His orbit had a perigee of 181km and apogee of 327km. Gagarin appears to have remained inside the spherical capsule for the landing, although the 5 subsequent Vostok cosmonauts all used ejection seats at 7,000m, presumably because of the danger of the spacecraft bumping too heavily when it hit the ground. It was 4yr before pictures of Vostok were released, and the secrecy surrounding the early Soviet flights, never fully relaxed, makes it difficult even today to establish with certainty precise details of the flight.

The handsome Gagarin, with an engaging personality, ready smile and quick sense of humour, had clearly been chosen with post-flight public relations in mind. His first Western visit was to England, where he lunched with the

Laika, a mongrel bitch, was the first living creature to be sent into space and experience weightlessness. She countered the transition by tossing her head. Her spacecraft, Sputnik 2, had no re-entry capsule and she died after 7 days. *(Novosti)*

Vostok 1 in the field in which it landed. This picture was not seen in the West until some years later. Note the heavy charring on the exterior of the capsule, caused by the searing heat of re-entry into the atmosphere.

Queen at a time when East-West relations were particularly strained, and he went on to make successful appearance in many other countries. He was only 34 when he was killed with another pilot in what seems to have been a routine training flight in a jet aircraft.

Vostok 2 The 1st 1-day flight. Gherman Titov, launched on 6 Aug 1961, completed 17 orbits in 25hr 18min. Apogee was 244km, perigee 183km and incl 65°; the orbit and inclination were similar for all the Vostok flights. A major factor in the Soviet decision to go straight from 1 orbit to 17 was the problem of a suitable landing site on Soviet territory if fewer orbits were done. Spacecraft improvements were said to include a more advanced air-conditioning system. Titov, who was 26, emerged with credit from man's 1st full day in space, and from the subsequent news conferences and hero's parade in Red Square. Only later was it disclosed that he had suffered serious disorientation during the flight and had inner ear trouble for some time afterwards.

Vostoks 3 and 4 The first group or double flight. Andrian Nikolayev, launched in Vostok 3 on 11 Aug 1962, completed 64 orbits in 94hr 27min. For the last 3 days he was accompanied in orbit by Pavel Popovich, L 12 Aug in Vostok 4, who completed 48 orbits in 70hr 29min. Although he became the 4th Soviet cosmonaut to fly, Popovich had in fact been the 1st to be appointed. At one time the 2 spacecraft were within 5km of one another. This has been attributed to the reliability of the Vostok launcher rather than to the sophistication of the spacecraft. Though at the time Western speculation was that the flight indicated that Russia was already moving towards rendezvous and docking techniques, there is no evidence that the Vostok spacecraft had any independent manoeuvring capability. The flight was notable for the 1st TV transmissions from space, and the 1st demonstrations of weightlessness seen by the public. Popovich was later said to have suffered some disorientation. The 2 cosmonauts, having used their ejection seats following simultaneous re-entries, finally landed only 193km and 6min apart.

Vostoks 5 and 6 The 2nd group flight. Valeri Bykovsky, L 14 Jun 1963 in Vostok 5, established a space record of nearly 5 days (119hr 6min) and 81 orbits which stood for more than 2yr until broken by Gemini 5. World attention was however focused on Vostok 6. Launched 2 days later, this carried the world's 1st spacewoman, Valentina Tereshkova, aged 26, who completed 48 orbits in 70hr 50min. According to some reports, Tereshkova was substituted at the last moment for a much more highly trained woman pilot who had become indisposed, and it is known that she suffered some disorientation and space-sickness. Tereshkova herself said soon after the flight that at the time of Titov's flight she was "not even dreaming of becoming a cosmonaut".

Five months after her flight she married Nikolayev in Moscow, with the Soviet Prime Minister, Mr Krushchev, leading the wedding festivities. The healthy daughter born to the 2 cosmonauts within a year was an added bonus for Soviet scientists, still much concerned with possible radiation damage as a result of spaceflight.

Zond

History When the series began in 1964 these spacecraft were described as "automatic interplanetary stations and deep spaceflight technology development tests". For Western observers this series has been the most puzzling of all Soviet programmes. President Keldysh of the Soviet Academy of Sciences confirmed in Jan 1969, following Russia's 1st manned docking, by Soyuz 4 and 5, that Zond spacecraft were "adapted for manned flight". However, he added a warning that such flights should not be expected "in the next 2 or 3 weeks". At that time, it seems, Russia was still undecided upon the division of space exploration between manned vehicles and automatic craft. Zond 5 was the world's 1st spacecraft to make a circumlunar flight and return safely to Earth, in Sep 1968; 2 months later the feat was repeated by Zond 6. Tortoises, insects and seeds carried on board showed no ill effects.

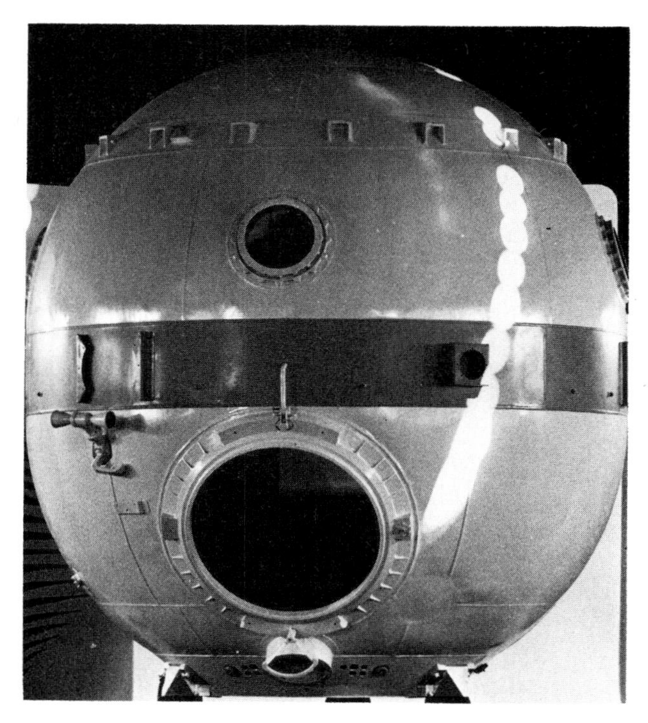

Vostok spacecraft, displayed years later at Paris Air Show. *(Graham Turnill)*

Spurred on by the fact that the Russians were apparently making final circumlunar tests with a manned spacecraft, America took a chance and sent Apollo 8 round the Moon at Christmas 1968. Though they did not land, the 1st manned flight around the Moon was an achievement ranking 2nd only to the 1st landing. Perhaps one day we shall learn why the Russians, so conscious of the prestige value of space firsts, failed to achieve this one when they clearly had the capability. Probably their rockets did not have sufficient lifting power for a fully manned Zond. One theory is that they could have sent a single cosmonaut but felt it was too risky to send him alone on the first circumlunar flight. If he became ill, or technical problems led to survival depending on manual control of the spacecraft, the workload might well have proved too much for him.

Cosmos 21 on 11 Nov 1963, C 27 on 27 Mar 1964, and C 379, 382, 398 and 434 in 1970 and 1971 were probably unsuccessful Zond missions. Nothing more was heard of the series.

Zond 1 L 2 Apr 1964 by A2e from Tyuratam. Wt 950kg. Believed to be a Venus probe, but communications were lost. It finally passed Venus at a distance of 99,779km and went into solar orbit after course corrections at 563,270km and 14 million km from Earth.

Zond 2 L 30 Nov 1964 by A2e from Tyuratam. Wt 950kg. Passed less than 1,609km from Mars but failed to return any data and went into solar orbit.

Zond 3 L 18 Jul 1965 by A2e from Tyuratam. Wt 950kg. Took 25 photographs of lunar far side at distances of 9,219-11,568km and transmitted them to Earth 9 days later from distance of 2,200,000km. Transmission was repeated from 31·5 million km, presumably in a test of photographic systems for Venus 2 and 3 later that year.

Zond 4 L 2 Mar 1968 by D1e (Proton) from Tyuratam. Wt 2,500kg. Flight-tested new systems "in distant regions of circumterrestrial space".

Zond 5 L 15 Sep 1968 by D1e (Proton) from Tyuratam. Wt 2,500kg. 1st spacecraft to circumnavigate the Moon and return to Earth, having been fired out of Earth parking orbit before completion of 1st orbit and into a free-return translunar trajectory. A Soyuz-like vehicle, it consisted of a heat-shielded re-entry module with camera, scientific package, radio and telemetry, re-entry control and parachute systems; and an instrument compartment with both major course-correction and low-thrust vernier engines, and solar-powered batteries. Tortoises, insects, plants and seeds were carried. After passing within 1,950km of the Moon it took Earth photos at a distance of

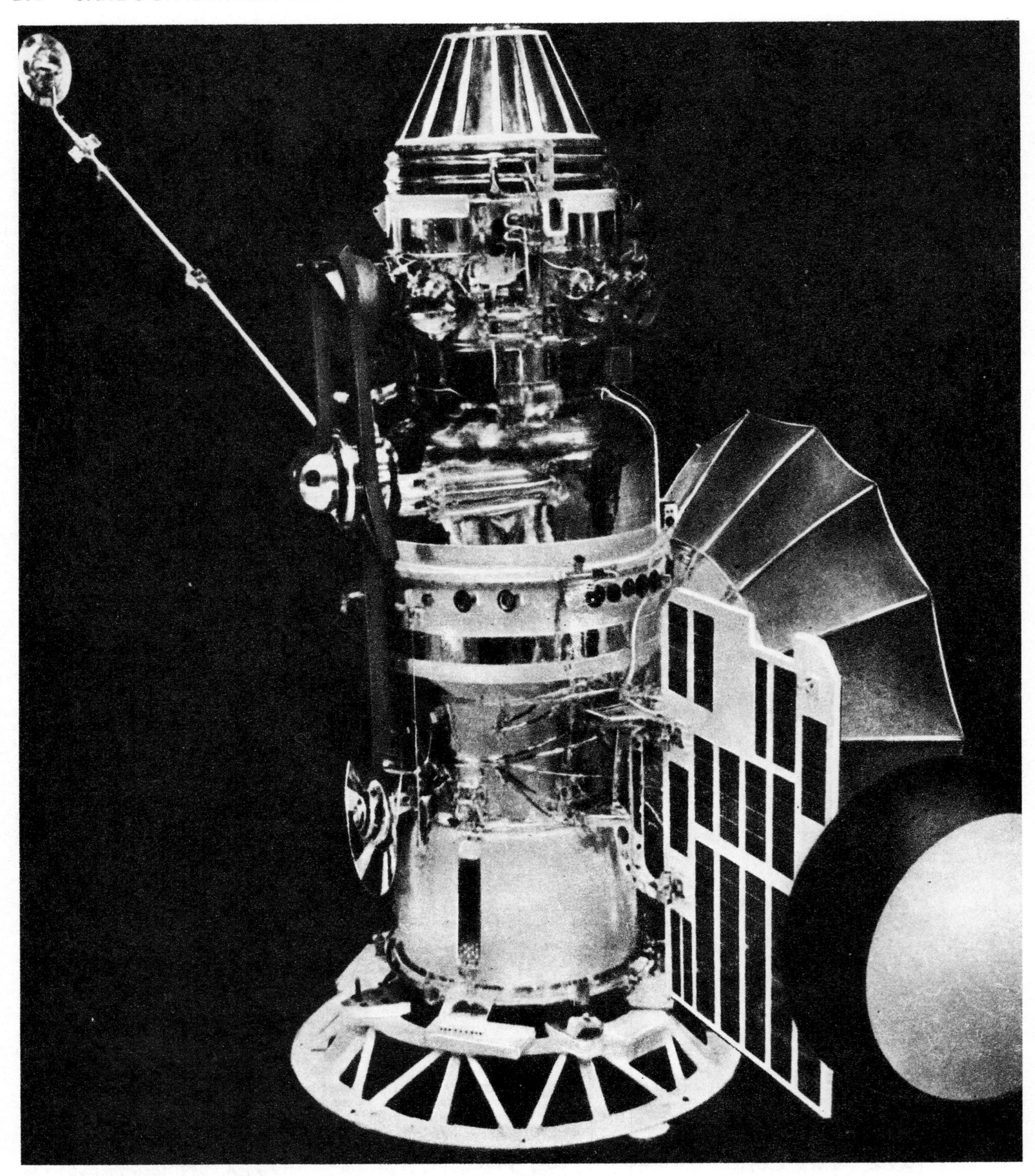

Zond 3, which sent back 25 photos of the Moon's far side.

90,000km (no Moon photos were mentioned). During the ballistic-trajectory return to Earth a tape-recorded Russian voice was heard calling out simulated instrument readings. Parachutes were deployed at 7km. Splashdown, at the end of a 7-day flight, was in the Indian Ocean; it was Russia's first sea recovery. The capsule was taken to Bombay and then flown back to the Soviet Union.

Zond 6 L 10 Nov 1968 by D1e (Proton) from Tyuratam. Wt 2,720kg. 2nd circumlunar flight, described by Tass as intended "to perfect the automatic functioning of a manned spaceship that will be sent to the Moon". The lunar far side was filmed as it passed around the Moon at a minimum distance of 2,420km. Re-entry was made by a skip-glide technique: the spacecraft's aerodynamic lift was used to bounce it off the atmosphere at a shallow angle, thus reducing speed and gravitational forces. Recovery, after another 7-day flight, was on land inside the Soviet Union.

Zond 7 L 7 Aug 1969 by D1e (Proton) from Tyuratam. Wt 2,720kg. 3rd circumlunar flight. Far side was photographed from distance of 2,000km, and colour pictures of both Earth and Moon were brought back. Re-entry was again carried out by the skip-glide technique, with recovery on 14 Feb.

Zond 8 L 20 Oct 1970 by D1e (Proton) from Tyuratam. Wt ? 4,000kg. 4th circumlunar flight, with more colour pictures of Earth and Moon. The spacecraft itself was photographed by an optical telescope at distances up to 277,000km from Earth; the telescope was pointed with the aid of an onboard laser. The main variation in this flight was re-entry from the northern hemisphere, as opposed to the normal southern hemisphere approach for touchdown on Soviet territory. The result was Russia's 2nd sea recovery, following splashdown on 27 Oct in the Indian Ocean. Photographs taken by Zonds 3, 6, 7 and 8 were used to compile a 3-volume atlas of the Moon's far side.

INTERNATIONAL SPACE PROGRAMMES

AMPTE

Active Magnetospheric Particle Tracer Explorers (AMPTE) are dealt with under the US Explorer programme. Germany and Britain are participating and providing 2 of the 3 satellites.

Apollo-Soyuz

History The plan for a joint US/Soviet spaceflight, later named the Apollo/Soyuz Test Project (ASTP), was included in an agreement on the peaceful exploration of outer space signed by President Nixon and President Kosygin on 24 May 1972. Target launch date was 15 Jul 1975, and 4yr of intense preparations were rewarded with on-time lift-offs from launchpads 16,000km apart. The crews were selected and announced 2yr before the flight; this was the 1st time that names of cosmonauts were known in the West before they reached orbit. Average age was 45: the occasion called for men of diplomacy as well as experience. US commander was Tom Stafford, with Docking Module pilot Donald K. Slayton making his 1st flight at 51, 16yr after being chosen as one of the original Mercury 7. Vance Brand was Command Module pilot, having replaced Jack Swigert following a NASA inquiry into the Apollo 15 and other philatelic deals. Soviet commander was Alexei Leonov, 1st man to walk in space, with Valeri Kubasov as flight engineer. The original plan – an Apollo and Soyuz docking with a Salyut – was dropped in favour of a direct link-up of an Apollo and Soyuz with the aid of a docking module 3·15m long, 1·4m dia and 2,012kg in weight. At first few in the West believed the flight would ever take place. But Russia, facing a period of setbacks and failures, badly needed to improve in such areas of management technique as quality control, about which she could learn much from America. For America, ASTP had immediate political benefits, gave her scientists their first opportunity to learn something about the Soviet space programme, and provided some much needed manned spaceflight activity during the long gap between completion of Skylab and the start of Space Shuttle flights. There was also a good deal of left-over Saturn/Apollo hardware which would otherwise be wasted.

Hardware Each country independently designed its docking system to meet the requirements of a jointly defined "common interface" which included structural rings, seals, latches and guide petals. The system did away with the docking probe which had caused problems throughout the Apollo programme (and finally on ASTP itself when used for the CSM/Docking Module link-up, carried out in the same way as the extraction of the Lunar Module from the Saturn 4B on the Moon flights). This was the one major technical advance achieved by the joint flight, since NASA decided that the new docking system would be used on future Shuttle missions. Designed and constructed in the US, the docking assembly was tested jointly by NASA and Soviet engineers. Both spacecraft can use it in either an active or passive mode for docking. Each half consists of an extendable guide ring with 3 petal-like plates, with a capture latch inside each plate. The spacecraft in the active mode advances with its guide ring thrust forward on 6 hydraulic attenuators, or retractable arms. As the spacecraft touch, the 6 capture latches (3 on each craft) engage on the guide ring. The active craft then retracts its guide ring, pulling the 2 spacecraft together against pressure seals. 8 structural latches then engage to hold them firmly and safely together.

Docking Module Length 3·15m, max dia 1·4m, wt 2,012kg. Basically an airlock with docking facilities at each end to permit crew transfer between the Apollo and Soyuz spacecraft. It contained control and display panels, VHF/FM transceiver, life-support systems and storage compartments. Equipment included oxygen masks, fire extinguisher, floodlights and handholds, junction ("J-Box") for linking Soyuz communications circuits to Apollo's, TV equipment, etc. Also housed in the DM was the Multipurpose Electric Furnace, which enabled the

Apollo-Soyuz: artist's concept depicting more clearly than the actual photos the only occasion when East and West have met in space. The original concept had been for Apollo to dock with a Salyut. *(NASA)*

crews to conduct 7 high-temperature experiments to investigate crystallisation and materials processing in weightlessness. 4 spherical tanks attached outside the DM contained 18·9kg of nitrogen and 21·7kg of oxygen for use during crew transfers. Because Apollo's normal atmosphere in orbit is 100% oxygen at 5lb/sq in and the Soyuz atmosphere an oxygen/nitrogen mix at 14·7lb/sq in, crew transfers provided a major problem. To avoid developing what divers call "the bends", the cosmonauts would have needed to spend a long time, possibly 2hr, in the DM airlock, pre-breathing oxygen to purge suspended nitrogen from their bloodstreams before passing into the Apollo. The problem was overcome, however, when the Russians agreed to lower the Soyuz pressure to 10lb/sq in during the joint exercise. Transfers were then possible in shirt-sleeves. The DM system was always operated by an Apollo crewman, with 2 men in the DM during a transfer. The rules were that there must always be one cosmonaut in the Soyuz and one astronaut in the Apollo; hatches at both ends of the DM, with pressure-equalisation valves, made it possible to transfer without disturbing the atmosphere in either spacecraft.

Mission summary The most complicated countdown in space history called for the co-ordination of 2 Soyuz and 1 Apollo at sites 16,000km apart. Russia's Soyuz 19 left the pad within 10sec of the planned time at 12.20 GMT on 15 Jul 1975; the Apollo followed within 3·8sec of its appointed time 7½hr later. Soyuz lift-off weight was approx 317,520kg, including 6,690kg for the spacecraft. Apollo/Saturn 1B lift-off weight (it was the 32nd and last Saturn, none the worse for being 7yr old) was 567,000kg; in-orbit weight of the Apollo/Docking Module was 14,679kg. Another Soviet first was provided by live TV of their launch. Both crews had technical troubles and were tired but still good-tempered when it came to the docking on 17 Jul. The Soyuz crew spent much time attempting to repair their TV system, which failed to provide any pictures of the Apollo throughout the mission, so that all the watching world saw was Soyuz from Apollo via the latter's excellent TV system. It had been agreed that if there were problems, attempts to achieve the first international link-up would continue for 3 orbits, followed by a rest and more attempts the next day. In practice the new docking system worked so smoothly that a soft docking was achieved 6min early over the Atlantic, at the start of the Soyuz's 36th orbit and the Apollo's 29th. The agreement that each crew would speak the other's language also worked well. (The Apollo crew were said to have spent over 1,000hr learning Russian.) It was 3hr after docking, with Stafford and Slayton sealed in the Docking Module, when the Soyuz hatch was finally opened and the handshakes in space took place. The cosmonauts read from prepared texts, Mr Brezhnev and President Ford joining in the mutual congratulations amid a good deal of confused communications.

Because of its superior manoeuvring ability and much larger supply of propellant (1,290kg compared with about 227kg on Soyuz), it had been agreed that the Apollo would play the major active role, carrying out both the 1st docking and a brief 2nd docking after separation. The Soyuz had in effect waited in orbit for the Americans to locate and dock with it 52hr later. There was much public confusion because the Soyuz was in the active mode for the 2nd docking, but the manoeuvres were in fact carried out by Slayton. The crews worked together, ate together and visited each other's craft, holding a TV news conference and carrying out joint experiments in the Docking Module, for 43hr. After some anxiety as to whether its pressurisation system had been damaged by what the Russians referred to as Slayton's "heavy" docking, the Soyuz re-entered on its 96th revolution and touched down amid live TV coverage 87km NE of Arkalyk in Kazakhstan to conclude a 5-day 22hr 31min flight. The Apollo crew continued with a busy programme of Earth-resources observations and experiments for another 3 days, on the 8th day holding a TV news conference at which for the 1st time newsmen could talk directly to them. Re-entry and splashdown on revolution 138, to conclude a 9-day 1hr 28min mission, was thought to have gone well but brought near-disaster. Crew switching errors during the descent led to poisonous RCS gases being drawn into the spacecraft, where they affected the crew's lungs. Doctors insisted that they remain in hospital at Honolulu (in intensive care at first), and wives and children were flown out to them. Stafford and Brand were then announced to be fully recovered, but a small shadow on one of Slayton's lungs, shown to have been there on pre-flight X-rays, led to an operation. It proved to be a benign lesion, with no cancer, and he was finally able to accompany Stafford and Brand on a much delayed visit to Russia in early Oct for a reunion with Leonov and Kubasov. Subsequent discussions of more joint projects to study the Moon and planets, meteorology and space medicine have brought some results over the years, particularly in the area of Venus exploration. But the prospect of more Soviet-American manned missions, including a joint journey to Mars, now seems further away than ever.

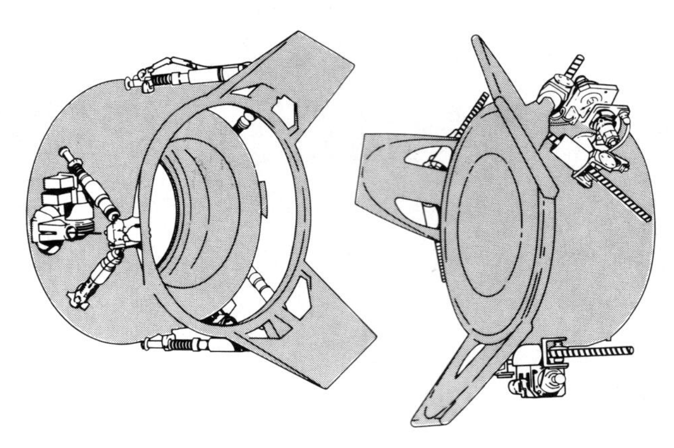

Complementary US and Soviet docking system. It proved to be a major advance on the earlier Apollo docking probe.

The actual handshake: Stafford greets Leonov through Hatch 3. The two men have remained close friends ever since. *(NASA)*

Arabsat

History The Arab Satellite Communications Organisation consists of 22 nations and organisations of the Arab League. In May 1981 a contract worth $133,350,000 was awarded to Aérospatiale of France and Ford Aerospace of the US for development and manufacture of 3 satellites - 2 flight models and 1 spare - capable of being launched either by Ariane/Sylda or Space Shuttle. With an in-orbit weight of 680kg, 21m width with solar panels deployed, and 7yr life, they will provide member nations with telecommunications and community TV. Later in the year, following Congressional protests that Arabsat included Libya, the Palestine Liberation Organisation and S Yemen, and that this back-door sale of satellite equipment through France had possible military implications, the State Dept withdrew its approval of the Ford sale. This was later restored, and in Nov 1982 Arabsat and NASA signed an agreement for a Shuttle/PAM-D launch at a cost of $20.1 million; an earlier Ariane launch had

Arabsat will be used by 22 member states and organisations of the Arab League. *(Aérospatiale)*

been negotiated for $23 million. By Mar 1984 Arabsat A had slipped to Ariane's 12th launch, in Nov 1984, and Arabsat B to Shuttle Mission 22 (51-G) in May 1985. Arabsat's operations centre will be in Riyadh, Saudi Arabia, with a back-up centre in Tunis.

Aureole

Aureole 1 L 27 Dec 1971 by C1 from Plesetsk. Wt ? 300kg. Orbit 410 × 2,500km. Incl 74°. The 1st Soviet-French satellite, launched under international project "Arcade", for "Arctic auroral density". It studied the aurora borealis, believed to be huge "plasma explosions" which change the ion composition of Earth's upper atmosphere and affect radio conditions. Instruments were designed and supplied by Russia's Space Research Institute and France's Toulouse Centre for Study of Space Radiation. Orbital life approx 70yr.

Aureole 2 L 26 Dec 1973 by C1 from Plesetsk. Wt ? 400kg. Orbit 400 × 1,975km. Incl 74°. Continued the previous satellite's work; little was heard of the results. Life approx 30yr.

Aureole 3 L 21 Sep 1981. Wt 1,000kg. Orbit 380 × 1,920km. Incl 82·6°. France's CNES said that this was the 1st Soviet spacecraft to be evaluated, developed and installed using joint Franco-Soviet facilities and staff. Designed for 6-month investigation of the upper latitudes of Earth's magnetosphere, carrying 4 French as well as Soviet experiments on arms and masts that included a 17m boom for the magnet system. France also supplied the on-board programming and telemetry unit; both countries received data direct. 60yr orbital life.

Biosatellites

History At least 6 international biosatellites have been launched by Russia in the Cosmos series, with America participating in the last 4. This development was partly a result of the collapse of America's own Biosatellite programme in 1969. Soviet and US scientists have used more than 50 biological specimens in this series, ranging from viruses to mammals, and including monkeys, dogs, rabbits, guinea pigs, rats, mice, tortoises, higher plants, fish, birds and insects. Such research began with the aim of protecting spacemen from the adverse effects of long-term spaceflight, and to discover the limits to their tolerance. Now there are many other aspects to the programme. The livers of successive generations of fruit flies hatched in space are being analysed to see what effects weightlessness has on ageing. France is particularly concerned about radiation. One particular US interest is spaceflight's effect on cancer cells, the growth of which can be slowed by an increase in gravitational force. At least 9 countries have participated. Launches are by A2 from Plesetsk into orbits of approx 220 × 400km with 62·8° incl; weight is approx 5,900kg. Despite the Aug 1983 furore over the shooting down by Russia of a South Korean airliner, NASA was told that it could continue with plans for a Soviet/US/French flight in Nov 1983 carrying monkeys and rats.

Cosmos 605 and 690 L 31 Oct 1973 and 22 Oct 1974. Based on the original Vostok spacecraft. Cosmos 605 contained several dozen rats, 6 boxes of tortoises, a mushroom bed, 4 beetles, and living bacterial spores. Recovered after 3 weeks, C 605 provided data on the reaction of mammal, reptile, insect, fungal and bacterial forms to prolonged weightlessness. Cosmos 690 contained specially trained albino rats. Biologists, physicians, physicists and engineers from Russia, Czechoslovakia and Romania subjected the rats by ground command to daily radiation doses from a gamma-ray source carried on board. When they were recovered 20½ days later many rats had developed lung trouble, and their blood and bone marrow showed that in space radiation had a much greater effect than in ground conditions. As a result of this mission, Russia invited America to participate in at least one of a further series.

Cosmos 782 L 25 Nov 1975. As a direct result of the ASTP collaboration, American experiments were flown on this Soviet spacecraft for the 1st time. Announced by Russia as the start of an international biological programme, it carried 14 experiments prepared and conducted by 7 countries. At the suggestion of US biologists C 782 was fitted with a centrifuge with revolving and fixed sections in which identical groups of animals, plants and cells could be compared; the subjects included white rats provided by Czechoslovakia. Each could move freely in a separate cage equipped with feeding troughs, water bowls, lamps and air vents. Tortoises were placed in holds in the centrifuge, which was used to simulate gravity. Biologists from both the US and Soviet Union studied the ageing effects of the flight on the liver of fruit flies. Plant tissues with cancerous growth grafted on to them were provided by Moscow and Colorado universities in order to test US observations that cancerous growths slow down if the force of gravity is increased. French research involved tobacco seeds and crayfish. Hungary, Poland and Romania took part in post-flight research. C 782 was successfully recovered on 15 Dec after 19½ days, the flight being cut short because of snowstorms in the Siberian recovery area. NASA later described the results as "very encouraging".

Cosmos 936 L 24 Aug 1977. This time 9 countries participated in a 19½-day mission, with the US providing 7 experiments, mainly as follow-ups to C 782. Czechoslovakia provided 30 bacteria-free white rats, distributed with other animals in 2 centrifuges, each with 5 pens for individuals. The Soviet Institute of Medical and Biological Problems carried out parallel work with NASA's Ames Research Centre, including studies of similar sets of animals on Earth. Bulgaria, Hungary, GDR, Poland and Romania also took part.

Cosmos 1129 L 26 Sep 1979. The same 9 countries participated, with Russia and America providing 13 experiments, including 38 white rats from Czechoslovakia. The 1st attempt to breed mammals in space, it proved unsuccessful. On the 2nd day in orbit 2 males were allowed to mingle with 5 females, but no litters resulted after the 18½-day flight. There was an equally negative result with an identical group housed in the same conditions on the ground and subjected to exactly the same vibration and re-entry forces, so that the one difference between the two groups was weightlessness. Fertilised quail eggs were also carried on the flight, but the incubator failed on the 13th day.

Cosmos 1514 L 14 Dec 1983 from Plesetsk. Orbit 226 × 288km. Incl 82°. This Soviet-US mission survived even the shooting-down of the Korean airliner, but was 4yr later than US scientists expected. NASA's Ames Research Centre participated in 3 animal experiments. Two 4kg Rhesus monkeys were flown implanted with sensors to permit monitoring of bloodflow to the head for space-sickness and biorhythm studies. 18 pregnant white rats were used for studies of the effects of zero g and radiation, and insects, guppy fish and plants were also flown. Recovered after 5 days, the animals, said to have "withstood the conditions of weightlessness well," were taken with the other specimens to the Institute of Medico-Biological Problems of the USSR Ministry of Health. In Apr 1984 Soviet scientists reported that the monkeys' blood circulation had slowed down instead of increasing early in the flight, suggesting that existing theories about the cause of spacesickness were incorrect.

Cosmos consortium

This is a consortium of 8 companies from 6 European countries linked to Ford Aerospace, California, which was awarded the Meteosat development contract by ESA in 1973. They are developing the Exosat astronomy satellite, with MBB of Germany as prime contractor, and contribute major components to the Ford Intelsat 5 spacecraft.

Cospas/Sarsat

This international search and rescue satellite system, operating on the 121·5 and 243MHz emergency location transmitter beacon frequencies, was started with the launch of Cosmos 1383 on 1 Jul 1982 from Plesetsk (orbit 1,004 × 1,041km, incl 83°). Within a month Cospas had saved 7 lives by making it possible to locate crashed aircraft in Canada and New Mexico, and a capsized catamaran off the New England coast. The US added Sarsat capability to NOAA-8, L 28 Mar 1983 (orbit 808 × 830km, incl 98·75°). The circular polar orbits, with a field of radiovision of 5,000-6,000km, permit the satellites to cover the whole planet. Norway, Britain, France and Canada are participating, and there are receiving terminals in Ottawa, Toulouse and Moscow. France has also given Boeing a $50,000 contract to evaluate the carriage of a combined Sarsat/Argos payload by the proposed Mesa satellite.

By the time C 1447 (Cospas 2), L 25 Mar 1983, had been added the Soviets were claiming that 16 people had been rescued, and this had risen to 145 by the end of 1983. By then the goal was to have 4 spacecraft operational at all times, managed by Inmarsat or a similar body. A major disadvantage of the new system, however, was the large number of false alarms from aircraft emergency beacons, which imposed a heavy burden on both military and civil authorities.

European Space Agency

Status By 1983, with the Ariane launcher at last declared operational, 17 satellites (including 8 operational) behind it, and at least 14 exciting new projects (including a free-flying Spacelab) under development, the 11-nation European Space Agency had reached maturity. A study of ESA history shows how successfully the Agency has stimulated both scientific space research and the development of applications satellites, with their now clearly apparent commercial prospects. Inevitably, however, ESA's troubles are not yet entirely over. Just as there was rejoicing that the 2½yr hiatus in ESA satellite

Ariane's successful 3rd launch, carrying Meteosat and Apple. *(Aérospatiale)*

launchings resulting from Ariane delays was over (and in any case had been largely nullified by similar delays with the US Shuttle), half of the 4 launches planned in 1982 were lost because of new problems with the satellites (see Launchers). By then ESA had handed over launch marketing to a newly created European company called Arianespace. ESA's pioneering work with communications satellites like OTS and ECS had led to the creation of Eutelsat to handle Europe's comsats, and helped to create Inmarsat to handle ship-to-shore traffic.

A sign of maturity was the 1980 decision to go it alone on a mission to intercept Halley's Comet in 1986. Early troubles with Spacelab were overcome with the completion and delivery of the first example to NASA for the STS-9 mission in Nov 1983, and a study to overcome commercial doubts about this versatile equipment has been started. In anticipation of future success in this area, 2 of ESA's astronauts have been upgraded from payload to mission specialist to enable them to participate in more advanced activities like spacewalking.

Anxieties, evident in 1979, that Germany and Britain would lose patience with the stresses and strains within ESA and follow France's lead by starting their own national space agencies lessened in the following 2yr. NASA's unilateral withdrawal from ISPM, the joint solar-polar mission, had an unexpectedly unifying effect as ESA members combined to protest with some success. NASA changed places with ESA as the unreliable partner in their relationship. France finally ratified the ESA convention in Oct 1980, at last giving the agency full legal and political status. It is hard to see what political advantage France had derived from delaying the signing so long, since she had been the most active member since ESA's inception 5yr earlier. National pressures within the ESA council have been responsible for frequent changes of director-general. The 1st, Roy Gibson of Britain, was replaced in May 1980 by Erik Quistgaard of Denmark; he in turn was replaced by Prof Reimar Lust of W Germany in May 1984. By then development of Ariane, Spacelab and the maritime satellites were all virtually complete. ESA began looking ahead to space station participation, the setting up of a European remote-sensing system, a joint solar-terrestrial programme (ISTPP) with the US and Japan, and a possible 24-spacecraft navsat system for civil navigation. It was also consulting member countries about the possibility of forming an international satellite group to monitor adherence to arms agreements.

Spacelab 1 in *Columbia*'s payload bay. The Z-shaped docking tunnel matched the levels of the *Columbia* and Spacelab hatches, and allowed longitudinal movements during launch and re-entry to avoid overstressing the tunnel. *(NASA)*

History Formation of a European Space Agency was agreed by 11 countries at a meeting in Brussels on 31 Jul 1973, and ESA came into operation in May 1975. The widely differing views of the biggest participating countries were resolved by agreement that Germany would be the biggest contributor to Spacelab, France to production of Ariane because she was the only country insisting on an independent European launcher, and Britain to production of Marots, a maritime communications satellite.

Because every member has a different currency and ESA regularly has to deal in at least 17, a universal currency of "Accounting Units" (AU) was created, with a value at that time of AU1 = $1·3 (now around $1·06). With constantly varying inflation rates and currency valuations, member states' percentage contributions also vary. By 1982 they were as follows: Belgium 3·0, Denmark 1·1, France 20·4, W Germany 15·6, Italy 8·9, Netherlands 2·8, Spain 2·1, Sweden 1·6, Switzerland 1·4, United Kingdom 10·3. Ireland, the 11th member state, together with Austria, Norway and Canada, which participate in experiments and other activities, added a total of 1·5%, and ESA derived 31·3% from other sources such as NASA and a wise investment policy. There are co-operation agreements with Japan (now also able to offer launch services), India, the US and the USSR. The budget for 1982 totalled 840·6 million AU, with Ariane taking the largest percentage at 20·2, science programmes 12·7, telecommunications programmes 9·9, and Spacelab 10·0. The general budget (administration) accounted for 14·9%.

With a staff of 1,400, of whom about 600 are engineers,

ESA's European Space Operations Centre at Darmstadt, West Germany, has 250 staff responsible for satellite control, tracking and data-processing. *(ESA)*

120 scientists and 22 technicians, ESA's headquarters is in Paris. Its largest establishment, ESTEC (European Space Research and Technology Centre) with over 800 staff, is at Noordwijk, Netherlands, and is responsible for the study, design, development and testing of both space components and complete vehicles. Next comes ESOC (European Space Operations Centre), at Darmstadt, Germany, with about 250 staff responsible for launch control, satellite tracking and data-processing, followed by the increasingly important SPICE (Spacelab Payload Integration and Co-ordination) at Porz-Wahn, Germany, which also provides a training centre for ESA's astronauts.

Like NASA, ESA is entirely non-military, its main aims being to develop a long-term European space programme covering research, technology and applications. To achieve this it must progressively Europeanise the parallel national space programmes on which many European nations still spend much of their meagre space budgets. In the early days of ELDO and ESRO (detailed below), there was much national conflict about the need to build European rockets capable of orbiting large satellites independently of either Russia or America. The scientists mostly preferred to use their limited money and resources to develop satellites, and to pay America for rockets and launch facilities. Technicians and some politicians took the view that America would ultimately be reluctant to provide launching facilities for TV, telephone and communications satellites which would compete with her own systems, particularly Intelsat, being developed for international use under American leadership. When the British Government finally ceased to support Blue Streak, originally developed as a British ICBM, ELDO withered away. But ESRO survived and between 1968 and 1972 8 scientific satellites were launched (7 successfully) on its behalf by America. Their major theme was the study of Sun-Earth relationships.

HEOS-1 and 2 (Highly Eccentric Orbit Satellites) are models of what can be achieved by genuine European collaboration. Their experiments were contributed by Imperial College, London; Rome University; the Danish Space Research Institute; the Netherlands' European Space Research and Technology Centre; the Centre d'Etudes Nucleaires, Saclay; Milan University; and the Max Planck Institutes in Garching and Heidelberg. They showed how the Sun continuously blows out a stream of plasma (the solar wind) at speeds of around 400km/sec. Where this meets Earth's magnetosphere it creates a permanent shockwave and squeezes the magnetic field into a truncated comet shape.

ELDO The 7-nation European Launcher Development Organisation was set up on 29 Feb 1964. The initiative came from Britain, anxious to find a use for Blue Streak, originally developed as a strategic missile, as the first stage of an independent European launcher system. Member countries were Belgium, France, W Germany, Italy, Netherlands and Britain, plus Australia; her government was equally anxious to keep the Woomera Test Range, N of Adelaide, in business. With France providing the Coralie 2nd stage, Germany the Astris 3rd stage, Italy the satellite test vehicles, Belgium the ground guidance station, and the Netherlands the telemetry links and other equipment, Europa 1 was built. By 12 Jun 1970 10 test firings had been carried out at Woomera. Despite a series of earlier failures there was confidence that the last, F-9, would place a test satellite in orbit. But loss of thrust on the 3rd stage, and failure to jettison the nose cone, prevented this. Despite growing doubts about the project and Britain's 1970 decision to withdraw, ELDO pressed ahead with development of the more advanced Europa 2, which had a French-built 4th-stage perigee motor able to place a satellite of up to 200kg in stationary orbit. The objective was to launch 2 Franco-German Symphonie telecommunications satellites in 1973 and 1974. Unfortunately the first test firing of Europa 2, from Kourou on 5 Nov 1971 with the world's space correspondents observing, ended in disaster. After 2½min the launcher deviated from its trajectory, became overstressed and exploded. Amid much bitter argument plans continued for the next test launch, F-12, and the vehicle was aboard a French ship en route to Kourou when on 27 Apr 1973 the project was finally abandoned in favour of creating a European Space Agency and building Spacelab for America's Space Shuttle system. Up to that time Europe had spent about $745 million on ELDO.

ESRO The 10-nation European Space Research Organisation was set up in 1964 to promote European collaboration in space research and technology. Member states were Belgium, Denmark, France, W Germany, Italy, Netherlands, Spain, Sweden, Switzerland and Britain. With headquarters in Paris and payroll of 1,100, its 1973 income was $127·5 million.

ESRO/ESA satellites

ESRO The early launches were out of sequence because of rocket faults and technical delays with the satellites. ESRO-2A, L 29 May 1967 by Scout from Vandenberg, weight 74kg, was intended to study solar astronomy and cosmic rays but failed to orbit because of 3rd-stage failure. ESRO-2B (Iris), L 17 May 1968 by Scout from Vandenberg (weight 164kg, orbit 325 × 957km), returned solar and cosmic data; dec 8 May 1971. ESRO-1A (Aurorae), L 3 Oct 1968 by Scout from Vandenberg (weight 84kg, orbit 252 × 1,231km), established "poleward" boundary of Van Allen radiation belts; dec 26 June 1970. ESRO-1B (Boreas), L 1 Oct 1969 by Scout from Vandenberg (weight 80kg, orbit 291 × 389km), was a duplicate of ESRO-1A but had only 52 days life as a result of booster malfunction. ESRO-4, L 22 Nov 1972 by Scout from C Canaveral (weight 130kg, orbit 245 × 1,173km), found that the temperature of the upper

Europa 2 on the Kourou launch platform. Its failure in front of the world's space correspondents led to the end of ELDO and creation of ESA. *(Guiana Space Centre)*

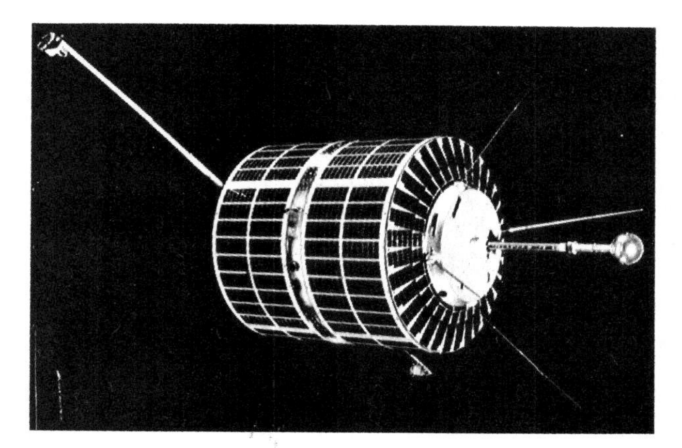

ESRO-1 established the boundary of the Van Allen radiation belts. *(ESA)*

atmosphere over the poles was higher instead of lower than that over the Equator; dec 15 Apr 1974.

HEOS-1 L 5 Dec 1968 by TAID from Cape Canaveral. Wt 108kg. Orbit 418 × 223,440km. Incl 28°. Re-entered on 28 Oct 1975, having operated perfectly for 7yr—6yr longer than planned. Studied interplanetary magnetic field conditions, and with HEOS-2 covered 7yr of the 11yr solar cycle. It not only measured solar particles in interplanetary space, but established in what strength and from which direction they arrived. By the time it re-entered after 542 revolutions, 90 scientific publications had resulted from correlating its results with data from other ESRO, European, American and Russian spacecraft.

HEOS-2 L 31 Jan 1972 by Delta from Cape Canaveral. Wt 117kg. Orbit 405 × 240,164km (raised in Jul 1973 to 5,442 × 235,589km). Incl 89·9° (later 88°). With HEOS-1 this discovered the plasma mantle. It also found that the very-high-energy particles emitted by the Sun, which reach Earth within a few minutes, travel through 3-dimensional, snake-like "tubes" in interplanetary space.

TD-1A L 12 Mar 1972 by Thrust Augmented Delta from Cape Canaveral. Wt 472kg. Orbit 524 × 551km. Incl 97·55°. Despite the failure of its tape recorder, a British/Belgian experiment surveyed 95% of the sky and examined 15,000 stars as seen through "wide-open" ultra-violet eyes. Dutch astronomers also used it to study individual stars in great detail, and found one, called Cygni, which was losing mass at a very high rate. Orbital life 20yr. A proposed TD-2 was abandoned.

HEOS-1 being assembled for what proved to be a 7yr mission studying the sun. *(ESA)*

COS-B Cosmic satellite L 9 Aug 1975 by Delta from Vandenberg. Wt 275kg. Orbit 337 × 99,067km. Incl 90·2°. (By 1980 orbit was 13,191 × 86,223km, incl 97·8°.) The 8th European and 1st ESA satellite, it was constructed by Belgium, Denmark, France, W Germany, Italy, Spain and Britain, and carried a single gamma-ray experiment provided by institutes in France, W Germany, Italy and Netherlands. Intended to spend 2yr studying the unexplained gamma-ray bursts detected by earlier US satellites, and shortly after launch was receiving radiation from the Crab Nebula. It was not switched off until 26 Apr 1982, after 6yr 8 months and over 1,000 highly elliptical orbits during which it had studied 63 different regions of the sky, searching for new radio pulsars. It carried out a complete survey of the galactic disc.

TD-1A surveyed 95% of the sky and 15,000 stars. *(ESA)*

GEOS: despite its failure to achieve stationary orbit, this spacecraft made a major contribution to the International Magnetosphere Study of 1976-79. ***(ESA)***

GEOS The world's first purely scientific satellite, intended to be the reference craft for IMS (International Magnetosphere Study, 1976-79), GEOS-1 (weight 574kg) was to have been launched from Kourou by Europe's abandoned Europa rocket. When it was launched (20 Apr 1977 by Delta from C Canaveral) a booster failure meant that geostationary orbit could not be achieved. However, ESOC successfully fired the apogee motor and achieved a "rescue orbit" of 2,050 × 38,000km. For 14 months its 7 experiments, integrated by the British Aircraft Corporation as main contractor for 15 companies in 10 countries, contributed significantly to IMS, making a detailed study of the magnetosphere's radial distribution of plasma, particles and waves in conjunction with ground-based observations in Scandinavia, Iceland, Antarctica and Alaska. It was reactivated in 1978 and 1979 so that 5 of its experiments could be used in co-ordination with 3 other magnetospheric craft, GEOS-2, Scatha and S-3-3. It also supported a campaign of lithium-release rocket launches from Alaska to study magnetospheric transport mechanisms. Total cost of GEOS-1 was $54 million.

GEOS-2, L 14 Jul 1978 by Delta from C Canaveral (weight 573kg, orbit 25,640 × 35,592km at 0·77° incl), was originally the back-up craft. It was successfully manoeuvred to 37°E and, despite a solar cell short-circuit a month later which degraded 3 of the 7 experiments, it provided 2yr of valuable operation and was then placed in a dormant mode until revived for 8 months in 1981 to help the EISCAT programme of upper-atmosphere motion measurements. In Jul 1979 GEOS-2 supported a programme in which ground-based rockets and 29 high-altitude balloons were launched from N Scandinavia in a series of auroral and magnetospheric experiments. It remained in use until the end of 1983. On 24 Jan 1984 the European Space Operations Centre (ESOC) boosted it 270km clear of geostationary orbit before its fuel ran out.

Marecs Three Marecs, a maritime version of the ECS satellite, are being produced by the Mesh consortium for ESA to lease to Inmarsat. This service will enable ships on the high seas to dial shore-based subscribers directly; merchant shipping will have reliable communications in the 4 and 6GHz bands vastly superior to the present congested HF circuits. To simplify ship and shore terminals Marecs is being equipped with a high-efficiency "shaped-beam" antenna, coupled with a transistorised L-band power amplifier. Marecs 1 and 2 have been leased to Inmarsat for 5yr for a total of $65 million. Marecs 1, L 20 Dec 1981 by Ariane L-04 (wt 582kg) and successfully placed in stationary orbit, finally became prime satellite for Atlantic shipping on 1 May after problems with its telemetry and command systems, caused by a build-up of electrostatic charges, had been overcome. These necessitated modifications to Marecs 2 and ECS-1, delaying the former's launch by 6 months. By Dec 1983 Marecs 1 was providing Inmarsat with 46 phone channels. Due to be stationed over the Pacific for a planned life of at least 7yr, Marecs 2 was lost on 10 Sep 1982 due to a 3rd-stage failure of Ariane L-05. Its cost had been given as $37·4 million plus $26·4 million for the Ariane launch. British Aerospace, as prime Mesh contractor, was given a £15 million contract in Oct 1979 for Marecs B2, now scheduled for launch from Kourou by Ariane in Sep 1984.

OTS The Orbital Test Satellite programme, ultimately to be so successful, started badly. OTS-1, intended to be ESA's 1st communications satellite, was lost on 14 Sept 1977 when the Delta 3914 exploded 54sec after lift-off from C Canaveral. Fortunately ESA, which had suffered losses and delays as a result of a series of US launch failures, had on this occasion insured the launch for a total of $29 million. OTS-2, the identical back-up craft, was successfully launched by Delta from C Canaveral on 12 May 1978 and placed in geostationary orbit at 10°E, its 6 SHF antennas covering W Europe, the Middle East, N Africa and Iceland. With 865kg weight (444kg in orbit), the 6-sided satellite, for which Hawker Siddeley (now part of British Aerospace) was prime contractor for the 10 Mesh countries, was built up from service, communications and antenna modules, enabling the later ECS and Marecs satellites to be readily developed from it. It was also one of the 1st comsats to operate in the 11/14GHz band,

Marecs: developed from ECS, 3 examples are being built for Inmarsat. As they become operational they provide ships with direct-dial services to shore-based telephones. ***(British Aerospace)***

particularly important for avoiding interference in Europe OTS has 4 wide-band communications channels able to accommodate up to 7,200 telephone circuits; alternatively, an individual channel can handle one or two TV transmissions. After 1yr of non-revenue experimental use, covering telephone, data transmission and TV (including transmissions between Europe, Egypt and Morocco), the commercial potential of OTS had been proved. By the end of 1980 the 1st experiments with videoconferencing between the German Bundespost and the British Post Office, using small Earth stations, had been successful, and TV programme exchanges on occasions like the Farnborough Air Show, using small terminals suitable for community reception, had aroused the interest of cable distribution companies. The ability of such systems to cope with emergencies was vividly demonstrated in Nov 1981 when the French PTT system at Lyons was destroyed by fire. Within hours the OTS satellite and a transportable Earth station had been brought into use to handle priority traffic. Britain started to use it in early 1983 for Project Universe, using 6 UK ground sites equipped with 3m dish antennas for high-speed communications between computers. Like GEOS-2, it was to be boosted clear of geostationary orbit when its useful life finally ended in 1984.

ISEE The International Sun-Earth Explorer programme was a 3-spacecraft mission, with ISEE-1 and 3 being provided by NASA and ISEE-2 by ESA as part of the International Magnetosphere Study (see GEOS and US Explorer 18, etc). ISEE-1 and 2 were launched together on 22 Oct 1977 by Thor-Delta from C Canaveral (weight 340kg and 166kg), orbit 340 × 137,900km, incl 28·9°. The spin-stabilised ISEE-2 could be manoeuvred to vary its separation from ISEE-1 from a few hundred to 5,000km to enable their matched experiments to distinguish between new and static magnetospheric phenomena. Apart from some minor problems (for instance, flashes of sunlight from the structure of ISEE-1 were entering a main telescope, making the data unusable), by 1980 most of the scientific objectives had been achieved. Phenomena better understood as a result of ISEE include the Earth's bow shock, the magnetosheath and magnetopause, boundary layers, interactions between the tail and aurorae, and particle populations and flows in the tail. During 1981 separation distances were successfully varied by 8 orbital manoeuvres. Operations were continued through 1983.

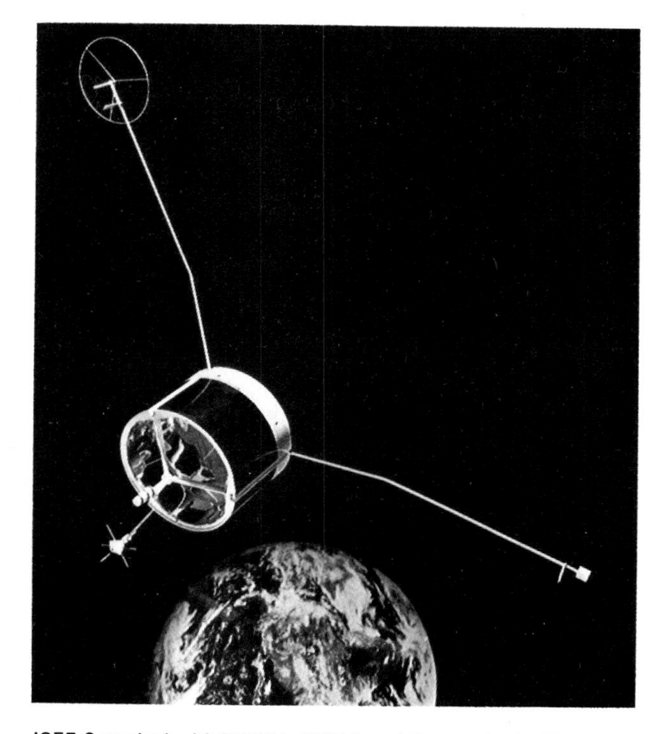

ISEE-2 worked with NASA's ISEE-1 and 3 to study the Earth's bow shock and magnetosphere during 1977-82. ***(ESA)***

Meteosat Although fully operational for only 2 of its 3yr planned life, Meteosat 1 (L 23 Nov 1977 by Delta from C Canaveral; wt 697kg, 300kg in orbit; geostationary at 0° above Gulf of Guinea) dramatically performed its role as one of 5 weather satellites girdling the Earth (see US - GOES). Its first pictures, taken less than a month after launch and covering a quarter of the Earth's surface, were among the best satellite images ever seen. They were obtained with a high-resolution radiometer with a ground definition of 2·5km in daylight and 5km in the infra-red; an additional channel observes water vapour, indicating its amount and altitude in the upper atmosphere. The radiometer is a 400mm-aperture telescope in which a series of mirrors is used to focus the optically collected signals onto 5 detectors, which convert the information to analogue electrical signals for transmission. Images are composed of 5,000 lines in visible light and 2,500 in the infra-red; line-scanning is achieved by the satellite's 100 rotations/min. Two images can be transmitted every 30min, 1 in the visible region, the 2nd either in the thermal infra-red or the water-vapour infra-red. The unprocessed images are good enough for immediate use, but to make the data compatible with international standards the image can be processed by computer before being sent back to the satellite for retransmission to weather stations. To its own information Meteosat adds data collected from Earth stations, buoys at sea and weather balloons to provide meteorologists, oceanographers, hydrologists and others with cloud heights and temperatures, wind velocities, Earth and sea surface temperatures, and much other information. Meteosat 1 also exchanged transmissions with its neighbouring satellites, GOES-1 over the Atlantic and GOES-3 over the Indian Ocean. In addition to demonstrating that the system could meet the requirements of Europe's meteorologists for accurate longer-range weather forecasting, Meteosat 1 contributed to 2 programmes set up by the World Meteorological Organisation (WMO): a continuous programme called World Weather Watch, and the Global Atmospheric Research Programme. (By Sep 1983, with funding from 8 countries, WMO was planning to add automatic weather data relay from airliners through Meteosat.)

Originally Russia was to have provided the Goms satellite to cover her own territory, with America providing 2 GOES satellites and Japan the GMS craft. When Russia pulled out of these projects America was able to fill the gap over the Indian Ocean by moving GOES-3, available as an in-orbit spare, into that position. The Global Weather Experiment, a major exercise involving virtually all the 147 member nations of WMO, was almost completed when, a day after its 2nd anniversary, Meteosat 1 suffered an apparent onboard overload. Despite efforts spread over some months its imaging capacity was never restored, although its data-collection function remain unimpaired. By then, however, more than 40,000 images had been processed and archived in digital form, and an outline proposal for a 5-satellite Meteosat programme for 1984-94 had been submitted to the European meteorological agencies.

Meteosat 2, L 19 Jun 1981 by Ariane L-03 from Kourou (weight 697kg full, 295kg empty; geostationary at 0°), was modified following Meteosat 1's radiometer problems, and repaired following damage during tests designed to assess its resistance to the additional vibration involved in an Ariane launch, as compared with Delta. It was soon sending back weather pictures for Europe, Africa and the Middle East that were even sharper than Meteosat 1's, but the system for collecting and relaying environmental data from Earth platforms could not be activated. Happily, however, the two satellites together provide a

Meteosat: an example of the images provided every 30min in 3 spectrum bands. *(ESA)*

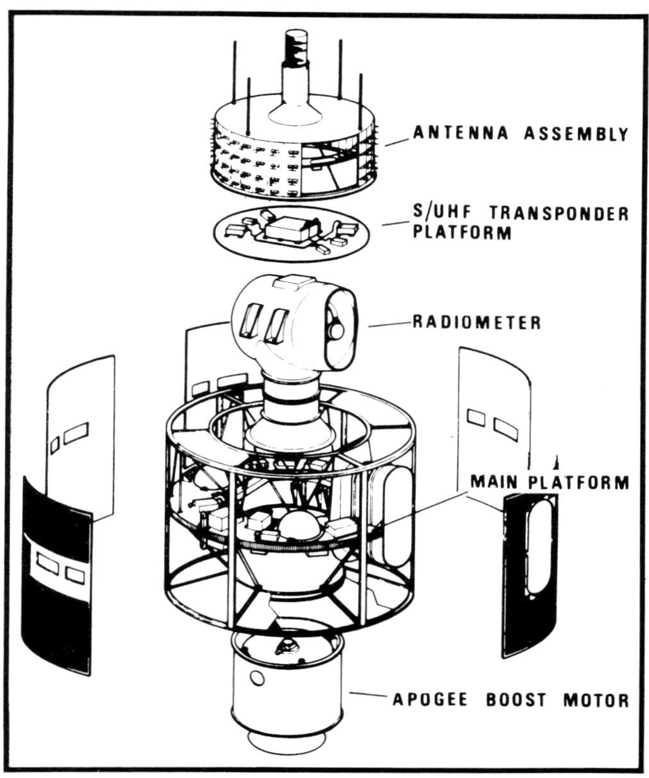

Meteosat details.

full service which is expected to continue until 1985-6. The University of Istanbul, Turkey, is examining whether the system can be used to predict natural disasters and earthquakes in time to permit alleviatory measures.

In Nov 1983 ESA approved development of a new version, Meteosat P2, to be launched in Mar 1986 by the 1st Ariane 4.

The Meteosats were built by the Cosmos consortium, with Aérospatiale of France as prime contractor. In early 1981 a meeting at ESA headquarters agreed to set up a European weather satellite organisation, Eumetsat, to be based on Meteosat 2 and its successors. Sweden has withdrawn, but other countries likely to become members of Eumetsat are Austria, Greece, Ireland, the Netherlands, Norway, Portugal, Spain, Turkey and Yugoslavia. ESA was entrusted with the conduct of the Meteosat operational programme pending the entry into force of the Eumetsat convention. In Jul 1983 ESA gave Arianespace a $140 million contract for the launch of 3 advanced Meteosats in mid-1987, mid-1988 and end-1990.

IUE International Ultraviolet Explorer, L 26 Jan 1978 by Delta from C Canaveral, was a joint NASA, ESA and UK venture. For details see US - Explorer.

Firewheel Designed by the Max Planck Institute of W Germany, with ESA/NASA support, this consisted of a main spacecraft and 4 subsatellites and was lost on the unsuccessful Ariane L-02 launch on 23 May 1980. The object was to study the effects of injecting barium and lithium into the magnetosphere. It is to be followed up by the German/UK/US AMPTE project.

Exosat ESA's 1st X-ray observer, Exosat was L 26 May 1983 by Delta 3914 from Vandenberg into a 356 × 191,581km orbit with 72° incl. Wt 510kg. The highly eccentric polar orbit will keep it in full sunlight for most of its 4yr operational life, maintaining the sensors at relatively stable temperatures and providing observation intervals of up to 80hr, necessary for precise location of cosmic X-ray sources. Following Ariane delays ESA switched the launch to Delta to avoid missing the window. Part of ESA's obligatory science programme, Exosat was developed by the Cosmos consortium (see Meteosat), with MBB of W Germany selected as prime contractor in 1976. Total cost will be about $170 million, including $26 million paid to NASA for the launch. Following up NASA's HEAO series, Exosat is being operated through the European Space Operations Centre (ESOC) at Darmstadt. Precise pointing allows its 2 imaging telescopes, proportional-counter array and gas scintillation spectrometer to study X-ray sources in the 0·04 - 80keV range. During the 1st 30 orbits Exosat detected the presence of highly ionised iron in clusters of galaxies such as Virgo, Coma and Perseus, yielding clues to the creation and evolution of some of the largest formations in the Universe. The hope was that Exosat would give European astronomers a leading role in the study of such exotic objects as neutron stars, black holes and quasars.

ECS 5 European Communications Satellites - developed from the successful OTS-2 programme and, at about 900kg in-orbit weight, twice the size - are to provide Europe with its own regional communications satellite system. Under a 10yr agreement ESA is providing the space segment for Eutelsat, which will become owner and manager of the satellites as soon as they are on station. Each able to carry 12,000 telephone circuits plus 2 TV channels, the spacecraft are being developed by the Mesh consortium, led by British Aerospace, to carry a "significant proportion" of future European, telephone and telex traffic, and to carry TV programmes between member countries of the European Broadcasting Union.

ECS-1 and 2, ordered in 1978 at a cost of $114 million plus Ariane launches at $26·4 million each, were delayed by Ariane failures and Marecs electrostatic problems. ECS-1, L 16 Jun 1983 by Ariane L-6 from Kourou, was placed at 10° E. With all 12 transponders operational, it was handed over to Eutelsat on 12 Oct 1983. ECS-2, due for launch in mid-1984, was to complete the Eutelsat system by adding telephony, data and business communications as well as EBU TV. ECS-3, 4 and 5, due to assure continuity of service into the 1990s, were ordered from British Aerospace in Oct 1979 under a £37 million contract. ECS-3 is due for launch in Aug 1985.

ESA projects

The scientific satellites are part of ESA's "obligatory" programme, funded on the basis of the member countries' percentage contributions. The applications satellite and other projects are mostly funded by the countries with the greatest interest (commercial or national) in them. Because each country expects to receive work in proportion to the amount it pays, funds cannot easily be switched from one project to another, making the task of ESA's administration very difficult.

Scientific satellites

Space Telescope ESA is contributing about 15% of the cost of this US project in return for 15% observation time for European astronomers. Europe's contribution covers provision of the Faint Object Camera with its associated photon-counting detector, the solar arrays, and support of the special NASA operations centre. In 1977 British Aerospace, as leader of a consortium of 1 US and 10 European companies, was awarded a contract worth £7·3 million to develop the photon-detector assembly. A 2nd BAe contract, worth £6 million, covers development of the 11·82m-long solar arrays, carrying 50,000 silicon solar cells and delivering at least 4kW of power after 2yr in orbit. The arrays must be retractable so that the Telescope can be brought back to Earth periodically by Shuttle for refurbishing (see also US national programmes).

ISPM The International Solar Polar Mission (formerly OOE, or Out-of-Ecliptic) is a joint ESA/NASA mission originally conceived as sending 2 spacecraft, one provided by each partner, in opposite directions over the Sun's poles. It was much changed and delayed by NASA budget cuts, and in Feb 1981 the latter announced that the US spacecraft would have to be cancelled. This was followed by protests at ambassador level by ESA, which pointed out that they had already spent about $70 million, half the cost, on the spacecraft and a further $30 million on experiments. Prime contractor for ESA is the Star consortium, led by Dornier of Germany, which was given a $59M contract in 1979 (see main ISPM entry for details).

Giotto Named after the Italian painter Giotto di Bondone, who depicted Halley's Comet's 1301 pass in a fresco on the Arena Chapel in Padua, this is an attempt to intercept the comet on 13 Mar 1986, on its next 76-yearly perihelion. A 750kg (430kg at encounter) spacecraft based

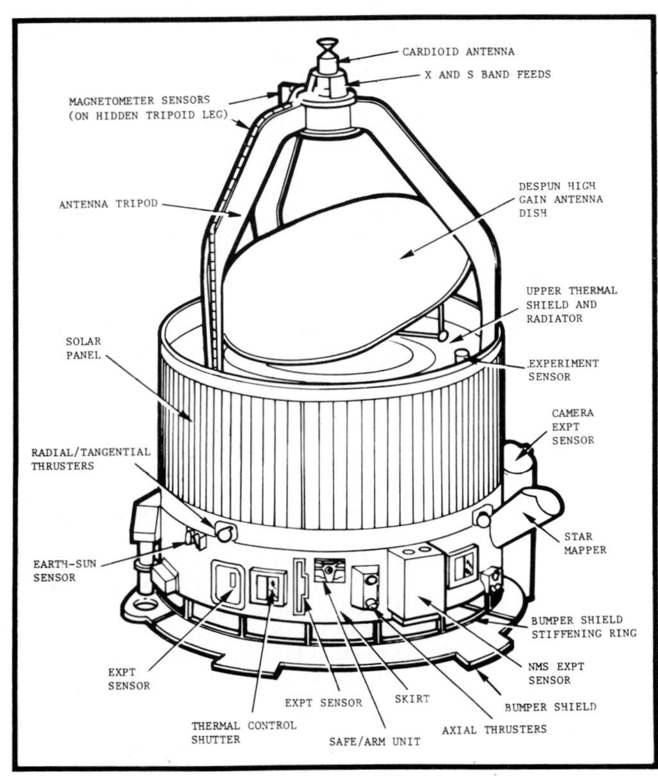

Giotto is due to intercept Halley's Comet on 13 March 1986. ***(British Aerospace)***

on the British Aerospace GEOS craft, it must be launched by Ariane 2 within a 2-week period in Jul 1985 to pass less than 500km from Halley's nucleus. 8 experiments, with a total weight of 53kg, will include instruments to detect the composition and send back colour pictures of the nucleus (if it exists), and assess the composition of the cometary gases and dust. The spacecraft will be struck by dust particles with velocities 5 times that of a bullet, making its survival uncertain, so data will be transmitted in real time for 4hr up to closest approach via Australia's 64m Parkes antenna. Control will be exercised from ESOC in Darmstadt, Germany, with trajectory data provided by the Space Telescope. Total cost will be $115 million at 1979 prices. Attempts to agree on a joint mission with NASA failed over differences about the launcher (NASA preferred a Delta, able to send a larger spacecraft) and NASA's wish to have leadership of the major experiments. Japan and Russia are also carrying out Halley-intercept missions.

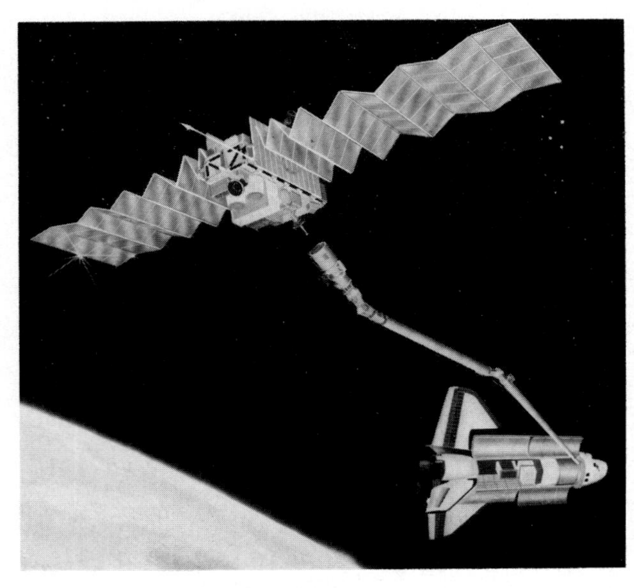

Eureca: ESA's free-flying, retrievable platform as it will be placed in orbit by the Shuttle in 1987. ESA hopes it will become commercially self-supporting. ***(MBB/ERNO)***

Eureca The European Retrieval Carrier is a Spacelab follow-on programme agreed in Apr 1982, to be launched by Shuttle in 1987 and retrieved 9 months later. Able to carry 1,100kg, with 3,500kg total weight, it would have an onboard propulsion unit to take it from 300km to a maximum altitude of 800km, and would be powered by solar arrays. Development cost was estimated at 156 million AU at mid-1980 prices. Eureca 1, with MBB/Erno as prime contractor and based on a double-SPAS body (Britain considered it should have been based on the BAe Spacelab Pallet), will be devoted to materials and life-sciences research in microgravity. ESA is seeking to make it commercially self-supporting, and is studying Eureca 2 for Earth observation, electrophoresis and general science. In addition to Germany, 7 other ESA members were originally due to take part - France, Belgium, Italy, Switzerland, Spain, Denmark and Britain - though with the rejection of the Pallet and Britain's withdrawal from the solar panel competition the latter's participation was looking increasingly unlikely. ESA has also been studying the possiblity of a free-flying platform launched to synchronous orbit by Ariane.

Disco Dual-Spectral Irradiance and Solar Constant Orbiter is being considered as a long-term solar observatory which would provide uninterrupted viewing of the Sun from the L1 Lagrangian point, 1·5 million km from Earth, where the gravitational fields of Earth and Sun balance. ESA was due to choose between this and 4 other projects (X-80, an X-ray observatory; ISO, an infra-red telescope; Magellan, an ultra-violet spectrograph; and Kepler, a Mars orbiter) in 1983.

Space Sled Operating on a 3·6m-long runway in Spacelab's Manned Module, this medical facility is intended to find ways of alleviating the spacesickness encountered by many astronauts. Sitting cross-legged on the Sled, with instrumentation on the head, an astronaut can be accelerated to 0·2g and subjected to a variety of motions. Originally intended as a major experiment on Spacelab 1, it was postponed to a later mission, partly because Spacelab 1 was overweight but mostly because of development and cost problems with the Sled itself. Finally completed by ESA in Mar 1981, 2 models are stored at SPICE and it has been agreed that 1 should be flown on Germany's Spacelab D1 mission in Apr 1985.

ISO Approved in Apr 1983 for launch in the early 1990s, the Infra-red Space Observatory will make detailed observations of selected galaxies and star formations within large clouds of gas and dust. It will complement studies in the visible range by the Space Telescope. Within the solar system, ISO will study the giant planets, asteroids and comets. Preliminary discussions with the USSR Academy of Sciences' new facility at Samarkand may lead to ISO's activities being co-ordinated with those of a proposed Soviet submillimetre observatory, possibly based on a Progress, to provide coverage in the 100-micron to 1mm range, compared with ESA's 5-200 microns.

Applications satellites

ERS The European Remote Sensing Satellite, also known as Earth Resources Satellite, is intended for coastal pollution, ocean and ice monitoring, and is funded to the extent of $430 million by 13 nations. Germany, with 25%, is providing Dornier as project leader. France (22%) has Matra supplying a Spot-type bus. Britain (16%) is in charge of the ground stations. Italy is paying 10·6% and Canada 9·1%. With ERS-1 due for launch by Ariane in late 1988 and ERS-2 2-3yr later, ESA is hoping that the project will be a "market-opener" for Europe in remote sensing. Its microwave sensors will see through cloud to provide 30m resolution from a Sun-synchronous 700km orbit.

Hipparcos This astrometry satellite is due for launch by Ariane from Kourou in late 1986, with the object of making accurate measurements of the trigonometric parallaxes, proper motions and positions of 100,000 selected stars, most of them brighter than magnitude 10. The Hipparcos telescope, with a launch weight of 835kg, will be placed in an almost geostationary orbit with less than 3° equatorial incl so that all operations can be conducted from a single existing European station. Named after the Greek astronomer who measured the length of the year to within 6min from his star observations, the mission is expected to take 2½yr and cost $104·7 million (139·3 million AU).

Olympus: a multi-purpose platform carrying a variety of direct-broadcast and other services. The 1st should be launched in 1986. *(British Aerospace)*

Olympus Studies of a large telecommunications satellite, known until 1983 as L-Sat, to act as a multi-purpose platform carrying a variety of facilities for direct and semi-direct TV and radio, inter-city telephones, etc, were started in 1979. Olympus 1 is to be launched by Ariane 3 or Shuttle in 1986. Weighing 2,300kg and generating 3,500W from its solar arrays, it will provide 1 direct TV channel for Italy and a 2nd "steerable" channel for experimental EBU use; there will be 6 more channels for mixed business using small 3m Earth antennas. Future Olympuses will grow to 3,500kg, generating 7,500W from 60m solar arrays and able to provide 5 direct TV channels to large countries like France or Spain. Satcom International, a British Aerospace/Matra partnership, was appointed prime contractor for the £150 million, 7-nation project and foresees a market for 150 Olympuses worth £5,000 million by the year 2000. Companies in Italy, Belgium, Netherlands and Britain are working on 20/30GHz transponder and beacon payloads.

Sirio 2 Part of ESA's meteorology programme, Sirio 2 (weight 419kg) was lost in the Ariane L-5 launch failure on 10 Sep 1982. It should have followed up Italy's Sirio 1. It was intended to enable various African meteorological centres to communicate directly via simple receive/transmit stations for 2yr using 36 channels. It also carried the 2nd LASSO (Laser Synchronisation from Synchronous Orbit), intended to enable the 11 participating countries to synchronise atomic clocks at low cost with 1 nanosec accuracy.

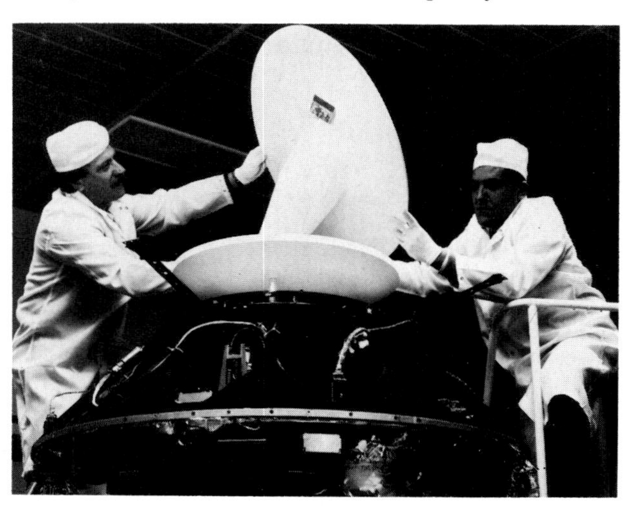

Sirio 2: preparing the despun antenna for testing before this meteorological satellite was lost with Ariane L-05 in 1982. *(ESA)*

Eumetsat

The 12yr operational European Meteorological Programme (Eumetsat) was finally agreed by 17 European countries (the 11 ESA member states plus Austria, Finland, Greece, Norway, Portugal and Turkey) in Mar 1983. The object is to establish, maintain and exploit European meteorological satellites, taking into account the recommendations of the World Meteorological Organisation. The programme will also distribute meteorological research results and weather-forecasting data to African and Middle East countries. The current Meteosat system will continue to be used and will be followed by 3 new, improved versions, MO-1, 2 and 3, to be launched by Ariane in May 1987, Aug 1988 and Nov 1990. Cost of the programme, to run until Nov 1995, is put at 400 million AU (in 1983 1 AU = $1). It will be run by an ESA management team, and the W German Government is pressing for the HQ to be sited in the Frankfurt area.

Eurosatellite

History Established in 1981 to develop Franco-German direct-broadcast TV satellites, this organisation aims to sell derivatives to many other countries, including China and Saudi Arabia. Pre-operational versions of TDF-1 for France and TV-Sat for Germany - with in-orbit weight of 1,190kg, 7yr life and 1984 launch - will relay 3 colour channels. There will also be 1 spare. Belgium and Sweden are also participating. Principal groups involved are AEG-Telefunken (Germany), Aérospatiale (France), ETCA (Belgium), MBB (Germany) and Thomson-CSF (France). Eurosatellite was further strengthened in Sep

1983 with the signing with Sweden of an agreement covering development and manufacture of the Scandinavian Tele-X direct-broadcast satellite.

Eutelsat

History Formed as "Interim Eutelsat" on 30 Jun 1977 and given "definitive status" in Sep 1982, the European Telecommunications Organisation is providing Europe with its own regional system. From early 1984 2 ECS-1 channels were supplied to Britain, 2 to W Germany, and 1 each to Belgium, France, Italy, Norway, the Netherlands and Switzerland. Financial participation on the basis of user requirements, similar to the Intelsat system, was initially shared as follows: Austria 1·97%, Belgium 4·92%, Cyprus 0·97%, Denmark 3·28%, Finland 2·73%, France 16·40%, W Germany 10·82%, Greece 3·19%, Ireland 0·22%, Italy 11·48%, Luxembourg 0·22%, Netherlands 5·47%, Norway 2·51%, Portugal 3·06%, Spain 4·64%, Sweden 5·47%, Switzerland 4·36%, Turkey 0·93%, Britain 16·40% and Yugoslavia 0·96%.

Eutelsat's HQ is in Paris. Under a 10yr agreement signed on 15 May 1979 ESA undertook to provide 5 ECS satellites, with Eutelsat becoming owner and manager as soon as they are on station and in working order. About 20 Earth stations are being built by the member countries' telecommunications administrations. Covering Europe from the Shetlands to Gibraltar and from Sweden to Greece, the system will route telephone and telex calls, transmit data and TV, and provide companies with teleconferencing and teletext. The European Broadcasting Union will lease 2 ECS transponders full-time for transmission of 2 high-quality TV programmes between EBU broadcasting authorities.

Studies of a 2nd generation of satellites to continue the service into the 1990s have begun. But Intelsat, which has agreed to the Eutelsat operation until 1988 because it does not yet have facilities to carry European business traffic, has given notice that by then it should have its own facilities for Europe and will oppose continuation of Eutelsat under its convention (signed by most Eutelsat members), which forbids rival commercial satellite systems if they pose an economic threat to Intelsat.

ECS-F1, also known as Eutelsat 1 F1, L 16 Jun 1983 by Ariane L-6 from Kourou and placed at 10° E, was transferred to Eutelsat ownership on 12 Oct 1983, providing cable TV within Europe. ECS-F2, due for launch in 1984, will extend the network to telephony, data transmission, EBU TV and business communications.

Inmarsat

History The International Maritime Satellite Organisation took over responsibility for maritime communications from the US-run Marisat on 1 Feb 1982. At that date 37 countries were members. By 1983 1,600 of the world's 74,000 ships of over 100 tonnes, plus oil rigs and other maritime platforms, had each been equipped with a $50,000 terminal providing direct access to high-quality telephone, telex and other communications services. The Inmarsat system includes dial telephones and an emergency button giving priority to distress and other urgent messages. The need for a global satellite system serving shipping in the same way that Intelsat links countries was first discussed by the Inter-Governmental Maritime Consultative Organisation (IMCO) in 1975. Agreement to establish Inmarsat, with headquarters in London, followed in 1979. Many countries were concerned to ensure that neither of the superpowers had such a large stake that it could dominate the organisation, and on inauguration day the investment shares were as follows: USA 23·37%, USSR 14·09%, Britain 9·89%, Norway 7·88% and Japan 7%, with other members' shares much less. Despite the USSR's large stake, only 10 Soviet ships had been equipped at that time with receiving stations (all US or Japanese-made), compared with 240 US ships. By Mar 1983, however, an Inmarsat station at Odessa had been brought into use and the Soviet space tracking ships were using the service.

Any ship can use the service, irrespective of whether its country of registration is a shareholder and provided that it is equipped with the necessary 1m-dia parabolic antenna inside a protective radome above deck; Panama, with 130 ships on inauguration day, is an example of a significant non-shareholding user. Shareholdings are adjusted periodically according to national use of the system, in the same way as Intelsat shares. Initially Inmarsat leased from Comsat General, the US corporation, the 10-telephone-circuit capacity of each of the existing Marisats, stationed over the Atlantic at 15°W, the Indian Ocean at 73°E and the Pacific at 176·5°E. These were to be followed by ESA's Marecs 1 (L 20 Dec 1981), offering an additional 40 telephone circuits over the already saturated Atlantic at 26° or 23°W. But Marecs 2, due to follow at 177·5°E over the Pacific, was lost on 10 Sep 1982 due to the Ariane L-05 failure. However, Intelsats 5F5 to 5F9 are carrying maritime packages for lease to Inmarsat, each carrying 30 phone calls to ships in the L-band (1·5-1·6GHz) and providing connections to Inmarsat shore stations through Intelsat's normal frequencies. Intelsat 5F5 was launched 28 Sep 1982; the plan

Inmarsat system.

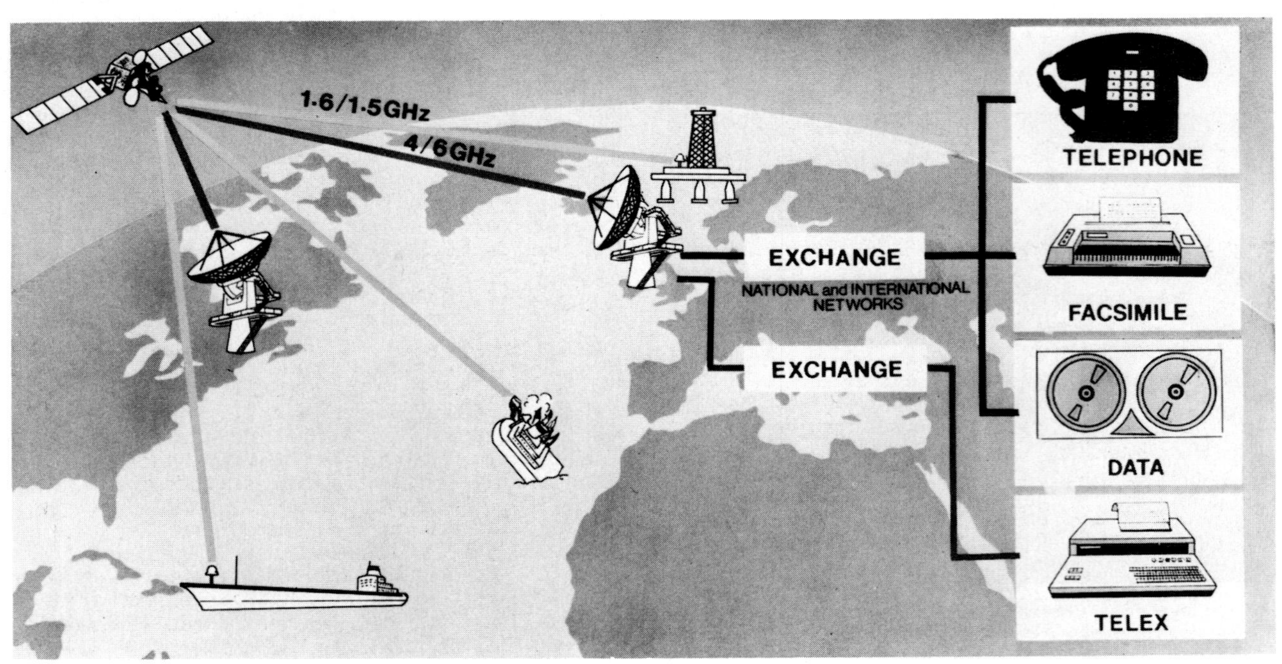

is to place one of these packages over the Atlantic, 2 over the Indian Ocean and a 4th, when available, over the Pacific.

Inmarsat's initial investment was about $250 million. The cost of calls varies from country to country, but at the start it was expected that a call from Britain to a ship in the Atlantic would fall from £5·50 to £3·50/min. Other member countries and their shareholdings on inauguration day were as follows: Italy 3·4, France 2·9, West Germany 2·9, Greece 2·9, Netherlands 2·9, Canada 2·6, Kuwait 2·0, Spain 2·0, Sweden 1·9, Australia 1·7, Brazil 1·7, Denmark 1·7, India 1·7, Poland 1·7, Singapore 1·7, People's Republic of China 1·2, Argentina 0·6, Belgium 0·6, Finland 0·6, New Zealand 0·4, Bulgaria 0·3, Portugal 0·2, Algeria, Chile, Egypt, Iraq, Liberia, Oman, Philippines 0·05 each, Sri Lanka to be decided. The United Arab Emirates became the 38th member in early 1983. (The initial investment cost to individual countries is illustrated by Britain's 1st payment of £13 million.)

In Aug 1983 Inmarsat invited 26 international satellite producers and space agencies to bid for its 2nd-generation network of 7-9 satellites. Estimated cost is $500 million, starting with 1st launch in 1988, and the system may include aviation communications and maritime distress facilities. Bids can include new or shared satellites. Launch vehicles can be Ariane, Atlas-Centaur, Proton, Thor-Delta, Titan or Space Shuttle. Satellites must be compatible with at least 2 launchers, one of which must be Ariane, Proton or Shuttle. Inmarsat also gave a $215,000 contract to Decca and Miller Communications to study the technical aspects of providing satellite/aircraft communications.

Intelsat

History By 1983 the International Telecommunications Satellite Consortium was providing two-thirds of the world's transoceanic communications (equivalent to 65,000 phone channels) to 165 countries or territories. 109 of them were member countries, with investments (minimum 0·05%) in the system in proportion to their use of it. By far the largest were the United States with 24·35% and Britain with 12·99%, followed by France (5·45%), Australia (3·45%), W Germany (3·29%), Japan (3·13%), Saudi Arabia (3·12%) and Brazil (3·04%). The 35 satellites launched by Intelsat over the 18yr to the end of 1982, with another 15 on order, represented an investment of $1,094 million. The system had brought clear and reliable TV, telephone and other communications to countries and areas which had little hope of enjoying good reception before the age of the satellite. An example of the benefits was that in Britain the cost of a transatlantic call had dropped from £3 to 75p/min, with the added advantage of direct dialling. Intelsat itself is the prime example of the way in which new technology both creates and satisfies new demands. Established by 14 countries in Aug 1964, since 1973 it has achieved an average 15·9% return on capital, shared proportionately by its investors. Charges, reduced in spite of inflation for 12 successive years, are now held at the 1981 level. The annual cost of a telephone circuit had come down from $64,000 in 1965 to $9,360 in 1981, and 1hr of TV transmission from $22,000 to less than $200. The benefits of that are indicated by the fact that in 1982 an estimated 1·3 billion people watched the final of the World Cup soccer tournament via Intelsat.

Since Intelsat 1 (the famous Early Bird) was launched in Apr 1965 successive generations of more advanced satellites have been deployed, with no end to the development process yet in sight; the 1st launch of a 6th series (Intelsat 5) took place in Dec 1980. Early Bird provided 240 high-quality voice circuits and made transatlantic TV possible for the 1st time when placed in synchronous orbit over the Atlantic. From 5 TV transmissions a month in 1965 via Early Bird, the average was 100 a month by 1970 and 1000 a month by 1980, representing well over 20,000hr per year of TV. The possibilities became apparent in 1972 during the Olympic Games at Munich, when the whole world not only saw the games live but shared in the horror of the massacre of the Israeli athletes.

The system played a major part in all 10 manned Apollo flights, by providing support communications and in making worldwide TV coverage possible. The manned flights created an enormous demand for live TV coverage, and as quality improved and colour was added there was a rapid growth in demand for satellite transmissions of news and sports events, both live and in the form of edited and recorded packages sent shortly after the event. Another landmark was in Nov 1975, when the Intelsat 4 system was used for the 1st time to relay facsimile pages of the *Wall Street Journal* from the paper's Massachusetts headquarters to Florida. They were used as the basis of

Intelsat: how the satellites have developed over the past 20 years. ***(Intelsat)***

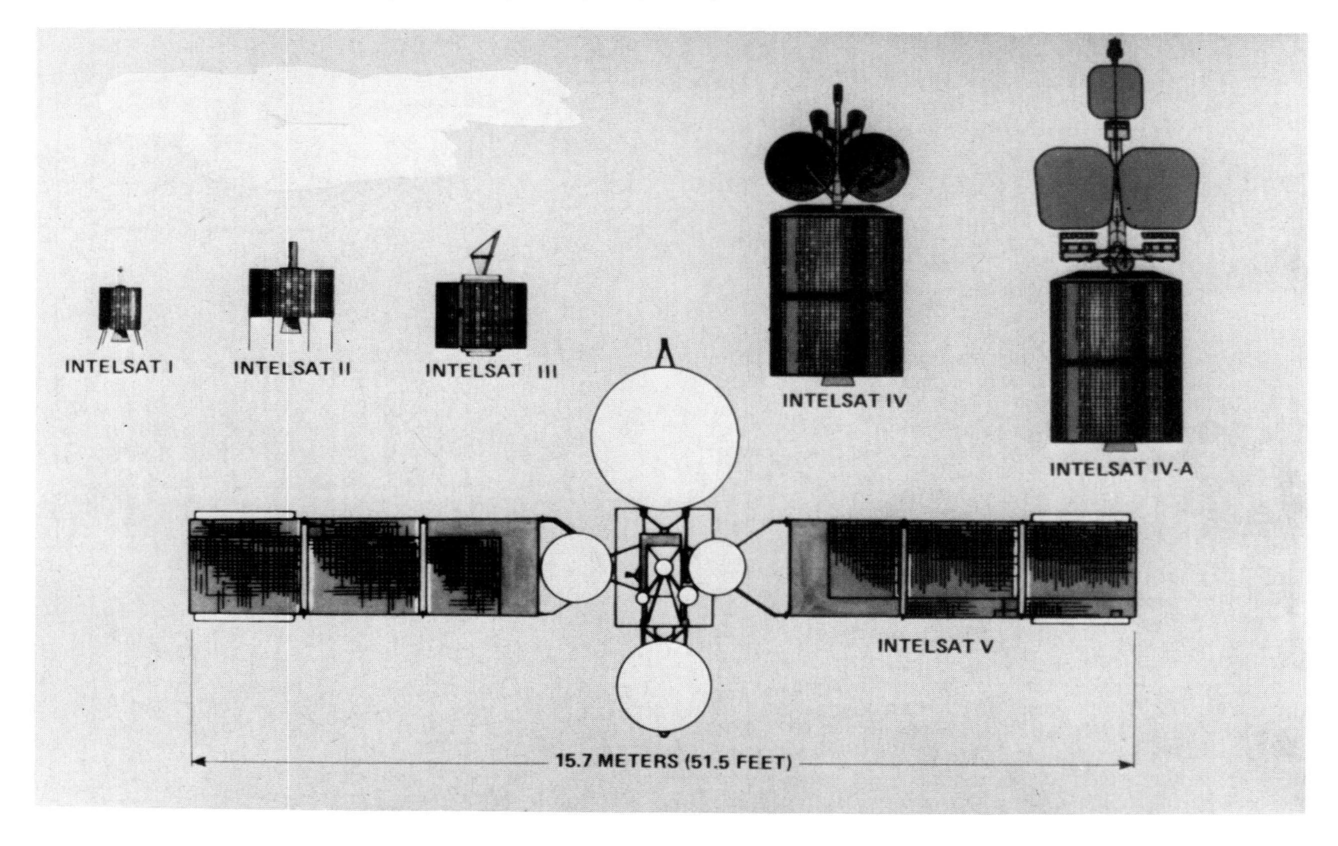

the 50,000-copy south-eastern edition. The parent company, Dow Jones, said that satellite transmission was 70% cheaper than conventional telephone landlines and microwave links, as well as being much quicker.

The speed of the technical advance, spurred on by the commercial demand, is best illustrated by the fact that just over 10yr after Early Bird had provided the organisation's 1st 240 circuits, Intelsat 4A was able to supply 6,250 2-way voice circuits plus 2 colour TV channels. Intelsat 5 in 1980 provided 12,250 circuits with 2 TV channels, and Intelsat 6 will offer up to 41,000 channels.

The fact that communications satellites were a commercial proposition was established as early as 1962, and Intelsat was set up two years later. The originally favoured name of "Glotelsat," short for "Global Telecommunications Satellites," was changed when the Japanese representative pointed out that the first syllable might suggest "grotesque" in Japanese. America's Comsat (Communications Satellite Corporation), representing the USA and its 63% stake at that time, managed Intelsat on behalf of the member nations. Their fears that the US might develop an international monopoly of the system proved groundless, with the US share diminishing steadily. Intelsat's 27-member board represents 94 of the 109 signatories. For 10yr until 1983 the director-general, chief executive and legal representative was Santiago Astrain of Chile (0·55% shareholding). He was succeeded by Richard Colino of the USA. An expression of Intelsat's independence was the decision to leave the Washington building it shares with Comsat, and the selection of an Australian architect's design for a new $51 million HQ being built in NW Washington DC.

Any country belonging to the International Telecommunications Union can join Intelsat, and in fact only 3 – North Korea, the German Democratic Republic and the Soviet Union – have not done so. The USSR sends representatives to Intelsat meetings but has not so far joined. She now seems unlikely to do so, since she is building up her own Statsionar system. This at first caused much concern to Intelsat because it seemed that the two systems might overlap, resulting in interference. The Soviet Union, claiming her own share of the limited number of geostationary slots despite being a latecomer to this essential orbit, remained unmoved and Intelsat's technicians soon solved the problem by shifting some of Intelsat 5's frequencies from the 6/4GHz to the 14/11GHz bands.

Intelsat's annual report is addressed to the Secretary-General of the United Nations. But UN proposals that it should be provided with a free service, particularly for disaster relief, have so far met with a negative response. However, in Oct 1983 Intelsat decided to lease a quarter transponder (9MHz) to the UN for peacekeeping and emergency relief activities on the same terms as domestic leased services.

1982 revenues of $315 million were expected to reach $400 million in 1983. But despite the continued confident investment in new satellite generations like Intelsat 6, there were signs that the years of unchallenged expansion were ending. The formation of Eutelsat and Arabsat, and the announced intention of Britain and France to use their domestic Unisat and Telecom systems for transatlantic services as well, threatened to erode Intelsat's monopolies; complicated negotiations on intersystem co-ordination have followed. More threats are looming as a result of applications by private companies like the Orion Satellite Corporation and International Satellite Inc for permission to set up business systems. Intelsat responded in Sep 1983 by offering the new Intelsat Business Service, involving $9 million-worth of modifications to Intelsats 5A F-13 to F-15 (to be known as 5ABs). The 1st should be in service in 1985-86.

In Oct 1983 the Intelsat Assembly unanimously confirmed its support for Intelsat's single global system, and expressed concern about the adverse impact of competing systems on its ability to continue providing high-quality, low-cost services to its members. Of the 80 countries attending, 50 spoke in favour of the resolution; apparently none of those developing the competing systems spoke against. Director-general Colino also warned the US Senate that any lessening of US support for the system would mean higher rates and adverse political, social and economic consequences, particularly in Third World countries, for many of which Intelsat was the only means of international communications.

Status By mid-1983 the Intelsat-owned space segment consisted of 12 operational satellites in synchronous orbit providing overlapping services in the Atlantic, Indian and Pacific Ocean regions. The ground segment, owned by the telecommunications organisations in the countries of location, consisted of 243 antennas at 173 station sites in 146 countries and territories. Standard A Earth stations, with 30m or larger dish antennas, operate in the 6/4GHz band. Standard B stations, with 11m antennas, also operate in 6/4GHz. Standard C stations, with 14m or larger antennas, operate in 14/11GHz. Provisional Standard E_1, E_2, and E_3 stations, with 3·5-7·0m antennas, are proposed for Intelsat's new digital business services. Standard Z, with selection of parameters available to the Earth station owner, is proposed for leased domestic services.

Hughes built Intelsats 1, 2, 4 and 4A; TRW built Intelsat 3; Ford Aerospace built Intelsats 5 and 5A; and Hughes is building Intelsat 6. Intelsat claims 99·8% continuity of service over the past 11yr, but interruptions do occur occasionally: in 1975 static discharges caused by solar storms twice affected the directional control systems of the antenna on a Pacific Intelsat 4, and once on an Atlantic Intelsat 4 for "relatively short" periods.

Intelsat 1 (Early Bird) Dia 70cm. Ht 60cm. Wt 68kg at launch, 38·5kg after apogee motor fire. Capacity 240 circuits or 1 TV channel. Early Bird, the only one in this series, was launched 6 Apr 1965 and became operational over the Atlantic at 325°E 28 Jun 1965. The world's 1st commercial communications satellite, it made transocean TV possible for the 1st time. Its antenna was focused N of the Equator to service N America and Europe. Design life was 18 months, but operated satisfactorily for 3½yr.

Intelsat 1: known as Early Bird, this was the world's 1st communications satellite. *(Comsat)*

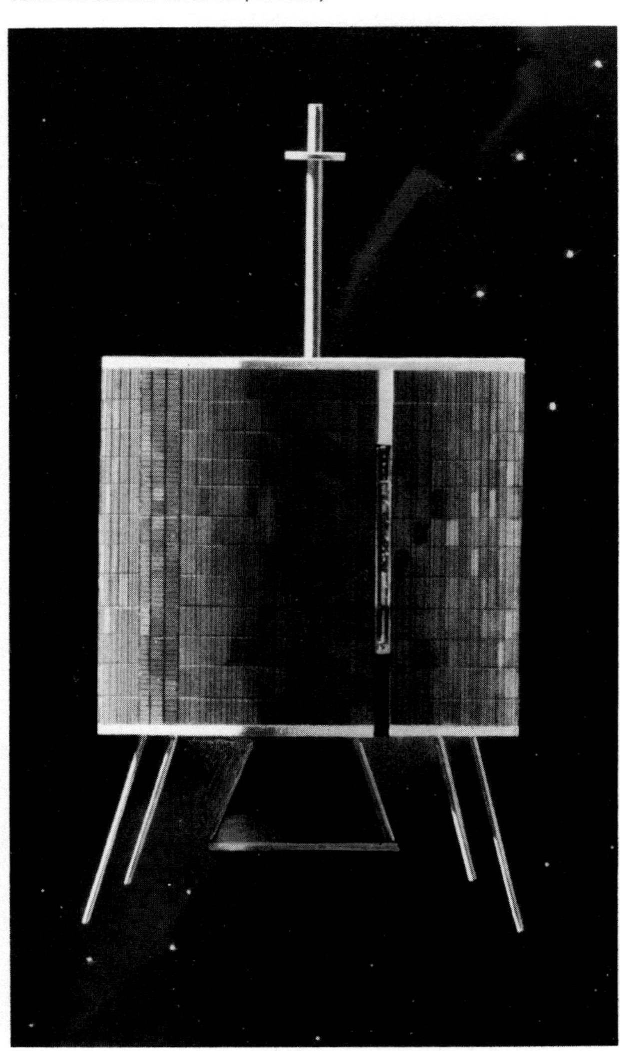

Intelsat 2: this series extended coverage to two-thirds of the world. *(Comsat)*

Intelsat 2 The series extended satellite coverage to two-thirds of the world, though only the last 2 achieved the planned 3yr life. Wt 162kg at launch, 86kg after apogee motor fire. F-1, L 26 Oct 1966, failed. F-2, L 11 Jan 1967, operational for 2yr over Pacific. F-3, L 22 Mar 1967, operational 3½yr over Atlantic. F-4, L 27 Sep 1967, operational 3½yr over Pacific.

Intelsat 3 Capacity increased to 1,200 circuits or 4 TV channels, with 5yr design life (which was not achieved). Wt 293kg, 151kg after apogee motor fire. F-1, L 18 Sep 1969, failed. F-2, L 18 Dec 1968, operational 1½yr over Atlantic. F-3, L 5 Feb 1969, partly operational 7yr over Indian Ocean. F-4, L 21 May 1969, operational 3yr over Pacific. F-5, L 25 Jul 1969, failed. F-6, L 14 Jan 1970, operational 2yr over Atlantic. F-7, L 22 Apr 1970, operational 1¾yr over Atlantic. F-8 failed. Starting with this series, satellites were boosted to higher orbit whenever possible at the end of their useful life.

Intelsat 3: despite increased capacity, the 5yr design life was not achieved and this was the least successful of the series so far. *(Comsat)*

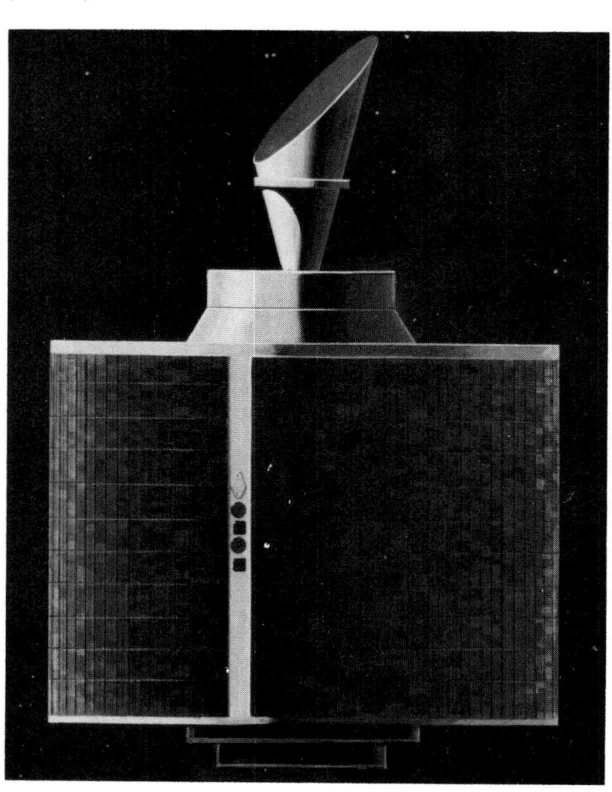

Intelsat 4 Capacity increased to 4,000 circuits and 2 TV channels, with 7yr design life. Wt 1,414kg, 730kg in-orbit. F2, L 25 Jan 1971, operational 12yr over Atlantic. Raised to higher orbit and deactivated 1983. F3, L 19 Dec 1971, primary Atlantic 3yr, operational 12yr+, placed in orbital storage 1983. F4, L 22 Jan 1972 (1st commercial satellite assembled outside US, by British Aircraft Corporation at Bristol; provided TV link between US and China for President Nixon's visit to Peking), operational over Pacific 7yr+, raised to higher orbit and deactivated 1983. F5, L 13 Jun 1972, operational over Indian Ocean 7yr+, raised to higher orbit and deactivated 1983. F7, L 23 Aug 1973, major-path satellite Atlantic 2yr, operational 6yr+, retired and placed in orbital storage 1983. F8, L 21 Nov 1974, operational Pacific from Dec 1974, in use for international and domestic lease 1983. F6, L 20 Feb 1975,

Intelsat 4: with 4,000 circuits each, some of these achieved a 12yr life. *(British Aerospace)*

failed to achieve transfer orbit. F1, L 22 May 1975 after 4yr storage, operational Indian Ocean 3yr, moved to Atlantic as major-path satellite 1979, domestic lease service 1983.

Intelsat 4A Capacity increased to 6,250 circuits plus 2 TV channels, with 7yr life. Wt 1,515kg, 825kg after apogee motor fire. F-1, L 25 Sep 1975, primary satellite Atlantic from Feb 1976. F-2, L 29 Jan 1976, major path Atlantic 2yr from 1976, then leased. F-4, L 26 May 1977, leased for Atlantic until Dec 1977, then major-path satellite. F-5, L 29 Sep 1977, failed due to launcher malfunction. F-3, L 6 Jan 1978, leased over Indian Ocean until 1979, then primary satellite.

Intelsat 5 9 ordered at cost of approx $300 million. Cube-shaped instead of spinning cylinder; 3-axis-stabilised using onboard momentum wheels. Capacity increased to 12,250 circuits plus 2 TV channels, with 7yr life; increase mainly due to dual-polarisation technique permitting reuse of 6/4GHz band, with addition of 14/11GHZ band. Wt 1,928kg, 1,012kg on station. Length with solar panels deployed 19·9m, height with antenna mast 6·4m. F5-F9 have maritime packages for lease to Inmarsat and capable of carrying the equivalent of 30 phone calls in the L-band (1·5-1·6GHz). Launches by Atlas-Centaur. F2 (1yr late due to manufacturing problems), L 6 Dec 1980, scheduled as primary over Atlantic. F1, L 23 May 1981, placed at 335·5°E and took over as primary Atlantic from 4A-F1. F3, L 15 Dec 1981, placed at 0·3°E. F4, L 5 Mar 1982 to take over as Indian Ocean primary from 4-F3. F5, L 28 Sep 1982 with 1st Inmarsat package for placing at 63°E. F6, L 19 May 1983, placed at 0·23°E. F7 and F8, L by Ariane L-7 and L-8 19 Oct 1983 and 5 Mar 1984 and placed at 60°E and 63°E, were delayed for modifications following interference problems caused by the F5 Inmarsat package. F9, L 9 Jun 1984 by Atlas-Centaur 62 from C Canaveral, was placed in a 178 × 1,111km orbit due, it is believed, to a Centaur failure. Later it was raised to 278 × 1,111km but written off by Intelsat, which had paid $10 million to insure it for $102 million.

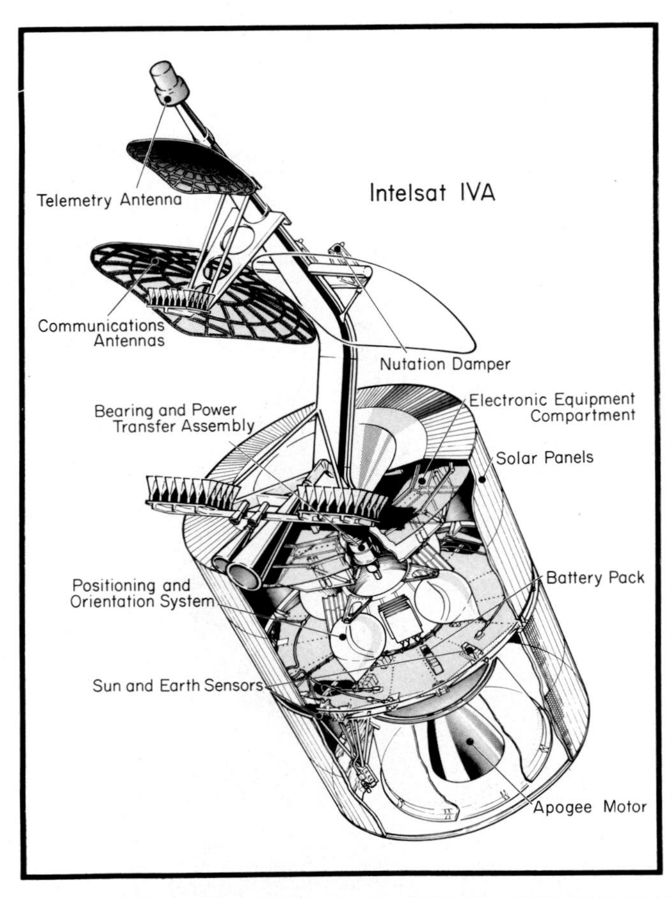

Intelsat 5A Improved Intelsat 5s with capacity increased to 15,000 circuits, achieved by orthogonal polarisation techniques through 6 additional transponders and incorporation of 2 fully steerable 4GHz spot beams. 6 ordered from Ford Aerospace from Jan 1981. Launches in 1984 and 1985, 4 by Atlas-Centaur, 2 by Ariane.

Intelsat 6 Industry proposals for a new generation, with 3 times the Intelsat 5 capacity, were called for in 1980. A commercial battle between international consortia led by Hughes and Ford ended in Apr 1982 with the award of a $700 million contract to Hughes for 5 satellites, with 1st launch in 1986, and options on another 11, which would bring the contract to $1·3 billion. Spin-stabilised, with 3,500kg weight, 3·6m dia and 11·8m height, and using both 6/4 and 11/14GHz frequency bands, they will offer 33,000 2-way phone circuits and 4 TV channels. Designed for launch by Shuttle or Ariane 4. Because of the cost and the smaller US stake in Intelsat, greater non-US participation in producing hardware will lead to Britain getting at least $100 million-worth of work, with other large subcontracts going to Germany, Canada, France, Italy and Japan. Hughes claims to have won the contract because of its plan for an integral propulsion unit, even though the Ford satellite design was preferred by Intelsat. The unit will reduce the cost of each launch by $16 million. Each Intelsat 6 is to be insured for $150 million, covering replacement cost and failure during the 1st 180 days in orbit.

Intelsat 5: providing 12,250 circuits each, 12 of this series have been built, with the last of them currently being orbited. Total span of the solar wings is over 15m. ***(Ford)***

Launchers Competition to provide launchers increased in 1981 when both the US Shuttle and the European Ariane became operational, taking on the well tried US rockets. Until then all Intelsats had been launched by NASA from Cape Canaveral, with Intelsat paying NASA for both the rocket and the launch service. Intelsats 1, 2 and 3 were launched by McDonnell Douglas Deltas; Intelsats 4 and 4A by General Dynamics/Convair Atlas-Centaurs; and for Intelsats 5 and 5A a mixture of Atlas-Centaur, Shuttle and Ariane was planned. Intelsat 4A missions cost $23·5 million for each spacecraft and $23 million for the Atlas-Centaur launch. By Intelsat 5 each satellite was costing $34 million, with an Atlas-Centaur launch being offered at $42 million, a Shuttle launch at $25·8 million and an Ariane launch at $27 million. In order to gain its 1st order the European Space Agency in Feb 1979 contracted to launch an Intelsat 5 (originally planned for Apr 1981) for $25·29 million, with an option on a second at $27·46 million. The Intelsat Board had to take into account questions of reliability, as well as the political need to share out its contracts, when deciding which launcher to use. An additional complication relating to Intelsat 6 was the fact that it would be possible to build larger and more advanced satellites if Intelsat could be sure that the Shuttle, with its much larger payload bay, would be available. Lack of confidence in both the readiness of the Shuttle and Ariane led to Intelsat's 23rd Atlas-Centaur being ordered for Intelsat 5-F9.

Launch technique This can be illustrated by a description of the launch procedure for Intelsat 4A. The 1,506kg satellite is mounted on the Atlas-Centaur launch vehicle, protected by a 3m fibreglass nose fairing. Once clear of Earth's atmosphere this is jettisoned during the first burn of the Centaur upper stage. Between them the Atlas booster and the Centaur place the satellite in a 185 × 644km parking orbit. After a 15min coast period the Centaur is again ignited. This accelerates the satellite so that, on separation from the Centaur, it coasts to an altitude of 35,680km above Earth's surface. Final synchronous equatorial orbit is achieved by firing the solid-fuelled apogee motor, which provides 5,670kg of thrust for 34sec. The speed of the spacecraft then matches the rotational speed of Earth, so that it remains over the same point on the surface. Small gas jets are then used so that the spacecraft drifts in orbit and is finally positioned accurately at any desired point on the equator. This process can take up to 2 months, and the spacecraft is not brought into operational use until it has been achieved. The jets are also used for 2 other vital manoeuvres. One is to orient the spacecraft so that it is correctly lined up with the Earth and its solar cells are in position to gather energy from the Sun to charge its batteries. The other is to set the spacecraft spinning at the rate of 60 revolutions per minute; this stabilises it in the same way as a gyroscope: 40% of the spacecraft is "despun" by means of a ball-bearing and power-transfer assembly so that the highly directional antennas can be kept pointed towards the Earth. The satellite carries 136kg of hydrazine propellant for these manoeuvres; once this is exhausted the spacecraft ceases to be operational despite its orbital life of 1 million yr.

Intercosmos

History A highly successful programme enabling 9 Soviet-bloc countries to participate in space acitivities, and which during 1978-81 enabled each to send cosmonauts on 7-day missions to Salyut space stations. Intercosmos was created in Apr 1967 by the Soviet-affiliated countries of CMEA (Committee for Mutual Economic Assistance). They were Bulgaria, Czechoslovakia, Cuba, GDR, Hungary, Mongolia, Poland, Romania and the Soviet Union; Vietnam was added in 1979. Areas of study were space physics, meteorology, communications, biology and medicine, with Earth resources being added in 1975. 1st spacecraft to be launched, Cosmos 261 on 20 Dec 1968 from Plesetsk, studied air density and polar auroras. Baikonur (Tyuratam) and Kapustin Yar have also been used for

Intercosmos 1: the start of a long series which has provided 9 Soviet-bloc countries, as well as France and Sweden, with opportunities to develop space technologies. *(Graham Turnill)*

subsequent launches, usually carried out at the rate of 2 per year, and by the end of 1981 Intercosmos 22 had been launched. Each country finances its own contributions, with the USSR supplying the basic satellite and the launch; scientific results are the common property of all Intercosmos members. The early flights overlapped both the Proton and Prognoz series, though Academician Boris Petrov said they were a continuation of earlier research, with improved instruments and greater capacity to handle information.

The first programme ran from 1967 to 1980, with Czechoslovakia winning much praise for its participation in every mission. This culminated in its provision on Intercosmos 18 of the Magion subsatellite, the first non-Soviet satellite in the series. The second programme, for 1980-85, saw a move from research to applications satellites, with the ocean-dynamics studies of Nos 20 and 21. Future plans also included Intershock, in which several craft will study shock waves in interplanetary space, a Mars mission and a Venus/Halley's Comet mission. The latter, first planned as a Franco-Soviet exercise with France undertaking the Venus observations and Russia the subsequent Halley's Comet interception, was later extended to include Intercosmos members. Hungary is to be responsible for the TV system monitoring the comet and transmitting the pictures to Earth.

Manned flights For details see Soyuz 28, 30, 31, 33, 36, 37, 38, 39 and 40, and Spacemen - Intercosmos.

Spacecraft description With the exception of No 6, which was a Vostok-type, and 15, which was an AUOS prototype, Nos 1-16 were light research satellites. Weighing 320 - 375kg and consisting of a cylinder and 2 hemispheres, they were mostly battery-powered, hence their short life. Their disadvantage was that they could not be accurately stabilised against the Sun. Czech scientists made contributions to all these satellites, and provided the orientation system for the more advanced, 422kg AUOS (Automated Unified Orbital Station), introduced in operational form with Intercosmos 17. This features a computer able to process data before transmission, is able to carry a 15kg subsatellite for jettison later in the mission, and is solar-powered.

Intercosmos 1 L 14 Oct 1969 by B1 from Plesetsk. Orbit 260 × 640km. Incl 48°. E Germany and Czechoslovakia contributed instruments, and launch was attended by representatives of 7 participating countries.

Intercosmos 3 being checked out before launch.

Studied Sun's ultra-violet and X-ray radiation, and its effects on Earth's upper atmosphere. Dec 2 Jan 1970.

Intercosmos 2 L 25 Dec 1969 by B1 from Kapustin Yar. Wt ?400kg. Orbit 206 × 1,102km. Incl 48°. Investigated Earth's ionosphere concentrations of electrons and positive ions. Dec 7 Jun 1970.

Intercosmos 3 L 7 Aug 1970 by B1 from Kapustin Yar. Orbit 207 × 1,320km. Incl 49°. Instrument and operational team jointly provided by USSR and Czechoslovakia. Studied composition and temporary variations of charged particles (protons, electrons, alpha particles); low-frequency electromagnetic waves; and Earth's magnetic field. Dec 6 Dec 1970.

Intercosmos 4 L 14 Oct 1970 by B1 from Kapustin Yar. Orbit 250 × 628km. Incl 43°. Continued studies of Sun's ultra-violet and X-rays and their effects on Earth's atmosphere. Found that amount of oxygen at 96km altitude was less than expected, and varied with time of day. Dec 17 Jan 1971.

Intercosmos 5 L 2 Dec 1971 by B1 from Kapustin Yar. Orbit 205 × 1,200km. Incl 48°. Equipped with Czech and Soviet instruments, and operated by a joint Soviet-Czech team, it continued the study of charged particles started by Intercosmos 3. Dec 7 Apr 1972.

Intercosmos 6 L 7 Apr 1972 by A1 from Tyuratam. Wt 4,082kg. Orbit 203 × 256km. Incl 51°. Modified Vostok; scientific payload, wt 1,070kg, brought back after 4 days. A nuclear photo-emulsion unit recorded several thousand collisions of cosmic-ray particles, among them a "space wanderer" with a pulse of 1,000 million million electron-volts. The photo-emulsion sheets, pasted on to glass, were divided for study among Polish, Hungarian, Czech, Romanian and Soviet scientists.

Intercosmos 7 L 30 Jun 1972 by B1 from Kapustin Yar. Orbit 267 × 568km. Incl 48°. With Czech instruments and controlled by a Soviet-GDR-Czech team, this continued the upper-atmosphere studies of Intercosmos 1 and 4. Ellipsoid, with 6 panels and an estimated 10-week life, it studied the short-wave solar irradiations which are absorbed by oxygen molecules and do not reach Earth's surface. Dec 5 Oct 1972.

Intercosmos 8 L 1 Dec 1972 by B1 from Plesetsk. Orbit 214 × 679km. Incl 71°. Instruments and operating team provided by Bulgaria, GDR, Czechoslovakia and USSR, to continue study of Earth's ionosphere by measuring concentration and temperature of electrons between the satellite and Earth's surface. This time an orbit more inclined to the equator was chosen so that the spacecraft would pass though the ionosphere at higher latitudes; in these regions Earth's magnetic field has "a less barring effect" on particles. Dec 2 Mar 1973.

Intercosmos 9 L 19 Apr 1973 by B1 from Kapustin Yar. Orbit 202 × 1,551km. Incl 48°. Soviet-Polish satellite, named Intercosmos Copernicus 500 to celebrate 500th anniversary of birth of Polish scientist. Elliptical orbit "stretched towards the Sun" enabled the instruments to record radio frequencies of solar radiations which fade out in the ionosphere before reaching Earth's surface. It also recorded the concentration of charged particles (electrons and ions) at the Earth's "space threshold". Dec 15 Oct 1973.

Intercosmos 10 L 30 Oct 1973 by C1 from Plesetsk. Orbit 265km × 1,477km. Incl 74°. With an orbital life of 3yr, it ceased transmitting after 2yr and was later said to have provided new data about the origin of powerful electric currents. It had been sent up to study electro-magnetic

Intercosmos 20 was similar in purpose to the US Seasat, supplying ocean-surface data which were used for the plotting of ice fields and the prediction of wave heights. *(Novosti)*

connections between the magnetosphere and ionosphere, and to measure magnetic field variations in some specific frequency ranges. During its flight meteorological rockets made in both the USSR and GDR were launched to provide comparative results. Dec 1 Jul 1977.

Intercosmos 11 L 17 May 1974 by C1 from Kapustin Yar. Orbit 484 × 526km. Incl 50°. Timed to mark the 250th anniversary of the USSR Academy of Sciences, and carrying instruments designed by the GDR, USSR and Czechoslovakia, it was intended to study solar ultra-violet and X-ray radiations, and Earth's upper atmosphere. Orbital life 6yr.

Intercosmos 12 L 31 Oct 1974 by C1 from Plesetsk. Orbit 243 × 707km. Incl 74°. This launch marked 5yr of space co-operation. Similar objectives, with Romanian instruments included for the first time. Dec after 8½ months.

Intercosmos 13 L 27 Mar 1975 by C1 from Plesetsk. Orbit 296 × 1,714km. Incl 83°. This continued studies by Intercosmos 3, 5 and 10 of cosmic radiation, spectral composition of protons in Earth's radiation belts, and low-energy electrons. It contained Soviet and Czech equipment and had a 5½yr orbital life.

Intercosmos 14 L 11 Dec 1975 by C1 from Plesetsk. Orbit 345 × 1,707km. Incl 74°. Intended to study low-frequency electromagnetic fluctuations in Earth's magnetosphere, and structure of the ionosphere. Orbital life 10yr.

Intercosmos 15 L 19 Jun 1976 by C1 from Plesetsk. Orbit 487 × 521km. Incl 74°. Prototype AUOS; its information could be received and processed outside the Soviet Union for the first time. Ground stations had already been built in Hungary, the GDR and Czechoslovakia; later stations in Bulgaria and Cuba would be added. With an orbital life of 6yr, dec 18 Nov 1979.

Intercosmos 16 L 27 Jul 1976 by C1 from Plesetsk. Orbit 465 × 523km. Incl 50°. Replaced Intercosmos joint solar studies mission with Sweden, which failed to reach orbit on 3 Jun 1975. Carried an ultra-violet spectrometer polarimeter jointly developed by Russia's Crimea Observatory and Sweden's Lund Observatory, and a small telescope with 100mm focal length and a mirror device to detect the direction of light-beam oscillations in the ultraviolet spectral region. Dec 10 Jul 1979.

Intercosmos 17 L 24 Sep 1977 by C1 from Plesetsk. Orbit 468 × 519km. Incl 83°. 1st operational AUOS. 6 Czech experiments out of 12 included instruments to measure electrons, the mass and speed of meteorites, and a laser reflector to measure satellite-Earth distances. Dec 11 Sep 1979.

Intercosmos 18 L 24 Oct 1978 by C1 from Plesetsk. Orbit 407 × 768km. Incl 83°. ESA provided electric field experiment in co-operation with Moscow Izmiran Institute for ionospheric and magnetospheric studies. Czechoslovakia's 1st satellite, Magion (acronym for Magnetospheric and Ionospheric) ejected on 14 Nov. Dec 13 May 1981.

Intercosmos 19 L 27 Feb 1979 by C1 from Plesetsk. Orbit 502 × 996km. Incl 74°. Carried 9 detectors to study ionosphere at peak of solar cycle. 50 countries participated, and 25 balloons carrying complementary equipment were released from Kiruna, Sweden, plus sounding rockets. 30yr life.

Intercosmos 20 and 21 L 1 Nov 1979 and 6 Feb 1981 into 470 × 520km orbits with 74° incl to monitor sea surface temperatures, wave formation processes and ice borders. Data from 21, with Cosmos 1151 data, enabled ocean surface maps to be compiled. 20 dec 3 Mar 1981.

Intercosmos 22 (Bulgaria 1300) L 7 Aug 1981 by A1 from Plesetsk. Wt ? 1,400kg. Orbit 799 × 893km. Incl 81·2°. Ionosphere and magnetosphere studies. 500yr life.

Intersputnik

History A Soviet-led competitor with the Western-controlled Intelsat, this was created in 1971 to improve and develop communications between Socialist countries. It offers international satellite transmissions - TV, radio, telex and telephones - to all interested nations. The first installations to go into operation were those of the Soviet Union and the Cuban Caribe ground station, used for a TV transmission (via a Molniya 1) of Mr Brezhnev's visit to Cuba in 1973. The system proper began operations in 1974, and by 1980 there were 8 ground stations and 11 member countries: Soviet Union, Poland, GDR, Hungary, Czechoslovakia, Mongolia, Vietnam, Yemen, Romania, Bulgaria and Cuba. In addition, the 9th Intersputnik Council meeting at Havana in Oct 1980 was attended by 3 observer countries, Nicaragua, Mozambique and Afghanistan, as well as 4 international communications organisations. A channel-distribution plan and co-operation protocol between Intersputnik and Intercosmos were approved during that meeting. The Statsionar 3 satellites now in use are to be followed by the more advanced Lutch series. By 1983 Afghanistan and Laos had become members, with ground stations completed. Iraq had also been provided with a ground station and Syria had applied for membership. According to Intelsat, by the end of 1983 Intersputnik had 14 member nations but only 150 telephone circuits.

IRAS

The international Infra-Red Astronomical Telescope, L 25 Jan 1983 by Delta from Vandenberg into a 900km polar orbit, is a 3-nation project to advance the techniques of deep space observations. JPL is managing the project,

IRAS: intense infra-red radiation is shown around the Orion constellation. In 10 months of operation IRAS found more than 40 stars within 75 light-years of Earth which have infra-red characteristics suggesting that they might have solar systems something like our own. *(NASA)*

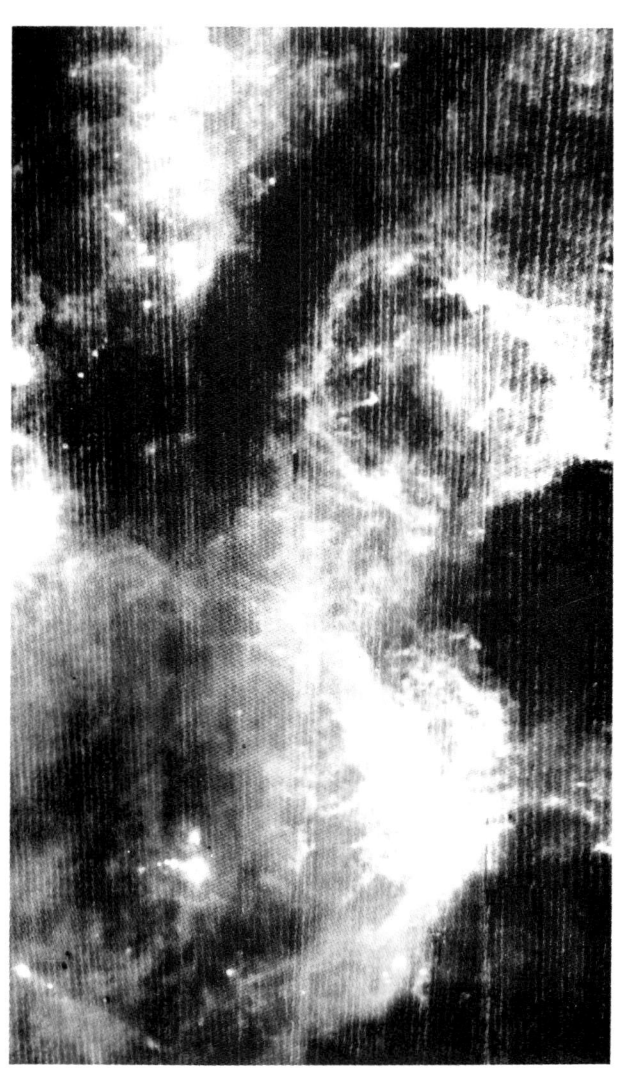

ISPM: ESA's solar polar satellite, which will be launched by NASA in 1985. The US agency will not now provide a complementary 2nd satellite. ***(ESA)***

with the Netherlands providing the spacecraft, Britain's Science Research Council installation at Chilton the control facilities, and Ball Aerospace of Boulder, Colorado, the telescope. Satellite weight is 834kg, of which the science payload is 499kg; the primary instrument, a 60cm Cassegrain infra-red telescope cooled to -271°C, weighs 27kg. The telescope cost $40 million to develop.

With 1,500 scientists involved, the primary mission of IRAS is to provide a complete infra-red map of the observable Universe, including perhaps 1 million new infra-red sources never observed before, and to study the expanding shell of infra-red radiation which exponents of the Big Bang theory say has been streaming outwards since the explosion that began the Universe 18 billion years ago. On its 1st day of operation, 10 Feb 1983, it identified more than 4,000 infra-red sources; this was equal to the total number previously identified, and included 20 hitherto unidentified galaxies.

The liquid helium used to cool the detectors to within 2°C of absolute zero ran out on 21 Nov 1983, a month earlier than estimated, but IRAS scientists claim that they had already achieved far more than they could have hoped. On 26 Apr, while scanning the Cygnus constellation, IRAS detected a moving infra-red source which was finally identified as Comet Iras-Araki-Alcock 1983D. Soon 4 more comets had been found. On 11 Aug much excitement was generated when solid objects were found to be orbiting Vega (also known as Alpha Lyrae), a star twice the size of the Sun and 26 light-years away in the Milky Way. This was claimed as the first direct evidence of the existence of another solar system. It was decided that, out of 9,000 candidate stars, 50 would receive greater study to see if they also had planets or material from which planets could be formed. By Nov 95% of the sky had been scanned 4 times, and 72% six times. In addition to a large number of previously unknown galaxies, vast new dust clouds had been found to be an important element of deep space. Another previously unknown object, designated 1983 TB, was believed to be the 1·2km-dia spent nucleus of a comet, and the parent body of the Geminid meteor shower which reaches Earth each Dec. Passing within the orbit of Mercury, it comes closer to the Sun than any other known object. By Dec 1983 scientists had been able to analyse only 1% of the data received, and while no evidence of a 10th planet had yet been found, they were confident that if such a body exists the data would reveal it. Project scientists in the US and Britain have been able to use their experience of processing IRAS data to help solve problems with the USAF Teal Ruby project.

ISPM

History The International Solar Polar Mission, formerly known as the Out-of-Ecliptic (OOE) Mission, is a joint NASA/ESA project now planned for launch by Shuttle/Centaur on 15 May 1986. Until 1981 it had been a 2-spacecraft mission to be launched by a single Shuttle and 3-stage IUS in Feb 1983. ESA was to provide 1 spacecraft and about half of the 16 experiments, with NASA providing the other spacecraft, the rest of the experiments and their nuclear power supplies, and the Shuttle launch. The 2 spacecraft would make a 15-month flight to Jupiter, 1 passing above and 1 below the planet at a distance of 450,000km so that both were thrown out of the ecliptic plane and into opposite solar orbits. 4yr after launch they would pass over opposite poles of the Sun at distances of about 250 million km, pass through the ecliptic plane, and then pass over the opposite poles. The 5yr mission would be the first to investigate the Sun from the 3rd dimension, providing an accurate assessment of the solar magnetic field and solar wind, and studying solar flares, coronal holes and loss of mass and energy from the Sun. Involving 200 investigators from 65 universities and 13 countries, it is seen by ESA and NASA as a major scientific project. However, NASA's decision in 1980 not to go ahead with the 3-stage IUS raised doubts about the feasibility of the double-launch plan. Then, under the US budget cuts of Feb 1981, NASA was obliged to save $250-$300 million on ISPM by cutting out its spacecraft. Shuttle delays and other problems had already delayed the launch to 1985. It was finally agreed that NASA would continue to provide launch and tracking services and a radioisotope thermoelectric generator, and that launch would be by wide-body Centaur in 1986.

ISTPP

The International Solar Terrestrial Physics Programme is a tri-nation project involving NASA, ESA and Japan's Institute of Space and Astronautical Science (ISAS). It envisages a detailed look at the Sun, the Earth's space environment and Sun-Earth interaction, with a 1986 start. ESA's planned solar observatory and cluster of 4 spacecraft to study basic plasma physics processes would become elements, and ISAS would develop and operate one of the craft. NASA would provide Shuttle launches and some instruments.

Marisat

History After several years' international controversy as to who should provide satellite communications for ships and aircraft, America's Communications Satellite Corporation (Comsat) set up a subsidiary called Comsat General. Marisats 1, 2 and 3 (L 19 Feb 1976, 10 Jun 1976 and 14 Oct 1976 by Delta from C Canaveral. Wt 362kg) were placed over the Atlantic, Pacific and Indian Oceans. Commercial services started in the summer of 1976, and complete global coverage was achieved in Nov 1978 with the installation of an Earth station in Japan. A slow commercial start was offset by 5yr of contracts for the supply of communications services to the US Navy; this brought in $138 million, about the same as the launch costs. As this work tailed off following the introduction of the US Navy's own satellites, Britain's Royal Navy leased capacity in the Atlantic region for 1981-84. (This proved useful during the Falklands conflict.) By 1980 there were over 300 commercial users, ranging from cargo ships and Esso oil tankers to Norwegian cruise liners, oil rigs, gas carriers and seismic ships searching the sea bottom for oil and minerals. The *Queen Elizabeth 2* was also fitted with the necessary antenna, and operator Cunard agreed to pay the minimum of $9,600 for a year's service. Charges soon came down to $4/min for telex messages and $30 for a 3min phone call. With the Marisats expected to remain operational until 1984 - 8yr instead of the planned 5 - Comsat General leased its services to Inmarsat when the new international maritime service took over on 1 Feb 1982. By then plans for Aerosat satellites for transoceanic aircraft had long since been dropped in the face of pilots' complaints that they had too many communications already.

Mesh consortium

This consortium of 14 industrial companies from 10 European countries (Belgium, Denmark, FRG, France, Italy, Netherlands, Spain, Sweden, Switzerland and Britain) was set up to compete for ESA and other space contracts. Mesh is an acronym of the initial letters of the 4 founder members: Matra (France), ERNO (Germany) Saab-Scania (Sweden) and Hawker Siddeley Dynamics, now British Aerospace. Mesh participates in Spacelab but its main activities concern the OTS, ECS and Marecs communications satellites.

Oscar

History *O*rbiting *S*atellite *C*arrying *A*mateur *R*adio is a series of small satellites launched piggyback with more important missions, initially to enable radio amateurs to obtain training in satellite tracking and for experiments in radio propagation. Transmitting low-powered signals, battery-operated and at first with only short lives of 2 months to 1yr, they have been a great success and have become much more sophisticated. The latest can serve school science groups, provide emergency communications for disaster relief and transmit pictures for display on domestic TVs.

Oscars 1-3 L by the US Defence Dept by Thor-Agena on 12 Dec 1961, 2 Jun 1962 and 9 Mar 1965.

Oscar 4 L 21 Dec 1965 by Titan 3E.

Oscar 5 Provided by Australia, L 23 Jan 1970 with ITOS-1. Wt 18kg. Orbit 1,435 × 1,481km. Incl 102°. Operated for 1½ months and Melbourne University compiled tracking reports from hundreds of stations in 27 countries.

Oscar 6 L 15 Oct 1972 with NOAA-2. Wt 16kg. Orbit 1,450 × 1,459km. Incl 101·7°.

Oscar 7 L 15 Nov 1974 with NOAA-4. Wt 29kg. Orbit 1,444 × 1,462km. Incl 101·7°. Built by a multinational team of radio hams under the direction of Amsat (Radio Amateur Satellite Corporation, PO Box 27, Washington DC 20044). With 4 radio masts mounted at 90° intervals on the base and 2 experimental repeater systems, it was able to receive and store messages, to be forwarded later as it moved around the world.

Oscar 8 L 5 Mar 1975 with Landsat 3. Wt 27kg. Orbit 903 × 917km. Incl 99°.

Oscar 9 (Uosat) L 6 Oct 1981 with SME. See Britain - Uosat for details.

Oscar 10 1st lost on the unsuccessful Ariane L-02 23 May 1980. 2nd, German-built, L 16 Jun 1983 by Ariane L-6. Struck twice when the 3rd stage caught up shortly after payload separation, damaging an antenna. Its boost motor also fired 80sec too long, so that the planned 1,500 × 35,800km orbit at 55° incl with 11hr period ended as 3,952 × 35,510km at 25·9°. With 90kg in-orbit wt, spin-stabilised and shaped like a 3-pointed star, Oscar 10 should have been available to most of the world's radio amateurs, but by Sep 1983 only a quarter of its radio relay coverage had been achieved. It was also to have provided emergency communications for disaster relief, transmissions of education programmes direct to classrooms, and 2-way long-distance communications for 17hr per day, transmitting on the 435MHz amateur radio satellite band with a 50W transponder; downlink is 145·9MHz.

Oscar 11 (Uosat 2) L 1 Mar 1984 with Landsat 5. Wt 60kg. Planned orbit Sun-synchronous at 700km. Like Oscar 8, built by Surrey University, England. With 3yr life, its experiments include transmission of Earth images to domestic TV screens.

Future projects Amsat is planning a 3-satellite global system (Syncart, for Synchronous Amateur Radio Transponder). Frequencies will be 1269MHz uplink and 436MHz downlink, known as Mode M. Most ambitious of these, to be launched on the 1st Ariane 4 in 1985, will attempt to look at asteroids. A low-budget project, it will travel free with Arsène, a French venture into amateur radio satellites supported by CNES and French colleges.

Radarsat

The Canadian-proposed Radarsat, based on Britain's L-Sat platform, with Shuttle launch and radar provided by the US, is intended to aid in exploitation of natural gas and oil reserves in the Arctic. British Aerospace is due to complete the definition study by 1985-86, with launch set for 1990.

Rosat

Rosat (*Roentgensatellit*) is a co-operative German/NASA satellite due for launch by Shuttle in Jul 1987 to study X-ray emissions from non-solar objects and to follow up Uhuru, HEAO-1 and 2 and Exosat. With Dornier as prime contractor and DM100 million funding from the W German Government, it is expected to detect more than 100,000 new X-ray sources with an accuracy of 1sec arc; results will be used to build up a celestial atlas. In addition to the launch into a 460km orbit, NASA will provide an electronic X-ray imager. Britain will participate in satellite development and provide the extreme-ultra-violet wide-field camera.

Satcom International

A joint-venture company linking British Aerospace and France's Matra, Satcom International was formed to bid for Inmarsat's proposed ship-to-shore satellites. It is currently developing the Eurostar platform for the French Athos satellite and the UK direct-broadcast satellite, due for Ariane launches in 1985 and 1986.

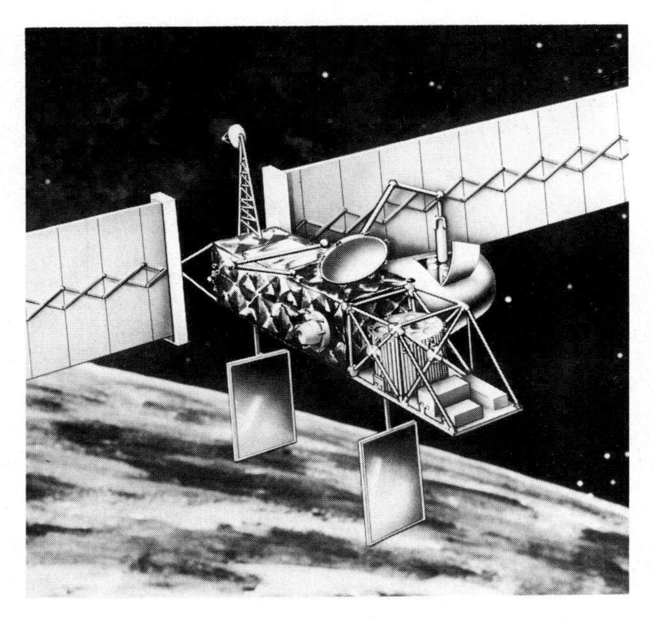

Solaris: a French proposal for a European materials-processing platform, automatically operated and able to deal with 2 tonnes of material. It could work alongside the US Space Station. *(Aérospatiale)*

Solaris

France's Aérospatiale has designed the proposed Station Orbital-Laboratoire Automatique, and Germany, Italy and Britain are participating in discussions of roles for it. 2 research platforms in equatorial and Sun-synchronous orbits are proposed. They would be able to process up to 2 tonnes of material in an orbital furnace working at 600-700°C, and to conduct Earth observations with a new type of synthetic-aperture radar. Automatic vehicles would carry payloads to and from the platforms twice a year. With a 15yr life and needing 10kW of power, they would be launched by Ariane 4 from Kourou, with recovery of re-entry vehicles off the French Guiana coastline. Costs have yet to be established.

Spacelab

History Spacelab is Europe's contribution to the US Space Shuttle system. When it went into orbit for the first time on STS-9 on 28 Nov 1983 it was 4yr later than planned. ESA's most important long-term project, it enables Europeans to participate in manned flight and share in the development of microgravity materials processing, with all its long-term commercial possibilities. When US Vice-President Bush took delivery of the first flight version of Spacelab on 5 Feb 1982, together with an engineering model, ground support equipment and some computer software, it represented an ESA contribution worth nearly $1 billion at no cost to NASA. A second Spacelab is being bought by NASA for approx $300 million.

This co-operative venture provides for the design, development and production of a versatile space laboratory which can be used either manned or unmanned. It should dramatically reduce costs, since it can be prepared for missions (about 50 are planned in a 10yr period) quite separately from the Space Shuttle Orbiters, which will be available for other flights during the preparation periods. It remains inside the payload bay throughout the mission, which usually lasts 7-9 days (the 1st was extended to 10 days), and is supported by a crew of 6 - pilots, mission specialists and payload specialists - working 12hr shifts in 2 teams. While these flights are a first step towards internationalising the US manned spaceflight programme, NASA will control all its activities. "European flight crew opportunities will be provided in conjunction with flight projects sponsored by ESA or by governments participating in the Spacelab programme," says NASA.

The discussions which gave rise to Spacelab started in 1969, when NASA invited Europe to participate in its post-Apollo programme. Agreements on Spacelab between NASA and ESA and between the US and 9 European governments followed in Sep 1973. Spacelab 1, 1st of 8 flights planned for the following 5yr, carried the 1st European to participate in the US manned programme. With Germany by far the largest contributor among the 10 ESA nations participating in Spacelab, it was inevitable that he should be a German. On that basis, on future flights Italy and France would also appear to have prior claims over Britain. STS-9/Spacelab 1 is described under Shuttle missions. Percentage contributions are: W Germany 54·94, Italy 15·57, France 10·29, Britain 6·51, Belgium 4·32, Spain 2·88, Netherlands 2·16, Denmark 1·54, Switzerland 1·00 and Austria 0·79.

Development An initial 6yr contract worth $226 million and covering design and development of Spacelab was awarded in Jun 1974 to ERNO of Bremen, whose major subcontractors were Bell Telephone in Belgium, Aeritalia in Italy and British Aerospace Dynamics (which made the Pallets). The industrial consortium finally

Spacelab 1: configuration for 1st mission. For centre-of-gravity reasons only 1 Pallet could be carried in *Columbia,* but almost all objectives were achieved. *(ESA)*

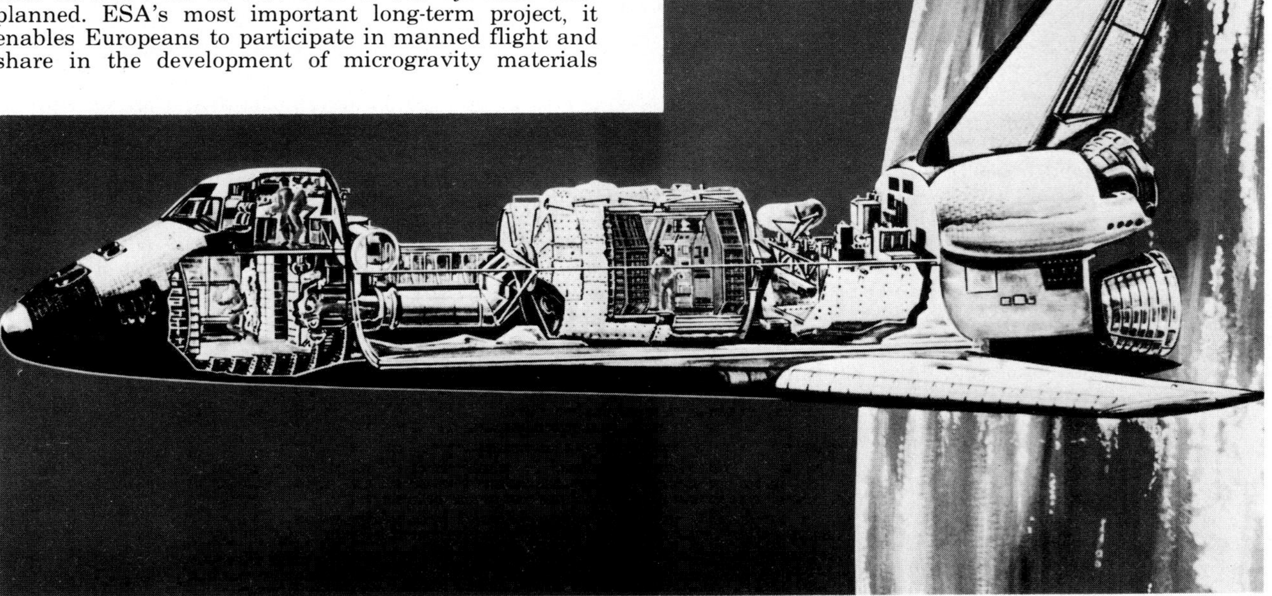

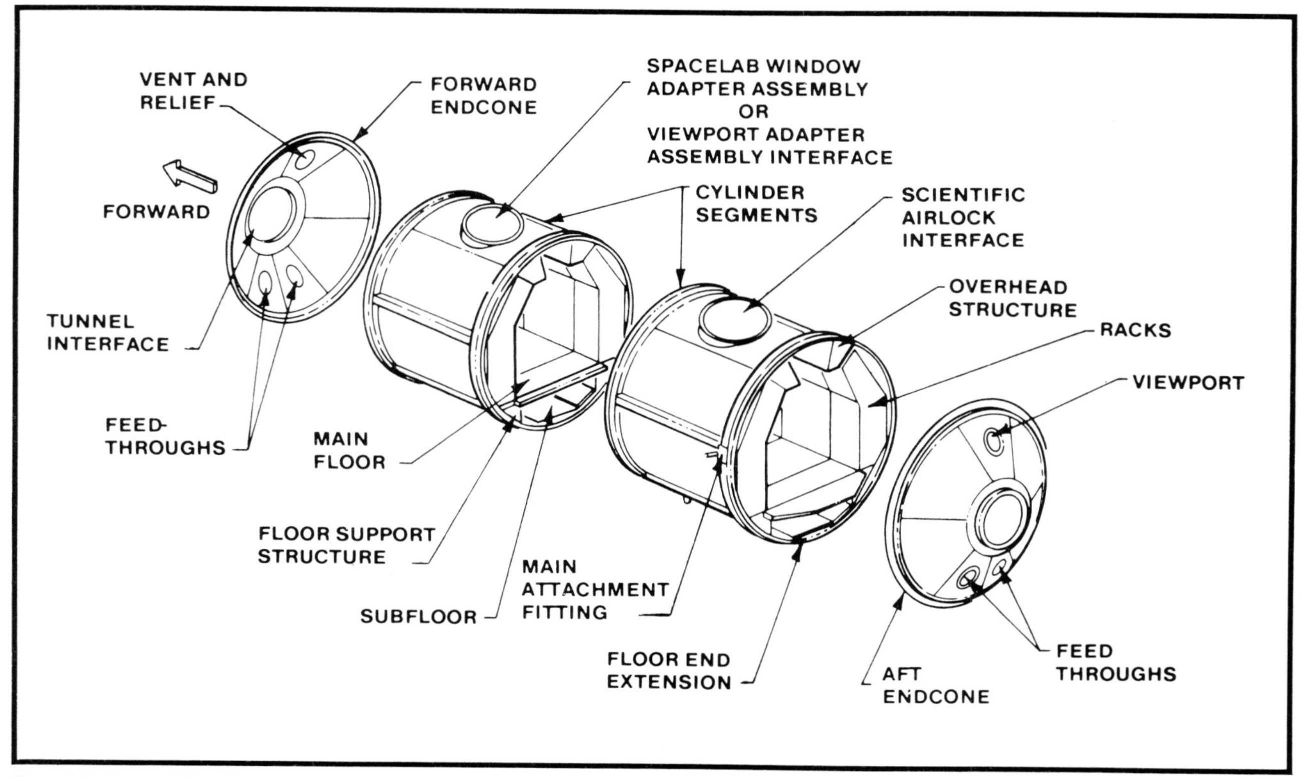

Spacelab Manned Module structure. *(NASA)*

embraced 50 firms in 10 European countries. Technical assistance to ERNO was provided by the US company TRW. While Spacelab never held up NASA, that was only because the Shuttle itself was delayed, and 1979 proved to be a testing year for both projects as well as for NASA/ESA relations. Spacelab integration tests brought to light numerous mismatches of hardware and incompatibilities between hardware and software. It became clear that costs would pass the 120% ceiling, and more funds had to be sought from the participating countries; by then the original estimated cost of $515·7 million had grown to $850 million. However, in Nov 1980, after over 6yr of development, the Spacelab engineering model, a non-flying prototype, was delivered to NASA at the Cape, where it was used at the Spacelab Flight Unit for interface verification and training. NASA undertook when the 1973 memorandum was signed not to develop its own Spacelab but to buy any similar units needed for its own projects from ESA. But it was only after much hard bargaining that NASA in 1980 ordered a 2nd Spacelab worth $144 million, and subsequently a 2nd Instrument Pointing Unit worth $18·4 million for delivery in 1983. NASA would have preferred a barter deal under which it would have provided flight facilities in return for the hardware. ESA hopes that a total of 3 Spacelab units will ultimately be needed seem likely to be disappointed, and British Aerospace's proposals for many different Pallet roles in future space platforms appear to have been overtaken by Germany's independently developed SPAS-01.

Spacelab development is likely to follow 3 lines. Priority is now being given to a free-flying Spacelab called Eureca (European Retrievable Carrier), which is being designed for European users wishing to pursue materials and life sciences research in microgravity (see ESA - Projects). Another follow-on element depends on NASA decisions to improve the Shuttle: ESA would like to extend Spacelab's mission duration to 20 days, which requires improved cooling capability and electrical power supply for experimenters. Finally, in the longer term, ESA wants to know how far it will be possible to integrate Spacelab with a NASA Space Operations Centre. Until 1984 Spacelab was expected to provide at least 1,500 jobs in European industry, as well as requiring 100 ESA staff.

General description Spacelab can be flown in at least 8 configurations, providing research opportunities in a broad range of disciplines. Biologists can study weightlessness effects on organisms ranging from simple bacteria to the human body. Astronomers can observe stars and planets with greater clarity and precision. Materials scientists can create new composite materials in its miniature furnaces and form stronger and purer metals and crystals; 10 experimental crystals were brought back from the 1st mission. Its versatility starts with the 4m-dia Manned Module, consisting of 2 segments each 2·7m long and designed to withstand meteoroid and radiation impact for 50 missions. Used together, the 2 segments are known as the Long Module and provide accommodation, including airlocks, for up to 4 experimenters and over 4,000kg of experiments carried in standardised racks which can be readily installed or removed. Behind the Long Module 1 or 2 Pallets, 3m long and 4m wide, can be added, carrying telescopes, antennas, sensors and other equipment requiring direct exposure to space, and controlled from inside. 3 Pallets can be carried with the short Manned Module. With no Manned Module, Spacelab can consist of a "train" of up to 5 Pallets, with an "igloo" or mini-laboratory providing automatic control under the supervision of mission and payload specialists on the Shuttle flight deck.

Total Spacelab weight must not exceed 14,525kg, since that is the maximum landing payload for the Orbiter. For a nominal 9-day mission module-only weight is 5,000kg, Pallet-only weight 9,100kg, and module-plus-Pallet 6,000kg. Maximum Pallet length is 15m. The pressurised module sections are designed to offer a maximum volume of 22 cu m and a minimum of 5 cu m. Centre-of-gravity requirements mean that Spacelab must be carried towards the rear of the Orbiter, necessitating a transfer tunnel; this was developed by McDonnell Douglas for NASA. The 6·1m-long version was used for Spacelab 1; there is also a 2·7m version. Design changes led to a decision that the Spacelab crew would live on the Shuttle mid-deck and not inside Spacelab. To save weight and cost it was also decided that the Shuttle would supply Spacelab's life-support and power requirements, with the result that it cannot at present be left in orbit as an independently operating spacecraft and recovered by the Shuttle on a later flight. Cost per dedicated flight is estimated at $82 million in 1982 dollars; an igloo with 2 Pallets would cost $46 million, and 1 Pallet $18 million.

Flights 1st-flight date became a source of friction between NASA and ESA, since a full return of scientific data was dependent upon the availability of 2 TDRS, which were delayed for a variety of reasons. The 1st was that the USAF, which will share the TDRS system,

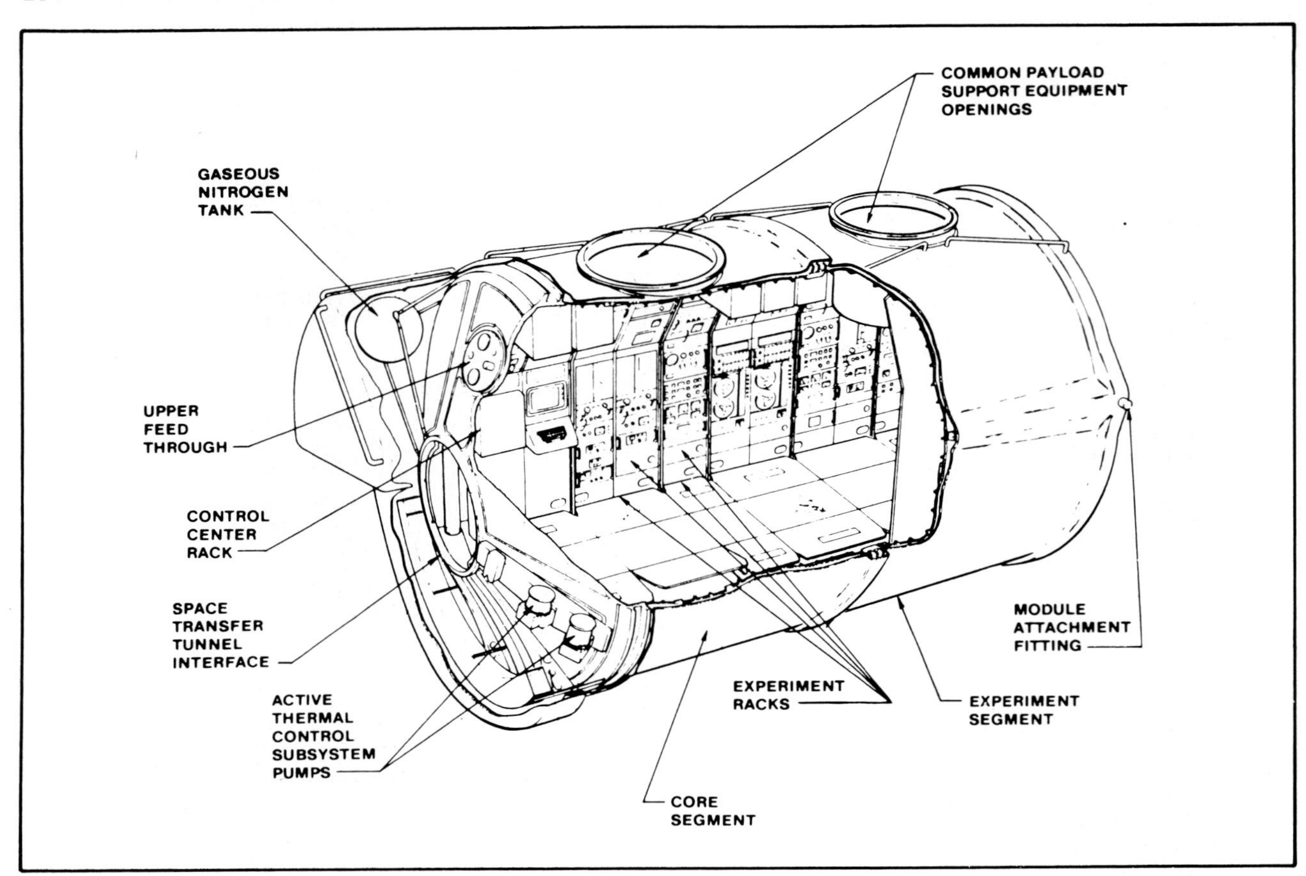

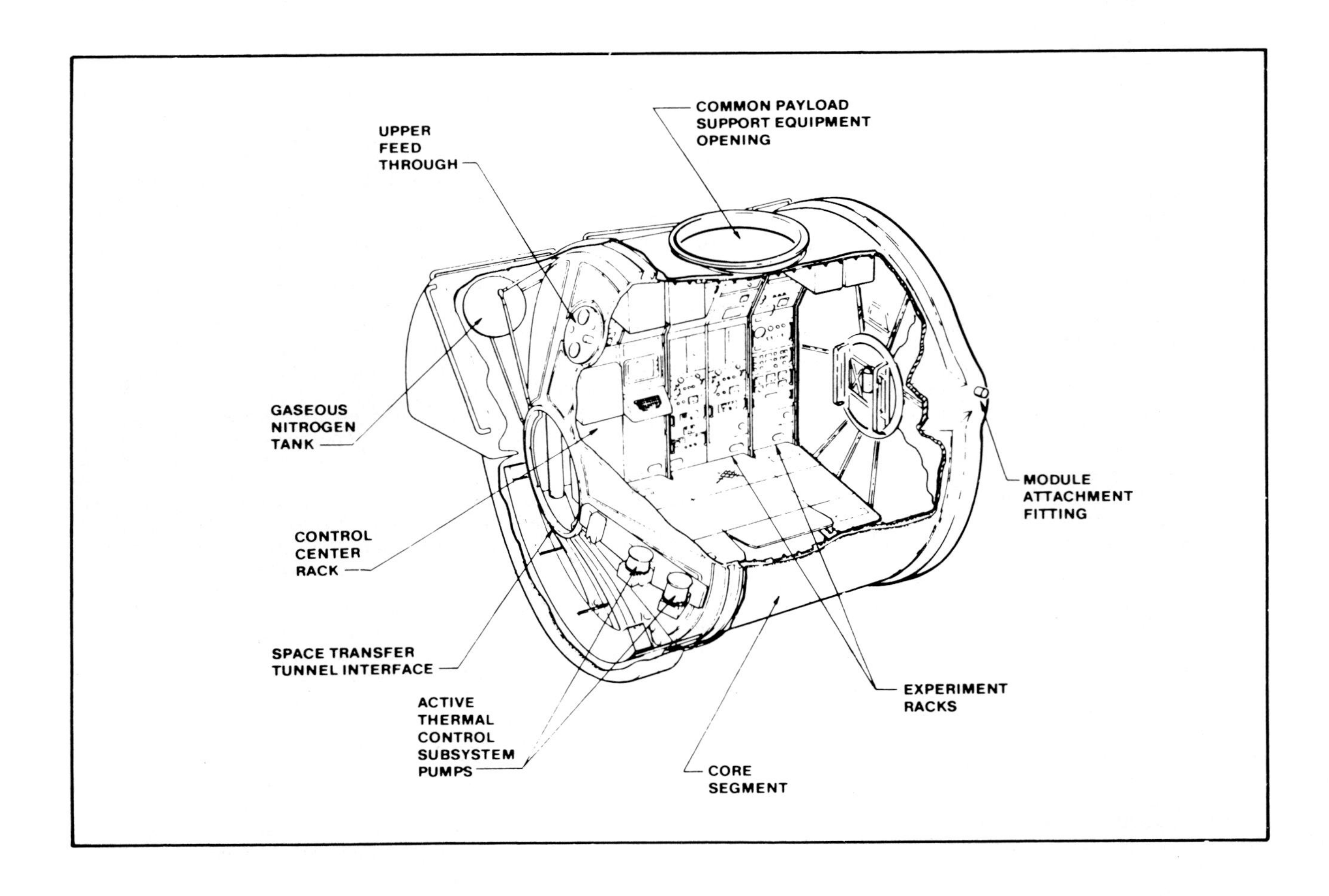

The Manned Module can be flown in either long or short form. *(NASA)*

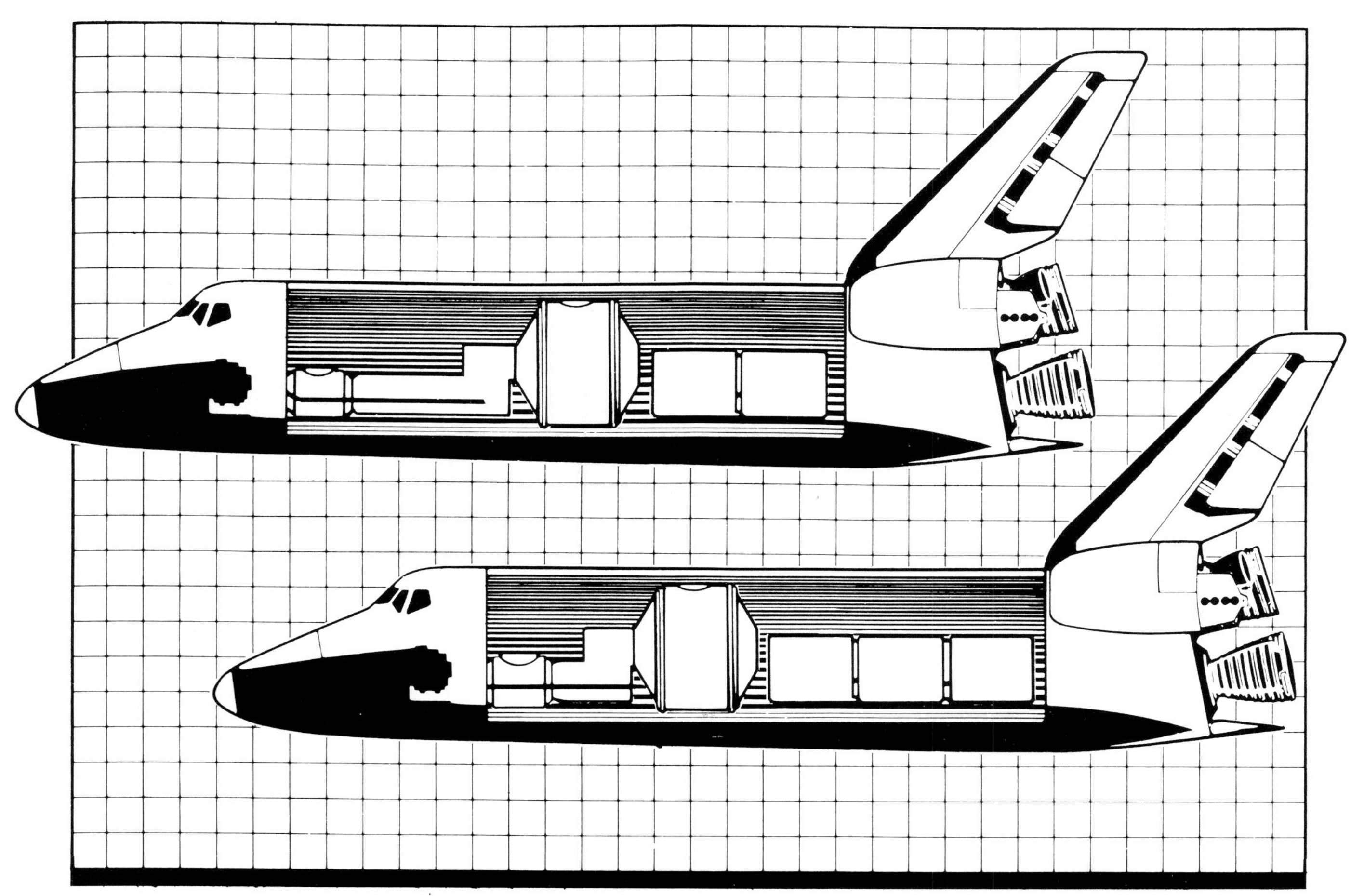

Up to 3 Pallets can be flown with the short Manned Module. *(NASA)*

wanted to improve their resistance to possible Soviet radio penetration. This plan was dropped at the end of 1981, largely for cost reasons, and the Spacelab 1 mission was swapped with the TDRS-2 launch so that the latter could be flown before Spacelab 1. However, the IUS problem on the TDRS-1 launch obliged both organisations to make concessions. Though Spacelab 1 was flown too late in the year to provide some investigators with maximum results, its success despite that provided all-round reassurance for future missions.

Spacelab 1 was a co-operative NASA/ESA mission primarily intended to test its capabilities as a space laboratory, and the two agencies sponsored 72 scientific investigations provided and supported by 11 European nations, the US, Canada and Japan. They covered 5 areas of research: life sciences, atmospheric physics and Earth observations, astronomy and solar physics, space plasma physics, and materials science and technology. There were many setbacks during the run-up to this mission. Centre-of-gravity requirements for landing meant that the Manned Module had to be carried at the rear of the payload bay, leaving room for only 1 Pallet. The landing had to be at Edwards AFB to ensure plenty of over-run capability in the event of a long landing with such a heavy payload. This meant that Spacelab, with its cargo of completed experiments, had to be retained in the payload bay for about a week until *Columbia* was flown back to KSC by the carrier 747. Several stops were needed en route for refuelling, causing anxiety among the scientists about possible disturbance to the experiments.

Spacelab 2 Due for launch on *Challenger* in Apr 1985 (after Spacelab 3), this will be an all-Pallet mission devoted to life sciences, plasma physics, infra-red astronomy, high-energy physics and solar physics. Sponsored entirely by the US, the 12 investigations will however include 2 from the UK. The 6 crew, who will include 2 payload specialists, will operate from the Orbiter's cabin.

Spacelab 3 Due for launch on *Discovery* in Jan 1985, the mission will cover 15 investigations (13 from the US, 1 from France and 1 from India) concentrating on materials-processing, space technology and life sciences; extremely low acceleration levels will be required. The major activity will be the deployment and retraction of a large solar array, which will also be used to study the dynamic response of a large structure to thruster-induced disturbances. The 6 crew members will have the use of a Long Module.

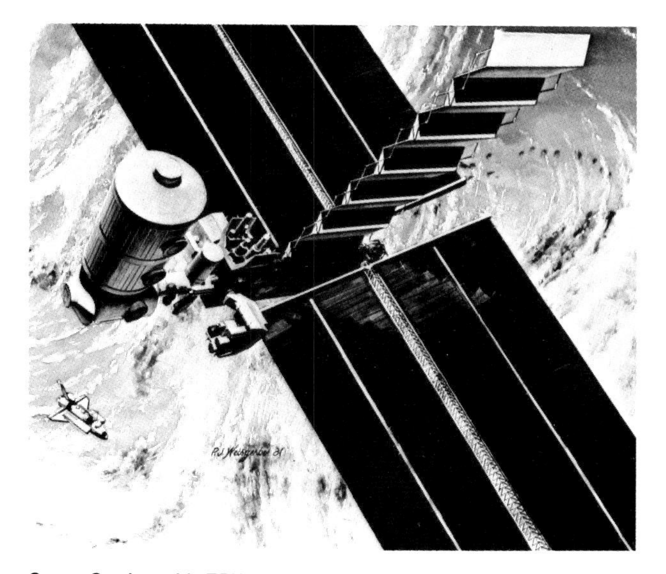

Space Station: this TRW proposal, based on Spacelab hardware, includes 2 solar arrays, a space radiator, 3 rotating Pallets for payloads, and an airlock joining 2 Manned Modules. *(NASA)*

Spacelab D1 Due for launch on *Columbia* in Oct 1985, this is a German-financed mission devoted to national research. It will include ESA's Space Sled, which will carry an astronaut back and forth on rails to study the effect of accelerations in weightlessness on vision and balance, spacesickness, etc.

Spacelab 4 Due for launch on *Challenger* in Jan 1986 with 25 experiments aimed at obtaining medical and

STS-9/Spacelab 1: left to right, Merbold, wearing headband to monitor space adaptation, working with Lichtenberg and Parker in the Manned Module. Pallet-mounted experiments are controlled from here. ***(NASA)***

biological data through carefully planned experiments on living organisms. A Long Module equipped with a biological laboratory will be operated by 2 mission and 2 payload specialists for 7 days.

European crew members In Dec 1977 ESA selected, from 2,000 applicants from 12 European countries, 4 Europeans for final training by NASA as payload specialists. They were Franco Malerba (Italy), Ulf Merbold (Germany), Claude Nicollier (Switzerland) and Wubbo Ockels (Netherlands). One woman was included in the final 12 candidates (as a result of political pressures) but eliminated with 7 others in the final tests. Later Malerba too was eliminated, and the surviving 3 given ESA staff jobs and training as payload specialists. As pressures grew among US mission specialists for the abolition of payload specialists on the grounds that MSs could do all that the PSs could do - plus EVAs, etc - ESA pressed for its astronauts to be upgraded to MS rank. Nicollier and Ockels were finally accepted for NASA MS training (though at ESA expense) but Merbold was disqualified on minor medical grounds. This did not however rule him out as a potential PS on Spacelab 1, and with Nicollier remaining at Houston for continued training after completing the basic MS course, the final choice for Spacelab 1 lay between Ockels and Merbold. It proved difficult for ESA to make the final choice until NASA began complaining that little time was left for final training. Then on 20 Sep 1982, Dr Merbold was named as prime crew member with Dr Ockels as back-up.

Star consortium

This consortium of 11 industrial companies in 10 European countries (Belgium, Denmark, FRG, France, Italy, Netherlands, Spain, Sweden, Switzerland and UK) specialises in *S*atellites for *T*elecommunications, *A*pplications and *R*esearch. 1st major contract was for ESA's GEOS, with British Aerospace as prime contractor and Hughes Aircraft, Calif, as consultant. Then followed ESA's ISEE-2 and ISPM, both with Dornier as prime contractor.

MILITARY SPACE

STS-2 crew Engle and Truly, seen here leaving *Columbia* at the end of the mission, had a strong military background: Engle earned his astronaut status flying the X-15, and Truly was originally recruited as a military astronaut. *(NASA)*

Introduction This section attempts to outline the history of the increasing military use of space, first in a general survey of developments, and then with as much detail of the rival US and USSR programmes as can be culled or deduced. Russia's military space activities have always been buried in her vast Cosmos series of launchings - the 1,500th was launched in Sep 1983 - making them difficult to disentangle from the many purely scientific satellites. And despite the fact that America's military launches are far fewer, her programmes too have become increasingly complicated, overlapping because new technologies encourage deployment of more advanced systems long before those already in orbit have completed their operational life.

Status In 1984 a much bigger issue than the US space station was whether the West - which really means America - should embark on defence against ballistic missiles (DABM). Dr Richard DeLauer, US Defence Department Under-Secretary for Research and Engineering, has warned that such a system would be "staggeringly expensive," but there seems to be an irresistible move towards it. In Oct 1983 NATO asked the US to continue development of a space-based anti-ballistic missile system.

The details of a 4-layered satellite defence system in a recently released report to the President from the US Government's Defensive Technologies Study Team also make staggering reading. But since the team is led by James Fletcher, former head of NASA, it must be taken seriously even if talk of such a multiple-layered defence system being operational by 2000 seems optimistic. Space-based battle stations would have a maximum of 300sec to destroy between 2,000 and 3,000 attacking missiles. They would do this in the 4 distinct phases of ICBM flight: boost, post-boost, mid-course and terminal. The theory is that each layer would detect and intercept 90% of the live warheads, with an overall "system leakage" of no more than 0·1%. That would imply no more than 2-3 nuclear warheads reaching their targets.

At least 8 US companies are working on the technology needed to produce space-based electromagnetic railguns firing hit-to-kill projectiles; pulsed and continuous-wave lasers able to "shock" the missiles into blowing up, burn holes in them, or destroy internal components are other candidate kill mechanisms.

DeLauer says that the "most fragile part" of the whole concept is the ability of the battle stations to survive a sudden direct attack upon them. America has not yet produced either the hardening they would need to protect them against attack, or the advanced computers needed to detect a massive missile launch, discard decoys and direct the destruction of the incoming warheads. Research and development alone could cost $18-27 billion in the next 5 years. NATO's N Atlantic Assembly has been told that the cost for the total system could be anything between $100 and $500 billion - and the Western European allies would have to contribute.

The decisions that emerge in 1984 now seem certain to be accompanied by a government edict imposing a unified space command upon the US services by 1985. The USAF set up its own Space Command, with HQ in Colorado Springs, in Sep 1982, thereby acknowledging that military contests in space had become a real possibility. By then, with "national security space spending" rising above $8 billion a year, the USAF was starting to lobby for its own Shuttle launchers, alongside studies of "a more capable Block II Shuttle". But the USAF proposal that its Space Command should become a tri-service organisation was countered by a US Navy announcement of the setting up of Naval Space Command (NavSpaceCom) with a 60-strong HQ at the Naval Surface Weapons Station at Dahlgren, Va, from 1 Oct 1983. NavSpaceCom was expected to include the Naval Space Surveillance System, the Naval Astronautics Group and Fltsatcom, and it was also intended to open career opportunities for Navy and Marine Corps astronauts at a time when the USN's 1st space engineering and operations specialists were graduating from courses started in Oct 1981. In addition to Fltsatcom, the Navy is planning a new ocean surveillance satellite, ITSS, and NavSpaceCom is expected to monitor and react to Soviet space activity. Capt Richard Truly, one of the original MOL astronauts, was appointed commander of NavSpaceCom immediately after completing the STS-8 mission.

Running parallel with all this is the anti-satellite (ASAT) controversy, dealt with later. This took a new turn in Jul 1983 when the Senate Foreign Relations Committee, supported by 102 Congressmen and 30 scientists, approved a resolution urging a delay in ASAT tests while negotiations on a ban were resumed with the Soviets. Senator Paul Tsongas described the Soviet system as a bulky, cumbersome weapon, launched from a fixed site and requiring extensive support facilities. It was launched by a ground-based SS-9 (F1) missile into an orbit close to the target satellite. It was limited to an altitude of 2,400km and could not attack for several orbits, and a full-scale attack on US satellites with such a system would take more than a day to complete. Once both sides had deployed ASAT systems, the superpowers would have moved to a launch-on-warning posture, greatly increasing the risk of accidental nuclear war. Within a few days of

ASAT: F-15 fighter carrying mock-up of the 2-stage hit-to-kill missile. 58 missiles have been ordered but there is political pressure to delay tests in the hope of encouraging Soviet-American talks on an ASAT ban.

this call for a suspension of the F-15 ASAT tests, President Andropov announced that the Soviet Union would suspend all ASAT development if the US did the same. A draft treaty placed before the UN also proposed prohibiting the use of manned spacecraft for military activities of any kind – an obvious move against the US Space Shuttle. Possibly to counter the Senate resolution, the US Joint Chiefs of Staff released more information about the Soviet ASAT system, which they said now included the ability to destroy target satellites by firing pellets at them; 5 launchpads were available at Tyuratam. The Soviets were also said to be developing a laser ASAT satellite, 6 of which could be operational within 7yr.

A final twist in this verbal space war came early in 1984 with the revelation that US intelligence has observed perhaps a dozen Soviet tests with tumbling re-entry vehicles in the last two years. The inference is that their object has been to test – perhaps to display – the Soviets' ability to spray lethal chemicals over heavily populated areas, missile ranges or space centres. To some this will seem to be the final example of Soviet cynicism, to others a final illustration of the utter futility of embarking upon a ruinously expensive game of technological leapfrog in space.

History The Soviet-American space race originally began in the 1950s not so much because of the international prestige involved but because each feared the other might achieve complete military dominance if it became the first to master space technology. Of 14 major programmes being conducted in that period for the USAF's Ballistic Missile Division, all aimed at searching out the possible advantages of space activity, perhaps the most effective was to be WS-117L. Under the sponsorship of the CIA contracts were offered for a "Strategic Satellite System" in Mar 1955. By Jun 1956 Lockheed had been given the contract to develop the Agena, an upper-stage rocket which would go into orbit and also serve as a satellite. Developed versions are still being launched, and Lockheed has been responsible for more satellites than any other company in the West, though the company is strictly forbidden to publicise this part of its activities.

The design and use of military satellites was thus far advanced long before the first 2 Sputniks were orbited in 1957. America's first photo-reconnaissance satellite was Tiros 1 in Apr 1960; Russia's was Cosmos 4 in Apr 1962. America's Projects Midas (Missile Defence Alarm System) and Samos (Satellite and Missile Observation System) were the 1st efforts to use space to double the warning time of an enemy missile attack given by the ground-based BMEWS (Ballistic Missile Early Warning System) by detecting the missiles the moment they were launched instead of when they rose above the horizon. Since 1980 these systems have been supplemented by the ground-based Pave Paws, a pair of phased-array radars sited on opposite coasts of the US at Otis, Massachusetts, and Beale, California. Their primary task is to watch for a sea-launched ballistic missile attack on the US. They also have the advantage for America that, linked with satellite systems, they make the US independent of foreign-based systems like BMEWS.

By the end of 1983 the comparative tables prepared by the late Dr Charles Sheldon and currently by Miss Marcia Smith for the US Government showed that US military launches numbered 351 compared with 1,075 by Russia, though another 96 US Defence Department launches had been classified as civil rather than military. Examples were the announced launches of small research satellites like Radcat and Radsat, launched 2 Oct 1972 from Vandenberg as radar calibration targets and to measure radiation effects on the lifetimes of instruments. In recent years the Soviet Union has launched at least 10 times as many military satellites as America, probably because of her more difficult northerly territory, her less advanced technology and a national tendency to believe that there is safety in numbers.

The late 1970s and the early 1980s have seen the gradual erosion and finally the collapse of the agreements of the 1950s and 1960s that neither East nor West would use space for offensive purposes. There was much alarm when in 1976 and 1977 Russia renewed tests of her anti-satellite system, launching satellites capable of inspecting and destroying suspect spacecraft. There were talks about a possible treaty to limit this type of activity, but by the end of 1981 Russia was considered to have an operational low-orbit ASAT system, with a high-altitude system under development.

American unease was also stirred in the late 1970s by reports – officially discounted by both sides – that Russia had used ground-based weapons to "blind" some of America's satellites. Following the invasion of Afghanistan in 1980 came reports that America was studying a $10 billion plan to enable the USAF to place about 250 Minuteman ballistic missiles, with single re-entry warheads, in parking orbit so that during a period of tension they could provide more time for the President's final decision on how great should be the response to a limited Soviet nuclear attack. This proposal aroused less interest in official circles in America at that time than it did in Russia, but public discussion of it in Feb 1980 was followed within 2 months by a resumption of Soviet ASAT tests.

By the end of 1981 the main US response had become "a high-technology interceptor": a SRAM missile launched from an F-15 aircraft. Details are given under the US ASAT entry; the circumstances in which it might be used are discussed under the Soviet OSS entry. In FY1982 DoD was spending $150 million on ASAT research and development, rising to $300 million in FY1983.

With no sign of any slackening in the 80-a-year pace of Soviet military satellite launchings, US defence experts have now become anxious about the vulnerability of their much thinner systems. The theory is that Russia might attempt to blind US defences by intercepting and destroying her early-warning satellites, and by putting the navigation and communications systems out of action by electronic jamming, either from the ground or from higher-altitude satellites. (A more sophisticated argument is that the early-warning satellites would be left untouched, so that America could see that an attack was on a limited scale and thus be tempted to limit or hesitate about her own response.) Given at the end of the US section are some details of moves to harden the warning satellites by shielding them against jamming (this may include protection against laser attack later on); and to provide multiple communications paths for others. "Cold birds" - satellites emitting no signals unless they are turned on to replace damaged or destroyed spacecraft - will also be launched. They could be placed in deceptive orbits and then quickly manoeuvred into the necessary paths to act as replacements as the need arose.

Despite all the media speculation about the imminence of laser war in space, all the evidence suggests that such a possibility is still a long way off. Pentagon reports that Soviet-based lasers might be able to threaten US synchronous satellites as early as 1983-88 were promptly discounted by civil and military scientists - though the 5yr spread of the forecast made it difficult to dismiss altogether. But the reports did elicit statements that pressure for costly laser programmes was premature when the technology had yet to be proven, and that much cheaper ASATs were available to deal with hostile satellites up to medium altitude at least. Whether ground-based or space-based, high-energy laser weapons are not likely to be operational until the mid-1990s. America's Defence Advanced Research Projects Agency (DARPA) continues to work on the 3 main components of a laser weapon, the so-called "space laser triad": pointing and tracking, beam generation, and conditioning of the beam through optics. Development is governed by the rate of funding, now about $150 million a year, and is unlikely to be speeded up unless there is any evidence that Russia's space laser capability is ahead.

US reports that pointing and tracking tests similar to those planned for the US space-based, high-energy laser weapon system had already been conducted by cosmonauts on both Salyuts 6 and 7 may have been aimed at obtaining additional Talon Gold funding rather than being based on real evidence of activities in the Soviet space stations at all. It is also interesting to note that there has been little public reference by US military men to the Soviets' announced details of materials-processing work aboard their space stations. A key technology in the development of a space-based radar capable of detecting bombers, ballistic and low-flying missiles, and space objects and debris is the production of solid-state transmit-and-receive modules no larger than a 25-cent piece. Experiments involve silicon on sapphire; gallium arsenide; and a possible hybrid module combining silicon and gallium arsenide into wafers. Compare this with the Soviets' own description of the activities in the furnaces of Salyuts 6 and 7, and it all makes much more sense!

	Objects in orbit	*Decayed objects*
Australia	1	1
Canada	12	0
Czechoslovakia	0	1
ESA	19	0
ESRO	0	10
France	25	56
France/FRG	2	0
FRG	5	9
India	12	4
Indonesia	3	0
Intelsat	30	1
Italy	1	4
Japan	51	21
NATO	5	0
Netherlands	0	4
People's Republic of China	3	39
Spain	1	0
UK	9	9
US	2,712	2,438
USSR	2,173	6,825
Totals	5,064	9,422

Monitoring space

The accompanying table, compiled and continuously updated by NASA's Goddard Space Flight Centre at Greenbelt, Maryland, shows that between 4 Oct 1957, when Sputnik 1 was launched, and 31 Dec 1983 over 14,000 man-made objects - satellites, rocket casings, discarded protective panels and other debris - have been placed in orbit. Their progress as they slowly fall back to Earth has been monitored by a series of overlapping and updated space detection and tracking systems. On 30 Jun 1982 North American Defence Command (Norad) was tracking 4,965 objects and quietly issuing "close approach" warnings through the USAF Satellite Control Facility when a primary communications satellite appeared likely to pass within 50km of another object. Increased tracking is started at 20km separation, and a collision avoidance manoeuvre considered when the 2 objects are within 5-8km.

At present there are about 80 active geostationary satellites, with another 120 uncontrolled pieces of debris drifting through this orbit. Miss distances are sometimes as low as 2·6km. So far it seems that serious collisions have been avoided at geostationary height, but it is believed that several catastrophic encounters have occurred in other orbits. There was the Pageos balloon which mysteriously disintegrated in 1976; the famous nuclear-powered Cosmos 954 was probably disabled by a collision in 1977; and Cosmos 1275 broke up 50 days after launch in 1981. The most serious incident involving a manned vehicle (though it might not have been occupied at the time) affected Salyut 6, which suffered torn vacuum insulation and splintered solar panel coverings.

More collisions are inevitable as space debris increases, even though they are likely to be few in number. The worst feature of this hazard is that there are now many thousands of pieces of debris too small to be tracked, mostly resulting from more than 70 explosions in space. At least 9 of these have been deliberate explosions, Soviet interceptor craft having been used as shrapnel to rehearse their ability to destroy target satellites. Another 11 have been US Delta rockets which have exploded up to 3yr after launch, a problem now apparently cured by a modification. At present Norad can track objects as small as 4cm up to 400km, though in geostationary orbit an object must be 2·5m in dia to be trackable. Against this, objects of only 1mm dia travelling at a relative velocity of 10km/sec can penetrate most satellites. According to the US Congressional Research Service, there is a population of up to 10,000 objects too small to be tracked at altitudes of 500-1,100km, with maximum density at 850km. An international agreement banning deliberate explosions in space, and measures to control and reduce other debris, are therefore becoming urgent. Collection of dead satellites should soon be possible, and it is now usually possible to remove geostationary satellites from that orbit with a final manoeuvre at the end of their active lives. This is estimated to average 7-8yr for US and 2½yr for Soviet satellites.

Meanwhile, the latest USAF system, due to be operational in 1983, is providing improved monitoring facilities. This is Ground-based Electro-Optical Deep Space Surveillance (GEODSS). The first of its 5 facilities spread around the world is already working. Their powerful telescopes are said to be able to identify objects "no bigger than a football" in orbits between 5,500 and 37,000km. Their photo-imaging tubes convert the pictures into electronic impulses which are fed into computers and relayed, with information from other monitoring systems, into the master computers of Norad's Command and Control System, buried deep inside Cheyenne Mountain, Colorado, where it is safe from nuclear attack.

There are more important reasons than collision hazards for such an unrelenting spacewatch. First, because re-entering space debris falling back to Earth could trigger a false alarm in the Western missile alarm system, giving the impression that Russia had launched an attack with nuclear-tipped missiles; and second, because under the 1967 United Nations Space Treaty each country is responsible for any damage caused by its returning space debris. With large objects like the 90-tonne Skylab, which re-entered in 1979, many tonnes survive the fiery fall through the Earth's atmosphere, and in a populated area could have the effect of a bomb. Happily, so much of the world's surface is either sea or unpopulated that the chances of damage are still small, and Skylab fell harmlessly in the Australian outback. But Russia had to pay compensation when its nuclear-powered Cosmos 954 fell in Canadian territory in 1978.

Development of America's spacewatch, operated by the US Air Force with help from Britain, Canada and other countries, started in the late 1950s to cover both missile firings and satellite launches. As Russia developed the ability to attack the West with intercontinental ballistic missiles (ICBMs), the West responded with Ballistic Missile Early Warning System (BMEWS), which still has an important role to play. It extended the then existing range of tracking radars from a few hundred km to about 5,000km. The most probable flightpath of a Soviet-launched ICBM attack was then over the Arctic, so 3 BMEWS radar stations were built at Thule in N Greenland, at Clear in Alaska and at Fylingdales, England. Each has antennas the size of a football field which are able to detect a missile as it rises above the horizon; trackers and computers then calculate its trajectory and indicate its impact point. A single missile suggests a satellite launch or test firing, and is regarded as having a low "threat value"; multiple launches would obviously have a high threat value. BMEWS trackers, even when they are satisfied that a Soviet launch is not hostile but a routine operation, still automatically feed the details to the Space Defence Centre in Cheyenne.

Many experts thought that by the early 1970s BMEWS would be completely out of date. In fact, valuable new roles have been found for it. The trackers, particularly in England, compute the precise location of any re-entering space derbis; the result is that America especially has been able to learn a great deal about Soviet metallurgy and space techniques by recovering and analysing sections of rockets and spacecraft which reach the surface. This has become a significant activity in both East and West, leading both sides to fit their spy satellites with self-destruct mechanisms so that they can be blown up or de-orbited at the end of these missions in such a way that any debris falls on home territory. It has been discovered too that by rolling the 3 Fylingdales trackers on their backs they can be made to look south instead of north, so providing some cover against missile attacks via the "back door", the South Pole. Astronomers and space scientists also look forward to the day when Fylingdales' military role does come to an end, since it has endless possibilities as a space observatory.

Detection and warning of submarine-launched ballistic missiles (SLBMs) started in 1968 with an interim system based on 8 radar sites located on the Atlantic, Pacific and Gulf coasts. They could detect the launch of a missile from a patrolling submarine and then track its course. It was replaced in 1980 by the USAF/Raytheon Pave Paws, a pair of solid-state radar systems on the US E and W coasts. With no mechanical parts to limit the speed of its radar scan, it embodies phased-array technology - thousands of small radar antennas co-ordinated by 2 large computers - allowing it to detect, track and predict the impact point and numbers of missiles. The two sites, at Otis, Massachusetts, and Beale, California, also feed display data on the position and velocity of satellites to Norad. Two more sites, in Georgia and Texas, are planned.

Ground-based detection was never considered adequate, however, since the fixed ground sites were vulnerable to attack. Midas and Samos satellites, with infra-red sensors able to detect a land or submarine-launched missile by comparing its hot plume with the cooler background, were launched from the early 1960s. But they had their limitations: sun shining off the clouds could sometimes look like the infra-red signature of a missile launch, and on one memorable occasion they misread beams from the rising Moon as a missile attack. At first they did not cover all possible Soviet launch sites, nor provide accurate impact-point predictions. They have been succeeded by the IMEWS and DSP satellites, whose capabilities are highly classified. No doubt they too have their limitations, but with the combination of Pave Paws, BMEWS and early-warning satellites, Soviet strategists must feel that there is little chance of launching an undetected pre-emptive strike.

It was the development of Pave Paws which led to the abandonment of over-the-horizon radar (OTHR), which went into operation briefly in Mar 1968. This was to meet the threat of the fractional orbit bombardment system (FOBS), Soviet missiles which could be launched into orbit via the South Pole, thus avoiding detection by BMEWS; they would then travel threequarters of the way round the world before re-entering to strike their target. OTHR was not really radar at all, consisting instead of a series of transmitters producing a continuous signal which bounced back and forth repeatedly between the ionosphere and the Earth's surface until received at another station several thousand miles away. A missile penetrating the ionosphere disturbed the signal and was thus detected. One of these stations operated at Orfordness on England's E coast for several years until it was closed down in 1974.

Then came SPADATS, which includes the US Navy's Space Surveillance System, made up of a line of radio transmitters and receivers strung across the US; Canada's Satellite Tracking Unit, which operates a telescopic camera able to photograph satellites; and Spacetrack, operated by the US Air Force. It was this system, with its worldwide network of cameras and radars, which a few years ago was making 18,000 satellite observations per day, logging their decay and predicting their re-entry to ensure that they would not be mistaken for attacking missiles. This system made the 1st use of the Eglin AFB phased-array radar, consisting of 2 fixed electronic arrays set in a 10-storey-high bank of concrete sloped at 45° to the horizon. One array, consisting of 5,184 transmitter modules, sends out an electronically aimed and steered beam which sweeps the area of coverage in milliseconds; the other array, 4,660 modules in a hexagonal shape, receives the signal as it bounces back from an orbiting satellite. Florida was chosen for the first of these arrays because most space objects pass within its coverage every day.

Spacetrack radar sensors can be used to reconstruct the exact shape and size of a space object. When Skylab was damaged during launch in 1973 correspondents like myself suspected that NASA had obtained a top-secret military picture of the space station which confirmed that a solar panel had failed to deploy. Over a year later this was officially admitted. That incident demonstrates how America can ascertain, within minutes of its launch, the size, shape and purpose of every Soviet spacecraft, whether its job is military or civil, and whether it is operational or damaged. If Russia does not yet have a similar ability to monitor American launches, she is no doubt working hard to obtain it.

With the development of Pave Paws and GEODSS, all these systems are now being pulled together in the interests of both efficiency and economy into a Consolidated Space Operations Centre (CSOC) covering both satellite and Shuttle operations. When completed in 1986 it will be manned by 1,800 military and civilian personnel. By then the military will also be sharing NASA's 4 Tracking and Data Relay Satellites, which will permit the closure of most of NASA's foreign-based space tracking stations. Thus the US will soon have little need for outside help with its space monitoring activities. As a first step, USAF officers are already being trained as mission controllers at NASA's Houston Centre.

Soviet monitoring The Soviet equivalents of the US BMEWS and Pave Paws consist of installations at Pechora, near the Arctic Circle, to detect an ICBM attack from the North; Kiev, to cover the Mediterranean and E Atlantic; and Komsomolsk-na-Amure, NE of Japan's Hokkaido Island, to cover the S and SE, and particularly SLBMs. In accordance with the ABM treaty of 1972, these stations are on the periphery of national territory and looking outwards. In late 1983 US spy satellites were examining a new radar detection system at Abalakova, central Siberia; this could be an ABM system protecting SS-18 missile fields, which would be a breach of the treaty.

In addition to satellite systems and these ground-based radars, the Soviets have a fleet of satellite communications vessels to compensate for their lack of round-the-world ground bases. Named after dead cosmonauts, the largest is *Kosmonaut Yuri Gagarin*. Another, referred to as "new" in 1979, is *Kosmonaut Pavel Belyayev*, which spent 200 days at sea maintaining communications with Salyut 6 crews. First of a new series announced at that time was *Kosmonaut Vladislav Volkov*; fitted with 4·5m dish antennas which track spacecraft automatically, they receive data from both manned and unmanned spacecraft, orbital stations and interplanetary probes. Data are automatically deciphered, computer-processed and transmitted either direct by short-wave or via a communications satellite to the appropriate flight control centre. Others in the *Volkov* class are *Kosmonaut Georgi Dobrovolsky* and *Kosmonaut Viktor Patsayev*.

United States

Contents

Space Shuttle

As East-West relations deterioriated the pressures to use the Shuttle for developing space-based weaponry steadily increased. US military men have for long expected that the Shuttle would enable them to continue to achieve more in space with far fewer launches than the Soviets. But few consider that the Shuttle should be used as a military spacecraft: visions of Orbiters manoeuvring to evade attacks by missiles and laser beams, or even by rival Soviet orbiters, are at least premature. Since a direct attack on an Orbiter would be "a huge act of war", such a move is unlikely unless hostilities had already broken out; much more importance is attached to protecting communications between the Orbiters and their command and control stations.

The Shuttle's routine military work will consist of delivering and servicing satellites, with testing of sensors and technology for new programmes building up later. Of 92 missions planned by Sep 1988, 37 will include DoD payloads; 18 of these flights will be dedicated entirely to military purposes. Military use of the Shuttle started inauspiciously: on STS-4 the lens cap on the infra-red telescope refused to budge, and STS-10, the 1st dedicated mission, had to be cancelled altogether because of the IUS failure on STS-6. Later all 3 military flights planned for 1984 were postponed, and a decision made to order 10 more Atlas or Titan launchers for use up to 1989. However, priority is likely to be claimed for a series of tests of infrared sensors, code-named Validator. This is the start of a campaign to move the US away from its long-established policy of deterrence by assured mutual destruction and towards an "assured survival" policy. Validator would start modestly, with 3 missions each needing only 1m of

3rd operational orbiter *Discovery*, seen here at KSC being prepared for its maiden flight, will be based at Vandenberg. Its schedule will include military missions in the polar orbits possible from this USAF site. *(Margaret Turnill)*

the cargo bay to start gathering the acquisition, pointing and tracking data needed for space-based surveillance and laser weapons.

The right to claim priority was established by a NASA/DoD memorandum of understanding signed in 1980. Military anxiety about the Shuttle's delayed development was assuaged by giving the USAF the right to claim priority, "upon declaration that national security was involved," over commercial cargoes, even when these had been ordered and paid for long in advance. This, incidentally, did much to help Ariane's order book.

Soviet criticism that the Shuttle is mainly a military vehicle is at least partly supported by the fact that the delta wings, giving a cross-range of over 2,000km, and the size of the cargo bay were insisted upon in anticipation of future military requirements. Ultimately the Shuttle should be able to carry twice the payload and 3 times the volume of the current largest booster, Titan 3, but initially it has serious limitations from a military point of view. Despite a determined programme to increase payload by reducing the weight of the ET and SRBs, it is likely that small boosters will have to be added to the SRBs if the planned 14,500kg payloads are to be launched into polar orbit from the 2nd launch site, at Vandenberg, California. Other improvements will be needed to enable the Shuttle to recover KH11 and other spy satellites for refurbishing and updating with the latest technology. A proposal to fly reconnaissance cameras in the Shuttle on three 20-day flights per year to supplement and improve upon the information provided by unmanned spy satellites was cut from the 1980 budget for financial reasons. But such flights are in any case beyond the Shuttle until endurance-increasing power modules become available.

1st use of the polar launch site at Vandenberg has slipped to 25 Oct 1985, when OV-103 (*Discovery*), with modifications for polar and military use, is due to inaugurate USAF-launched flights. The delay was due to problems with assembling the cluster on the launchpad in high winds. The USAF also says that it will need 2 operational Orbiters for "a significant period" in 1984-5 for use in checks of the new launch facilities.

With a rapidly expanding manned spaceflight support effort, and selection well under way long before any official announcement of its own exclusive cadre of military astronauts, the USAF is moving enthusiastically into the Space Shuttle programme. A policy of direct participation in Shuttle operations started with the formation of Manned Spaceflight Support Group (MSFSG) in 1979, followed by the appointment of 13 officers to train as flight controllers. From STS-1 on, USAF personnel parallelled and gradually began taking over many of the launch and mission control activities. NASA's traditional "open" policy, with newsmen entitled to know everything that was going on, was gradually obscured by "measures to assure protection of information and resources for DoD activities on Shuttle flights". Classified control rooms are being provided for USAF use at both Launch Control at Kennedy and Mission Control at Johnson. The latest NASA flight assignment manifest omits all details of DoD flights; mission length, crew numbers, inclination and altitude are all left blank. Initially military astronauts will work on the aft flight deck, being provided with their own control panel and joystick control to operate and point equipment mounted in the payload bay. One of the first tasks for a new unit called Space Test Program is to develop a standard panel so that the Shuttle can be used for many experiments not needing an individual satellite. An example of "carry on" experiments to be conducted at short notice was a hand-held camera for obtaining infra-red images of cloud build-up which the USAF Air Weather Service wanted to try out on Mission 14. Free-flying military satellites to be launched from the Shuttle and later recovered include one which will carry Teal Ruby's staring-mosaic focal-plane arrays, designed to detect aircraft from space. In addition to all this, there seems little doubt that as these activities advance the military astronauts will soon be pioneering all sorts of exciting EVA work. Since they will not be involved in flying the Shuttle itself, it seems likely that the 6 remaining Manned Orbiting Laboratory (MOL) astronauts transferred to NASA from the USAF in 1969 will be given preference on military missions. It was no coincidence that another former MOL astronaut, Gen James Abrahamson, was briefly placed in charge of the Shuttle during its vital development period before being transferred back to the USAF to take charge of the so-called "Star Wars" programme. For this he has been given $26 billion to spend in the next 5 years on investigating what military activities are possible in space. His much less rigid personality may help to lessen the conflict between NASA and USAF personnel while the latter seek to carry out their orders: "Prepare for the eventual command and control of DoD Shuttle flights from a Consolidated Space Operations Centre". This is to be built at Colorado Springs in the mid to late 1980s, and more than 900 military and civilian specialists will be trained for Shuttle operations alone.

By then it seems likely that production of a small, "quick-reaction" shuttle will be under way. Boeing and Rockwell have both done preliminary designs, the former under a $100,000 a year contract. The vehicle would have a 3,629-4,536kg payload, compared with Shuttle's 29,500kg. With a single-stage-to-orbit capability (though Boeing proposes to assist take-off with a rocket-powered sled), it would look much more like a conventional winged aircraft and be capable of up to 1,000 missions, or 10 times the Shuttle's designed mission life. Under both concepts the Shuttle main engine would be used to achieve orbital velocity, and both would use metallic heatshields, probably a combination of titanium, nickel and coated columbian. One of the jobs of such a "pop-up" spacecraft would be to rotate the crews of permanent space platforms in times of crisis or war.

Boeing's latest concept is the Air-Launched Sortie Vehicle (ALSV), an unmanned mini-shuttle using a small version of the External Tank and weighing a total of 125 tonnes. Launched from the back of a Boeing 747, it would be fitted with an SSME plus liquid oxygen and hydrogen tanks to feed it, and would be able to reach any part of the world within 100min of take-off. Rockwell is experimenting with the 7·6m-long Manoeuvrable Re-entry Research Vehicle (MRRV), which could be launched from under the wing of a B-52 bomber and could return to its point of origin for rapid re-use.

ASAT

1st of 12 test firings of a 2-stage hit-to-kill missile from a modified McDonnell Douglas F-15 fighter took place on 21 Jan 1984 over the Western Test Range. The USAF's 1st ASAT should be "initially operational" in 1987 and fully operational in 1989. A full-size mock-up of the ASAT was fitted to an F-15 in Sep 1981; it was 5·4m long and 50·8cm wide, with a launch weight of 1,179kg. Boeing is responsible for design and development of the 1st (Starr) stage, a modified version of the company's SRAM missile, and also for system control; Vought, which had by then received contracts worth $328·5 million, is responsible for the 2nd (Altair) stage and the interceptor vehicle. The latter is reported to be a simple infra-red seeker spinning at 1,200rpm and guided by a laser-gyro roll sensor. The direct-ascent trajectory is intended to reduce the time available for the Soviets to manoeuvre the satellite target, which the ASAT destroys by collision; no explosive warhead is used. Earlier assumptions that this weapon was to intercept Soviet ASAT satellites have proved wrong. Its weaknesses, recently emphasised, are that it cannot attack Soviet ASATs and is dependent on space-based surveillance systems which cannot yet adequately track manoeuvring satellites. 58 missiles have been ordered for delivery by 1989. Development and procurement costs of even this low-altitude system have now risen to $3·85 billion. An all-altitude capability, not fully operational until at least 1999, would cost $15 billion. This might be a Spartan ABM missile armed with pellet warheads, or a low-powered space-based laser.

In the meantime, it has been decided to base F-15 ASAT squadrons at Langley AFB, Va, and McChord AFB, Wash, on account of the orbital inclinations of Soviet target spacecraft and the need to have booster debris from tests fall into the Atlantic or Pacific. ASAT command and control is conducted from a facility developed since 1979 in Norad HQ. Its responsibilities include monitoring vulnerable ground facilities which might be sabotaged, and watching for radio interference and physical anti-satellite or high-energy weapon attack against spacecraft.

Spy satellites

More respectably known as "reconnaissance satellites", these spacecraft are generally divided into the Area Survey and Close Look categories, though their tasks often overlap and they are sometimes supplemented by "piggyback ferret" electronic surveillance satellites. It was the Discoverer series that established that both area-survey pictures, transmitted by radio as well as more detailed close-look photographs, dropped off in recoverable capsules, would be necessary.

Close Look First Agena B Close Look satellite with recoverable film capsule was launched by Atlas from Vandenberg on 7 Mar 1962 into a 251 × 676km orbit with 90° incl. Later orbits were around 135 × 400km. By 1976, when 93 launches had been made, there had been 3 generations, with Titan 3B taking over to launch the Agena 3Ds, which had increased in weight from 2,000 to 3,000kg. The restartable Agena engine permitted perigees as low as 135km to be maintained for over 50 days, so that annual launch rate could be reduced from 9 in the early years to 2 or 3 in the 1970s. Versatility of the USAF launch system became apparent when the failure of a launch on 5 Jun 1974, apparently intended to observe the results of India's first nuclear tests, was followed by a successful launch the next day. When Big Bird came into use in mid-

Agena target vehicle photographed from Gemini 8 by David Scott in 1966. Much-modified versions of Agena are still used as spy satellites. *(NASA)*

1972 these satellites were continued, apparently to fill the gaps in Big Bird coverage. In 1964-66 this series was used for 8 piggyback launches of ferret satellites.

Samos The Satellite and Missile Observation System, formerly called Sentry, was able to photograph all parts of the world from polar orbits, tape-recording TV pictures over potentially hostile territory and later transmitting them when passing over US territory. Samos 1, weight 1,860kg and L 11 Oct 1960 by Atlas-Agena A from Pt Arguello, failed to achieve orbit. Samos 2, L 31 Jan 1961 into a 475 × 554km orbit with 97° incl, operated for 1 month and proved the value of the system. Its 150kg of instruments sent back pictures which established that the US had greatly overestimated the number of Soviet ICBMs and the extent of the "missile gap". Samos 3 exploded on the launchpad on 9 Sep 1961. Subsequent launches came after the decision that details of military satellites should be classified, and the name disappeared from the log tables. However, there were 23 launches, with the orbits getting lower as techniques improved, up to 7 Jan 1963. Attempts to introduce a 2nd generation, carrying more film and consumables and launched by Thrust-Augmented Thor-Agena D (TAT-Agena D), began with failures on 28 Feb 1963 and 18 Mar 1963, but the programme was soon achieving 90% reliability. Last launch in the 13yr programme, which ended with the introduction of Big Bird, was on 25 May 1972; the satellite went into a 158 × 305km orbit with 96° incl. By then they were carrying infra-red scanners in addition to cameras, making night photographs possible, and a 1·5mm antenna giving a much higher rate of transmission. 19 of the series also carried 60lb ferret satellites. Lifetime averaged about 20 days at first and around 80 days at the end.

Big Bird A large Titan-launched USAF multi-function satellite designed to perform both the area-survey and close-look functions, which required 2 different spacecraft until it came into operation in 1970. Also known as Low Altitude Surveillance Platform (LASP). Weighing over 13,000kg in orbit, it probably consists of a modified Agena rocket casing, 15·2m long and 3·05m in dia, fitted with a high-resolution Perkin-Elmer camera capable of identifying objects as small as 0·3m across from heights of more than 160km. Operational techniques are similar to those employed by the ERTS satellites: they are placed in Sun-synchronous orbits so that they pass regularly over the targets at the same time of day. A series of pictures with identical Sun angles is thus obtained, and changes occurring, such as the construction of new missile sites and the number and types of missile being installed, are easily read. Film is processed on board, scanned by a laser device, and converted into electronic signals for transmission to at least 7 receiving stations at the USAF's global bases. The drag encountered by such a large vehicle at such low altitude would normally cause it to re-enter in 7–10 days. The Agena engine is therefore fired periodically to raise the orbit and extend its life. Big Bird is almost certainly capable of carrying out some electronic intelligence (elint) as well, and sometimes carries with it a small 60kg piggyback capsule which is placed in a higher orbit for such "ferret" operations. Because both East and West collect and examine each other's space debris whenever possible, the Americans adopted Soviet policy with Big Bird, blowing them up in orbit to ensure debris fell in their own area or in the sea, rather than allowing them to decay naturally.

Big Bird: night launch from Vandenberg by Titan 3. *(NASA)*

Big Bird production ended in 1981, with a number held in store for annual launches until later systems made them unnecessary. KH11, using direct TV imaging transmitted via relay satellite, had become more practical for routine use. The twice-yearly launches in the mid-1970s had become occasional launches by the end of the decade, reflecting the need to supplement KH11 intelligence during periods of crisis such as Middle East conflicts. By the end of 1983 a total of 18 Big Birds had been identified.

Big Bird 1, L 15 Jun 1971 from Vandenberg on the 1st known Titan 3D/Agena, weighed 11,400kg and was placed in a 114 × 186km orbit with 96° incl; re-entered after 52 days. By the end of 1973 a regular pattern of launches had developed, with slightly higher orbits and lifetimes extended to 90 days. There was a 3-month gap between Big Bird 8 (L 10 Apr 1974), which re-entered on 28 Jul, and Big Bird 9 (L 29 Oct 1974; orbit 162 × 271km). This was probably due to modifications to its sensors to counter Soviet efforts to camouflage missile silo and control centre construction. Big Bird 9 provided surveillance at a time of increased tension in the Middle East and took pictures of 16 Soviet ships unloading crated materials, believed to include SA-6 spare parts and components for Scud surface-to-surface missiles. Big Bird 9, L 29 Oct 1974, operated for 141 days; Big Bird 10, L 8 Jun 1975, 150 days; Big Bird 11, L 4 Dec 1975, 119 days; Big Bird 12, L 8 Jul 1976, 158 days; Big Bird 13, L 16 Mar 1978, 179 days; Big Bird 14, L 16 Mar 1979, 190 days.

Big Bird 15 (L 18 Jun 1980, wt ? 13,300kg, orbit 169 × 265km at 96·5°), supplemented KH11-3 with film returns for about 180 days, observing the Iran-Iraq war. The US satellites were matched by Russia's Cosmos 1210, a 2wk imaging vehicle, and Cosmos 1208, a 3-day recoverable craft which was also capable of returning film pods. With Big Bird 15 a new practice began: with it a 60kg electronic intelligence-gathering satellite was launched into a 1,331 × 1,333km orbit at 96·6° giving a 5,000yr orbital life.

Big Bird 16 (L 25 Feb 1981, orbit 138 × 336km, incl 96·4°) operated for 112 days but carried no electronic satellite. Big Bird 17 (L 11 May 1982, orbit 174 × 258km) operated for 208 days and was probably linked with the Falklands War. It carried an electronic satellite (wt 60kg, orbit 701 × 707km) with 70yr orbital life. Big Bird 18 (L 20 Jun 1983, orbit 169 × 229km) was accompanied by an electronic satellite which entered a 1,289 × 1,291km orbit at 96·7° (5,000yr orbital life). Big Bird probably identified the new Soviet missile radar at Abalakova, central Siberia, which was then studied in more detail by KH9 to check for a possible breach of the ABM treaty.

KH8 and KH9 First public report of the "keyhole satellite" KH9, a low-altitude film-return spacecraft with limited lifetime and used only to photograph highest-priority intelligence targets in the Soviet Union and elsewhere, came in Apr 1983. With lower orbit than either Big Bird or KH11, one was launched on 15 Apr 1983 by Titan 3B from Vandenberg into a 136 × 297km orbit incl at 96·53°; weight was believed to be 3,000kg. Another, launched 31 Jul 1983 by Titan 3B, was reportedly intended to take pictures of the new Soviet ballistic missile radar at Abalakova, central Siberia, believed to be a violation of the ABM treaty. KH8 had not been publicly discussed until Jan 1984, when *Aviation Week* revealed that production of both types had been ended because of cost over-runs and the need to keep KH11 alive.

KH11 Keyhole satellites are placed in higher orbits (300 × 500km) than Big Bird, weigh about 13,500kg and measure 19·5m in length and 2m in width. Built by TRW for the USAF, officially designated Program 1010 and launched by Titan 3D from Vandenberg into 97° incl, KH11 has a manoeuvring engine able to restore its original orbit every 3 months, giving a 1-2yr lifetime. Its sensing systems and high-resolution cameras enable it to distinguish military from civilian personnel, while its infra-red and multi-spectral sensing devices can locate missiles, trains and launchers by day or night, and distinguish camouflage and artificial vegetation from real plants and trees. Its sideways-looking radar can see through cloud cover. First publication of the designation resulted from the sale in 1978 by a CIA employee of a system manual to a KGB agent in Greece. In early 1980 a KH11 was used to watch construction of a large shed at the Severodvinsk naval yard on the White Sea, from which finally emerged a submarine even larger than the US Trident boats. In Sep 1980 the Pentagon let it be known that a KH11 had observed that the Russians had placed SS-20 and SS-16 missiles side by side. It was concluded that this had been done so that Soviet reconnaissance spacecraft could take pictures which would then be used to increase the external similarity of the missiles, making it impossible for America to check whether Russia was observing the Salt II agreement on ballistic missile limitation.

By the end of 1983, however, the KH programme appeared to be in both financial and technical trouble. Development costs had over-run by $1 billion, and the Soviets had introduced an elaborate camouflage system which made it much more difficult to obtain detailed pictures of new construction. More advanced satellites were needed, and in-orbit refuelling exercises planned for early Shuttle flights were aimed at extending the life and manoeuvrability of future KH craft in an effort to reduce their cost.

KH11-1, L 19 Dec 1976, re-entered on command 28 Jan 1979. KH11-2, L 14 Jun 1978, was still operational at the end of 1980. KH11-3, L 7 Feb 1980, observed the Iran-Iraq conflict with Big Bird 16, in competition with Cosmos 1210 and Cosmos 1208, and was probably still operational at the launch of KH11-4 on 3 Sep 1981. KH11-5 L 17 Nov 1982.

NOSS The US Navy's Ocean Surveillance Satellites were developed as a result of the growth of Soviet naval power in the late 1960s, and the increasing difficulty of maintaining aircraft surveillance of all the new missile cruisers and submarines, attack submarines and helicopter carriers. From 1971 the Navy relied upon USAF spy satellites until the long years of research culminated in NOSS-1, launched 30 Apr 1976 by Atlas from Vandenberg into a 1,092 × 1,128km orbit with 63° incl. No weight is available, but in orbit NOSS-1 and its cylinder-shaped successors released 3 box-shaped subsatellites into similar orbits, with a lifetime of about 1,600yr. Developed by the Naval Research Laboratory and Fairchild Industries, NOSS is believed to carry millimetre-wave radar, able to track surface ships in all weathers and with radio-frequency facilities for listening in on their radar and communications. It may also be able to track submerged nuclear submarines, detecting their warm-water wakes, and low-flying missiles. NOSS-2 was launched 8 Dec 1977, followed by NOSS-3 on 3 Mar 1980, suggesting a 3-5yr operational life. In 1981 Goddard was given project responsibility for a heavier, high-inclination NOSS to meet both civil and defence needs; its data would be used to improve weather forecasts and contribute to climate research, sea-ice forecasts and ocean acoustic propagation predictions.

Ferrets

Elint A general name for electronic intelligence or "ferret" satellites, now believed to be highly effective after 20 years' development. Usually launched into circular 500km orbits – higher than photographic satellites – they are electronic ears, recording radio and radar transmissions from areas of military activity. When replayed to ground stations the "radar signatures" – characteristics such as pulse-repetition frequency, pulsewidth, transmitter frequency and modulation – enable the likely function and method of operation of a particular centre to be identified. The number and type of electronic systems at a particular site, and subsequent changes in the signals, give a valuable indication of its purpose and capability. The ability to intercept and decode an enemy's satellite and ground-based communications, and to interfere with them by rival satellite activities, is likely to be a decisive factor in any large-scale future hostilities.

Heavy ferrets Starting on 15 May 1962 with a Thor-Agena B launch from Vandenberg into a 305 × 634km orbit (approx 1,500kg wt, 82° incl, life 560 days), 17 large satellites, launched at a rate of 2-3 per year, were identified as probable ferrets and characterised as "heavy" to distinguish them from the smaller subsatellite ferrets. As they were developed, Thrust-Augmented Thors and Agena Ds were used, then Long-Tank Thrust-Augmented Thors and Agena Ds, the last being launched 16 Jul 1971 into a 488 × 508km orbit with 94·5° incl and 7yr life. Following conviction in 1977 of 2 Americans for selling documents to Soviet agents, it was learned that satellites code-named Rhyolite had taken over this work, the 1st, weighing approx 700kg, being launched on 6 Mar 1973 by Atlas-Agena D into synchronous orbit at 0·2°. They are being followed by Project Aquacade, and because of Shuttle delays had to be modified for launch by Titan/IUS.

Subsatellite ferrets It is believed that a series of small subsatellites, launched piggyback by the Close Look and Area Survey satellites, were used to identify possible targets for detailed examination by the heavy ferrets. 1st of these, weighing 50kg and launched 29 Aug 1963 from Vandenberg by Thor-Agena D into a 310 × 431km orbit and 82° incl, had a 30-day life. By 8 Jun 1975 34 had been identified; Big Bird took over as their launcher from 20 Jan 1972. There were 5 interesting variations on 12 Dec 1968, 5 Feb 1969, 10 Oct 1972, 10 Nov 1973 and 8 Jun 1975, when the ferrets were fired into much higher, 1,400km orbits. Philip Klass, in *Secret Sentries in Space*, suggested that they were intended to monitor Russia's Sary Shagan centre at the time of Soviet ABM tests.

Early-warning satellites

Midas Equipped with infra-red sensors capable of detecting the exhaust plume of hot gases from a ballistic missile against the background of the Earth, this system was planned in the 1950s. Ground-based BMEWS could detect missiles once they rose above the horizon and give the US 15min warning of an attack. With 12-15 carefully spaced Midas satellites in polar orbit the theory was that this warning could be increased to 30min. But Midas 1 (L 25 Feb 1960) failed to reach orbit; Midas 2 (L 24 May 1960) reached a 484 × 511km orbit but suffered telemetry failure. Midas 3 (L 12 Jul 1961) was successfully launched from Vandenberg into a circular polar orbit of 3,428km and 91° incl. Orbital life was 100,000yr. Though it was known to be fully operational, there is no record of how long it lasted. Midas 4 (L 21 Oct 1961) and Midas 6 (L 9 May 1963) aroused an international furore when details of an experiment called Project West Ford became known. This involved the ejection of a 35kg canister containing 350 million hair-like copper dipoles each 21mm long. The idea was that after separation the spinning canister would slowly dispense the dipoles in an orbital belt 3,220km high, 8km wide and 40km deep to test whether they could act as passive reflectors for relaying military communications. For over a year the project was violently attacked by the world's scientists, particularly in Britain and Russia, because they felt the dipoles might interfere with astronomical observations, especially with radar telescopes. The late Professor Keldysh, president of the Soviet Academy of Sciences, said the experiment could result in "serious contamination of near-terrestrial space and greatly hamper both manned spaceflights and astronomical observations." Midas 4 successfully ejected its canister but the dipoles failed to disperse. Despite the protests, the US Air Force insisted on repeating the experiment with Midas 6, which was said to be successful. After that, however, talk of dipoles was dropped. Apart from the dipole incident, Midas 4 was credited with detecting the launch of a Titan missile from C Canaveral 90sec after lift-off – yet another leak which added to growing demands for greater secrecy for US military space experiments. A much better reason for secrecy at that time was the large number of false alarms given by the Midas system, which at first read sun reflections off cloud tops as missile launches. The name was dropped and Midas became Program 461. However, by 1963 President Johnson, apparently referring to the 7th and 9th launches (9 May and 19 Jul into circular 3,600km orbits with 88° incl), was able to claim that both liquid and solid-fuelled ICBM launches had been detected. After a 3yr gap 2 launches (19 Aug and 5 Oct 1966) apparently tested infra-red detectors directing TV cameras with telephoto lenses on a more advanced Agena D satellite in similar orbits. The 1st of 4 synchronous-orbit satellites of this type was launched on 6 Aug 1968 into 10° incl so that it traced a figure of 8 over W Russia. The last in this series, launched 1 Sep 1970 into a 31,947 × 39,855km orbit with 10·3° incl, was reported to have been involved in the "laser blinding" incident of 1975.

Missile Defence Alarm System (Midas) being launched by Atlas-Agena to give warning of Soviet launches. *(Lockheed)*

IMEWS Integrated Missile Early Warning Satellites – built by TRW, powered by cruciform solar panels spanning about 7m, with at least a 5yr life, and launched by Titan 3C – took over from the Midas series. "Integrated" referred to the fact that this series was also equipped with sensors to take over nuclear test detection from Vela. It is believed that they provided immediate information when Russia tested missiles carrying 3 separate warheads which spread out to make interception more difficult. IMEWS-1 (L 6 Nov 1970 from C Canaveral, wt 820kg, orbit 26,070 × 36,050km at 7·8° incl) ran out of fuel before it could observe Chinese missile tests, but IMEWS-2 and 3 (L 5 May 1971 and 1 Mar 1972 into synchronous orbits at 0·1 and 0·2° incl) were declared operational. The 1st was stationed over the Panama Canal to watch both the Atlantic and E Pacific oceans for submarine-launched missiles, the 2nd over the Indian Ocean to warn of an attack by Soviet land-based missiles as well as to provide details of Soviet and Chinese missile tests. IMEWS-4 and 5 followed on 12 Jun 1973 and 14 Dec 1975.

DSP Defense Support Program (also known as TRW Block 647 EW) satellites continue and probably overlap IMEWS. 3 are deployed in geostationary orbit over the E and W hemispheres to cover Soviet ICBM and SLBM launch areas and to give early warning of a possible ballistic missile attack. Some details were released as confidence in the operational ability of DSP grew. With in-orbit weight of 1,100kg, each is a 2·91m-long, 2·78m-dia cylinder covered in solar cells which are supplemented by 4 solar panels. It looks down with a 3·63m-long Schmidt telescope with an aperture of 0·91m. This collects infra-red energy emitted by missiles within 90sec of launch. Orientation in orbit is maintained by spinning about the Earth-pointing axis at 5-7rpm. The telescope's axis is offset by about 7½°, producing a conical scanning pattern as the vehicle rotates. From synchronous orbit each of the infra-red sensor's 2,000 lead sulphide cells views a region on Earth less than 3km across. In Feb 1981, following Soviet tests of a ground-based high-energy laser system against re-entry vehicles, it was decided that future DSPs must be hardened against thermal energy by using ablative materials, though they could not be deployed until at least the early 1990s.

The USAF plans to buy 4 additional modified satellites at a rate of 1 per year from 1982-86, thus saving money; they will be modified for launch by either Titan 34D/IUS or Shuttle/IUS. However, DSP is considered by some (according to USAF Space Division commander Gen Henry) to be "flawed" by its inability to distinguish between empty and loaded Soviet missile silos; to overcome this a DSP Upgrade, followed by a DSP-2 programme, is planned. DSP-2 would be equipped with a "staring mosaic" sensor which would see instantly anything worth observing, as opposed to current sensors, which sweep like a searchlight and are slower to react.

Nuclear detection

Vela/ERS Planned as a way of policing a nuclear test ban treaty before America's 1st satellite had been launched, 6 pairs of Velas were launched, starting on 17 Oct 1963. They were able to give instant warning of any violation of the treaty of that year prohibiting nuclear weapons tests either in the atmosphere or distant space. They could have detected nuclear explosions as far away as Mars and Venus. The initial pair, launched by Atlas-Agena D into circular 100,000km orbits with 38° incl, were placed on opposite sides of the Earth, well beyond the Van Allen radiation belts. Velas 3 and 4 followed on 17 Jul 1964, 5 and 6 on 20 Jul 1965, 7 and 8 on 28 Apr 1967, 9 and 10 on 23 May 1969, and 11 and 12 on 8 Apr 1970. The last 2 pairs, with 263kg weight, were launched by Titan 3C. Transmission life was about 3yr, with orbital life 1 million yr. Later versions also made solar flare and other observations and the launcher usually carried small piggyback payloads, notably Environmental Research Satellites (ERS) ranging in weight from 0·7 to 20kg and numbering 29 by the end of 1972. The nuclear detection role was taken over by IMEWS in the 1970s.

Communications

Development A military satellite operating in synchronous orbit overlooks 163 million sq km of Earth's surface, compared with an aircraft 8km high, which can see 284,900 sq km. The development of these satellites for military communications and navigation has therefore formed a major part of the US space effort.

History The world's 1st communications satellite, the 21st in the world log to be orbited, was **Score**, launched 18 Dec 1958 by Atlas B from C Canaveral into a 185 × 1,470km orbit with 32° incl; wt 70kg. It transmitted taped messages for 13 days and re-entered 34 days later. Next came **Courier**, launched 4 Oct 1960. The 1st active-repeater comsat, it operated for 17 days. Lack of rocket power, together with political and economic argument, delayed further developments until 16 Jun 1966, when a Titan 3C successfully orbited 8 satellites, including the first 7 of the Initial Defence Satellite Communications System (**IDSCS**). These 45kg, 26-sided polygons were 86cm in dia, covered with solar cells, and had no moving parts. Dispensed over a period of 6hr at slightly different orbital velocities to give them global coverage, they were placed just below synchronous altitude, at 33,915km. Drifting about 30° relative to Earth, each stayed in view of an equatorial station for 4½ days so that even if one malfunctioned there was always another drifting into position. Spin-stabilised, with a service life of 3yr, they shut off automatically at the end of 6yr. The system, totalling 26 satellites, was completed by 3 more Titan 3C launches, the last on 13 Jun 1968. The satellites were capable of linking ground points 16,090km apart, and from 1967 provided a S Vietnam—Hawaii—Washington link for transmitting, among other things, high-quality reconnaissance photographs. Designed to last 18 months, 5 were still operational after 8yr and 2 after 10yr in orbit.

Titan 34D/IUS carries 2 DSCS military communications satellites into stationary orbit in a midnight launch from Cape Canaveral on 30 October 1982. ***(USAF)***

NATO Based on IDSCS with the addition of a rocket engine to place them in geostationary orbit, the NATO series handle military communications between the US and the 13 other NATO countries. NATO-1 and 2, with 129kg in-orbit weight, were launched by Thor-Delta on 20 Mar 1970 and 2 Feb 1971. Stationed 5,950km apart at 18 and 26° W, they covered the northern hemisphere from Ankara, Turkey, to Virginia in the US. In Feb 1973 Ford Aerospace was given a $27·7 million contract to build 3 NATO-3 satellites, with 720kg launch and 310kg in-orbit weight and 7yr life. Each can handle simultaneously hundreds of communications of various types and has 3 channels for receiving, translating frequencies, amplifying and re-transmitting voice, telegraph, facsimile and wideband data to and from ground stations in the multi-nation network. NATO-3A (L 28 Apr 1976 by Delta from C Canaveral) is stationed midway between Africa and S America; NATO-3B (L 28 Jan 1977) is off the US W coast above the E Pacific, and in 1979 was augmenting the US DSCS system, but has probably ceased to be opertional; NATO-3C (L 19 Nov 1978) was positioned at 50°W between Africa and S America as a spare. A 4th has been ordered, with an option on a 5th to extend system life into the late 1980s.

Tacsat Tactical communications satellites (Tacsat) need much greater onboard power than conventional comsats so that they can work with the small ground terminals carried by ships, tanks, jeeps and aircraft. LES-5 (Lincoln Experiment Satellite), launched 1 Jul 1967 from C Kennedy as part of a 6-satellite payload (which included IDSCS-16, 17 and 18), was America's 1st. 2 days after it had been manoeuvred into a 33,360km, near-synchronous orbit, the 1st satellite communications between US aircraft, a US Navy submarine and surface vessel, and Army ground units had been carried out. The follow-on Tacsat 1 (L 9 Feb 1969, 7·6m tall, 2·7m dia, wt 725kg) was gyro-stabilised so that the antennas and telescopes could be continuously pointed while the major part of the satellite spun within them. It was designed to communicate with tiny land-based receivers using aerials 0·3m in dia.

DSCS Phase II of the Defence Satellite Communications System, prime contractor TRW, comprises about 16 satellites designed to operate into the 1990s. With 590kg fuelled weight, they are 2·75m in dia and 3·95m tall, and the extended solar arrays provide 535W of power. Capable

Two DSCS II satellites being prepared by TRW technicians for their dual launch by the USAF Space and Missile Systems Organisation. They serve the whole of DoD. ***(TRW)***

of handling 1,300 voice channels or 100 megabits/sec of data, they have steerable narrow-beam antennas which can focus a portion of the satellite's energy on to areas 1,600km in dia, enabling small ground terminals to be used. Beams can be steered to different locations within minutes, and the satellites moved within a few days to new synchronous positions to meet defence contingencies in any part of the world. They are launched in pairs by Titan 3C from C Canaveral. DSCS-1 and 2 (L 3 Nov 1971) had ceased to operate by 1978; of DSCS-3 and 4 (L 14 Dec 1973), one was then still operating over the W Pacific. DSCS-5 and 6 (L 20 May 1965), which would have completed world coverage except for polar areas, fell into the Pacific after 6 days, due to a Titan guidance failure. A second set of 6, with a 5yr life, began with DSCS-7 and 8 (L 12 May 1977), but system completion was again frustrated when DSCS-9 and 10 were lost on 25 Mar 1978 following a Titan 2nd-stage failure. A full complement of 4 operational satellites was at last achieved with the successful launch of DSCS-11 and 12 on 14 Dec 1978, and they were placed over the E and W Pacific. DSCS-13 and 14 followed on 21 Nov 1979.

DSCS Phase III satellites, with General Electric as prime contractor, have a 10yr life and are 3-axis instead of spin-stabilised. Their multiple-beam antennas (19 transmit, 61 receive) are able to produce "steerable nulls," permitting them to recognise and tune out attempts at jamming while handling 1,300 2-way telephone conversations for both tactical and strategic users. The 1st operational satellite in this series, weighing 1,042kg, was launched in tandem with DSCS-15 by Titan 34D/IUS on 30 Oct 1982 from C Canaveral to join the 6 then operational. It was stated at that time that the Phase II and III systems would each consist of 6 satellites: 4 active and 2 spares. With both operating on 7-8GHz, they would provide secure strategic and tactical voice and data transmission, military command and control, and ground mobile communications. They would be used by the National Command Authority, the White House Communications Agency and the Diplomatic Telecommunications Service. By 1984 they were costing $150 million each.

Fltsatcoms provide global communications for command and control of US nuclear forces, and will be in use until the turn of the century. *(TRW)*

The 4·3m-wide Leasat, 1st satellite to take advantage of the extra size permitted by Shuttle launch. 1st of 5 was to be launched by the new Frisbee technique on *Discovery*'s maiden flight in 1984. The US Navy is leasing them for tactical communications. *(Hughes)*

Fltsatcom/Afsatcom High-priority UHF communications between the President and commanders in the USA and their remotest units are provided by the TRW-built and US Navy-sponsored Fltsatcom (FSC) system. Fully operational from Jan 1981, it provides the communications for command and control of US nuclear forces around the world. There are relay links between 900 ships, submarines and aircraft of the Navy and selected fleet ground stations; between 1,000 USAF aircraft and air-to-ground terminals; and for Strategic Air Command. Only the polar regions are not covered. The spacecraft, of which 4 are operational and 1 spare, have 1,005kg in orbit weight and 13·19m width with solar panels extended; there is enough fuel for at least 10yr of normal operations. They were launched into synchronous orbits by Atlas-Centaur from C Canaveral over a 3yr period. Each provides 9 25kHz channels and 12 5kHz channels for small mobile users, plus a 25kHz fleet broadcast channel and a 500kHz DoD channel. Until Fltsatcom became operational a UHF naval service, Gapfiller, had been provided by a leased portion of Marisat. Fltsatcoms also carry special transponders for the Afsatcom service: UHF data communications for the command and control of USAF nuclear forces. 2 on-board processors, one each for the Navy and Air Force, protect against uplink signal jamming. FSC-1, launched 9 Feb 1978, was stationed at 100°W over Midway Island to give coverage across the US to the Azores and Atlantic. FSC-2, launched 4 May 1979, was stationed at 71·5°E to cover the Indian Ocean from Africa to the Philippines. FSC-3, launched 17 Jan 1980 to 23°W over the mid-US, covers the Atlantic and Mediterranean. FSC-4, launched 30 Oct 1980 to 172°E over the Pacific, provides coverge from E Asia to the US W Coast. FSC-5, launched 6 Aug 1981, was to be placed at 93°W near FSC-1, the busiest area, to act as an instantly available spare, but was damaged during launch. TRW was then given a $47 million contract to procure long-lead components for 3 more satellites for launch between mid-1985 and early 1987.

SDS Satellite Data System is the code name for satellites periodically placed in highly elliptical orbits similar to those of Russia's Molniyas. Few details have been announced, but these spacecraft are used to communicate with the US nuclear and other forces in the polar regions uncovered by geosynchronous satellites. They also relay image data from KH11 and, probably, other spy satellites. It is suspected that the recent "temporary anomalies" from which these satellites have suffered while over Soviet territory may have been caused by ground-based lasers illuminating them from the Sary Shagan test facility, known to have 3 different types of laser device. Weighing approx 700kg, they are launched by Titan 3B/Agena D from Vandenberg into orbits ranging from 390 × 33,800km to 460 × 39,296km with 63° incl and 10-100yr life. SDS-1 launched 21 Mar 1971, SDS-2 21 Aug 1973, SDS-3 6 Aug 1976, SDS-4 25 Feb 1978, SDS-5 5 Aug 1978, SDS-6 13 Dec 1980, SDS-7 24 Apr 1981 and SDS-8 31 Jul 1983.

Navigation

Transit The 1st navigation satellite, this was developed primarily to provide Polaris missile submarines with the

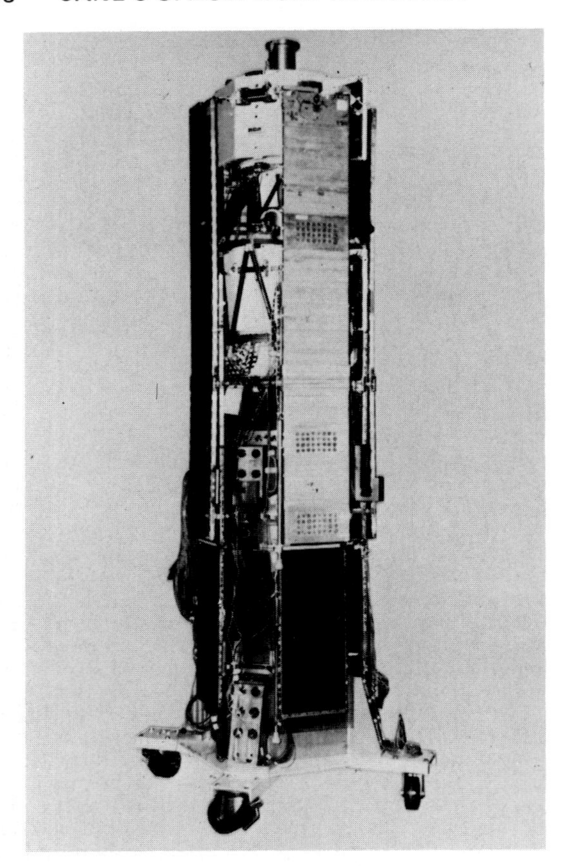

A pair of USN Transit navigation satellites stacked one on top of the other. Eventually 8 will be dual-launched in this way by Scout rockets. *(RCA)*

ability to fix their positions within 150m. It soon became evident that all-weather navigation could be provided in this way for all types of shipping. After an initial launch failure, Transit 1B, launched 13 Apr 1960 into a 373 × 745km orbit at 51° incl, transmitted information including time signals for 3 months. By 1968 3 Transit series totalling 23 satellites had been placed in circular orbits of about 805km. Some operated for more than 5yr; one (L 14 Apr 1967 into a 1,049 × 1,077km orbit with 90° incl) was still operating 11yr later. Later the system became known as Navy Navigation Satellite System (NNSS). The last shown in the Satellite Table was NNS-20 (L 30 Oct 1973 by Scout from Pt Arguello; wt 58kg; orbit 895 × 1,149kg; incl 90°). By then 5 of these polar satellites, needing only cheap battery-operated receivers, were allowing ships' masters to determine their position to about 1m.

In Sep 1983 it was announced that of 15 Transits built by RCA in the 1970s 12 had been kept in store because the 3 launched had far exceeded their 3yr design life. RCA was given a $9·9 million contract to modify 8 for dual launches by Scout from 1985 in a program called Stacked Oscar on Scout (SOOS). The intention is to store them in orbit so that they are more readily available if needed. It is not clear how far, if at all, this programme is linked to Navstar.

Nova Still further improved versions of Transit, designed by Johns Hopkins University and built by RCA, these include a more powerful transmitter, better reference clock, greater computer capacity and onboard ability to compensate for orbital disturbances. First of 3, Nova 1 (L 14 May 1981 by Scout from Vandenberg; wt 136kg; orbit 1,170 × 1,187km at 109°) had a 7·6m-long gravity-gradient boom for stabilisation. Orbital life 3,000yr.

Timation Time Navigation began as a series of experiments with multiple satellites in high polar orbits of about 16,000 × 20,000km. The only named launch, Timation 2 on 30 Sep 1969, placed 6 Timations together with 2 other satellites in 906 × 940km orbits with 70° incl. An unnamed launch on 14 Dec 1971 placed 4 more satellites in similar orbits. By the time Timation 3 was

Nova, an improved version of Transit, has a 7·6 m gravity-gradient boom and onboard ability to compensate for orbital disturbances. *(RCA)*

12 Transits stored for immediate use. One is always kept in "orbital storage". Commercial ships as well as the USN can use them to obtain their exact position. Inset, a Transit inspection by technicians. *(RCA)*

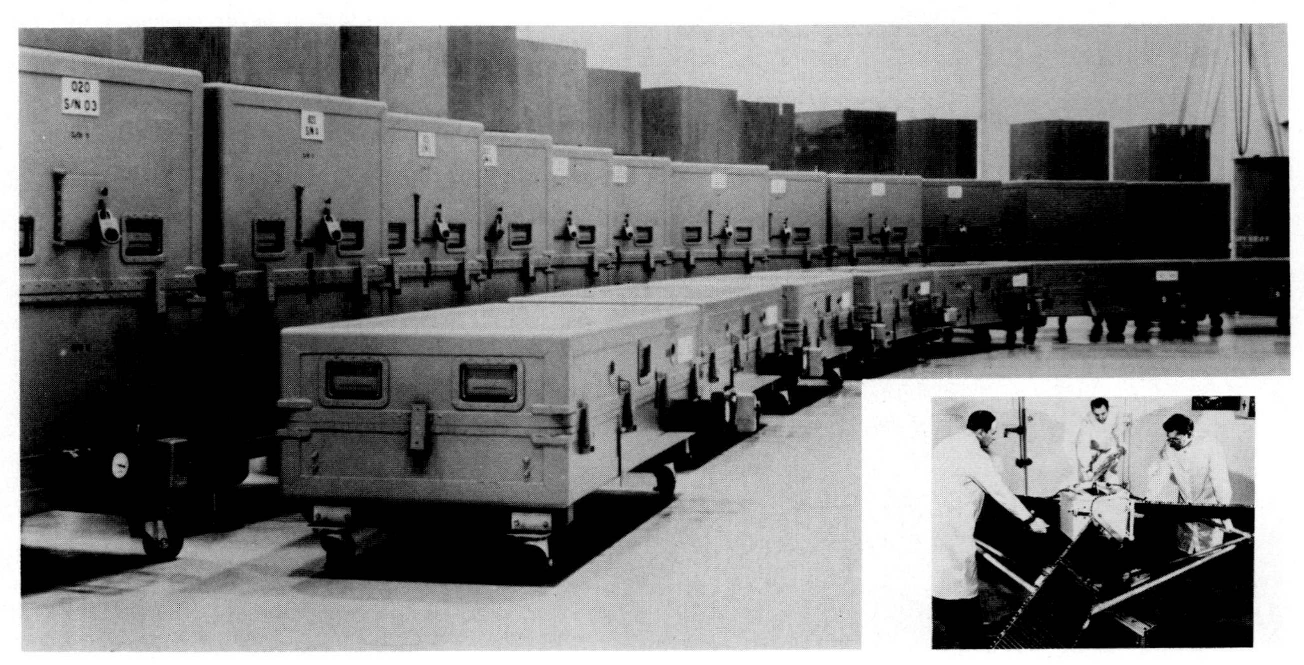

launched (14 Jul 1974; wt 293kg; orbit 13,445 × 13,767km; incl 125°), the programme had been merged with Navstar and Timation 3 was renamed NTS-1 (Navigation Technology Satellite).

Navstar The Navstar Global Positioning System, being designed by Rockwell International for the USAF at a 10yr cost of $2 billion, is intended to meet the future precise navigation needs of all the US services, entering full operation in 1984. The name is short for Navigation System using Timing and Ranging, with the key being 3 highly accurate atomic clocks built into each of the 433kg satellites; the clocks will gain or lose only 1sec in 36,000yr. The system should enable its users to fix any point on or near Earth with a horizontal and vertical accuracy of 10m, and since it will be able to measure velocity (of ICBMs, for instance) to an accuracy of 0·03m/sec, its potential applications include weapon delivery. Originally planned as consisting of 24 satellites in three 8-satellite rings in sub-synchronous 12hr orbits at 20,200km, the system was later cut for financial reasons to 18 satellites, lowering position-fixing accuracy to 16m. Users, ranging from foot soldiers carrying 5·4kg manpacks to ships, aircraft, spacecraft and ballistic missiles, will remain passive so that they cannot be electronically located. Their sets comprise a combined radio receiver and computer which locks on to the Navstar signals from the satellites most favourably located at that moment and transforms the signal's time and range into navigational data. Extra funding was added in 1979 for development of onboard sensors able to report whether the satellite had been illuminated by laser energy or touched by another spacecraft. In an example of Navstar versatility and accuracy reported in Feb 1980, a USAF F-4 was able to rendezvous with a C-141, playing the role of aerial tanker, without either radiating a radio signal. But by 1982 the Air Force was complaining that Navstar had suffered so many funding cuts that the programme was "barely alive". There were also Congressional reservations on the grounds that Soviet military systems could "plug in" to Navstar, and that there were so many benefits to foreign and civilian users that they should share the costs. The USAF agreed that Soviet use of the system was possible, but to siphon off high-resolution data, such as those required for all-weather bombing, would need sophisticated terminals and knowledge of the code structure which would not be available to Russia.

A design review completed in mid-1982 resulted in a Block 2 Navstar design one-third larger, and presumably eliminating some of the weaknesses. Its life was increased from 5 to 7½yr, and the approx 450kg weight was increased to 787kg. Larger vertical solar panels replacing the horizontal panels increased power from 400W to 700W. It can pass nuclear detection data via other satellites if necessary. Rockwell has completed 11 Block 1 Navstars; the Block 2 version can be launched by Shuttle.

Navstars 1-6 were launched with great precision by Atlas F from Vandenberg into their 20,000km circular orbits at 63° with perhaps 5-10yr operational life on 22 Feb 1978, 13 May 1978, 7 Oct 1978, 11 Dec 1978, 9 Feb 1980 and 26 Apr 1980. Navstar 6 was the 1st to take over the detection of nuclear explosions. Navstar 7 was lost on 18 Dec 1981 due to an Atlas launch failure, by the end of 1982 Navstar 2 was out of commission, and Navstar 8 was launched 14 Jul 1983. By early 1984 the Navstar schedule had been speeded up by about 1·5yr with the issuing of a $1·2 billion long-term contract to procure 28 satellites, and the programme was said to be in "excellent shape ".

Navstar satellites can fix the position of a soldier, aircraft or ship within 10m with the aid of atomic clocks which will remain accurate for 36,000yr. *(Rockwell)*

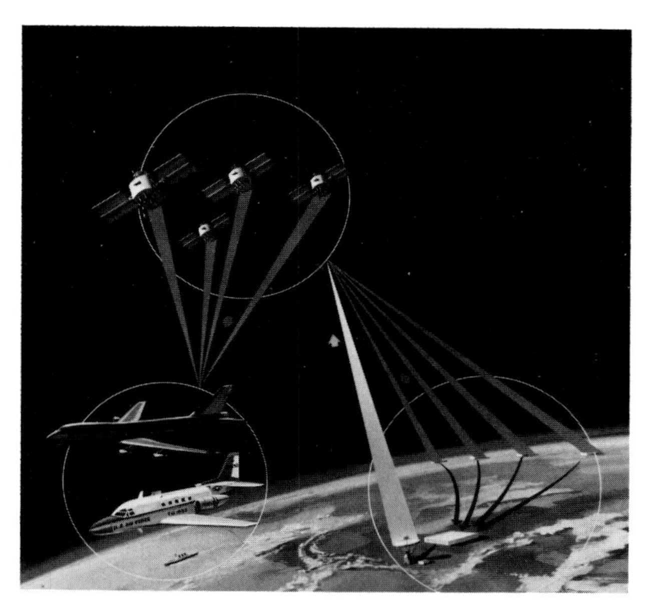

Navsat satellites operate in groups spaced around the globe. *(Rockwell)*

Geosat Due for launch by Atlas from Vandenberg, this is designed to provide gravitational data about the S Hemisphere and N Pacific which are expected to contribute to a 10% improvement in the accuracy of Trident 2 SLBMs. Developed by Johns Hopkins University at a cost of $42 million, it is intended to provide the data originally expected from the short-lived Seasat.

Meteorology

DMSP This is now known as the Defence Meteorological Satellite Programme after working through a whole series of classified names and numbers. Military weather forecasters need to observe weather conditions throughout the world in order – among other reasons – to prevent spy satellites from wasting film (and payload) when their targets are covered by cloud. In a joint USAF/USN operation the tiny experimental Calspheres 1 and 2 were launched on 6 Oct 1964, followed by 5 series numbered Block 4A, Block 4B and Blocks 5A, 5B and 5C. Some, weighing less than 1kg, were launched piggyback to test tracking radars. By 1980, with RCA Corporation as prime contractor, more than 26 had been launched. Currently the aim is to keep 2 satellites in near-polar Sun-synchronous orbits providing both ground and shipborne stations of the USN and USAF with warning of thunderstorms, typhoons and hurricanes, as well as routine data and 3-dimensional cloud analyses which are the basis of computer simulations of weather conditions.

The fitting of side-looking radar was considered but ruled out when it was found that a ship's pitching and rolling movements destroyed the necessary phase relationships. Current Block 5D-1 (Advanced Meteorological Satellites), placed in circular 850km orbits with 98° incl and 1½yr operational life by Thor-Burner 2, were not very successful; 1 failed after launch and another was not fully operational. 4 were launched: AMS-1, 11 Sep 1976; AMS-2, 5 Jun 1977; AMS-3, 1 May 1978; AMS-4, 6 Jun 1979. Block 5D-2 satellites were given extra redundant systems and 3yr operational life, adding 155kg to the 513kg weight and necessitating an Atlas launcher. Weather data had to be received from NOAA, with which the DMSP service is

DMSP 5D-2, the latest Defence Meteorological Satellite. Military weather forecasters use them to prevent spy satellites wasting film when targets are cloud-covered. *(RCA)*

RCA technicians performing pre-launch systems tests on DMSP 5D-2. Mission life was doubled by comparison with the previous version, to 3yr. *(RCA)*

shared. 1st Block 5D-2 was AMS-5, launched 20 Dec 1982 by Atlas E from Vandenberg into 724km Sun-synchronous orbit; weight? 751kg. This series takes infra-red and daylight images with 0·3-mile resolution at least twice a day. Weather information is stored before transfer via US readout stations to the US Air Force Global Weather Centre. Data are available to tactical commanders through transportable terminals and to civilian users worldwide.

Future projects

The Defence Advanced Research Projects Agency (DARPA) has sponsored a number of ambitious missions intended to be Shuttle-launched and handed over to the USAF or USN when operational. By 1979 it was working out methods of improving both the security of space transmissions and their ability to resist jamming. Sire (see below) is an example of this work. Advances in ground-based satellite surveillance are also necessary; see GEODSS.

CRRES Combined Release and Radiation Effects Satellite, due for deployment on Shuttle Mission 39 in Jul 1986, is a joint USAF/NASA spacecraft. NASA will use it for 45 days in low orbit for chemical-release experiments. It will then be boosted to synchronous orbit for DoD trials of high-efficiency, radiation-resistant solar cells; a secondary task will be to map the radiation belts more accurately.

GEODSS Ground-based Electro-Optical Deep Space Surveillance, being developed by TRW for USAF Strategic Air Command (and due to be handed over to Space Command in 1983), consists of 5 installations which should be fully operational by 1987. By 1983 3 sites at White Sands, NM, Taegu, Korea, and Maui, Hawaii, were already operational, with equipment for a 4th being shipped to Diego Garcia in the Indian Ocean and a 5th site still to be selected "along the zero longitude in the E Atlantic". Due to supersede the current Norad satellite surveillance network, based on the Baker-Nunn system developed in the mid-1950s, GEODSS provides detailed data on man-made objects 5,500km from Earth and beyond. Each site has 3 telescopes on a mount able to move to a new field of view in only 1sec, with an accuracy of 1·5 arc sec. If a satellite is not at its plotted co-ordinates, a search for it is automatically triggered. The Norad Resident Space Object Catalogue currently identifies about 800 man-made objects in deep-space orbit, a figure which is expected to increase to 1,500 by 1985.

MilStar Military Strategic-Tactical and Relay is a survivable 2-way communications service started in 1983 and intended to become operational in the second half of the decade. An all-service, high-priority programme, it will presumably complement and ultimately supersede Fltsatcom/Afsatcom. It will consist of at least 4 satellites in stationary orbit over the Indian, E and W Pacific and Atlantic oceans, with 3 more in highly elliptical polar orbits, all backed up by an unspecified number of spares parked in supersynchronous orbits up to 177,000km and capable of being commanded down to synchronous height if needed. Ultimately all MilStars might be in supersynchronous orbits to increase their ability to survive ground-launched interceptions. In addition to being hardened against nuclear attack and radiation from directed-energy weapons, MilStars would be coupled by laser or other data links. Service agreement has been reached that they will operate uniformly in the 44GHz (uplink) and 20Hz (downlink) ranges. In Jul 1983 Lockheed was awarded a 5yr $1·05 billion USAF contract; Hughes and TRW are subcontractors for the payload.

Scatha L 30 Jan 1979 by Thor-Delta from C Canaveral. Wt 357kg. Orbit 27,543 × 41,775km. Incl 27·3°. Designated P78-2, Scatha was intended primarily to help designers to control the effects of electrostatic charging. In such high orbits solar radiation and the influx of particles can lead to the build-up of an unsymmetrical charge on the spacecraft until a high-voltage arc discharges between its components and damages them. Built by Martin Marietta, with 1·8m height and 1·8m dia, Scatha is made of glass-fibre epoxy, magnesium, titanium and aluminium honeycomb. Experiments include particle-beam systems to control spacecraft charging by means of an electron gun and xenon-ion gun, electric field detectors with a 91m-long antenna to measure electric fields, plus instruments to measure damage to the satellite surface caused by charging. Designed for 1yr operation.

Teal Ruby Potentially a revolutionary method of detecting and tracking strategic aircraft and cruise missiles by following the fuselage rather than the heat plume against Earth's difficult "background clutter," this uses staring-mosaic focal-plane arrays. Technical problems, now apparently overcome, led to its planned launch in 1981 being delayed to about 1986. Intended to demonstrate the feasibility of maintaining surveillance of air-breathing vehicles from high-altitude space platforms, Teal Ruby will now be flown in a polar 740km orbit aboard the AFP 888 USAF satellite (previously designated P80-1). About 100 target missions, with aircraft flying across the sensor's field of view, are planned for Teal Ruby's 1yr operational life. Like IRAS, it is cooled by superfluid helium, which limits its life; but unlike IRAS, which was designed to look away from Earth and scan, Teal Ruby will point towards Earth and stare. It has a 3·3m sensor barrel and shield to protect it from Earth's radiated heat. Developed by Rockwell for DARPA, the project has been helped by JPL and Britain's Rutherford Appleton Laboratory, which are experienced in processing IRAS data. Programme costs have risen from $110 million to about $500 milion.

SBSS/Sire The USAF's proposed Space Based Surveillance System, consisting of 4 satellites capable of detecting and tracking both low and high-altitude objects worldwide, would replace the now cancelled Satellite Infra-Red Experiment (Sire). Developed by Hughes, Sire was to use a long-wavelength infra-red sensor to detect the thermal radiation of objects against the coldness of space. It was cancelled following the success of the international IRAS, which carried similar sensors. But a much longer-duration active refrigerator would be needed to give SBSS satellites the required 5yr duration. Because a major objective would be the monitoring of covert attempts to interfere with US space systems, the go-ahead has been delayed for political reasons.

Laser battle stations Formation of a US Space Force, creating a 4th arm of the US military forces, seemed increasingly likely by mid-1982 following a General Accounting Office report to Congress calling for accelerated development of space-based laser weapons. There remains however some doubt as to whether the existing programme known as the "Space Laser Triad" (see below), which is responsible for developing the 3 key technologies involved, can be speeded up. Nevertheless, it is already virtually certain that directed-energy weapons, comprising devices for generating power and controlling laser, particle and microwave beams, will reverse existing military strategy. In the West that has been based for the past 40 years on the concept of deterrence - the ability to reply to any initial attack with a shower of nuclear-tipped ballistic missiles. Laser battle stations, as they are somewhat misleadingly called, offer a genuine defence against the initial attack by making it possible for chemical lasers to destroy attacking ballistic missiles in the first 4min of their flight so that their nuclear debris would fall back upon the territory of the attacker. Under this concept 24 laser battle stations with a range of 5,000km would be orbited in 3 polar rings at an altitude of 1,200km. Such a system would have the enormous task of dealing with 1,000 hardened ICBMs, but once that capability had been achieved the laser battle stations would also be able to intercept SLBMs and bomber, surveillance and early-warning aircraft. The Space Laser Triad comprises 3 complementary programmes: Talon Gold, Alpha and Lode.

Talon Gold Improvements in fire control and precision beam direction are needed before laser weapons can become useful for space defence. Talon Gold, using a low-power laser, will be tested against both high-altitude aircraft and space targets to obtain the fundamental information about fire-control and other features needed to design high-energy laser weapons based in space. It continues laser tracking research carried out at the MIT Lincoln Laboratory.

Alpha This programme involves demonstration and development, on the ground at first, of a chemical laser suitable for space operation. It concentrates on the technologies needed to generate extremely high-power beams - about 10MW - expanding power by adding generator modules. TRW is building the device, with 1st tests planned for the mid-1980s.

Lode Large Optics Demonstration Experiment involves development and testing of the 4m-dia primary mirror and associated beam-control system needed for space laser weapons. DARPA says it will "integrate significant advances in large mirrors, high-bandwidth fine tracking and beam stabilisation, and advanced structures into an ultra-high-performance electro-optical system." Mirrors that are light but so big that the Shuttle will probably have to launch them in segments will be needed if laser energy is to be focused on targets thousands of miles away. The mirror and beam technology is so advanced that a prototype will probably not be available before the 1990s.

Strategic laser communications Blue-green laser pulses can in theory penetrate clouds and water, enabling a satellite or space-based relay mirror to transmit communications to a deeply submerged submarine. This would eliminate the need for Poseidon and Trident missile submarines to rise close to the surface to receive instructions, thus revealing their location to Soviet sensors. As a first step towards launching a laser communications satellite, DARPA is carrying out trials in which blue-green laser beams are directed from high-flying aircraft through clouds to submerged submarines.

Advanced space sensors 2nd-generation infra-red sensor technology is to be built into a 1988 geosynchronous satellite carrying a telescope with optical filters which stares rather than scans over a wide field of view. This will convert focal-plane data into target track information for transmission to small ground terminals. The aim is to be able by the year 2000 to track selected targets anywhere in the world for strategic and tactical air war missions as well as fleet defence. Alongside this is a development, started in 1979, of a space-based phased-array radar capable of detecting and tracking hostile aircraft and ships. Equipped with specialised computers performing a variety of other tasks and able automatically to tune out attempts to jam it, this would be so large that it would have to be delivered by a series of Shuttle flights and assembled in orbit.

USSR

ASAT
FOBS
Early-warning satellites
Spy satellites
Ocean-surveillance satellites
Ferrets
Radsats
Navigation
Communications
Unknown

ASAT

History By the end of 1982, 20 tests had enabled Russia to develop an operational system for intercepting and destroying enemy satellites up to altitudes of 1,500km. Interceptors, as they were first called (ASATs, or anti-satellites, becoming the accepted designation later), began to appear in Oct 1968, with two forms of interception taking place as the series progressed. In one form the interceptor co-orbited with the target, allowing time for close inspection before demonstrating its ability to blow up itself and the target. In the other the interceptor swooped in upon the target at a closing speed of about 400m/sec, using an elliptical orbit. This allowed time for only a quick inspection before destruction could be achieved with an explosion as close as 30m to the target. Until 1976 no target had in fact been destroyed, the Russians being content with simulations in which the interceptor was detonated far enough away to avoid actual destruction of the target. When tests started interception was usually achieved about 3½hr after launch, although at least twice, in Apr 1971 and Apr 1976, the ability to intercept in less than 1hr and on the 1st orbit was demonstrated. The latest interceptors "pop up" from below the target, enabling an interception to be made within half an hour of the ASAT's lift-off. They would however take at least 6hr to reach the all-important targets in synchronous orbit, and higher-powered launchers would be necessary. Moreover, there so far appears to have been no demonstration of the ability of interceptors to chase manoeuvring targets. By 1982 US sources were saying that Russia was testing a spacebased anti-satellite homing vehicle which could operate from a stationary orbit. Some US service chiefs believed that a rocket able to destroy target satellites in synchronous orbit was under development, but since such a rocket would have to be of Saturn 5 size it would be "a pretty vulnerable target in itself". As such a vehicle should be instantly detected by DSP and other sensors, there would be time to take evasive action.

General description Interceptions tend to take place over E Europe or the USSR in the early hours, local time, apparently to allow optical tracking of the ASAT and the target during the closing manoeuvres. Western observers were able to study the new technique during 20-30 Oct 1970 by observing Cosmos 373, 374 and 375. C 373 was probably sent up to play the part of an enemy satellite. About 3·9m long, with 2·1m dia, it was probably able to report back on miss distances. C 374 went up 3 days later, passed very close on its 2nd orbit and then exploded into over 16 pieces. C 375 was launched on 30 Oct and also passed very close to C 373 on its 2nd orbit, 230min after launch; it then exploded into 30 pieces. With 12 interceptions by the end of 1971, Russia had succeeded in developing the ability to intercept and destroy satellites at relatively low level. This is much more difficult than at high level, because as the altitude decreases the target satellite moves faster in relation to the ground. C 462 exploded on 3 Dec 1971 when approached by C 459; during earlier interception tests target satellites had been destroyed at altitudes between 579 and 885km. On this occasion the target was destroyed below 257km. There was a 4yr gap in ASAT tests between C 462 and C 803 in Feb 1976, which may have been connected with the Salt talks. US observers regarded Russia as having demonstrated that its ASAT was operational following C 1243, launched 2 Feb 1981 into a 296 × 1,015km orbit. At the end of its 2nd orbit it passed within 15km of C 1241, launched 21 Jan 1981 into a 976 × 1,010km orbit. The 2 craft were 1,010km above Leningrad when C 1243 manoeuvred, probably taking it much closer to the target and also causing it to re-enter. Had it not been a simulation, the attacking craft would have exploded, destroying the target as well as itself. It was probably the 17th exercise since the programme began. In Apr 1980 C 1174 intercepted C 1171, but the miss distance on the 2nd orbit was 60km and it was thought at first to be a failure. However, C 1174 later made 2 more close passes, one within 20km, at a time when C 1167, an ocean ferret, was directly below and able to report on what happened. C 1174 apparently used optical-thermal detection of targets rather than radar, and this led to speculation that the attack satellite might have been trying out a laser as a means of disabling the target. There were 2 tests in 1981, 1 in 1982 and none in 1983.

FOBS

Fractional Orbit Bombardment System was demonstrated in 18 tests between 1966 and 1971. They caused much concern in the West when they began with Cosmos U1 and U2 in Sep and Nov 1966. These "space bombs" can be fired into an orbit of 160km and then slowed down by retro-rockets so that they re-enter, their nuclear warheads (though none was ever carried during the tests) falling on the target before the completion of the 1st orbit. This makes it possible to attack Western targets by the "back door," travelling threequarters of the way round the world via the S Pole, instead of by the shorter, more obvious N Pole route, which is monitored by BMEWS. In the tests the satellite was launched on a path avoiding the US, with the test warhead being called down on to Soviet territory just before the end of the 1st orbit. From the Soviet point of view FOBS had the advantage of forcing the West to invest in more complicated defences such as over-the-horizon radar (OTH); but the disadvantages of the system were reduced accuracy and payload.

Launches were by F1 (Scarp) rockets from Tyuratam, with a retro-rocket stage fitted to the warhead. The 1966 tests, unannounced by Russia and given "U" designations by RAE Farnborough, were followed up by 9 more in 1967: Cosmos 139, 160, 169, 170, 171, 178, 179, 183 and 187. There were 2 in each of the following 3 years, and the last, C 433, was on 8 Aug 1971. Opinion is divided as to whether the lack of tests in the last 13yr means that the Soviets consider the system operational or that it has been abandoned.

Early-warning satellites

History Russia has developed a system of 9 satellites in highly elliptical (approx 600 × 40,000km) semi-synchronous orbits spaced at 40° intervals around the equator to provide early warning of any launch of ballistic missiles against her territory. They also monitor routine launches from the US and other countries. With high points over the N Atlantic and W Pacific, and overlapping coverage, each can view ICBM sites in the US for 5-6hr on each revolution (12hr per day) and transmit data to Soviet bases at the same time. First to be identified was Cosmos 520, launched 19 Sep 1972; weight 1,250kg, orbit 750 × 39,470km, incl 62·8°, period 715min, orbital life about 15yr. C 606, with similar statistics, followed in 1973; C 665

1974; C 706 1975; C 862 1976; C 903, 917 and 931 in 1977; C 1024 and 1030 in 1978; and C 1109 and 1124 in 1979. In 1980 6 launches - C 1164 12 Feb, C 1172 12 Apr, C 1188 14 Jun, C 1191 2 Jul, C 1217 24 Oct, and C 1223 27 Nov - led to suggestions that there had been many failures. However, Robert Christy considers it more likely that Russia was at last filling the empty slots and that the large number needed was due to the gravitational "dips" around the equator, which cause them to drift off station and back again; constant use of an onboard motor to correct the drift restricts satellite life to a few months. Until 1980 only 5 of the 9 planes had been used. There were no launches in 1981 and it was not until Cosmos 1367 was launched on 20 May 1982 (orbit 581 × 39,624km; incl 62°; period 707min) that the 9-strong system was finally completed. In 1983 3 early-warning satellites were launched; 1 failed after 2 days and another after 4 months.

General description Similar to Molniya in design and weight (now probably up to 2,000kg), and windmill-shaped with 6 vanes. Probably launched by A2e from Plesetsk.

Spy satellites

General description The activities of Soviet spy satellites reached a new peak in 1983. One US observer estimated that film-return spacecraft observing the Iran-Iraq war, the Lebanon and Middle East, and the US invasion of Grenada were in use for over 800 mission days. Photographic reconnaissance by satellite to gather high-resolution pictures of military installations and activities was so obviously going to be of value to both East and West that its development was one of the main incentives in the early years of the post-war space race. Russia's 1st - her 18th launch - was Cosmos 4 in 1962. The last 2 of the 4 generations of spysats described below are currently being launched at the rate of about 35 per year. The main reason why the Soviet Union launches so many more than the US is that Russia still relies heavily upon photo-reconnaissance rather than data-imaging. Photography provides higher resolution - in Russia's case it is believed that objects as small as 0·2m can be seen - even though the need to recover film capsules makes it much slower. In the early days both high and low-resolution pictures were taken from orbits of about 200 × 250km. From about 1974 high-resolution quality was increased by lowering perigees to about 170km. Generation 4 spysats nowadays manoeuvre to 150km or less over areas of military or geographical significance at local midday to ensure good lighting conditions. Weather satellites check whether the target is cloud-free to avoid waste of film and manoeuvring propellant.

It was a Soviet spy satellite - C 932, launched 20 Jul 1977, orbit 180 × 342km, incl 65° - which detected that S Africa was about to explode a nuclear device in the atmosphere, with the result that joint East-West pressure was able to stop it.

Current Generation 3 low-resolution spysats are manoeuvred up to about 350 × 420km at the start of their missions, and in 11 days have covered the whole world in strips at 2° intervals of longitude. These missions are usually flown at 73° inclination from Plesetsk and 70° from Tyuratam. Nowadays some of the 82° Plesetsk flights are announced as having an Earth-resources mission. It is possible that this refers to the supplementary package carried on Generation 3 which is part of Russia's programme to photograph and recover film of the whole of its territory and that of some other Soviet-bloc countries.

US intelligence believes that beam-splitter mirrors have been used on some Soviet spysats to photograph US ICBM sites in laser light while simultaneously photographing the satellites' star background. This method enables the sites to be pinpointed with an accuracy of 15-30m. The detonation of some of the main craft at the end of their missions could be designed to ensure that debris does not fall on populated areas or, more likely, that none of it can be recovered and examined in the West.

Recent conflicts have provided evidence of Russia's quick-response launch capability. In 1980, early in the Iraq-Iran war, C 1210, launched 19 Sep into a 187 × 320km orbit with 82° incl, passed over the battlegrounds at noon each day and was recovered on the 14th day, 3 Oct. It was overlapped by C 1211, launched 23 Sep into a similar 216 × 242km orbit with the same incl, and recovered on the 11th day, 4 Oct. C 1208, launched 26 Aug into a 173 × 339km orbit with 67° incl, was already available on station to send back film pods until its mission was completed on 24 Sep.

During the Anglo-Argentine conflict at least 6 Soviet spacecraft were able to monitor events in the area, and there is little doubt that some of them were specially launched for that purpose. 2 imaging reconnaissance spacecraft, C 1347 and C 1352, launched 2 and 21 Apr 1982 into 173 × 340km and 209 × 361km orbits with 70° incl, were able to film relative positions during clear weather and were recovered 50 and 14 days later. A 3rd, C 1368, launched 23 May into a 218 × 365km orbit with 70° incl, was sent up 2 days after the British landing on the islands. An Eorsat ocean-surveillance ferret, C 1365, launched 29 Apr into a 425 × 443km orbit at 65° incl, was followed by a pair of nuclear-powered Rorsats, C 1365 and C 1373, launched 15 May and 2 Jun (orbits 252 × 264km and 246 × 270km at 65° incl).

Generation 1 This started with C 4, launched 26 Apr 1962 by A1 from Tyuratam; wt ? 4,750kg; orbit 298 × 330km; incl 65°. Believed to be an unmanned Vostok, it was recovered after 3 days. The 5 launches in 1962 averaged 4 days in orbit, increasing by 1968 to 8 days (sometimes more, sometimes less) before being brought back for their film to be processed. Last of the series was C 153, launched 4 Apr 1967; orbit 199 × 279km; wt ? 4,750kg; recovered on 8th day. Both Plesetsk and Tyuratam were used for launches.

Generation 2 Started with C 22, launched 16 Nov 1963 by A2; wt ? 5,530kg; orbit 192 × 381km; incl 65°; recovered after 6 days. This larger series had higher-resolution cameras, and mission durations were usually 8 days. Usual launch pattern for 1st 2 generations was 2 per month, often overlapping, with a noticeable increase in numbers during periods of international tension. Last of this series was C 355, launched 7 Aug 1970; wt ? 2,500kg; orbit 191 × 304km; incl 82°; recovered on 8th day. Launches were from Plesetsk to 65°, 72° and 81°, Tyuratam to 52° and 65°.

Generation 3 This series, which carried a supplementary payload of a military or scientific nature on the forward end, started with C 208, launched 21 Mar 1968; wt ? 5,900kg; orbit 208 × 274km; incl 65°; recovered on 7th day. Starting with C 228 - launched 21 Jun 1968; wt ? 5,900kg; orbit 199 × 252km; incl 52° - they began to stay in orbit for 12 days or longer before recovery. The limited manoeuvring ability observed on C 251 in 1968 has steadily improved so that either high or low-resolution cameras can now be directed very accurately on to the same region over a period of days to check progress. Mission durations are now 13-14 days, with launches usually by A2 into 65° and 70° incl from Tyuratam and 73°, 81° and 82° incl from Plesetsk.

Generation 4 Started with C 758, launched 5 Sep 1975; wt ? 6,700kg; orbit 174 × 326km; incl 67°; exploded after 20 days. This series may use some Soyuz hardware and is believed to carry 6 recoverable film capsules which are returned periodically during the flight. Highly manoeuvrable, it can look at several targets during its mission, at the end of which the main craft is either detonated in orbit or burns up on re-entry. Typical missions last 1 month. C 1347, launched 2 Apr 1982; wt ? 6,700kg; orbit 173 × 340km; incl 70°; established new long- duration (50 days) pattern. They are probably powered by solar panels, not fitted on earlier generations. Launches are by A2 into 67° incl from Plesetsk and 65 and 71° incl from Tyuratam. C 1426, L 28 Dec 1982, made 10 major manoeuvres in orbits ranging over 200 × 300-400km during a record 67 days. C 1446 (L 16 Mar 1983, orbit 222 × 242km, incl 67°) spent 2 weeks observing the Iran-Iraq war. C 1454 (L 22 Apr 1983, orbit 171 × 343km, incl 67°) and C 1457 (L 26 Apr 1983, orbit 171 × 349km, incl 70°) between them watched the Middle East until the end of May. C 1504 (L 14 Oct 1983, orbit 173 × 305km, incl 65°) watched Grenada during its 53-day life.

Ocean-surveillance satellites

General description Known more specifically as Radar-Equipped Ocean Reconnaissance Satellites (Rorsats) and Elint Ocean Reconnaissance Satellites (Eorsats), both series work in pairs and their data can be combined to obtain a very detailed picture of what is going on below. They have low-thrust engines to maintain height and the correct distance between the pairs.

Rorsats Easily identified by their powerful pulsed radar signals, these are used to locate naval ships in all weathers. They can identify formations of ships and their direction and speed. Targeting data can be downlinked within 1 orbit to ships or other weapon platforms carrying long-range anti-ship missiles, giving ships of the Western powers very little time to take evasive action. The series began with C 198, launched 27 Dec 1967 (orbit 894 × 952km, incl 65°), and in the 1970s developed into the controversial nuclear-powered pairs which were being launched within a few days of each other by F1m from Tyuratam into 270km circular orbits. About 16m long with 2m dia, they were powered by a radioisotope thermal generator containing 50kg of slightly enriched U235. Normally at the end of the 60-70-day mission the 6m-long reactor was separated and boosted to a 900km orbit with a 500-600yr life. (Since the half-life of U235 is 700 million yr, raising the orbit merely postpones the problem of its ultimate return to Earth. The practice is however possibly based on the hope that the reactors could be retrieved and even reused in later years.) Inevitably the day came when things went wrong: C 954, launched 18 Sep 1977 into a 630 × 641km orbit with 64·98° incl, lost pressurisation and started to tumble on 1 Nov. The Soviet controllers were unable to boost its nuclear power pack to a safe high orbit. An international furore broke out while re-entry was awaited, with apprehensive speculation as to whose territory would ultimately receive the radioactive debris. It proved to be Canada: 92 days after launch, on 24 Jan 1978, C 954 re-entered, with the main debris falling W and S of Yellowknife along an 800km strip of the NW Territories. The Nuclear Emergency Search Team was set up to carry out Operation Morning Light. Weeks later, tiny globules which had formed as the molten uranium cooled and solidified were located and collected from the snow by giant bulldozers. Elaborately protected, they were flown with other debris in large jet transports to Canada's Atomic Energy Board in Manitoba. US protests about C 954 were somewhat muted, since it emerged that 14yr earlier, on 21 Apr 1964, a US Navy satellite launched from Vandenberg and powered by a Snap-9 nuclear generator with plutonium 238 had failed to reach orbit, broken up, and released radioactive material over the Indian Ocean. The Canadian Government estimated that the C 954 search had cost $6 million, and not until Dec 1980 was it reported that the Soviet Union had agreed to pay Canada $3 million.

In Apr 1979 the Soviet Union used a pair of non-nuclear satellites for ocean survey during a review of this programme. When C 1176 (launched 29 Apr 1980; orbit 250 × 266km at 65° incl, raised on 10 Sep 1980 to 870 × 966km with 600yr orbital life) marked the resumption of the use of nuclear power, the US State Department issued a "statement of regret". Despite this, the programme continued in 1981 with C 1249, launched 5 Mar into a 252 × 265km orbit; its nuclear power pack was boosted to 898 × 985km 5 months later. C 1266, launched 2 Apr 1981, was used for only 8 days before its manoeuvring rocket and platform were jettisoned and raised to an 891 × 965km orbit with 600yr life. Assuming that this was a result of a malfunction, a similar fate befell C 1299, launched 24 Aug 1981 into a 248 × 269km orbit and boosted to 910 × 984km after only 12 days.

For the 1st time since the C 954 drama a pair of Rorsats, C 1365 and C 1372, were launched on 15 May and 1 Jun 1982 and operated respectively for 136 and 71 days before being boosted into higher orbit. They were thus able to work together during Jun, Jul and early Aug, surveying the S Atlantic and, presumably, the Falklands war. They were followed by another pair, C 1402 and C 1412, launched 30 Aug and 2 Oct to operate for 118 and 39 days. Thus they were able to pick up the S Atlantic watch together during Oct and early Nov. Then, exactly 5yr after C 954, C 1402 repeated that drama. Between orbits 1,925 and 1,926 the Kettering Space Group in England noted that the attempt at separation and boosting of the nuclear unit to higher orbit had gone wrong. The reactor section's propulsion motor had apparently failed to fire, having possibly been damaged or blocked during separation. Russia at first denied that there was any problem, saying it would burn up safely during re-entry. In fact, with Norad issuing daily reports and safety teams on standby in many countries, 1 section burned up harmlessly over the Indian Ocean on 23 Jan 1983; another section containing the radioactive core followed uneventfully over the S Atlantic on 7 Feb. Meanwhile, C 1412 had been successfully boosted out of harm's way to 901km altitude. During coverage of these events it emerged that the USAF considers these satellites to be a warfighting system. They would therefore be a prime target for ASATs if war broke out, and simulated attack trajectories against them had already been run by Norad. No Rorsats were launched in 1983; modifications were possibly under way following the C 1402 mishap.

Eorsats These identify the type and role of naval ships by noting transmission frequencies of both radio and radar, and the pulse patterns of the latter. During the British Task Force's journey to the Falklands during the Anglo-Argentine conflict of 1982 the commander was asked how he could avoid being tracked by Soviet satellites. His reply was: "We know where they are and can take measures". At the time there were 3 Eorsats in orbit: C 1286, launched 4 Aug 1981, orbit 432 × 434km, incl 93° (probably out of action); C 1306, launched 14 Sep 1981, orbit 409 × 462km, incl 93° (operational); and C 1337, launched 11 Feb 1982, orbit 429 × 447km, 93° incl (apparently inactive at the time). All the British commander had to do was to maintain radio and radar silence while they passed overhead. There was no Rorsat launch until after the fleet had arrived at the Falklands.

Ferrets

There are several varieties of Soviet ferret, also known as elint (electronic intelligence) satellites, and they are difficult to disentangle, especially as some of their orbits are also used for radar calibration vehicles. One system comprises 6 large (2,500kg) Cosmos satellites spaced at 60° intervals in a 630km orbit at 81° incl. Their task is to identify and pinpoint military radio and radar stations. First was C 389, launched 18 Dec 1970 by Al from Plesetsk (orbit 642 × 687km). They averaged 2 per year until 1976, when the number increased to 4 per year, with an average operational life of 18 months (orbital life 60yr). The orbit suggests that they may be based on Meteor, and when the series started they were at first thought to be Meteor failures.

Starting with C 1340, launched 19 Feb 1982 into 626 × 654km orbit, replacements began to be placed in orbit planes 12-13° further west, as shown by C 1346, launched 31 Mar 1982 into 622 × 661km, and C 1356, launched 5 May 1982 into 632 × 671km. These have orbit spacings of 45°, and it is suggested that a "handover" to an improved version is taking place.

Larger craft in co-planar orbits of 440km with 65° incl operate over ocean areas, probably collecting data on naval radars. These could be transmitted to weapons platforms for targeting purposes. Starting with C 699, launched 24 Dec 1974 by F1m from Tyuratam, 1 has been flown annually between pairs of nuclear-powered ocean-surveillance craft. In 1983 there were 6 ferret launches.

Radsats

Radar calibration satellites are believed to be used for calibrating ground-based early-warning radars, and possibly for testing the ground radar installations before ASAT launches. The latter Cosmos launches are from Plesetsk by C1 into 66° incl. Other launches from Plesetsk go to 74°, while Cosmos 1311, launched 28 Sep 1981 (orbit

463 × 519km, 3yr orbital life), went to 83°. Other launches are from Kapustin Yar, also by C1, and generally go to 400-500km orbits with 50·6° incl and 90-95min period. Some of these satellites, particularly from Kapustin Yar, release clusters of small objects more than once in their lifetime, and it is thought that these are used to simulate an incoming multiple-warhead ICBM attack.

Navigation

Glonass The Soviet Union informed the International Telecommunications Union that the Global Navigation Satellite System - apparently almost identical to the US Navstar - would enter service in 1982. While indications are that it is basically a military system, analysts believe that Russia hopes eventually to offer it as an international standard for civil aviation. An orbital height of 20,000km will give a period of 12hr, and there will be 3 constellations of 3-4 satellites at 60° intervals and 63° incl, making a total of 9-12 compared with Navstar's 18. Since such a system would be incomplete, there has been speculation that Glonass might be designed to receive and process signals from Navstar. With transmissions at 1·2 and 1·6GHz to compensate for ionospheric distortion, the Soviets say it is intended for "worldwide aircraft radio navigation". It could also be used by Soviet ships. First launch was probably Cosmos 1413/4/5 on 12 Oct 1982 by Proton into 51·6° parking orbit, before being manoeuvred into elliptical orbit with 19,100km apogee and 64·8° incl. By the end of 1983 9 Glonass satellites were in place and the system was almost operational.

Communications

Tactical Direct communications between ships, aircraft and bases are provided by an average of 3 launches per year of clusters of 8 small satellites by a single C1 from Plesetsk into approx 1,500km circular orbits at 74° incl with 115min periods. Officially announced as having "scientific equipment, radio systems for the precise measurement of orbital elements, and radio telemetric systems for transmitting data to Earth," they are actually used for medium-range contacts between forces in the field, and possibly at sea in the VHF and UHF bands. 1st launch in this series was C 336-43, on 25 Apr 1970; the 30th was C 1357-64, launched 6 May 1982 into orbits of 1,403/78 × 1,480/1,526km. A minimum of 24 of these 100cm spheres, each weighing 41kg, is needed to provide global coverage. Soviet military chiefs apparently prefer a total of 38-48 to ensure plenty of redundancy and to make complete jamming impossible. Each has an operational life of about 2yr, but since their orbital life is 7-10,000yr, with a 20,000yr life for the rocket stages, their orbits are becoming increasingly cluttered with dead ironmongery already amounting to more than 220 satellites and rocket stages. Collecting this debris may one day become an important task for a Soviet space shuttle.

Store Dump Longer-range military communications are provided by about 3 Cosmos launches per year by C1 from Plesetsk into an 800km orbit at 74° incl with 101min period. The satellite stores messages as it passes over the transmitter for playback later as it passes over a receiving station in another part of the world. Its design is probably similar to that of the navigation satellites.

Unknown

More than 90 small Cosmos launches, identified only by the fact that the B1 launcher was used, started in 1964 and apparently ended with Cosmos 919 in 1977. They are classified as military because they were not announced as scientific and no results of any work they might have done have been published. One series, totalling 62, was launched into approx 280 × 500km orbits with 49° incl from Kapustin Yar until 1967-68, when Plesetsk was opened; they were then switched to that launch site and 71° incl. Another series, with approx 200 × 1,100-1,500km orbits at 49°, started from Kapustin Yar in 1967, with a switch to Plesetsk in 1968. A third series was launched from Kapustin Yar into 250 × 2,100km orbits at 49°, and a fourth from Plesetsk into 280 × 840km orbits at 71°. The constant replacement of one satellite by another suggested that continuous observations were made. Another 40 Cosmos satellites using similar orbits and including the early Intercosmos vehicles were identified as scientific through the publication of their measurements and findings.

Landsat 1 view of Tyuratam (Baikonur) area taken from altitude of 905km shows road system linking launchpads. *(NASA)*

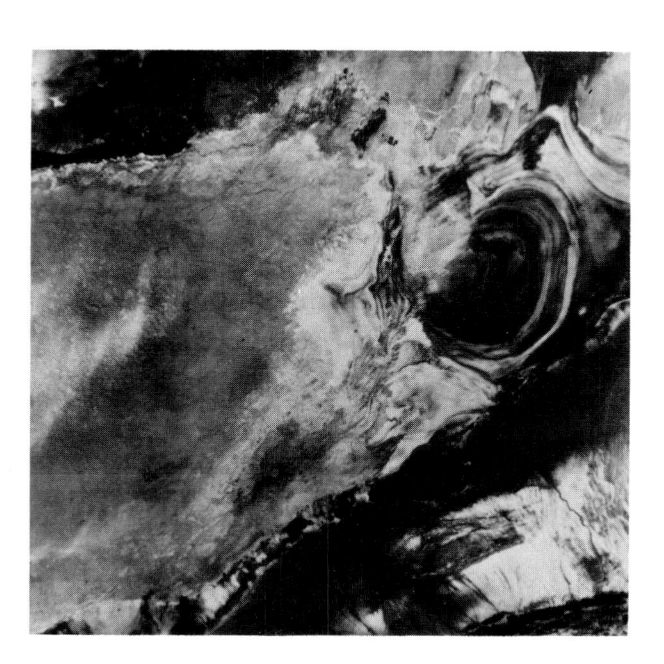

Landsat 1 view of China's Lop Nor launch area. Circular feature is a dry lake bed. China has since developed a more advanced centre for launches to geostationary orbit. *(NASA)*

LAUNCHERS

Introduction The difficulties experienced by Europe's Ariane appear to have been the main stimulus for a whole series of new launch programmes in the US, aimed at capturing a share of the Western market for commercial satellite launchings, which until 1982 everyone had expected would be shared between the Space Shuttle and Ariane.

Since 1977 NASA has expressed rocket thrust in newtons instead of pounds. While the general rule in *Jane's Spaceflight Directory* is to use only the metric system, in the case of thrusts we have usually given both the newton measurement and the value in pounds rather than kilogrammes, since all those who have followed space developments are familiar with thrust expressed in lb. Soviet launch thrusts are usually given in estimated tonnes, since exact figures are seldom available.

China

China's first 2 satellites were launched by CSL-1 (Long March 1), based on the CSS-2 IRBM, which has a range of over 1,500km. The next 6 were launched by the FB-1, designated CSL-2 in the West and based on the CSS-3 ICBM with a range of up to 5,000km. Its launch weight is 190 tonnes, height 32·77m, and payload 2 tonnes in a 200km, 69° orbit. It is built at the Shanghai Xinxin Machine Factory.

Presumably the same factory is developing the CSL-X3, called Long March 3; 43m in height, comparable with Thor, Atlas and Ariane, it is targeted for use in 1985. During the UN conference on peaceful uses of space in Aug 1982 it was also stated that a 3-stage, liquid-propellant vehicle, CZ-3, was being developed for placing "large-scale spacecraft" in low orbits or smaller ones in geostationary orbits.

Europe

Ariane Planned as a series of increasingly heavy launchers since being declared operational on 25 Jan 1982, Ariane was developed by 10 European nations as a result of French insistence that Europe should have the ability to launch its own satellites, and not rely upon either America or Russia. A new European company, Arianespace, was formed in 1980 to take over production and launch operations as soon as ESA had completed 4 test and 6 production launches, a position likely to be reached in 1984. Following the collapse of ELDO and the failure of the Europa rocket based upon Blue Streak, France launched the Ariane programme in 1972. Development cost was estimated in Jul 1975 at Fr 3,046 million. With France paying 62·5%, the programme was managed by the Centre National d'Etudes Spatiales (CNES). Other contributors are: Belgium 5%, Denmark 0·5%, W Germany 20·12%, Netherlands 2%, Italy 1·74%, Spain 2%, Sweden 1·1%, Switzerland 1·2%, Britain 2·47%, others 1·37%. All launches are from the French Guiana Space Centre, Kourou.

By 1981, the situation was that Ariane was ESA-financed and managed by CNES with industrial facilities provided by Aérospatiale; final production, marketing and launching were the responsiblity of Arianespace.

Ariane's 1st launch, on 24 Dec 1979. After many delays, it went perfectly. Had this failed, instead of the 2nd launch, commercial success would have been imperilled. *(CNES)*

Once Ariane had been declared operational ESA's programme board began planning future developments with more confidence, unaware of a whole series of new setbacks that lay ahead. All 1982 flights had to be postponed until Sep due to the need for satellite modifications and an inability to find alternative payloads. Then the Sep flight itself failed, losing the payloads. Before then it had been decided to build a 2nd launch pad at Kourou at a cost of 102 million AU to permit a launch rate of 12 per year by 1985. Both the Ariane 2 and 3 developments were to be available by mid-1983 (now mid-1984), with Ariane 4 coming in 1986. One task for Ariane 4 is to launch Solaris, an automated space station capable

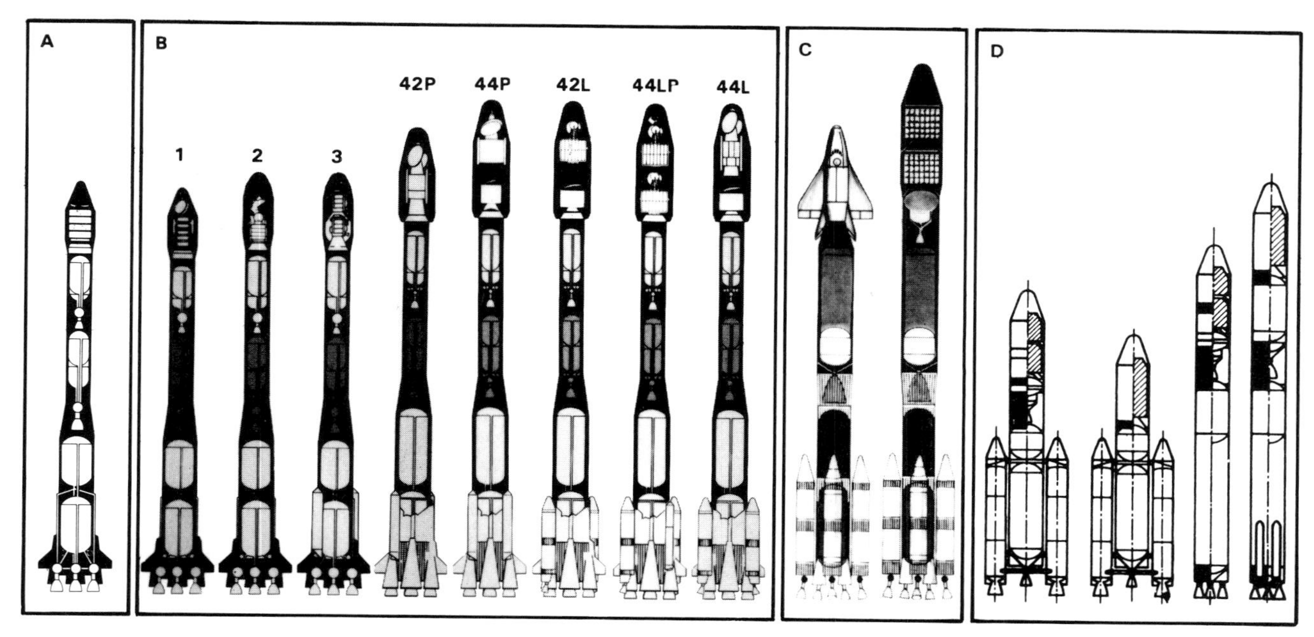

A: Ariane 1 has a maximum lift-off weight of 210 tonnes and can place a 1-tonne payload in stationary orbit.

B: Arianes 1 to 4 are in service, in production or under development. Due to enter service in 1986, the Ariane 4 range will offer twice as much geostationary payload as Ariane 2, as well as a variety of liquid and solid-propellant strap-on boosters.

C: Ariane 5 as originally conceived. On the left is the configuration that would have been used to launch the proposed Hermes mini-shuttle.

D: A complete rethink of Ariane 5 has resulted in these new configurations. Left to right: AR 5P (solid rocket boosters) with 2 geostationary satellites; AR 5P with heavy low-Earth orbit payload; AR 5C (all-cryogenic) with 2 geostationary satellites; AR 5C with heavy LEO payload. *(Air et Cosmos)*

Note: Only Arianes 1-4 are to common scale.

Assembly of Ariane's Sylda double-launch system.

of processing 2 tonnes of material in its ovens. Ariane 4 development costs were estimated at $241 million at 1981 prices.

Meanwhile, steps towards reusability included plans for parachute recovery of Ariane's 1st stage and studies of fly-back systems to avoid ocean recovery. ESA is studying a Future Launch System (FLS), called Ariane 5 by CNES. While FLS is thought of as a heavier unmanned launcher, CNES is more interested in a 4-crew shuttle vehicle called Hermes and is hoping that ESA will take over development of a proposed 100-tonne-thrust rocket engine, the HM60. Hermes is aimed at the mid-1990s.

General description Ariane is a 210-tonne vehicle with the ability to place 1,700kg in transfer orbit (950kg in stationary orbit). More powerful versions are in an advanced stage of development, and there are even hopes that it will ultimately be possible to use it for launching a manned mini-shuttle. Ariane 2 and 3, capable respectively of placing 2,065kg and 2,470kg in geostationary transfer orbits, were due to supersede Ariane 1 in 1983. Ariane 3 will embody a pair of 7-tonne-weight solid-propellant 1st-stage boosters each giving 70 tonnes thrust, as well as offering improved thrust in all 3 stages. Ariane 4, due for test in 1985, is intended to be fully competitive with the US Space Shuttle and to be available in 3 main versions: Ariane 40, with no additional booster; Ariane 42, with 2 additional boosters, will double Ariane's original capability; and Ariane 44 will be available with 4 boosters and enlarged main tanks. Ariane 5, intended to enable Europe to send men into orbit in the Hermes mini-shuttle, is being studied for use "before the end of the century". *Système de Lancement Double Ariane* (Sylda) is an egg-shaped carbon-fibre container enabling 2 600-700kg satellites to be placed in stationary orbit simultaneously.

Launchings L-01 was successfully launched on 24 Dec 1979 after being postponed from Jun 1979 following a ground test malfunction, and after 2 unsuccessful countdowns on 15 and 23 Dec. The orbit of 202 × 35,996km, with 17·55° incl, was close to nominal. A technological capsule was carried in place of a satellite to monitor the rocket's behaviour. L-02, launched 23 May 1980, broke up at 108sec as a result of a high-frequency combustion problem in one of the 1st-stage Viking engines. It fell into the Atlantic Ocean with its 2 German-designed satellites, Firewheel for investigating the magnetosphere, and Oscar 9 for radio amateurs. They were the first of 5 satellites, chosen from 93 proposals in 1975, carried in the Ariane Passenger Experiments (Apex) programme, under which payloads instead of ballast were carried on the last 3 test flights. This made for more satisfactory test flights and gave the users a chance to gamble on a free launch. After months of investigation, during which L-02's engines were recovered from the sea 5km S of the Isle of Salut, the suspected problem was confirmed and modifications for the engine injector components developed. The consequent delays meant the postponement of planned launches for at least 13 customer payloads. Ariane L-03 was successfully launched on 19 Jun 1981, injecting 3 satellites totalling 1,635kg into a 200 × 35,800km transfer orbit with 10° incl; they were the CAT-3 technology capsule, ESA's Meteosat 2 and India's Apple. The last 2 were successfully transferred into stationary orbit, though neither was fully operational. Ariane L-04, launched 20 Dec 1981, was again completely successful, injecting ESA's Marecs 1 (weight 1,014kg) into the required 200 × 36,038km orbit with 10·5° incl; a technological capsule (weight 157kg) included a 16-day experiment developed by a Parisian science club for young people to record profiles of electron density in the ionosphere. This mission enabled ESA to declare the launcher operational.

Ariane L-5, carrying Inmarsat's Marecs B and ESA's Sirio 2, was lost 560sec after launch on 10 Sep 1982 as a result of a 3rd-stage turbopump failure. Due to be the first operational launch and the first use of the Sylda double-payload capability, it was another major setback for the programme as well as a severe financial loss to ESA, which had been able to afford insurance cover of only $20 million. Cost of this failure was later estimated by the French Government to be at least $78·6 million, including replacement/relaunch costs of $30·3 million and $18 million for the Marecs and Sirio 2 satellites.

Ariane L-6, with lift-off weight of 209,794kg, was successfully launched from Kourou amid much tension on 16 Jun 1983. Sylda worked perfectly and ESA's ECS-1 and Germany's Amsat 3B (Amateur Satellite, Phase B), which became Oscar 10, were placed in transfer orbit 19min later. ESA was quick to claim that the flawless performance confirmed that the L-5 problems had been overcome. Later however it emerged that Oscar 10's attitude and spin rate after separation were not as expected, and it was thought that the Ariane's 3rd stage, with a higher than expected residual thrust because of LOX tank venting, had caught up with the Oscar and damaged the antenna used for initial orbital operations. ESA said that the problem could be avoided in future launches.

Ariane L-7, L 18 Oct 1983, delayed 1 month at Intelsat's request, placed Intelsat 5-F7 in a 183 × 36,158km orbit at 8·58° incl compared with the planned 185·4 × 35,987km at 8·50°. (See Addenda for Arianes L-8 and L-9.)

Arianespace consortium With a share capital of about Fr 120 million, Arianespace was formed on 26 May 1980 by 36 European aerospace companies and 13 banks, together with CNES, representing nearly 60% of the capital, to take over from ESA the production of operational Ariane launch vehicles and their marketing, launching and financing. ESA remains responsible for development of the planned new versions of Ariane, which will only be taken over by Arianespace when operational. CNES continued to act for ESA in handling the 4 test flights and first 6 production launches, with Arianespace taking over from the 9th launch, in May 1984. The consortium expected to make a profit if there were 30 launches between 1983 and 1990; at least 40 out of a world total of about 200 were anticipated, necessitating production of 4-6 Arianes per year. Ariane assembly is integrated at Arianespace's Les Mureaux facility near Paris; from there the vehicles are sent by barge and ship to the Kourou Launch Centre in French Guiana, for which Arianespace also became responsible. In Jun 1981 Arianespace decided to offer piggyback launches of small scientific and research satellites of 300-700kg weight on a platform designed by Boeing for Sweden's Viking; this facility fits into an otherwise unused space beneath Ariane's normal payloads.

By mid-1982 the early competition to obtain payloads at uneconomic prices appeared to have settled down to a more realistic level. 2 RCS GStars were to be launched with other payloads by Ariane 3 in May and Aug 1984 at a total cost of $50 million. Arianespace offered to launch 2 Brazilian comsats in Jul 1985 for $58 million, or $29 million each, while NASA was offering a Shuttle/PAM-D launch at $34 million each. By Jun 1984 Arianespace was quoting $25-30 million for a dual launch by Ariane 3.

The Ariane L-5 failure led to some losses of orders, notably Exosat, which was transferred to a Delta. Confidence was restored by the L-6 and L-7 successes, and by May 1984, with the addition of firm contracts for ESA's Olympus (by Ariane 3 in late 1986), Eumetsat's 3 Meteosats (dual launches by Ariane 4 in 1987-90), and a dedicated Intelsat 6 launch in 1987 (for $53 million in 1983), 28 launches worth $813 million were guaranteed – 40% of them for non-European customers. There were also 19 options. But the likelihood was that technical problems with Ariane 3 would mean only 4 launches in 1984.

Flight	*Date*	*Payload*
	1984	
L-8	Mar	Intelsat 5-F8
L-9	May	Spacenet 1
L-10	Aug	ECS-2 + Telecom 1A (1st Ariane 3)
L-11	Sep	Marecs B2 + GStar 1A
L-12	Nov	Arabsat 1 + Spacenet 2
	1985	
L-13	Jan	Telecom 1B or SBTS-1 + GStar 1B
L-14	Mar	SBTS-1 or Telecom 1B + Spacenet 3
L-15	May	Intelsat 5 or Spot + Viking
L-16	Jul	Giotto
L-17	Aug	SBTS-2 + ECS-3
L-18	Sep	TV-Sat
L-19	Oct	Intelsat 5 or Spot + Viking
L-20	Nov	TDF-1
	1986	
L-21	Jan	Intelsat 5A-F15
L-22	Mar	Athos + Meteosat 3 + Arsène + Amsat
L-23	May	Intelsat 5A-F13
L-24	Jun	Payload opportunity
L-25	Aug	Unisat 1(R) + Payload opportunity
L-26	Nov	STC(R) + Payload opportunity
L-27	Dec	Intelsat 6 (N)
	1987	
L-28	Feb	Tele-X + Unisat 2(R)
L-29	Mar	DBSC-1(R)
L-30	Apr	Intelsat 6
L-31	May	TDF-2(R)
L-32	Jun	Anik + MOP-1 + DFS-1(R)
L-33	Jul	L-Sat (Olympus)
L-34	Aug	Intelsat 6(N)
L-35	Sep	DBSC-2(R)
L-36	Oct	Italsat(R) + Rainbow(R)
L-37	Dec	Spot 2

R = Reservation; N = Under negotiation.

Members Belgium 4·40%, Britain 2·40, Denmark 0·70, France 59·25, Germany 19·60, Ireland 0·25, Italy 3·60, Netherlands 2·20, Spain 2·50, Sweden 2·40, Switzerland 2·70.

Future launches The forecast at 18 Feb 1984 is given. ESA is responsible only for L-17, Arianespace for the rest. Ariane 1 was to be used on L-9, L-15 (or Ariane 2), L-16 and L-19 (or Ariane 2). Ariane 2 or 3 will be used for the other flights up to L-22, due to be the Ariane 4 test flight.

German Federal Republic

Otrag Orbital Transport und Raketen AG, with headquarters and production facilities at Garching, near Munich, is developing as a private venture low-cost launch vehicles and sounding rockets. Tank-plus-engine modules are clustered in increasing numbers to form larger vehicles, and it is claimed that a 600-module vehicle could place a 10,000kg payload in low Earth orbit or 2,000kg in synchronous orbit.

Each module consists of a 2-tank/2-engine unit, one tank containing white fuming nitric acid and the other kerosene; each engine develops 29·36 kN (6,600lb) and can be throttled. 2 modules form the smallest flight vehicle. Guidance is by an inertial strap-down platform and computer in the payload section.

A basic 2-module vehicle was successfully test-launched from North Shaba in Zaire on 17 May 1977 and reached an altitude of 12,000m. On 20 May 1978 a 4-module vehicle reached 30,000m. On 5 Jun 1978 there was another 4-module test.

Otrag then moved from Zaire to a new test range in the Sahara, 800km S of Tripoli, Libya. On 1 Mar 1981 the 1st suborbital test was carried out, and it was stated that the propulsion and guidance systems had worked perfectly.

Following these basic flight tests, Otrag was required to return to Germany and has now re-established good relations with the government aerospace agency, DFVLR. Through that organisation Otrag organised two sounding-rocket launches in late 1983, and is still working towards production of a series of efficient, inexpensive orbital launchers and scientific sounding rockets. 1st orbital flight, using a 65-module vehicle, is planned for late 1984; it is hoped to launch a vehicle of many hundreds of modules by 1986.

India

SLV-3 Comparable in performance to the US Scout, this is a 4-stage, solid-propellant vehicle with a lift-off weight of 16,900kg, 22·77m height, 1m max dia and 1st-stage thrust of 422kN (95,000lb). It orbited a 45·3kg satellite in Jul 1980. An Augmented SLV, costing about £12 million and with 2 strap-on solid-propellant boosters and augmented 4th-stage motor, can place 150kg in low Earth orbit. Polar SLV, costing £190 million, will have 4 stages, the 1st with 125 tonnes of solid propellant plus 6 strap-ons, and will place 1,000kg in 1,000km polar, Sun-synchronous orbit from 1987-88.

Italy

Iris The Italian Research Interim Stage is a solid spinning upper stage designed to carry 600kg satellites from low to synchronous transfer orbits. Aeritalia has prime responsibility. 1st use is likely to be for launching Nasa's Lageos 2 from the Shuttle in 1986; 2nd use will be with Sax, Italy's astronomy satellite, in 1987.

Japan

Lambda Developed by Tokyo University and the Aeronautical and Space Division of Nissan Motor Co Ltd from the family of Kappa sounding rockets, the 4-stage, solid-fuel Lambda-4S, used for the 1st Japanese satellite, consisted of 1st-stage core with 36,970kg thrust, augmented by 2 13,150kg strap-on boosters; 2nd stage providing 11,800kg thrust; 3rd stage with 6,580kg thrust; and 4th stage with 816kg thrust. Height 16m, weight 9,390kg. First 3 stages were unguided; 4th stage was provided with attitude control only. Between Sep 1966 and Sep 1969 there were 4 failures to orbit a test satellite with versions of the Lambda-S; the 5th attempt resulted in the brief life of Osumi. This led to hopes that the technologies for the much-delayed and more advanced Mu-4S had been proved. The 1st attempt, on 25 Sep 1970, failed when the 4th-stage motor did not ignite, but was followed by a success 5 months later.

M Developed by the Aeronautical and Space Division of Nissan Motor Co Ltd from 1970, the 23·56m-high, 43·6-tonne M-4S was used for Japan's 2nd, 3rd and 4th satellites. A combination of 4 different solid-propellant rockets, it had no sophisticated guidance system. Controlled only by radio commands, it was capable of putting a 120kg payload into a 500km circular orbit. 1st stage, M-10, had an average thrust of 85 tonnes; M-20 2nd stage 33·1 tonnes; M-30 3rd stage 13·2 tonnes; 4th stage, a spherical motor, 2·7 tonnes. The improved, 41·6-tonne M-3C, with only 3 solid stages and liquid-injection thrust-

First launch of N-2, in February 1981. With 9 strap-on boosters, this will be Japan's main launch vehicle until the mid-1980s. ***(NASDA)***

Japan's N-1 launcher, based on the McDonnell Douglas Thor, is seen here launching the Engineering Test Satellite. ***(NASDA)***

vector control on the 2nd stage, was used for the 5th and 6th satellites. Then came the M-3H and the M-3S with further improved 2nd and 3rd stages and 2 large strap-on boosters for the 1st stage. The latter can place a 700kg payload in a 750km circular orbit, and will be used to launch Japan's first interplanetary spacecraft to Halley's Comet in 1984.

N-1 A Thor-based 3-stage radio-guided vehicle, N-1 was developed for NASDA by Mitsubishi Heavy Industries under licence from McDonnell Douglas and could launch 130kg into geostationary orbit. After delays and problems with earlier national launchers, it was decided to utilise American technology. With total length of 33m, 90 tonnes weight and max dia 2·4m, N-1 has 2 liquid-fuelled stages and a solid 3rd stage. 1st stage for ETS-1 launch used a Rocketdyne MB-3 engine with 3 strap-on solid motors to provide lift-off thrust of 150 tonnes. 2nd stage, designed by Japan's NASDA with Rocketdyne assistance, was powered by a storable liquid rocket, 3rd stage by a solid rocket. The 7th and last N-1 orbited ETS-3 in Sep 1982.

N-2 With longer 1st-stage fuel tanks and 9 solid strap-on boosters, an Aerojet AJ10-118FJ 2nd-stage engine and inertial guidance, this can put 350kg into geostationary orbit and will be Japan's main vehicle until the mid-1980s. 6 launches are scheduled by 1986.

H-1 Designed to place 550kg in geostationary orbit and replace N-2 as the main vehicle from the mid-1980s, this will use Japan's own liquid oxygen/liquid hydrogen engine, 2nd-stage inertial guidance system and 3rd-stage solid rocket motor. 1st launch of H-1 2-stage test vehicle is planned for 1986, test flight of 3-stage H-1 in 1987.

H-2 Development of this vehicle, which would be able to launch a mini-shuttle and compete with Ariane for international launchings, depends on the Nippon Telephone and Telegraph Corp and its choice of launcher for the 4-tonne stationary communications satellite planned for the 1990s. Cost is estimated at $1-3 billion.

USA

History The wide range of US launch rockets preceding the Shuttle has enabled NASA and the Department of Defence to place in orbit payloads as light as 9kg and as massive as Saturn 5's 136,000kg. The manned Mercury spacecraft launched by Atlas rockets weighed about 1,360kg – the same orbital lift capacity as Russia's original ICBM, which was put to work in the Soviet space programme in 1957. Gemini, weighing about 3,630kg, was launched by Titan 2. The first manned Apollo flight, launched by Saturn 1B, was about 20,410kg; the lunar Apollos, launched by Saturn 5, carried about 52,600kg to the Moon. With the phasing out of the Saturns in 1975, America faced a 6yr wait until the Space Shuttle became operational, during which its largest launch vehicle was the Titan 3E-Centaur, with a maximum payload into Earth orbit of 13,600kg. That vehicle, in company with Atlas-Agena and Saturns 1B and 5, is no longer in service. Scout, Delta and Atlas-Centaur are NASA's current expendable launchers, with the USAF relying on versions of Titan for its heaviest, polar, satellites.

But the long-term aim of using the Shuttle for all civil launches – and augmenting its thrust by 1985 to enable it to place military payloads of up to 14,515kg in 98° polar orbits, thus finally eliminating the need for the heaviest current expendable launch vehicle, Titan 34D – has been virtually abandoned. Continuing delays with development of both the Shuttle and Europe's Ariane, coupled with the growing worldwide demand for satellite launchings, have given rise to half a dozen programmes aimed at taking over and continuing production of well proven expendable launchers for commercial use. This trend was encouraged first by DoD, increasingly concerned about the vulnerability of the Shuttle and the need for a back-up launch system, and latterly, with the Shuttle heavily booked until 1989, by NASA too.

Challenger **lifting off in April 1984 on the first satellite rescue mission. But despite the Shuttle's success, NASA's hopes that all other US launchers could be phased out are unlikely to be realised. *(NASA)***

Agena

History The versatile Agena upper stage, produced by Lockheed Missiles and Space Co (LMSC), has been in use since 1959 and is much more than a mere rocket. It has been used as a satellite or booster on more than 300 flights – about half of all US space missions. Most have been classified, and a high proportion of the unidentified US satellites in the annual space logs can be assumed to be Agena payloads of various kinds. The current Big Bird and KH11 spy satellites are derivatives of Agena. Launched as the 2nd stage of an Atlas-Agena (a configuration no longer used), it counted among its early missions the Mariner probes to Mars and Venus, the Ranger photographic flights to the Moon, OAO and ATS.

Its most exciting early use was as a docking target on 5 of the Gemini flights. On Gemini 8 the target for history's 1st space docking, achieved by Armstrong and Scott, was an Agena. It was selected as target vehicle because of its record of achieving precise predetermined orbits and the fact that once in orbit it could be successfully stabilised and then pointed accurately towards the Earth. More recently it formed the basis of the Seasat ocean survey satellite.

Agena D Normally used as the upper of 2 stages, in combination with Atlas, Thor or Titan 3B. The current Agena D version consists of a cylindrical body containing a Bell Aerosystems Model 8096 (YLR81-BA-11) restartable liquid-propellant rocket engine (71·2kN, 16,000lb) and propellant tanks, telemetry, instrumentation, guidance and attitude-control systems. It has carried most types of power supply, including a nuclear reactor, and an ion engine. The payload section (nosecone) can accommodate a wide variety of Earth-orbiting and space probes weighing up to several hundred pounds. The Agena system and its attached payload have functioned for more than six months in some missions for the US Air Force.

Agena D differs from earlier versions in being able to accept a variety of payloads, whereas its predecessors had integrated payloads. The restartable engine permits orbit changes.

Dimensions	
Length (typical)	7·09m
Diameter	1·52m
Weights (typical)	
Propellant	6,148kg
Vehicle, empty	673kg
In orbit, less payload	579kg

Atlas

History It was on top of a General Dynamics Atlas D that on 20 Feb 1962 John Glenn became the 1st American in orbit. There were 10 Atlas D flights in Project Mercury, the pioneering US manned programme: 5 unmanned test flights were followed by MA-5, with a chimpanzee, and 4 manned flights starting with Glenn. All were successful despite initial doubts as to whether Atlas would ever be safe enough for manned flight. It has now been used in more than 25 space programmes, having launched Score, the world's 1st communications satellite, in 1958 and sent Rangers to the Moon and Mariners to Mars and Venus. All US planetary craft have so far been launched either by Atlas or its stablemate, Centaur. Originally developed as the first US ICBM, it became operational in that role in Sep 1959. A total of 159, with a range of 8,050km, were at one time deployed at sites across the US; later versions had a range of 12,875km. As they were withdrawn from military service they were refurbished at Vandenberg as space launchers. By 1981 23 remained in store, having been assigned for use in the GPS, DMS, Geosat and NOAA programmes. 8 Atlas-Centaur launches are scheduled up to 1986, and 4 companies have told NASA that they are interested in continuing production.

Atlas SLV Two versions of the Atlas Standardised Launch Vehicle are currently in service: SLV-3A for use with Agena, and SLV-3D for use with Centaur D-1A. These vehicles differ mainly from their immediate predecessor, SLV-3 having in increased tank length and engine thrust.

Atlas is a "stage-and-a-half" vehicle consisting of side booster and central sustainer sections. The sustainer section includes the propellant tanks and a single rocket engine. The booster engines receive fuel from the sustainer tanks and are jettisoned midway into the flight. The main propellant tanks are currently being lengthened to provide more usable propellant for Intelsat 5A missions.

The engine system is the Rocketdyne MA-5, using liquid oxygen and RP-1 propellants. Total thrust developed is 1,917·4kN (431,040lb), including 1,646kN (370,000lb) total from the two boosters, 266·9kN (60,000lb) from the sustainer, and 4·6kN (1,040lb) total axial thrust from the two vernier rockets. All engines are ignited at lift-off.

Atlas-Centaur launching Mariner 7 to Mars in 1969. By mid-1984 this combination had been used 61 times, with only 6 failures.

Most of the electronic command and control functions for SLV-3D are generated by its Centaur D-1A upper-stage electronics system. SLV-3A has its own systems, including radio guidance.

Dimensions	
Diameter	3·05m
Length	
SLV-3A	24·0m
SLV-3A/Agena	36·0m
SLV-3D	21·2m
SLV-3D/Centaur	39·9m
Performance	
Atlas-Centaur	See Centaur
Atlas SLV-3A/Agena	3,856kg into 185km orbit 1,238kg into synchronous transfer orbit

Centaur

History This was the 1st US high-energy upper stage, and the 1st to use liquid hydrogen as a propellant. It has the advantage of being restartable several times during its 7½min burn time. Produced by the Convair Division of General Dynamics, it was first used in 1963 for the Surveyor lunar landing missions, and by Oct 1982 had made 66 flights, the last 40 flawless. In Apr 1973 the first Atlas-Centaur D-1A launched Pioneer 11 on its Jupiter/Saturn flyby, and other successes have included Mariner, Intelsat, HEAO and Intelsat. Because of Shuttle delays, such use will continue until 1985 at least. The Centaur mission total includes 7 as a Titan 3E upper stage on flights which included Viking and Voyager. A so-called "wide-body" version, under development for use in the Shuttle for the Galileo and Solar Polar missions, was cancelled early in 1982 because both NASA and the USAF preferred a completely new, liquid-propellant upper stage which could also be used as a space tug and possibly be manned as well. Congress reinstated the wide-body Centaur G 6 months later on the grounds that it was cheaper and would be ready sooner. Development contracts worth $250 million to General Dynamics and Pratt

Development of Centaur G, formerly known as Wide-Body Centaur, will cost over $500 million if modifications to Orbiters *Challenger* and *Atlantis* and to KSC launchpads are included. *(General Dynamics)*

& Whitney were expected to follow. Space Services Inc and General Dynamics themselves are both proposing commercial operations with Atlas-Centaur, and the proposed dual-launch capability will compete with Ariane/Sylda. By early 1984 the USAF had decided that IUS had a limited future, and was supporting Centaur G as a follow-on (see Titan).

Centaur D-1A This retains the propulsion and structural features of its predecessor, Centaur D, with stainless steel pressurised tanks and two 73·4kN (16,500lb) Pratt & Whitney RL10A liquid oxygen/liquid hydrogen rocket engines. Specific impulse with this propellant combination is 445sec, equalling that of the advanced main engines on the Space Shuttle Orbiter. Total propellant weight is 13,950kg (30,750lb). Attitude control is achieved by gimballing the two main engines or by clusters of small hydrazine rocket motors.

Several of the electronics components were redesigned for Centaur D-1A. The most significant addition is a 16,000-word-capacity Teledyne digital computer. Navigation, guidance, vehicle stability, tank pressurisation, propellant management, telemetry formats and transmission, and event initiation are all controlled by the computer. Guidance, control and sequencing for the Atlas booster are provided by the Centaur D-1A electronics system.

Payloads are carried on adapters mounted on the forward end of the Centaur. A 3·05m (10ft) diameter fairing protects payloads for Centaur D-1A.

Dimensions

Centaur length	9·14m
Centaur diameter	3·05m
Atlas-Centaur length	39·9m

Performance

Atlas-Centaur	5,942kg into 185km circular orbit
	2,222kg into synchronous transfer orbit
	1,179kg to near planet (with kick motor)

Centaur G The new wide-body version for DoD - 2 have been ordered - will cost $280 million to develop and will boost 4,500kg from Shuttle to geostationary orbit. Centaur G-Prime, the NASA version, will cost another $88 million, plus $80 million for modifications to Orbiters *Challenger* and *Atlantis* and another $80 million for changes to the KSC launchpad. G-Primes for Galileo and ISPM for use in 1986 will cost another $190 million.

Conestoga

Two private companies, Space Services Inc of Houston, Texas, and GHC Inc of Sunnyvale, Calif, combined to attempt to provide a private launch service able to compete with NASA and Europe's Arianespace at one-third the cost. Percheron (French for "workhorse"), with a liquid-fuelled 1st stage, exploded on its 1st launch attempt, from Matagorda Island off the Texas coast, on 12 Aug 1981. The company later switched its efforts to a multi-stage solid rocket called Conestoga (with waggon-train connotations) and capable of placing 11,000kg in low Earth orbit and 2,200kg in synchronous orbit. Conestoga 1, using a 206·8kN-thrust Minuteman 1 2nd-stage motor, was successfully launched on a 10·5min, 313km-altitude sub-orbital flight on 9 Sep 1982 from Matagorda, with successful separation of the 490kg mock payload. Mission director was former astronaut Deke Slayton. In Sep 1983 Slayton announced that Conestoga 2 would be used to launch 3 remote-sensing satellites for the newly formed Space America in 1986, 1988 and 1990.

Constellation

Starstruck Inc, formerly Arc Technologies of Redwood City, Calif, is developing Constellation as a commercial launcher with the help of tests by the 9,000kg sub-orbital Dolphin rocket. This was due to be taken by ship 300km from the S Californian cost and launched in early 1984 while floating in a vertical position. To date 30 static tests of prototype engines have been conducted in preparation for design of a multi-stage orbital vehicle.

Delta

History Delta was derived by McDonnell Douglas from the original Thor ICBM, and since its 1st firing on 13 May 1960 has undergone a profusion of extensions and developments. By Jan 1984 it had launched 173 satellites, with 12 failures. In that time its lifting capacity had grown from 90kg to 1,815kg, and missions included Explorer, Pioneer, Nimbus and Landsat. It can be used as 2 or 3 stages, augmented with 3, 6 or 9 solid-fuel strap-on 1st-stage motors, to lift payloads to synchronous orbits. Its latest use is with PAM, and production was maintained, especially for use from Vandenberg, as a back-up during the early operational Shuttle flights. In 1983 9 launches were scheduled, including ESA's Exosat, transferred from Ariane. However, the makers' hopes that its use would continue into the 1990s received a setback in Mar 1983 when NASA issued a stop-work order following production of the 178th vehicle in 1984. NASA said it had no objection to a commercial concern taking over the project if it was technically and financially able to do so. TransSpace Carriers will be allowed to do this if it can find customers by Oct 1984. Pad 17B at C Canaveral has been modified to handle a launch every 5 weeks, in addition to the Vandenberg polar launches. By Mar 1982 cost of a Delta launch was $25 million, and apart from the Shuttle it was the main competitor with ESA's Ariane for commercial payloads.

DSV-3P Extended Long Tank Delta (Also known as "Straight-Eight" or "2000 Series Delta") This vehicle has a constant 2·44m diameter from the base of the boat-tail to the conical nose section of the shroud; this provides an enlarged payload volume. The 1st stage, with a TRW LMDE engine, is suspended within the 2·44m-diameter barrel section. 1st-stage length is increased by 3·05m by comparison with earlier long-tank versions, providing an increase of 13,600kg in propellant capacity. In addition, the MB-3-III main engine is replaced by a Rocketdyne RS-27 of 911·84kN (205,000lb). As an alternative to the TE-364-3 motor, the higher-performing TE-364-4 is available as a 3rd stage. The 2-stage capability is increased to 1,880kg into a 370km circular orbit. The 3-stage capability is increased to 700kg into a synchronous transfer orbit.

A Delta 3910 carrying out one of its 155 launches, IRAS in January 1983. Delta's future is in doubt. *(NASA)*

DSV-3P (Delta 3914) First launched on 12 Dec 1975, this uprated version of the "Straight-Eight" Delta utilises nine Castor IV solid-propellant strap-on motors in place of the nine Castor IIs used on the Delta 2914. Dimensions are the same as for the standard DSV-3P Delta 2914, but firing weight is increased to 191,400kg. Delta 3914 is capable of putting a 930kg payload into a geosynchronous transfer orbit. The payload is increased to 1,091kg with PAM.

DSV-3P (Delta 3920) This version, development of which was authorised by NASA in March 1980, has an upgraded 2nd-stage propulsion assembly produced by Aerojet. This new stage has a larger propellant tank than the present 2nd stage, extending the burn time by 30%, and uses an Aerojet AJ10-118K-ITIP (Improved Transtage Injector Programme). Delta 3920 with PAM is designed to place 1,250kg (2,750lb) in a geosynchronous transfer orbit.

Dimensions

Length overall	35·15m
Body diameter	2·44m

Weights

Firing weight of DSV-3P (2000 Series)	
3 solid motors	104,330kg
6 solid motors	117,930kg
9 solid motors	131,540kg

Dolphin

Being developed by ARC Technologies of Redwood City, Calif, for ocean launch using hybrid fuel, this is intended initially for suborbital use, taking 230-900kg payloads up to 500km. An orbital version may be produced later. 1st test, 400km SW of San Diego on 6 Feb 1984, had to be hurriedly abandoned following a nitrogen leak, but up to 24 tests are planned for the next 3yr.

HEUS

High Energy Upper Stage was advocated by DoD as necessary for both USAF and NASA to follow the IUS in the late 1980s for delivering spacecraft into high-energy

COS-B being launched for ESA by an earlier Delta in 1975. Derived from the original Thor ICBM, it is fitted with 9 solid-fuel strap-ons. *(McDonnell Douglas)*

orbits from the Shuttle. At the time of writing efforts continued to have Wide-Body Centaur cancelled once more in favour of HEUS, not only for delivering military spacecraft but for use later as a NASA Orbital Transfer Vehicle, thus avoiding the need for a 2nd major development.

IUS

What is now called the Inertial Upper Stage was known as the Interim Upper Stage when conceived in the mid-1970s to fill the gap before a manned Space Tug became available to convey Shuttle payloads from low Earth orbit to their final destination. The name was changed when Space Tug plans were shelved indefinitely. In 1976 Boeing was selected by the USAF to develop the IUS for use with Shuttle and as an upper stage for the USAF's Titan 34D; a one-third-larger version for NASA use would put 2,268kg into synchronous orbit. As IUS became more expensive and was slower than expected in development, the NASA version was cancelled in favour of PAM. The basic IUS 2-stage vehicle is 5·0m long and 2·9m in diameter. It consists of an aft skirt, a 9,707kg aft-stage solid rocket motor generating an average of 189·5kN (42,600lb) of thrust, an interstage, a 2,722kg forward-stage solid rocket motor generating an average of 77·5kN (17,430lb) of thrust, and an equipment support structure which contains electronics for guidance and navigation, re-action control subsystem and electrical power. Its weight is 14,515kg. Chemical Systems Division of United Technologies designs and tests all solid motors.

1st use of Titan 34D/IUS on 30 Oct 1982 was successful, but 1st Shuttle/IUS use, on STS-6 in Apr 1983 and intended to deliver TDRS-1, was marred by a 2nd-stage malfunction. This left TDRS-1 14,000km below the desired synchronous transfer orbit and with damaged thruster nozzles when the 2 vehicles bumped during separation. An inquiry into the IUS malfunction revealed software shortcomings and led to the postponement of the Shuttle/TDRS-2/IUS launch in Aug 1983. The USAF had followed the original $700 million order for 8 development IUS with a $277·3 million contract for 6 production versions for delivery by Apr 1985. But by Mar 1984 the service was talking of IUS having "a limited future" and was supporting Centaur as a follow-on upper stage.

Juno

See Explorer.

Jupiter

See Explorer.

Liberty

Satellite Propulsion Inc of Beverley Hills, Calif, is considering development of the liquid-propelled Liberty to compete with Atlas-Centaur. It would deliver 2,900kg to low Earth orbit (due E) or 2,040kg to geostationary transfer orbit.

IUS: artist's concept of its use for launching TDRS from the Shuttle. The IUS failure with TDRS-1 in Apr 1983 has delayed that programme for about 2yr. ***(Boeing)***

PAM/SSUS

The Payload Assist Module (PAM) is a new high-altitude space vehicle designed by McDonnell Douglas to improve the load-carrying capability of the Delta and Atlas launchers, and for use on Shuttle missions as the Spinning Solid Upper Stage (SSUS). Its first use was on 15 Nov 1980, when as the 3rd stage of a Delta 3910 it placed the SBS-1 communications satellite in synchronous orbit. PAM is powered by a Thiokol Corporation solid-propellant Star 48 rocket motor, which has more thrust and a longer burn time than the standard TE-364-4 Delta 3rd stage. PAM is spin-stabilised; small rockets on the structure that joins it to the Delta 2nd stage set the PAM and its payload rotating before the PAM motor is fired. The version for Shuttle launchers is identical to the Delta PAM except that the rocket exhaust nozzle will be slightly shorter, and the stabilising rotation will be imparted by a pair of small electric motors. As many as 4 satellites with PAM stages attached could be launched during a single Shuttle mission; it will carry payloads of up to 1,110kg into synchronous orbit. By 1985 PAM-D2 will increase this to 1,886kg, and the USAF plans to use 28 of them for Navstar launches from the Shuttle. Firm and "potential" orders for all versions total 75, worth $210 million. PAM-A, a larger version being developed for use from the Shuttle, will be able to lift payloads up to 2,000kg into transfer orbits; by May 1983, however, there were still no customers even though 8 had been built. PAM-Ds were successfully used on 11 Nov 1982 to launch SBS-3 and Anik C3 from STS-5, the Shuttle's 1st operational mission, on STS-7 to launch Anik C2 and Palapa B1, and on STS-8 to launch Insat 2B. By the 10th mission (41-B), in Feb 1984, successful firings totalled 16, so it was a shock when PAM-Ds on both the Westar 6 and Palapa B2 satellites malfunctioned, placing them in unusable orbits. PAM unit cost is $5-10 million, depending upon the version.

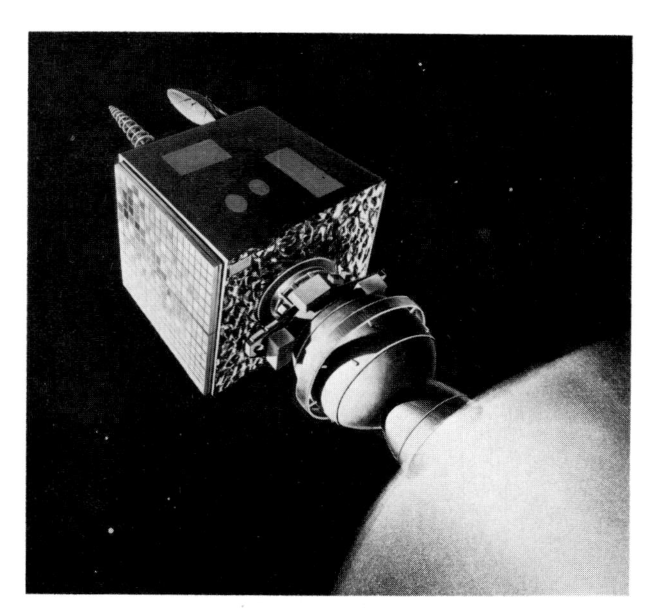

PAM: double failure of this upper stage on the 10th Shuttle mission after 16 successful firings was still delaying many satellites as we went to press. ***(McDonnell Douglas)***

Phoenix

A reusable launcher proposed by Pacific American Launch Systems Inc for manned or unmanned use, this would be a single-stage vehicle powered by high-efficiency liquid oxygen, hydrogen and propane engines and capable of missions to low Earth orbit, geostationary orbit or even the Moon. It could place 2,300-4,600kg in geostationary orbit without an upper stage. Its promoter claims that in 1986 a dedicated launch would cost $50 million compared with the Shuttle's $70 million.

Redstone

History Redstone was used to launch the 2 suborbital flights which inaugurated America's manned space programme. It was first developed by Dr Wernher von Braun for the US Army from his original V2 missile. With a 965km range, it was deployed in Europe in 1958 and became known as "Old Reliable". Some 800 engineering changes were made to transform it into a booster for the 1st experimental manned flights. In addition to major engine improvements, 20sec was added to the burn time by lengthening the tank section by 1·83m. This, with the spacecraft and escape tower, added 2,265kg to the original 27,670kg missile weight. Telemetry provided 65 measurements covering attitude, vibration, acceleration, temperature, thrust level, etc. Redstone became established as a space launcher in 1958 when used as the 1st stage of the Jupiter C rocket which put Explorer 1, America's 1st satellite, into Earth orbit.

6 Mercury-Redstone flights were made, starting in Nov 1960. The first 4 were test firings. Then came the historic 15min "lob" which made Alan Shepard America's 1st man in space, on 5 May 1961, followed by Virgil Grissom on 21 Jul 1961. For the manned flights lift-off weight was 29,935kg, thrust 346,972N (78,000lb), height 25·30m, dia 1·77m, speed at burn-out 7,080km/hr and range 320km.

Saturn

History The Wernher von Braun organisation, then working with the Army Ballistic Missile Agency, first proposed the need for a launcher of 680,400kg thrust, able to place between 10 and 20 tonnes in Earth orbit or send 3-10 tonnes on escape missions, in Apr 1957. By 1959 the project had been named Saturn. This was because Saturn was the next planet after Jupiter in the solar system, and the Saturn rocket was the next von Braun project following completion of Jupiter missile development.

Saturn 1 With engines and tanks in clusters in order to make use of equipment already developed, Saturn 1 was a 2-stage vehicle. The first stage had 8 H-1 engines burning RP-1 kerosene and liquid oxygen and each generating 82,250kg thrust. The 2nd stage (designated S4) had 6 liquid oxygen/liquid hydrogen RL-10 A-3 engines, each generating 6,800kg thrust. 10 Saturn 1s were fired between Oct 1961 and Jul 1965, with an unprecedented 100% success record. The 5th placed a 17,100kg payload in Earth orbit, and the 6th and 7th each placed unmanned boilerplate models of Apollo spacecraft in Earth orbit. The 9th orbited a Pegasus meteoroid technology satellite. Meanwhile, it had been decided that elements of Saturn 1 and of the planned much larger Saturn 5 should be combined to form a new mid-range vehicle, Saturn 1B. This would have 50% more capability than Saturn 1, enabling complete Apollo spacecraft to be tested in manned Earth-orbital flights 1yr earlier than would be possible with Saturn 5.

Saturn 1B 9 firings maintained Saturn 1's record of success. The 1st launch, which was also the 1st flight test of a powered (unmanned) Apollo spacecraft, was on 26 Feb 1966. After 3 more test flights the 1st manned flight of the series, Apollo 7, was successfully launched on 11 Oct 1968. 3 Saturn 1Bs were used to send 3 crews to the Skylab space station in 1973, followed by the equally successful ASTP launch in 1975. Of the 12 originally built, 3 were unused. The Saturn 1 programme cost $767 million, Saturn 1B an additional $1,127 million.

Saturn 5 Also designed and developed at NASA's Marshall Space Flight Centre under Dr Wernher von Braun, this continued the success record from the start of development in Jan 1962. Its 1st flight, on 9 Nov 1967, was the unmanned Apollo 4 mission, which successfully tested both the rocket and the spacecraft. It was first used for a manned flight on Apollo 8, on 21 Dec 1968, when Frank Borman commanded man's 1st flight around the Moon. Its flexibility was demonstrated when it survived a lightning strike during the Apollo 12 launch. On Apollo 13, when the centre engine of Stage 2 shut down 2min early the other 4 engines automatically burned to depletion to make up the lost thrust and the 3rd stage automatically burned for an extra 10sec to complete the task. There was still ample fuel left to complete the translunar injection. Structural weight reductions, with improved engine performance and operational techniques, enabled the Saturn 5s used in the final Apollo missions to place payloads of 152 tonnes in Earth orbit and to send 53 tonnes to the Moon.

General description Saturn 5's 1st stage (S1C) was built by Boeing at NASA's Michoud Assembly Facility, New Orleans. Its 5 F-1 engines consumed kerosene and liquid oxygen at a rate of 13,319kg/sec and boosted the vehicle to approx 8,530km/hr and a height of 61km in 2½min. The 2nd stage (S2) was built by North American Rockwell at Seal Beach, Calif. Its 5 J-2 engines developed a thrust of 1,160,200lb, burning liquid hydrogen and liquid oxygen, to raise speed to 24,625km/hr and height to 183·5km in 6min. The 3rd stage (S4B), built by McDonnell Douglas at Huntingdon Beach, Calif, and also powered by hydrogen and oxygen, was restartable and produced 230,000lb thrust. After a 2½min burn to place the vehicle in Earth parking orbit a 5min firing, usually on the 2nd orbit, injected the vehicle into translunar orbit at 39,270km/hr. The Apollo Command Module then separated, docked at the opposite end of the S4B to withdraw the Lunar Module, then separated again, leaving the S4B to be sent into solar orbit or impacted on the lunar surface.

Lift-off weight (Apollo 11)	2,938,312kg
1st-stage thrust (Apollo 11)	7,553,227lb
Height	
(with Apollo)	111m
(with Skylab)	109m
Diameter	
1st and 2nd stages	10·06m
3rd stage	6·60m

Scout

History Designed by Vought as a low-cost launcher for sub-orbital, orbital and re-entry research, Scout was the 1st US vehicle to use solid fuel. First launch was Explorer 9 on 16 Feb 1961. When launches totalled 100 with the British Ariel 6 on 2 Jun 1979, 78 payloads had been for NASA and DoD, and 22 for European nations, with a 95% success rate. Many developments have taken the payload from 59kg to 204kg. By Jan 1984 launches totalled 104.

Scout (XRM-91) The 1st stage motor is an Algol III, manufactured by Chemical Systems Division of UTC and providing a total thrust of 484kN (109,000lb). 2nd stage is the 285·2kN (64,000lb) Castor II by Thiokol. 3rd stage is the 83·1kN (18,700lb) Antares III by Thiokol. The Altair III 4th stage, by Thiokol, develops 25·5kN (5,700lb) Honeywell provides the simplified gyro guidance system. Spin stabilisation of the 4th stage is by Vought.

In this standard 4-stage form Scout is able to put a 193kg (425lb) payload into a 500km (310-mile) easterly orbit. A 5-stage version offers increased hypersonic re-entry performance and makes possible highly elliptical deep-space orbits.

Dimensions	
Height overall	22·92m
Max body diameter	1·14m
Weight	
Launch	21,400kg

Space Van

A reusable mini-shuttle for manned or unmanned use proposed by Transpace Inc of Washington DC, Space Van would be launched from a 747 and be powered by Pratt & Whitney RL-10 cryogenic engines. In the cargo mode, when an expendable upper stge would be used, 3,000kg could be placed in a 450km orbit. 800kg could be placed in synchronous orbit, but this could be increased to 3,000kg by using low-orbit refuelling. The promoters claim that cost could be one-fifth that of a Shuttle launch.

SRB-X

Boeing has a NASA contract to study Shuttle-derived cargo vehicles for use when an Orbiter is either unnecessary or not available. An SRB-based vehicle could place 30,000kg in LEO, or 5,500kg in synchronous orbit; a vehicle using the SRBs and ET could orbit 36-64,000kg. Other "mixed fleet" concepts being studied by Martin Marietta include an Orbiter and ET to place 45,000kg in LEO, and a cargo canister to replace the Orbiter and put 64,000kg in LEO.

Thor

Produced by Douglas (later McDonnell Douglas) as the 1st Intermediate Range Ballistic Missile (IRBM) - 60 of them were stationed in Britain during 1957-63 - Thor was fired as a missile or space launcher over 500 times and was developed into the Delta. The original Thor, with 1st-stage thrust of 666kN (150,000lb), was developed by combining Jupiter and Vanguard stages, and as Thor Ablestar was first used for a successful orbital launch in 1959. Thor Ablestar was flown 19 times between 1960 and 1965, placing 12 Transits, Courier 1B and Anna 1B in orbit, with 5 failures. At the same time McDonnell Douglas was developing the Thor-Delta improved version, which achieved 33 successful orbital flights out of 36 up to 1969, with payloads rising to 430kg in Earth orbit. Many succeeding variants included Thor-Agena, Thrust-Augmented Thor-Agena (TAT Agena), Thrust-Augmen-

Thor: this version is Thrust Augmented Improved Delta (TAID), seen here launching ESA's HEOS-1. Delta has now taken over from Thor. *(ESA)*

ted Thor-Delta (TAD), Improved Thor-Delta (TAID) and Long-Tank Thrust-Augmented Thor-Delta (Thorad). TAD, built by Japan under licence, provided that country with the basis of its current N launcher. The 1st communications satellites, Telstar and Early Bird, were orbited by Thor. By 1976, 20 years after the 1st delivery, 364 Thors had been launched, of which 350 had been successful. About 10 are still in store and available for future use.

Titan

History Titan 3, America's most powerful unmanned booster, is used for both military and civil purposes and can launch payloads ranging from 15,875kg in Earth orbit to 3,175kg for planetary missions. It had its origins in the Titan 1 and 2 ICBMs, test flights of which began in Feb 1959; 52 Titan 2s, currently deployed in Arizona, Kansas and Arkansas, are still the most powerful missiles in the US arsenal. A modified version of Titan 2 was successfully used to launch all 10 of the 2-man Gemini spacecraft in 1965-6.

Martin Marietta builds the airframe and liquid-propellant stages, supplies the flight-control system, and is integrating contractor. Aerojet-General produces the liquid-propellant engines. UTC's Chemical Systems Divion supplies the solid-propellant boosters used in the more powerful models. Guidance systems for Titan 3C and D are built by General Motors Corporation's Delco Division and Western Electric respectively. Titan 3's core section consists of 2 booster stages and an upper stage, known as Transtage, that can function both in the boost phase of flight and as a restartable space tug. All stages use storable liquid propellants and have gimbal-mounted thrust chambers for vehicle control. There are usually 7 Titan launches per year, 5 from Vandenberg and 2 from C Canaveral. In early 1984 about 5 Titan 3Bs and 10 Titan 34Ds were in store for future use. An agreement between Martin Marietta and SpaceTran to market 34D as "Commercial Titan" looked like coming to nothing because of high costs, but an earlier USAF decision to end Titan production was about to be reversed. The USAF was considering ordering 10 Titan-Centaurs as a Shuttle back-up from 1988; the Centaurs could be used with either Shuttle or Titan.

Titan 3B Basically the first 2 stages of the core section, accommodating a variety of specialist upper stages. First launched 1966, by the end of 1979 it had been used 58 times (including 4 failures) from Vandenberg with Agena D as 3rd stage and carrying classified USAF payloads: 51 were directed at low orbits, suggesting observation, and 7 at Molniya-type orbits, probably to provide communications coverage over the Arctic more effectively than geosynchronous satellites over the Equator.

Titan 3C Consists of the core section, including the Transtage upper stage, with the addition of 2 solid-propellant rockets attached to each side to function as a booster stage before ignition of the main engines. Payloads include USAF and NASA scientific and communications satellites, and Titan 3Cs have been responsible for about 80% of US synchronous launches. From 1965-1979 there were 33 flights, including 3 Vela nuclear-detection payloads in circular 112,650km orbits (well beyond synchronous altitude), 10 military payloads possibly designed to give early warning of missile launches, and NASA's ATS-6. So far flown only from C Canaveral.

Titan 3D Similar to 3C but without Transtage, and with radio guidance instead of the standard inertial. By 1981 there had been 20 launches, all from Vandenberg. 16 of them were Big Birds, starting on 15 Jun 1971; 4 were believed to be KH11s.

Titan 3E A special NASA version of 3D for deep-space flights, using Centaur instead of Transtage. 6 successful launches comprise 2 Helios, 2 Vikings and 2 Voyagers.

General description The 1st stage of the core section is 22·25m long and 3·05m in dia. Its engines, which use a blend of hydrazine and unsymmetrical dimethylhydrazine (UDMH) for fuel and nitrogen tetroxide as an oxidiser, have a 15:1 expansion ratio and are ignited at an altitude where efficiency is maximised, giving a thrust of 2,339·6kN (526,000lb) in vacuum. The second stage is 7·10m long and 3·05m in dia. Its engine uses the same propellants as the 1st stage and develops 453·7kN (102,000lb).

Transtage is 4·57m long and 3·05m in diameter and also uses UDMH/hydrazine and nitrogen tetroxide. The twin-chamber engine produces 71·17kN (16,000lb) and is capable of multiple restarts in space, which permits a wide variety of manoeuvres, including change of plane, change of orbit, and transfer to deep-space trajectory. Transtage also houses the control module for the entire vehicle, including the guidance system and segments of the flight-control and vehicle safety systems.

Titan 3C/D's solid-propellant booster motors are each 25·91m long and 3·05m in dia. Each motor is built in 5 segments and develops more than 5,115kN (1,150,000lb). The booster stage is steered by a thrust-vector control system which injects nitrogen tetroxide into the engine nozzle.

Titan 34D The latest Titan 3, redesignated because instead of Transtage it uses the Boeing IUS developed for the Shuttle. With 1st stage lengthened by 1·75m to contain more propellants, these are being used for some primary launches as well as acting as Shuttle back-ups. 1st flight, due mid-1981, was delayed more than a year by IUS problems. Able to place 1·9 tonnes in stationary orbit or 12·5 tonnes in low Earth orbit. 14 have been ordered by the USAF: 8 for launch from C Canaveral, the rest from Vandenberg. 1st use of Titan 34D/IUS was finally achieved on 30 Oct 1982, placing 2 DSCS satellites in synchronous orbit.

TOS

Development of the Transfer Orbit Stage (ie, 4th stage) was announced by NASA, Orbital Systems Corp and Martin Marietta in May 1983. With fully loaded weight of 10,930kg, 3·2m length and 2·8m dia, it is designed to send spacecraft of 2,700-5,900kg from the Shuttle to transfer orbit and is intended to fill a gap between the smaller PAM and the more expensive Centaur and IUS. At $20

Titan-Centaur launching Viking 1 to Mars in 1975. Titan 34D, derived from the original Titan ICBM, is currently America's most powerful unmanned booster. ***(Martin Marietta)***

million it should cost half as much as an IUS. The $50 million development, at no cost to NASA, is considered an important step in the commercialisation of space activity and is intended to strengthen the competitiveness of US launch systems with Europe's Ariane. Also compatible with Titan, TOS will be spring-launched like IUS from the Shuttle in 55° orbit. TOS will be available from late 1986, and OSC hopes to sell 34 by 1992.

Vanguard

See Vanguard satellites.

USSR

Status A whole family of new Soviet launchers seemed to be imminent as we went to press. According to some US sources the launch site for the largest of these, a Saturn 5-class booster, is still incomplete, and 1st firing may still be as far away as 1986. But other pads, capable of accommodating vehicles providing much greater lifting power than current launchers, appeared to be ready. A 100m-tall booster, camouflaged in nets to foil spy satellites, was said to be awaiting flight test at Tyuratam. Some confirmation was offered by cosmonaut training chief Gen Vladimir Shatalov in Jan 1984, when he spoke of "heavy transport ships able to lift cargo weighing more than the aggregate on existing stations". Other pads are apparently in preparation for launching the spaceplane, weighing around 12,000kg, and the larger space shuttle.

History In 25yr of spaceflight the Soviet Union has revealed few details of its launch vehicles. It is therefore surprising to find how much detailed information has been painstakingly assembled by Western observers. Their task has been simplified by the fact that there have been few completely new Soviet launchers, especially in recent years. Developments have consisted of additions and improvements to the original military missiles. This compares with the much faster development of rocketry techniques resulting from the more varied launchers produced by the competing aerospace companies of the United States. This Directory uses the identification system originated by the late Dr Charles Sheldon of the US Library of Congress, and combines information provided by him with explanations from Dr J. A. Pilkington, former director of the Scarborough Planetarium.

By 1969 6 basic Soviet launch vehicles had been identified and designated A, B, C, D, F and G. Dr Sheldon labelled added upper stages "1" or "2," and when there was doubt whether the stage used was 1 or 2, labelled it "1/2". It should be noted however that the "1" in A1, B1 and C1 usually refers to a different upper stage in each case. The escape rocket, often a 4th stage, is labelled "e"; manoeuvrable stages thus become "m" and re-entry vehicles "r".

G is "Webb's Giant," named after the former NASA Administrator who said in 1967 that Russia was developing a huge new launcher bigger than Saturn 5. Reports that 3 test firings, beginning in 1969, had ended in disastrous explosions appear to be true, and work on it was suspended for some years. Recently the author learned from US sources that the final explosion, causing much loss of life, was on on 3 Jul 1969, only 13 days before launch of Apollo 11 – thus finally disposing of any remaining doubts as to whether there really was a Moon race. (There is, however, no evidence that there were any cosmonauts aboard at the time of the explosion.)

Current Soviet rockets go back to the early 1950s and the first ICBMs, and the remarkable thing about Soviet spaceflight, particularly manned flight, is the small amount of rocket development which appears to have taken place between the first Vostok mission and the Soyuz series. This contrasts with the 1,632kN thrust of the Atlas rocket used for the 1st US orbital flight and the 33,343kN of the Saturn 5 used for Apollo. The cluster design of the original Vostok rocket used to place Yuri Gagarin in orbit, with its central unit and clip-on boosters, was still clearly recognisable in the pictures of the launch of Svetlana Savitskaya, the 2nd space woman, on Soyuz T-7.

In 1983 the Soviets offered Proton launches to Inmarsat at a cost (according to Inmarsat) of $24 million for a dedicated launch, compared with $22 million for a shared Shuttle launch and about $29 million for Ariane or a US expendable. The Soviets refused to disclose Proton's success rate or say how many firings had been made. They offered, however, to provide a 2nd launch at half cost if the 1st failed. Conditions were that the satellite be flown from Moscow to Baikonur by Aeroflot and mated horizontally. According to Inmarsat, launch vibration levels were mid-way between those of the relatively smooth Ariane and the "brutal" Shuttle.

Here is Robert Christy's analysis of Soviet launchers:

A The original Sapwood (SS-6) ICBM was used to launch Sputniks 1-3. A core vehicle contains a single Lox/kerosene-burning RD-107 engine with 4 main thrust chambers and 4 verniers. The core is supplemented with 4 strap-on boosters containing RD-108 engines, each with 4 main chambers and 2 verniers. At lift-off the 32 thrust chambers provide a total of over 500 tonnes thrust.

SOVIET LAUNCH VEHICLES

Designation	Derivation	Max Payload to LEO (kg)	Typical payloads (kg)
A	SS-6 (Sapwood)	2,000	Sputnik 1 to 3, 84-1,327
A1	SS-6 + small stage	5,000	Luna 1 to 3, 360 Vostok, 4,700 Meteor 2, 2,500
A2	SS-6 + large stage	7,500	Voskhod, 5,700 Soyuz, 6,600 Cosmos biosat, 6,000
A2e	SS-6 + large stage + probe rocket	n/a	Venera/Mars, 900-1,100 Luna, 1,400-1,800 Molniya 1, 1,750 Prognoz, 900
B1	SS-4 (Sandal) + stage	500	Cosmos/Intercosmos, 450
C1	SS-5 (Skean) + restart stage	1,000	Cosmos/Intercosmos, 550 Cosmos navsat, 700
D } D1 }	Original	20,000	Salyut, 19,000 Proton, 12,200-17,000
D1e	Original + probe rocket	n/a	Luna 15-24, 6,000 Venera/Mars, 5,000 Statsionar (incl apogee motor), 6,500
F1	SS-9 (Scarp) + stage	4,500	FOBS, 4,500 Rorsat, 4,000
F2	SS-9 + 2 stages or SS-9 + large stage	7,000	Cosmos reconsat, 6,500
G	Original	100,000+	Large space station

LEO = Low Earth orbit. n/a = Not applicable; in LEO some upper stages have still to be fired.

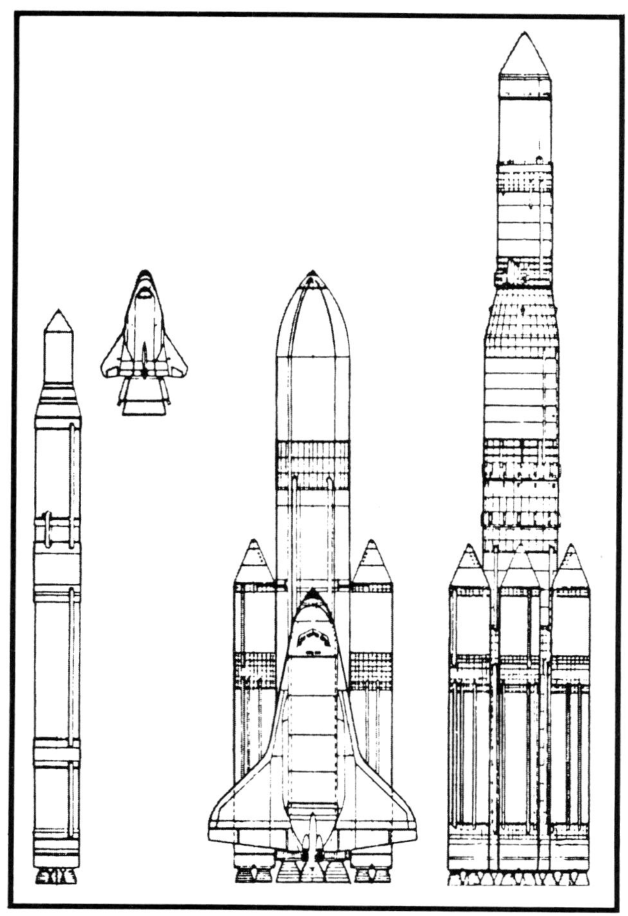

US concepts of the forthcoming generation of Soviet launchers and spacecraft. Left to right, Titan-class booster; spaceplane (a possible Soyuz T replacement); shuttle; and the long-awaited big booster. *(DoD)*

Vostok with Soyuz 18 goes by rail to the launchpad. It takes only 6min to raise the cluster to the vertical position, and nowadays this is not done until 2 days before launch. *(Novosti)*

Vostok: not shown publicly until 6yr after placing Gagarin in orbit, its cluster configuration is still in use. The original Vostok remains clearly recognisable in current pictures of Soyuz T launches. *(Reginald Turnill)*

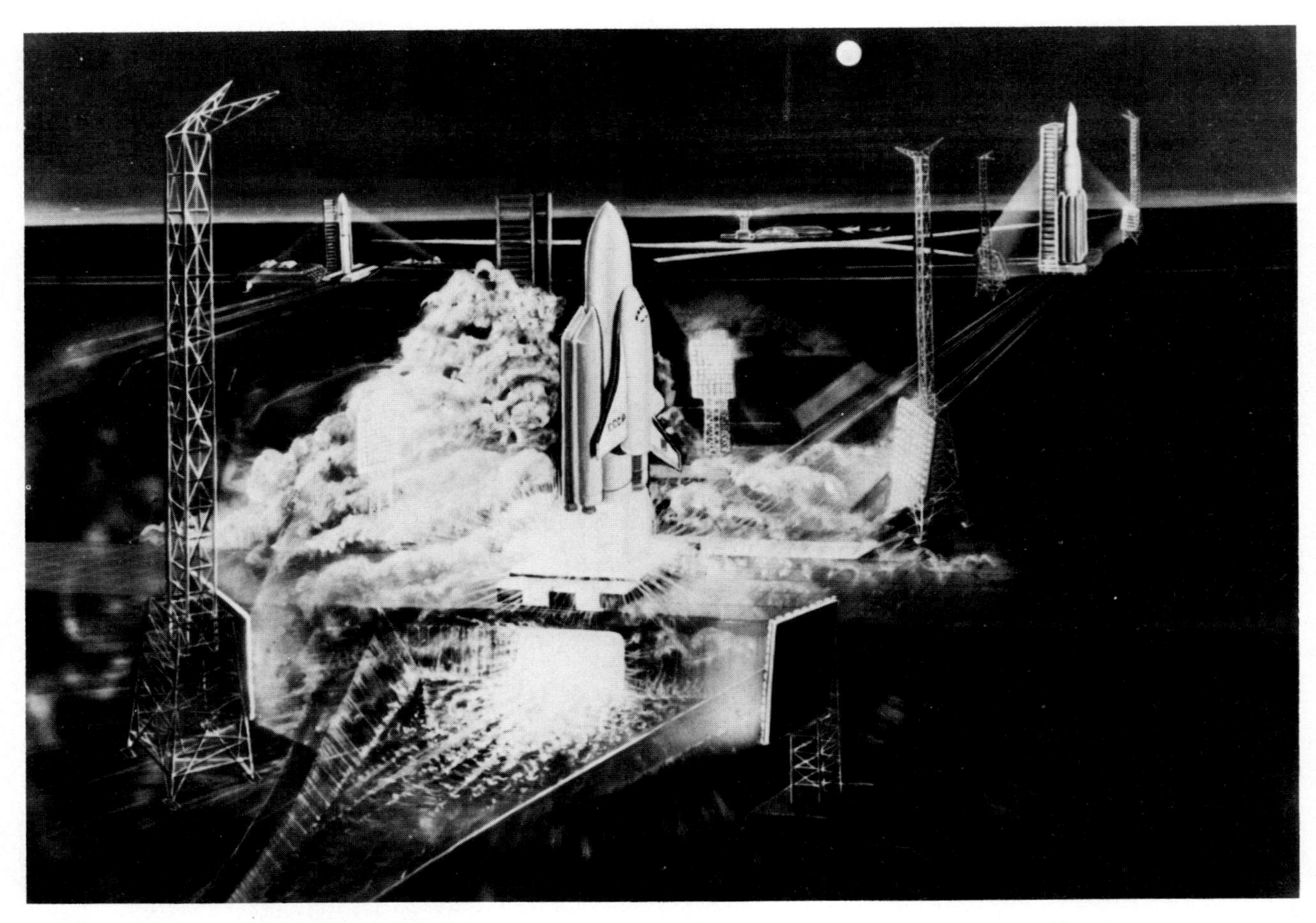

Artist's concept of Soviet shuttle lift-off. The most important difference is believed to be that the main engines are on the external tank instead of on the orbiter. This relieves the orbiter of much unnecessary weight, but means that the engines are lost unless the Soviet ET is recoverable. *(DoD)*

Largest payload launched was Sputnik 3 (1,327kg), to 226 × 1,881km orbit. Overall length 28m (excluding payload).

A1 First used in 1959; one version still in use. With Sapwood as 1st stage, it has a 3·1m-long × 2·6m-dia upper stage attached by a trelliswork structure. A single-chamber engine gives approx 10 tonnes thrust. Early spacecraft launched were 1959 Luna probes, Vostok and 1st-generation recoverable Cosmos. Currently used for Meteor weather satellites and electronic ferrets at 81° incl. With overall length of 34m (including payload shroud), it can place nearly 5 tonnes in LEO from Tyuratam and Plesetsk.

A2 Another Sapwood development with a more powerful upper stage; used today for launching Soyuz and Progress as well as 3rd and 4th-generation recoverable Cosmos observation satellites. It also launched the Voskhod manned vehicles and the 2nd-generation reconsats. Upper stage is 8m long × 2·6m dia, with 4-chamber engine giving approx 30 tonnes thrust. Overall length of the Soyuz version with launch escape system is 49·3m. Soviets have said payload capacity to LEO is 7·5 tonnes, though its largest known payload is Progress at 7 tonnes. More than 700 A2s have been used, about 100 having been A2es.

A2e A still more powerful version of Sapwood, with an additional stage on top of A2. First flown in 1961 for Venus attempts; A2 itself was not used until 1962. The "e" stage fits under the shroud and appears to be a version of the A1 upper stage; shroud is similar to that of Vostok. A2e has been used for interplanetary, Luna and other deep-space probes such as Prognoz. Current use is for Molniya and early-warning satellites. Payload for this type of mission is approx 2 tonnes.

B1 Based on Sandal with added upper stage, this has been launched from Kapustin Yar and Plesetsk. Kapustin Yar launches have been from silos, probably relating to its original development as a missile. The 1st-stage engine has been identified by Soviets as the 72-tonne-thrust RD-214; upper stage is RD-119 with 11 tonnes thrust. Max payload is about 600kg. Overall length 32m, dia 1·65m. Last B1 launch was Cosmos 919 in 1977 from Plesetsk. Has been used in Cosmos and Intercosmos programmes.

C1 Based on Skean (SS-5), this is also 2-staged. Upper stage is restartable for placing small payloads in high circular orbits from the apogee of an elliptical transfer orbit. First used from Plesetsk 1967, now also launches occasional payloads from Kapustin Yar. Has taken over from B1 as launcher of small scientific satellites, eg the Intercosmos programme. Is also used for several military programmes, including navigation and store/dump communications satellites. Length about 32m, overall dia 2·5m; payload up to 1½ tonnes to LEO.

D Basic design follows that of the A vehicle, with central core and 6 strap-on boosters. Appears to be the only Soviet launcher so far not directly adapted from a military missile. Basic form may have been used without upper stages in its first 3 launches, the 1965-66 Protons. Overall length about 40m; total thrust about 1,500 tonnes. Proton 1 weighed 12·2 tonnes.

D1 This version had an added upper stage with basic core and strap-on 1st stage. Used in 1968 for 17-tonne Proton 4, and may also have launched the earlier Protons. Currently used for the 19-tonne Salyuts; maximum capacity to LEO said to be over 22 tonnes.

D1e Further development of D. Additional "e" stage enabled it to launch Zond circumlunar series, starting as Cosmos test flights in 1967. Now used to launch geostationary comsats. For this an apogee motor, which could be the old A1 upper stage, is also carried. Planetary-flight payload nearly 5 tonnes.

F1 Based on the Scarp (SS-9) ICBM, this is believed to be a 2-stage launcher with capacity almost as high as that of A1. Used mostly for military purposes, firstly for FOBS, then as a satellite interceptor and for launching radar and electronic ocean-reconnaissance satellites. FOBS version often referred to as "F1r", with the "r" standing for a retro-rocket stage which is actually part of the payload. Similarly, the satellite interceptor and ocean-reconnaissance launchers are referred to as "F1m," the "m" standing for manoeuvrable stage, which is again actually part of the payload. Either F1 or F2 (see below) may be in use for launching some electronic ferrets, oceanographic satellites and Meteors from Plesetsk. Apart from these, F1 is Tyuratam-based, offering LEO payloads of up to approx 4 tones.

F2 The latest F version. Some recent 3rd-generation spy satellites have been launched from Plesetsk with a rocket other than the usual A2. This could be a Scarp with a new upper stage, or F1 with an additional stage.

G Tests of the original "Webb's Giant" failed in Jun 1969, Jun 1971 and Nov 1972, and after that it was thought that the big booster had been abandoned. By mid-1980, however, US experts were able to produce a diagram of the revised giant booster, and US Defence Secretary Caspar Weinberger's 1983 report said that this Saturn 5-class booster had a core vehicle about 95m tall, with 2 or 3 large liquid-propellant strap-on boosters each about 35m tall. Lift-off thrust, at one time estimated at 10-14 million lb, was then estimated at 8-9 million lb, compared with Saturn 5's 7·5 million lb. Payload capability to 180km would be 130-150,000kg, compared with Saturn 5's 127,000kg.

THE SOLAR SYSTEM

Note on escape velocities The approx speeds which must be attained by a spacecraft, manned or unmanned, intended to escape from the gravitational pull of Earth, Moon or one of the major planets to travel to any other are as follows:

	km/hr
Mercury	15,000
Venus	37,000
Earth	40,000
Moon	8,500
Mars	18,000
Jupiter	214,000

Planetary priorities

The first 2 decades of planetary exploration, largely carried out by the United States and Soviet Union, have yielded so much information that the scientists and astronomers have had difficulty in keeping pace with it. Perhaps this explains why, for the past 6 years, Russia has concentrated on Venus and NASA has been unable to obtain funds for new programmes. By the end of the 3rd decade, however, all the major planets except Pluto should have been visited, while the 1980s will see the first planetary probes launched by Europe and Japan. Current planetary plans are as follows:

Sun ESA/NASA Solar Polar mission, due for Shuttle/Centaur launch in May 1986 to explore interplanetary space out of the plane of the orbits of the planets and above the solar poles.

Venus Soviet flyby and landers in 1985, en route to Halley's Comet. NASA's Venus Radar Mapper will be launched by Shuttle/Centaur in Apr 1988.

Mars ESA is studying an orbiter called Kepler. Russia is planning a 1986 orbiter to formate with Phobos.

Jupiter NASA's Galileo, due for Shuttle/Centaur launch in May 1986, will orbit Jupiter and drop an atmospheric probe.

Uranus Voyager 2 due to fly by in Jan 1986.

Neptune Voyager 2, if still working, arrives Sep 1989.

Tenth Planet Pioneers 10 and 11 may establish within a few years whether a 10th planet, dark star or even black hole exists at about 80 billion km beyond Pluto; IRAS data could establish this in 1984.

Asteroids

The Asteroid Belt was discovered at the end of the 18th century by astronomers who thought there might be an undiscovered planet between Mars and Jupiter since they were separated by such a disproportionate distance. Instead they found thousands of tiny planets and fragments, estimated to contain enough material to make a planet about one-thousandth the size of Earth. Astronomers have so far calculated the orbits of over 2,400, all of which have been officially numbered and many named. There may be a total of over 50,000, ranging from the largest, Ceres (955km dia), Pallas (558km dia) and Vesta (503km dia), to bodies of only 2km dia, and there are hundreds of thousands of fragments below that size.

Jupiter and its 4 Galilean moons, photographed by Voyager 1, are assembled here in their relative positions. ***(NASA)***

While the largest are spherical, many have been battered by collisions into strange shapes: 624 Hektor is believed to be a 210km-long cylinder with 50km dia, probably made of iron.

The belt of debris forms a doughnut-shaped region. In its centre the asteroids orbit the Sun at about 61,200 km/hr. The orbit is elliptical, ranging over 300-545 million km from the Sun, and the Belt is about 245 million km wide. Many meteorites which penetrate Earth's atmosphere and reach the surface are believed to be made of asteroidal material. Most are stony but some are iron; some contain large amounts of carbon and inorganic chemicals, and occasionally diamonds.

It is now thought that meteorites are collisional fragments from several parent bodies, so that each represents a different class of asteroid. A rendezvous with about 6 asteroids is needed to ensure that each class is covered.

Pioneers 10 and 11 in 1973 and Voyagers 1 and 2 in 1979 passed safely through the Asteroid Belt, disproving fears that they would be destroyed by collisions; Pioneer 10 took a few glancing blows from tiny fragments but was undamaged. The debris is very widely dispersed and Soviet scientists have calculated that a collision with a large body is likely only once in 600yr. In Nov 1975 the Russians identified an asteroid with a 10km dia and named it after the late Sergei Korolev.

NASA is already giving high priority to an unmanned asteroid-inspection mission soon after the Space Station becomes operational in 1992. Asteroids are a ready-made source of material for space colonies. It might be possible for men to select, say, an asteriod of pure nickel, attach low-powered ion engines to it and leave it to make its way over a period of about 5yr into a suitable near-Earth orbit. There it could be mined for its materials, also providing a convenient base while that was being done.

ESA has also considered (but not yet approved) a proposal to launch a 750kg asteroid probe (Asterex) by

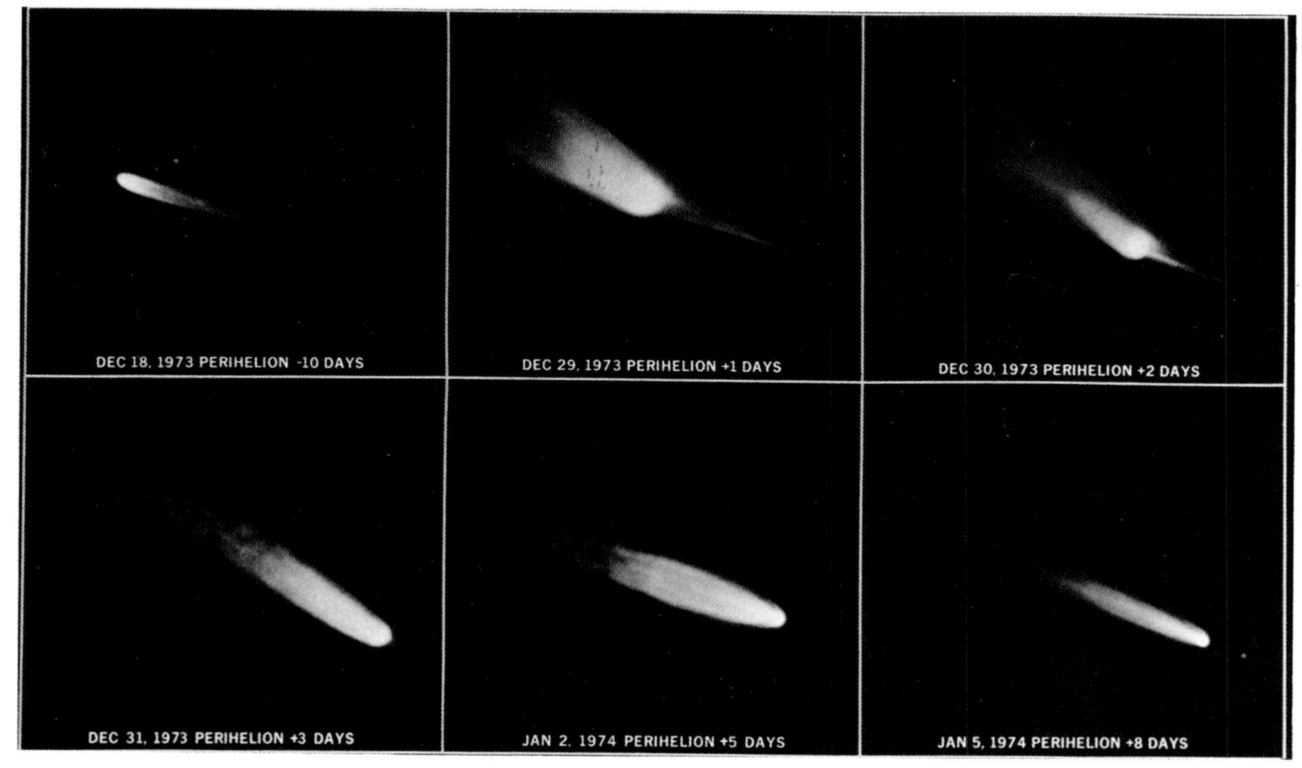

Comet Kohoutek as sketched from observations by Skylab 4 crewmen. Satellite inspection of comets will be a worldwide preoccupation for astronomers during 1985-87. ***(NASA)***

Ariane 4 on 15 Apr 1987 on a trajectory which would result in encounters with 5 different asteroids, including Ceres on 16 Sep 1989. N. D. Hulkower and D. J. Ross have suggested that man's first close-up views of an asteroid could be obtained by launching a modified Tiros N by Titan 34D/IUS in either 1985 or 1987 to rendezvous with 1943 Anteros at its aphelion, 1·4 times Earth's distance from the Sun. RCA Astro-Electronics was given a $43,000 contract to study the project. NASA is now recommending that Mariner 11 be launched by Shuttle in Jul 1990 to pass the main-belt asteroids Namaqua and Lucia, and then rendezvous with Comet Kopff, keeping station with it for several years and sometimes as close as 10km. In Jun 1983 Soviet scientists saw evidence of an atmosphere around Pallas, thought to be a temporary result of meteorite bombardment or other factors.

Comets

The most active but least known members of the solar system, the comets are estimated to number some billions, orbiting the Sun at distances of up to 50,000 Astronomical Units (AU) and with a total mass about equal to Earth's. Occasionally some are deflected within the inner planets. So far 360 different comets have been identified, with 5-10 new ones being added each year. They are believed to consist of a solid nucleus a few km in diameter and composed of a mixture of ice, condensed gases and solid particles - often described as "a dirty snowball". As the comet approaches the Sun the surface of the nucleus is heated and gas and dust expand to form the "tail" visible from Earth. The sweeping effect of the solar wind ensures that the tail is always opposite the Sun as the comet orbits. An encounter with a comet is one of the last purely exploratory missions left for man to tackle inside the solar system, which explains why Russia, Europe and Japan are all planning to intercept Halley's Comet in 1986 on the next close pass of its 76yr orbit around the Sun. A new comet on its first orbit would be preferable, but the forward planning needed for an interception means that the orbit must be well known, and Halley's visits have been observed and recorded 29 times over a 2,000yr period. Another disadvantage of a Halley interception is the fact that since its orbit is retrograde (opposite in direction to Earth's) the spacecraft's flyby velocity will be about 68km/sec or about 245,000km/hr. The Americans abandoned their Halley project on cost grounds, but decided to manoeuvre ISEE-3 in its solar orbit for a flyby of Comet Giacobini-Zinner in 1985 (see US programmes - Explorer).

Halley encounter dates and flyby distances are expected to be as follows:

6 Mar 1986 Soviet Vega 1 (10,000km)
7 Mar 1986 Japan's Planet A (2000,000km)
8 Mar 1986 Japan's MS-T5 (1 million km)
9 Mar 1986 Soviet Vega 2 (3,000km)
13 Mar 1986 ESA's Giotto (500km)

Earth

3rd planet from the Sun, at an average distance from it of 149·6 million km, Earth has an equatorial dia of 12,723km; the polar dia is 42km less. It is not therefore a true sphere, but is flattened at the poles. Equatorial circumference is 40,054km. Earth is believed to be about 4.5 billion yr old, with man himself having evolved about 1 million yr ago. Its interior is least known but probably consists of a solid inner core with a 1,206km radius and a temperature of about 10,000°F, surrounded by a liquid outer core of nickel-iron. Of the surface, 70·8% is covered with water to an average depth of 3,810m. Average land elevation is 820m. The average annual temperature on the surface is 60°F. The atmosphere consists of 78% nitrogen and 20·9% oxygen (the balance being made up of minute quantities of various other gases) for 100km from the surface, and above that there are successive layers of atomic oxygen, helium and hydrogen. Planet Jupiter probably has the same type of reducing atmosphere that Earth did when life originated here. Travelling at 30km/sec, Earth orbits the Sun in 365 days 5hr 48min 45·51sec; it completes 1 rotation on its axis in 23hr 56min 4·09sec.

Earth's Moon With an elliptical 356,334 × 406,610km orbit, the Moon takes 27 days 7hr 43min to revolve around Earth. Because its rotation about its axis is equal to its period of revolution, the Moon always shows the same face to Earth, although 59% of the lunar surface is visible at different times. With a dia of 3,475km, the Moon has a

Study of lunar rocks obtained during the Apollo landings has established that water and oxygen can be obtained for manned bases. ***(NASA)***

mass 1/81 of Earth's. Man's conquest of the Moon can be traced by studying America's Lunar Orbiter, Ranger and Surveyor projects, which were succeeded by the manned Apollo programme. Geologists are still studying the 385kg of lunar rocks and core samples brought back; only one-third of the 16 cores have yet been examined. In the case of Russia, much was learned from the Luna and Lunokhod projects. What is still needed is a polar orbiting mission, something which the European Space Agency is considering. Russia has proposed that such a mission, possibly manned, should be carried out on an international basis in the late 1980s.

Jupiter

JUPITER SPACECRAFT

Sequence	*Date*	*Spacecraft*	*Notes*
1	03.03.72	Pioneer 10	1st to Jupiter, passing at 130,000km on 4 Dec 1973. Sent pictures and data
2	05.04.73	Pioneer 11	Passed at 42,800km on 2 Dec 1974. Pictures included S Pole
3	11.09.77	Voyager 1	Passed at 278,000km on 5 Mar 1979. 19,000 pictures
4	20.08.77	Voyager 1	Passed at 650,000km on 9 Jul 1979. 15,000 pictures

By far the largest of the 9 planets, Jupiter contains 70% of the material in the solar system apart from the Sun. Its nearest approach to Earth is 627 million km. There are 16 known moons, of which the 4 Galilean satellites are the size of small planets. Accurate measurements and details of its atmosphere were finally provided by Pioneers 10 and 11, which flew past in Dec 1973 and Dec 1974. Voyagers 1 and 2, which followed in Mar and Jul 1979, discovered that Jupiter had a ring, added 3 more to the list of moons given below, and identified 8 active volcanoes on Io. Jupiter has an equatorial dia of 142,800km and a polar dia of 133,100km, the flattening being caused by the rapid rotation period of 9hr 55min. (The cloud tops at the equator have a speed of 45,600km/hr.) It orbits the Sun in 11·9 Earth years at a distance from the Sun of 778 million km, or 5·2AU.

The Great Red Spot in the southern hemisphere, visible through telescopes for several centuries, was photographed by Pioneer 10 and studied in detail by the following spacecraft. A mass of whirling clouds 20,000km wide and towering about 8km above the surrounding cloud deck, it could easily swallow 3 Earths. The Pioneer pictures revealed that there were other, smaller red spots in the Jovian atmosphere. Pioneer-inspired theories that it was the vortex of an intense hurricane which had been in progress for centuries were discarded following the Voyager flybys, when it was agreed that the spot was an anti-cyclonic formation with a flow quite different from that of Earth hurricanes.

Jupiter's atmosphere circulates much less rapidly at the poles than at the equator. As well as spectacular thunderstorms, dwarfing anything known on Earth, there is "blue sky" visible at the poles. Analysis of the seething, coloured cloud bands surrounding the planet suggests that the grey-white zones are cloud ridges of rising atmosphere - crystals of frozen ammonia - about 19km above the darker, orange-brown belts, which are troughs of descending atmosphere 19km deep and thought to consist of icy particles of ammonium hydrosulphide. As sunlight reaches this material it turns yellow, then orange, then brown, providing a partial explanation of much of the colour we see on Jupiter. Below that a 3rd cloud layer consists of liquid-water droplets suspended in hydrogen-helium. It is in this area, in the relatively stagnant polar regions, with pressures and temperatures within the limits that terrestrial organisms can tolerate, that some form of life may one day be found. It contains hydrogen, methane and ammonia - the ingredients which probably gave rise to life on Earth thousands of millions of years ago. Further down, however, about 200km below the uppermost cloud layers, temperatures reach 800°F and 100 times Earth atmospheric pressure at sea level. Beneath this, hydrogen is probably squeezed into a dense, hot fluid. About 2,900km down, pressure has reached 100,000 atmospheres and temperatures 12,000°F higher than on the Sun's surface. At 24,000km the liquid hydrogen becomes metallic, a form not known on Earth. Jupiter may have a small rocky core with temperatures rising to 54,000°F and the pressure to 40 million atmospheres.

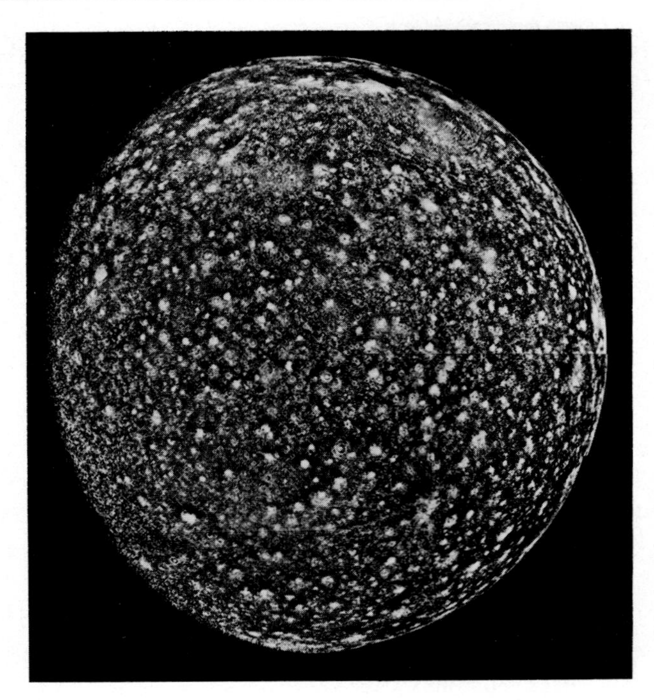

Callisto, 2nd largest of Jupiter's moons, was revealed by Voyager 1 to be uniformly cratered. The bright rays show that many of the craters are "very young". ***(NASA)***

Unlike any other planet, Jupiter radiates 2·7 times the energy it receives from the Sun, thus behaving more like a star than a planet. This heat is probably "left over" from that generated by gravitational contraction after the planets condensed some 4·6 billion yr ago.

Dr R.C. Parkinson has suggested a 5-6yr manned mission to Jupiter next century. It would concentrate on the Galilean satellites, use water tanks as radiation shielding, and set up a propellant factory making use of Callisto's water ice. Voyager pictures suggesting that Europa has small, temporary warm spots and a deep water ocean under its global ice sheet, in which some form of life is possible, add to the attractions of such a mission.

Clouds streaming counter-clockwise around Jupiter's Red Spot are shown in this Voyager 1 picture. Much remains to be learned on the Galileo mission to Jupiter. ***(NASA)***

JUPITER'S RINGS AND MOONS

Name	*Distance from planet centre (km)*	*Diameter (km)*	*Year of discovery*
Faint ring	71,900–123,700	—	1979
Bright ring	123,700–130,100	—	1979
Metis	128,000	40	1980
Adrastea	128,400	40	1979
Amalthea	181,300	155 × 270	1892
Thebe	225,000	80	1980
Io	421,600	3,632	1610
Europa	670,900	3,126	1610
Ganymede	1,070,000	5,276	1610
Callisto	1,883,000	4,820	1610
Leda	11,100,000	10?	1974
Himalia	11,470,000	170	1904
Lysithea	11,710,000	20?	1938
Elara	11,743,000	80	1905
Ananke	20,700,000	20?	1951
Carme	22,350,000	20?	1938
Pasiphae	23,300,000	20?	1908
Sinope	23,700,000	20?	1914

Mars

The Red Planet, named by the Romans after their god of war because of its blood-red hue, visible even to the unaided eye, is much further from Earth than Venus. On average, Mars is 228 million km from the Sun, travelling round it at 77,000km/hr. In its larger orbit Mars moves more slowly than Earth, so that a Martian year is 687 days 23hr. With a dia of 6,786km, Mars is about twice the size of the Moon and half the size of Earth. Maximum distance between the 2 planets, when they are on opposite sides of the Sun, is 398,887,000km, but every 2nd year the 2 planets come within about 55,784,000km of each other, providing launch windows approximately every 25 months. Depending on the initial velocity given to the spacecraft, Mars can be reached in periods varying from 259 to 105 days. Faster journeys mean higher fuel consumption by the launch rocket in order to achieve the greater velocity, and this of course can only be achieved by reducing the weight or payload of the spacecraft. It was Galileo's invention of the telescope in 1608 which enabled man to begin his study of the surface. In 1659 Christian Huygens in 1659 first sketched the dark region, Syrtis Major ("giant quicksands"), and from that established that Mars rotated on a north-south axis like Earth, with a day 30min longer than ours. The dust storms – one of which in 1971 was to ruin Russia's Mars 2 and 3 missions – were identified in the 1700s. The discovery that Mars had seasons and polar ice caps, just like Earth, inevitably led to the belief that Mars was inhabited, a conviction that was strengthened by the observations of Lowell and Schiaparelli between 1877 and 1920. Their discovery of channels or "canali" (mistranslated as "canals"), which

JPL has proposed this unmanned Mars soil-recovery mission. The return vehicle would orbit at 500km while the lander collected the soil. The ascent stage, seen at top returning to the orbiter, would be left behind after transferring the soil. ***(NASA)***

MARS SPACECRAFT

Sequence	*Date*	*Spacecraft*	*Notes*
1	10.10.60	USSR(U)	Failed to reach Earth orbit
2	14.10.60	USSR(U)	Failed to reach Earth orbit
3	24.10.62	USSR(U)	Exploded in Earth orbit
4	11.11.62	Mars 1	Passed Mars at 200,000km after communications failed
5	14.11.62	USSR(U)	Failed to leave Earth orbit
6	05.11.64	Mariner 3	No communications when shroud failed to jettison, throwing it off course
7	28.11.64	Mariner 4	Returned Mars pictures and data, passing at 9,844km. Still active
8	30.11.64	Zond 2	Communications failed, but passed Mars at 1,500km
9	18.07.65	Zond 3	Engineering test towards orbit of Mars; returned lunar far side pictures en route
10	24.02.69	Mariner 6	Returned Mars pictures; passed planet on 31 Jul 1969 at 3,215km
11	27.03.69	Mariner 7	Returned Mars pictures and data; passed planet on 5 Aug 1969 at 3,516km
12	08.05.71	Mariner 8	Failed to achieve orbit
13	10.05.71	Cosmos 419	Failed to leave Earth orbit
14	19.05.71	Mars 2	Gathered data in Martian orbit; lander, carrying the hammer-and-sickle emblem, crashed on surface 27 Nov 1971
15	28.05.71	Mars 3	Gathered data in Martian orbit; lander survived, touched down, but TV transmissions from surface on 2 Dec 1971 ended after 20sec
16	30.05.71	Mariner 9	Entered Martian orbit on 13 Nov 1971; sent over 7,000 TV pictures of planet and its moons
17	21.07.73	Mars 4	Failed to orbit
18	25.07.73	Mars 5	Operated in Mars orbit for a few days and sent some good pictures of S Hemisphere
19	05.08.73	Mars 6	Lander transmitted 1st atmospheric data but crashed on surface
20	09.08.73	Mars 7	Lander missed Mars
21	20.08.75	Viking 1	Successfully orbited; after 2 weeks' delay lander sent back 1st Mars surface pictures
22	09.09.75	Viking 2	2nd successful orbiter and lander

U = Unannounced by Soviet Union.

Arsia Mons, 19km high and 120km across, is one of Mars' 3 large Tharsis volcanoes. They are among the youngest of its surface features. This is a mosaic of 5 pictures taken by Viking Orbiter 1. ***(NASA)***

Lowell thought were the work of "intelligent creatures", led to speculation that there was life on Mars.

Such prospects were progressively diminished by the 16 spacecraft launched since 1962 by Russia in her Mars series, and by America in the Mariner and Viking series. The Mariner flybys, which disclosed details of the cratered surface with its gigantic volcanoes, established that there were no man-made channels or canals, and the Viking landers proved that there can be no advanced life forms there. But following the discovery in 1978 of living organisms inside rocks in the Antarctic, the possibility of microbial life still cannot be ruled out. It is now known that Mars is geologically active, and in a manner different from both Earth and Moon. Details of the boulder-strewn dusty landscape and pink sky are described under Mariner and Viking, as are the 2 cratered, potato-shaped moons, Phobos (dia 19km) and Deimos (dia 10km), which orbit Mars at distances of 6,100km and 19,900km. The planet's magnetic field, polar ice caps (now known to contain much water), thin atmosphere (on the surface no denser than Earth's at 30,000m) and dried-up river beds are also discussed in more detail under the entries which record their discovery. When the 1st Apollo men reached the Moon there was optimistic talk of a 12-man expedition to Mars departing on 12 Nov 1981, arriving at the planet on 9 Aug 1982, and arriving back on Earth on 19 Aug 1983. With the cutback soon afterwards in America's space programme that project was abandoned indefinitely. However, Dr R. Parkinson, speaking at a British Interplanetary Society symposium in 1981, proposed a 5-man mission to Mars starting in 1995 and lasting 560 days. It would consist of three 132-tonne vehicles docking in Earth orbit after being launched by heavy-lift Shuttle. An Apollo-like lander would take 3 crewmen to the Martian surface, and a visit to Phobos was possible during the 45-day stay, with a Venus flyby on the return trip to Earth. More immediate possibilites are a Soviet Martian orbiter planned for 1986 which will fly in formation within a few thousand metres of Phobos to determine whether it is in fact a captured asteroid; a pair of NASA/ESA orbiters launched by Shuttle/Centaur for geochemical mapping and atmospheric studies; and an unmanned sample-return mission launched from NASA's proposed space station in the 1990s, with sample examination taking place in the station as a safeguard against contaminating Earth with possible Martian micro-organisms.

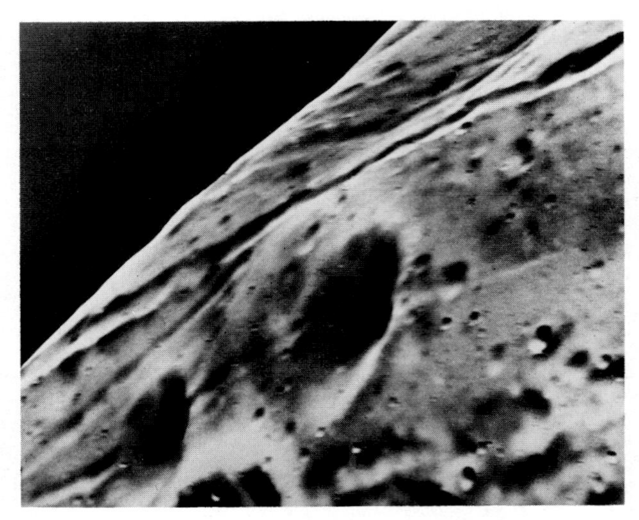

Phobos: Viking Orbiter 1 also took this close-up of Mars' satellite Phobos from only 120km. The striations, or grooves, are 100-200m wide and tens of kilometres long. ***(NASA)***

Mercury

MERCURY SPACECRAFT

Sequence	Date	Spacecraft	Notes
1	03.11.73	Mariner 10	1st to Mercury, passing at 271km on 29 Mar 1974; 2,300 pictures. 2nd and 3rd passes in Sep 74 and Mar 75. Total 10,000 pictures

Little was known about Mercury, the nearest planet to the Sun, until more than 10,000 pictures were provided by America's Mariner 10 in 11 remarkable days as the spacecraft flew past within 431km at the end of May 1975. The smallest of the planets, with a 4,878km dia, it had always been difficult for astronomers to study because of its proximity to the Sun. It was not until 1965 that radar measurements by Cornell University established that Mercury's eccentric orbit took it, at its closest, to within 46,259,000km of the Sun; it then retreats to 70,393,000km. The long-held view that Mercury always keeps the same face to the Sun proved to be wrong. In fact it rotates on its axis in 58·6 days; the combination of that and its orbit makes a Mercury day equal to 175 days on Earth. It is a heavy planet for its size, like the Moon outside but similar to Earth inside, with a density of 5·5 times that of water, the same as Earth's. Unexpectedly, it was found to have a magnetic field, though 100 times weaker than Earth's.

The Mariner 10 pictures revealed a heavily cratered, dusty surface like the Moon's, while the instruments indicated a large, heavy core of iron like Earth's. Like both Mars and the Moon, Mercury is "two-faced"; the spacecraft showed that the "in-coming" side (furthest from the Sun at that time), was heavily cratered, like the lunar highlands. Then, as it passed to the far side (nearer the Sun at that time), a less heavily cratered area was revealed, with extensive volcanic "flooding" similar to the lunar seas. Giant basins, the result of terrible meteorite impacts when the planet was young, include at least 18 which are more than 193km in dia. Most prominent of these, given the name Caloris, is 1,280km across; the shock wave appears to have passed right through the planet to shatter the surface opposite, creating an area named "Weird Terrain". The basins show no sign of erosion from water or wind, leading US geologists to conclude that Mercury has had no appreciable atmosphere for at least 4,000 million yr.

The condition of Mercury, coupled with recent knowledge of the Moon and Mars, also suggests that all the planets from Mars inwards suffered heavy bombardment at the same time. On Earth, similar cratering has been largely erased by erosion, volcanic action and movement of the crust. Other interesting features identified on the Mercurian surface include towering, curving cliffs, often 3·2km high, running for hundreds of km and cutting across crater walls and floors.

Mercury has 2 "hot poles" resulting from its elliptical orbit around the Sun and its rotation. To a man on the surface the Sun would appear to stop at Mercury's perihelion; it would then go back more than 1° before resuming its forward course. This erratic action takes about 1 Earth week, and during it the surface directly beneath the Sun gets baked longer. One of these poles is near Caloris; on them the temperature is believed to range from 800°F to -300°F - an incredible 1,100°F variation, greater than on any other planet.

Like Venus, Mercury has no moons. Dr R. C. Parkinson has suggested a manned mission to Mercury using a Venus swing-by manoeuvre, possibly starting in 1997 and lasting 421 days with 136 days in Mercury orbit. Alternatively, the return vehicle would not go into orbit but would rendezvous with the excursion module on one of the repeated approaches of the type made by Mariner 10.

Mercury: photomosaic of 18 pictures taken at 42sec intervals by Mariner 10 from 210,000km. N Pole is at top. Bright rayed craters are prominent. *(NASA)*

Neptune

The 8th planet, Neptune orbits the Sun every 164·8yr at a distance of 4,497 million km. Finally identified in 1846 as a result of deductions from irregularities in the orbit of Uranus, it is slightly smaller at 49,500km equatorial dia. Consisting largely of hydrogen and helium, it has methane in its dense atmosphere, giving it a blue-green colour. It rotates every 15hr 48min. Voyager 2 is due to reach Neptune in Aug 1989.

Neptune's moons Triton, discovered in 1846 and with a dia of over 5,000km, could prove to be the solar system's largest satellite. Orbiting at only 355,000km, it seems in danger of being torn apart by Neptune's gravity. Nereid, discovered in 1949 and with 500km dia, has a highly elliptical orbit of 1·28 × 9·6 million km.

Pluto

The outermost of the 9 planets so far discovered, Pluto orbits the Sun at an average distance of 5,900 million km, with estimates of its equatorial dia varying between 3,500 and 6,400km due to its reflective (possible methane ice) surface. It was discovered in 1930 as a result of a systematic search based on the sort of predictions that led to the discovery of Uranus. Pluto takes 247yr to orbit the Sun in an elliptical path with brings it inside Neptune's orbit; perihelion in 1989 will bring it within 4,000 million km of Earth, its closest point in 250yr. It will continue to travel inside the orbit of Neptune until 1999. This strange orbit, inclined 17° to the solar system plane, has led to speculation that Pluto might be an escaped moon of Neptune. The mystery was deepened in 1978 when astronomers at the US Naval Observatory announced the discovery of a satellite, which they named Charon; its dia was originally given as about 850km and since revised to about 2,000km. The two bodies may have a "dual planet" relationship similar to that of Earth and Moon, though with a separation distance of only about 19,000km. With no spacecraft due to visit Pluto in the foreseeable future, it is hoped that the Space Telescope may provide further information. Some Soviet astronomers have calculated that there are likely to be 2 more "transplutonian planets" beyond Pluto.

Saturn

SATURN SPACECRAFT

Sequence	Date	Spacecraft	Notes
1	05.04.73	Pioneer 11	1st to Saturn, passing at 20,800km on 1 Sep 1979. 1st close-up pictures
2	11.09.77	Voyager 1	Closest approach 138,000km on 12 Nov 1980; 17,000 pictures
3	20.08.77	Voyager 2	Closest approach 650,000km on 9 Jul 1979; 15,000 pictures

The most spectacular of the 9 planets, Saturn is 6th-furthest from the Sun at 1,425 million km compared with Earth's 150 million km. With a dia of 120,000km it is 2nd in size only to Jupiter, and since the Voyager 1 and 2 encounters in Nov 1980 and Aug 1981 it is known to have at least 23 moons in addition to its fascinating ring systems. Saturn is composed almost entirely of hydrogen and helium; rotating in only 10hr 39min 24sec, it is even more flattened than Jupiter. The cloud bands include spots like Jupiter's, and Voyager 2 saw one 7,000 × 5,000 km white oval with 100m/sec (360km/hr) circumferential winds. Near the equator the Voyagers found 1,800km/hr winds blowing primarily in an easterly direction. The astonishing complexity of the ring system is still being studied, but Voyager 1 found that the A, B and C-rings consist of hundreds of rings or ringlets, a few of which are elliptical in shape. The F-ring consists of 3 apparently interwoven rings bounded by 2 "shepherding" satellites, S13 and S14. The D (inside the C) and E (outside the F) rings were also observed, and along with F were found to have a large population of very small particles (1/10,000in). It was thought that the spokes or fingers in the B-ring might be due to fine electrically charged particles, perhaps resulting from lightning discharges within the rings. *Not* found were the predicted moonlets in the rings, which were expected to explain the gaps between them. The latest list of rings and moons, compiled by Jonathan McDowell, appears below. Titan, once thought to be the largest satellite in the solar system, is now known to be smaller than Jupiter's Ganymede, though both are larger than the planet Mercury. The only satellite known to have a dense atmosphere, Titan may be going through the same chemical processes as Earth several billion years ago. Mimas, Enceladus, Tethys, Dione and Rhea are spherical and appear to be composed mostly of water ice. Enceladus, the most active of the group, reflects almost 100% of the sunlight that strikes it. Tethys and Mimas each have an enormous impact crater one-third their diameter. Phoebe, red in colour, is unlike Saturn's other satellites in that it does not always show the same face to the planet and rotates in a retrograde direction; it may be a captured asteroid. Hyperion, hamburger-shaped, appears to have the oldest surface in the system.

Voyager 2 found that Saturn's rings were much brighter, due to the higher Sun angle, than when Voyager 1 passed 10 months earlier. ***(NASA)***

SATURN'S RINGS AND MOONS

Official name	Distance from centre of Saturn (km)	Diameter (km)	Discovery year
(Saturn radius)	60,330	—	—
D-ring	67,000–74,400	—	1980
C-ring	74,400–91,900	—	1850
B-ring	91,900–117,400	—	1659
Cassini Division	117,400–121,900	—	1675
A-ring	121,900–136,600	—	1659
Atlas	137,670	40 × 20	1980
1980 S27	139,350	140 × 100 × 80	1980
F-ring	140,300	—	1979
1980 S26	141,700	110 × 90 × 70	1980
Epimetheus	151,422	140 × 120 × 100	1980
Janus	151,472	220 × 200 × 160	1980
G-ring	170,000	—	1980
Mimas	185,540	392	1789
1981 S12	186,000?	10?	1981
Enceladus	238,040	510	1789
E-ring	180,000?–480,000?	—	1980
Tethys	294,670	1,060	1684
Telesto	294,670	34 × 28 × 26	1980
Calypso	294,670	34 × 22 × 22	1980
1981 S6	294,670	20?	1982
1981 S10	350,000	15?	1982
Dione	377,420	1,120	1684
Dione B	378,060	36 × 32 × 30	1980
1981 S7	377,500	20?	1982
(?) 1981 S9	470,000?	20?	1982
Rhea	527,100	1,530	1672
Titan	1,221,860	5,150	1655
Hyperion	1,481,000	410 × 260 × 220	1848
Iapetus	3,560,800	1,460	1671
Phoebe	12,954,000	220	1898

Note Satellites recently discovered have "provisional designations" – eg, 1981 S6 – which will later be replaced by permanent numbers and names assigned by the International Astronomical Union

Sun

The Sun has been likened to a very slow-burning hydrogen bomb. On average it is 149·6 million km from Earth, and scientists call this distance an Astronomical Unit (AU) and use it for measuring the solar system. Because of Earth's slightly elliptical orbit, the Sun's actual distance varies from 147 million km to 151 million km. A gaseous body, twisting on its axis so that the equatorial regions rotate faster than the polar caps, it has a dia of 1,390,176km compared with Earth's 12,732km. Since the Sun's mass is 330,000 times that of Earth, a man on its surface (if that were possible) would weigh 2 tonnes. Light takes 8min to reach Earth from the Sun's surface; every second 4 million tonnes of solar hydrogen transforms itself into radiant energy which eventually floods into space. The Sun has probably been consuming itself at this rate for 5,000 million yr, and current thinking is that ultimately the thermo-nuclear reactions will move outwards as they consume unused hydrogen and the Sun will become a monstrous ball of red-hot gas, big enough to engulf and destroy its 4 nearest planets, Mercury, Venus, Earth and Mars. This will probably not occur for another 5,000 million yr, however. It will then cool and shrink to a white dwarf no bigger than Earth but weighing several tonnes per cubic cm. The main body of the Sun is

Relative sizes of planet Mercury and the largest satellites in the Solar System.

surrounded by a thin shell of gas, about 322km thick and called the photosphere; most of our light comes from this. Outside this is a layer of flamelike outbursts of gas called the chromosphere; outside that again is an outer atmosphere called the corona. Men have been observing the Sun for more than 4,000yr, and as a result we know that its eruptions reach a peak every 11yr, slowly quietening down and then increasing again between the peaks.

Solar outbursts (also called solar flares) send out streams of invisible radiation and clouds of solar gas. On Earth man is protected from the worst effects because our atmosphere acts as a shield, but men in spacecraft would have to be brought home during a big solar storm because the invisible particles could penetrate their craft and damage and even destroy human cells. The biggest eruption ever recorded was on 12 Nov 1960, when American astronomers detected a huge explosion on the face of the Sun. 6hr later a gigantic cloud of hydrogen gas, 16 million km across and travelling at 6,400km/sec, collided with Earth. Its effect on Earth's atmosphere was to start worldwide electrical storms and black out all long-distance radio communications for hours, interfering with the navigation systems of ships and aircraft.

In only 3 days the Sun delivers to us as much heat and light as would be produced by burning Earth's entire oil, coal and wood reserves. Now that those reserves are running short, an increasing number of spacecraft, both manned and unmanned, are being used to develop ways of collecting that energy in outer space and beaming it down to Earth. Major advances in our understanding of the Sun were achieved by astronauts aboard America's Skylab space station in 1973-4, Soviet cosmonauts in Salyut 4 in 1975, and America's OSO and SMM satellites.

Uranus

Uranus is 3rd largest planet after Jupiter and Saturn, and at a mean distance of 2,870 million km 7th from the Sun. Its pale-green colour suggests a large amount of methane in the atmosphere, with less hydrogen and helium than Jupiter and Saturn, and a larger proportion of heavy elements. With an equatorial dia of 51,800km, Uranus spins rapidly on its side, giving it a day of about 16hr. But the year lasts 84 Earth years, and the unique tilt means that the Sun shines directly on each pole in turn for 42yr at a time. Voyager 2, if it is still working, should provide more accurate details when it arrives in Jan 1986.

Uranus's moons The planet itself was discovered in 1781, and its 5 moons, with their year of discovery and estimated distances, are as follows:

Oberon (1787) 585,600km, dia 1,690km
Titania (1787) 438,000km, dia 1,600km
Umbriel (1851) 266,900km, dia 644km
Ariel (1851) 191,600km, dia 960km
Miranda (1948) 130,000km, dia 240km

Uranus's rings Discovered in 1977 by an airborne Cornell University research team, the rings circle the planet from about 18,000-25,000km above the visible cloudtops. There are now believed to be 9 rings.

Venus

The mysteries of Venus have been penetrated within the last 10yr, principally by Russia's Venera 9 and 10 in 1975 and America's Pioneer-Venus missions in 1978. Its surface has been vividly radar-mapped and its atmosphere and racing cloud belts analysed, even if scientists are still far from understanding how these conditions came about. Similar in size and mass to Earth, Venus has very different atmospheric and surface conditions which have been a source of fascination to astronomers since Galileo first studied the planet through his newly invented telescope in 1610. Of the 9 planets, it is 2nd nearest the Sun, at a distance of 108 million km compared with Earth's 150 million km. Dia is 12,400km, compared with Earth's 12,900km. Venus has no detectable magnetic field and, like Mercury and Pluto, no known satellites. It is often so bright in a clear night sky that when appearing just above the horizon it has frequently started flying saucer scares in England, such is the effect of the sunlight reflecting from its heavy cloud layers. Mariner 10's ultraviolet pictures revealed in 1974 that the cloud layers move around at different speeds. Comparison with data from the Pioneer-Venus Probes and Orbiter 4yr later showed that the wind patterns change dramatically over such periods. Wind speed, only 3-18km/hr from the surface up to 10km altitude, rises to 192km/hr at 50km and

VENUS SPACECRAFT

Sequence	Date	Spacecraft	Notes
1	04.02.61	Sputnik 7	Earth-orbiting platform, failed to launch probe
2	12.02.61	Venus 1	Communications failed, but passed Venus at 100,000km
3	22.07.62	Mariner 1	Destroyed by range safety officer at 161km altitude
4	25.08.62	USSR(U)	Failed to leave Earth orbit
5	27.08.62	Mariner 2	Returned data, passing Venus at 34,830km
6	01.09.62	USSR(U)	Failed to leave Earth orbit
7	12.09.62	USSR(U)	Failed to leave Earth orbit
8	11.11.63	Cosmos 21	Engineering test only; failed to leave Earth orbit
9	27.03.64	Cosmos 27	Failed to leave Earth orbit
10	02.04.64	Zond 1	Communications failed; passed Venus at 100,000km
11	12.11.65	Venus 2	Communications failed; passed Venus at 23,810km
12	16.11.65	Venus 3	Communications failed; impacted on Venus
13	23.11.65	Cosmos 96	Failed to leave Earth orbit
14	12.06.67	Venus 4	Returned atmospheric data on 18 Oct 1967 during descent and impact
15	14.06.67	Mariner 5	Returned data during flyby at 3,990km on 19 Oct 1967
16	17.06.67	Cosmos 167	Failed to leave Earth orbit
17	05.01.69	Venus 5	Returned atmospheric data during descent and impact on 16 May 1969
18	10.01.69	Venus 6	Returned atmospheric data during descent and impact on 17 May 1969
19	17.08.70	Venus 7	Reached surface and returned data for 23min on 15 Dec 1970
20	22.08.70	Cosmos 359	Insufficient velocity to leave Earth orbit
21	27.03.72	Venus 8	Reached Venusian surface and returned data for 50min on 22 Jul 1972
22	31.03.72	Cosmos 482	Failed to leave Earth orbit
23	03.11.73	Mariner 10	Venus/Mercury flyby. 6,800 pictures of Venus, Mercury, Earth and Moon; 1st pictures of Venus; 1st mission to Mercury
24	08.06.75	Venus 9	1st pictures from Venus surface; operated 53min after landing
25	14.06.75	Venus 10	Lander survived 65min on surface; sent more pictures
26	20.05.78	Pioneer Venus 1	Orbiter; radar-mapped 93% of surface; continues to 1985
27	08.08.78	Pioneer Venus 2	5 probes; confirmed greenhouse theory
28	09.09.78	Venera 11	Analysed clouds and atmosphere; no pictures
29	14.09.78	Venera 12	As Venera 11
30	30.10.81	Venera 13	1st surface soil analysis and colour pictures
31	04.11.81	Venera 14	2nd soil analysis

U = Unannounced by Soviet Union.

Venus: artist's concept of the Ishtar Terra continent-sized highland region is based on topography measurements by Pioneer Venus Orbiter. The mountain on the right, higher than Everest at 10·6km, has been named Maxwell Montes. ***(NASA)***

360km/hr at 65km above the equator. The latter figure is about 60 times as fast as the planet's own very slow rotation around its axis. Still unexplained is the mystery, discovered by American radar measurements in 1962, of why it is that, whereas Earth spins on its axis once a day, Venus turns lazily in the opposite direction once every 243 Earth days. At the same time it revolves around the Sun once every 225 Earth days, resulting in a solar day (from one sunrise to the next) of 117 Earth days. The Sun rises in the west and sets in the east. When Venus is closest, only the dark side faces us.

With major volcanic regions at Beta Regio and Aphrodite Terra, Venus is now considered to be the most active body in the solar system after the Earth and Io. The main clouds consist of sulphuric acid particles 2 microns in diameter, with the particles in the haze layer only a quarter of that size. The haze layer, which can be 30km thick, has been observed to appear and disappear over a several-year period. It is this which acts as the "sealer" for the Venus greenhouse effect, now thought to be proved. The 96% carbon dioxide atmosphere plays the greatest role in trapping the heat, now measured on the surface at 482°C (900°F). For a period it was thought that the thick clouds would result in a "gloomy greenhouse" effect on the surface, like the dullest winter day on Earth. For this reason, Russia equipped her Venus 9 and 10 spacecraft with searchlights to illuminate their landing areas for the benefit of the TV cameras. In fact, the surface proved to be "as bright on Venus as in Moscow on a cloudy June day". Details could be clearly seen at a distance of 50–100m. Neither Venus 9 nor Venus 10, when they landed 2,200km apart in Oct 1975, raised a dust cloud, but the 90kg Pioneer Day Probe, landing 3yr later in the southern hemisphere at 35km/hr, did so, and it took 3min to settle.

Venera 10 recorded a pressure of 92 Earth atmospheres (1,352lb/sq m). The surface, as radar-mapped by Pioneer Venus Orbiter, is 60% flat, rolling plain but includes 2 continent-sized highland areas, one of which has a mountain massif higher than Everest. There is also a 2,250km-long rift valley. Soviet and US scientists continue to explore and map Venus and have been exchanging information despite poor political relations between their governments; the Russians changed the site for their Venera 13 and 14 landers after studying the Pioneer Venus radar images.

WORLD SPACE CENTRES

SUMMARY OF WORLD-WIDE SUCCESSFUL SPACE LAUNCHES BY SITE 1957-1983

Launch Site	*Earth orbit*		*Lunar or Escape*		*Cumulative*
	1983	*1957-1982*	*1983*	*1957-1982*	*Total*
Plesetsk (USSR)	62	869	0	0	931
Tyuratam (USSR)	30	541	2	53	626
Vandenberg AFB (USA)	11	441	0	0	452
Cape Canaveral (USA)	11	275	0	58	344
Kapustin Yar (USSR)	4	75	0	0	79
Wallops Island (USA)	0	18	0	0	18
Uchinoura (Japan)	1	12	0	0	13
Shuang Cheng Tse (China)	1	10	0	0	11
Kourou (French Guiana)	2	9	0	0	11
Tanegashima (Japan)	2	9	0	0	11
Indian Ocean Platform (Kenya)	0	8	0	0	8
Hammaguir (Algeria)	0	4	0	0	4
Sriharikota (India)	1	2	0	0	3
Woomera (Australia)	0	2	0	0	2
Totals	**125**	**2,275**	**2**	**111**	**2,513**

Australia

Woomera About 430km N of Adelaide and running eastwards for 2,000km across the desert, the Woomera missile and rocket range was started jointly by Britain and Australia in 1946 and cost them well over £200 million. Australian hopes that what was originally a test range for ballistic missiles and sounding rockets, and for shooting down pilotless aircraft, would have a future as a launch centre for satellites were not fulfilled. During the period when Britain's Blue Streak was being developed as the main stage of ELDO's Europa rocket, a thriving township of 4,500, with over 500 houses supplied with power and water across 160km of desert, sprang up. But Woomera was not suitable for equatorial launches and the French preferred to develop their own centre at Kourou. After successive cuts in its contributions to the range, Britain announced that there would be no more work for Woomera after 1976. NASA also decided to close down its Carnarvon tracking station, which had played a major part in manned and unmanned space flights.A small test satellite, launched into polar orbit by an American Redstone in 1967, and Britain's Prospero in 1971 are likely to be the only 2 satellites ever orbited from Woomera.

Brazil

Barreira do Inferno (Barrier of Hell) Launch site, located near Natal, for Brazil's Sonda sounding rockets and future orbital launcher. Otrag, the private German company, is interested in the use of this equatorial site for its modular launcher.

China

Jiuquan This site is known in the West as Shuang Cheng Tse, from its position in central China approx 1,600km W of Peking and 800km E of Lop Nor. The Chinese Academy of Space has recently given its official name and some details, describing it as "a fairly large launch site" in Gansu Province, NW China, with tracking and telemetry equipment for monitoring and controlling launch vehicles and satellites during their ascent. It was added that the main control centre for the nationwide telemetry tracking and command stations was located in Weinan, Shaanxi Province, central China. There is also an oceangoing telemetry/tracking/command (TTC) vessel capable not only of tracking and monitoring China's satellites, but of controlling re-entry at precisely designated areas.

Landsat 2 pictures showed what was originally identified as Shuang Cheng Tse as a small facility, with little expansion in progress. Orbital missions are launched SE, with military IRBMs fired into the Lop Nor impact site. A gap of nearly 4yr in launches between Jan 1978 and Sep 1981 coincided with reports that a new launch centre was being built further away from the Soviet border. This gap was no doubt also a result of the recently announced explosion at "Launchpad No 5" on 28 Jan 1978. The existence of a new site was confirmed by the China 14 and 15 launches in Jan and Apr 1984. It lies about 1,280km S of Jiuquan, at or near 31°N, and is more suitable for attaining synchronous orbits.

France

Kourou In 1964 France decided to transfer her national launch centre from Hammaguir in the Sahara Desert to Kourou, French Guiana. Though hot and wet, its position 2°N of the Equator makes it ideal for launching synchronous satellites and gives it a 17% payload advantage over C Canaveral for eastward launches. Both E and N, vehicles can travel 3,000km without passing over land. 1st firing was the 1st Veronique sounding rocket on 9 Apr 1968, with 1st satellite launch in Mar 1970. With the 1966 decision to transfer ELDO (later ESA) activities from Woomera, Australia, to the more favourable Kourou site, development into a space centre somewhat comparable to C Canaveral began. Installations now extend over a 30km strip of waterfront and include 3 launchpads. One is for sounding rockets. The 2nd, disused since 1976, handled Diamant, which placed 7

Kourou: only 2°N of the Equator, the French launch centre has a 17% payload advantage over Cape Canaveral for eastward launches. *(CNES)*

Kourou launch centre, showing monitoring and command consoles. *(ESA)*

satellites in orbit. The 3rd, originally built for the abandoned Europa 2, was converted for Ariane and used for the 4 test flights between Dec 1979 and Dec 1981, 3 of which were successful. The 1st successful operational launch took place in Jun 1983. A 2nd Ariane launch site (ELA-2), due for completion in 1984, is 500m S of ELA-1, will be able to accommodate the more advanced Ariane 4 and will permit 12 launches per year. Radar and telemetry stations are located on a hill 20km away at Montagne de Peres, with a tracking radar on Ile Royale, one of 3 islands forming the disused penal colony known as Devil's Island.

Hammaguir This military base at Colomb Bechar in Algeria's Sahara Desert was used as a French rocket testing and development site from the mid-1950s until she was required to evacuate the base in 1967 following Algerian independence. It had an IRBM test corridor stretching 3,000km SE towards Fort Lamy. About 300 rockets were fired in upper-atmosphere and missile tests, some of them carrying high-explosive charges to the upper atmosphere. France developed the liquid-propellant boosters Emeraude, Saphir and Diamant at Hammaguir, and 4 small test satellites were orbited during 1965-1967.

India

Sriharikota Located in the Bay of Bengal, the Shar Centre, operated by the Indian Space Research Organisation, carried out its first successful SLV-3 launch on 18 Jul 1980. It has facilities for launching and flight-testing multi-stage sounding rockets as well as satellites. The main Indian ground station for tracking and commanding ISRO satellites and receiving their data is also located here.

Italy

San Marco Owned and operated by the Italian Government, this consists of 2 offshore platforms in Formosa Bay 4·8km off the coast of Kenya and 2·9°S of the Equator. It became operational in 1966 and by mid-1976 8 satellites had been launched: 4 Italian, 1 British and 3 American. The launch platform has 20 steel legs sunk into the sandy seabed, and is linked by 23 cables to the Santa Rita control platform 920km away. This is a modified oil rig. Explorer 42, the Small Astronomy Satellite launched by an Italian crew on 12 Dec 1970, was the 1st US satellite to be launched by a foreign country. An example of the benefits of this equatorial site is that it was possible to

Italy's San Marco launch platforms, located 4·8km off the Kenyan coast, were due to be used in 1984 for the 1st time since 1976. *(NASA)*

launch the 195kg Explorer 43 with a Scout rocket instead of the much larger booster needed to achieve the same orbit from Cape Canaveral. In an addition to the launch ramp for Scouts, there is a ramp for US Nike Apache and Nike Tomahawk sounding rockets. San Marco was due to be brought back into use in 1984, 8yr after the last orbital launch, for 2 joint NASA/Italian spacecraft.

Japan

Kagoshima, Uchinoura Japan's first 6 satellites were launched from these levelled hilltops facing the Pacific Ocean on the S tip of Kyushu Island. Construction of a site for launching of sounding rockets began in Feb 1962, with extensions for scientific satellite launchings by M rockets completed in 1966. The first Lambda 3 rocket was fired in 1964, but it was 6yr later before the first successful orbital launch was achieved. The launch pads are 220m above sea level, but local fishermen complained about the noise and hazards associated with the launches over their fishing grounds. The government decided that improved and safer launch facilities should be provided at Tanegashima for the bigger N rockets, designed to launch applications satellites, although Kagoshima remained available to Tokyo University for launching scientific satellites.

Tanegashima The 1st phase of the Tanegashima Space Centre was completed in 1974 in time for the launch of Japan's 7th satellite, ETS-1, by the new N rocket. Tanegashima is a 58km-long island about 100km S of Kagoshima. Appropriately, Tanegashima became the first Japanese word for "firearm" because it was there that Fernao Mendes Pinto landed in 1542 carrying a musket. NASDA's applications satellites are launched from the main Osaki site by N rockets, the first N-2 firing being Kiku 3 in Feb 1981. The nearby Takesaki site is used for sounding rockets, and since 1978 has conducted a series of materials-processing tests in preparation for Japan's participation in US Space Shuttle flights. But here too fishermen's objections have resulted in launches being limited, to Feb and Aug each year.

Tanegashima: Japan's new space centre, completed in 1974, is idyllically sited on a sparsely populated island. But fishermen, protecting their traditional areas, have succeeded in restricting launches to February and August. *(NASDA)*

Sweden

Esrange, Kiruna Operated by the Swedish Space Corporation in the far N of Sweden. The Swedish-German Microgravity Sounding Rocket Programme provides ESA members with 6min of microgravity conditions per flight in preparation for Spacelab and Eureca (orbital platform) missions.

USA

CSOC The Consolidated Space Operations Centre (CSOC), being built for USAF Space Command at Peterson AFB, Colorado, will become the focal point of military space activity and will take over Shuttle operations in emergencies such as an unscheduled landing. One element, due to become operational in 1986, will be a satellite control facility both duplicating and interoperable with Sunnyvale. Its functions will include control of new systems such as MilStar, Navstar and DSC-3. The 2nd main element will duplicate the Johnson Space Centre at Houston; it should have Shuttle planning capability in 1987, and Shuttle control capability in 1990.

Dryden Flight Research Centre This is located 100km N of Los Angeles, on the edge of an 88-sq km dry lake bed in the Mojave Desert and on part of Edwards Air Force Base, famous for the 1950s test flights of the X-series research aircraft and lifting-bodies. It proved ideal for the Shuttle Approach and Landing Tests in 1978, with plenty of safe overrun room if the 8-17km runways proved insufficient. 3 of the 4 orbital flight tests also landed successfully here; STS-3 in Mar 1982 had to be diverted to White Sands because Edwards had for once been flooded by heavy rain. All 1st flights of Orbiters will terminate at Edwards.

Goddard Space Flight Centre Located 24km NE of Washington DC and built in 1958, this is mainly known for its Satellite Situation Reports, based on analysis of information from the Spacecraft Tracking and Data Network. It is also responsible for research and development of meteorological and applications satellites.

Johnson Space Centre, Houston Control of all manned spaceflights passes from Launch Control Centre (LCC) at C Canaveral to Johnson Space Centre (formerly Manned Spacecraft Centre), 32km SE of Houston, 10sec after lift-off. Until 1961 it was undeveloped cattle country; in 3yr it became the world's most sophisticated scientific

Orbiter *Columbia* making its 1st landing, at NASA's Dryden Flight Centre, Edwards Air Force Base. Chief Astronaut John Young is among an influential group who think Dryden is preferable as prime landing site because of Kennedy's unpredictable weather. *(NASA)*

JSC's Mission Operations Control Room (MOCR) studies a replay of STS-8. Here astronauts complete their training as spacecraft communicators: left to right, O'Connor, Hoffman, Gardner, William Fisher. *(NASA)*

centre. Its focal point, Mission Control Centre, has been the control point for all NASA's manned flights since Gemini 4 in June 1965. Flight controllers at 171 consoles can check, at a touch of a button, the progress of the mission against the flightplan; 133 TV cameras and 557 receivers are used. JSC is also responsible for design, development and testing of manned spacecraft and their systems, as well as the selection and training of astronauts, who live nearby in the Clear Lake and Key Largo areas It has now been named as the lead centre for Space Station development.

Kennedy Space Centre, Cape Canaveral A strip of sandy jungle on the Florida coastline known as Cape Canaveral was originally set aside as a development area for ICBMs in 1947. What became the USAF's Eastern Test Range, now known as the Eastern Space and Missile Centre, was joined in 1963 by NASA's Kennedy Space Centre on nearby Merritt Island. By 1983 344 launchings, including all US manned flights, had been made from the 2 sites. C Canaveral was renamed C Kennedy following the late President's assassination, but after several years of local protests and presentations to Congress the Cape's original name was restored, though the NASA centre continues to be named after the President. About 40 launch complexes have been built, all except Pads 39A and B by the USAF; until Apollo the USAF conducted NASA's manned launches from its own pads. The Merritt Island "Moonport," however, was developed and operated exclusively by NASA for the Apollo flights, the Skylab missions and the Space Shuttle test flights. That situation is changing now that Shuttle missions with DoD payloads are starting from Pad 39.

Bus tours are operated by NASA, and on Sundays visitors are allowed to drive their own cars to the launch sites.

Launch complexes involved in manned spaceflight are as follows:

5-6 Site of first 2 US suborbital manned flights, Mercury-Redstone 3 and 4. Complex now a museum piece.

14 Used for Mercury-Atlas 6-9, the four orbital flights, starting with John Glenn in Feb 1962. Blown up in 1976 when corrosion made it dangerous.

19 All 10 Gemini-Titan flights launched from in 1965-6. Site now inactive.

34 Site of 1st Saturn 1 launch. Altogether 7 Saturn 1 and 1B launches took place here, including Apollo 7, in Oct 1968, the project's first manned flight. It was also the scene of the Apollo disaster; Grissom, White and Chaffee died when fire broke out in their spacecraft during what should have been a final rehearsal for the Apollo 7 launch. Site now inactive and preserved as a memorial to them.

37 Site of 8 Saturn 1 and 1B launches, but no manned flights. Inactive.

39 Site of all manned Apollo flights except the 1st (Apollo 7), of Skylab launches and ASTP. Its construction reversed the fixed launch concept of earlier missions, under which assembly, checkout and launch were all carried out on the pad. With the mobile concept employed on Complex 39, assembly is carried out in the huge Vehicle Assembly Building (VAB), which is 160m tall and so large that 4 buildings the size of the United Nations in New York would fit inside. The Launch Control Centre (LCC) nearby has 4 firing rooms, each with 470 sets of control and monitoring equipment, plus conference and display rooms, etc. During launch 62 TV cameras at the pad supply 100 monitor screens. The Mobile Launcher, 136m high and 4,808 tonnes in weight, serves as an assembly platform within the VAB and as a launch platform and umbilical tower at the launch site 5·6km away. The Mobile Launcher is moved along a special roadway by the Crawler-Transporter, a double-tracked vehicle the size of a football field, 40m long and 35m wide. The assembled Apollo-Saturn 5 vehicle which it carried stood 36 storeys high and the total load weighed 7,711 tonnes.

Access to both the Apollo spacecraft and Saturn vehicle at the pad was provided by the Mobile Service Structure (MSS). 125m high, this had 2 lifts and provided 5 platforms, the top 3 for spacecraft access and the lower 2 for Saturn 5. About 11hr before Apollo launches the MSS was moved back 2,134m from Pad A. Pads A and B are almost identical, although Pad B was used only once, for Apollo 10, up to the end of the Apollo project. They are 2,657m apart and roughly octagonal in shape. Below them are flame trenches 13m deep and 137m long. For emergency, a 61m escape tube led from the bottom of the lifts to a blast-resistant room 12m below the pad. For Space Shuttle launches this has been replaced with a system of 5 slidewires leading to bunkers 3-400m to the W of the pad.

Following the conclusion of the Skylab and ASTP flights major modifications were carried out on Pad 39A and in the VAB to adapt them for Shuttle assembly and launch, and a 4,500m runway was laid down among Merritt Island's orange and grapefruit groves. Modifications and improvements to Pad 39B, incorporating the lessons of the test flights, are due to be completed by Oct 1986 in time for the Galileo and ISPM launches. The aim is to support a possible 40 launches per year by the end of 1986.

Cape Canaveral: Landsat 3 photo from 917km altitude shows: A: Apollo/Shuttle launchpads; B: VAB; C: Shuttle runway; D: Unmanned launchpads; E: NASA HQ building. It also demonstrates how detailed the much better spysat pictures must be. *(NASA)*

Complex 14 at Cape Canaveral, from which Glenn became the 1st US astronaut in orbit in 1963, had to be blown up in 1976 because of dangerous corrosion caused by salt winds. *(UPI)*

Marshall Space Flight Centre Established as the Army Ballistic Missile Agency at Huntsville, Alabama, in 1955, this is best known for its development of the Saturn 5 rocket which sent men to the Moon. The Saturns were developed here by Dr Wernher von Braun, who arrived with his team of 118 Germans at Ft Bliss, Texas, in 1945 ("Operation Paperclip"). They were transferred to Huntsville in 1954. Marshall also developed the Lunar Roving Vehicle. It is now in charge of the Space Telescope; other work includes space platforms and solar power concepts.

Space Technology Centre Established by the USAF at Kirtland AFB in Oct 1982, this is responsible for develoments in laser research, spacecraft survival, propulsion and the military use of the space environment. It manages and focuses the space technology work by Air Force laboratories on weapons at Kirtland, geophysics at Hanscom, Mass, and rocket propulsion at Edwards AFB, Calif. Most urgent studies at present concern the need to protect microprocessor electronics from natural radiation at a time when smaller and lighter components tend to become more instead of less susceptible to radiation; improvements, through experiments on a DSCS-3 satellite, to improve the ability of spacecraft to switch automatically to back-up systems when threatened or damaged; and the speeding up of development of more efficient gallium arsenide solar cells and higher-energy nickel-hydrogen batteries.

Sunnyvale This USAF satellite control centre, located in California, came into the news during STS-4, when the Shuttle carried its first DoD payload. Classified communications and crew instructions were passed through NASA at Houston. The USAF Consolidated Space Operations Centre (CSOC) at Peterson AFB,Colorado, now being built, is expected to be additional.

Vandenberg Air Force Base From Oct 1985 this California location will provide the Space Shuttle with a much overdue polar launch site for use by NASA as well as the US Air Force. About 160km N of Los Angeles, it had seen 441 launches by the end of 1982. Since Vandenberg is also perhaps the most important US base for operational ICBMs, sunk deep in protective underground silos, it is inevitably the most secret of America's 3 launch centres. The base has been created along 40km of coastal desert and scrub, similar to the C Canaveral area. From here are launched the Big Bird and KH11 spy satellites. From here too ICBMs are test-fired under operational conditions over the Western Test Range,extending 8,000km across the Pacific to the tip of S America. About 10,000 service personnel and their families live in the isolated area behind the base. From its southern tip, Point Arguello, spacecraft are launched into polar orbits for both NASA and the USAF; polar launches from C Canaveral are not permitted, since this would involve overflight of the US mainland.

Vandenberg's Space Launch Complex 6 (SLC-6) was

Kennedy's 4,500m runway has to be checked for wandering alligators and bobcats before being used. But it was a mixture of "conservative operation" and weather conditions that resulted in only 1 Shuttle landing there in the 1st 12 missions. *(NASA)*

Vandenberg: a rare view of the Western Test Range, with a simultaneous test firing of 2 Minuteman missiles. When Shuttle launches from VAFB start at the end of 1985 manned spaceflight will at last have access to polar orbits. *(USAF)*

originally built in the early 1960s for the Gemini based Manned Orbiting Laboratory, cancelled in Jun 1969. SLC-6 is being converted at a cost of $2·5 billion for Space Shuttle use. Delays increased to over 3yr when it was found that the high winds and low temperatures encountered at Vandenberg would not permit the planned assembly of SRBs, ET and Orbiter on the launchpad. At the end of 1981 a 71m-high, $40 million movable assembly building, to protect the Shuttle cluster and technicians during launch preparations was approved. Lengthening of the existing 2,438m runway, located 25·7km N of the launchpad, to 4,572m for Shuttle landings was completed by the end of 1983. The journey from there to the pad will take 3-4hr. After launches the SRBs will be recovered from the Pacific about 200km off Vandenberg and towed to Port Hueneme, 125km S, for cleaning and disassembly before being shipped back to the manufacturer. Eventually Vandenberg will have capacity for about 20 launches per year into high-inclination polar-type orbits. 8 launches between Oct 1985 and Sep 1987 are currently planned, but because the maintenance and checkout facilities will not be complete until 1987 the Orbiter will have to be ferried to Vandenberg from C Canaveral for the first 3 launches. At lift-off control will pass from Vandenberg to the new CSOC in the same way as Kennedy passes control to Johnson.

Wallops Flight Centre Off Virginia's E shore and connected to the mainland by a causeway, Wallops Island was the 3rd US launch site after C Canaveral and Vandenberg. Established in 1945, it was the scene of many early launch vehicle tests, including the Little Joe series in support of Project Mercury. 19 scientific satellites have been orbited by Scouts from Wallops, the last of them Britain's Ariel 6 in Jun 1979. Future satellite launches look unlikely, but since 1945 more than 8,000 sounding rockets have been launched, including 54 on one occasion in 1974 in an elaborate upper-atmosphere experiment; 6 launch areas and 24 launch installations were used.

White Sands Missile Range Test site in New Mexico for sounding rockets and short-range vertical firings, White Sands has also been designated as either an abort-once-around or end-of-mission landing site for the Space Shuttle. Its Northrup Strip, 1,200m above sea level and with 2 gypsum-sand runways each about 11km long, was used for the STS-3 landing on 30 Mar 1982 after Edwards became waterlogged just before the mission started; the landing at Northrup had to be delayed one day because of a dust storm. The Strip is regularly used as a training site for astronauts, and the STS-3 crew were said to have made 800 practice landings there.

USSR

Kaliningrad The Soviet Union's Manned Spaceflight Control Centre, N of Moscow at Kaliningrad, employs about 2,000 people. The main control room, not unlike its counterpart at Houston in its array of 2 display screens and consoles, is responsible for controlling Salyut space stations and their supporting Soyuz and Progress missions. A recently admitted *Aviation Week* reporter said that its map display showed the coverage area of Russia's ship-based tracking station near Cuba. A 2nd, smaller control room is responsible for Soyuz and Progress flights to and from Salyuts, while a "psychological support" room prepares radio programmes, music and messages for transmission to Salyut crews.

Kapustin Yar This was Russia's 1st rocket development centre, preceding the much more important centres at Tyuratam and Plesetsk. Its first orbital launch was Cosmos 1 in 1962. In 1975 a Landsat 2 picture revealed that its facilities extend over a 96 × 72km area NE of the River Volga and 965km SE of Moscow. During its early years Kapustin Yar was used for testing captured V2 rockets and for sounding-rocket experiments carrying dogs and other research animals to heights of 500km. By 1980 there had been 70 orbital launches, mostly small scientific Cosmos satellites, but as work was switched to Plesetsk the launch rate fell to an average of only 1 per year. Recent activities have involved anti-ballistic missiles, which are fired through the atmosphere so that they fall back towards the Sary Shagan missile base 2,000km to the E, near the Chinese border. There was a surprise in Jun 1982 when Cosmos 1374, believed to be the first test of a re-usable Shuttle, was launched from Kapustin Yar.

Plesetsk This is Russia's most important military launch site, equivalent to America's Vandenberg. It is located 170km S of Archangel, a northerly position which enables communications and spy satellites to be placed in polar and highly elliptical orbits. Far more payloads - over 800 by 1982 - have been launched from here than from any other space centre in the world. Its existence was first established in the West by Geoffrey Perry of Kettering Grammar School following its first orbital launch, Cosmos 112 on 17 Mar 1966. Perry and his science students noted that the satellite could not have been launched from either Tyuratam or Kapustin Yar. Landsat pictures taken in 1973 showed it to be about 100km long, with at least 4 launch complexes and heavily defended by surface-to-air missiles. The scale of its activities can be judged from the fact that in 1980, a typical year, Tyuratam had 24 launches compared with Plesetsk's 68, nearly all of them military.

Space ships Because her spacecraft are within direct visibility of Soviet territory for only 9 out of every 24hr, Russia has compensated for her lack of land bases by building up a flotilla of space ships. Since 1960 more than 10 have been brought into service, operating under the flag of the USSR Academy of Sciences. The flagship, *Kosmonaut Yuri Gagarin* (45,000 tonnes), has 75 computer-controlled aerials and 86 laboratories. Communications can be maintained simultaneously with 2 or more satellites, manned craft or orbital stations; signals can be picked up with precision even during Force 7 gales. It is claimed that the ships enable space vehicles to be brought down and, if necessary, recovered almost anywhere in the world. The ships' activities are closely monitored by Western navies, since their positioning provides advance warning of manned and other important launchings.

Star City The Gagarin Cosmonaut Training Centre was established at Star City, 30km NE of Moscow, in 1960. It has flight simulators for both Salyut 7 and Soyuz

T space vehicles, and in 1982 about 50 cosmonauts were training here under the commander, Gen Beregovoi. Foreign cosmonauts are also trained here in a course lasting up to 2yr. Facilities include classrooms, laboratories, scale models of spacecraft and orbital stations, a sports training complex, centrifuges and thermal chambers. Recent additions include a new pool, 23m in dia and 12m deep and containing scale models of both Soyuz and Salyut, which can be used by cosmonauts and engineers to work out EVAs and many other operations.

Tyuratam (Baikonur) The Soviet equivalent of C Canaveral, with about 80 pads from which all manned flights and unmanned planetary missions are launched. Missile and rocket tests began in the area in the early 1950s, with construction of the pad from which both Sputnik 1 and Yuri Gagarin were launched starting in 1955. A 2nd pad with manned launch capability was completed about 10yr later about 30km from the 1st. The location of the site was pinpointed by a Japanese astronomer in 1957 as being near Tyuratam in Kazakhstan, E of the Aral Sea and about 370km SW of Baikonur. But despite this the Russians continued for at least 17yr to give its latitude and longitude as that of the town of Baikonur, though during the ASTP mission in 1975 it was finally admitted that "the highway [leading to the site] passes near the Tyuratam railway station". Though Tyuratam is much busier than C Canaveral (see launch table on page 281), Plesetsk, the main military site, is busier still. Not until Soyuz T-6 in Jun 1982 was a Western journalist admitted to Tyuratam, but no doubt America's Big Bird spy satellites have provided detailed information for the military, while Landsat photographs have given the public some idea of the vast sprawl of launchpads. The Novosti press agency once said that it could not boast a mild climate, being hot in summer and suffering violent snowstorms and 40°C frosts in winter. However, the cosmonauts have comfortable accommodation in cottages, and for pre-launch checks and activities there is a hotel complex with swimming pool and sports centre. The pad used for Yuri Gagarin's flight is about 32km from Leninsk; 2 small houses used by Gagarin and spacecraft designer Sergei Korolev, "who preferred to stay there during his frequent visits to Baikonur rather than at the hotel," are now memorials. It is presumably these houses which are traditionally visited by cosmonaut teams before their missions. Another tradition is that cosmonauts plant trees in the park before their flights; those planted by Gagarin and Titov are now 25yr old. The tradition was followed by the 3 US astronauts on their brief visit before the 1975 ASTP flight. Leninsk, 2,100km or a 3hr jet flight from Moscow, now has a population of over 50,000. Three railway tracks lead to the manned launchpads: 1 for the launcher and spacecraft, which are moved horizontally instead of vertically as in the case of America's Apollo and Space Shuttle, and 2 for the rocket fuels. Launch equipment is "housed in several storeys beneath the concrete pad". Adjacent is the command bunker, with periscopes giving it a submarine-like appearance. Though no Soviet reference to them has ever been made, it is known that 2 massive launch complexes have been built for the SL-X (designated "G" in America) giant booster, which is said to be more powerful than Saturn 5. A pad had to be rebuilt when an SL-X exploded during a fuelling test in 1969, and 2 launch failures followed in 1971 and 1972. In mid-1976 testing was again under way in a 122m static rig, but 8yr later there had still been no launch.

SPACEMEN

American astronauts

History 10 groups of pilots and scientists, totalling 144, were selected by NASA between Apr 1959 and May 1984 as follows:

Group	*Date*	*Total*	*Details*
1	Apr 1959	7	The Mercury astronauts
2	Sep 1962	9	Test pilots
3	Oct 1963	14	Pilots
4	Jun 1965	6	Scientists
5	Apr 1966	19	Pilots
6	Aug 1967	11	Scientists
7	Aug 1969	7	USAF pilots transferred to NASA when the Manned Orbiting Laboratory (MOL), a military version of Gemini, was cancelled
8	Jan 1978	35	20 mission specialists (incl 6 women); 15 pilots
9	May 1980	19	11 mission specialists (incl 2 women); 8 pilots
10	May 1984	17	10 MS; 7 pilots

Of the first 3 groups, only John Young remained on flight status long enough to fly on the Shuttle. Deke Slayton, last survivor of the famous "Mercury Seven," retired in 1982 when his contract was not renewed.

By the end of Jun 1984 76 were on flight status; 35 of those were pilots and 41 were mission specialists. Now that the Space Shuttle is operational the duty of pilots is to operate the Shuttle Orbiters, manoeuvring them in Earth orbit and flying them back for a runway landing. Mission specialists have responsibility for co-ordinating Shuttle crew activity, consumables usage and experiment operations. Their duties include EVAs and use of the RMS. A 3rd category - payload specialists, who are selected for individual missions - is dealt with separately. A new group of military astronauts - likely to be known as mission specialists because they will be engineers and scientists rather than test pilots, and with the task of handling classified payloads - was being quietly selected by DoD in 1982. Their numbers may rise to more than 50, and the 1st should have flown on the since cancelled STS-10. With a number of spare seats available in the late 1980s, NASA established a task force in late 1982 to study the possibility of using them for "citizens' flights". These will not be just for journalists, it is emphasised, though *Aviation Week*'s Craig Covault appears to be under training for the 1st.

Age is proving no bar to spaceflight: Slayton was 51 when he made his 1st flight, and Young was the same age when he commanded the 1st Shuttle mission.

NASA now has 3 married couples available for flights: Drs William and Anna Fisher were joined in May 1981 by Robert Gibson and Rhea Seddon, who married 3yr after selection, and in Jul 1982 by Sally Ride and Steve Hawley.

Of the 49 astronauts no longer on the active list, 9 are dead: 3 were lost in the Apollo fire during a ground rehearsal, but not one in space. By the end of Shuttle mission 41-B 68 astronauts had flown. 1 MOL astronaut, selected in Aug 1967 but not included in the list above since he elected to remain with the USAF instead of transferring to the NASA astronaut pool, is now Lt-Gen James Abrahamson, who became Associate Administrator for Space Flight and head of Shuttle development in Nov 1981.

The alphabetical list gives each astronaut's name, date of birth, resignation date or flight status, missions flown or assigned, if any, details of career following resignation,

Deke Slayton in 1975. (NASA)

Rhea Seddon. ***(NASA)***

US ASTRONAUTS IN ORDER OF 1ST MISSION

No	Name	Total flights	Total duration (days, hr, min)	Missions
1	A. Shepard (5th man on Moon)	2	09.00.17	M3, A14
2	V. Grissom	2	00.05.08	M4, G3
3	J. Glenn	1	00.04.55	M6
4	S. Carpenter	1	00.04.56	M7
5	W. Schirra	3	12.07.13	M8, G6, A7
6	G. Cooper	2	09.09.16	M9, G5
7	J. Young (9th)	6	34.19.41	G3, G10, A10, A16, STS-1, STS-9
8	J. McDivitt	2	14.02.57	G4, A9
9	E. White	1	04.01.56	G4
10	C. Conrad (3rd)	4	49.03.38	G5, G11, A12, SL2
11	F. Borman	2	19.21.36	G7, A8
12	J. Lovell	4	29.19.05	G7, G12, A8, A13
13	T. Stafford	4	21.03.43	G6, G9, A10, AS
14	N. Armstrong (1st)	2	08.14.00	G8, A11
15	D. Scott (7th)	3	22.18.54	G8, A9, A15
16	E. Cernan (11th)	3	23.14.15	G9, A10, A17
17	M. Collins	2	11.02.05	G10, A11
18	R. Gordon	2	13.03.53	G11, A12
19	E. Aldrin (2nd)	2	12.01.53	G12, A11
20	D. Eisele	1	10.20.09	A7
21	W. Cunningham	1	10.20.09	A7
22	W. Anders	1	06.03.00	A8
23	R. Schweickart	1	10.01.01	A9
24	A. Bean (4th)	2	69.15.45	A12, SL3
25	J. Swigert	1	05.22.55	A13
26	F. Haise	1	05.22.55	A13
27	S. Roosa	1	09.00.42	A14
28	E. Mitchell (6th)	1	09.00.42	A14
29	J. Irwin (8th)	1	12.07.12	A15
30	A. Worden	1	12.07.12	A15
31	T. Mattingly	2	18.03.01	A16, STS-4
32	C. Duke (10th)	1	11.01.51	A16
33	R. Evans	1	12.13.51	A17
34	H. Schmitt (12th)	1	12.13.51	A17
35	J. Kerwin	1	28.00.50	SL2
36	P. Weitz	2	33.01.13	SL2, STS-6
37	O. Garriott	2	69.18.56	SL3, STS-9
38	J. Lousma	2	67.11.14	SL3, STS-3
39	G. Carr	1	84.01.15	SL4
40	E. Gibson	1	84.01.15	SL4
41	W. Pogue	1	84.01.15	SL4
42	V. Brand	3	22.02.59	AS, STS-5, 41-B
43	D. Slayton	1	09.01.28	AS
44	R. Crippen	3	15.18.25	STS-1, 7, 41-C
45	J. Engle	1	02.06.13	STS-2
46	R. Truly	2	08.07.22	STS-2, STS-8
47	C. Fullerton	1	08.00.05	STS-3
48	H. Hartsfield	1	07.01.10	STS-4
49	R. Overmyer	1	05.02.14	STS-5
50	J. Allen	1	05.02.14	STS-5
51	W. Lenoir	1	05.02.14	STS-5
52	K. Bobko	1	05.00.23	STS-6
53	D. Peterson	1	05.00.23	STS-6
54	S. Musgrave	1	05.00.23	STS-6
55	F. Hauck	1	06.02.24	STS-7
56	S. Ride	1	06.02.24	STS-7
57	J. Fabian	1	06.02.24	STS-7
58	N. Thagard	1	06.02.24	STS-7
59	D. Brandenstein	1	06.01.09	STS-8
60	D. Gardner	1	06.01.09	STS-8
61	G. Bluford	1	06.01.09	STS-8
62	W. Thornton	1	06.01.09	STS-8
63	B. Shaw	1	10.07.47	STS-9
64	R. Parker	1	10.07.47	STS-9
65	R. Gibson	1	07.23.17	41-B
66	B. McCandless	1	07.23.17	41-B
67	R. Stewart	1	07.23.17	41-B
68	R. McNair	1	07.23.17	41-B
69	F. Scobee	1	06.23.40	41-C
70	T. Hart	1	06.23.40	41-C
71	J. van Hoften	1	06.23.40	41-C
72	G. Nelson	1	06.23.40	41-C
	Totals (to 41-C)	**111**	**1,198.20.43**	**42**

Notes Payload specialist flights (2 totalling 20.15.34) not included. Abbr: M=Mercury, G=Gemini, A=Apollo, AS=Apollo-Soyuz, SL=Skylab, STS=Space Transportation System (Shuttle).

US FLIGHT SUMMARY (TO 41-C)

Astronauts		Flights
1	man has made 6 flights	6
3	men have made 4 flights	12
5	men have made 3 flights	15
15	men have made 2 flights	30
47	men have made 1 flight	47
1	woman has made 1 flight	1
72	**Totals**	**111**

and marital status. Back-up and support roles are not given.

Not included here are 6 of the 8 X-15 pilots who officially qualified as astronauts by flying higher than 80km; the other 2, Neil Armstrong and Joe Engle, became NASA astronauts later. Also excluded is Jerrie Cobb, who nearly became the world's first woman in space under a 1959 program called Women in Space Soonest (WISS). Considered at the time to be "a unique and exceptional astronaut candidate" by reason of her speed and altitude records as a pilot, she began training but NASA soon afterwards dropped the idea of sending women into space.

ALDRIN, Edwin E. (20 Jan 1930). Col USAF (Ret). Res Jul 1971. Gp 3. Plt G12, Nov 1966, 5½hr EVAs. 2nd man on Moon as LMP A11, Jul 1969. Pres, Research and Engineering Consultants Inc, LA. M 3 ch; div and Re-M.

ALLEN, Joseph P. (27 Jun 1937). Civ, PhD (Physics). Flight, MS. Gp 6. NASA A/Admin Legislative Affairs, 1975-78. MS STS-5, Nov 1982. Selected for 41-H, originally scheduled for Sep 1984. M 2 ch.

ANDERS, William A. (17 Oct 1933). Col USAF (Reserve). Res Sep 1969. Gp 3. LMP A8, Dec 1968, 1st manned circumlunar flight. Atomic Energy Comm, 1973; Nuclear Regulatory Comm, 1975; US Ambassador, Norway, 1976-7; General manager, GEC Aircraft Equipment, Jan 1980. M 6 ch.

ARMSTRONG, Neil A. (5 Aug 1930). Civ. Res 1971. Gp 2. Cdr G8, Mar 1966, performed 1st space docking with Agena. Cdr A11 and 1st man on the Moon, Jul 1969. Prof of Engineering, Cincinnati U, 1971-80; Cardwell Intl Ltd. M 2 ch.

BAGIAN, James P. (22 Feb 1952). Civ, MD. Flight, MS. Gp 9. Selected for 61-D, Jan 1986.

BASSETT, Charles A. (30 Dec 1931). Maj USAF. Dead. Gp 3. Killed with Elliott See in T-38 jet crash at St Louis in Feb 1966 while training as G9 pilot.

BEAN, Alan L. (15 Mar 1932). Capt USN (Ret). Res Jun 1981. Gp 3. LMP A12, Nov 1969, 4th man on Moon. Cdr Skylab 3, Jul 1973. Head, Shuttle Operations and Training. Full-time painter. M 2 ch; Div.

BLAHA, John E. (26 Aug 1942). Col USAF. Flight, plt. Gp 9.

BLUFORD, Guion S. (22 Nov 1942). Lt-Col USAF. Flight, MS. Gp 8. MS1, STS-8, Aug 1983. M 2 ch.

Left to right: Lt-Col Guion Bluford, Dr Ronald McNair, Col Frederick Gregory and Maj Charles Bolden Jr. *(NASA)*

BOBKO, Karol J. (23 Dec 1937). Col USAF. Flight, plt. Gp 7. Plt, STS-6, Apr 1983. Selected as cdr 41-F, Aug 1984. M 2 ch.

BOLDEN, Charles F. (19 Aug 1946). Maj. USMC. Flight, plt. Gp 9.

BORMAN, Frank (14 Mar 1928). Col USAF (Ret). Res Jul 1970. Gp 2. Cdr G7, Dec 1965, 1st RV with G6. Cdr A8, Dec 1968, 1st circumlunar flt. Pres, Eastern Airlines, 1975. M 2 ch.

BRAND, Vance D. (9 May 1931). Civ. Flight, plt. Gp 5. CMP ASTP, Jul 1975. Cdr STS-5, Nov 1982. Cdr 41-B, Feb 1984. M 4 ch; Div and Re-M 1 ch.

BRANDENSTEIN, Daniel C. (17 Jan 1943). Cdr USN. Flight, plt. Gp 8. Plt STS-8, Aug 1983. Selected as cdr 51-A, Oct 1984. M1 ch.

BRIDGES, Roy D. (19 Jul 1943). Lt-Col USAF. Flight, plt. Gp 9.

BUCHLI, James F. (20 Jun 1945). Maj USMC. Flight, MS. Gp 8. Selected for 51-C, Dec 1984. M 2 ch.

BULL, John S. (25 Sep 1934). Lt-Cdr USN (Ret). Res Jul 1968 due to pulmonary disease. Gp 5. Research scientist, aircraft navigation and guidance, Ames. M 2 ch.

CARPENTER, M. Scott (1 May 1925). Cdr USN (Ret). Res Aug 1967. Gp 1. Plt M7, May 1962. Pres, Sea Sciences Corp. M 4 ch; Div and Re-M.

CARR, Gerald P. (22 Aug 1932). Col USMC (Ret). Res Jun 1977. Gp 5. Cdr Skylab 4, Nov 1973, at 84 days longest US flight. V-p, Bovay Engineers. M 6 ch.

CERNAN, Eugene A. (14 Mar 1934). Capt USN (Ret). Res Jul 1976. Gp 3. Plt G9, Jun 1966, 2hr EVA. LMP A10, May 1969. Cdr A17, Dec 1972, 11th man on Moon. Exec V-p, Intl Coral Petroleum. M 1 ch.

CHAFFEE, Roger B. (15 Feb 1935). Lt-Cdr USN. Dead. Gp 3. Died in Apollo 1 fire on launchpad, Jan 1967.

CHANG, Franklin R. (5 Apr 1950). Civ, PhD (plasma physics). Flight, MS. Gp 9.

CHAPMAN, Philip K. (5 Mar 1935). Civ, ScD. Res Jul 1972. Gp 6. M 2 ch; Div.

CLEAVE, Mary L. (5 Feb 1947). Civ, PhD. Flight, MS. Gp 9. Selected for 51-D, Feb 1985.

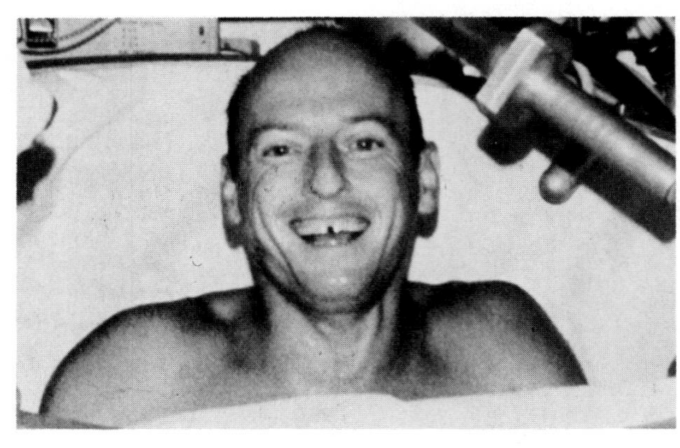

Pete Conrad takes a shower during Skylab 2. ***(NASA)***

COATS, Michael L. (16 Jan 1945). Cdr USN. Flight, plt. Gp 8. Selected as plt 41-D, Jun 1984. M 1 ch.

COLLINS, Michael (31 Oct 1930). Maj-Gen USAF (Reserve). Res Jan 1970. Gp 3. Plt G10, Jul 1966. CMP A11, Jul 1969, in orbit during 1st Moon landing. Dir, Nat Air and Space Mus, 1971-78; V-p, Vought Corp. M 3 ch.

CONRAD, Charles "Pete" (2 Jun 1930). Capt USN (Ret). Res Feb 1974. Gp 2. Plt G5, Aug 1965. Cdr G11, Sep 1966. Cdr A12, Nov 1969, 3rd man on Moon. Head Skylab project and Cdr Skylab 2, May 1973. V-p marketing, Douglas Aircraft Corp. M 4 ch.

COOPER, L. Gordon (6 Mar 1927). Col USAF (Ret). Res Jul 1970. Gp 1. Plt M9, May 1963. Cdr G5, Aug 1965. V-p, WED Enterprises. M 2 ch; Div and Re-M.

COVEY, Richard O. (1 Aug 1946). Maj USAF. Flight, plt. Gp 8. Selected as plt 51-C, Dec 1984. M 2 ch.

CREIGHTON, John O. (28 Apr 1943). Cdr USN. Flight, plt. Gp 8. Selected as plt 51-A, Oct 1984.

CRIPPEN, Robert L. (11 Sep 1937). Capt USN. Flight, plt. Gp 7. Plt *Enterprise* ALTs 2 and 4, Sep and Oct 1977. Plt STS-1, Apr 1981. Cdr STS-7, Jun 1983. Cdr 41-C, Apr 1984. Selected as cdr 41-G, Oct 1984. M 3 ch.

US EVA RECORD

		In-flight		*Lunar exploration*		*Lunar-stay time*	
Mission	*Launch date*	*Astronaut*	*hr, min*	*Astronaut*	*hr, min*	*hr, min*	*Site*
Gemini 4	03.06.65	White	00.23				
Gemini 9	03.06.66	Cernan	02.10				
Gemini 10	18.07.66	Collins	01.27				
Gemini 11	12.09.66	Gordon	02.44				
Gemini 12	11.11.66	Aldrin	05.30				
Apollo 9	03.03.69	Schweickart	00.46				
Apollo 11	16.07.69			Armstrong	02.40	21.36	Sea of
				Aldrin	02.15		Tranquillity
Apollo 12	14.11.69			Conrad	07.45	31.31	Ocean of
				Bean	07.45		Storms
Apollo 14	31.01.71			Shepard	09.17	33.31	Fra Mauro
				Mitchell	09.17		
Apollo 15	26.07.71	Worden	00.38	Scott	18.35	66.54	Hadley
				Irwin	18.35		Apennine
Apollo 16	16.04.72	Mattingly	01.13	Young	20.15	71.14	Descartes
				Duke	20.15		
Apollo 17	07.12.72	Evans	01.06	Cernan	22.04	74.59	Taurus-
				Schmitt	22.04		Littrow
Skylab 2	25.05.73	Conrad	06.34				
		Kerwin	03.58				
		Weitz	02.11				
Skylab 3	28.07.73	Bean	02.41				
		Garriott	13.42				
		Lousma	11.01				
Skylab 4	16.11.73	Carr	15.48				
		Gibson	15.17				
		Pogue	13.31				
STS-6	07.04.83	Musgrave	04.10				
		Peterson	04.10				
STS 41-B	07.02.84	McCandless	12.12				
		Stewart	12.12				
STS-41-C	09.04.84	Nelson	10.04				
			(MMU 00.35)				
		van Hoften	10.04				
			(MMU 00.50)				
Totals		**24**	**153.32**	**12**	**82.51**	**299.45**	

Ronald Evans (right), Dr Harrison Schmitt (left) and Eugene Cernan (front), crew of Apollo 17. Schmitt was the only scientist to get to the Moon. *(NASA)*

CUNNINGHAM, Walter (16 Mar 1932). Civ. Res Aug 1971. Gp 3. LMP A7, Oct 1968. Sr V-p, The Capital Gp, 1979. M 2 ch.

DUKE, Charles M. (3 Oct 1935). Brig-Gen USAF (Ret). Res Jan 1976. Gp 5. LMP A16, Apr 1972, 10th man on Moon. Pres, Southwest Wilderness Art Inc. M 2 ch.

DUNBAR, Bonnie J. (3 Mar 1949). Civ. Flight, MS. Gp 9.

EISELE, Donn F. (23 Jun 1930). Col USAF (Ret). Res Jul 1972. Gp 3. CMP A7, Oct 1968. Eastern mgr, Marion Power Shovel Co. M 5 ch.

ENGLAND, Anthony W. (15 May 1942). Civ, PhD (geology). Flight, MS. Gp 6. Res Aug 1972 to join US Geological Survey, rejoined NASA 1979. Selected for 51-F/Spacelab 2, Mar 1985. M 2 ch.

ENGLE, Joe H. (26 Aug 1932). Col USAF. Flight, plt. Gp 5. Cdr *Enterprise* ALTs 2 and 4, Sep and Oct 1977. Cdr STS-2, Nov 1981. Selected as cdr 51-C, Dec 1984. M 2 ch.

EVANS, Ronald E. (10 Nov 1933). Capt USN (Ret). Res Mar 1977. Gp 5. CMP A17, Dec 1972. Mgr Space Products, Sperry Flt Systems. M 2 ch.

FABIAN, John M. (28 Jan 1938). Lt-Col USAF. Flight, MS. Gp 8. MS STS-7, Jun 1983. Selected as MS 51-A, Oct 1984, and 2nd plt 61-D, Jan 1986. M 2 ch.

FISHER, Anna L. (24 Aug 1949). Civ, MD. Flight, MS. Gp 8. Selected for 41-H, Sep 1984. M (to astronaut William Fisher).

FISHER, William F. (1 Apr 1946). Civ, MD. Flight, MS. Gp 9. Selected for 51-C, Dec 1984. M (to astronaut Anna Fisher).

FREEMAN, Theodore C. (18 Feb 1930). Capt USAF. Dead. Gp 3. Killed when T-38 collided with snow goose at Houston, Oct 1964.

FULLERTON, C. Gordon (11 Oct 1936). Col USAF. Flight, plt. Gp 7. Plt *Enterprise* ALTs 1, 3 and 5, Aug, Sep and Oct 1977. Plt STS-3, Mar 1982. Selected as cdr 51F/Spacelab 2, Mar 1985. M 2 ch.

Left to right: James Lovell, William Anders and Frank Borman, 1st men around the Moon, on Apollo 8. *(NASA)*

GARDNER, Dale A. (8 Nov 1948). Lt-Cdr USN. Flight, MS. Gp 8. MS2 STS-8, Aug 1983. Selected for 41-H, Sep 1984. M 1 ch.

GARDNER, Guy S. (6 Jun 1948). Maj USAF. Flight, plt. Gp 9. M.

GARRIOTT, Owen K. (22 Nov 1930). Civ, PhD (electrical engineering). Flight, MS. Gp 4. Sc plt Skylab 3, Jul 1973. Dir Space and Life Sciences, JSC, 1977. MS STS-9/Spacelab 1, Nov 1983. M 4 ch.

GIBSON, Edward G. (8 Nov 1936). Civ, PhD (engineering and physics). Gp 4. Res Nov 1974, rejoined NASA Mar 1977, res Oct 1980. Sc plt, Skylab 4, Nov 1973, at 84 days longest US flight. M 4 ch.

GIBSON, Robert L. (30 Oct 1946). Lt-Cdr USN. Flight, plt. Gp 8. Plt 41-B, Feb 1984. M 1 ch; Div and Re-M (to astronaut Rhea Seddon, May 1981).

GIVENS, Edward G. (5 Jan 1930). Maj USAF. Dead. Gp 5. Killed in car accident near Houston, 1967.

GLENN, John H. (19 Jul 1921). Col USMC (Ret). Res Jan 1964. Gp 1. M6, Feb 1962, 1st US man in orbit. Sen (Dem) Ohio, 1974, and Presidential candidate. M 2 ch.

GORDON, Richard F. (5 Oct 1928). Capt USN (Ret). Res Jan 1972. Gp 3. Plt G11, Sep 1966. CMP A12, Nov 1969. Pres, EDCO Inc. M 6 ch.

GRABE, Ronald J. (13 Jun 1945). Maj USAF. Flight, plt. Gp 9.

GRAVELINE, Duane E. (2 Mar 1931). Civ, MD. Res Aug 1965. Gp 4. Resigned for personal reasons; in practice in Colchester, Vermont. M 5 ch.

GREGORY, Frederick D. (7 Jan 1941). Lt-Col USAF. Flight, plt. Gp 8. Selected as plt 51-B, Nov 1984. M 2 ch.

GRIGGS, Stanley D. (7 Sep 1939). Civ. Flight, plt. Gp 8. Selected for 41-F, Aug 1984. M 2 ch.

GRISSOM, Virgil I. (3 Apr 1926). Lt-Col USAF. Dead. Gp 1. Plt M4, Jul 1961. Cdr G3, Mar 1965, 1st US 2-man craft. Died in Apollo launchpad fire, 27 Jan 1967.

HAISE, Fred W. (14 Nov 1933). Civ. Res Jun 1979. Gp 5. LMP A13, Apr 1970. Cdr *Enterprise* ALTs 1, 3 and 5, Aug, Sep and Oct 1977. V-p Space Progs, Grumman Aerospace Corp. M 4 ch.

HART, Terry J. (27 Oct 1946). Civ. Flight, MS. Gp 8. As MS1 on 41-C, Apr 1984, used RMS to make world's 1st satellite recovery. Res Jun 1984 to join Bell Labs. M 1 ch.

HARTSFIELD, Henry W. (21 Nov 1933). Col USAF (Ret). Flight, plt. Gp 7. Plt STS-4, Jul 1982. Selected as cdr 41-D, Jun 1984. M 2 ch.

HAUCK, Frederick B. (11 Apr 1941). Capt USN. Flight, plt. Gp 8. Plt STS-7, Jun 1983. Selected as cdr 41-H, Sep 1984. M 2 ch.

HAWLEY, Steven A. (12 Dec 1951). Civ, PhD (astronomy and astrophysics). Flight, MS. Gp 8. Selected for 41-D, Jun 1984. M (to astronaut Sally Ride, Jul 1982).

HENIZE, Karl G. (17 Oct 1926). Civ, PhD (astronomy). Flight, MS. Gp 6. Selected for 51-F/Spacelab 2, Mar 1985. M 4 ch.

HILMERS, David C. (28 Jan 1950). Capt USMC. Flight, MS. Gp 9.

HOFFMAN, Jeffrey A. (2 Nov 1944). Civ, PhD (astrophysics). Flight, MS. Gp 8. Selected for 41-F, Aug 1984.

Dr Anna Fisher and her husband Dr William Fisher. *(NASA)*

Left: Gus Grissom. Right: Dr William Lenoir.

Left: Jim Lovell. The Apollo 13 explosion deprived him of his day on the Moon. Right: Judith Resnik, 2nd US woman in space.

HOLMQUEST, Donald L. (7 Apr 1939). Civ, MD, PhD (physiology). Res Sep 1973. Gp 6. Leave of absence May 1971 to become A/Prof, Radiology and Physiology, Baylor Col of Med, Houston; now has own practice at Nuclear Lab, Navasota, Tex. M 1 ch; Div and Re-M.

IRWIN, James B. (17 Mar 1930). Col USAF (Ret). Res Aug 1972. Gp 5. LMP A15, Jul 1971, 8th man on Moon. Left NASA to found and become Pres, High Flight Foundation, religious organisation, Colorado Springs. M 4 ch.

KERWIN, Joseph P. (19 Feb 1932). Capt USN, MD. Flight, MS. Gp 4. Sc plt SL2, May 1973. Head Shuttle MS, 1976. NASA Rep, Australia, 1982. M 3 ch.

LEESTMA, David C. (6 May 1949). Lt-Cdr USN. Flight, MS. Gp 9. Selected for 41-G, Aug 1984.

LENOIR, William B. (14 Mar 1939). Civ, PhD (electrical engineering). Flight, MS. Gp 6. MS STS-5, Nov 1982. M 2 ch.

LIND, Don L. (18 May 1930). Civ, PhD (physics). Flight, plt. Gp 5. Shuttle development. Selected as MS 51-B, Nov 1984. M 7 ch.

LLEWELLYN, John A. (22 Apr 1933). Civ, PhD (chemistry). Res Aug 1968. Gp 6. Born Cardiff, became US citizen 1966. Resigned for personal reasons. Prof Energy Conversion, U of S Fla. M 3 ch.

LOUNGE, John M. (28 Jun 1948). Civ, PhD. Flight, MS. Gp 9. Selected for 51-C, Dec 1984.

LOUSMA, Jack R. (29 Feb 1936). Col USMC. Res 1983. Gp 5. Plt SL3, Jul 1973. Cdr STS-3, Mar 1982. Senate candidate (R) for Michigan. M 3 ch.

LOVELL, James A. (25 Mar 1928). Capt USN (Ret). Res 1973. Gp 2. 1st man to fly 4 missions, 715hr 5min in space. Plt G7, Dec 1965. Cdr G12, Nov 1966. Cdr A8, Dec 1968, 1st circumlunar flt. Cdr A13, Apr 1970, aborted Moon mission. D/Dir Sc and Applications, JSC Houston, 1971; Pres, Fisk Telephone Systems, Houston. M 4 ch.

LUCID, Shannon W. (14 Jan 1943). Civ, PhD (biochemistry). Flight, MS. Gp 8. Selected for 51-A, Oct 1984. M 3 ch.

MATTINGLY, Thomas K. (17 Mar 1936). Capt USN. Flight, plt. Gp 5. Assigned CMP A13 but replaced following German measles contact. CMP A16, Apr 1972. Cdr STS-4, Jul 1982. Selected as cdr 41-E (cancelled). M 1 ch.

McBRIDE, Jon A. (14 Aug 1943). Cdr USN. Flight, plt. Gp 8. Selected as plt 41-G, Aug 1984. M 3 ch.

McCANDLESS, Bruce (8 Jun 1937). Capt USN. Flight, plt. Gp 5. Shuttle EVA specialist. MS3 41-B, Feb 1984, world's 1st untethered EVA. M 2 ch.

McDIVITT, James A. (10 Jun 1929). Brig-Gen USAF (Ret). Res Sep 1972. Gp 2. Cdr G4, Jun 1965. Cdr A9, Mar 1969, 1st LM flight. Manager, Apollo Prog, 1969. Pres, Pullman Standard Co, Chicago. M 4 ch.

McNAIR, Ronald E. (21 Oct 1950). Civ, PhD (physics). Flight, MS. Gp 8. MS1 41-B, Feb 1984. M 1 ch.

MICHEL, F. Curtis (5 Jun 1934). Civ, PhD (physics). Res Aug 1969. Gp 4. Prof Physics, 1969, then Chmn, Space Physics and Astronomy, Rice U. M 2 ch.

MITCHELL, Edgar D. (17 Sep 1930). Capt USN (Ret). Res Oct 1972. Gp 5. LMP A14, Feb 1971, 6th man on Moon. Founder and Chmn, Inst of Noetic Science, Palo Alto, Ca; and Pres, Edgar Mitchell Corp, Palm Bea, Fla. M 2 ch; Div and Re-M.

MULLANE, Richard M. (10 Sep 1945). Lt-Col USAF. Flight, MS. Gp 8. Selected for 41-D, Jun 1984. M 3 ch.

MUSGRAVE, F. Story (19 Aug 1935). Civ, MD, PhD (physiology). Flight, MS. Gp 6. MS STS-6, Apr 1983, 1st *Challenger* flight. Selected for 51-F, Mar 1985. M 5 ch.

NAGEL, Steven R. (27 Oct 1946). Maj USAF. Flight, plt. Gp 8. Selected for 51-A, Oct 1984. M.

NELSON, George D. (13 Jul 1950). Civ, PhD (astronomy). Flight, MS. Gp 8. MS3 41-C, Apr 1984. M 2 ch.

O'CONNOR, Bryan D. (6 Sep 1946). Maj USMC. Flight, plt. Gp 9. Former Harrier plt. Selected as plt 51-D, Feb 1985.

O'LEARY, Brian T. (27 Jan 1949). Civ, PhD (astronomy). Res Apr 1968 for personal reasons. Gp 6. Nuclear Energy Conslt, House of Rep; research member of Dept of Physics, Princeton U, NJ. M 2 ch.

ONIZUKA, Ellison S. (24 Jun 1945). Capt USAF. Flight, MS. Gp 8. Selected for STS-10 (cancelled). M 3 ch.

OVERMYER, Robert F. (14 Jul 1936). Col USMC. Flight, plt. Gp 7. Plt STS-5, Nov 1982, 1st operational Shuttle. Selected as cdr 51-B, Nov 1984. M 3 ch.

PARKER, Robert A. (14 Dec 1936). Civ, PhD (astronomy). Flight, MS. Gp 6. MS1 STS-9/Spacelab 1, Nov 1983. M 2 ch.

PETERSON, Donald H. (22 Oct 1933). Col USAF (Ret). Flight, MS. Gp 7. MS STS-6, Apr 1983, 1st *Challenger* flight. M 3 ch.

POGUE, William R. (23 Jan 1930). Col USAF (Ret). Res 1979. Gp 5. Plt SL4, Nov 1973, longest US flight at 84 days. Res NASA Sep 1975 to become V-p, High Flight Foundation (see Irwin); returned to NASA Apr 1976 to work on ER payloads. M 3 ch.

RESNIK, Judith A. (5 Apr 1949). Civ, PhD (electrical engineering). Flight, MS. Gp 8. Selected for 41-D, Jun 1984.

RICHARDS, Richard N. (24 Aug 1946). Lt-Cdr USN. Flight, plt. Gp 9.

RIDE, Sally K. (26 May 1951). Civ, PhD (physics). Flight, MS. Gp 8. MS STS-7, Jun 1983, 1st US woman. Selected for 41-G, Aug 1984. M (to astronaut Steven Hawley, Jul 1982).

ROOSA, Stuart A. (15 Aug 1933). Col USAF (Ret). Res Feb 1976. Gp 5. CMP A14, Jan 1971. Pres, Jet Industries, Austin, Tex, 1977. M 4 ch.

ROSS, Jerry L. (20 Jan 1948). Capt USAF. Flight, MS. Gp 9. Selected for 51-D, Feb 1985.

SCHIRRA, Walter M. (12 Mar 1923). Capt USN (Ret). Res Jul 1969. Gp 1. Plt M8, Oct 1962. Cdr G6, Dec 1965, 1st RV with G7. Cdr A7, Oct 1968, 1st Apollo flight. Only man to fly in M, G and A. V-p Development, Goodwin Coys, Middleton, Co. M 2 ch.

SCHMITT, Harrison H. "Jack" (3 Jul 1935). Civ, PhD (geology). Res Aug 1975. Gp 4. LMP A17, Dec 1972, only scientist and 12th man on Moon. NASA A/Admin Energy, 1974. US Sen (R) for New Mexico, 1976-82.

SCHWEICKART, Russell L. (25 Oct 1935). Civ. Res Aug 1979. Gp 3. LMP A9, Mar 1969, 1st Apollo EVA. Asst, Payload Ops, 1976; Sc Adviser, Calif Governor, 1977. M 5 ch.

SCOBEE, Francis R. (19 May 1939). Maj USAF (Ret). Flight, plt. Gp 8. Plt 41-C, Apr 1984. M 2 ch.

Left: Alan B. Shepard, 1st US man in space and the only Apollo astronaut to go to the Moon. Right: Lt-Gen Tom Stafford, who commanded the Apollo-Soyuz mission.

Left: Lt-Col Edward White. Right: Charles Walker, the Shuttle's 1st paying passenger. His task on Mission 12 was to produce the first usable space medicine.

SCOTT, David R. (6 Jun 1932). Col USAF (Ret). Res Oct 1977. Gp 3. Plt G8, Mar 1966, world's 1st docking. CMP A9, Mar 1969, 1st CSM/LM docking. Cdr A15, Jul 1971, 7th man on Moon. D/Dir, Dryden, 1973; Dir 1975. Pres, Scott Sc & Tech, Los Angeles. M 2 ch.

SEDDON, Margaret R. "Rhea" (8 Nov 1947). Civ, MD. Flight, MS. Gp 8. Selected for 61-D/Spacelab 4, Jan 1986. M (to astronaut Robert Gibson, May 1981) 1 ch.

SEE, Elliott M. (23 Jul 1927). Civ. Dead. Gp 2. Training as cdr G9 when killed with Bassett in T-38 crash at St Louis, Feb 1966.

SHAW, Brewster H. (16 May 1945). Maj USAF. Flight, plt. Gp 8. Plt STS-9/Spacelab 1, Nov 1983. Selected as cdr 51-D, Feb 1985. M 3 ch.

SHEPARD, Alan B. (18 Nov 1923). R-Adm USN (Ret). Res Aug 1974. Gp 1. M3, May 1961, 1st US man in space. Lost flight status due to ear trouble. Chief of Astronaut Off, 1963. Restored to flight status, 1969. Cdr A14, Feb 1971, 5th man and only Mercury astronaut on Moon. Chmn, Windward Co, Deer Pk, Tex. M 2 ch.

SHRIVER, Loren J. (23 Sep 1944). Maj USAF. Flight, plt. Gp 8. Selected as plt, STS-10 (cancelled). M 4 ch.

SLAYTON, Donald K. "Deke" (1 Mar 1921). Maj USAF (Ret). Res Mar 1982. Gp 1. Replaced as cdr M7 due to heart murmur. Dir Flt Crew Ops until 1974. Flight status restored 1972. DMP ASTP, Jul 1975. Mgr, Shuttle Orbital Flt Tests until Mar 1982, when contract not renewed. Mission Dir, Conestoga, Jul 1982. M 1 ch.

SMITH, Michael J. (30 Apr 1945). Cdr USN. Flight, plt. Gp 9.

SPRING, Sherwood C. (3 Sep 1944). Maj USAF. Flight, MS. Gp 9. Selected for 51-D, Feb 1985.

SPRINGER, Robert C. (21 May 1942). Lt-Col USMC. Flight, MS. Gp 9.

STAFFORD, Thomas P. (17 Sep 1930). Lt-Gen USAF (Ret). Res Nov 1975. Gp 2. Plt G6, Dec 1965, world's 1st RV. Cdr G9, Jun 1966. Cdr A10, May 1969, descending to 14·5km from lunar surface in final landing rehearsal. Cdr ASTP, Jul 1975. Cdr, Edwards AFB, 1975. USAF Dep Ch of Staff for R&D, 1978. American Farm Lines, Oklahoma. M 2 ch.

STEWART, Robert L. (13 Aug 1942). Lt-Col USAF. Flight, MS. Gp 8. MS2 41-B, Feb 1984, 2nd untethered spacewalk. M 2 ch.

SULLIVAN, Kathryn D. (3 Oct 1951). Civ, PhD (geology). Flight, MS. Gp 8. Selected for 41-G, Aug 1984.

SWIGERT, John L. (30 Aug 1931). Civ, plt. Dead. Gp 5. CMP A13, Apr 1970, aborted Moon landing. Leave of absence to become Staff Dir, Cttee on Science and Astronautics, Hse of Reps, 1973. Congressman (R, Colo) 1982. Died 28 Dec 1982, 6 days before taking his seat.

THAGARD, Norman E. (3 Jul 1943). Civ, MD. Flight, MS. Gp 8. MS3 STS-7, May 1983. Selected for 51-B, Nov 1984. M 2 ch.

THORNTON, William P. (14 Apr 1929). Civ, MD. Flight, MS. Gp 6. MS3 STS-8, Aug 1983. Selected for 51-B, Nov 1984. M 2 ch.

TRULY, Richard H. (12 Nov 1937). Capt USN. Res Sep 1983. Gp 7. Plt *Enterprise* ALTs 2 and 4, Sep and Oct 1977. Plt STS-2, Nov 1981. Cdr STS-8, Aug 1983. Cdr, NavSpaceCom, Sep 1983. M 3 ch.

van HOFTEN, James (11 Jun 1944). Civ, PhD (fluid mechanics). Flight, MS. Gp 8. MS2 and EV2 41-C, Apr 1984. M 2 ch.

WALKER, David M. (20 May 1944). Cdr USN. Flight, plt. Gp 8. Selected as plt 41-H, Sep 1984. M 2 ch.

WEITZ, Paul J. (25 Jul 1932). Capt USN (Ret). Flight, plt. Gp 5. Plt SL2, May 1973. Cdr STS-6, Apr 1983, 1st flt of *Challenger*. M 2 ch.

WHITE, Edward H. (14 Nov 1930). Lt-Col USAF. Dead. Gp 2. Plt G3, Jun 1965, 1st US spacewalk. Died during Apollo launchpad rehearsal fire, 27 Jan 1967.

WILLIAMS, Clifton C. (26 Sep 1932). Maj USMC. Dead. Gp 3. Killed in T-38 crash near Tallahassee, Fla, Oct 1967.

WILLIAMS, Donald E. (13 Feb 1942). Cdr USN. Flight, plt. Gp 8. Selected for 41-F, Aug 1984. M 2 ch.

WORDEN, Alfred E. (7 Feb 1932). Col USAF (Ret). Res Sep 1975. Gp 5. CMP A15, Jul 1971. Chief, System Studies Div, Ames, 1972. V-p, High Flight Foundation (see Irwin); and Dir, Energy Management Progs, Palm Bea, Fla. M 2 ch; Div and Re-M.

YOUNG, John W. (24 Sep 1930). Capt USN (Ret). Flight, plt. Gp 2. 1st man to make 5 and 6 flights. Plt G3, Mar 1965. Cdr G10, Jul 1966. CMP A10, May 1969. Cdr A16, Apr 1972, 9th man on Moon. Chief, Astronaut Office, since 1975. Cdr STS-1, Apr 1981, 1st Shuttle flight. Cdr STS-9/Spacelab 1, Nov 1983. M 2 ch; Div and Re-M.

US payload specialists

Spacelab 1 Dr Byron K. Lichtenberg (19 Feb 1948). Biomedical engineer and pilot, on MIT research staff. PS STS-9/Spacelab 1, Nov 1983. M 2 ch.
Dr Michael L. Lampton (1941). Physicist and space research scientist, U of Calif, Berkeley. Selected as back-up and for ground operations.

Spacelab 2 Selection will be made from: Loren Acton (1936), solar scientist; John-David Bartoe (1944), physicist studying solar ultra-violet radiation; Dianne Prinz (1938), solar terrestrial physicist; and George Simon (1934), Chief, Solar Research, AF Geophysics Lab, NM.

Spacelab 3 Selection from: Drs Eugene Trinh and Taylor Wang of JPL, specialists in liquid-drop dynamics; and Drs Mary Johnston (1945) of Marshall and Lodewijk van den Berg of EG & G Inc, Goleta, Calif, specialists in materials science.

Spacelab 4 Selection from: Drs Millie Fulford, U of Calif, San Francisco; Francis Gaffney, U of Texas, Dallas; Robert Phillips, Colorado State U, Ft Collins; and Bill Williams, Environmental Protection Agency, Corvallis, Ore. The 1st Spacelab mission dedicated to life sciences, this is scheduled for 61-D in Jan 1986.

Shuttle mission 41-D Charles D. Walker, McDonnell Douglas project engineer on the electrophoresis project since 1978 and who has trained the astronauts to operate this equipment in orbit, was announced in Jun 1983 as the first commercial PS. After 100hr of pre-flight "orientation" training in spaceflight hygiene, food-preparation techniques, systems, instrumentation, timelines and safety, he will fly as 6th crew member on 41-D to operate the electrophoresis apparatus throughout the mission. This follows the success of earlier experiments on STS-4, 5, 7 and 8. NASA and MDAC are sharing the cost of his training. Paul D. Scully-Power (40), civilian oceanographer with the USN, was later added as 7th crew member.

British payload specialists

Britain has accepted a NASA invitation to send a payload specialist on each of the Skynet 4 launches by Shuttle in Jan and Jun 1986. They will be selected from Maj Richard Farrimond (Army), Christopher Holmes (MoD), 33, Cdr Peter Longhurst (RN), 41, and Sq Ldr Nigel Wood (RAF), 34.

Canadian payload specialists

Canada has been allotted flights for 3 payload specialists in 1984, 1985 and 1986. 1st to fly, with Telesat 8 in Oct 1984, will be Cdr Marc Garneau (1948), selected by the National Research Council of Canada from among Roberta Bondar, Garneau, Steve MacLean, Ken Money, Bob Thirsk and Bjarni Tryggvason.

European astronauts

History In Dec 1977 the European Space Agency nominated 4 out of 53 candidates from 12 European countries as potential payload specialists on Spacelab missions. Later NASA agreed, in recognition of ESA's substantial contribution to the Space Shuttle, to include Europeans in its astronaut training programme, and in Jul 1980 2 of the 4 began training at Houston as mission specialists. Final selection of Ulf Merbold of W Germany as payload specialist on Spacelab 1 was announced in Sep 1982.

MALERBA, Franco (1946). Italian electronics engineer/physicist, Digital Equipment Corp, Milan.

MERBOLD, Ulf (20 Jun 1941). German research scientist, worked on crystal lattice defects at Max-Planck Inst, Stuttgart. Rejected by NASA for MS training on medical grounds, but flew as PS on STS-9/Spacelab 1, Nov 1983. Back-up for Spacelab D1, Jun 1985. M 2 ch.

NICOLLIER, Claude (1944). Swiss astronomer and former Swissair pilot. Participant airborne Spacelab simulation 1977. Completed MS training 1981. Named as PS to operate EOM-1 aboard 51-H, June 1985. M 1 ch.

OCKELS, Wubbo (28 Mar 1946). Netherlands physicist, carried out nuclear research at Gröningen Univ. Completed MS training 1981. Selected as back-up PS for Spacelab 1, Sep 1983, and prime PS for Spacelab D1, Jun 1985. M 2 ch.

French spationautes

2 French "spationautes" - apparently so called to distinguish them from both astronauts and cosmonauts - were selected by CNES in Jun 1980 from 400 applicants. They had to be aged 25-45 and not more than 82kg in weight or 95cm tall when seated. They started training at Star City near Moscow in Sep 1980, and 1 was chosen for a Soyuz/Salyut flight in 1982.

BAUDRY, Patrick (6 Mar 1946). Maj Armée de l'Air. Jaguar fighter pilot. Graduate of British Empire Test Pilots School. Selected for Soyuz T-6 back-up crew, Dec 1981. Selected to fly aboard Space Shuttle in 1985. M 1 ch.

Left: Dr Ulf Merbold of West Germany, ESA's 1st payload specialist. He flew on Spacelab 1. Right: Jean-Loup Chrétien.

CHRÉTIEN, Jean-Loup (20 Aug 1939). Lt-Col Armée de l'Air. Former chief military pilot, Mirage F.1. Test-researcher Soyuz T-6/Salyut 7, Jun 1982. Back-up to Baudry for 1985 Shuttle flight. M 4 ch.

German payload specialists

Prof Reinhard Furrer and Ernst Messerschmid have been named as payload specialists in training for Germany's Spacelab D1 mission in Sep 1985.

Indian cosmonauts

SHARMA, Rakesh (13 Jan 1949). Sqn Ldr Indian Air Force. Cosmonaut-researcher Soyuz T-11/Salyut 7, Apr 1984. 1st Indian in space; experiments included yoga exercises. Back-up was Wg Cdr Ravish Malhotra.

Intercosmos cosmonauts

History Russia's undertaking in Jul 1976 to train and fly a cosmonaut from each of the 9 Intercosmos member countries on a week-long mission to Salyut 6 was completed with Soyuz 40 in May 1981, 2yr early. By then India and France had been promised similar collaborative flights.

FARKAS, Bertalan (2 Aug 1949). Lt-Col, Hungarian AF. Cosmonaut-researcher Soyuz 36, May 1980. 5th Intercosmos cosmonaut, selected 1978.

GURRAGCHA, Jugderdemidiyn (5 Dec 1947). Capt, Mongolian aircraft engineer. Cosmonaut-researcher Soyuz 39, Mar 1981. 8th Intercosmos cosmonaut, selected 1978.

HERMASZEWSKI, Miroslaw (15 Sep 1941). Sqn Ldr Polish Air Force. Cosmonaut-researcher Soyuz 30, Jun 1978. 2nd Intercosmos cosmonaut, selected 1976.

IVANOV, Georgi (Jul 1940). Maj Bulgarian AF. Cosmonaut-researcher Soyuz 33, Apr 1979, failed to dock with Salyut 6. 4th Intercosmos cosmonaut.

JÄHN, Sigmund (13 Feb 1937). Lt-Col GDR AF. Cosmonaut-researcher Soyuz 31/29, Aug 1978. 3rd Intercosmos cosmonaut.

MENDEZ, Arnaldo Tamayo (1942). Pilot, Cuban AF. Cosmonaut-researcher Soyuz 38, Sep 1980. 7th Intercosmos cosmonaut; 1st black spaceman.

Left: Patrick Baudry. Right: Vladimir Remek.

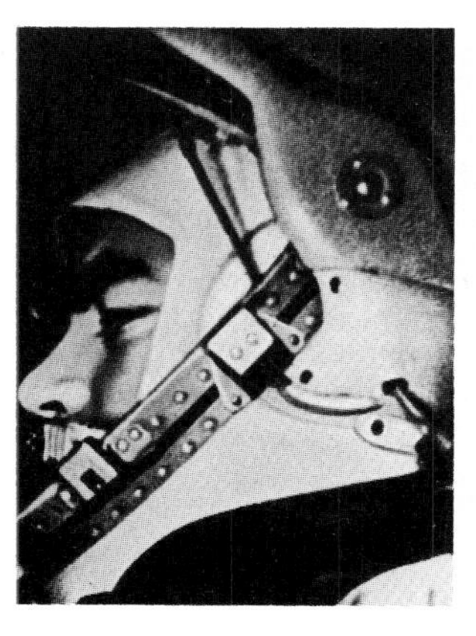

Left: Andrian Nikolayev, Cosmonaut No 3, took part in the 1st joint flight, with Valentina Tereshkova, and later married her. Right: Svetlana Savitskaya, 2nd woman in space.

PRUNARIU, Dumitru (27 Sep 1952). Snr Lt (Eng) Romanian AF. Cosmonaut-researcher Soyuz 40, May 1981. 9th and last Intercosmos cosmonaut.

REMEK, Vladimir (26 Sep 1948). Capt Plt Czech People's Army. Cosmonaut-researcher Soyuz 28, Mar 1978. 1st non-Soviet/US spaceman.

TUAN, Pham (14 Feb 1947). Lt-Col Vietnam AF. Cosmonaut-researcher Soyuz 37, Jul 1980. 6th Intercosmos cosmonaut.

Soviet cosmonauts

History With the flight of Soyuz T-10B in Feb 1984 57 different cosmonauts, including 2 women, had flown on Russia's 55 manned missions (excluding the T-10A abort). It has always been Russia's policy not to name cosmonauts until they are launched on their 1st mission, the only exceptions having been Apollo-Soyuz in 1975 and the Franco-Soviet flight in 1982. We can only guess how many of those who have flown are no longer active, and how many new ones are under training. Of the early cosmonauts, 6 are dead: Yuri Gagarin, 1st man in space, was killed in a military aircraft accident; 4 died in two different re-entry accidents; and another, Belyayev, died from natural causes. We know now that Russia's original cosmonaut group numbered 20, of whom 12 flew. The active list, it was at last revealed after Soyuz T-9, now numbers 65. In recent years Soviet cosmonauts and space chiefs have repeatedly said that no women were under training, but this changed, perhaps as a result of a political directive, within a week or two of the US announcement that Dr Sally Ride would fly on STS-7. Gen Nikolayev, who was married to Valentina Tereshkova, the world's 1st spacewoman, had only a few months earlier explained the original ban on women. After listing all the qualities needed by a cosmonaut, he continued: "Of course, a woman can possess all these qualities, but the mission programme makes big demands on her, especially if she is married. So nowadays we keep our women here on Earth. We love our women very much; we spare them as much as possible. However, in the future they will surely work on board space stations, but as specialists - as doctors, as geologists, as astronomers - and of course as stewardesses." After that it came as a surprise, following America's announcement about Sally Ride, when we learned that 2 women aged about 30 and both married were under training. When Svetlana Savitskaya flew on Soyuz T-7 it was stated that she had undergone a full course of training, but it was noticeable that no date was given as to when she first joined the cosmonaut group. She became Cosmonaut No 53.

Savitskaya's prestige flight was only a brief diversion from Soviet concentration on developing long-duration missions. Special tests have been worked out to establish the psychological compatibility of crew members, the better to enable them to withstand long periods in the confined quarters of Salyut space stations. But Gen Beregovoi, director of the Gagarin Training Centre, revealed in a radio interview in May 1978 that competition is used to keep cosmonauts at their sharpest during training. The selection of which crew would actually fly was announced only just before a flight "in the interests of morale". They had found it was unwise to decide on a team too soon: "As soon as the 2nd team finds out that they are only reserves, they cannot prepare for the task with the same enthusiasm."

AKSYONOV, Vladimir (1 Feb 1935). Flt eng Soy 22, Sep 1976. Flt eng T-2/Sal 6, Jun 1980. Specialist, systems and crew development. Selected 1973. Cosmonaut No 36. M 2 ch.

ALEXANDROV, Alexander (20 Feb 1943). Flt eng T-9/Sal 7, Jun 1983 (149 days). Designer, control systems. Selected 1978. Cosmonaut No 55. M 1 ch.

ANDREYEV, Boris (6 Oct 1940). Named in 1973 as 2nd member, 1st back-up crew for ASTP. Worked in design bureau from 1965. Selected for cosmonaut training 1970. M 2 ch.

ARTYUKHIN, Yuri (22 Jul 1930). Lt-Col AF. Flt eng Soy 14/Sal 3, Jul 1974. Selected 1973. Cosmonaut No 30. M 2 ch.

ATKOV, Oleg (9 May 1949). Dr (cardiologist). Cosmonaut-researcher Soy T-10B/Sal 7, Feb 1984. Invented portable ultrasound cardiograph used for 211-day flight and later for ambulance crews. Selected 1977. Cosmonaut No 57.

BELYAYEV, Pavel I. (26 June 1925). Cdr Voskhod 2, Mar 1965, from which Leonov made world's 1st spacewalk. Also poet and painter. Died from internal trouble, 11 Jan 1970. Cosmonaut No 10.

BEREGOVOI, Georgi T. (15 Apr 1921). Cdr Soy 3, Oct 1968, RV flight with Soy 2. Head Cosmonaut Training Centre, 1975. Selected 1964. Cosmonaut No 12. M 2 ch.

BEREZOVOI, Anatoli (11 Apr 1942). AF plt. Cdr Soy T-5, May 1982, 1st 200-day+ flight. Selected 1970. Cosmonaut No 51.

BYKOVSKY, Valeri F. (2 Aug 1934). Col. Plt Vostok 5, Jun 1963, 2nd group flight, with Tereshkova. Training officer for ASTP, Jul 1975. Cdr Soy 22, Sep 1976. Cdr Soy 31/29/Sal 6, Aug 1978. Cosmonaut No 5. M 1 ch.

DEMIN, Lev (11 Jan 1926). Col. Flt eng Soy 15, Aug 1974. Wartime student with Komarov at AF school, Moscow. 1st grandfather in space. Selected 1963. Cosmonaut No 32. M 2 ch.

DOBROVOLSKY, Georgi (1 Jun 1928). Cdr Soy 11/Sal 1, Jun 1971, new 24-day endurance record. Killed following pressurisation failure during re-entry. Selected 1961. Cosmonaut No 24.

DZHANIBEKOV, Vladimir (13 May 1942). Col. Named 1973 as back-up ASTP cdr. Cdr Soy 27, Jan 1978, 1st double docking with Sal 6. Cdr Soy 39/Sal 6, Mar 1981. Cdr Soy T-6/Sal 7, Jun 1982. Selected 1970. Cosmonaut No 43. M 2 ch.

FEOKTISTOV, Konstantin P. (26 Feb 1926). Scientist-cosmonaut Voskhod 1, Oct 1964, 1st 3-man craft. Accepted as cosmonaut despite wartime injuries (shot and left for dead when acting as scout in German-occupied territory). Cosmonaut No 8. M 1 ch.

FILIPCHENKO, Anatoli (26 Feb 1928). Lt-Col. Cdr Soy 7, Oct 1969, group flight with Soy 6 and 8. Cdr Soy 16, Dec 1974, ASTP rehearsal. Cosmonaut No 19. M 2 ch.
GAGARIN, Yuri A. (9 Mar 1934). Col. Vostok 1, 12 Apr 1961, 1st man in space. Subsequently Cdr Soviet Cosmonauts' Detachment and made worldwide public appearances. Komarov's back-up for Soy 1, then while training for Soy 3 killed with Col Seryogin during test flight of military jet, 27 Mar 1968. Ashes placed in Kremlin Wall; lunar far-side crater named after him. Cosmonaut No 1.
GLAZKOV, Yuri (2 Oct 1939). Lt-Col. Flt eng Soy 24/Sal 5, Feb 1977. Selected 1965. Cosmonaut No 39.

Left: Pavel Belyayev commanded Voskhod 2, during which Leonov made the world's 1st spacewalk. Belyayev died 5yr later. Right: Yuri Gagarin, 1st man in space, was killed in a jet crash 7yr later.

COSMONAUTS IN ORDER OF 1ST FLIGHT

No	*Name*	*No of flights*	*Total duration (days, hr, min)*	*Missions*
1	Y. Gagarin	1	00.01.48	V1
2	G. Titov	1	01.01.18	V2
3	A. Nikolayev	2	21.15.25	V3, S9
4	P. Popovich	2	18.16.29	V4, S14
5	V. Bykovsky	3	20.17.49	V5, S22, S31
6	V. Tereshkova	1	02.22.50	V6
7	V. Komarov	2	02.02.54	Vd1, S1
8	K. Feoktistov	1	01.00.17	Vd1
9	B. Yegorov	1	01.00.17	Vd1
10	P. Belyayev	1	01.02.02	Vd2
11	A. Leonov	2	07.00.33	Vd2, S19
12	G. Beregovoi	1	03.22.51	S3
13	V. Shatalov	3	09.21.59	S4, S8, S10
14	B. Volynov	2	52.07.20	S5, S21
15	A. Yeliseyev	3	08.22.24	S5, S8, S10
16	Y. Khrunov	1	01.23.49	S5
17	G. Shonin	1	04.22.42	S6
18	V. Kubasov	3	18.17.19	S6, S19, S36
19	A. Filipchenko	2	10.21.05	S7, S16
20	V. Volkov	2	28.17.03	S7, S11
21	V. Gorbatko	3	30.12.31	S7, S24, S37
22	V. Sevastyanov	2	80.16.19	S9, S18B
23	N. Rukavishnikov	3	09.22.10	S10, S16, S33
24	G. Dobrovolsky	1	23.18.22	S11
25	V. Patsayev	1	23.18.22	S11
26	V. Lazarev	2	01.23.16	S12, S18A
27	O. Makarov	4	20.18.28	S12, S18A, S27, T-3
28	P. Klimuk	3	78.18.19	S13, S18B, S30
29	V. Lebedev	2	219.05.00	S13, T-5
30	Y. Artyukhin	1	15.17.30	S14
31	G. Sarafanov	1	02.00.12	S15
32	L. Demin	1	02.00.12	S15
33	A. Gubarev	2	37.11.37	S17, S28
34	G. Grechko	2	125.23.20	S17, S26
35	V. Zholobov	1	49.06.24	S21
36	V. Aksyonov	2	11.20.35	S22, T-2
37	V. Zudov	1	02.00.06	S23
38	V. Rozhdestvensky	1	02.00.06	S23
39	Y. Glazkov	1	17.17.23	S24
40	V. Kovalyonok	3	216.10.17	S25, S29, T-4
41	V. Ryumin	3	361.20.24	S25, S32, S35
42	Y. Romanenko	2	104.06.34	S26, S38
43	V. Dzhanibekov	3	21.17.33	S27, S39, T-6
44	A. Ivanchenkov	2	147.12.39	S29, T-6
45	V. Lyakhov	2	324.10.22	S32, T-9
46	L. Popov	3	199.01.32	S35, S30, T-7
47	Y. Malyshev	2	11.20.32	T-2, T-11
48	L. Kizim	2	12.19.08	T-3, T-10B
49	G. Strekalov	3	22.17.17	T-3, T-8, T-10A, T-11
50	V. Savinykh	1	74.18.43	T-4
51	A. Berezovoi	1	211.08.05	T-5
52	A. Serebrov	2	09.22.10	T-7, T-8
53	S. Savitskaya	1	07.21.52	T-7
54	G. Titov	1	02.00.18	T-8, T-10A
55	A. Alexandrov	1	149.09.46	T-9
56	V. Solovyev	1		T-10B
57	O. Atkov	1		T-10B
	Totals	**103**	**2,851.05.38**	**56 missions**

Notes T-10A was a launchpad abort and is not included in totals. T-10B time not included because mission not finished at going to press.
Abbr V =Vostok, S = Soyuz, Vd = Voskhod, T = Soyuz T.

GORBATKO, Viktor (3 Dec 1934). Lt-Col. Research engineer Soy 7, Oct 1969, group flight with Soy 6 and 8. Cdr Soy 24/Sal 5, Feb 1977. Cdr Soy 37/36/Sal 6, Jul 1980. Selected with Gagarin, 1960. Cosmonaut No 21. M 2 ch.
GRECHKO, Georgi (25 May 1932). Flt eng Soy 17/Sal 4, Jan 1975. Flt eng Soy 26/27/Sal 6, Dec 1977 (126 days). Worked in Korolev's design team on automatic lunar landing systems. Selected 1966. Cosmonaut No 34. M 2 ch.
GUBAREV, Alexei (29 Apr 1932). Lt-Col. Cdr Soy 17/Sal 4, Jan 1975. Cdr Soy 28/Sal 6, Mar 1978, 1st international flight. Selected 1963. M 1 ch.
ILLARIONOV, Valeri. No details known. He was one of 9 Soviet specialists on Soyuz spacecraft at Houston Mission Control for ASTP, 1975.
IVANCHENKOV, Alexander (28 Sep 1940). Named 1973 as back-up flt eng ASTP. Flt eng Soy 29/Sal 6, Jun 1978 (1st 100-day flight). Flt eng Soviet/French T-6/Sal 7, Jun 1982. Spacecraft designer. Selected 1970. Cosmonaut No 44. M 1 ch.
KHRUNOV, Yevgeni V. (10 Sep 1933). Lt-Col. Flt eng Soy 5/4, Jan 1969, 1st docking of 2 manned craft, spacewalked with Yeliseyev to return in Soy 4. Selected 1960. Cosmonaut No 16. M 1 ch.
KIZIM, Leonid (5 Aug 1941). Col (military test pilot). Cdr Soy T-3/Sal 6, Nov 1980, resumption of 3-crew flights. Cdr Soy T-10B/Sal 7, Feb 1984, 1st triple long-stay crew. Selected 1965. Cosmonaut No 48. M.
KLIMUK, Pyotr (10 Jul 1942). Lt-Col. Cdr Soy 13, Dec 1973. Cdr Soy 18/Sal 4, May 1975 (63 days). Cdr Soy 30/Sal 6, Jun 1978. Selected 1965. Cosmonaut No 28. M 1 ch.

SOVIET FLIGHT SUMMARY (TO SOYUZ T-11)

Cosmonauts		*Flights*
1	man has made 4 flights	4
12	men have made 3 flights	36
19	men have made 2 flights	38
23	men have made 1 flight	23
2	women have made 1 flight	2
57	**Totals**	**103**

SOVIET EVA RECORD

Mission	*Launch date*	*Cosmonaut*	*Duration (hr, min)*
Voskhod 2	18.03.65	Leonov	00.24
Soyuz 4	14.01.69	Yeliseyev	01.00
Soyuz 5	15.01.69	Khrunov	01.00
Soyuz 26	10.12.77	Grechko	01.28
		Romanenko*	
Soyuz 29	23.09.78	Kovalyonok	02.05
		Ivanchenkov	02.05
Soyuz 32	25.02.79	Lyakhov	01.23
		Ryumin	01.23
Soyuz T-5	13.05.82	Berezovoi	02.33
		Lebedev	02.33
Soyuz T-9	01.11.83	Lyakhov	05.45
		Alexandrov	05.45
Soyuz T-10B	08.02.84	Kizim	17.50
		Solovyev	17.50
	Totals	**15**	**63.04**

*Unauthorised, length unknown.

Left: Vladimir Komarov became the 1st man to be killed in space when the landing parachute snarled at the end of his 2nd flight. Right: Alexei Gubarev.

Gen Alexei Leonov. *(NASA)* Oleg Makarov.

KOMAROV, Vladimir M. (16 Mar 1927). Col. Cdr Voskhod 1, Oct 1964, 1st 3-man craft. Flew twice despite heart condition similar to Slayton's. 1st man to be killed in space, when landing parachute snarled at end of Soy 1, 23 Apr 1967. Cosmonaut No 7. M 2 ch.
KOVALYONOK, Vladimir (3 Mar 1942). Col. Cdr Soy 25, Oct 1977, failed to dock with Sal 6. Cdr Soy 29/31/Sal 6, Jun 1978, 1st 100-day+ flight. Cdr Soy T-4/Sal 6, Mar 1981. Former paratroop instructor. Selected 1967. Cosmonaut No 40. M 1 ch.
KUBASOV, Valeri (7 Jan 1935). Flt eng Soy 6, Oct 1969, manoeuvred with Soy 7 and 8. Kubasov did 1st space welding. Flt eng Soy 19 (ASTP), Jul 1975. Cdr Soy 36/35/Sal 6, May 1980. Selected 1966. Cosmonaut No 18.
LAZAREV, Vasili (23 Feb 1928). Lt-Col. Cdr Soy 12, Sep 1973. Cdr Soy 18A, Apr 1975, aborted during launch. Test pilot and physician. Selected ? 1966. Cosmonaut No 26. M 1 ch.
LEBEDEV, Valentin (14 Apr 1942). Flt eng Soy 13, Dec 1973. Replaced by Ryumin for record Soy 35 flight following leg injury. Flt eng Soy T-5/Sal 7, May 1982, 1st 200-day+ flight. Selected 1972. Cosmonaut No 29. M 1 ch.
LEONOV, Alexei (20 May 1934). Gen. Flt eng Voskhod 2, Mar 1965, 1st man to walk in space. Later produced vivid impressionistic paintings. Cdr ASTP (Soy 19), Jul 1975. Dep Head, Gagarin Training Centre. Selected 1960. Cosmonaut No 11. M 2 ch.
LYAKHOV, Vladimir (20 Jul 1941). Col. Cdr Soy 32/34/Sal 6, Feb 1979 (175 days). Cdr T-9/Sal 7, Jun 1983 (149 days). Test pilot. Selected 1967. Cosmonaut No 45. M 2 ch.
MAKAROV, Oleg (6 Jan 1933). Flt eng Soy 12, Sep 1973. Flt eng Soy 18A, Apr 1975, aborted during launch. Flt eng Soy 27/26, Jan 1978, 1st double linkup with Sal 6. Flt eng Soy T-3/Sal 6, Nov 1980. Design specialist, worked under Komarov on Vostoks. Selected 1966. Cosmonaut No 27. M 1 ch.
MALYSHEV. Yuri V. (27 Aug 1941). Lt-Col. Cdr Soy T-2/Sal 6, Jun 1980. Named Dec 1981 as back-up for Soviet/French mission, 1982. Cdr Soy T-11, Apr 1984. Selected 1967. Cosmonaut No 47. M 2 ch.
NIKOLAYEV, Andrian G. (5 Sep 1929). Gen. Plt Vostok 3, Aug 1962, 1st joint flight, with Vostok 4. Cdr Soy 9, Jun 1970. Selected 1960. Cosmonaut No 3. M (Valentina Tereshkova, 1st woman in space, 1963). 2 ch.

Left: Valentina Tereshkova, 1st woman in space. Right: Pavel Popovich.

PATSAYEV, Viktor (19 Jun 1933). Test eng Soy 11/Sal 1, Jun 1971. After 23 days in Salyut killed by pressurisation failure during re-entry. Selected 1969. Cosmonaut No 25.
POPOV, Leonid (31 Aug 1935). Lt-Col. Cdr Soy 35/37/Sal 6, Apr 1980. Cdr Soy 40/Sal 6, May 1981. Cdr Soy T-7/T-5/Sal 7, Aug 1982. Fighter pilot, specialises in Earth resources. Selected 1970. Cosmonaut No 46. M 1 ch.
POPOVICH, Pavel R. (5 Oct 1930). Col. Plt Vostok 4, Aug 1962, 1st double flight, with Vostok 3. Cdr Soy 14/Sal 3, Jul 1974. Selected 1960, believed to be 1st member of cosmonaut team. Cosmonaut No 4. M 2 ch.
ROMANENKO, Yuri (1 Aug 1944). Lt-Col. Named as back-up crew member for ASTP, 1975. Cdr Soy 26/27/Sal 6, Dec 1977. Cdr Soy 38/Sal 6, Sep 1980. Selected 1970. Cosmonaut No 42. M 1 ch.
ROZHDESTVENSKY, Valeri (13 Feb 1939). Lt-Col. Flt eng Soy 23, Oct 1976, failed to dock with Sal 5 and made 1st Soviet splashdown in emergency return. Former Cdr, Baltic Deep Sea Divers' Rescue Service. Selected 1965. Cosmonaut No 38. M 1 ch.
RUKAVISHNIKOV, Nikolai (18 Sep 1932). Test eng Soy 10/Sal 1, Apr 1971. Flt eng Soy 16, Dec 1974, ASTP rehearsal. Cdr Soy 33, Apr 1979, failed to dock with Sal 6. Spacecraft designer under Korolev. Replaced on Soy T-11 due to illness. Selected 1967. Cosmonaut No 23. M 1 ch.
RYUMIN, Valeri (16 Aug 1939). Flt eng Soy 25, Oct 1977, failed to dock with Sal 6. Flt eng Soy 32/34/Sal 6. Feb 1979, setting new world record of 175 days. Flt eng 35/37/Sal 6, Apr 1980, increasing record to 185 days and Ryumin's total to 360 days. Electronics engineer and equipment designer. At 184cm (over 6ft) probably tallest cosmonaut. Selected 1973. Cosmonaut No 41. M 2 ch.
SARAFANOV, Gennadi (1 Jan 1942). Lt-Col. Cdr Soy 15, Aug 1974, failed to dock with Sal 3. Selected 1965. Cosmonaut No 31. M 2 ch.
SAVINYKH, Viktor (7 Mar 1940). Flt eng Soy T-3/Sal 6, Mar 1981. 100th person in space. Former railway engineer, specialist in aerial photography and cartography. Selected 1978. Cosmonaut No 50. M 1 ch.
SAVITSKAYA, Svetlana (8 Aug 1948). Aerobatic pilot and parachutist with 18 world records. Researcher-cosmonaut T-7/T-5/Sal 7, Aug 1982. 2nd woman in space. No selection date given. Cosmonaut No 53. M.
SEREBROV, Alexander (15 Feb 1944). Flt eng T-7/T-5/Sal 7, Aug 1982. Researcher-cosmonaut T-8, Apr 1983, failed to dock with Sal 7. No selection date given. Cosmonaut No 52.
SEVASTYANOV, Vitali (8 Jul 1935). Flt eng Soy 9, Jun 1970. Flt eng Soy 18B/Sal 4 (63 days), May 1975. Selected 1967. Cosmonaut No 22. M 1 ch.
SHATALOV, Vladimir A. (8 Dec 1927). Lt-Gen. Cdr Soy 4, Jan 1969, 1st docking, with Soy 5, of 2 manned craft. Overall cdr in Soy 8 of 7 crew in Soy 6, 7 and 8 group flight, Oct 1969. Cdr Soy 10, Apr 1971, docked with but did not enter Sal 1. Later Dir, Cosmonaut Training Centre. Selected 1963. Cosmonaut No 13. M 2 ch.
SHONIN, Georgi (3 Aug 1935). Lt-Col. Cdr Soy 6, Oct 1969, group flight with Soy 7 and 8. Selected 1960. Cosmonaut No 17. M 2 ch.
SOLOVYEV, Vladimir (11 Nov 1946). Space technology designer. Named as back-up flt eng for Soviet-French T-6/Sal 7, Jun 1982. Flt eng Soy T-10B/Sal 7, Feb 1984. Selected 1978. Cosmonaut No 56.

STREKALOV, Gennadi (28 Oct 1940). Res eng T-3(Sal 6, Nov 1980, resumption of 3-crew flights. Flt eng T-8, Apr 1983, failed to dock with Sal 7. Flt eng T-10A, Sep 1983, launchpad abort. Flt eng T-11, Apr 1984 (replacing Rukavishnikov at short notice). Spacecraft designer. Selected 1973. Cosmonaut No 49.

TERESHKOVA, Valentina V. (6 Mar 1937). Plt Vostok 6, 16 Jun 1963, 1st woman in space, in group flight with Vostok 5. She said that when Titov flew she "was not dreaming of becoming a cosmonaut," but was in orbit 21 months later. Member Supreme Soviet for Yaroslavl, 1967; Praesidium, Supreme Soviet, 1974. Cosmonaut No 6. M (cosmonaut Nikolayev, 1963) 2 ch.

TITOV, Gherman S. (11 Sep 1935). Gen. Plt Vostok 2, Aug 1961, 2nd man on orbit, 1st to spend 1 day there. Suffered from spacesickness during and after flight; concealed damaged wrist, which probably disqualified him for further flight. Head of Cosmonauts from 1964. Cosmonaut No 2. M 2 ch.

TITOV, Vladimir (1 Jan 1947). Test pilot, Lt-Col. Cdr T-8, Apr 1983, failed to dock with Sal 7. Cdr T-10A, Sep 1983, launchpad abort. Selected 1976. Cosmonaut No 54.

VOLKOV, Vladislav (23 Nov 1935). Flt eng Soy 7, Oct 1969, gp flt with Soy 6 and 8. Flt eng Soy 11/Sal 1 (24 days). Killed by pressurisation failure during re-entry. Selected 1966. Cosmonaut No 20.

VOLYNOV, Boris V. (18 Dec 1934). Col. Cdr Soy 5, Jan 1969, 1st docking of 2 manned craft, with Soy 4. Cdr Soy 21/Sal 5, Jul 1976 (48 days). Selected 1960. Cosmonaut No 14. M 2 ch.

YEGOROV, Boris B. (26 Nov 1937). Dr Voskhod 1, Oct 1964, 1st doctor in space. Cosmonaut No 9. M 1 ch.

YELISEYEV, Alexei S. (13 Jul 1934). Flt eng Soy 5, Jan 1969, 1st docking of 2 manned craft, with Soy 4. Spacewalked to Soy 4 for return. Flt eng Soy 8, Oct 1969, group flight with Soy 6 and 7. Flt eng Soy 10, Apr 1971. Flt dir ASTP. Selected 1966. Cosmonaut No 15. M 1 ch.

Boris Yegorov.

ZHOLOBOV, Vitali (18 Jun 1937). Lt-Col, Soviet Army. Flt eng Soy 21/Sal 5, Jul 1976 (48 days). Selected 1963. Cosmonaut No 35. M 1 ch.

ZUDOV, Vyacheslav (8 Jan 1942). Lt-Col. Cdr Soy 23, Oct 1976, failed to dock with Sal 5 and made 1st Soviet splashdown in emergency return. Selected 1965. Cosmonaut No 37. M 2 ch.

SPACE CONTRACTORS

Note The Editor is grateful to space companies from many countries for supplying information without which this book would not be possible. In contrast with the majority of Jane's reference books, this information is carried under the project name rather than that of the company. For cross-reference purposes, companies providing information are listed below with the names of their current projects.

Britain

BAe British Aerospace Public Limited Company, Dynamics Group, Six Hills Way, Stevenage, Herts SG1 2DA. ECS, Marecs, ISPM, L-Sat, Giotto, Skynet, Unisat.

MSDS Marconi Space and Defence Systems Ltd, The Grove, Stanmore, Middlesex HA7 4LY. Exosat, Marecs, Ariane.

Canada

Spar Aerospace Ltd Royal Bank Plaza, S Tower, PO Box 83, Toronto, Ontario M5J 2J2. Tel (416) 865 0480. Anik, Shuttle RMS, L-Sat solar array.

China

Chinese Academy of Space Technology PO Box 2417, 31 Bai Shi Qiao, Beijing, China.

France

Aérospatiale Société Nationale Industrielle Aérospatiale, Division Systèmes Balistiques et Spatiaux, Route de Verneuil, BP 96, 78130 Les Mureaux. Tel 33 (1) 474 7213. Ariane, Sonate, Sylda, TDF-1, Arabsat.

Matra SA Matra, Space Branch, 37 Ave Louis Breguet, 78140 Vélizy. Tel 946 96 00. Telecom 1, Spot.

German Federal Republic

Dornier Dornier System GmbH, Postbox 1360, 7990 Friedrichshafen. Tel (07545) 81. ISEE-B, GEOS, Faint Object Camera (FOC), ISPM, Spacelab, Azur, Aeros A/B, Rosat.

Erno Erno Raumfahrttechnik GmbH, Hünefeldstrasse 1-5, POB 10 59 09, D-2800 Bremen 1. Tel (0421) 5391. Spacelab, ECS, Marecs, Telecom 1, TV-Sat.

MBB Messerschmitt-Bölkow-Blohm GmbH, Space Division, 8012 Ottobrunn bei München. Tel (089) 60 00 25 90. Helios, Symphonie, Exosat, Intelsat 5, Galileo, SPAS, TV-Sat/TDF-1.

Otrag Orbital Transport und Raketen AG, Schleissheimer Str 59, 8046 Garching, Munich. Satellite launchers and sounding rockets

India

ISRO Indian Space Research Organisation, F Block, Cauvery Bhavan, Gowda Road, Bangalore 560 009. Tel 27371/76. SLV-3 launch vehicle, Bhaskara, Insat, Rohini.

International

Cosmos This consortium, formed in Nov 1970, consists of: Etudes Techniques et Constructions Aérospatiale SA (ETCA), BP 97, 6000 Charleroi 1, Belgium; Société Nationale Industrielle Aérospatiale, BP 96, 78130 Les Mureaux, France; Société Anonyme Télécommunications (SAT), 41 rue Cantagrel, 75624 Paris Cedex 13, France; Messerschmitt-Bolkow-Blohm GmbH (MBB), 8 München 80, Postfach 801 169, Federal Republic of Germany; Construcciones Aeronauticas SA (CASA), Division Espacial, Getafe, Madrid, Spain; Marconi, Space and Defence Systems Ltd (MSDS), The Grove, Stanmore, Middlesex HA7 4LY, England; and Selenia SpA, CP 7083 00100 Rome, Italy. Programmes include: Exosat, Meteosat.

Mesh This consortium, formed in Oct 1966, consists of: Aeritalia SpA, Piazzale V. Tecchio 51, 80125 Naples, Italy; British Aerospace Public Limited Company, Dynamics Group, Space and Communications Division, Gunnels Wood Road, Stevenage, Herts SG1 2AS, England; Erno Raumfahrttechnik GmbH, 28 Bremen 1, Hunefeldstrasse 15, Postfach 1199, Federal Republic of Germany; Fokker BV, PO Box 1065, 1000BB Amsterdam, Netherlands; Inta, Departamento de Equipo Armamento, Paseo del Pintor Rosales 34, Torrejon de Ardoz, Madrid 8, Spain; SA Matra, BP No 1, 78140 Vélizy, France; and Saab-Scania AB, S-581 88 Linköping, Sweden. Programmes include: ECS, Marecs, OTS.

STAR (Satellites for Telecommunications, Applications & Research) This consortium consists of: British Aerospace Public Limited Company, Dynamics Group, Space and Communications Division, Gunnels Wood Road, Stevenage, Herts SG1 2AS, England; Contraves AG, Schaffhauserstrasse 580, CH-8052 Zurich 11, Switzerland; CGE-Fiar, Via G.B. Grassi 93, 20157 Milan, Italy; Dornier System GmbH, Postfach 648, 799 Friedrichshafen/Bodensee, Federal Republic of Germany; Telefonaktiebolaget L.M. Ericsson, 126 25 Stockholm, Sweden; Montedel (Montecatini Edison Eletronica SpA), Via E. Bassini 15, 20133 Milan, Italy; Sener SA, Guzman el Bueno 133, Madrid 3, Spain; Société Européenne de Propulsion, Tour Roussel-Nobel, 92080 Paris Défense Cedex 3, France; and Thomson-CSF, 173 boulevard Haussmann, 75360 Paris Cedex 08, France. Programmes include: ESPM, Giotto.

Italy

Aeritalia Aeritalia-Societa Aerospaziale Italiana p.A, Space and Alternative Energies Gp, Corso Marche 41, 10146 Turin. Tel (011) 33321. Iris, Italsat, Tethered Satellite.

CNS Compagnia Nazionale Satelliti per Telecommunicazioni SpA, Via Salaria Km 9.3, 00138 Rome. (06) 840 2021. Sirio 1 and 2, MDD, Lasso.

Japan

Mitsubishi Mitsubishi Jukogyo Kabushiki Kaisha (Mitsubishi Heavy Industries Ltd), Aircraft and Space Vehicle Dept, 5-1 Marunouchi, 2-chome, Chiyoda-ku, Tokyo 100. N-2 and H-1 launchers, GS-1.
Mitsubishi Electric Corporation (Melco) 2-3 Marunouchi, 2-chome, Chiyoda-ku, Tokyo 100. ECS.
NASDA National Space Development Agency of Japan, 2-4-1 Hamamatsu-cho, Minato-ku, Tokyo 105. Tel 03-435 6111. N-2 and H-1 launch vehicles, CS-2a and CS-2b, BS-2a and BS-2b, GMS-3, MOS-1.
Nissan Nissan Jidosha Kabushiki Kaisha (Nissan Motor Co Ltd), Aeronautical & Space Division, 5-1, 3-chome, Momoi, Suginami-ku, Tokyo. Tel Tokyo (390) 1111. Mu launcher.

Sweden

Saab Saab-Scania AB, S-581 88 Linköping. Tel 46 13 18 13 97. Viking.

United States

Boeing The Boeing Company, Seattle, Washington 98124. IUS (Inertial Upper Stage).
Fairchild Fairchild Space and Electronics Co, Germantown, Md 20874. Tel (301) 428 6000. MMS, Leasecraft.
Ford Ford Aerospace and Communications Corporation, 20th Floor, Renaissance Center, PO Box 43342, Detroit, Michigan 48243. Tel (313) 568-7718. Intelsat 5.
General Dynamics General Dynamics Corporation, Convair Division, 5001 Kearny Villa Road, PO Box 80847, San Diego, Calif 92138 . Tel (714) 277 8900. Atlas, Centaur, Orbiter mid-fuselage.
General Electric General Electric Company Space Division, Valley Forge Space Centre, PO Box 8555, Philadelphia, Penn 19101. Landsat, Nimbus.
Hughes Hughes Aircraft Company, Space and Communications Group, PO Box 92919, Los Angeles, Calif 90009. Tel (213) 648 0884. Leasat, Palapa B, SBS, Telstar 3, HS 376, Intelsat 6.
LMSC Lockheed Missiles and Space Company Inc (subsidiary of Lockheed Corporation), 1111 Lockheed Way, Sunnyvale, Calif 94068. Tel (408) 742 6688. Agena D, Big Bird, Space Telescope.
Martin Marietta Martin Marietta Corporation, 6801 Rockledge Drive, Bethesda, Maryland 20034. Tel (301) 897 6000. Titan 3, Titan 34D.
MDAC McDonnell Douglas Astronautics Company, 5301 Bolsa Ave, Huntingdon Beach, Calif 92647. Tel (714) 896 1301. Delta, PAM/SSUS, MDTSCO.
NASA National Aeronautics and Space Administration, Washington DC 20546. Galileo, ISPM, Space Shuttle, Space Telescope, Voyager, etc.
RCA RCA Corporation, Government Systems Division, Cherry Hill Offices, Camden, New Jersey 08358. Tel (609) 234 1234. Block 5D DMSP, Nova, RCA Satcom, Tiros, Spacenet, GStar.
Rockwell International Rockwell International Corporation, Space Operations, 12214 Lakewood Boulevard, Downey, Calif 90241. Tel (213) 922 2111. Navstar, Space Shuttle Orbiter.
TRW TRW Defence and Space Systems Group, 1 Space Park, Redondo Beach, Calif 90278. Tel (213) 535 4321. DSCS II, Fltsatcom, HEAO, IMEWS (USAF 647).
Vought Vought Corporation (subsidiary of LTV Corp), PO Box 225097, Dallas, Texas 75265. Scout (XRM-91), ASAT.

NOTES

Electrophoresis The programme known as Electrophoresis Operations in Space (EOS) is being carried out by McDonnell Douglas and the Ortho Pharmaceutical Group (a division of Johnson & Johnson) with NASA help. It is the 1st of a series of collaborative projects aimed at giving private enterprise an opportunity to develop commercial applications. McDonnell Douglas says that EOS "may lead to dramatic breakthroughs in the treatment of a number of diseases like diabetes, emphysema, dwarfism, thrombosis and viral infection." In a process called continuous-flow electrophoresis biological materials, continuously injected into a buffer solution flowing through a thin, rectangular chamber, are pulled apart by an electrical field into separate streams which can be collected. By STS-7, after 3 Shuttle flight tests, by comparison with similar operations on Earth over 700 times more material with 4 times the purity had been separated. This led to the decision to send a prototype production unit on Shuttle mission 41-D, accompanied by McDonnell Douglas payload specialist Charles Walker to operate it continuously. A first production plant is to be carried on Spacelab 2 in 1985 to produce quantities sufficient for commercial sales. A free-flying production unit, mounted on a SPAS-type satellite and visited every 6 months by Shuttle crews to deliver raw materials and collect units of separated material, is expected by 1987.

Insurance In 1982 the maximum insurance value of a satellite was $130 million, the average $74 million and minimum $55 million. Total value insured was $1·26 billion and premiums $100 million. Average premium was 8% of insured value. Maximum in-orbit cover available was 3yr. The insurance industry was complaining that to that date it had paid out twice as much as it had received in premiums.

Nuclear space power 10yr of friction between NASA, DoD and the US Energy Department over development of nuclear space power-generation technology was ended with a tri-agency agreement early in 1983. DoD's Defence Advanced Research Projects Agency will direct the SP-100 programme for 2–3yr, after which it will be taken over by the Energy Department; technical aspects will be directed by NASA's Jet Propulsion Laboratory. The aim is to develop a 100kW ground demonstration system in 1989, with a prototype in orbit by 1995. The DoD lead is explained by the fact that a nuclear energy system in the 100kW range, with lower weight, size and cost than advanced solar arrays, would make future satellites less vulnerable to attack and provide the power, currently lacking, for laser and particle-beam weapons. So far NASA has limited its use of nuclear powerplants to lunar and interplanetary spacecraft operating so far from the sun that solar arrays are impractical. The Soviets use nuclear power in low Earth orbit, however.

Spacesickness About 45% of astronauts have experienced some nausea, vomiting and general malaise during the early stages of adjusting to weightlessness. NASA renamed this "space adaptation syndrome" in 1982 and introduced new rules under which journalists would not be told about individuals suffering from its effects unless their sickness necessitated changes to the flightplan. This followed complaints from astronauts over a long period about public discussion of their "private medical conferences," which they saw as a breach of the doctor-patient relationship. In Nov 1983 NASA established a Space Biomedical Research Institute at JSC in an attempt to speed up the search for countermeasures and perhaps identify individuals who might be more susceptible to spacesickness. (It was the astronauts' fears that even a mild attack would affect their selection for future missions that led to their successful campaign for personal privacy.) On Spacelab 1 in Dec 1983 50% of crew time was devoted to "life science" investigations, many of which were related to this problem.

Synchronous satellites It was in 1945 that Arthur C. Clarke, the space writer, suggested that a satellite placed in the equatorial plane 35,680km from Earth would orbit at such a speed as to appear to be hanging in space above one point on the surface. Such a "synchronous" or "stationary" satellite would always be in position to relay radio, TV and telephone signals, and no elaborate tracking devices would be needed. Compared with conventional underwater cables, satellite communications offered flexibility and limitless capacity. Clarke admits to being surprised that his vision became reality within his lifetime, and wishes now that he had patented the idea. It must however be some consolation that in tribute to his foresight geostationary orbits are now known as "Clarke orbits".

The first synchronous satellite, **Syncom 1**, was successfully placed in orbit in Feb 1963, but its radio equipment failed to work. **Syncom 2,** launched in Jul 1963, was then placed in a 35,567km orbit, and during the following 3 weeks manoeuvred into position over Brazil by the firing of small onboard hydrogen peroxide thrusters. On Sep 13 **Syncom 2** and **Relay 1** were used to link Rio de Janeiro, New Jersey and Lagos, Nigeria, in a 3-continent conversation. But Syncom 2, not quite in the plane of the equator, appeared to describe a figure 8 as the Earth turned beneath it. **Syncom 3,** weighing 38·5kg, was placed in true equatorial orbit, with no N-S swing, on 19 Aug 1964: perigee 35,670km, apogee 35,700km, period 1,436·2min, incl 0·1°. Drifting over the Pacific Ocean near the International Dateline on Oct 10, it telecast the opening-day ceremonies of the Olympic Games in Tokyo.

In 21yr up to 1983 72 synchronous satellites had been launched, with another 80 scheduled to follow in the next 3 yr. Western hemisphere slots as allotted by the US Federal Communications Commission on 9 May 1983 were as follows:

SATELLITE ORBITAL SLOTS

Orbital position (W long)	*User*	*Frequency band*
143°	Satcom 5	C
141°	Unassigned	C
139°	Satcom 1R	C
137°	Unassigned	C
134°	Galaxy 1	C
132°	Rainbow	K_u
131°	Satcom 3R	C
130°	ABCI	K_u
128°	American Satellite	Hybrid
126°	RCA	K_u
125°	Telstar/Comstar	C
124°	SBS	K_u
122°	Spacenet I	Hybrid
120°	Usat	K_u
119·5°	Westar 5	C
117·5°	Canada	K_u
116·5°	Mexico	Hybrid
113·5°	Mexico	Hybrid
112·5°	Canada	K_u
111·5°	Canada	C
110°	Canada	K_u
108°	Canada	C
107·5°	Canada	K_u

SATELLITE ORBITAL SLOTS

Orbital position (W long)	*User*	*Frequency band*
105°	GStar	K_u
104·5°	Canada	C
103°	GStar	K_u
101°	Unassigned	Hybrid
99°	SBS	K_u
98·5°	Westar 4	C
97°	SBS	K_u
96°	Telstar	C
95°	SBS	K_u
93·5°	Galaxy 3	C
93°	Unassigned	K_u
91°	Spacenet 3	Hybrid
89°	SBS	K_u
88·5°	Telstar	C
87°	RCA	K_u
86°	Westar	C
85°	Usat	K_u
83·5°	Satcom 4	C
83°	ABCI	K_u
81°	American Satellite	Hybrid
79°	Rainbow	K_u
78·5°	Westar	C
77°	RCA	K_u
76°	Telstar	C
75°	Unassigned	K_u
74°	Galaxy 2	C
73°	Unassigned	K_u
72°	Satcom	C
71°	Unassigned	K_u
69°	Spacenet 2	Hybrid
67°	Satcom	C

In Aug 1983, however, the decision to establish 2° spacing immediately for Ku-band spacecraft, with 2° spacing in C-band by the end of the century, started a reshuffling procedure, with special consideration being given to systems which had established large ground-station networks.

ADDENDA

This section covers developments which were too late for inclusion in the main body of the book.

INTRODUCTION

Like all previous giant leaps, the US attempt to establish a permanent manned space station as a micro-gravity manufacturing base, and as the next logical step towards manned lunar and planetary exploration, is becoming a battle between the pioneers and the doubters.

Neil Hutchinson, who achieved fame as an Apollo and Space Shuttle flight director and has now been appointed manager of the Space Station project, has led his team of 250 enthusiasts out of the confines of the Mission Control complex at Houston and into the independence of a rented building. Already his team has selected 3 possible rigid framework designs for the Station, each with 5 modules attached to solar arrays, radiator panels and railway-mounted robot arms. Space contractors all over the world are bidding for the development contracts being issued in the next year. They are trying not just to share in the $8 billion being made available, but to ensure their future in these expanding technologies.

Their combined task is to get the whole thing well under way before the doubters succeed in having it stopped. Already a Congressional committee report is questioning the concept and suggesting that something much cheaper and simpler would do just as well. This will be well received in countries like Britain, where the first TV programmes describing the Station as nonsense will comfort politicians anxious to justify their lack of interest despite government and industry announcements in France, West Germany and Italy that these countries will play a major part in it.

Meanwhile, the Soviets continue to man their Salyut 7 space station with both long-stay and short-visit crews, and to press on with their own Space Shuttle, spaceplane and new generation of heavy launchers. No doubt it was a briefing about this programme that led President Reagan to surprise NASA with a proposal for a simulated US Space Shuttle/Soviet Salyut rescue mission. This imaginative plan might - like the only previous joint manned mission, the Apollo-Soyuz link-up in 1975 - be the ideal way to improve East-West relations. (No one has yet thought out whether the mission would involve docking the two vehicles, or just practising the transfer of crews between them as they flew alongside.)

Hopes that US and Soviet scientists will persuade their governments that this is a good idea have been raised by the elaborate International Halley Watch, which will call for Soviet, US, Japanese and European scientists to work closely together for the next 3 years as Halley's Comet makes its 76-yearly approach to Earth on its way around the Sun. Such collaboration may also help to warm the atmosphere for Soviet/US talks aimed at limiting military space activities.

Meanwhile, with their usual skill at beating the US to the "firsts" in the record books, the Soviets in Jul 1984 sent up Svetlana Savitskaya to make the first spacewalk by a woman - 3 months before Kathryn Sullivan, scheduled to make an EVA on STS 41-G, in Oct 1984, when NASA will carry a record 7 crew members, including a Canadian payload specialist and a US oceanographer.

NATIONAL SPACE PROGRAMMES

Canada

General Another success for Canada's space industry came in Mar 1984 with the signing of a $20 million contract to supply China with 26 satellite Earth stations for its developing domestic system.

ISIS In 1984 both ISIS satellites, having completed their work for Canada, were donated to Japan for polar orbit communications research.

China

China 15 L 8 Apr 1984 by CZ-3 from new launch site. Wt 900kg (in-orbit 420kg). Orbit 35,520 × 36,383km. China's 1st synchronous satellite, it was placed at 125°E on 16 Apr and began experimental transmissions on 1 TV and 15 radio channels in Cantonese, Amoy, Japanese, Spanish, Russian, Burmese and Tagalog to 5 ground stations. In Jun 1984 tests were going well, using 2 C-band Earth terminals installed by Scientific-Atlanta Inc near Beijing and Chengdu.

Following the successful placing of China 15 in synchronous orbit in Apr 1984, many details of events leading up to it emerged, including a vivid description of "a roaring explosion on No 5 pad during a fuel flow test on 28 Jan 1978". "A dozen or more" people were injured, 7 of them seriously. During preparations to send the 1st synchronous satellite into orbit, "some scientific and technical workers unselfishly laid down their precious lives". One such was Ma Jingyang, an engineer who went into the experiment hall of the Atomic Energy Research Institute 31 times to obtain data about the effects of solar ions on the satellite. Though aware of the harmful effects of exposure to atomic radiation, he was unable to receive even minimum safety protection because "this happened during the chaotic time of the Gang of Four". Jingyang, who was in his 30s, had difficulty breathing for 6 or 7 years but continued work between hospital visits until he died.

The director of research was stated to be Zhu Senyuan, with a woman, Wang Zhiren, as deputy director. The Defence Minister, Zhang Aipang, said that while the Chinese would seek independence in space, they were still behind in space technology and were seeking co-operation with the US, Federal Germany, France, Italy and other countries.

Indonesia

Palapa B2 recovery Initial Indonesian Government protests at delays by the insurers in reimbursing them for the Palapa B2 loss have subsided following assurances that a recovery mission could have the satellite on station by mid-1985 compared with a replacement in Feb 1986.

Indonesia has agreed to $75 million of the insurance money being paid into an "escrow account" to fund retrieval and refurbishment, and, in the event of failure, a replacement. After negotiations with the insurers, Hughes and NASA, in Jun 1984 Indonesia signed a $40 million contract with Hughes for an HS376 satellite which could be returned to the production line if it was not needed. NASA would charge $4·8 million for the Palapa retrieval but only a total of $5·5 million if both Palapa B2 and Westar 6 were recovered on the same mission. The insurers are therefore pressing Western Union to take part in the recovery, since a joint mission would save them up to $40 million, compared with less than half that for Palapa-only. As we went to press, astronauts Joe Allen and Dale Gardner were rehearsing a recovery plan calling for MMU activity to attach grapple fixtures to the satellites, followed by recapture with the robot arm for return to Earth, refurbishment and relaunch. With the viability of the recovery mission depending upon speed, more changes were made to the Shuttle's already ravaged 1984 schedule to make it possible before the end of the year.

Japan

Status A revised space development policy reduced the number of national satellites to be orbited by the end of the century to about 50 but gave the go-ahead for the H-2 rocket, which will be able to place 2-tonne satellites in stationary orbit in the 1990s. It was also decided to use the US Space Shuttle for life-science and materials-processing activities, and to participate in the Space Station as actively as possible. Concern about the high cost of Japanese satellites compared with buying them from the US – CS-3 was expected to cost $145 million, or 2-3 times more than an American communications satellite – led to a move in early 1984 to lift government restrictions on such space technology imports. This however was affected by political friction following the failure of the BS-2 TV satellite due to malfunctioning US components which Japanese scientists had not been allowed to inspect for quality-control purposes. This situation was made worse by the failure of the US-built GMS-1 and 2 weather satellites, the 2nd long before its 7yr life had expired.

United States

DBSC In Apr 1984 the US Direct Broadcast Satellite Corp gave Ford Aerospace a $177 million contract to build DBSC-1 and 2 for Ariane 3 launches in Mar and Sep 1987. The contract includes launch costs and an optional 3rd satellite. With 2,380kg lift-off wt, 6 200W amplifiers giving broad-beam coverage of the US and 12 45W channels for spot beams, the craft will be used for direct broadcast of commercial TV. They will have a 10yr life.

LDEF The Long Duration Exposure Facility, L 7 Apr 1984 by STS 41-C into a 463km orbit with 28·5° incl, carried 57 science and technology experiments involving investigators from the US and 9 other countries and covering materials, coatings and thermal systems; power and propulsion; science; and electronics and optics. Due to fly every 18 months, LDEF was due to be recovered at the end of its 1st mission by STS 51-D in Feb 1985.

Built by Langley Research Centre, LDEF is a 12-sided open-grid structure made of aluminium rings and longerons, and measuring 9·14m long, 4·26m in dia and 3,629kg in weight (9,980kg loaded). Entirely free-flying, with no central power or data systems, it is placed in a gravity-gradient-stabilised attitude by the Shuttle's RMS and can accommodate 86 experiment trays, 72 around the circumference and 14 on the ends. Experiments on the 1st mission included 4 for the USAF Space Test Program – testing the durability in space of thermal control materials, electrical components, mirror coatings and fibre-optics – and 1 from the Naval Research Laboratory, studying the population of heavy nuclei in space and their effects on electronics and wiring.

STS 41-D The shutdown of *Discovery*'s engines within 4sec of lift-off on the Orbiter's maiden launch on 26 June 1984 (following a 24hr delay due to a computer fault), and the need to use water jets to douse a small fire in the tail, brought more concern both for NASA programmers and insurers. So close did NASA come to a launchpad fire that KSC's chief test director recommended emergency evacuation of the 6 crew members (cdr Henry Hartsfield; plt Michael Coats; MSs Judith Resnik, Steven Hawley and Richard Mullane; and Charles Walker, 1st commercial payload specialist) down the slidewire. This was not deemed necessary and 40min later the crew were finally hauled out looking rather shaken – starting with Judith Resnik, who should have become the 2nd US woman in space. The fault was a main fuel valve failure in No 3 engine.

2 weeks later NASA decided to amalgamate Missions 41-D and 41-F (Mission 41-E having been cancelled earlier) so that the 1985 flight schedule could be maintained. Another important factor was the need to achieve the November launch window for retrieval of the lost Palapa and Westar satellites. It all added up to a late-August flight, with SBS-4 and Telstar 3C accompanying the original US Navy Leasat in the cargo bay. Since both SBS and Telstar were dependent upon PAM upper stages (which failed on 41-B) to get them to GTO, the insurers were keeping their fingers crossed. The Hartsfield crew was given the amalgamated mission, with the Bobko crew selected for 41-F once more awaiting reassignment.

Shuttle status, Jul 1984 With the programme going through its worst period since orbital flights started in Apr 1981, the transfer of the dynamic Lt-Gen James Abrahamson from his position as Shuttle head to become director of the new Strategic Defense Initiative ("Star Wars") Organisation was a double loss.

If the Shuttle system is indeed to "pay off handsomely" in the future, as Administrator James Beggs predicted in his budget request of $3,166 million for FY 1985 (which runs Oct-Sep), it is essential to improve flight frequency. In the unlikely event of the 11 flights planned for that financial year being achieved, their cost would still average $287·8 million. In FY 1984, with a possible maximum of 6 flights (starting with 41-A in 1983), the average cost would be $583 million. Even if the planned yearly average of 24 is achieved by FY 1989, it might still prove difficult to get mission cost down to $100 million.

The heavily revised May 1984 manifest lists 110 flights up to Sep 1989 (the end of FY 1989), carrying 325 payloads. While 38 of those are for DoD, many of them are quite small, and the payload total falls far below NASA's expectation that the US military would use one-third of Shuttle capacity. DoD's decision to order 10 more expendable launch vehicles for use at the rate of 2 a year has been another severe blow – coming on top of the growing Ariane competition – to NASA's efforts to make the Shuttle fulfil its promise of providing cheap and easy access to space. Another revelation in the May manifest is that between Oct 1985 and Sep 1989 only 12 flights from Vandenberg are scheduled: 4 of them are dedicated missions for DoD, 2 of them "reflight opportunities" and the others frequently include "payload opportunities".

For commercial customers, shopping around for the cheapest launcher for their communications satellites, it looks like a buyer's market. This was apparent in the petition by Transpace Carriers Inc (a private company intending to market McDonnell Douglas Deltas) to the Office of US Trade accusing ESA and Arianespace of "predatory pricing" – a reference to the European decision of several years ago to offer launches to US customers at 25-30% lower than the rate to ESA member states. Inevitably, ESA replied sharply that while it was ready to discuss ways of creating healthy competition, Ariane pricing policy had been based on the rates for the heavily subsidised Shuttle. In any case, Arianespace must ultimately recover the full cost from all users.

Space Station On 28 Jun 1984 NASA announced the sharing of Space Station design and definition among its centres as follows:

Johnson, already named as lead centre, would select the configuration and have responsibility for overall systems engineering and details such as the RMS and structural framework, attitude and thermal control, communications and data management, and equipment of a module for the crew's living quarters.

Marshall would be responsible for a "common module" for use as living quarters and laboratories, the Station's propulsion system and environment, and an orbital manoeuvring vehicle (OMV), or space tug.
Lewis, responsible for power, would examine alternative systems for electric power generation, conditioning and storage, as well as large solar arrays.
Goddard would be responsible for definition of free-flying platforms to accompany the Station, and for servicing and maintaining all types of free-flying platform.
Kennedy was mentioned in passing as "of course" being responsible for pre-flight and launch operations and logistic support.

Contracts to industry were expected to be awarded in early 1985, with negotiations continuing with several nations "to determine means of international co-operation".

Westar 6 L 4 Feb 1984 by STS 41-B, this was originally booked for Ariane launch at $22 million compared with $30 million for a Delta launch. Following Ariane delays, Western Union negotiated a price with NASA of $18 million each for Westars 6 and 7, paying a penalty of about $100,000 to ESA for the transfer. But the PAM-D upper stage, provided by McDonnell Douglas in less than 1yr, malfunctioned, placing Westar 6 in a useless 307 × 1,218km orbit instead of in the planned stationary orbit at 91°. (That slot had been allotted only 1 day before launch - an indication of the pressure on this overcrowded orbit.) Westar 6 cost $75 million, but was insured for $105 million to cover loss of business. To ensure that it did not re-enter pending decisions about its future, the apogee kick motor was successfully fired on 12 May 1984 to raise its orbit to 1,060 × 1,666km. Its hydrazine thrusters could be used to lower it from there for a possible Shuttle retrieval mission. However, with the back-up Westar 7 due for launch in Dec 1985, Western Union appeared little interested in the complicated finances of a rescue mission. There was however pressure from the insurers to agree to a joint Palapa/Westar recovery mission in Nov 1984.

USSR

Orbital reflectors The Moscow Institute of Avionics has designed an experimental 200kg satellite reflector with a 110sqm working area capable of reflecting sunlight to illuminate large regions in different time zones at night. Such reflectors, it is claimed, would provide 7 times as much light as a full moon over areas of 10km dia, and should be able to prolong daylight for several hours in cities like Moscow. Their cost would be repaid in 4-5yr by electricity savings. They could be switched from one group of towns to another at short notice, providing a useful facility when night work was needed; they could also be used to combat night frost. They might be in use "as early as the next decade".

Soyuz T-11 L 3 Apr 1984, this was the Soviets' 11th international flight, carrying India's Maj Rakesh Sharma, 35, as cosmonaut-researcher. Commander was Col Yuri Malyshev, 42, making his 2nd flight, with the previously unlucky Gennadi Strekalov, 43, making his 3rd (his 4th including T-10A) as flight engineer. Docking, apparently uneventful, was at the end of the 16th orbit, and the crew were greeted in Salyut 7 - and "given the best sleeping places" - by Kizim, Solovyev and Atkov who were in their 55th day. A complicated programme during their 7-day stay included medical experiments with yoga exercises by Sharma, metal smelting (specimens of silver-germanium alloy had been prepared by Indian scientists) and studies of Indian territory. Particular attention was paid to the Himalayan glaciers and their water reserves, the Arabian Sea, the Brahmaputra and Ganges deltas, and the New Delhi and Agra areas. While filming 2,000 sequences, the crew urgently reported a 50 sq km forest fire in Burma. Shift-change news conferences and TV hookups with Moscow and New Delhi, enabling Prime Minister Gandhi and others to talk to the crew, paralleled US spaceflight procedures. Before undocking on 11 Apr the visiting crew carried out on their own initiative an additional experiment involving crystallisation of a gallium-bismuth alloy. They returned in Soyuz T-10B, landing at the end of a 7-day 21hr 51min flight 46km E of Arkalyk - 1min earlier and 1km further away than predicted. Launch, docking and landing times had all been given in advance - but Indian journalists who had expected to travel to Tyuratam for the launch never did get their visas.

Soyuz T-12 L 17 Jul 1984, this mission enabled Svetlana Savitskaya, 35, to become the 1st woman to walk in space, beating America's Kathryn Sullivan to this prestigious 1st by about 3 months. Savitskaya, making her 2nd flight, had been promoted to flight engineer. Her commander was Vladimir Dzhanibekov, who became only the 2nd cosmonaut to make 4 flights, with Research Cosmonaut Igor Volk making his 1st. They docked with Salyut 7 25hr 36min later, joining Kizim, Solovyev and Atkov, in their 24th week. The EVA was on 25 Jul; Dzhanibekov and Savitskaya spent 3·5hr working outside, Savitskaya being assisted by Dzhanibekov as she tried out a bulky welding tool with an earth weight of 30kg. She also sprayed a silver coating on an aluminium plate. Finally the pair recovered panels attached to Salyut 7's exterior to test the effects of long-term space exposure, and took them back inside for return to earth. Leonov said Savitskaya had worked "brilliantly, with pinpoint precision". This was the 6th spacewalk since the Soyuz T-10B crew started their long duration flight in Feb 1984, more than doubling the Soviets' EVA time to a total of 70hr. Flight duration was an unusual 13 days.

INTERNATIONAL SPACE PROGRAMMES

ESA

Space Station By mid-1984 France and Italy were leading the ESA nations in formulating ambitious plans for European participation in the US Space Station, with Britain taking little interest. With NASA badly needing European participation, Administrator James Beggs was accepting European arguments that such participation must include continued right of access to Space Station facilities. He envisaged the possibility of Europe docking its own module and using the Station's power and services under a leasing arrangement similar to that governing ownership of an apartment in a US condominium.

Ariane/Columbus NASA's launcher setbacks with Shuttle/IUS, Shuttle/PAM and Atlas-Centaur upper stages made France's continued advocacy of Ariane development, first through ESA and then commercially through Arianespace, look much more viable. France was pushing for funding of the HM60 large cryogenic engine to ensure that follow-on Arianes would be fully competitive with all possible US launchers and be capable of giving Europe independent access to manned spaceflight. With France, through CNES, willing to go on providing the greater part of Ariane funding, Germany and Italy appeared ready to support launcher development in return for parallel French support for Columbus, the now well advanced proposal for a free-flying space laboratory which could be launched either by Ariane or Shuttle. It would probably fly in formation with the US Space Station, making servicing and recovery of processed materials relatively cheap and easy. Alternatively, Columbus, based on Spacelab/Eureca/Spas technology, could act as the docked module discussed above.

Hermes Feasibility studies by French companies under contracts from CNES were given impetus by NASA's acceptance that the mini-shuttle could well be used to service a NASA/ESA space station. Latest configuration proposals show it as up to 18m long with 10m wingspan and 4·9-tonne payload, carrying 2 pilots and 2 passengers. 1st flight could be in about 1996.

Tethered Satellite

A joint US/Italian project, agreed in 1984, the Tethered Satellite System consists of an Aeritalia-built spherical satellite weighing 550kg and 4·5m in dia which can be reeled out from the Shuttle cargo bay on a 2mm-thick Kevlar rope up to 100km long. NASA is developing the deployment system for a 1st mission in Dec 1987. This calls for a 20km upward deployment, with an insulated copper wire incorporated in the tether, to study electrodynamic and other scientific phenomena. Downward deployment, with the help of 4 nitrogen thrusters on the satellite, will "troll" the Earth's atmosphere at heights of 130-150km - normally inaccessible because of drag - for magnetospheric, atmospheric and gravitational data. The satellite and its tether can become a generator, much as a copper coil moving within a magnet on Earth can produce a flow of electricity. By drawing off the energy from the tether, scientists will be able to study the magnetic lines of flux that surround the Earth. They also hope to simulate the electrodynamic conditions that exist between Jupiter and its satellite Io. Development is estimated to cost NASA and Italy's National Research Council $56 million each, with an additional $10 million for the 1st mission.

LAUNCHERS

Europe

Ariane L-8 L 5 Mar 1984, successfully placed Intelsat 5-F8 (launch wt 1,928kg) in GTO.

Ariane L-9 L 22 May 1984, successfully placed Spacenet 1 (launch wt 1,195kg) in GTO. 1st commercial launch by Arianespace, for $27 million fee.

United States

Centaur Atlas-Centaur No 62, a stretched version offering an increased geosynchronous payload of 2,320kg, failed on 9 Jun when attempting to launch Intelsat 5-F9. The Centaur's 2nd burn lasted 3·5sec instead of 90sec, placing the satellite, insured for $102 million, in a 178 × 1,111km orbit (later raised to a safer 278 × 1,111km). Following 58 successful launches, it was the 1st ELV failure since 1977. With the causes of the IUS and PAM malfunctions still not fully established, NASA set up a failure board to investigate Atlas-Centaur as well.

OTV In Jul 1984 Boeing was awarded a $1 million contract to study a reusable Orbital Transfer Vehicle, or space tug, capable of being maintained and refuelled at a space station. With a 10-tonne capacity, it could take a manned vehicle to stationary orbit and back, as well as being able to launch planetary observation missions. It could be operational by the mid-1990s. Martin Marietta has a similar contract (see TOS in main text).

THE SOLAR SYSTEM

Saturn, the asteroids and Mars Three joint US/European missions to Saturn, the asteroids and Mars were proposed by a European Planetary Science Symposium, in Apr 1984:

Titan Probe and Saturn Orbiter to explore Saturn, its ring system and satellites, and to drop a probe onto Titan to explore its lightning and wind dynamics as well as the surface.

Multiple Asteroid Orbiter with Solar Electric Propulsion. The latest consensus is that the asteroids are collisional fragments of a small number of parent bodies, each parent now being represented by a class of asteroids. It is therefore planned to rendezvous with 4-6 asteroids, to ensure investigating one of each class. The spacecraft, which would need a new type of rocket propulsion, would go into orbit around each asteroid to explore their composition, structure and surface.

Mars Surface Rover Mission. The most challenging of the missions, this would involve placing 2 or 3 rovers on the surface. Each would explore a few thousand km, for 1-2 years, studying Mars' chemical, isotopical and mineral composition.

Comets The path of Pioneer Venus, which has been orbiting the planet since 1978, was tilted 37° in Apr 1984 to allow the spacecraft to look across the solar system at Comet Encke. As a result, it was discovered that the comet was losing water 3 times faster than expected. After this success it was decided to use half Pioneer's remaining fuel to observe Halley's Comet at perihelion on 9 Feb 1986, when both Venus and the comet will be on the far side of the Sun. US scientists, disappointed at being refused funding for their own interceptor spacecraft, will now match or better Soviet, European and Japanese studies of Halley's Comet with observations from the Shuttle, the Pioneer, ICE, SMM and IUE satellites already in orbit, and sounding rockets.

SPACEMEN

American astronauts

A 10th group of astronauts - officially "astronaut candidates" for their first year - was announced in May 1984. 7 were pilots, of whom 1 was Hispanic; 10, including 3 women, were mission specialists. They were selected from 4,934 applicants, of whom 128 were interviewed.

ADAMSON, James A. (3 Mar 1946). Maj US Army; aeronautical/mechanical engineer; flight controller, JSC.

BROWN, Mark H. (18 Nov 1951). Capt USAF; pilot, McDonnell Douglas F-4 Replacement Training Unit, Homestead AFB, Fla.

CAMERON, Kenneth D. (29 Nov 1949). Maj USMC; Project Officer, Marine Aviation Detachment, Patuxent River Naval Air Test Centre, Md.

CARTER, Manley L. (15 Aug 1947). Cdr USN; test pilot under instruction, Patuxent River Naval Air Test Centre.

CASPER, John H. (9 Jul 1943). Lt-Col USAF; Deputy Chief, AF Special Projects Office, Pentagon.

CULBERTSON, Frank L. (15 May 1949). Lt-Cdr USN; pilot, Grumman F-14 Replacement Air Group, Naval Air Station Oceana, Va.

GUTIERREZ, Sidney M. (27 Jun 1951). Test Pilot, Edwards Flight Test Centre, Calif.

HAMMOND, Lloyd B. (16 Jan 1952). Instructor Pilot, USAF Test Pilot School, Edwards.

IVINS, Marsha S. (15 Apr 1951). Space Shuttle training aircraft flight simulation engineer, JSC.

LEE, Mark C. (14 Aug 1952). Capt USAF; General Dynamics F-16 pilot, 388th TFW, Hill AFB, Utah.

LOW, George D. (19 Feb 1956). Spacecraft systems engineer, JPL.

McCULLEY, Michael J. (4 Aug 1943). Lt-Cdr USN. Operations officer, VA-35, USS **Nimitz**.

SHEPHERD, William M. (26 Jul 1949). Lt-Cdr USN; Co, Special Boat Unit 20, Naval Amphibious Base, Little Creek, Va.

SHULMAN, Ellen L. (27 Apr 1953). MD; medical officer for Medical Sciences Division, JSC.

THORNTON, Kathryn C. (17 Aug 1952). PhD; physicist with US Army Engineering Science and Technology Centre, Charlottesville, Va.

VEACH, Charles L. (18 Sep 1944). Pilot, JSC Aircraft Operations Division.

WETHERBEE, James D. (27 Nov 1952). Lt USN; test pilot Naval Air Station Lemoore, Cal.

INDEX